AF883464

SPACE RESEARCH XIV

Organized by

THE COMMITTEE ON SPACE RESEARCH — COSPAR

and

THE "DEUTSCHE FORSCHUNGSGEMEINSCHAFT"
OF THE FEDERAL REPUBLIC OF GERMANY

Open Meetings of Working Groups on Physical Sciences
Sponsored by COSPAR

The Symposium
on Noctilucent Clouds and Interplanetary Dust

Sponsored by

THE INTERNATIONAL ASSOCIATION OF METEOROLOGY
AND ATMOSPHERIC PHYSICS OF
THE INTERNATIONAL UNION OF GEODESY AND GEOPHYSICS —
IAMAP/IUGG

and

THE COMMITTEE ON SPACE RESEARCH — COSPAR

COSPAR

SPACE RESEARCH XIV

Proceedings of Open Meetings of Working Groups of the Sixteenth Plenary Meeting of COSPAR

Constance, F.R.G. — 23 May—5 June 1973

and

Resumés of the
Symposium on Noctilucent Clouds and Interplanetary Dust

Constance, F.R.G. —24 and 25 May 1973

Edited by

M. J. RYCROFT
R. D. REASENBERG

AKADEMIE-VERLAG · BERLIN

1974

Executive Editor:
Dr. A. C. Stickland

Library of Congress Catalog Card Number 60-50 878

© Akademie-Verlag, Berlin, 1974
All Rights Reserved. No part of this publication may be reproduced, stored in a retrieval system, or transmitted, in any form or by any means, electronic, mechanical, photo copying, recording or otherwise, without the prior permission of the Copyright owner.
202 · 100/472/74
Gesamtherstellung: VEB Druckhaus „Maxim Gorki", 74 Altenburg
Bestellnummer: 761 940 9 (3059/XIV) · LSV: 1435,1495
Printed in GDR

FOREWORD

The sixteenth Plenary Meeting of COSPAR and associated scientific meetings took place in Constance, a town that has been for so many centuries a centre of cultural life and international traffic, and a communication link between South and North, East and West. The town and the meetings lived up to expectations perfectly: COSPAR was happy to see in Constance a gathering of scientists from nearly all the countries with institutions who are members of this Organization. The spirit of the many interesting sessions, often with intense scientific discussions, is reflected in these Proceedings.

In most of the preceding years it has been the practice to have the Proceedings split into "Space Research" and "Life Sciences and Space Research". This year it was felt to be necessary to have two more Volumes, viz. one containing the Proceedings of Symposium A on Approaches to Earth Survey Problems through the Use of Space Techniques, and another for Symposium B on Methods of Measurements and Results of Lower Ionosphere Structure.

The Proceedings of Symposium A contain also the papers presented during the Open Meetings of Working Group 6, while as usual those presented during the Open Meetings of Working Group 5 are included in the "Life Sciences" Volume. Hence, the 130 or so papers in *Space Research XIV* are the Proceedings of the Open Meetings of Working Groups 1, 2, 3, 4 and 7, and the abstracted Proceedings of Symposium C on Noctilucent Clouds and Interplanetary Dust.

We are happy to have found some excellent scientists willing to serve as Scientific Editors of our various publications: Space Research XIV: Dr. M. J. Rycroft and Dr. R. D. Reasenberg, Life Sciences and Space Research XII: Prof. P. H. A. Sneath. Proceedings of Symposium A: Prof. P. Bock. Proceedings of Symposium B: Prof. K. Rawer. We thank them all for their cooperation. At the same time it is our very pleasant duty to thank the Executive Editor of these Volumes, Dr. A. C. Stickland, who has served COSPAR well for a number of years now.

C. de Jager
President of COSPAR

PREFACE

"So the sun sits as upon a royal throne ruling the children his planets which circle round him"

Nicolaus Copernicus (1473–1543), in De Revolutionibus Orbium Coelestium; see R. G. Walgate, Physics Bulletin, 24, 284, 1973.

In the five hundredth anniversary year of the birth of Copernicus, it is fitting to preface the proceedings of the 1973 COSPAR Meeting with a quotation which succinctly summarizes his outstanding contribution to man's understanding of the solar system. This is so because, nowadays, scientists can study the sun and the planets, and in particular the earth and its environs, using space techniques.

It was this broad subject that formed the basis of the Sixteenth Plenary Meeting of COSPAR. This meeting was held, from the twenty-third of May to the fifth of June 1973, in the ancient city of Constance, Federal Republic of Germany. 630 scientists from 36 countries participated in the Meeting, attending one of the three specialized symposia and/or some of the meetings of the seven Working Groups. Two sessions in the Annual Reviews on Space Research series were also held.

In this volume are published some of the review papers and some of the papers presented during open meetings of all the COSPAR Working Groups except numbers 5 and 6. Résumés of the Symposium on Noctilucent Clouds and Interplanetary Dust are also given here. The book is comprised of ten main sections:

Satellite geodesy
The earth's neutral atmosphere
The ionosphere
The magnetosphere
The sun
Astronomy
Comets
The moon
Venus
Symposium on noctilucent clouds and interplanetary dust

Three companion volumes contain papers presented at:

a) the Symposium on Approaches to Earth Survey Problems through the Use of Space Techniques and open meetings of Working Group 6, on Applications of Space Research to Meteorology and Earth Surveys,

b) the Symposium on Methods of Measurements and Results of Lower Ionosphere Structure,

c) the Open Meetings of Working Group 5 on Space Biology.

In organizing the material for publication we, as Scientific Editors, are most grateful to the Chairmen and Reporters of the different Sessions of the Working Groups for their thorough evaluation of the papers presented and their recommendations concerning publication. For their considerable help we also thank Mr. Z. Niemirowicz, Executive Secretary of the COSPAR Secretariat, and other members of the Secretariat. In particular we take this opportunity of thanking Miss Connie O'Dea and Miss Beth Reynen (secretary to Prof. C. de Jager) who most ably filled the breach in the Editorial Office left by the absence, due to an accident, of Dr. A. C. Stickland, the Executive Editor. We are very grateful to Dr. Stickland, who nevertheless tackled the huge task of collating and editing all the material for the publishers, Akademie-Verlag, Berlin.

M. J. Rycroft
R. D. Reasenberg

CONTENTS LIST

Satellite Geodesy

The Earth's Neutral Atmosphere

Stratospheric and Mesospheric Temperatures and Winds

Solar UV Absorption; O, O_2 and O_3 Concentration Profiles

Atmospheric Models

Fluctuations of Satellite Brightness

The Ionosphere

Electon Content

Ionosphere and Neutral Winds

Equatorial Ionosphere

Composition

Sporadic E

The Magnetosphere

Energetic Charged Particles

High Latitude Phenomena

Very Low Frequency Radiophysics

Magnetic Fields

The Sun

Solar X-ray Flares

Solar Gamma Rays

Solar Particles

Astronomy

X-ray Astronomy

Ultraviolet Astronomy

Heliosphere

Primary Cosmic Rays

Comets

The Moon

General

Geology and Chemistry

Magnetism, Gravity, Radioactivity

Venus

Symposium on Noctilucent Clouds and Interplanetary Dust

Noctilucent Clouds and Interrelated Phenomena

Lunar Surface Micrometeoroid Studies

Observations of Dust from Space Vehicles

Satellite Geodesy

Space Research XIV — Akademie-Verlag, Berlin 1974

SATELLITE GEODESY AND THE INTERNAL STRUCTURE OF THE EARTH

G. Barta

Department of Geophysics, Eötvös University, Budapest, Hungary

The geoid obtained by means of the methods of satellite geodesy for low degree harmonics is the sum of two rotation-symmetrical forms. The axes of the components are directed towards Australia and India. From this it can be inferred that (i) the earth's core is an asymmetrical structure (at the centre of the earth, this being a singularity point of the gravity potential field, this is possible according to hydrostatic equilibrium); (ii) gravity and magnetic fields of the earth appear to be connected (the magnetic dipole is displaced towards Australia, while India is an important centre of the magnetic secular variation).

According to this concept the magnetic secular variation is in correspondence with a large scale mass motion, which may produce also a secular change of the earth's figure and of the gravity field. This motion of the material of the earth's core is obviously closely related to the internal dynamics of our globe and with its surface tectonics.

The large amount of information material involved in the earth figure investigations contains many data concerning the internal structure of the globe and its processes. From the scientific viewpoint it is of great importance to make a possibly complete use of this knowledge.

It is well known that the oblateness of the ellipsoid of the earth, being in hydrostatic equilibrium, has the value of about 1/300 [1] and more precisely 1/299.67 [2], while the value observed by the satellite geodesy amounts to 1/298.255; the difference is small, yet it is real, i.e. the polar oblateness is greater by about 0.5% than would follow from the condition of hydrostatic equilibrium. According to Munk and McDonald [3] this phenomenon is a remainder of a past higher rotational speed of the earth in geological times and along with the steady diminution of this speed the oblateness of the earth is, though with some lag, decreasing. The observed higher value of the flattening is due to this lag.

A further problem is presented by the deviation of the two hemispheres of the earth, i.e. the well known "pear-shaped" form of the globe. In the case of hydrostatic equilibrium, such a difference cannot be permanent. We know that against large scale global effects the earth behaves as a fluid. Also tidal phenomena contribute to the development of hydrostatic equilibrium. At the same time we have difficulties in interpreting geoidal anomalies. Determinations made by different methods at different times lead to the same anomaly systems, therefore their reality cannot be questioned. Their dimensions are, however, greater than could be supported by the rigidity of the globe. Geoidal anomalies represent a distortion of the gravity field of our earth. Also the terrestrial magnetic field shows distortions and the distribution of continents and oceans is also asymmetric. Some workers have studied the correlation between the geoidal figure and anomalies of the magnetic field: Hide and Malin have found a correlation between the low order harmonics of the magnetic and gravity fields if the former is rotated by 160° in longitude [4]. Others compared the surface distribution of continental tables and oceanic basins, i.e. the large scale topographic features, with those of the geoid.

When entering into investigations of this nature one has to proceed very carefully. The gravity and the magnetic fields of the earth are basically different regarding their physical nature: (i) the former is unipolar, the latter bipolar; (ii) different physical laws are valid for them, e.g. isostasy and magnetic induction, and (iii) their origin is due to different parts of the globe (to its whole mass and its core respectively). Thus it is evident that their anomaly features must be basically different one from another.

Comparison of surface topography with the geoidal figure presents also great difficulties, topography being, independently of its dimensions, a pronouncedly local phenomenon, while the shape of the geoid is a result of a combined effect of all gravitating and inertial masses of the globe. Thus the connections are so complex that a direct comparison seems not to be promising.

Good progress of the study is probably to be hoped for through a recognition of the physical backgrounds manifesting themselves in the phenomena as well as through their comparison. Such a physical background may perhaps be found in the general inhomogeneity of the globe and its asymmetrical structure.

It is well known that the dipole representing the geomagnetic field is significantly displaced towards Australia. The interpretation of this phenomenon inevitably suggests the necessity of modifying the spherically symmetric model of the earth. Since the permanent magnetic field furnishes information about the core of the globe, we can suppose that the eccentricity of the dipole corresponds to an eccentricity of the inner core of our globe. According to the magnetic observations made for centuries the dipole is moving slowly towards WNW. If the dipole can be identified with the inner core of the globe, then it is the inner core which is displaced in this direction. The asymmetry of the earth's core must of course also manifest itself in the figure of the earth, and that is where a connection between geodesy and terrestrial magnetism develops.

The earth's figure as determined by means of artificial satellites shows its greatest positive anomaly in the direction of the magnetic eccentricity, while its highest negative anomaly coincides with the large centre of symmetry of the magnetic secular variation in India. Nevertheless, a major theoretical difficulty arises with the interpretation of four further anomaly belts lying at middle latitudes in the northern and southern parts of the Atlantic and Pacific.

The displacement of the inner core from the centre of the earth leads to an elliptical geoid in the equatorial plane. However, this is not what is required. The observed equatorial geoid can well be approximated by two mathematically symmetrical forms shown diagrammatically in Fig. 1. Since these forms have rotational symmetry, we can compute the whole low degree harmonic geoid using only the equatorial data (Fig. 2) [5].

With this construction of the theoretical geoid we have used only the equatorial profile of the observed geoid. Thus the agreement between the two geoid figures is not a result of an approximation, but it is a proof, independently of the method applied, of the fact that the geoid is the sum of two rotation-symmetrical forms.

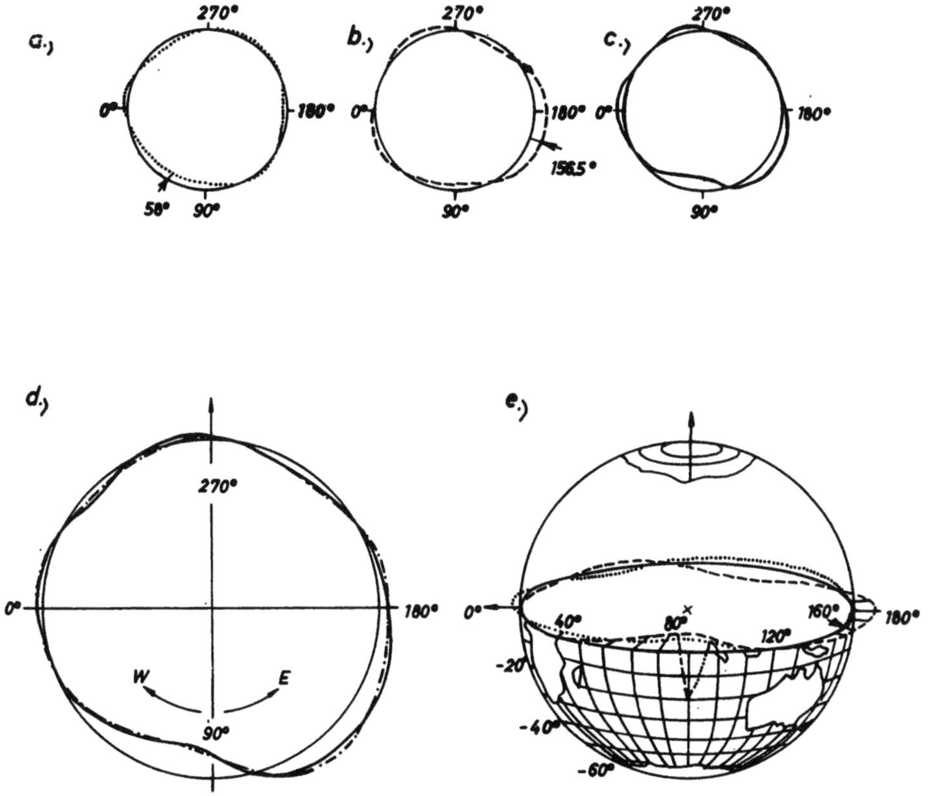

Fig. 1. a, b, symmetrical approximating forms; c, equatorial profile of the geoid; d, check of the approximation: —— measured; —·—· approximating form; e, method of constructing the theoretical geoid on the basis of assumption of rotational symmetry.

The anomalies of the middle latitudes have no physical background of their own; they are only superpositions of the two main anomalies from the directions of India and Australia [5]. Also the distribution of oceans and continents as well as of the superficial isostatic masses must have no major influence upon the satellite geoid. The low degree harmonic geoid is formed by agents working at great depth, possibly in the core.

Besides the high degree of similarity, the theoretical and the measured geoid patterns also show certain characteristic differences. The axes of the positive and negative anomaly systems of the northern and southern hemispheres are, as measured on the map, not perpendicular to the equator. Consequently, when determining the position of the agent more accurately, one has to consider symmetry about a plane which is displaced slightly from the equatorial plane. In this way, an approximation still better than the present one can be obtained. It may be possible to give an explanation also for a considerable part of the known polar asymmetry of the earth. The internal mass asymmetry causing the equatorial ellipticity of the globe also increases the oblateness, thus contributing to the explanation of the 0.5% excess in the value of the polar flattening.

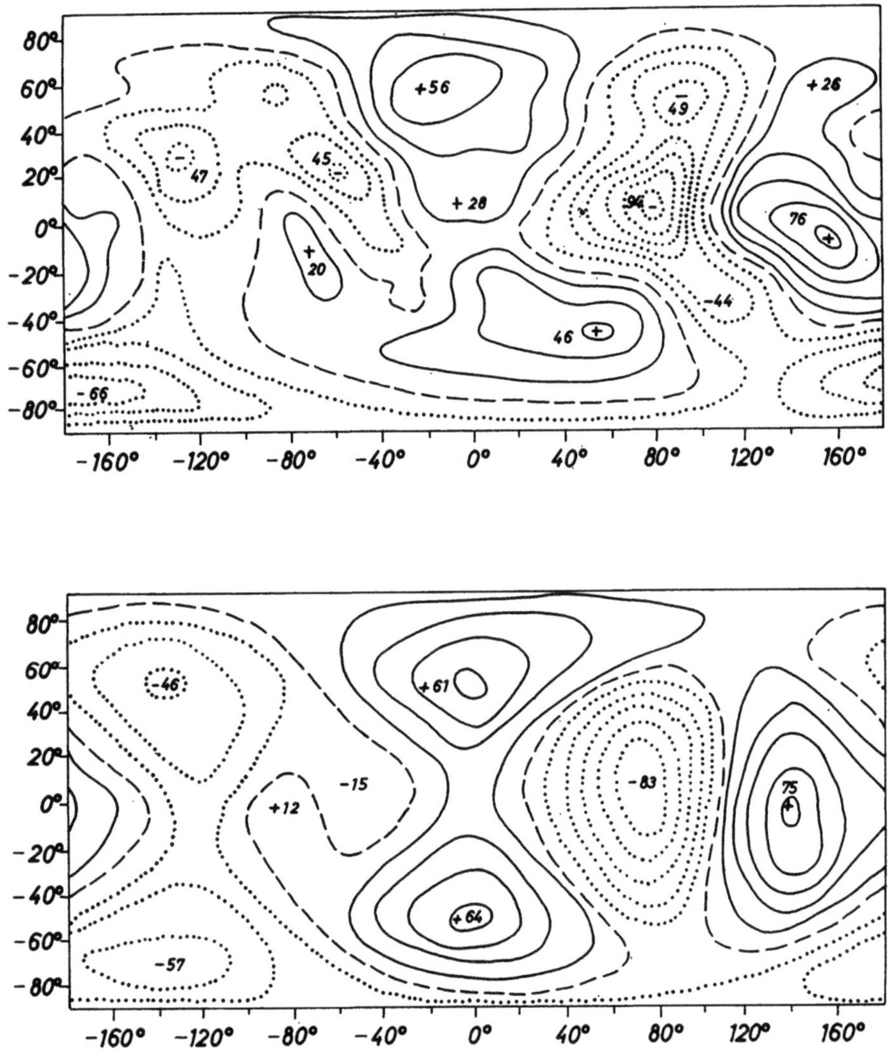

Fig. 2. Above, the "Smithsonian 1966" geoid; below, geoid figure computed from the two rotation-symmetrical anomaly systems. Intervals between lines 15 m.

To consider why the inner core of highest density does not sink in the liquid plasma material of the outer core to the centre of the globe is erroneous. This is because the more dense rigid body and the less dense fluid form the field of force of the outer and inner cores themselves. The inner core of the globe does not know where the geometrical centre of the earth happens to be, where it has to sink to. The pressure centre will be formed by the gravity field of the core itself. In other words, the centre of pressure is not determined by geometrical, but by physical conditions.

From the aspect of potential theory the inner core of the globe is a singularity of the gravity field of the earth's body. Therefore, it is not allowable to extrapolate usual hydrostatic views to phenomena of the earth's core and to fix the inner core of the globe at the geometrical centre. It is therefore necessary to investigate the position assumed by the inner core, rotating with the earth, in the case of hydrostatic equilibrium.

The inner core of the globe, influencing with its gravitational field the plasmatic material of the outer core, can only be in equilibrium if on the whole surface we

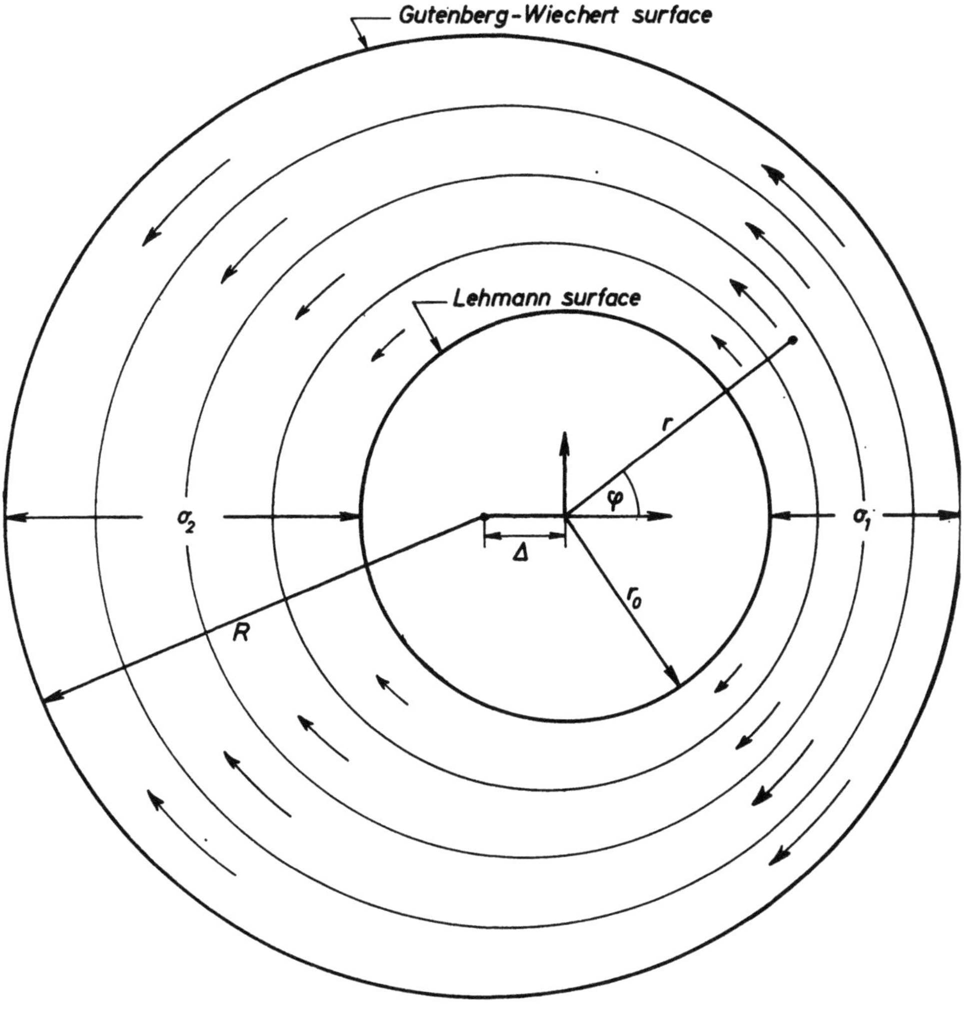

Fig. 3. Variation of the equipotential surfaces and of density in the outer core in the case of an eccentric inner core.

have a constant pressure. Assuming the material of the crust and mantle to be a homogeneous spherical shell, as a first approximation, then these masses have no internal gravitational effect. Their pressure manifests itself in what follows by an integration constant.

The centrifugal force appearing in the case of a small amount of asymmetry tends to move the inner core away from the axis of rotation. Thus the material of the plasmatic outer core flows away from the direction of the shifting and its density changes. We can easily calculate the density distribution for which, at a given asymmetry, the pressure remains constant over the whole surface of the inner core (Fig. 3).

Let us carry out an approximate calculation for the determination of these effects. Let Δ denote the distance of the centre of the inner core from the geometrical centre. The pressure must be constant on the surface of the inner core, i.e.

$$p_1 = \int_{r_0}^{R-\Delta} \sigma_1 \frac{m}{r^2}\, dr + C = p_2 = \int_{r_0}^{R+\Delta} \sigma_2 \frac{m}{r^2}\, dr + C$$

where m is the mass of the inner core; hence

$$\frac{\sigma_2}{\sigma_1} = \frac{(R+\Delta)(r_0 - R + \Delta)}{(R-\Delta)(r - R - \Delta)}.$$

Let the radius of the inner core be $r_0 = 1250$ km, that of the outer core $R = 3500$ km and its density $\sigma = 10$ g cm^{-3}, then in case of different eccentricity values the variation of the density is as follows:

$\Delta =$	$\sigma_1 =$	$\sigma_2 =$
0 km	10.00 g cm^{-3}	10.00 g cm^{-3}
100	10.17	9.83
200	10.35	9.65
300	10.52	9.48
400	10.70	9.30
500	10.87	9.13

Thus, when assuming a suitable density distribution, any eccentricity is possible. We therefore have to look for some restricting factors. When computing the pressure we have taken into account the gravitational interaction of the materials of the inner and outer cores, but the inhomogeneity of the outer core leads to further gravitational effects, thus giving limits to the possible values of eccentricity.

Let us make an estimate for this, by introducing the following considerations. In the central case the mass of the outer core should be divided by a plane dividing it into two equal parts: let their mass be denoted by M. The two spherical shells exert a mutual gravitational force on one another which is proportional to M^2. If the inner core is shifted by Δ, then the initial equal state breaks down and the mass of one side increases by m, that of the other diminishing by the same amount. The gravity interaction is proportional to the product of the two masses

M^2 and m^2. Thus, the change of the symmetry position is proportional to Δ^2. This is the force which fixes the limits for the possible values of the eccentricity.

The mass difference between the two sides can be computed by means of integration. We have also to take into account that the limits of the density as well as the upper limit of the integral will change with r and φ (see Fig. 3). A rough computation (neglecting, e.g., the elliptical form of the level surfaces) leads to the result that this force, being proportional to the square of the eccentricity, is equal and opposite to the centrifugal force which increases linearly with Δ at a distance of the order of 100 km from the axis of rotation. Therefore the two forces are in equilibrium at a distance of a few hundred kilometres from the axis of rotation, i.e. they compensate each other there.

Thus the supposition of an eccentricity is not absurd, but rather it is a consequence of a correctly interpreted hydrostatic equilibrium state at the centre of the globe, as a singularity point of the system. The inner core of any rotating body of similar material structure and dimension assumes such an asymmetric state.

It is to be expected that the position of the inner core is a natural consequence of the dynamic conditions in the interior of the earth; the direction of the eccentricity, however, remains undetermined. The axis of rotation is surrounded by

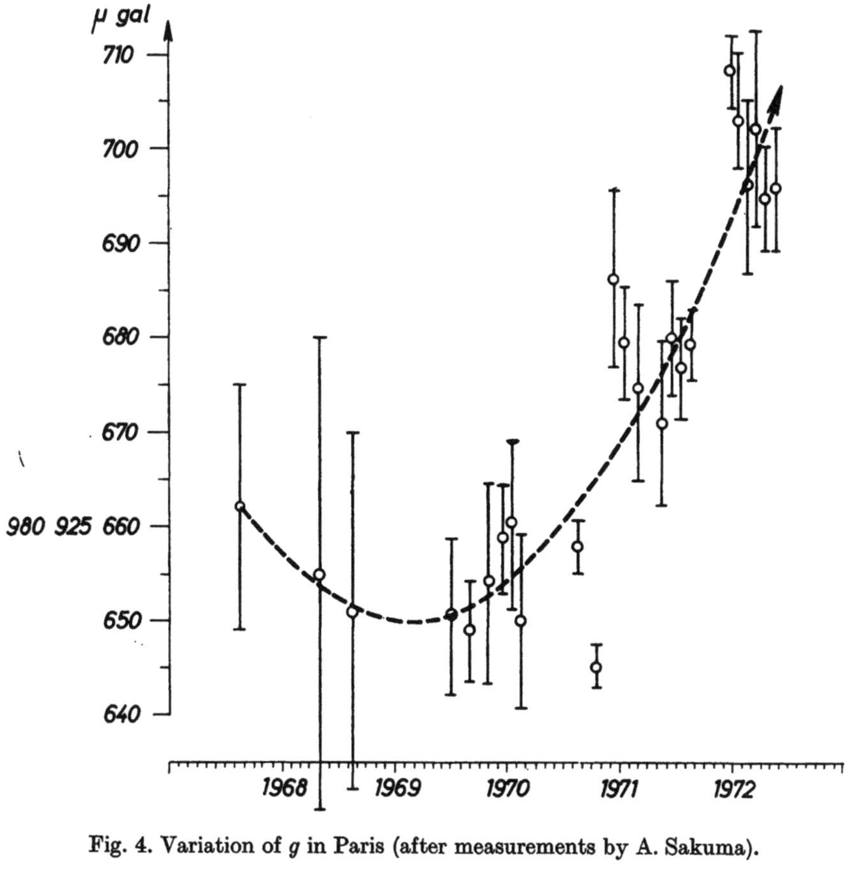

Fig. 4. Variation of g in Paris (after measurements by A. Sakuma).

a potential valley in which the inner core can assume its position at any place. Owing to the tidal effect of sun and moon the inner core can shift in a westerly direction, thus bringing about the magnetic secular variation. The shifting of the inner core can cause, as a matter of fact, a rearrangement and flow of the material of the outer core too. Thus on the basis of a comparison of the deformations of the magnetic and gravity fields and of the speed of the processes of variation we can tentatively draw conclusions concerning the physical characteristics of the material of the core (e.g. its compressibility and viscosity).

In connection with the shifting of the core, however, the gravity field must also show some changes. It seems that Sakuma [6, 7] succeeded in detecting a secular variation of the gravity field in the last five years by means of its observation to a microgal accuracy. According to him g had a minimum at Paris in 1969, while decreasing previously and increasing at present at a rate of 20 microgal per year (Fig. 4). This is not a proof, but a positive indication, of the correctness of our hypothesis.

A most accurate measurement of the variations of our gravity field is important not only from the point of view of the earth sciences, but it can contribute to the solution of questions of physical principles concerning the bases of the physical sciences (e.g. the "constancy" of the gravitational constant, fine structure of the gravity field etc.).

From all these results, what conclusions can be drawn for the features of the globe and what further investigations are suggested?

(i) In the investigations carried out as yet, we approximated the equatorial profile of the geoid by two symmetrical forms. It would be desirable to lift this restriction and to improve the approximation by going out of the plane of the equator. By solving this problem we could perhaps separate the primary source at great depth from the secondary one lying at a higher level.

(ii) On the basis of phenomena observed at the surface, conclusions can be drawn concerning physical characteristics of the materials of the outer core. From the speed of movements and from the effects of forces we can compute the viscosity of the material of the outer core and its compressibility.

(iii) Seismological investigations are generally carried out on the assumption of a radially structured earth. Recent seismic investigations have shown that the inner core is a solid of well defined radius and that around the boundary of the inner and outer cores a transition zone of about 500 km width is present. It seems that these ideas are contradictory.

(iv) C. A. Wagner recently, by investigating the paths of geostationary satellites, placed an upper limit of 0.02 degree of longitude per year on the westward drift of the gravitational field [8]. This observation is in contradiction to the hypothesis of a connection between the magnetic and gravity fields, and to the supposition of a secular change of the gravity field. Further research in this direction is needed.

(v) Investigations concerning the interaction of core and mantle materials could lead to a better understanding of tectonism of our globe. Through the correct interpretation of the precise geoid figure the location and dimension of the deviations from hydrostatic equilibrium could furnish important information.

Answers to these questions concern not only the earth sciences but basic physics also. Thus, efforts towards obtaining a possibly accurate geoid figure represent a very important task not only from a practical, but also from a purely scientific viewpoint.

References

[1] K. Lambeck, Nederlands Geodetisch Tijdschr. Mar. 1972, No. 3, 41 (1972).
[2] H. Jeffreys, Geophys. J. Roy. Astron. Soc. **8**, 196 (1963).
[3] W. H. Munk and G. F. J. MacDonald, J. Geophys. Res. **65**, 2169 (1960).
[4] R. Hide and S. R. C. Malin, Nature **225**, 605 (1970).
[5] G. Barta, Nature **243**, 156 (1973).
[6] A. Sakuma, paper presented at 15th Gen. Assembly of IUGG, Moscow 1971.
[7] A. Sakuma, extract from Procès Verbaux du Comité International des Poids et Mesures 1973, and private communication 1972.
[8] C. A. Wagner, J. Geophys. Res. **78**, 470 (1973).

Space Research XIV — Akademie-Verlag, Berlin 1974

THE "ARCTIC-ANTARCTIC" PROGRAMME: PRELIMINARY RESULTS

S. K. TATEVYAN

The Astronomical Council of the USSR Academy of Sciences, Moscow, USSR

Results of the satellite observations obtained during the Arctic-Antarctic programme up to March 1973 are considered. The astronomical reduction of the simultaneous observations and the perspective of the programme are discussed.

The principal object of the "Geodetic Vector Traverse Arctic-Antarctic" programme is the determination of the lengths and directions of a chain of lines, connecting eight principal sites: Mirny (Antarctic), Kerguelen Island, Afgoy (Somaly), Khartum (Sudan), Helwan (Egypt), Uzhgorod (USSR), Zvenigorod (USSR), Barentsburg (Spitzbergen).

The directions of all these lines are supposed to be determined with the highest accuracy from no less than 50 simultaneous pairs of observations, and the length of each line is to be determined from special laser measurements. The lasers are sited at no less than three basic stations, serving two directions at a time (southern and northern). The mean length of each line is about 3000 km, and the total length of all lines is 16900 km. The geodetic distance between Mirny and Barentsburg is about 12400 km, that is, almost equal to the diameter of the earth. Assuming an accuracy of the measured distances of about 1 m, and directional accuracy of 1^s one may roughly estimate the relative error of each line as 1:500000, and the error of the total length as 1:1000000.

Precisely reduced lengths and directions of each line may be considered as a base-line for a new geodetic survey and used for an improvement of the ground survey in adjoining districts. After that an overall adjustment of all these individual geodetic nets will be possible. It would be very interesting to obtain a connection of this traverse with a similar one for the other hemisphere.

During 1972—1973 the stations listed in Table 1 participated in observations on the Arctic-Antartic Programm; all these stations are equipped with Soviet AFU-75 cameras except Potsdam (camera SBG) and Poznan (Poznan II). Table 1 shows the total number of simultaneous observations obtained during 1972 and early 1973 (according to the stations' monthly reports).

Simultaneous observations, obtained in connection with the "Arctic-Antarctic" programme and combined with previously obtained simultaneous observations give an opportunity to determine the directions of several lines of traverse:

Helwan—Afgoy	116 plates	Helwan—Riga	361 plates
Helwan—Zvenigorod	128 plates	Uzhgorod—Riga	436 plates

Uzhgorod—Zvenigorod	187 plates	Riga—Afgoy	13 plates
Uzhgorod—Helwan	150 plates	Riga—Zvenigorod	462 plates

In 1972 (May—October) a French laser was mounted at the Uzhgorod station (in the south of USSR). During this period several simultaneous laser-optical observations were obtained (see Table 2).

An astrometrical reduction of the simultaneous observations obtained in 1971 at the Soviet stations has been made, and two volumes of the Catalogue involving more than 500 reduced plates of ISAGEX have been prepared. In 1973—1974 photographic observations along this traverse will be continued, mainly on the southern lines. Laser observations at Helwan station are scheduled to start in the autumn of 1973.

Table 1

Arctica—Antarctica, Simultaneous Observations: Number of Plates
January 1972 to March 1973

Station	Pageos 66056-1	Midas-4 61028-1	Geos A 65089-1	Geos B 68002-1	Explorer 19 63053-1	Explorer 39 68066-1
Riga	607	136	43	8	—	—
Zvenigorod	326	101	51	2	3	—
Uzhgorod	191	105	144	7	—	—
Helwan (Egypt)	457	155	8	—	11	15
Khartum (Sudan)	168	64	2	—	7	2
Fort-Lamy (Chad)	—	3	7	—	7	—
Afgoy (Somaly)	88	32	—	1	—	4
Kerguelen	—	—	—	—	—	—
Mirny	—	—	—	—	—	—
Baija (Hungary)	120	7	—	3	—	—
Poznan	12	—	—	—	—	—
Ondrejov	—	3	6	—	5	—
Potsdam	—	—	—	—	—	—
Bucharest	—	—	—	—	4	—
Plana (Bulgaria)	8	11	—	—	—	—

Table 2

Number of Plates

Station	Geos A	Geos B	BEB	BEC
Uzhgorod	208	14	1	10
Riga	48	—	—	—
Zvenigorod	32	—	—	—
Sofia	7	—	—	—
Helwan	5	—	—	—
Ondrejov	32	3	—	—

The Earth's Neutral Atmosphere

Space Research XIV — Akademie-Verlag, Berlin 1974

HIGH-LEVEL CIRCULATION STUDIES BASED ON RAWINSONDE, ROCKETSONDE AND SATELLITE OBSERVATIONS

F. G. Finger, M. E. Gelman and R. M. McInturff

National Meteorological Center, NWS, NOAA, Hillcrest Heights, Md, USA

While analyses of radiosonde and rocketsonde observations have provided the basis for many stratospheric investigations for well over a decade, satellite information now serves to complement the *in situ* observations. Attention here is focused on use of the various analyses between 15 and 55 km (100 and 0.4 mb) to delineate stratospheric warmings and circulation changes. Warmings since 1952 are reviewed, including comparisons of area affected, altitude and movement. Details are presented of the stratospheric warming event of early 1973. Latest methods of stratospheric monitoring are also noted. Mention is made of the several types of synoptic and dynamic investigations under way toward determining the causes and effects of warmings, especially with regard to tropospheric behavior.

The recent formation of the Eastern and Western Meridional Rocket Networks allows description of synoptic aspects of the high-level circulation in both northern and southern hemispheres. Cross-section analyses along these Networks are presented for selected days of 1972, illustrating the autumn and spring transition periods in the stratosphere and mesosphere. Inferences are drawn concerning the relative symmetry in thermal structure and circulation patterns of both hemispheres.

1. Introduction

Among the many problems regarding the behavior of the stratosphere and higher levels, perhaps the most important has to do with the genesis and effects upon the entire atmosphere of the wintertime stratospheric warming phenomenon. Another important problem involves the similarities and differences between the northern and southern stratospheres, especially during the respective winters and winter-summer (summer-winter) changeover periods. The tropical stratosphere is also of concern because it represents an area of possible interaction between hemispheres. Certainly the known tropical oscillations (the quasi-biennial, the annual and the semiannual) need more clarification. On the basis of past experience, it seems realistic to expect that further knowledge gained from the study of any of these problems will throw light on the others.

In spite of the many data problems, such as defining radiation error affecting high level *in situ* type observations, we now have as a background for studies at least two series of high-level constant-pressure charts. The northern hemisphere series published by the Free University of Berlin, including daily maps to the 10-mb surface and above, has been used by many researches for well over a

Table 1

Details of Stratospheric

Winter years	Warming intensity	Warming began	Warming moved	Warming peaked	Warming period (days)	Max. 10-mb Temp.
1951–52	Major	unk	unk	unk	unk	unk
1955–56	Minor	9 Jan	unk	20 Jan	unk	unk
1956–57	Major	19 Jan	E	6 Feb	27	−3°C
1957–58	Major	22 Jan	NW	30 Jan	21	−14°C
1958–59	Minor	23 Jan	NE	28 Jan	9	−32°C
1959–60	Minor	21 Feb	N	24 Feb	10	−19°C
1960–61	Minor	25 Jan	N	28 Jan	6	−16°C
1961–62	Minor	15 Feb	NW	17 Feb	5	−28°C
1962–63	Major	18 Jan	NW	27 Jan	19	0°C
1963–64	Minor	27 Jan	N	30 Jan	9	−23°C
1964–65	Minor	30 Dec	NE	6 Jan	13	−25°C
1965–66	Major	19 Jan	NW	31 Jan	43	−15°C
1966–67	Minor	29 Dec	N	5 Jan	25	−16°C
1967–68	Major	17 Dec	NW	1 Jan	29	−15°C
1968–69	Minor	29 Nov	N	14 Dec	23	−16°C
1969–70	Major	17 Dec	NE	2 Jan	49	−6°C
1970–71	Major	7 Dec	NE	7 Jan	48	−8°C
1971–72	Minor	8 Feb	N	15 Feb	24	−18°C
1972–73	Major	4 Jan	N	6 Feb	33	−12°C

Definitions are given in the text.
unk = unknown

decade. A series of objectively analyzed northern hemisphere charts at levels between 100 mb and 10 mb has been produced by the National Meteorological Center, NOAA, since 1964. In addition, the Upper Air Branch of the National Meteorological Center has published a five year series of 5-, 2- and 0.4-mb charts (35, 42 and 55 km) for a portion of the northern hemisphere [1]. Beginning in 1972, charts for the entire northern hemisphere have been analyzed primarily from rocketsonde observations, but also with the aid of satellite sounder measurements.

Although the stratospheric constant-pressure charts have necessarily been confined to the northern hemisphere, meridional rocket networks now allow some investigation of the tropics and southern hemisphere. The Experimental Inter-American Meteorological Rocket Network (EXAMETNET) has been in operation for a number of years [2]. Recently, this network has become part of the Western Meridional Rocket Network which extends from the Arctic region to middle latitudes of South America. The Eastern Meridional Rocket Network provides complementary information nominally from pole to pole in the eastern hemisphere.

In this paper we have elected to concentrate on the application of constant-pressure analyses to stratospheric warming studies and the use of cross section and time section analyses for inter-hemisphere comparisons and tropical latitude studies.

Warmings, 1952—1973

Polar temperatures			Strongest	Max. temp.	Springtime
Begin date	Peak polar	ΔT °C (pole)	E wind at 10 mb (major warm)	5 mb/2 mb (rocket data)	changeover
unk	unk	unk	unk	unk	unk
unk	unk	unk	—	unk	unk
unk	unk	unk	unk	unk	unk
−70	−30	+40	unk	unk	Mid May
−65	−50	+15	—	unk	Early May
−65	−55	+10	—	unk	Late Apr
−60	−50	+10	—	unk	Late Apr
−45	−35	+10	—	unk/−6°C	Mid May
−78	−15	+63	100 KTS	unk	Mid May
−78	−68	+10	—	−7°C/−5°C	Early May
−75	−55	+20	—	unk/−5°C	Early May
−81	−33	+48	120 KTS	+14°C/+14°C	Mid May
−82	−66	+16	—	−8°C/+10°C	Late Apr
−75	−19	+56	120 KTS	+3°C/+43°C	Mid May
−57	−49	+8	—	unk/+36°C	Late Apr
−77	−31	+46	140 KTS	+16°C/+38°C	Mid May
−68	−26	+42	120 KTS	+25°C/+35°C	Mid May
−68	−35	+33	—	+8°C/+2°C	Early May
−69	−31	+38	90 KTS	+11°C/+11°C	—

2. Stratospheric Warmings

The synoptic aspects of stratospheric warmings have been the focus of numerous studies. Details of these events were rather scarce during the 1950s since radiosondes seldom penetrated as high as the 10-mb level (30 km). However, with the development of the meteorological rocketsonde as an observational tool in the 1960s came a significant increase in the information on the warming phenomenon. For the first time the full vertical extent could be studied.

The information contained in Table 1 presents some details of northern hemisphere stratospheric warmings since the initial observation in 1952. For each winter, only that warming which was considered most significant is represented. The 10-mb level was used as a basis for categorizing the types of warmings because it is the highest level for which a long series of maps is available. Accordingly, major warmings are defined as those in which the 10-mb temperature approached or exceeded −15 °C over a significant area and for more than a brief period. They were also accompanied by a circulation reversal (easterly winds over the region of the North Pole). Minor warmings include events in which temperature changes are more limited, with temperatures at 10 mb exceeding at least −35 °C. These were associated with only limited circulation effects. It should be noted that warmings are observed to altitudes of at least 50 km and, in many cases, strong high altitude events may not be classified as major at 10 mb. However, all major warmings at 10 mb have been preceded, or accompanied, by strong warmings above that level. Thus, at least the entire stratospheric layer is significantly involved in major warming events.

Rather arbitrary criteria were used for definition of "warming began" and "warming moved". The former is based on the approximate date when 10-mb temperatures exceeded −35 °C in a cell which later became a major or minor warming, the latter on movement during the initial warming stages (approximately the first three days). In addition, the period of the phenomenon included only days when 10-mb temperatures exceeded −35 °C.

The peak temperature and temperature changes over the polar area provide a good index of the warming of a particular event, even though these values may have been exceeded in various other areas. Also of interest are the maximum temperatures at 5 and 2 mb during the warmings. This information from the rocketsonde network is based on considerably fewer observations than at 10 mb, where a great amount of rawinsonde data and indeed daily analyses are available. The 2-mb level temperature of +43 °C, which was measured by a rocketsonde at West Geirinish, Scotland, on 21 December 1967, appears to be the highest recorded by this type of observation. A wind of 182 m s^{-1} was recorded at a slightly higher level a week before (the strongest known wind of 198 m s^{-1} was measured at 39 km over Heiss Island on 1 February 1966). Unfortunately, very little rocketsonde temperature information is available for the warming of early 1963, which appears to be the most pronounced on record.

3. Stratospheric Warming: January—February 1973

As can be seen from Table 1 a major type stratospheric warming took place in early 1973. This event was traced continuously with the aid of charts constructed from radiosonde, satellite and rocketsonde data. A convenient method of monitoring the stratospheric thermal fields on a day-to-day basis is by using the observed radiances from the various satellite temperature profiler instruments.

The analyses in Fig. 1 depict the radiance patterns derived from the Selective Chopper Radiometer (SCR) and the Infrared Temperature Profile Radiometer (ITPR) instrumentation on board the Nimbus 5 satellite. The SCR data from this channel represent the weighted mean temperature in a layer centered at about 2 mb (42 km) and the ITPR, a layer centered near 30 mb (24 km).

It may be visualized that during a non-warming winter-time situation the higher SCR and lower ITPR regions would show low (cold) radiance values in the polar region, with segmented high (warm) centers near the lower middle latitudes (about 30 °N). There is a suggestion of this pattern in the lower (ITPR) field on 10 January. In contrast, the configuration of the upper SCR field indicates relatively warm air nearer the higher latitudes. Radiance charts such as these were used to monitor the mean stratospheric thermal fields so that the movement of the various thermal perturbations could be followed throughout the winter period.

Synoptic constant-pressure charts play an important role in monitoring events at particular pressure levels. The effect of the warming at levels throughout the stratosphere can be seen from Fig. 2. The 500-mb tropospheric surface is also

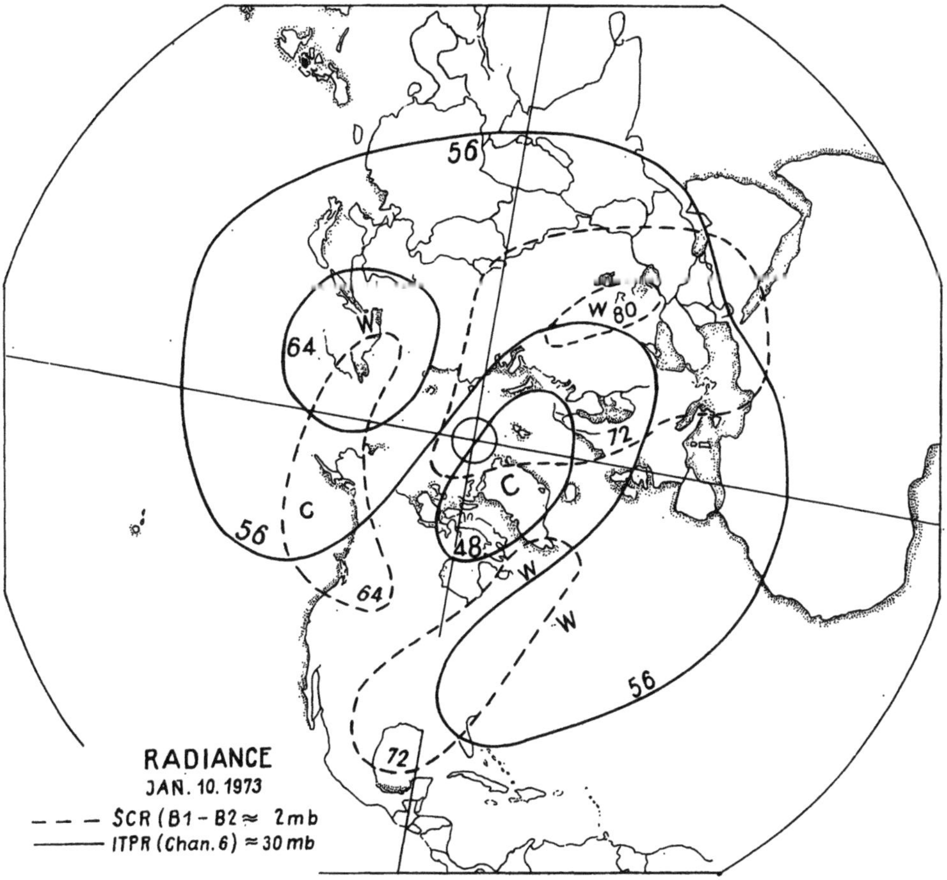

Fig. 1. Radiance fields for 10 January 1973 as measured by the Selective Chopper Radiometer (SCR) and the Infrared Temperature Profile Radiometer (ITPR). Peaks of the transmittance weighting functions are at 2 mb and 30 mb respectively. Contours are marked in mW (m sr $cm^{-1})^{-1}$. W = warm center; C = cold center.

included in the representation. Latitudinal mean temperatures were computed from each of three days of northern hemisphere constant-pressure charts. The most significant feature is the behavior of the temperature field with time and height. Warming at the higher altitudes is obvious from the 10 January 1973 curves at 0.4 and 2 mb and to a large extent at 5 mb. However, the major thrust of the warm air in the 100- to 10-mb layer occurred nearly one month later. It may be noted that the mean temperatures at 30° N throughout the 100- to 10-mb layer reached minimum values when the polar area was warmest. This effect has been noted during previous warmings. As the warm air spread over the polar region, the colder air was displaced southward. Hemispheric (30°—80° N) mean temperature profiles are also shown in Fig. 2. These values have been cosine-weighted for latitudinal distribution. The profiles clearly show the warmer air

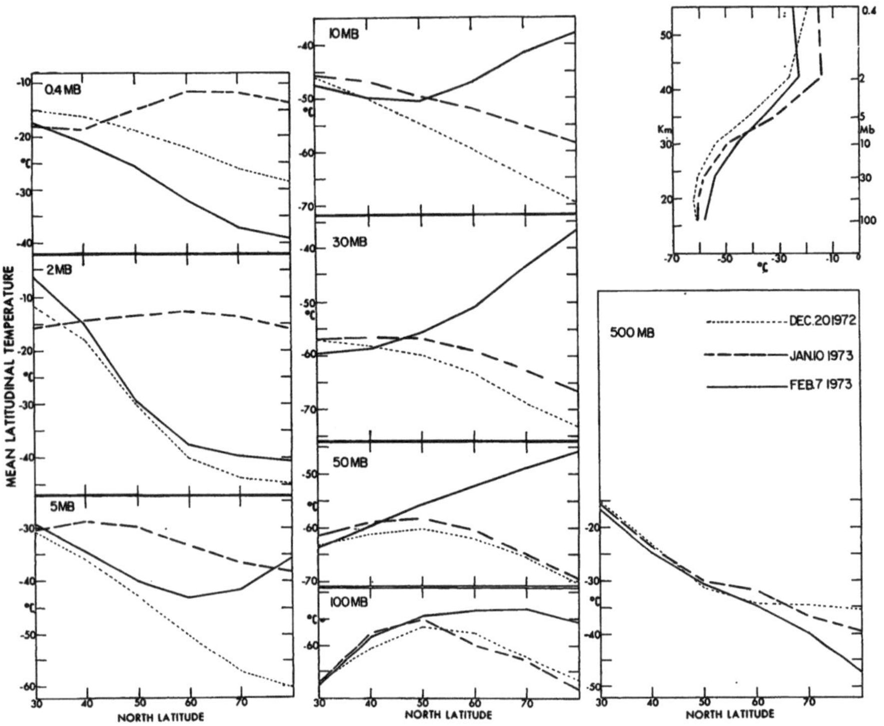

Fig. 2. Mean latitudinal temperatures in °C as functions of pressure surfaces, latitude and date. Upper right: mean northern hemisphere temperatures (30° N—80° N; cosine-weighted for latitude) in °C as a function of pressure or height.

at higher levels above 10 mb on 10 January, before large-scale warming took place below 10 mb.

Warming was not evident at the tropospheric level of 500 mb during the course of events in the stratosphere. In fact, on 10 February, when temperatures were highest at 100 mb, the reverse was true at 500 mb. Whether the cold air over northern latitudes at 500 mb is associated with the warming phenomenon, or due to independent tropospheric behavior, is a problem that requires more investigation.

Additional information on the behavior of the warm air with height, time and latitude is included in Fig. 3. These height cross sections are oriented along 60° W longitude across the Pole and down 120° E longitude; generally they coincide with the major axis of the warm air within the 100- to 10-mb layer. Contours in this diagram show differences between analyzed temperatures at the various constant pressure surfaces and the US Standard Atmosphere Supplements, 1966, and are thus indicators of warming intensity.

A significant area of warm air was noted above 10 mb on 27 December. By 3 January the center was located over Siberia and considerable intensification occurred during one week. The warming at high altitudes reached its climax within the next week as indicated by the 10 January panel. At the same time,

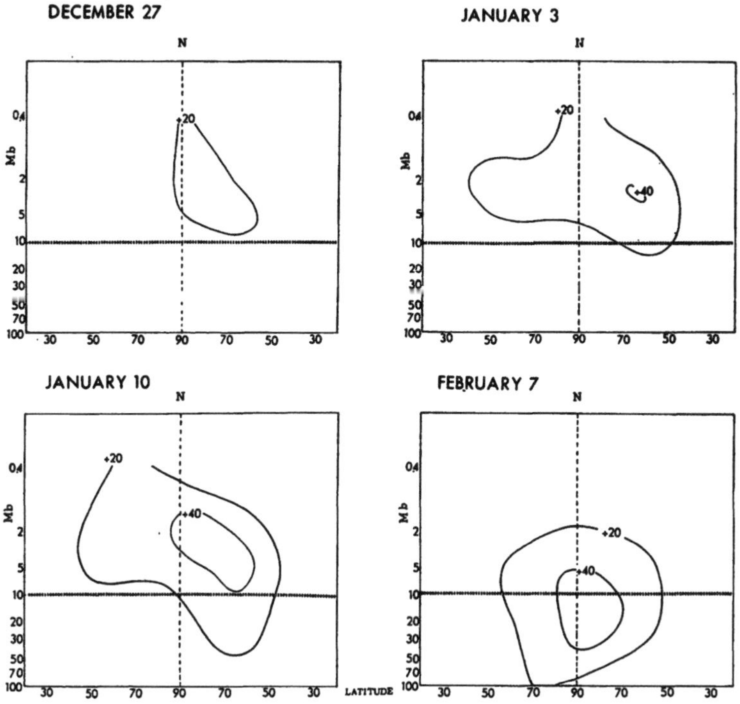

Fig. 3. Behavior of warming in northern hemisphere for selected days as a function of pressure surface (height) and latitude. Cross sections are oriented along 60° W longitude (left of dashed line), across the North Pole and along 120° E longitude (right of dashed line). Contours represent differences between analyzed and winter model (US Standard Atmosphere Supplements) temperatures.

levels below 10 mb were first becoming affected by significant changes. Thus, sometime near this date the vertical depth of the warm air included practically the entire stratosphere. By 7 February, peak temperatures had been reached in the lower stratosphere and cooling at higher altitudes had returned temperatures to near pre-warming levels.

The behavior of the geopotential height fields at various surfaces can also provide valuable information on the vertical extent of significant atmospheric perturbations, since the differences or similarities in these fields are functions of mean layer temperatures. Mean hemispheric geopotential height (cosine weighted from 20° N to 90° N) on a daily basis from November 1972 through February 1973 is shown in Fig. 4. The curves are restricted in the stratosphere to the 10-mb surface since this is the highest level for which computerized maps are available for computational purposes. Most evident during the autumn season are the lowering heights at all levels due to the intensification of the winter-time polar cyclone, with most rapid hemispheric height changes occurring at 10 mb. These falls were abruptly reversed in late November as the polar low began to lose strength at stratospheric levels. The only suggestion of a relationship of the earliest perturbation with events at lower levels is a slight decrease in the rate of height fall in the troposphere. Mean daily temperatures changed significantly

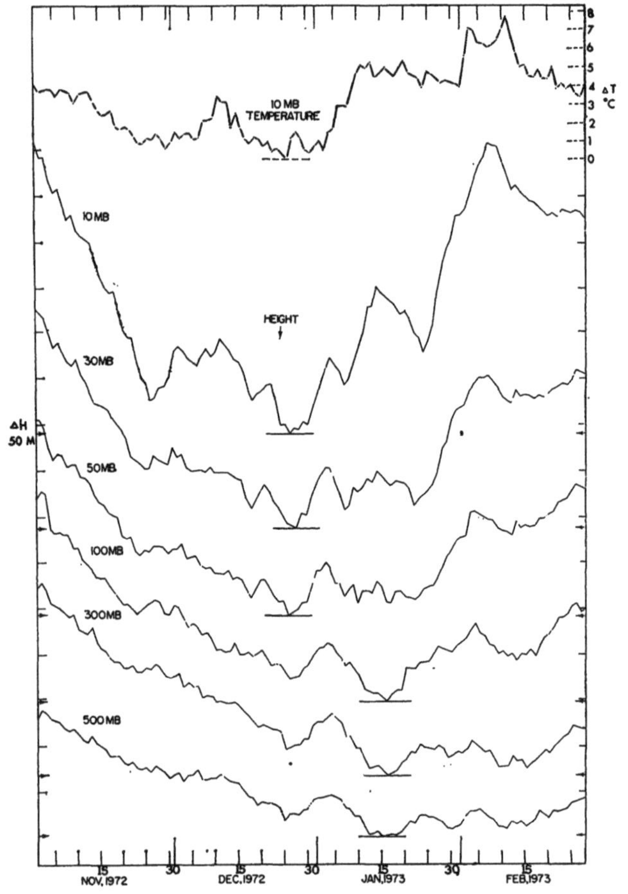

Fig. 4. Daily mean northern hemisphere heights (20° N—90° N; cosine-weighted for latitude) at various pressure surfaces. Horizontal line segments and arrows indicate minimum values. Ordinate scale at 50-m geopotential intervals allows calculation of height range. Daily mean hemispheric 10-mb temperatures are shown at top.

during this period. This is shown by the broken curve at the top the diagram. An inspection of temperature computations for other constant-pressure surfaces indicated that this perturbation decreased rapidly with decreasing height. Thus, it appears that the stratospheric change is not an obvious direct reflection of tropospheric events. This is not to say that tropospheric events did not precipitate the stratospheric changes, but the type of analysis under discussion here is certainly not capable of deciding this issue.

The pulse that occurred at all levels in late December is most interesting. One can conclude from hydrostatic considerations that at least part of the variation at higher altitudes was forced from below. This is a specific case where the entire atmosphere appears to be operating in concert. It is difficult to deduce a lag in the trend at any level, but there is some evidence that height rises in the troposphere led those in the stratosphere by a few days. The December pulse at 10 mb was clearly the first of a series that was associated with the major stratospheric warming. The relationship between the various levels during the later stages of the event is very complex, especially since the height fields even as high as 100 mb reached minimum winter-time mean values.

The relatively large amount of data available for the winter of 1972–1973 should allow us to determine the sequence of events before, during and after the warming event with more precision than heretofore possible. The data shown here combined with other computations, including those for specific latitude bands, should allow a rather complete synoptic description. This combined with planned dynamic studies [3, 4] should result in a significant increase in our knowledge of the cause and effects of stratospheric warmings.

4. Hemispheric Circulation as shown by Meridional Cross Sections

Since January 1972 rocketsonde data have become available from the Eastern and Western Meridional Networks. These networks with stations in both the northern and southern hemispheres are located near the 70° E and 70° W meridians. Rocket observations are scheduled for each Wednesday although reports for other days are sometimes also available. One of the primary functions of the networks is to provide data which allow comparisons of the stratospheric behavior of the northern and southern hemispheres. Information that meteorologists require from both hemispheres would allow answers to questions regarding the differences in the intensities of the thermal and circulation fields, the differences in the preferred areas of winter-time warm and cold air centers and, of course, the differences in modes of stratospheric warmings. Cross section analysis has already contributed to the climatology of the stratosphere [5]. We will concentrate here on depicting the transition in the stratosphere and lower mesosphere from winter-type to summer-type circulation regimes in the northern hemisphere, and on the summer-type to winter-type changeover in the southern hemisphere. The latter depiction, however, is restricted since the networks still have not attained full pole to pole coverage. The analysis of the meridional cross sections during a portion of the 1972 period included consideration of the large amount of satellite sounding data from the Nimbus 4 satellite. Northern hemisphere charts at 5, 2 and 0.4 mb which were also based in part on satellite information were checked for consistency with the cross sections [6].

The cross section for 15 March (Fig. 5) shows conditions characteristic of this time of year, with fairly strong westerlies of the winter circulation prevailing in the northern hemisphere, and with weak easterlies typical of late summer in the southern hemisphere at Mar Chiquita, between 20 and 40 km height. The quasi-biennial oscillation is known to have been in its easterly phase at this time, accounting for the easterly winds over Thumba, Ascension Island and Fort Sherman in the lower stratosphere. As is to be expected, the stratopause is warmest near the equator, in both western and eastern hemispheres, with maximum temperatures near 0 °C at 45 km.

Spring and autumn have advanced in the northern and southern hemisphere respectively by the beginning of April (Fig. 6). There are now winds with weak easterly components at high and middle latitudes of the northern hemisphere extending to altitudes greater than 50 km, while Molodezhnaya, the USSR station in Antarctica, indicates moderately strong westerlies in the polar area. The warm cell near 60° N, between Volgograd and Heiss Island should be noted,

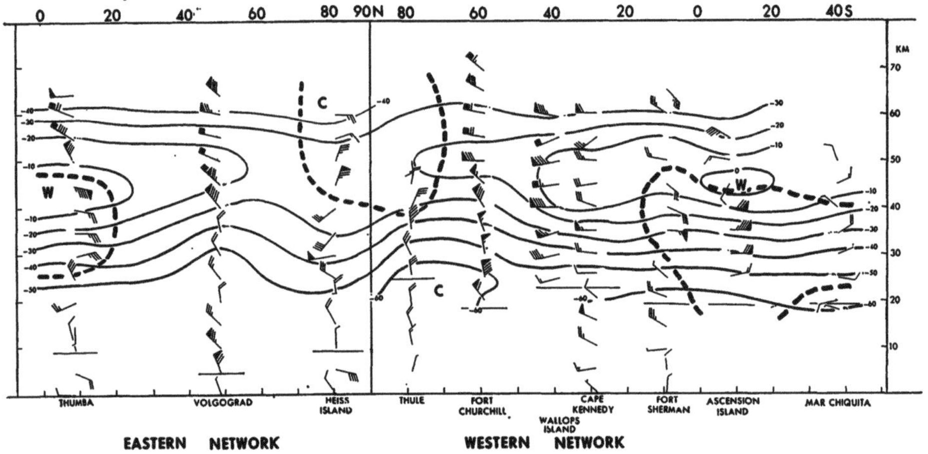

Fig. 5. Height cross section for Eastern and Western Meridional Networks for 15 March 1972. Solid lines are isotherms labelled in °C and heavy dashed lines separate winds with easterly and westerly components. Northerly winds blow from top of diagram: square flag 100 knots, pennant 50 knots, full barb 10 knots. Horizontal line segments show separation between radiosonde and rocketsonde data.

because typically the warm cells at the stratopause occur near the equator at the equinoxes and then, as the season advances, they appear in the vicinity of the summer Pole.

Finally, by the end of May (Fig. 7) the transition from winter to summer in the northern hemisphere and from summer to winter in the southern hemisphere is complete. Warm cells appear quite close to the North Pole, and easterly winds prevail throughout the stratosphere and lower mesosphere of the northern hemisphere. Mar Chiquita, Argentina, shows very strong westerly winds; its 60-km wind is in excess of 300 knots. It should be added that the wind shear reported

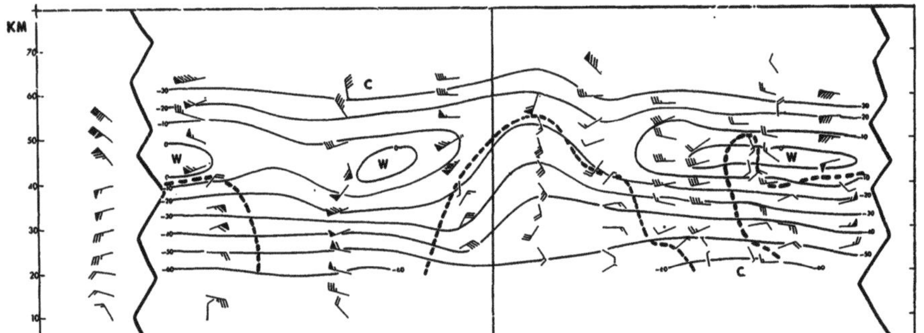

Fig. 6. Height cross sections for Eastern and Western Meridional Networks for 5 April 1972. Explanation as for Fig. 5. Vertical jagged lines indicate gaps in latitudinal representation.

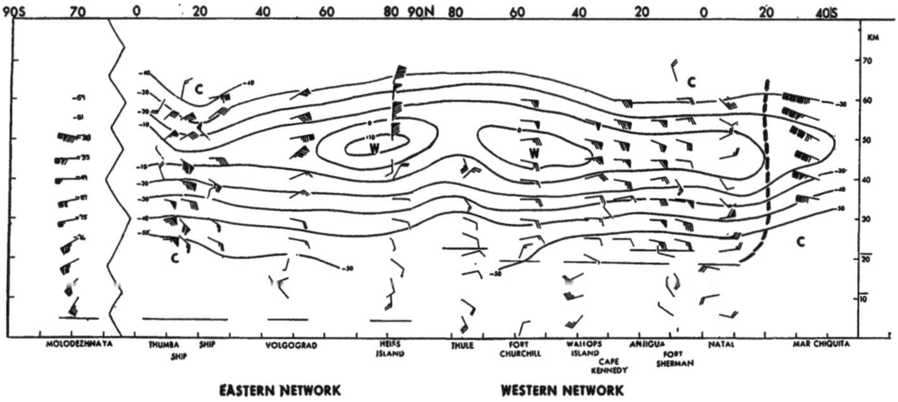

Fig. 7. Height cross section for Eastern and Western Meridional Networks for 31 May 1972. Explanation as for Figs. 5 and 6.

over Mar Chiquita on this date was checked against the wind shear calculated from satellite radiance measurements by means of the thermal wind equation and the two were found to be in general agreement.

Another transition period for which analyses are available extends roughly from 15 September to 30 November 1972. Thus, additional information has been provided on the relative symmetry between phenomena of the northern and southern hemispheres. For example, the springtime transition from westerlies to easterlies outside the tropics appeared to take place in a more regular fashion in the southern hemisphere than in the northern hemisphere. This sort of pattern has been indicated over the past ten years by means of different types of analyses [7], and is no doubt related to differences in dynamic and radiational warming patterns between the two hemispheres. These differences are probably related to differences in topographical characteristics [8]. In both hemispheres there is a marked tendency for the westerlies to replace the easterlies before the autumnal equinox, and to persist after the vernal equinox. Both winter-time westerlies and summertime easterlies appear to be stronger on the average between 20 and 60 km over the southern hemisphere than over the northern hemisphere, though more data are needed to confirm this.

5. Tropical Oscillations

It is well known that the tropical stratosphere is dominated by fairly regular oscillations with different periods [9]. Perhaps the most discussed is the so-called quasi-biennial oscillation. Others are the semiannual and annual cycles. Fig. 8 includes a time-height section of monthly mean zonal winds over Ascension Island, a rocket station located in the Atlantic Ocean at latitude 8° S.

The quasi-biennial oscillation appears prominently between 20 and 40 km, but gives way rather abruptly at high levels to the semiannual oscillation which has a maximum somewhere around 48 km. The semiannual is in turn replaced by a stronger annual cycle at 60 km. It is noteworthy that, if this figure is considered

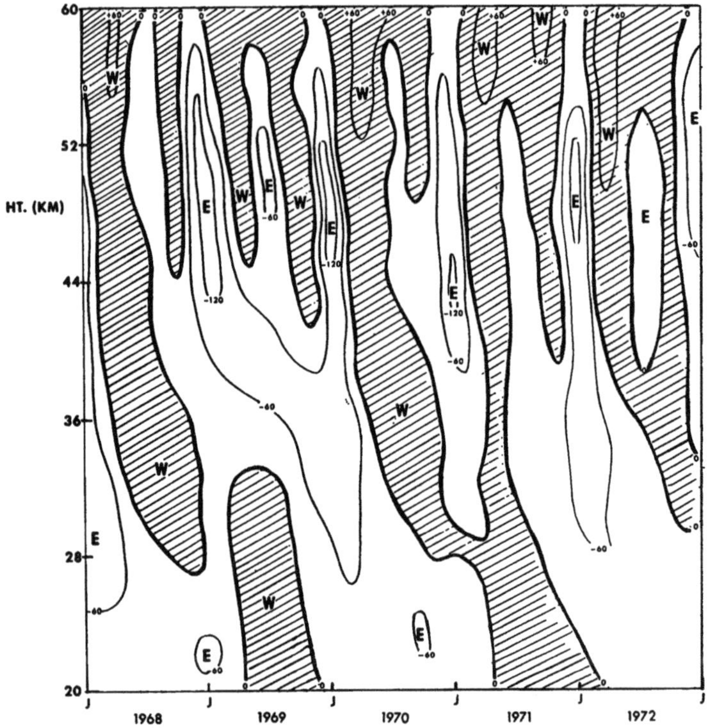

Fig. 8. Time height section of monthly mean zonal winds over Ascension Island, 1968—1972. W = westerly, E = easterly; areas of westerlies are shaded. Isotachs are given at 60 knot intervals with positive values for westerlies and negative values for easterlies.

representative of equatorial conditions (and all available evidence suggests that it is), then easterlies prevail about 70% of the time within the 20- to 40-km layer of the stratosphere. Near the stratopause, easterlies occur about 50%, and westerlies 50% of the time and near 60 km westerlies occur more than 70% of the time. This time section, together with the cross sections shown earlier, illustrates several features of the high atmosphere. For example, in the stratosphere beyond tropical latitudes, westerlies prevail most of the time and are stronger than the easterlies when they do occur. At equatorial latitudes easterlies tend to dominate the stratospheric circulation regime and are stronger than the less frequent westerlies. In the lower mesosphere, however, westerlies appear to be dominant at all latitudes.

It must be stressed that the semiannual, annual and quasi-biennial cycles are present at all times and probably at all levels. Thus, it is meaningful to speak of the occurrences of maxima and minima in these oscillations, while it is not meaningful to refer to the presence of one to the exclusion of others. In particular, it is noteworthy that the annual cycle has not been removed in Fig. 8. If it were, both the quasi-biennial and semiannual cycles would show up even more clearly. The advantage of this particular representation, in which the only smoothing used is due to the taking of monthly averages, is that the contribution of each of the three cycles to the resultant wind field is simultaneously exhibited. A disadvantage is that it gives only limited quantitative information, and, in order to obtain more, one must resort to harmonic or spectral analysis. However, it can readily be seen that such time sections are very useful in the analysis of meridional cross sections, since for the latter it is important to know the phase of each of the regular oscillations.

References

[1] Staff, Upper Air Branch, National Meteorological Center, NOAA Technical Reports, National Weather Service, Nos. 2, 3, 9, 12 and 14 (1967—1971).

[2] EXAMETNET Executive Committee, NASA SP-293, 1972.

[3] T. Matsuno, J. Atmos. Sci. **24**, 387 (1971).

[4] A. J. Miller, J. A. Brown and K. A. Campana, Quart. J. Roy. Met. Soc. **98**, 730 (1972).

[5] R. J. Murgatroyd, in: The Global Circulation of the Atmosphere, Roy. Met. Soc., London 1969 (p. 159).

[6] F. G. Finger and M. E. Gelman, Space Research XIII, 147 (1973).

[7] F. G. Finger and H. M. Woolf, J. Atmos. Sci. **24**, 387 (1967).

[8] S. Fritz and R. M. McInturff, Mon. Weath. Rev. **100**, 1 (1972).

[9] J. K. Angell and J. Korshover, J. Geophys. Res. **75**, 543 (1970).

Space Research XIV — Akademie-Verlag, Berlin 1974

OPERATIONAL RADIANCE MAPS OF THE STRATOSPHERE, WITH PRELIMINARY DETAILS OF A MAJOR STRATOSPHERIC WARMING

R. S. QUIROZ[a], M. P. WEINREB[b] and D. Q. WARK[b]

[a]National Meteorological Center, NWS, NOAA, Washington, D.C., USA
[b]National Environmental Satellite Service, NOAA, Washington, D.C., USA

Direct use of radiances measured by the Vertical Temperature Profile Radiometers (VTPR) on the NOAA 2 satellite is being made to monitor the structure and circulation of the stratosphere to altitudes above 30 km. For this purpose, daily radiance maps for the northern hemisphere are being produced for the VTPR stratospheric channels (668 cm^{-1} and 678 cm^{-1}). The development of a major stratospheric warming was observed during January 1973. A significant phase in its development was the merging of an eastward-travelling thermal cell with a quasi-stationary thermal wave in eastern Siberia. Thereafter, poleward migration of a major warm-air system resulted in a reversal of the meridional temperature gradient in the lower and middle stratosphere. Thicknesses of the 100 to 2 mb layer derived from the radiation data were employed to deduce the circulation state in the upper stratosphere. An anticyclonic circulation prevailed in the Arctic during early February in the middle and upper stratosphere.

1. Introduction

The NOAA 2 satellite [1] carries two Vertical Temperature Profile Radiometers (VTPR) with nearly identical measuring properties. In addition to the routine generation of discrete-level temperature data up to the 10-mb level (about 30 km), daily radiance maps are being produced operationally for spectral CO_2 channels with transmittance weighting functions, $d\tau/dx(p)$, whose peaks lie in the stratosphere. These are channels 1 (668.5 cm^{-1}) and 2 (678.5 cm^{-1}), with maximum weighting near 30 and 65 mb, respectively.

Reduced weighting above the 30-mb level in the uppermost stratospheric channel imposes a limitation on the altitude to which temperature profiles can be "retrieved" reliably by inverse solution of the radiative transfer equation. Thus special techniques (§ 2) are used to infer the temperature and circulation state of the upper stratosphere directly from the radiation measurements.

2. Direct Interpretation of Radiance Data

Recent investigations have shown a high correlation ($0.85 < r < 1.00$) between the radiance measured in individual channels and both the temperature at an altitude near the peak of the weighting function, and the mean layer temperature over the approximate transmittance range 0.1—0.9 [2]. Because the weighting functions for the VTPR and SIRS stratospheric channels are very similar, it is possible to use VTPR radiances in regression relationships developed earlier using SIRS transmittance data. In the special case of an isothermal stratosphere, the radiances are equivalent, and the Planck equation can be used directly to estimate the stratospheric temperature. Finally, the ratio of the time rates of change in these two channels has been shown [3] to be useful for indicating the altitude and amplitude of warming in an explosive warming situation.

3. Radiance Data Processing

As described in [1], VTPR spectral radiances are mapped from 23 scan positions distributed symmetrically about the twelfth nadir position. Before mapping, all radiance measurements are normalized to the nadir by linear least-squares regression. The average of the four highest radiances within a "box", approximately 600 km square, of about 60 observations yields values to represent the entire box, while eliminating the effects of high clouds, which, experience demonstrates, depress individual radiance measurements. The representative radiances are plotted on polar stereographic maps of the northern hemisphere,

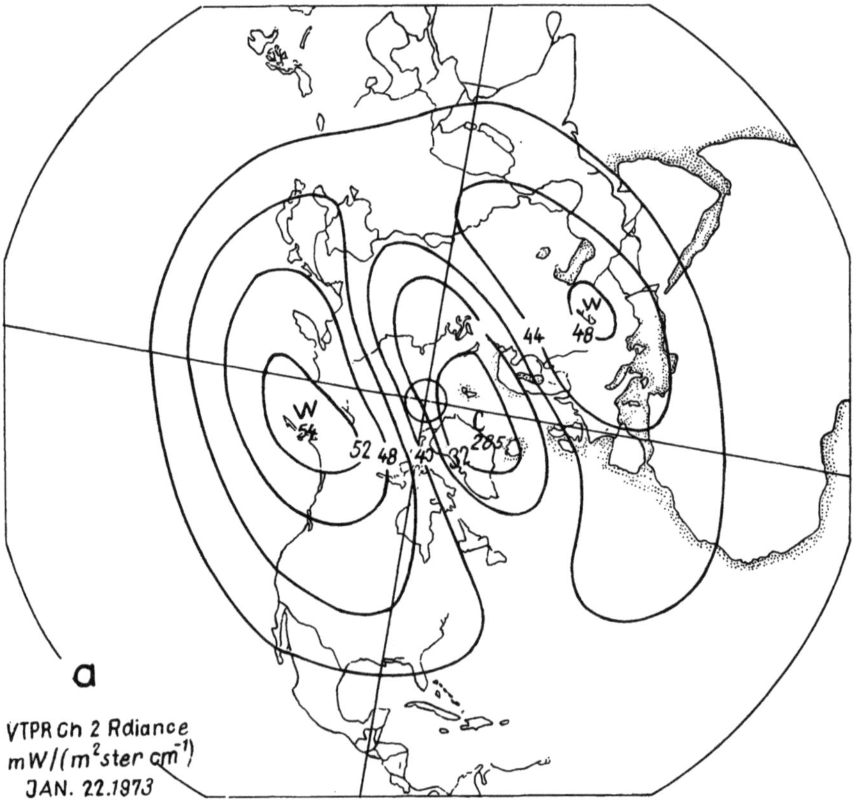

Fig. 1a–c. VTPR Channel 2 (678 cm^{-1}) radiance contour maps for selected dates, showing warm (W) and cold (C) regions. Contours are marked in mW (m^2 sr cm^{-1})$^{-1}$.

one map for each of the two channels. All entries on a single pair of maps are derived from observations in one 24-hour period centered at 0000 GMT. Each day's maps are produced before 2100 GMT, within nine hours of the latest measurements.

4. The Stratospheric Warming of January—February 1973

Radiance maps for the stratospheric channels were produced from mid-December 1972. A significant radiance increase occurred, mainly in middle latitudes, from 27 December to 8 January; but the greatest change in the radiance fields began in late January. Fig. 1 shows the Channel 2 radiance fields on 22 January, 29 January and 6 February. On 22 January an eastward-moving high-radiance system is located over southern Europe, and a quasi-stationary system is found over the Aleutian chain. Interaction of the traveling wave with the standing wave resulted in a single high radiance system on 27 January in northeastern Siberia, which then migrated poleward, with the system recorded by both Channels 1 and 2 centered near 90° N on 6 February.

The development and movement of high and low radiance systems was basically similar in the two channels. Fig. 2 shows the trajectory of the Channel 1 high

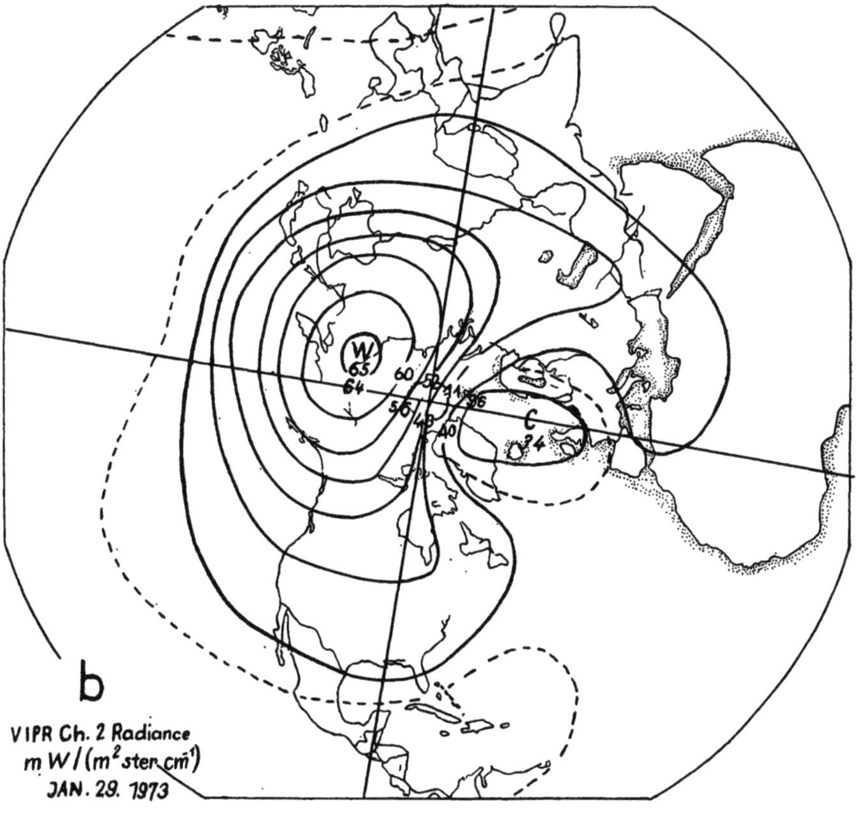

Fig. 1b

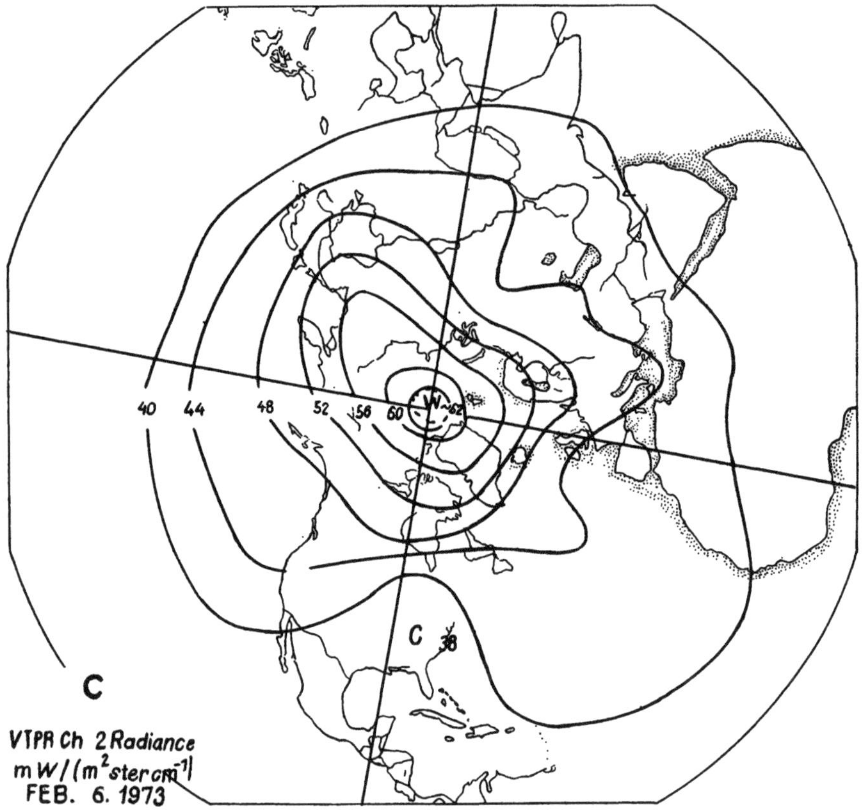

Fig. 1c

centers during several weeks. Note the curious northwestward excursion of the eastward-traveling system, from Africa to Newfoundland, in early January; during the same period the Far East system underwent a moderate northward displacement from southern Japan. In late January, the North Atlantic system accelerated eastward. The merging of the two cells (note their positions on 26 January) was completed by 27 January. The poleward displacement of the single high-radiance system then proceeded at a small fraction of the speed (5–10 m s^{-1}) of the arctic stratospheric flow ($\sim$50 m s^{-1}), yet in the general direction of this flow. Thus, while advection by the horizontal wind was a significant factor in this warming sequence, other factors such as vertical motion must be taken into account in explaining local temperature increases.

Wave interaction as a precursor to sudden or "explosive" warming has previously been depicted with satellite radiance data for the major warming of December 1969 [3]. Such evidence indicates the dominance of a wavenumber 1 planetary thermal pattern in the "sudden" phase of recent major events. In the 1973 warming, the sudden phase may be said to begin about 26 January and it applies primarily to high latitudes (see [4] in reference to the 1969 event).

At the North Pole, maximum increases were 22 mW (m^2 sr $cm^{-1})^{-1}$ in Channel 1, 25 January–1 February; and 29 mW (m^2 sr $cm^{-1})^{-1}$ in Channel 2, 22 January to

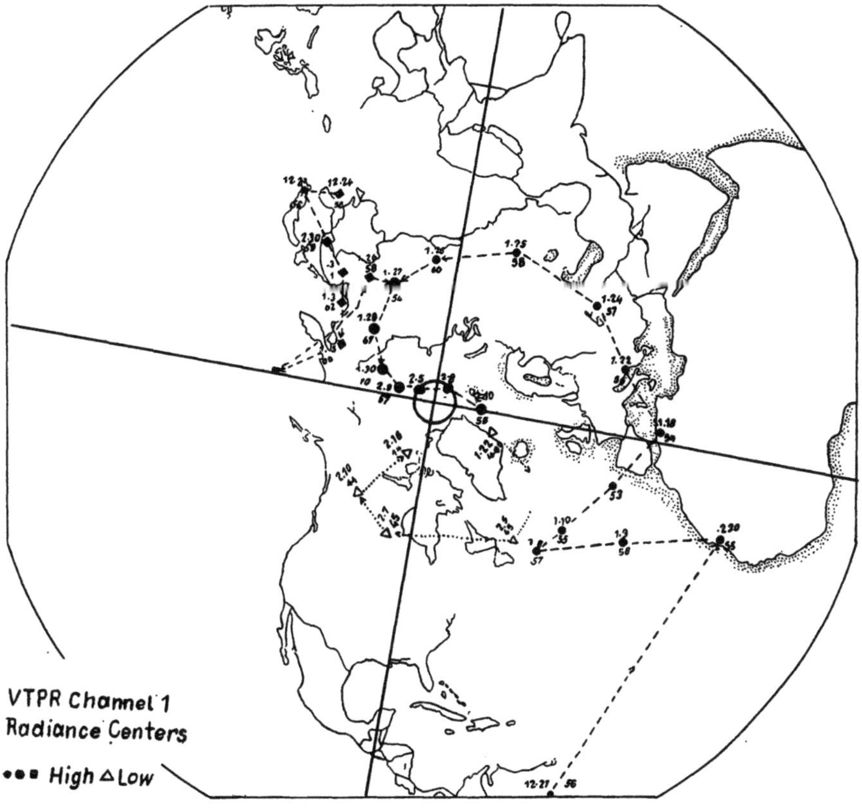

Fig. 2. Trajectories of Channel 1 (668 cm^{-1}) radiance centers. For each point the upper numbers show the month and day, and the lower number radiance values in mW (m^2 sr cm^{-1})$^{-1}$.

5 February. These changes correspond approximately to changes in the mean temperature of the stratosphere of +27 °C from 100 mb to 2 mb, and +38 °C from 100 to 5 mb, respectively. The temperature changes in more restricted regions of the stratosphere were, however, larger. Use of the radiance-change ratio (§ 2) indicates only slight warming above 40 km in the polar area at the end of January. This does not preclude the possibility of significant mesospheric warming [5]. Ratios computed over various time intervals from 25 January to 5 February indicate that at the North Pole and at the Soviet station, Heiss Island (81° N, 58° E), there was warming by less than 40 °C near 40 km to about 55 °C at 20 to 30 km.

The sudden phase of the warming, beginning about 26 January, is depicted for Heiss Island in Fig. 3. The upper time section shows the actual temperature, determined from weekly rocket observations together with more frequent 2-mb level temperatures estimated from the radiance data, and radiosonde data below 30 km. The lower diagram shows the departure from the December 1972 mean temperature. Three points are noteworthy: (i) minor warming of the upper stratosphere in early January; (ii) maximum positive departure near 25 km, about 60–65 °C, by 5 February, with weak upper stratosphere warming leading by

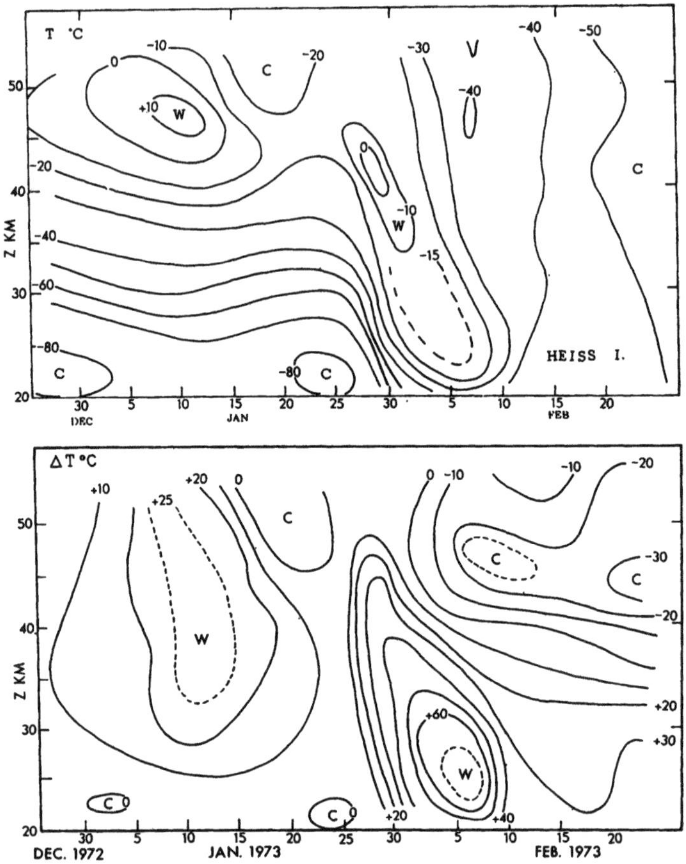

Fig. 3. Time sections of temperature (upper diagram) and temperature departure from December 1972 mean (lower diagram) at Heiss Is., based on combined rocket and radiosonde data and 2-mb temperatures estimated from VTPR radiances.

several days; and (iii) a quasi-isothermal structure after the sudden warming phase (upper diagram).

This time section provides qualitative verification of the altitude and amplitude of warming as specified from the radiance data, and moreover highlights the isothermal structure after the sudden warming. The apparent downward propagation of warming is evidently due largely to the poleward horizontal migration of the warm-air system.

The minor warming in early January was related to the northward excursion of the standing wave in eastern Siberia, 27 December—10 January (Fig. 2). At Ft. Churchill, Canada (59° N, 94° W) an increase of 40 °C at 45 km, indicated by rocket observations, can be shown to be related to the northward excursion of the high-radiance cell in the North Atlantic, in the same period.

The occurrence of maximum warming in late January in the *lower* stratosphere (in contrast to the event of December 1969 and a warming in December 1970, which were marked by maximum warming in the *upper* stratosphere), had a profound effect on the atmospheric circulation. Maximum warming at low stratospheric altitudes tends to lead to quasi-isothermal temperature profiles; accordingly, a near-equivalence of Channel 1 and 2 radiances was observed over a substantial area of the Arctic. Daily constant-pressure maps prepared by the National Meteorological Center indicate that, in accordance with the low-level warming, anticyclonic flow was established in the polar area as early as 4—5 February at the 10-mb level.

Graphs of the height of the 2-mb surface, constructed from thickness data derived directly from the radiance fields, as a function of latitude along selected meridians have been studied. Interpreted with the aid of the geostrophic wind equation, these indicate that in the polar region an anticyclonic circulation was established at the end of January at the 2-mb level, being symmetric about the pole and much stronger than at 10 mb on 5 February. By 11 February a weak cyclonic flow set in at 10 mb, and a stronger cyclonic flow at 2 mb. A strong cyclonic polar vortex was not observed at 10 mb until early March.

Acknowledgment

The authors thank Mr. H. M. Woolf for the use of his FORTRAN code for plotting polar stereographic maps.

References

[1] D. Q. Wark et al., in: COSPAR Symposium on Approaches to Earth Survey Problems through the use of Space Techniques, Akademie Verlag, Berlin 1974 (p. 463).
[2] R. S. Quiroz and M. E. Gelman, Mon. Weath. Rev. **100**, 788 (1972).
[3] R. S. Quiroz, J. Appl. Meteorol. **10**, 555 (1970).
[4] S. Fritz and S. D. Soules, Mon. Weath. Rev. **100**, 582 (1972).
[5] K. Labitzke, J. Atmos. Sci. **29**, 756 (1972).

Space Research XIV — Akademie-Verlag, Berlin 1974

THE STRATOSPHERIC-MESOSPHERIC CIRCULATION OVER THE NORTH PACIFIC OCEAN

I. Shimizu and T. Suzuki

Japan Meteorological Agency, Tokyo, Japan

Stratospheric-mesospheric circulation over the North Pacific Ocean in the winters of 1970/71, 1971/72 and 1972/73 are described based on rocketsonde data. Time sequences of temperature and wind at Ryori, Japan, show a close relation between the stratopause height and the wind variation. The circulation patterns are closely related to the features of the Aleutian High.

1. Introduction

Three winters have passed since the Japan Meteorological Agency commenced routine observation by means of the meteorological rocket MT-135P at Ryori (39°02′ N, 141°50′ E), Japan, on 5 August 1970 [1]. Successive observations thereafter were made every Wednesday (every other Wednesday in 1970). The launching was sometimes postponed to Thursday or interrupted by bad weather or other trouble. 112 launchings have been made and 86 observations obtained up to 6 May 1973. Contour maps of 1 mb, 2 mb and 5 mb, with isotherms, are constructed every week from rocketsonde data at Ryori and many other stations received in ROCOBs. The purpose of this paper is to present some features of the stratospheric-mesospheric circulation over the North Pacific Ocean based on the above-mentioned observations.

2. Time Sequence of Temperature and Wind at Ryori

Time sequences of temperature and the east-west component of wind in the winter of 1970/71 are shown in Fig. 1. The stratopause is shown by double dashed lines. The stratopause is higher in early December 1970 and strong westerlies prevail. A similar pattern is observed in early February 1971. On the contrary the stratopause is lower in early January 1971 and there are strong easterlies. The stratopause height decreases rapidly in early December 1971 and increases gradually thereafter. Westerly and easterly winds, which are not so strong in this winter, alternate with a period of about one month as shown in Fig. 2. The variation of the stratopause height is not so closely related to that of the wind in this winter.

In the winter of 1972/73 the higher stratopause corresponds to the stronger westerly and the lower stratopause to the easterly, as shown in Fig. 3, but features of their variation are somewhat different from those in the winter of 1970/71.

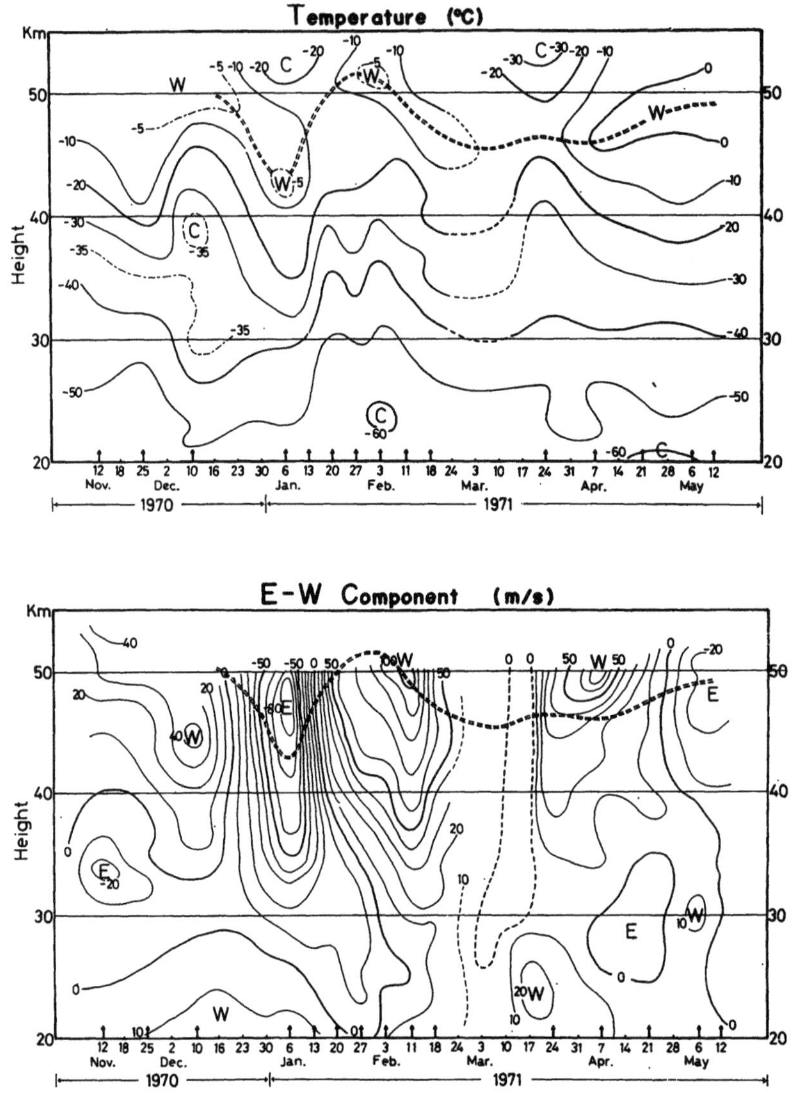

Fig. 1. Time sequence of temperature and east-west component of wind in the winter of 1970/71 observed by means of MT-135P rocket at Ryori, Japan. Arrows on the abscissa, observation dates; dashed lines, estimated values; double dashed lines, stratopause.

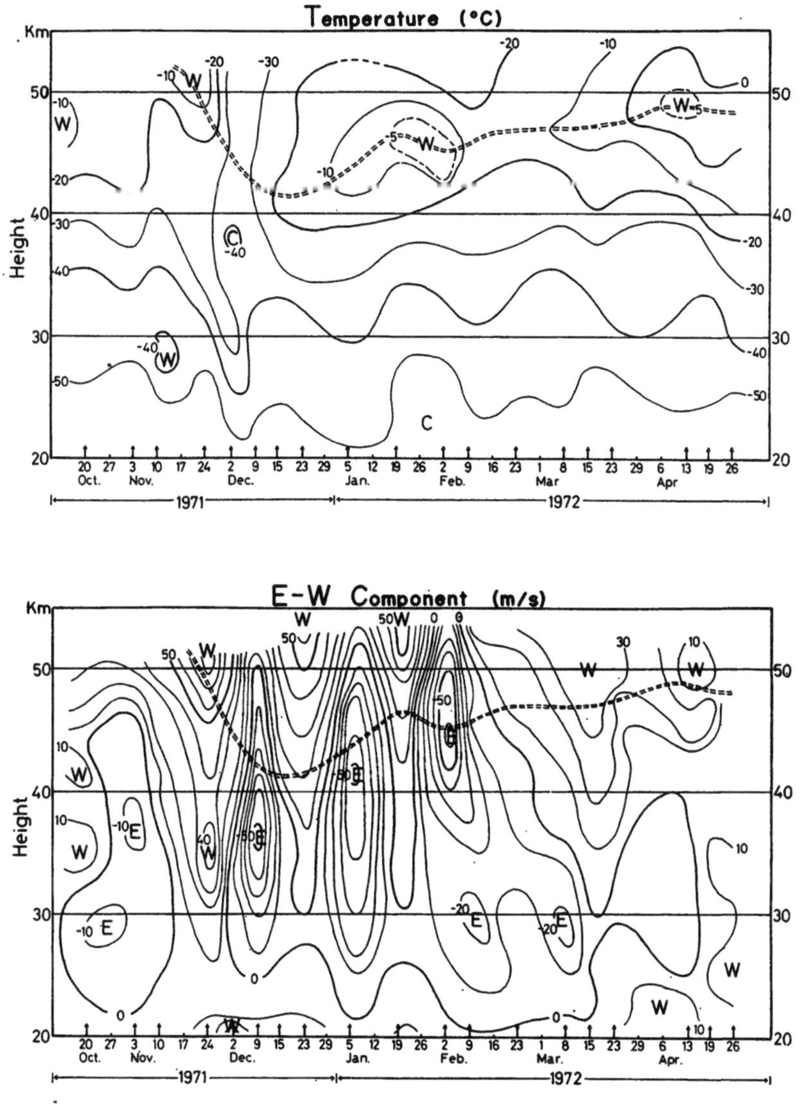

Fig. 2. Time sequence of temperature and east-west component of wind in the winter of 1971/72 observed by means of MT-135P rocket at Ryori, Japan. Key as Fig. 1.

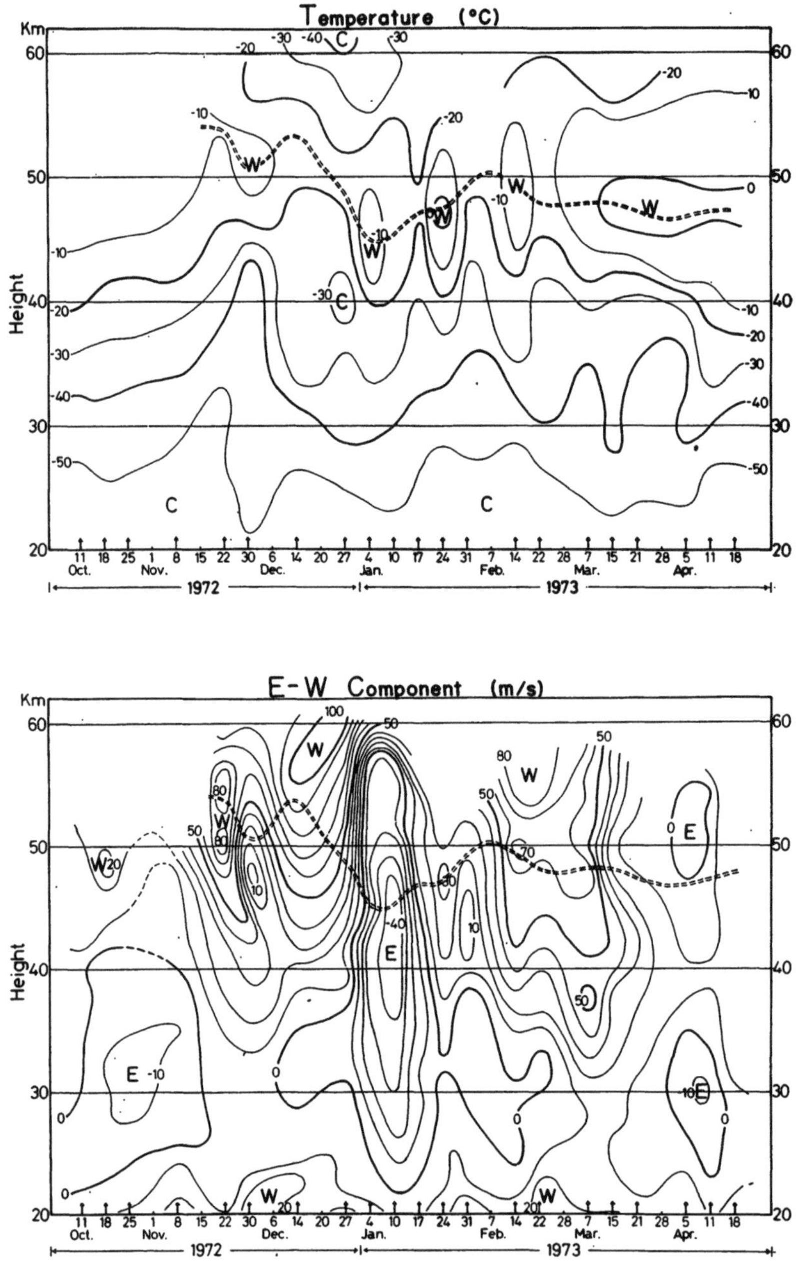

Fig. 3. Time sequence of temperature and east-west component of wind in the winter of 1972/73 observed by means of MT-135P rocket at Ryori, Japan. Key as Fig. 1.

3. Contour Maps and their Vertical Cross Section

Contour maps of 1 mb, 2 mb and 5 mb for the second week in December 1970 and their vertical cross section along 180° E (W) longitude are shown in Fig. 4*. Isotherms are shown by the dashed lines. The standard heights of the 45° N January model, derived from the US Standard Atmosphere Supplements, 1966 [2], are shown in Fig. 4d. The percentage departures from standard heights are also plotted.Height differences in the levels of 10 mb, 30 mb, 50 mb, 100 mb and 300 mb are taken from the weather maps sent from Institut für Meteorologie der Freien Universität Berlin. The circulation pattern in this week corresponds to the stage 0 of Labitzke [3]. The polar vortex prevails and the Aleutian High does not appear. A slight sign of the High is seen over the western part of Japan. The higher stratopause and westerly winds are observed at Ryori. Contour maps of 1 mb, 2 mb and 5 mb with isotherms for the first week in January 1971 corresponding to stage 2 of Labitzke have also been studied. The centre of the Aleutian High is over about 150° E and north of Ryori, which has a lower stratopause and a strong easterly as shown in Fig. 1. The vertical cross section shows that the higher level has maximum height in the higher latitudes.

Contour maps of 1 mb, 2 mb and 5 mb with isotherms for the fifth week in January 1973 and their vertical cross section along 180° E (W) longitude are shown in Fig. 5*. The Aleutian High is found to be over the polar region. The maximum height in a level above 5 mb appears near the pole. In this case the minimum height of the 500 mb level is found to be just at the latitude of the maximum height of the 1 mb level.

4. Remarks

Observation data at Ryori and some examples of contour maps are presented to show features of the stratospheric-mesospheric circulation over the North Pacific Ocean. Further studies are necessary to understand them.

Acknowledgments

We wish to express our thanks to Dr. Karin Labitzke for her invitation to present this paper at an open meeting of the COSPAR XVI, to Mr. F. G. Finger for providing us with ROCOB data, and to Mr. K. Saotome and Mr. S. Okimasa for preparing figures.

References

[1] I. Shimizu, Space Research XII, 607 (1972).
[2] US Standard Atmosphere Supplements, 1966, US Government Printing Office, Washington, D.C. 1966.
[3] K. Labitzke, J. Atmos. Sci. **29**, 757 (1972).

* Figs. 4 and 5 are on following pages.

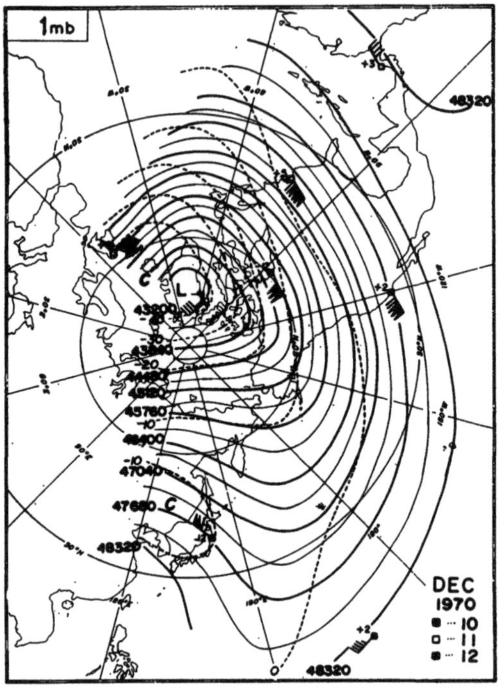

Fig. 4a. 1 mb map for the second week in December 1970. Solid line, height in geopotential metres; dashed line, temperature in C.

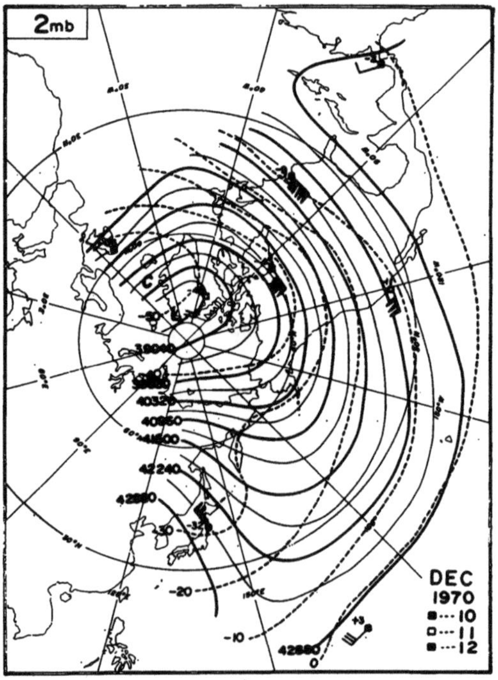

Fig. 4b. 2 mb map for the second week in December 1970.

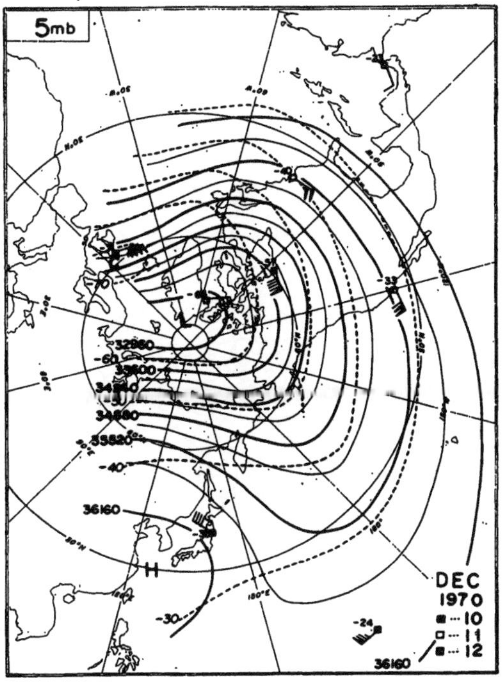

Fig. 4c. 5 mb map for the second week in December 1970.

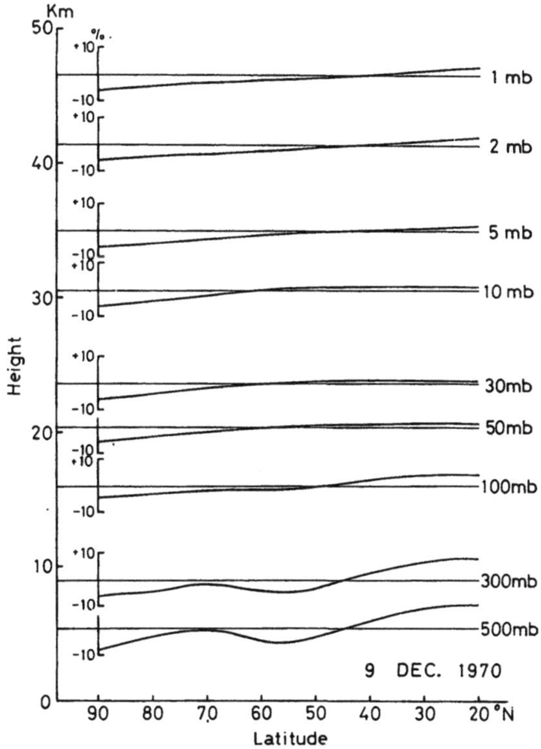

Fig. 4d. Vertical cross section of standard levels along 180° E (W) for the second week in December 1970. Height differences of the standard level are shown as a percentage with respect to the standard height of the standard level at 45° N in January derived from US Standard Atmosphere Supplements, 1966 [2].

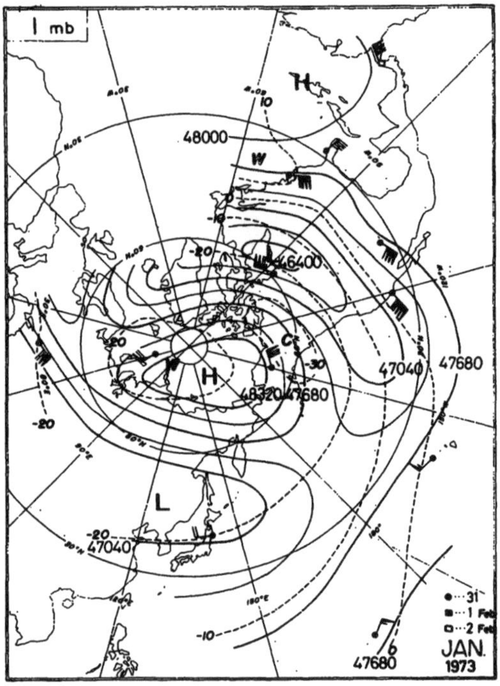

Fig. 5a. 1 mb map for the fifth week in January 1973.

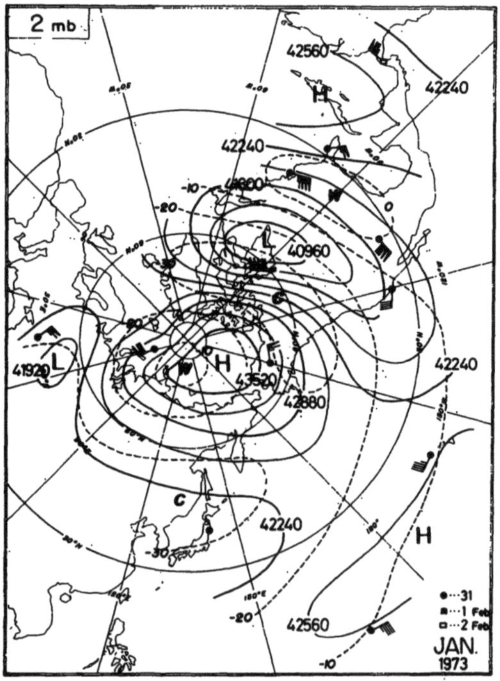

Fig. 5b. 2 mb map for the fifth week in January 1973.

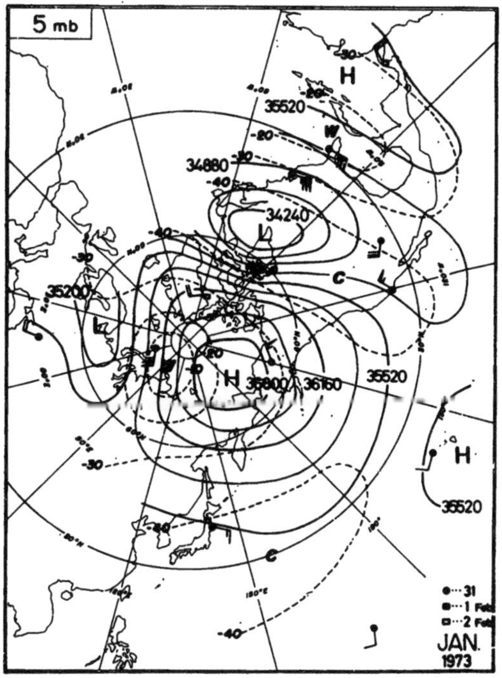

Fig. 5c. 5 mb map for the fifth week in January 1973.

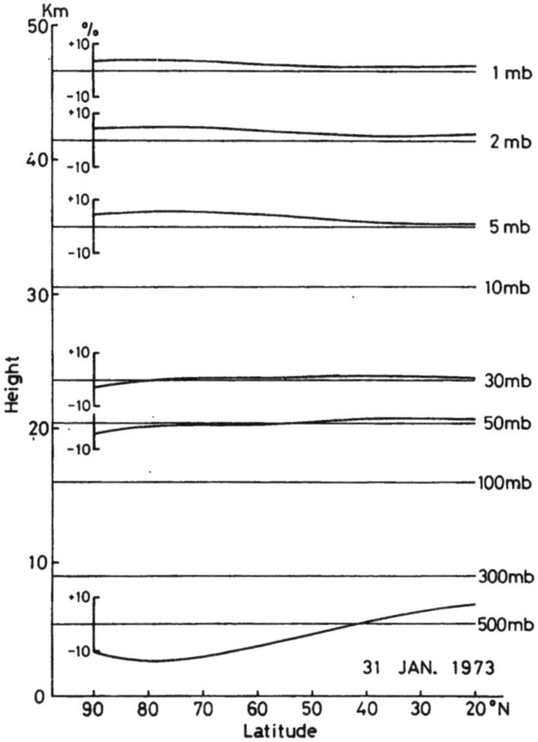

Fig. 5d. Vertical cross section of standard levels along 180° E (W) for the fifth week in January 1973. Height differences of the standard level as percentage, as in Fig. 4d.

Space Research XIV — Akademie-Verlag, Berlin 1974

DIFFERENCES IN CIRCULATION OF THE UPPER ATMOSPHERE IN LOW LATITUDES OF THE SOUTHERN AND NORTHERN HEMISPHERES

Yu. P. Koshelkov

USSR

The annual temperature variation in the tropical stratosphere is known to be in phase north and south of the equator. Differences in the magnitudes of the thermal and pressure gradients and the winds in the two hemispheres are discussed.

The in-phase annual temperature variations in the tropical stratosphere in the northern and southern hemispheres [1, 2] results in similar variations of density [1] or pressure. Thus, in low latitudes of the southern hemisphere, pressure values in winter are higher than in summer up to 55 or 60 km (Fig. 1), with a maximum amplitude between 40 and 50 km. As a result of this, annual variations of thermo-

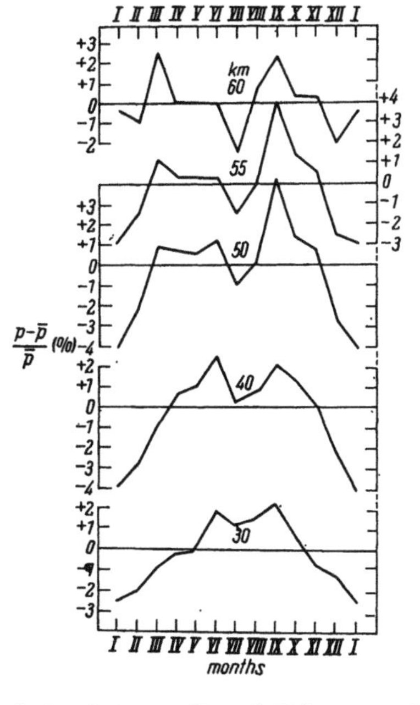

Fig. 1. Seasonal pressure variation between 30 and 60 km over Ascension Is. in 1965—1971, in terms of percentage deviation from the annual mean.

dynamic parameters in the low and high latitudes of the southern hemisphere appear to be out of phase. This in turn may have a marked effect on the magnitude of the horizontal temperature and pressure gradients. The anomalous rise of temperature and pressure occurring south of the equator in July and their lowering in January tend to smooth meridional temperature and pressure gradients in low latitudes of both hemispheres in July, and to strengthen them in January.

Calculations based on mean rawin data for the middle stratospheric levels indicate [3] that in low latitudes of the southern hemisphere in summer (January) temperature gradients are in excess of those in the northern hemisphere in the appropriate season (July). The reverse is true in the winter season. In summer, easterly winds in the stratosphere must grow faster with altitude south of the equator than north of it. On the contrary, in winter the growth of the easterlies with altitude must be faster in the northern hemisphere than in the southern hemisphere. This is illustrated in Fig. 2 showing the difference between thermal winds at the 30 mb level in the southern and northern hemispheres.

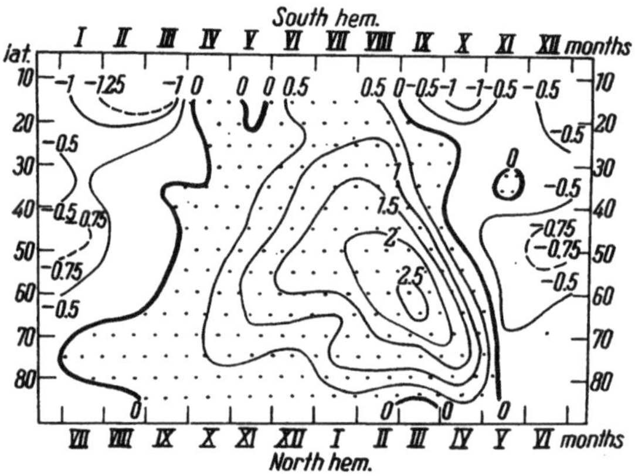

Fig. 2. Contours of thermal wind difference between the southern and northern hemispheres, marked in m s^{-1} per km, for the 30 mb pressure surface. The dotted region shows stronger westerly winds in the southern hemisphere.

This fact agrees well with the meridional pressure distribution in the upper stratosphere based on rocket sounding data. The number of rocket data are known to be substantially higher for the northern hemisphere than for the southern one where the bulk of rocket data have been collected at Ascension Island, in Argentina and by Soviet research vessels in the Indian and Pacific Oceans. Pressure values at 40 km have been calculated with the help of mean rocket temperature profiles and mean heights of the 30 mb surface [4, 5].

It may be seen that in summer the values of the meridional pressure gradients are similar in the two hemispheres in middle latitudes, while in low latitudes the magnitude of the gradients is higher in the southern hemisphere than in the northern. (Unfortunately, for the winter period the comparison of gradients in low latitudes is hampered by lack of reliable rocket data in the southern hemisphere and great scatter of results in the northern hemisphere.) Hence, the summer maximum of easterlies in the upper stratosphere in low latitudes of the southern hemisphere must be higher than that in the northern hemisphere while the winter tropical easterlies are expected to be weaker than in the northern hemisphere. Actually, in summer the wind speed in the upper stratosphere is found to be considerably higher south of the equator than north of it. It may be supposed that the discrepancy is not local but reflects differences in circulation between the two hemispheres.

Only scant information has been gained so far in middle latitudes of the southern hemisphere. One may note, however, gradual reduction with latitude of differences between hemispheres in the summertime easterly circulation which agrees well with the meridional pressure distribution represented in Fig. 3. The differences can still be traced at 30° latitude, while wind profiles in higher latitudes of the two hemispheres are in good agreement. Thus in summer the speed of the easterly stratospheric flow of the southern hemisphere does not differ much from that of the northern hemisphere in middle latitudes but exceeds the latter in tropical regions. Strong winds observed in low latitudes of the southern hemisphere determine the shift of the axis of the easterly flow towards lower latitudes in the southern hemisphere as compared with the northern hemisphere. So it seems not to be

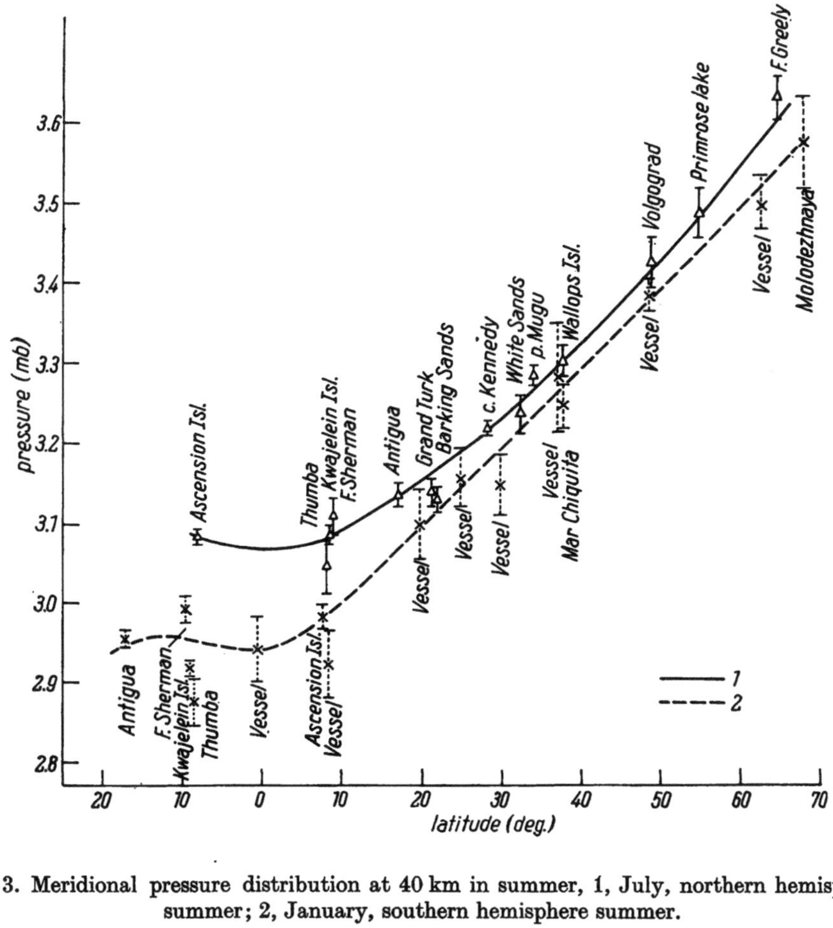

Fig. 3. Meridional pressure distribution at 40 km in summer, 1, July, northern hemisphere summer; 2, January, southern hemisphere summer.

justifiable to use rocket data obtained in one hemisphere for describing the wind pattern in the opposite hemisphere. The magnitude of the wind differences between hemispheres in summer, amounting to 20 or 30 m s^{-1} at 10° and 20° lat., makes it reasonable to take them into consideration when compiling atmospheric wind models. Values of wind speed deviations, in m s^{-1}, in the stratosphere of the southern hemisphere from the northern hemisphere data are given in Table 1 for the summer period (January and July, respectively).

Table 1

Height (km)	Latitude (deg.) 10	20	30	40	50
50	−30	−25	−15	−10	0
45	−30	−25	−10	−10	0
40	−15	−10	−10	−10	0
35	−10	− 5	−10	− 5	0
30	− 5	0	0	− 5	0

In winter, easterly wind speeds at 10° S are lower in the southern hemisphere at heights of 40 and 50 km than in the northern hemisphere. The comparison for 20° lat. is hampered by lack of data in the southern hemisphere. In higher latitudes the westerlies are known to be more intense in the southern hemisphere. Table 2 summarizes differences in zonal wind (m s^{-1}) between the southern and northern hemispheres in winter.

It follows from Tables 1 and 2 that in low latitudes maximum differences in wind speed between hemispheres are observed at a height of 40—50 km; this height corresponds to the level of maximum range of the anomalous annual pressure variation in low latitudes of the southern hemisphere (Fig. 1). This fact confirms the role of this variation in the development of wind differences between hemispheres.

When looking for reasons for the in-phase annual variations of meteorological elements in the stratosphere north and south of the equator, one should obviously agree with the conclusions [2, 6] about seasonal variations of vertical motions

Table 2

Height (km)	Latitude (deg.) 10	20	30	40	50
50	5	20	25	15	40
45	10	20	25	10	40
40	5	15	20	10	40
35	5	5	10	10	40
30	0	5	10	10	35

related to the Hadley cell, though direct evaluation of meridional and vertical circulations in the stratosphere on the basis of rawin observations is difficult [7]. The origin of differences between hemispheres in the intensity of Hadley circulation and possibly in the annual variation of meteorological parameters in the stratosphere lies in the underlying surface characteristics [6]. However there may also be a relation between the intensity of vertical circulation in the tropical stratosphere and the development of large-scale processes at higher latitudes.

The main differences between hemispheres in the intensity of these processes in the stratosphere are known to occur in the winter period. Frequent break-up of mid-winter zonal circulation accompanied by warmings have been observed in the stratosphere of the northern hemisphere, while in the southern hemisphere these processes are neither so intense nor so regular. Descending motions are commonly assumed to be a direct cause of stratospheric warmings. These may be accompanied by ascending motions in the tropical stratosphere of the northern hemisphere [8]. According to recent satellite data analyses [9, 10], cooling of the stratosphere has usually been observed in the tropical regions and in the summer southern hemisphere at least to 30° S simultaneously with the stratospheric warming in the northern hemisphere. This fact may be a reflection of the development of ascending motions in the tropical stratosphere of both hemispheres during the period of warming in middle and high latitudes of the northern hemisphere. As warmings are more frequent and intense during the winter period of the northern hemisphere than during that of the southern hemisphere, it may be supposed that ascending motions in low latitudes should attain greater intensity in January or February than in July or August. The resulting effect would be decrease of temperature over the equator at the beginning (and end) of the calendar year and increase in the middle of it. Hence the annual cycle caused by radiation will be strengthened north of the equator and radically changed south of it, which leads to the anomalous annual variation observed in this region.

References

[1] A. E. Cole, Space Research VIII, 823 (1968).
[2] R. J. Reed and Ch. L. Vlcek, J. Atmos. Sci. **26**, No. 1 (1969).
[3] Yu. P. Koshelkov, Trudi Tsent. Aerolog. Obs. **99**, 155 (1971).
[4] Yu. P. Koshelkov and N. I. Lisogufsky, in: Antarctica, Nauka, Moscow 1971 (p. 117).
[5] R. Scherhag et al., Meteor. Abh. Inst. Met. Geophys. **100**, H. 1 (1969).
[6] R. E. Dickinson, Mon. Weath. Rev. **99**, No. 6 (1971).
[7] A. H. Oort and E. M. Rasmusson, Mon. Weath. Rev. **98**, No. 6 (1970).
[8] J. S. Perry, J. Atmos. Sci. **24**, No. 5 (1967).
[9] S. Fritz and S. D. Soules, J. Atmos. Sci. **27**, No. 7 (1970).
[10] K. Labitzke, paper presented at XV COSPAR Plenary Meeting, Madrid 1972.

Space Research XIV — Akademie-Verlag, Berlin 1974

ATMOSPHERIC MERIDIONAL CROSS SECTIONS BASED ON DATA FROM EASTERN AND WESTERN ROCKET NETWORKS

S. S. Gaigerov, D. A. Tarasenko, V. V. Fedorov, M. Ya. Kalikhman, A. M. Kolomiitseva and B. P. Zaichikov

Academy of Sciences of the USSR, Moscow, USSR

Some results of the analysis of meridional cross sections of the upper atmosphere and upper contour charts based on meteorological rocket data of the western and eastern networks are presented. It is confirmed that the winter polar vortex and summer anticyclone spread up to the mesosphere while tropical anticyclones are confined to the stratosphere.

It is noted also that large-scale disturbances of the thermal field and circulation caused by winter warmings of planetary scale appear most significant in the mesosphere. Global cross sections of the atmosphere during the period of the disturbed circulations in the mesosphere and stratosphere are presented as examples.

Global meridional cross sections and charts of the northern hemisphere indicative of large temperature and wind variabilities in the mesosphere of the winter hemisphere are presented.

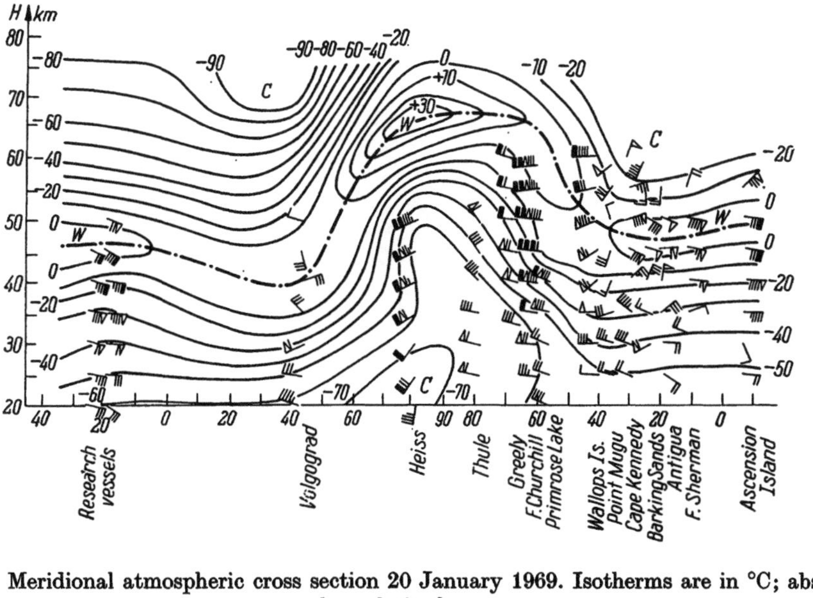

Fig. 1. Meridional atmospheric cross section 20 January 1969. Isotherms are in °C; abscissa shows latitude.

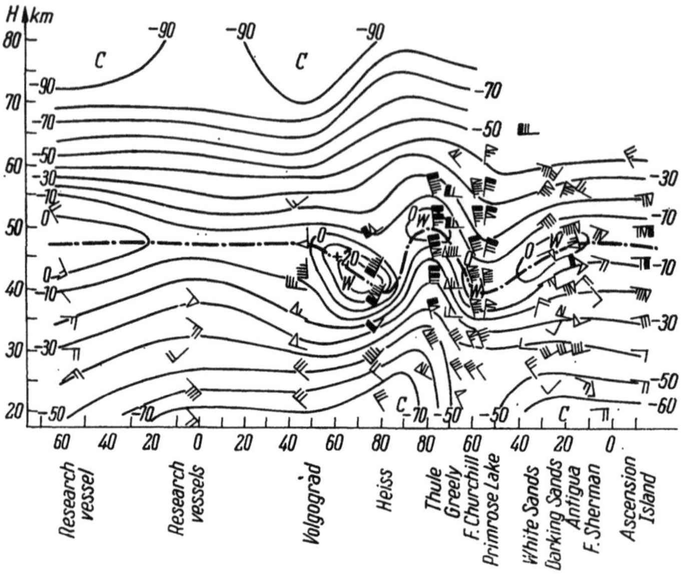

Fig. 2. Meridional atmospheric cross section 5 February 1969. Isotherms are in °C; abscissa shows latitude.

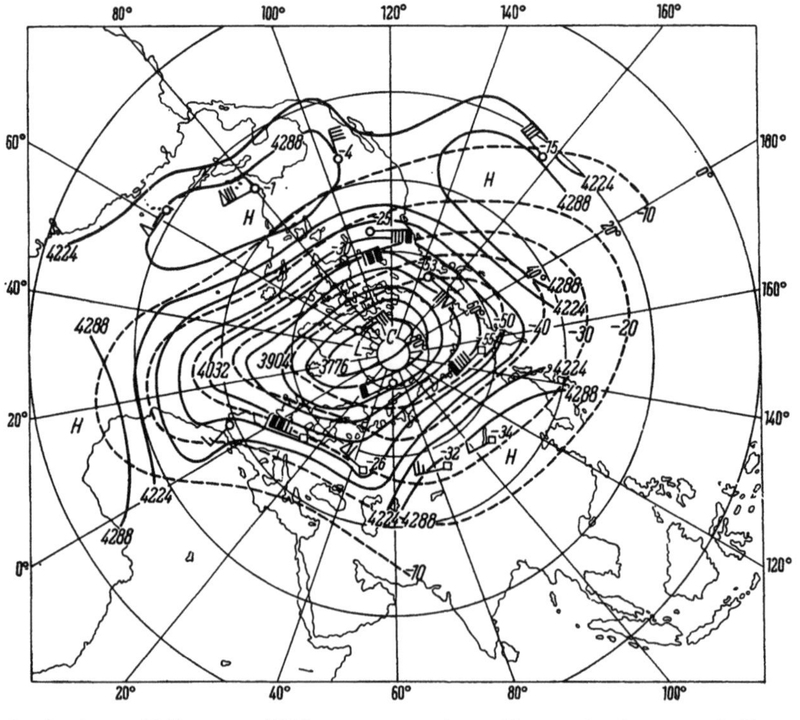

Fig. 3. 2 mb chart, 20 January 1969. ——— contours (decametres); – – – isotherms (°C).

As is known, orographic differences cause appreciable longitudinal differences in the temperature and circulation patterns of the stratosphere and lower mesosphere in the northern hemisphere. Studies of global processes in the upper stratosphere and lower mesosphere have been carried out by Labitzke [1], Finger and Gelman [2]. Using promptly exchanged rocket sounding data we are making a systematic analysis of charts for the upper stratosphere and lower mesosphere of the northern hemisphere as well as an analysis of global meridional cross sections, covering atmospheric layers as high as possible.

In Figs. 1 and 2 meridional atmospheric cross sections for 20 January and 5 February 1969 are presented. Besides data from the ground-based network of rocket stations, sounding by M-100 rockets from research vessels are used. The cross sections presented provide a spectacular illustration of the sharp temperature variations which may occur in the stratosphere and mesosphere over the Arctic region. It should be noted that the strong stratospheric warming observed in February 1969, had been discussed previously [3]. The important feature of Fig. 1 is the fact that the mesospheric warming, with its maximum over the Central Arctic, is global in character. Strong cooling in the stratosphere and lower mesosphere was also global in character. This is confirmed by the 2 mb chart for 20 January (Fig. 3), from which it can be seen that the centre of the winter cyclonic vortex practically coincides with the centre of the cold region over the pole, and is slightly shifted towards the Canadian Archipelago. In the subsequent period, the development of a stratospheric anticyclone over East Asia and its joining with

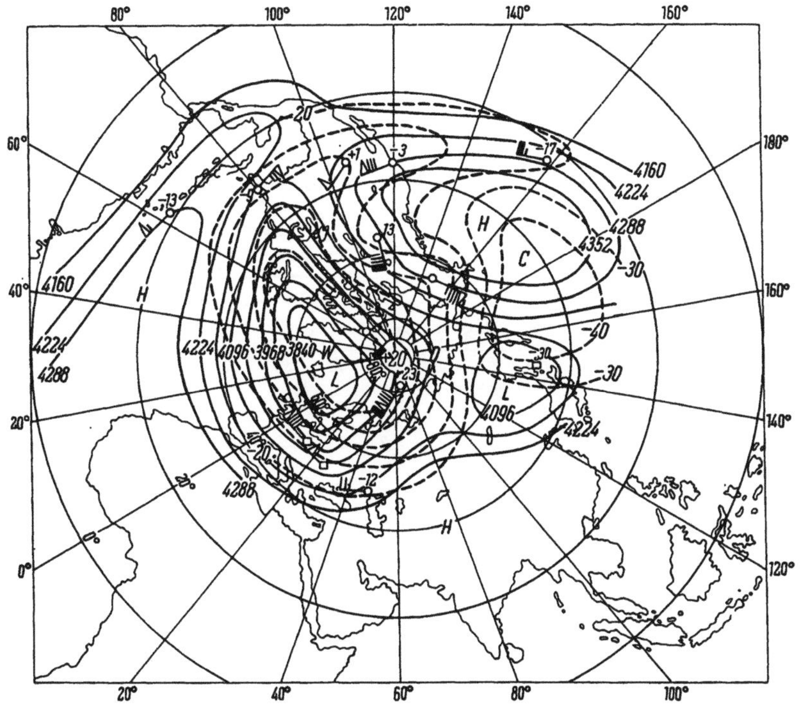

Fig. 4. 2 mb chart 5 February 1969. ——— contours (decametres); – – – isotherms (°C).

the high pressure region over the north Pacific occurred (Fig. 4). There was intensive warm advection on the periphery of the anticyclone, which resulted in the formation of a warm region in the Central Arctic stratosphere and in a temperature increase of up to +20 °C at the height of 45 km (Fig. 2). Thus, over the period from 20 January to 5 February 1969 a cold region over the pole changed to a warm region. From the time cross section for Heiss Island for this period, the warm region descending from mesosphere to stratosphere as well as variations of the stratopause altitude related to this process can be seen [3].

As another illustration of the high variability of the temperature field in the stratosphere and mesosphere in winter, the cross section for December 1972 (Fig. 5) may be given. This contains wind data up to altitudes of 75—80 km (at Molodezhnaya, Heiss Island and Cape Kennedy) which make the temperature field analysis at these levels much easier. Moreover, an attempt has been made to extend this cross section to the Antarctic. This seems to be possible under conditions of undisturbed summer temperature and wind patterns in this hemisphere.

For more detailed studies of large-scale processes in the mesosphere it is desirable to provide temperature and wind soundings up to heights of 80—100 km at all stations of the rocket network. For the best matching of rocket data from eastern hemisphere networks rocket-grenade experiment data are necessary.

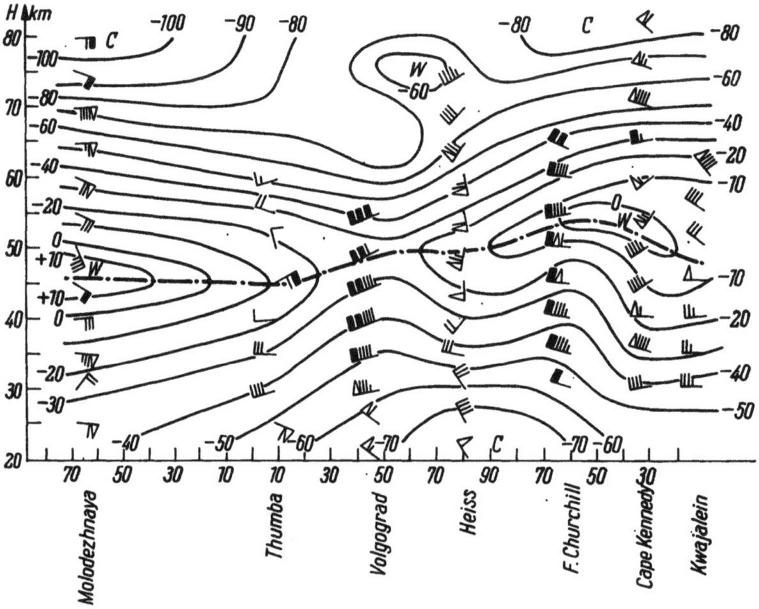

Fig. 5. Meridional atmospheric cross section 27 December 1972. Isotherms are in °C.

References

[1] K. Labitzke, paper presented at XV COSPAR Meeting, Madrid, May 1972.

[2] F. G. Finger and M. E. Gelman, Space Research XIII, 147 (1973).

[3] S. S. Gaigerov, L. M. Kolomiitseva and D. A. Tarasenko, Space Research XII, 637 (1972).

Space Research XIV — Akademie-Verlag, Berlin 1974

TEMPERATURE AND WIND VELOCITY VARIATIONS IN THE WINTER POLAR MESOSPHERE

S. S. GAIGEROV, V. V. FEDOROV, M. YA. KALIKHMAN and B. P. ZAICHIKOV

Academy of Sciences of the USSR, Moscow, USSR

Meteorological rocket data obtained in the polar regions reveal large temperature and wind variability in the winter mesosphere. The maximal range of temperature fluctuations amounts to about 40—80°C in 7 to 10 days in the mesosphere. Temperature fluctuations gradually diminish in the stratosphere towards a minimum in the lower stratosphere. Warmings (and coolings) in the upper atmosphere precede those in the lower layers. Warm and cold regions can be seen overlaying each other.

Rocket observations at the Molodezhnaya station show that the intensification of the westerly flow in the upper and middle mesosphere is usually accompanied by a warming, while during the decrease of westerly winds, as well as the growth of meridional wind components and development of easterly flow, cooling of the upper atmosphere has been recorded.

Meteorological rocket data obtained in the Arctic show large variability of temperature and wind in the winter mesosphere. Results of frequent observations made in January—February 1969 are the best indications of such variability [1]. Similar variations of temperature were also observed in subsequent years although they were not so large as in 1969 (Fig. 1).

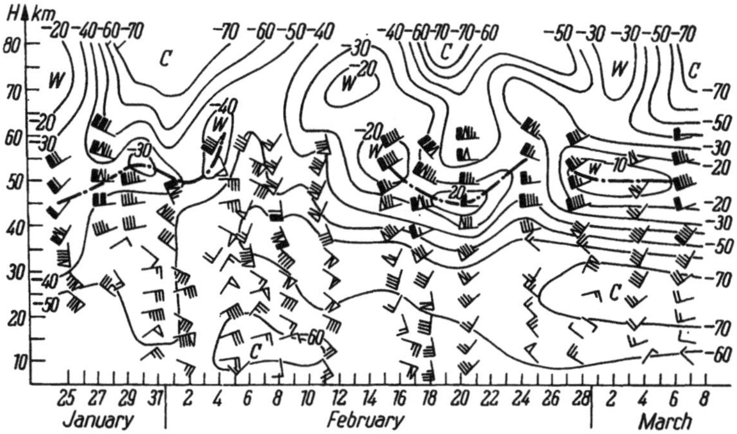

Fig. 1. Atmospheric time cross section for Heiss Island, Jan.—Feb. 1970. Temperatures in °C; W = warm, C = cold region.

Up to now studies of temperature and wind patterns in the polar mesosphere have only been made in the northern hemisphere. Temperature and wind rocket soundings made at the USA Antarctic McMurdo station were limited to heights of about 50—60 km [2]. In the middle of 1969, regular stratospheric and mesospheric sounding by M-100 meteorological rockets was started at the Antarctic station Molodezhnaya (68° S 45° E). Until May 1971 results of soundings consisted of temperatures for heights up to 80 km and wind velocity up to 50—55 km. Since May 1971, due to tracking of chaff drift, the height of wind soundings has increased up to 80—90 km.

As seen in Fig. 2, in the winter of 1971 in the Antarctic mesosphere significant disturbances of temperature and wind velocity distribution were observed, exceeding similar variations at lower levels. The period of oscillations at heights of 60—70 km (as far as they can be detected from relatively rare rocket launchings) ranged from one to three weeks, and changes of temperature were from 20 to 70 °C.

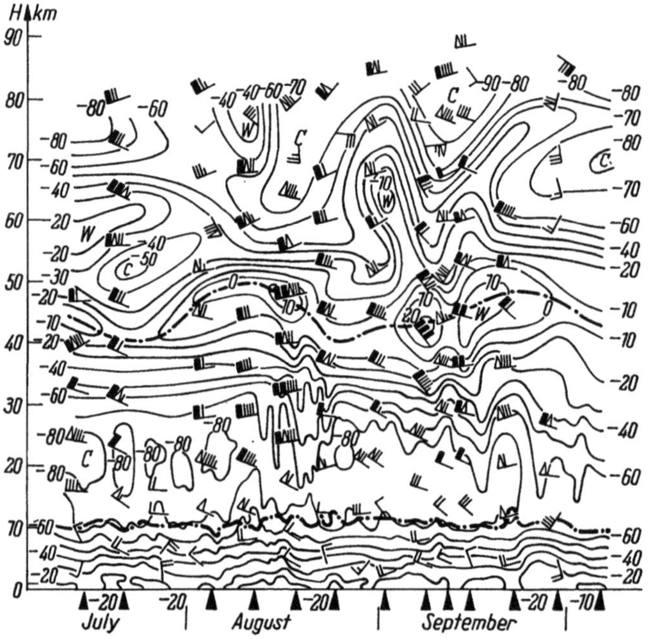

Fig. 2. Atmospheric time cross section for Molodezhnaya station (Antarctic), July—Sept. 1971. Temperatures in °C; W = warm, C = cold region.

The strengthening of westerly flows in the middle and upper mesosphere as well as in the stratopause layer were accompanied by warming, while the weakening of a westerly flow, the increase of meridional components and the appearance of easterly components coincided with cooling in this part of the atmosphere. Also, strong cooling in the upper mesosphere was accompanied by warming in the stratopause layer.

The above-mentioned features of temperature and wind variations enable one to conclude that, along with the relative stability of the southern hemisphere winter polar cyclone within the stratosphere and lower mesosphere, its intensity and centre position apparently vary in the middle and upper mesosphere. Again, from rather stable south-easterly flows that occurred in May—June 1971 in the upper mesosphere an assumption can be made about a disturbance of the cyclonic vortex and possible change to anticyclonic circulation. It should be noted that large temperature and wind variations occurring in the winter Antarctic mesosphere are also typical of northern high latitudes. This is evidenced by results of rocket soundings in Fort Churchill in January—February 1966 [3] and at Heiss Island in the winter of 1969 [1]. Upper-air charts for the southern hemisphere

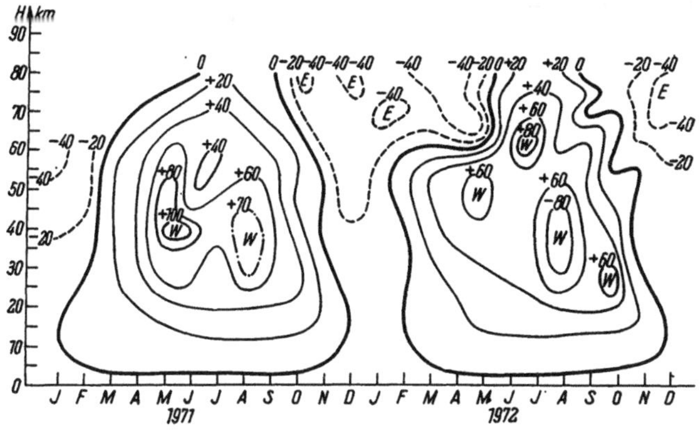

Fig. 3. Mean montly values of zonal wind component for Molodezhnaya, Jan. 1971—Dec. 1972

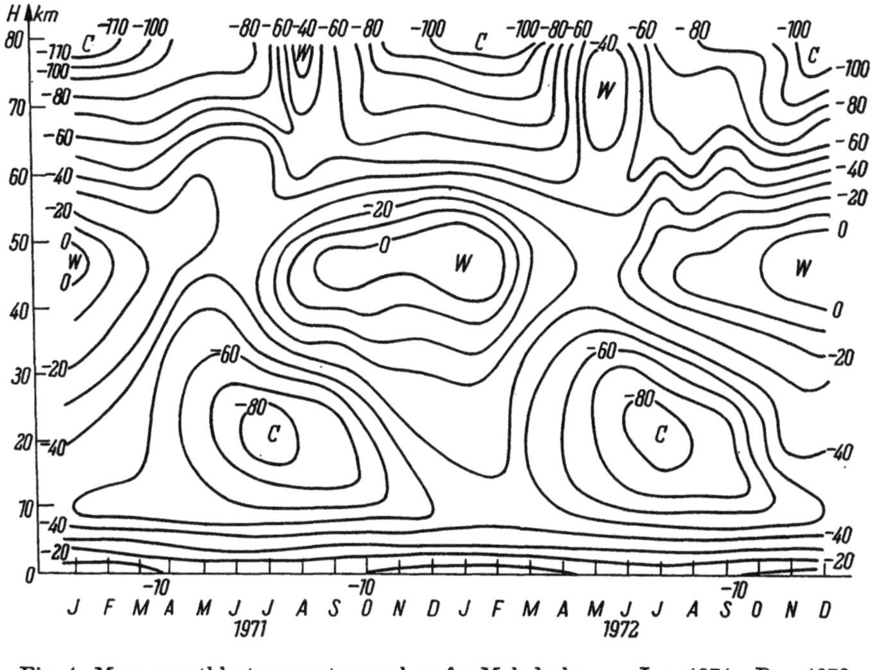

Fig. 4. Mean monthly temperature values for Molodezhnaya, Jan. 1971—Dec. 1972.

upper mesosphere are not available as yet. However as far as can be concluded from the preliminary attempts to construct 0.1 and 0.001 mb constant pressure charts for the winter of the northern hemisphere [4], in the upper mesosphere within the cyclone area the distances between contour lines prove to be much larger than in the stratosphere, and a high pressure area appears which is indicative of the possible mesospheric change of circulation in polar regions.

Features mentioned above are also supported by mean monthly temperature and wind values for the period from May 1971 to December 1972 (Figs. 3 and 4). A rather interesting fact is that in the upper mesosphere, as seen in Fig. 3, easterly wind components are predominant for 8—9 months. As can be seen in Fig. 4 these periods of easterly winds correspond to a period of low temperatures. Periods of westerly flow correspond to short duration mid-winter warmings in the upper mesosphere.

References

[1] S. S. GAIGEROV, L. M. KOLOMIITSEVA and D. A. TARASENKO, Space Research XII, 637 (1972).
[2] R. S. BRIGGS, J. Appl. Met., 4, No. 2 (1965).
[3] R. S. QUIROZ, Mon. Weath. Rev. 97, No. 8 (1969).
[4] S. S. GAIGEROV, M. YA. KALIKHMAN, V. E. SEDOV, E. G. SHVIDKOVSKY, D. A. TARASENKO and B. P. ZAICHIKOV, Space Research XI, 799 (1971).

Space Research XIV — Akademie-Verlag, Berlin 1974

STRUCTURE VARIATIONS IN THE WINTER POLAR ATMOSPHERE

A. C. FAIRE and E. A. MURPHY

Air Force Cambridge Research Laboratories, Bedford, Mass., USA

Neutral armospheric density and temperature results are presented that were obtained from two rocketborne 7-inch falling spheres launched at Churchill Research Range, Canada, on 10 December 1971 at 0846 and 1200 LT. Except for minor wave-like structure variations the successive sphere measurements are in very good agreement with each other. While the wavelengths of the fluctuations observed in each measurement remain approximately the same, a phase shift of about 5 km in the height profile is observed between the two. The sphere data are also in very good agreement with the 60° N January (cold) model from the US Standard Atmosphere Supplements 1966 up to 80 km where the model terminates. This is a significant result since it implies that the cold model may be related to quiet geomagnetic conditions.

1. Introduction

Two 7-inch rigid falling spheres instrumented with omnidirectional accelerometers were launched onboard Nike Iroquois rocket vehicles to determine atmospheric density and temperature over Churchill Research Range, Canada (lat. ~59° N) on 10 December 1971. The rockets were launched at 0846 and 1200 LT. The data appear to represent a relatively undisturbed atmosphere and are particularly significant since they imply that the cold January model from the US Standard Atmospheric Supplements, 1966 (USSAS-66) [1] may be related to quiet geomagnetic conditions.

Density results have been analyzed with reference to departures from the US Standard Atmosphere, 1962 (USSA-62) [2] and are discussed briefly in relation to similar measurements made at Churchill during 9—11 December 1965 [3]. Details of the experimental technique and instrumentation have been reported earlier [4—6].

2. Results

Densities. In Fig. 1 the sphere density data are plotted as percentage departures relative to the USSA-62 atmospheric model. Also plotted for comparison are the 60° N January (mean) and the 60° N January (cold) models from the USSAS-66 model. In addition, the sphere results are compared with those obtained from meteorological rocketsondes (metrocks) in the altitude region between 40 and 50 km where the measurements overlap. The sphere and meteorological rocket data of 10 December are in very good agreement.

Small scale variations in structure are observed in the stratosphere, mesosphere and lower thermosphere. Density departures from the USSA-62 model for successive sphere measurements lie within 0 to 8% of each other in the altitude range 45—90 km and within 0 to 14% in the region 90—98 km. Densities derived

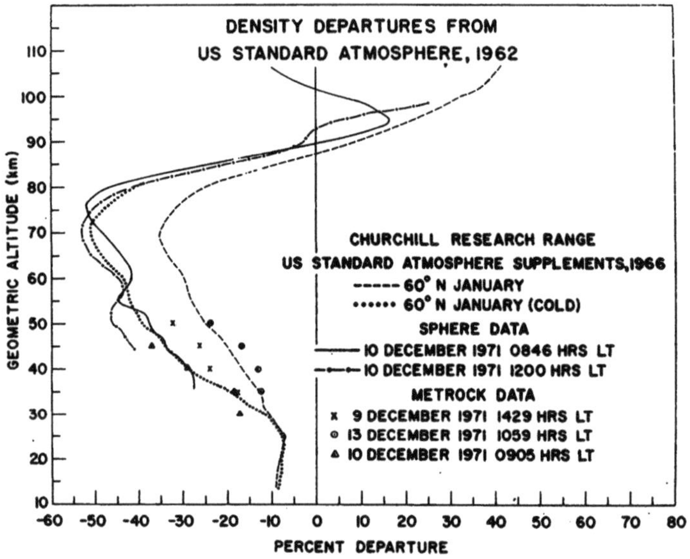

Fig. 1. Density departures from US Standard Atmosphere, 1962, 10 December 1971.

from both spheres are in very good agreement with those of the 60° N January cold model from USSAS-66 which terminates at 80 km. The wavelengths of the fluctuations observed in both sets of data are approximately the same except for a vertical phase shift of about 5 km between the waves of the two measurements. In the vicinity of 95 km the curve for the 0846 hours data lies close to that of the mean January model before sharply reversing its direction to attain a maximum departure of about 52% relative to the model at 105 km. Also, near 95 km the 1200 hours sphere curve intersects that of 0846 hours, but then continues close to that of the mean January model. The difference between the results of the two measurements may be due to atmospheric dynamics.

Temperatures. In Fig. 2 are plotted molecular-scale temperatures as a function of geometric altitude. It is not surprising to note that the temperature-height profiles exhibit more structure than those of density. This results from the fact that the temperature is quite sensitive to the slope of the density curve from which it is derived [5]. Except for fluctuations over small altitude increments temperatures derived from the 1200 hours density measurement are in good agreement with the cold model. Below the stratopause, temperatures from both spheres lie within 0 to 13 K of cold model values. In the lower mesosphere noon-time sphere temperatures are in somewhat better agreement with the cold model than those

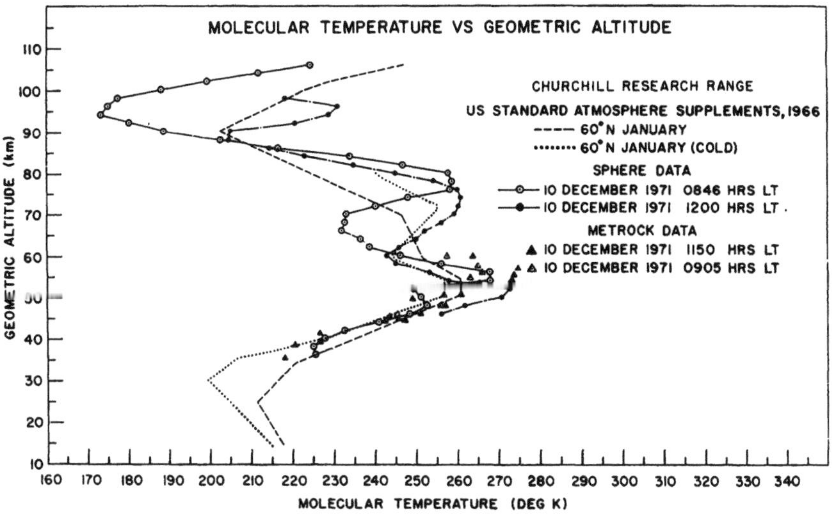

Fig. 2. Molecular temperature vs geometric altitude, 10 December 1971.

obtained in the morning. The 1200 hours sphere temperatures are within 0 to 5 K of the cold model values from 53 to 72 km and within 0 to 14 K from 72 to 80 km. In the lower mesosphere the 0846 hours sphere temperatures lie within 0 to 20 K of cold model values. In the region of the lower thermosphere (~95 to 100 km) the 0846 hours observed temperatures are as much as 30 K cooler than corresponding values of the normal January model. This large difference is attributable to the density perturbation discussed above.

3. Discussion and Conclusions

In general, the data presented in this paper seem typical of quiet or undisturbed atmospheric conditions, a result that is consistent with the relatively low level of geomagnetic activity (average $Kp = 2^-$) reported for the 12-hour period preceding the measurements. The agreement between the sphere results and the 60° N January cold model is remarkably good. These observations tend to support the tentative conclusion that the cold January model may apply to quiet geomagnetic conditions. The very good agreement between results obtained from two spheres launched three hours apart again demonstrates the reliability of the measurements. The data further provide information on the extent of short-term variability of the polar atmosphere under quiet geomagnetic conditions.

Variations similar to those discussed in this paper have been observed in the same altitude regime (80—100 km) at both high and mid-latitudes [7]. Referenced for comparison [3] is a set of falling sphere density measurements obtained at Churchill on 9 and 11 December 1965. These data lie between the normal and warm models from about 40 to 60 km and between the normal and cold models up to 80 km. The measurement made on 9 December at 1433 LT, shows a perturbation near 95 km that is less pronounced but similar to that discussed in this paper, whereas the two measurements made on 11 December 1965 at 1255 and

2255 LT show negligible perturbations in this region. This comparison is made to point out that both short-term and long-term perturbations are common in this altitude regime.

Acknowledgments

The authors gratefully acknowledge the valuable technical assistance provided by Gustave Stromberg, Paul Mundis, Milton Baratz and Fred Fritzler of Wentworth Institute, Boston, Mass., and Kenneth McGee of AFCRL. Also we wish to thank Dr. K. S. W. Champion, Chief of the Atmospheric Structure Branch, Aeronomy Laboratory, AFCRL for valuable and stimulating discussions and for reviewing the manuscript.

References

[1] COESA, US Standard Atmosphere Supplements, 1966, Government Printer, Washington, D. C.
[2] COESA, US Standard Atmosphere, 1962, Government Printer, Washington, D. C.
[3] COESA, US Standard Atmosphere Supplements, 1966, Government Printer, Washington, D. C. (pp. 29, 33).
[4] A. C. FAIRE, 1966 National Telemetering Conf. Proc., Boston, Mass. 1970 (p. 70).
[5] A. C. FAIRE and K. S. W. CHAMPION, Space Research V, 1039 (1965).
[6] A. C. FAIRE and K. S. W. CHAMPION, Space Research VI, 1048 (1966).
[7] A. C. FAIRE and E. A. MURPHY, Space Research XI, 821 (1971).

Space Research XIV — Akademie-Verlag, Berlin 1974

ANNUAL VARIATIONS OF DENSITY, TEMPERATURE AND WIND BETWEEN 30 AND 97 KM AT WOOMERA SHORTLY AFTER DUSK, 1962–1972

P. H. O. Pearson

Department of Supply, Australian Defence Scientific Service, Adelaide, South Australia

Over a number of years falling sphere experiments have been conducted at Woomera on a regular basis. Most of these firings, which have released two metre diameter inflatable spheres, have been made shortly after local sunset. Sinusoidal curves representing annual variations have been fitted to 53 of these firings and models produced of density and of temperature between 97 and 40 km, and wind values from 80 to 30 km, and compared with the US Standard values given in the 1966 Supplements. The density results showed an isopycnic level at 96 km with below standard values between 85 and 60 km and amplitudes in excess of the US Standard. The temperature results showed no significant annual variation above 46 km with annual amplitudes of 12 °C below this height. The zonal wind results gave a westerly component 30 m s^{-1} higher than given in the US Standard for 30° latitude while in the meridional results a positive 7 km thick annual stream centred at 76 km with an amplitude of 18 m s^{-1} was evident.

1. Introduction

For a number of years falling sphere experiments have been conducted at Woomera on a regular basis. The large majority of these experiments have been at a time shortly after sunset when the earth's shadow height has been approximately 50 km. The purpose of this paper is to collect together these results and examine the annual variation by finding the best fit sine curve with a fixed period of 365.25 days.

The distribution of the 53 firings over the years is not even and was divided into two groups, one from 1962 to 1966 and the other from 1970 to 1972. However, for this study the more important distribution is the pattern between the seasons. As can be seen in Fig. 1 this is relatively even with a maximum difference of only two firings between any pair of three month periods. All of these firings were conducted shortly after sunset, which should minimize any diurnal effect and thus produce a true annual variation. The condition to produce a true annual effect will be if, and only if, the phase of any diurnal effects is such that a null point occurs at the time of the firing. If, however, the diurnal effect has a phase shift from this condition, a bias error will be produced in the results which might vary during the course of a year. Thus it is possible to confuse a mean annual variation with an annual variation in the diurnal maxima unless care is taken in selecting the time of launch.

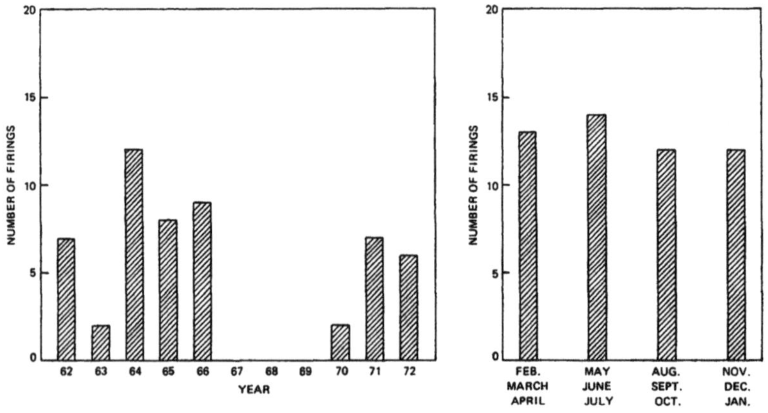

Fig. 1. Temporal distribution of falling sphere firings conducted at Woomera in dusk conditions.

2. Data Reduction Method

The atmospheric parameters were produced by processing the data obtained from an AN FPS 16 radar which tracked the sphere from ejection until it reached a height which overlapped with a ground released meteorological sonde. The data were then processed in the manner outlined by Pearson and Bissell [1]. The density, temperature and wind values are punched on cards for each kilometre height level. A second computer programme takes the data for each kilometre height level and determines the existence of sinusoidal components and produces coefficients representing the best least squares fit to the experimental data for increasing periods up to 4000 days. The information is plotted as a graph of the correlation coefficient and amplitude against period, thus allowing the significant periods to be selected. The phase shift relative to a fixed datum is also printed thus allowing study of the phase relationship between different height levels of the atmosphere. In the case of density and zonal wind and, for certain height levels, in the temperature and meridional wind values, the annual period is dominant. Other variations are clearly evident, such as 6, 8 and 24—26 month oscillations. This paper examines the annual component as this has, in general, the largest amplitude.

3. Density

Fig. 2 shows the Woomera mean annual density variation with height together with the amplitude of the sine wave corresponding to the summer and winter conditions. For comparison, the US Standard Atmosphere Supplement 1966 [2] values for 30° N in the northern hemisphere are shown.

The correlation coefficient between zero and one, which has a 5% significance value of 0.4, is large, starting at 0.5 at 40 km and increasing to 0.8 at 52 km, and then varying between 0.7 and 0.8 up to 92 km when there is a decrease to 0.5 as the low amplitude variation associated with the isopycnic level becomes evident. There are three noticeable features:

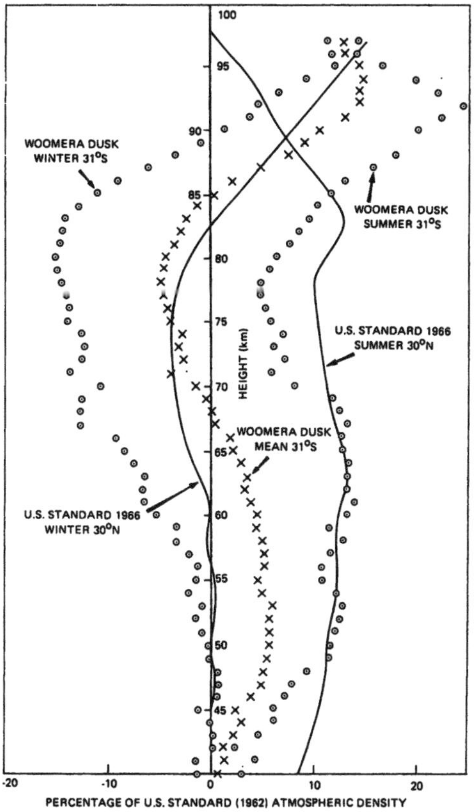

Fig. 2. Annual variation of atmospheric density as a function of height for Woomera at dusk.

(a) The amplitudes at the lower levels (40—48 km) are considerably less than those predicted by the US Standard. One explanation might be that when compiling the standard atmosphere there was a bias toward results with a significant annual variation in diurnal maxima, which could not be separated from the true annual variation. It seems unlikely that this difference could be attributed to a southern hemisphere effect.

(b) The apparent lower density values between 65 km and 85 km could be attributed to drag coefficients used in the data reduction. The values used were the latest available from Bailey and Hiatt [3] and have been adopted as standard for all Australian falling sphere experiments. However, other work has indicated that drag coefficient errors would not account for a shift of 10% and it is therefore concluded that the density values are realistic.

(c) The height of the upper isopycnic level appears to be at about 96 km which is about 6 km higher than suggested in the US Standard Atmosphere Supplements 1966 or by Cole [4] for high latitudes.

An annual variation with zero phase lag would be a sine wave passing through the mean value on 22 March and a positive lag indicates that this point occurs after 22 March. It appears that the lag has a maximum value of approximately

13 days just below the mesopause at 75 km and again below the stratopause at around 45 km. There is no lag at around 57 km in the lower mesosphere but above this height the lag increases at the rate of 0.7 day km^{-1}. The mesopause has a reasonably constant lag of about seven days with an increasing lag below this region of 1.2 day km^{-1} until at 68 km a lag of 12 days is evident. The high lag above 90 km represents the start of the complete phase change associated with the upper isopycnic level.

4. Temperature

Analysis of the temperature values gave some interesting results which are shown in Fig. 3. Firstly, it is surprising that there appeared to be no high correlation for annual variations in the mesopause region. In general, correlation coefficients were reasonably low at 0.4, with the values of lag most erratic, even though the US Standard Atmosphere predicts summer to winter variations of 25 °K. Secondly, the lack of any significant annual oscillations continued down through the entire mesosphere until the middle of the stratopause when a reasonably strong

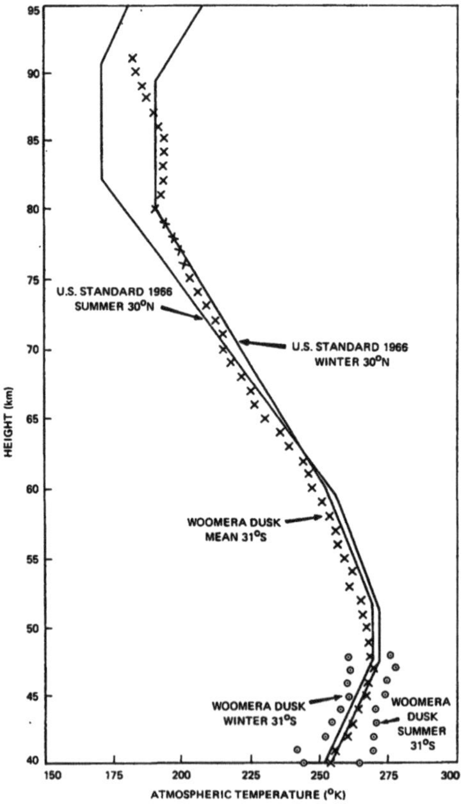

Fig. 3. Annual variation of atmospheric temperature as a function of height for Woomera at dusk.

correlation suddenly appeared, with correlation coefficient varying between 0.6 and 0.8 combined with a stabilized lag with an average value of +4.5 days. The peak to peak amplitude of about 15 °C was about five times greater than that given by the US Standard models. It is interesting to note that the 5% significance values of annual temperature variations stop at approximately the height of the earth's shadow for these dusk firings. The mean values of temperature follow the US Standard values reasonably well although they follow the "winter values" more closely than the mean values for all points above the stratopause and especially in the mesopause. The lowest measured point of the mesopause coincides with the height predicted for the "winter values" of the US Standard Atmosphere.

5. Winds

The annual zonal wind model is shown in Fig. 4 together with the profile for the 30—45° latitude annual profiles give in the US Standard Atmosphere. The summer values followed the US Standard from 30 km up to 65 km and the showed a progressively reducing easterly wind up to 80 km where there was no easterly

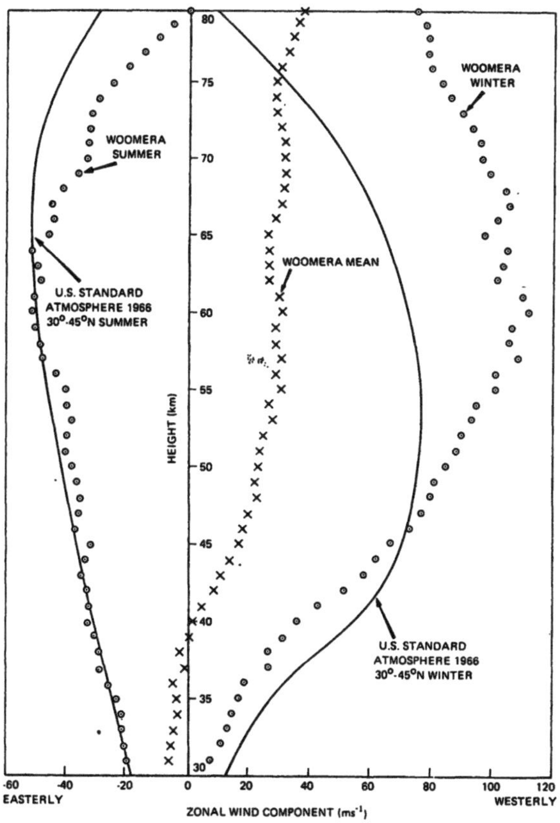

Fig. 4. Annual variation of zonal wind as a function of height for Woomera at dusk.

component throughout the entire year. The winter values showed a marked difference from the US Standard especially above 50 km where the winter peak showed westerly winds of 80—110 m s^{-1} from 50 km up to 80 km while the US Standard model decreased progressively to below 10 m s^{-1} at 80 km. Thus the amplitude at 80 km of the annual variation of zonal wind at Woomera at dusk was twice that predicted by the US Standard and was reduced with height until by 30 km the amplitude was only 0.75 of the US Standard value.

The phase lag showed the largest lag, about 14 days, to occur as a near constant value between 42 and 58 km; then through the stratopause the lag was reduced above 58 km at about 0.7 day km^{-1} until zero lag was found at 80 km. Below 42 km the lag was reduced at 3.0 day km^{-1} until zero lag occurred at 38 km and then continued until an advance of approximately nine days occurred at 36 km; this remained reasonably constant down to 30 km. The correlation coefficients were 0.6 at 30 km and quickly rose to between 0.85 and 0.95 by 42 km and remained at this level until 76 km where they decreased to 0.7 at 80 km.

The meridional wind profile (Fig. 5) showed only small values of mean wind below 70 km although a southerly wind up to 5 m s^{-1} was evident between 50 and 60 km. Below 70 km there was no significant correlation with an annual period, but between 74 and 78 km the annual period was detected with a peak

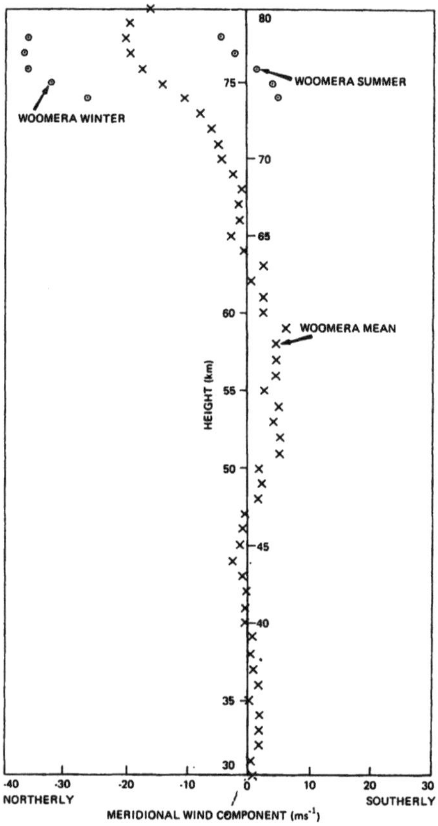

Fig. 5. Annual variation of meridional wind as a function of height for Woomera at dusk.

correlation coefficient of 0.55 at 76 km falling to 0.47 at 74 and 78 km. Over this region the phase changed linearly between a lag of 35 days at 74 km to an advance of 7 days at 78 km showing a slow downward propagation of about 10 day km^{-1}. This annual variation results in negligible mean meridional wind in the summer but a relatively strong stream of up to 35 m s^{-1} northerly during the winter.

6. General Conclusions

The amount of data is such that a reliable model can be constructed for latitude 31° S at a time shortly after dusk. The method of reduction has remained constant for all of the results, using the latest available sphere drag coefficient values [3] which should give the lowest bias errors to date. The time chosen for the firings minimizes the diurnal effects which might account for some of the differences between this model and the US Standard values. It is possible that the differences, especially in the annual temperature variations, are explained by the fact that the US Standard model was constructed from experiments conducted during daylight hours. Some of the differences might be attributed to a southern hemisphere effect, although it cannot be stated that this is true without a considerably greater number of controlled measurements in both northern and southern hemispheres.

The existence of a meridional wind stream centred at 76 km is considered to have been demonstrated quite positively in view of the methods of data reduction and errors that can occur. It is noted that if results had only been computed at 5 km intervals, as is often the custom, only one point would have emerged at 75 km which might well have been neglected as a bad point by the experimenter.

These dusk firings are continuing and a study is in hand to isolate other periodic components and produce, by use of a multiple correlation programme, a model to include all significant periods.

References

[1] P. H. O. Pearson and R. A. Bissell, COSPAR Technique Manual Series (Ed. L. M. Jones). COSPAR, Paris 1967.

[2] NASA, ESSA and USAF, U. S. Standard Atmosphere Supplements, Government Printer, Washington, D.C. 1966.

[3] A. B. Bailey and J. Hiatt, AEDC Tech. Rep. No. 70-291 (1968).

[4] A. E. Cole, J. Geophys. Res. 66, 2773 (1961).

Space Research XIV — Akademie-Verlag, Berlin 1974

RECENT STRATOSPHERIC TEMPERATURE MEASUREMENT COMPATIBILITY TESTS AT WALLOPS ISLAND

F. J. Schmidlin

NOAA, NWS, Weather Service Support Facility, Wallops Island, Va, USA

Several tests have been conducted from NASA Wallops Station in a continuing effort to determine the accuracy and compatibility of various rocketsondes. Tests indicate that the commonly used US rocketsonde gives quite repeatable temperature values with an rms difference of ~1 °C. Other tests showed that differences exist between rocketsonde and satellite measurements. An international intercomparison of rocketsondes, carried out in March 1972, indicated that above 45 km differences exist between the rocketsondes of different countries. Finally, a recent day-night test is discussed; the solar radiation error on the Datasonde instrument may be ~7–8 °C at 60 km and an additional adjustment may be required to compensate for the diurnal effect.

1. Introduction

The *in situ* technique used to measure stratospheric-mesospheric temperatures between 20 and 60 km in the United States is the meteorological sounding rocket using the bead thermistor as the sensing element. While this method provides representative temperatures below about 45 km, the uncertainty of the measurement error above this altitude is of considerable concern. To investigate the accuracy obtainable from this rocketsonde and its compatibility with other instruments, a number of tests were conducted at the NASA Wallops Island launch facility.

2. Description of Tests

2.1. Repeatability Test

Occasionally inconsistencies in the measured temperatures are found that cannot be resolved by any correction scheme. These discrepancies are of particular concern since it is not known whether the cause is true atmospheric variability or an instrument peculiarity. To resolve this problem an experiment was designed to determine the reliability and repeatability of measurements made with the United States' Datasonde instrument [1]. This test, conducted during June when temperature gradients are expected to be small, called for Datasondes to be launched in pairs one hour apart with the instruments of each pair launched within 5 minutes of each other. Since differences between the diameters, coatings, and shapes of bead thermistors tend to affect the measurements [2], the condition of the sensors and how they were paired in the experiment was noted and recorded. Approximately 100 thermistors were inspected at the manufacturer's plant and 12 were selected for the experiment. These were classified as being either average or below average based on the criterion that those sensors not similar in appearance were considered to be below average. Four pairs with average sensors and two with below average sensors were launched.

Fig. 1 reveals that reasonably good agreement between the temperature measurements exists. The rms differences of five pairs were found to be between 0.7 °C and 1.4 °C, with a mean of 1.1 °C. One below average pair at 1835—1840 GMT exhibited an rms value of 2.8 °C; it is significant that of the thermistors examined only one had a cracked coating and this was launched as part of this pair. The other below average pair at 1735—1740 GMT showed consistent measurements, and in fact, had a smaller rms than some of the average pairs. Based on the results of this experiment it seems that the Datasonde instrument is quite repeatable in operation and that the present manufacturing and inspection methods are adequate.

2.2. Rocketsonde-Satellite Comparisons

An experiment was carried out [3] in which data obtained from various rocketsonde techniques (grenades, pitot probes, Arcasondes, and Datasondes) were compared with data obtained from various satellite sounders (SIRS-A and SIRS-B, IRIS, and SCR) on Nimbus 3 and 4 satellites. The rocketsonde launchings were scheduled to occur within ± 1 hour of the time the satellites passed within ± 5 degrees latitude of Wallops. This test showed that the several rocketsondes in use can provide comparable layer-mean temperatures though detailed profiles may not agree. Furthermore, radiances obtained from the various satellite sounders completely agreed neither with each other nor with the rocketsondes, nor can the sounders provide fine temperature structure in the vertical. Rocketsondes can and should be used as a check on the calibration of satellite instruments.

2.3. International Rocketsonde Intercomparison

Routine meteorological rocket observations to 60 km and higher are being made by a number of countries and the data are exchanged. In order to resolve inconsistencies noted in these data, the Commission for Instruments and Methods of Observations (CIMO) suggested a comparison of the different rocket systems used by the member countries. The intercomparisons were to be held in two phases. Participants of the first phase, held at Wallops Island, Va., during March 1972, were France, Japan, and the United States; Brazil and India sent observers. A total of 29 rocketsondes were launched, 12 USA, 7 French, and 10 Japanese. The second phase is planned to be held at Kourou, French Guiana, during September 1973. Day and night launchings were scheduled with five comparisons made during the day and two at night between the US Loki-Datasonde and Japanese MT-135P borne Echosonde systems, three day and two night comparisons between the US and French DMN systems; three day and one night comparison were obtained with all systems. Preliminary results indicate that all systems agreed within ~ 5 °C up to about 45 km; above this level the French system gave colder temperatures than the US and Japanese systems, both day and night.

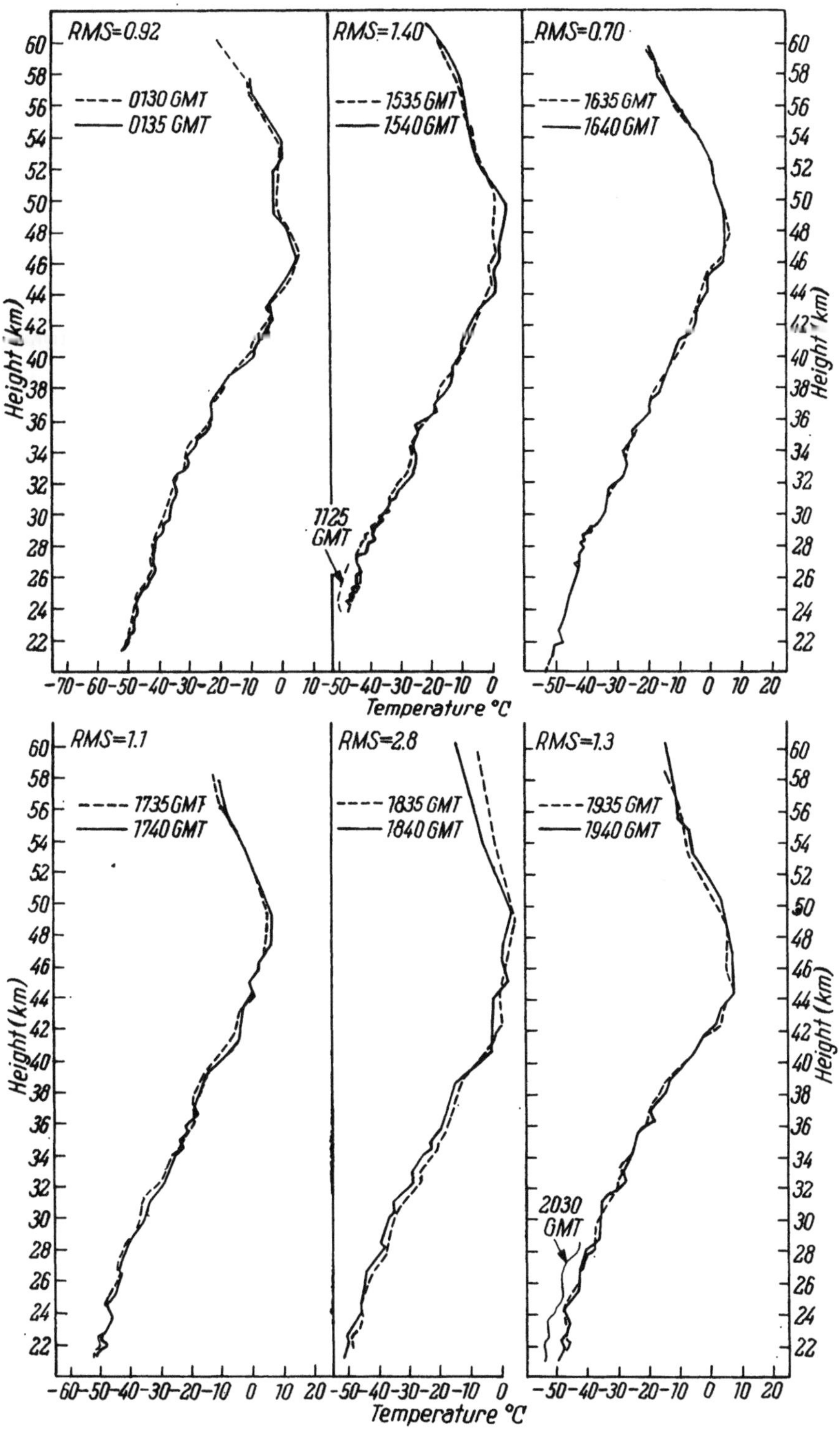

Fig. 1. Temperature profiles obtained with Datasonde instruments launched 5 minutes apart. RMS differences between the two measurements are shown in °C.

2.4. Day-Night Differences

With the establishment of the Eastern and Western Meridional Networks early in 1972, an increasingly large number of rocketsonde data became available. Hemispheric analysis using these data suggests that a bias exists between the instrument types; the Eastern Meridional Network predominantly uses the USSR M-100 system and the Western Network uses the US Datasonde. It is noted that the stratosphere above the Eastern Network is consistently colder above 45 km than that of the Western Network. Temperature differences are estimated to be ~15 °C at 50 km and ~30 °C at 60 km, with the Datasonde being warmer. A number of possibilities for the cause of this temperature difference have been considered: The Western Network generally launches during the day and the Eastern Network generally at night. In addition, the Western Network data are not adjusted for dynamic heating, lag, etc.

Until recently, corrections being developed for the US Datasonde instrument were based on laboratory studies and some flight experiments which generally applied to the Arcasonde 1A instrument. Henry [4] suggested a correction technique for the flat-plate, film-mounted thermistor, where the heat inputs were calculated separately for the thermistor and film and then combined into a single correction value for the sensor. This correction technique was applied to the loop mount geometry actually used and new heat input quantities were calculated [5]. Since the corrected Datasonde temperatures are still much higher than those of the M-100 system the correction values obtained may be too small.

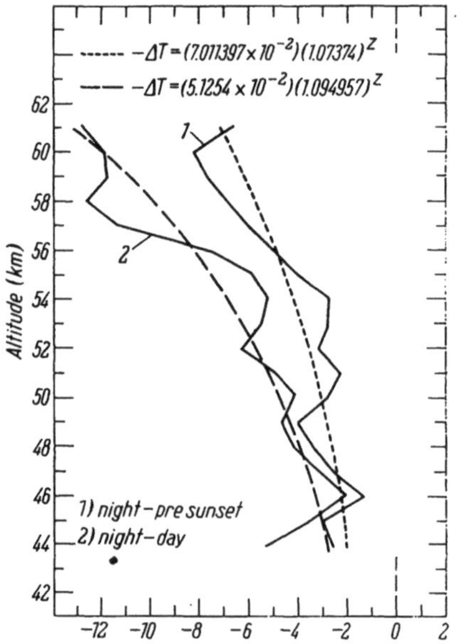

Fig. 2. Mean temperature differences obtained from 12 sets of data. Curve 1, mean difference between night and pre-sunset launch observations, dotted curve represents a least squares fit, according to equation shown. Curve 2, mean difference between night and day observations, dashed line represents a least squares fit according to equation shown. Here Z is the geometric altitude in km.

In an effort to determine an adjustment which would bring the Western Network temperatures into agreement with those of the Eastern Network, Datasondes were launched during daylight and darkness. The experiment was conducted during October and November 1972, with rocketsondes launched near 1400 hours LST, about 1 hour before sunset, and immediately after sunset (at 60 km). At sunset (at 60 km) the only temperature change observed would be that due to the removal of solar radiation; any diurnal change between these launch times is considered to be negligible. Twelve sets of data were obtained.

Because of weather influences that might affect the measurements a least squares fit to the mean temperature differences (Fig. 2) for the day-night data (Curve 2) and solar radiation (pre-sunset-night) data (Curve 1), were calculated. The portion of the diurnal variations affecting the measurement is simply the difference between the smoothed curves 1 and 2. The values obtained are ~5.2 °C at 60 km and ~2 °C at 50 km. Previous studies [6, 7] suggest that the maximum of the diurnal variation should occur near 1200—1400 local time, and the minimum near local midnight; ΔT values might be of the order of 9—12 °C at 60 km and 6—8 °C at 50 km. Adjustment values for aerodynamic heating, emissivity, thermal lag, long-wave radiation, and electrical heating [5] were applied to the observed radiation effect (Curve 1), together with the expected difference due to the diurnal variation, to produce an estimated adjustment value of ~—20°C at 60 km and ~—7 °C at 50 km. If the full range [6, 7] of the diurnal variation is considered this adjustment would be somewhat larger, but would still leave a few degrees difference between the Eastern and Western Network temperatures to be explained.

Final Remarks

These tests show that measurements obtained with the Datasonde are quite repeatable and that the Datasonde is capable of providing results compatible with those of instruments of different countries. The adjustments suggested for the Datasonde instrument as a result of the day-night tests are, obviously, gross values at the very least. Although an adjustment for a part of the diurnal variation was included in the day-night test, such adjustments are best left to the user of the data. Additional work is being carried out to determine a more suitable correction. Certainly, direct comparison between the M-100 and the Datasonde instruments is desirable and would probably help to explain more conclusively the reason for such wide temperature differences.

References

[1] A. J. Miller and F.J. Schmidlin, J. Appl. Met. **10**, 330 (1971).
[2] D. C. Thompson, J. Appl. Met. **7**, 306 (1968).
[3] National Aeronautics and Space Administration Spec. Rep. NASA SP-296 (1972).
[4] R. M. Henry, paper presented at Amer. Met. Soc. Conf. on High-Altitude Meteorology and Space Weather, Houston, Texas, March 1967.
[5] M. V. Krumins and W. C. Lyons, Naval Ordnance Lab. Tech. Rep. NOLTR 72-152 (1972).
[6] N. J. Beyers and B. T. Miers, J. Atmos. Sci. **25**, 155 (1968).
[7] L. R. Hoxit and R. M. Henry, in: Int. Conf. on Aerospace and Aeronautical Meteorology, AMS, May 1972 (p. 149).

Space Research XIV — Akademie-Verlag, Berlin 1974

THE ALADDIN II EXPERIMENT: PART I, DYNAMICS

S. P. Zimmerman[a], N. W. Rosenberg[a], A. C. Faire[a], D. Golomb[a], E. A. Murphy[a], W. K. Vickery[a], C. A. Trowbridge[b] and D. Rees[c]

[a]Air Force Cambridge Research Laboratories, Bedford, Mass., USA
[b]PhotoMetrics, Inc., Lexington, Mass., USA
[c]University College London, London, UK

A co-ordinated through-the-night rocket experiment was carried out on 12—13 April 1972 at Eglin AFB, Florida (30° N). The program measured vertical profiles between 50 and 150 km altitude of total neutrals, individual neutral species, temperature, diffusion coefficients, both turbulent and molecular, and turbulent structure. Results of a photochemical-transport computer model using these parameters were compared with the observed profiles. Vertical wind fields were measured at spaced time intervals from sunset to before sunrise. Zonal and meridional sequential wind patterns identified the major wind feature as the diurnal solar tide. Total densities derived from falling sphere data from 50 to ~110 km and kinetic temperatures measured from fluorescence of chemically released AlO also displayed vertical oscillations indicative of tidal motions. Vertical shears and temperature profiles have been included in a Richardson number stability criterion to obtain a through-the-night correlation between predicted and observed turbulent and laminar regions. Values for the minimum vertical turbulent diffusivities over this time period are also presented.

1. Introduction

The simultaneous observation of transport parameters and total mass density and temperature in the upper atmosphere makes it possible to more realistically compute the local distribution of minor neutral and ionic species. These measurements, spaced in time, give the contribution of the tidal and higher frequency motions to wind, turbulence, and temperature fields as well as offering some explanations of the large deviations from diffusive equilibrium profiles observed in the minor neutral species measurements above the turbopause at ~108 km altitude.

The Aladdin II experiment, which was carried out from 1800 CST on 12 April to 0430 CST on 13 April 1972 at Eglin AFB, Florida (30° N), was designed to examine the temporal and spatial variations of different atmospheric parameters and their effects upon minor and major neutral species, also measured and discussed by Philbrick et al. [1]. The program resulted in measured vertical profiles (between 90 and 170 km altitude) of the concentration of total neutrals and of individual species (O, O_2, N_2 and Ar), temperature, density, diffusion coefficients, horizontal winds, and turbulent structure. The density and temperature measurements were extended down from 90 km to 32 km by inflatable and rigid sphere measurements [2].

2. Wind and Shears

Five chemical trails spaced over a ten-hour period provided wind profiles from ~90 to 160 kilometers, with a 1 kilometer height resolution. These have been interpolated to provide a height/time history of wind (Fig. 1) covering the period 1843 to 0430 local time. The downward phase progression of velocity with time (Fig. 2) which has been observed in several other wind and time series measurements was again present above an altitude of 105 km, but is not in evidence below this altitude. Phase hodographs at specific altitudes demonstrate clearly the dominance of the diurnal tide from about 90 km to ~140 km. Above this altitude there appears to be a possible higher frequency mode; however the lack of data at 0230 hampers the analysis here. Wind phase comparison with the Lindzen and Blake [3] theory (Fig. 3) shows large disagreement between ~110 km and 150 km, the divergence beginning at ~100 km. While it is not possible here

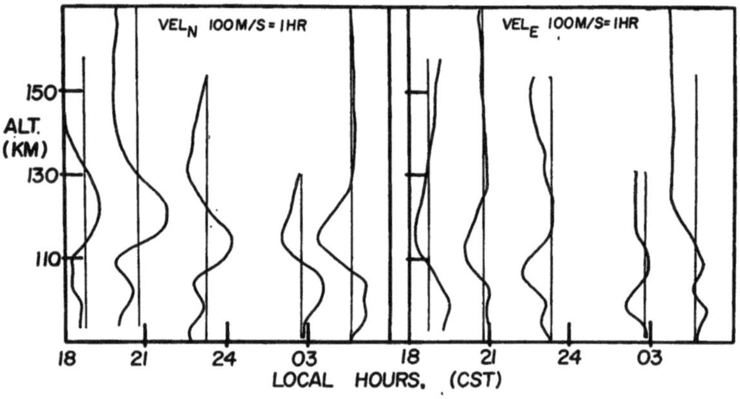

Fig. 1. Wind velocities to the north and east for the time period 1800 to ~0430 (CST).

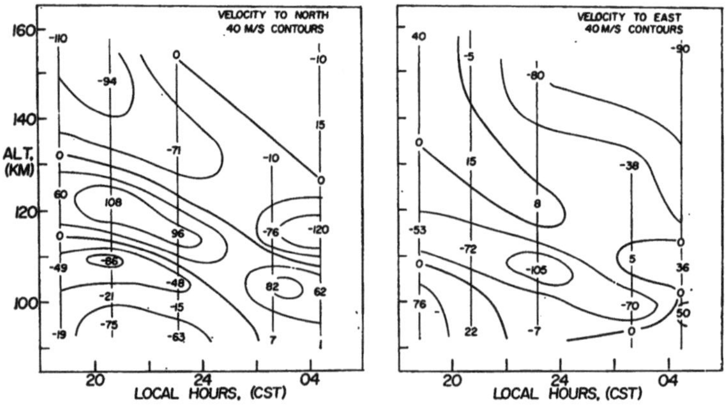

Fig. 2. Contour maps of the easterly and northerly wind components. The downward phase progression is quite discernible.

to explain completely the reasons for the phase shift, it is suggested that the difference between the turbulent diffusivity used by Lindzen and Blake (10^5 cm^2 s^{-1}) and the minimum measured (§ 4) vertical turbulent diffusivities determined here (10^6—10^8 cm^2 s^{-1}) is the cause.

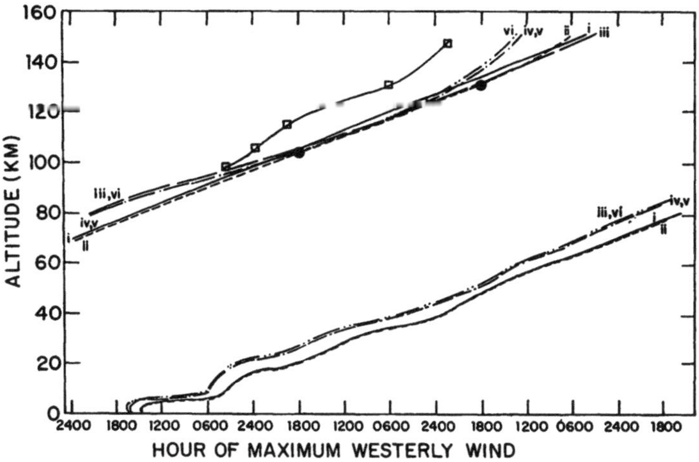

Fig. 3. The hour of the maximum westerly component of the diurnal solar tide compared with the theoretical calculations of Lindzen and Blake [3].
□ Aladdin II; ● Aladdin I.

3. Temperature

Two independent measurements of AlO clouds were performed (CRL and UCL) to compare temperatures and diffusion coefficients over the altitude region 118 to 175 km for two times (1845 and 0430 LST). The details of the experiment are to be discussed more fully by Rees et al. [4], who also show the measured diffusion coefficients and derived densities. However, the probable dynamical effect on the temperature field (Fig. 4) is evident above the scatter and experimental errors. These data appear to confirm, although in an extremely qualitative manner, the diurnal period evident in the ~180° phase difference of the oscillatory components in the two temperature measurements taken about ten hours apart.

4. Atmospheric Instabilities

The simultaneous measurements of winds and temperature resulted in the determination of the Richardson number where the measurements overlapped:

$$Ri = \frac{g}{T}\left[\frac{\partial T}{\partial z} + \Gamma\right] \Big/ \left(\frac{\partial V}{\partial z}\right)^2 \qquad (1)$$

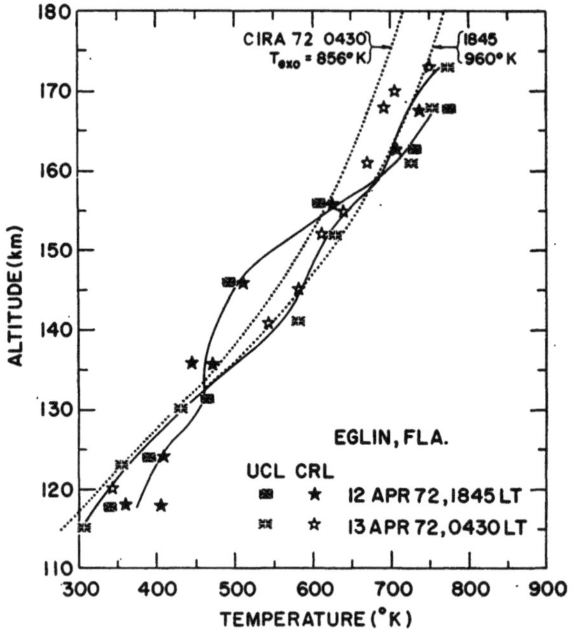

Fig. 4. Temperature measurements from the AlO radiance at 1845 and 0430.

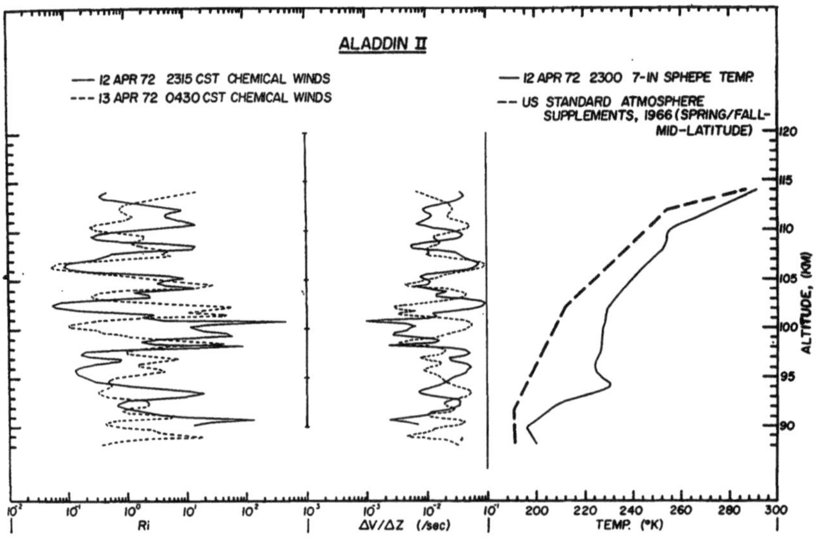

Fig. 5. Richardson numbers, total vertical shears for the times 2300 and 0430 CST, and 2300 CST temperature measurement. Also shown (dashed line) is the 1966 Supplements [6] model temperature profile.

where g is the acceleration of gravity, T is the kinetic temperature, $\partial T/\partial z$ is the vertical temperature gradient, Γ is the adiabatic lapse rate ($= +10\,°K\,km^{-1}$), and $\partial V \partial z$ is the vertical shear of the horizontal winds. The temperature used for the altitude region above 88 km was derived from the falling sphere density measurement at 2300 LST [2]. Unfortunately this was the only temperature measurement available, as the other sphere measurement of 2045 went only to 85 km. Thus, the only accurate Richardson number would be for the time at 2300 LST, where there was an altitude (90—115 km) overlap with the wind measurement, at 2315 LST. These data are shown in Fig. 5. Also shown is the estimated Richardson number profile for the time 0430, which is not a correct Richardson number, but only reflects the effects of the wind shears measured at these times since we were forced to use the 2300 temperature profile in the calculations.

While there were no direct wind measurements by chemical trail below 88 km, winds were derived from the 1905 LST inflatable sphere [2] drag data. The Richardson number was calculated using these winds and the temperature profiles of both the 2045 and 2300 rigid spheres. The results show no indication of turbulence until ~73 km, and that only for the 2045 temperature profile which demonstrates large lapse rates at 73 and 76 km. Above ~85 km the wind data obtained by this technique become highly uncertain.

5. Turbulence Spectra, Rates of Dissipation and Vertical Diffusivities

Zimmerman and Trowbridge [5] demonstrated that the rate of turbulent dissipation of wind energy ε may be derived from the Fourier analysis of the fluctuations in density observed in the chemical releases below the turbopause. When the spectrum shows an inertial sub-range, $P(f) \sim f^{-5/3}$ where $P(f)$ is the spectral energy density and f is the spatial frequency, the rate of dissipation ε may be determined from the expression

$$\varepsilon = \frac{(C_1 \varepsilon_\gamma)^3}{4\pi^2 P(f)^3 f^5}. \tag{2}$$

ε_γ [$= 6D\overline{(\partial n_i'/\partial x_i)^2}$] is the rate of decay by molecular diffusion, of the turbulent density fluctuations n', exhibited in the chemical trail, D is the molecular diffusion coefficient and $\partial n'/\partial x$ is the longitudinal gradient of the turbulent density fluctuations. Fig. 6 shows the results of a measurement at 107 km where the solid straight line is representative of the "inertial" sub-range, and the rate of dissipation derived from the measurement is $\sim 2 \times 10^5$ ergs $g^{-1}\,s^{-1}$. This is an exceedingly large energy deposition rate; however, the local shears ($\sim 0.03\,s^{-1}$) would transfer energy through molecular kinematic viscosity at a rate of $\sim 10^3$ ergs $g^{-1}\,s^{-1}$. Thus, for such large shears, the turbulent dissipation rate is not excessive.

To arrive at estimates of the vertical turbulent diffusivities, we must use the one-dimensional turbulent energy equation

$$K_m \left(\frac{\partial V}{\partial z}\right)^2 - K_h N^2 = \varepsilon \tag{3}$$

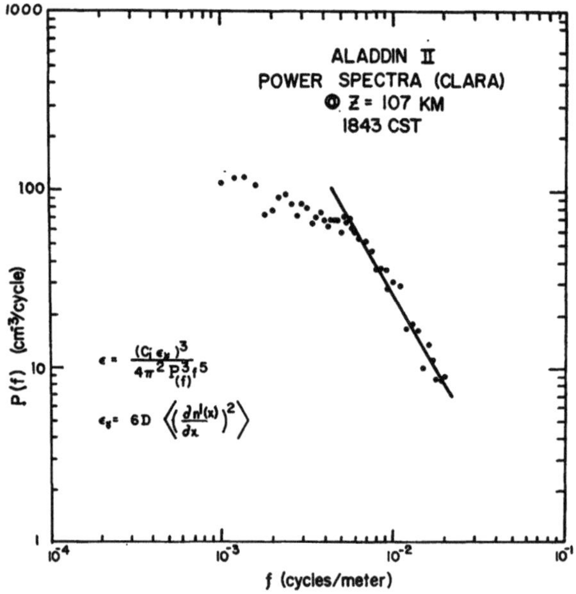

Fig. 6. The measured turbulent scalar density spectrum. This shows a significant extent of the "inertial" (—5/3) spectrum.

where: K_{m} is the momentum transfer coefficient, K_{h} is the heat transfer coefficient, $\partial V/\partial z$ is the vertical shear of the horizontal wind,

$$N^2 = \frac{g}{T}\left[\frac{\partial T}{\partial z} + \Gamma\right]$$

is the square of the Brunt-Väisälä frequency. For the condition of

$$R_i = \frac{N^2}{(\partial V/\partial z)^2} \ll \frac{1}{4}$$

and where we assume $K_{\mathrm{h}} \approx K_{\mathrm{m}}$ (a fairly good assumption),

$$K_{\mathrm{m}} \approx \frac{\varepsilon}{(\partial V/\partial z)^2} = K(z)\,. \tag{4}$$

The results of such an analysis for the release at 1845 are shown in Fig. 7. The squares are the vertical diffusivities for the altitudes 107, 102 and 95 km. The minimum at ~100 km is where the Richardson number is very large (~100) and we would not expect turbulence at this altitude. The altitude region below the dashed line is where we cannot yet make turbulence measurements; this is why we make two estimates of the amplitude of the turbulent diffusivities. The molecular diffusion coefficient calculated for atomic oxygen in a nitrogen atmosphere is also shown for reference.

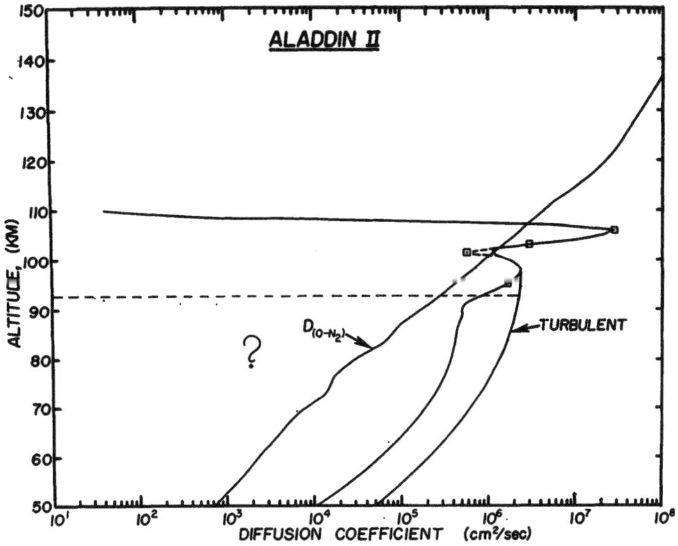

Fig. 7. The minimum vertical turbulent diffusion coefficient determined from the turbulent energy equation and the *in situ* determination of the rate of viscous dissipation.

6. Summary

In summary, we have shown that the simultaneous measurements of the transport parameters of the upper atmosphere in the Aladdin II program provide necessary information for the calculation of model atmospheres, and also information on wind, temperature and turbulence fields necessary for comparison with existing theories of winds and minor constituents in order to assess their validity.

References

[1] C. R. Philbrick et al., Space Research XIV, 89 (1974).
[2] A. C. Faire, E. A. Murphy and R. Olsen, Space Research XIV, 97 (1974).
[3] R. S. Lindzen and D. Blake, Geophys. Fluid Dynam. 2, 31 (1970).
[4] D. Rees et al., Space Research XIV, 103 (1974).
[5] S. P. Zimmerman and C. A. Trowbridge, Space Research XIII, 203 (1973).
[6] US Standard Atmosphere Supplements, 1966, US Government Printing Office, Washington, D. C. 1966.

Space Research XIV — Akademie-Verlag, Berlin 1974

THE ALADDIN II EXPERIMENT: PART II, COMPOSITION

C. R. PHILBRICK[a], D. GOLOMB[a], S. P. ZIMMERMAN[a], T. J. KENESHEA[a], M. A. MACLEOD[a], R. E. GOOD[a], B. S. DANDEKAR[a] and B. W. REINISCH[b]

[a]Air Force Cambridge Research Laboratories, Bedford, Mass., USA
[b]Lowell Technological Institute Research Foundation, Lowell, Mass., USA

During the Aladdin II (Atmospheric Layering And Density Distribution of Ions and Neutrals) experiment, parameters relating to the structure and dynamics of the atmosphere were measured. Vertical profiles of N_2, O_2, O, Ar, CO_2 and O_3 concentration and of total neutral density and temperature were obtained. The atomic oxygen profile was measured using three different techniques, mass spectrometer, green-line photometer, and NO chemical release. Chemical trail measurements of dynamic parameters were used in a numerical model for the E region to predict the regions of ion layering throughout the night. The predicted regions of ion layers are compared with ionosonde measurements. The turbulent and molecular diffusion coefficients, total mass density, and temperature measured during the period of the Aladdin II experiment were used as input data to a numerical model which calculates vertical time-dependent distributions of atmospheric constituents. The computed distributions are compared with the measured profiles.

1. Introduction

The Aladdin II (Atmosphere Layering And Density Distribution of Ions and Neutrals) experiment was performed on the night of 12—13 April 1972 at Eglin AFB, Florida. The experiment consisted of five chemical trails spaced through the night together with two 7-inch falling spheres, one instrumented inflatable sphere, one passive inflatable sphere, one nitric oxide headglow measurement, two mass spectrometer measurements, an one green-line photometer measurement. In the companion paper [1], a summary of the launch sequence together with the measurements of the wind, shear, turbulence, density and temperature structure have been presented. This paper describes preliminary results relating to the structure, composition and modeling of the experiment; it is divided into three sections which include the ionospheric measurements and calculations, the neutral composition measurements, and the neutral atmosphere model [2—7].

2. Ionosphere Measurements

A Digisonde situated 2 km from the launch pad continuously monitored the ionosphere during the ten days surrounding the launches. No unusual ionospheric activity was observed during this time except at periods following certain of the evening chemical releases. Strong Es activity developed during the night after two of the six chemical releases while all other nights showed no significant Es occurrence. Similar observations, with *f*oEs values reaching 10 MHz in one case, were made at Eglin after twilight releases in November 1969, November 1970, and May 1971 [8].

Previous experiments [9] have demonstrated that Al^+ ions are produced following trimethylaluminum (TMA) releases, probably by photo-ionization and charge exchange processes. Assuming Al^+ to be continuously produced in these TMA clouds, the motion of the ions has been calculated using sporadic E theory [10] and the observed neutral winds. The ions moved away from the field of view of the Digisonde for the 1845 LT release on 12 April 1972 (which showed no enhancement of Es) and towards it for the 2045 LT release (which was followed by a large enhancement of Es), with the calculated ranges corresponding approximately to the observed values.

Ion density profiles were calculated for the afternoon and early evening phases of the Aladdin II experiment 12 April 1972 in a manner similar to that done previously [11]. The coupled continuity equations were solved as a function of altitude and time for the molecular ions and a representative metal ion, assumed of mass 28 amu. After turning on the model sun at noon, photochemical profiles were calculated through the afternoon until sunset (1806 LT). At sunset the transport terms were included in the equations and the metal ions were introduced

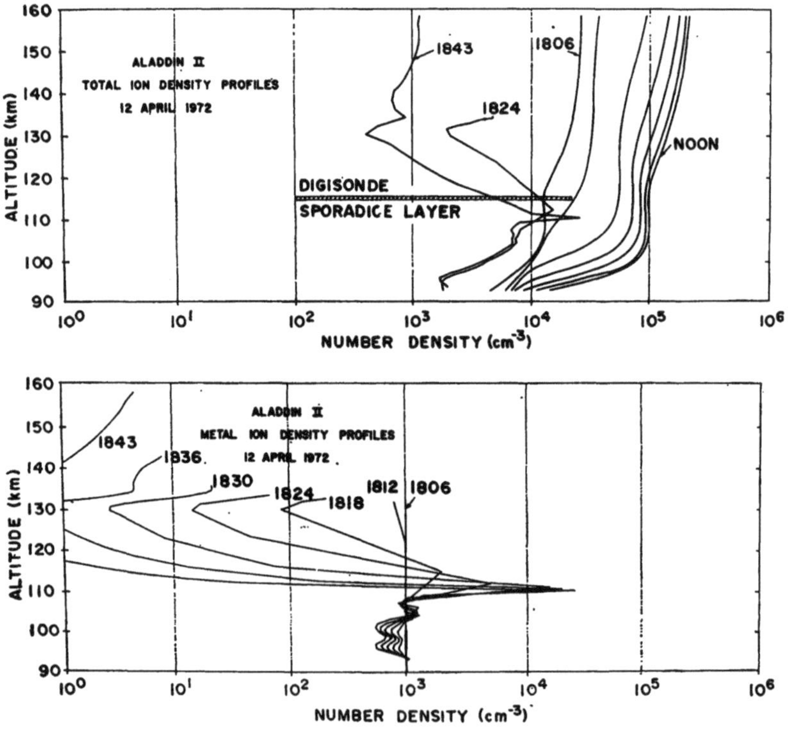

Fig. 1. Total ion density profiles (above) and metal ion density profiles (below) obtained from E-region model for the conditions on the day of the experiment, 12 April 1972.

with an intially constant density of 1000 cm^{-3} (see Fig. 1). As time progressed to the first experiment time (1843 LT), the metal ions were rapidly compressed into a sporadic E layer which appears at 109 km, which the Digisonde picks up at 116 km. As can be seen from the figure, this transport compression of the metal ions easily builds the sporadic E layer in this half-hour period.

3. Neutral Atmosphere Measurements

The neutral composition was measured by a cryosorption pumped RF quadrupole mass spectrometer [12, 13]. Fig. 2 shows the altitude profiles obtained for N_2, O_2, Ar, and O number densities. The N_2 profile above 105 km was obtained by a free molecular flow calculation using the laboratory calibration of the instrument sensitivity. The mass spectrometer and falling sphere techniques were in agreement in the overlap region near 105 km. The profile below 105 km was obtained from the mass density measurement of the falling sphere. The solid line shown is the N_2 density profile of the appropriate USSAS 1966 model. The O_2 profile is in ground level mixing ratio at the 90—95 km region and departs from ground level ratio above this altitude, decreasing to about 30% of its ground level ratio at 120 km. The Ar profile is also in ground-level mixing ratio in the region from 90—95 km and rapidly separates above this altitude. The Ar profile however shows a departure from the normal diffusive separation in the vicinity of 110 km where a strong turbulent layer is observed in the chemical trails.

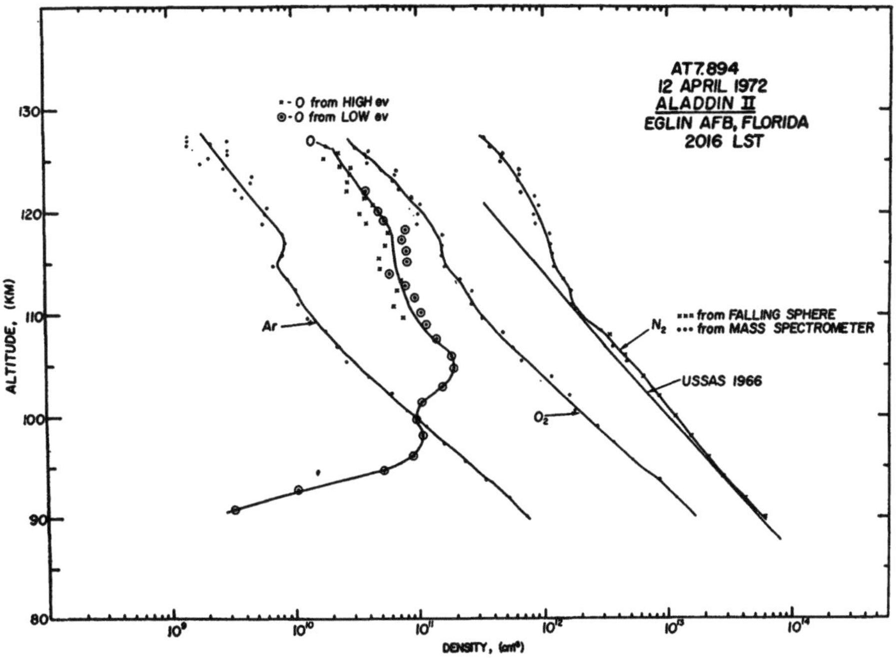

Fig. 2. Profiles of N_2, O, O_2 and Ar determined from the mass spectrometer ascent measurements. Measurements were made using 19 and 85 eV ionizing electrons.

One goal of the Aladdin II experiment was to compare atomic oxygen density profiles ,with different techniques. In addition to the mass spectrometry, a 5577 Å green-line photometer was flown on the same payload and a nitric oxide headglow measurement was performed with a payload about 20 minutes later. The nitric oxide release obtained measurements from 110 to 150 km and the green-line photometer obtained measurements in the region from 94 to 110 km. The results from the three techniques are shown in Fig. 3.

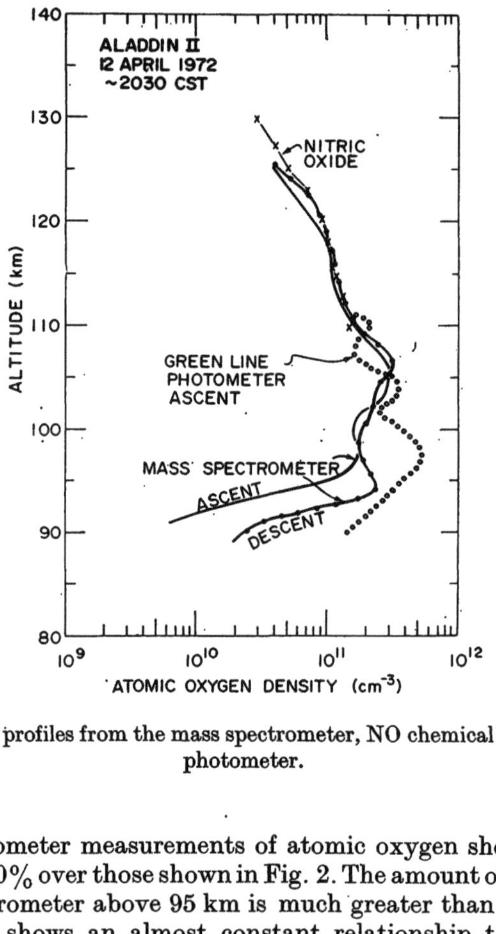

Fig. 3. Atomic oxygen profiles from the mass spectrometer, NO chemical release and green-line photometer.

The mass spectrometer measurements of atomic oxygen shown in Fig. 3 have been increased by 60% over those shown in Fig. 2. The amount of mass 44 measured by the mass spectrometer above 95 km is much greater than atmospheric abundance of CO_2 and shows an almost constant relationship to the O profile. If the mass 44 abundance is assumed to be due to the loss of two atomic oxygen atoms per molecule, then an increase in the O profile of 60% would be necessary. The downleg profile shown in Fig. 3 is preferred since atomic oxygen densities are higher below 97 km than on the upleg; also a large flux of atomic oxygen has been encountered by that time which could have stabilized initial surface loss processes. The magnitude of the error bars for this profile has not yet been determined and thus the profile is considered as preliminary.

The nitric oxide headglow technique, which has been described previously [14, 15], consists of releasing a jet of a high pressure gas mixture of 75% N_2 and

25% NO through a sonic orifice. In the hemispherical expansion ahead of the rocket, the chemiluminescent reaction between ambient atomic oxygen and the nitric oxide results in a headglow. The headglow has a radius of about 10 m at 90 km and 150 m at 150 km. Due to the adiabatic expansion of the jet, some of the nitric oxide forms dimers and larger clusters. The chemiluminescent reaction between NO dimers and O has been shown to be very fast, accounting for the high intensity of the headglow [16]. A photometer situated near the sonic orifice views the nose portion of the headglow from its center. In wind tunnel tests it has been shown that the headglow radiance, is proportional to the flux of atomic oxygen impinging on to the nose of the jet, $B = q[O]v_\infty$, where B is in rayleighs (10^6 photons cm^{-2} s^{-1}), [O] is ambient atomic oxygen number density (cm^{-3}), v_∞ is the free stream velocity equal to the vehicle velocity. The factor q is the efficiency at which the impinging O flux is converted into a photon flux. For the 75% N_2—25% NO mixture, $q = 0.004$ with an uncertainty factor of two.

The 5577 Å OI line emission of atomic oxygen was measured with a photometer which consisted of two narrow band interference filters at 5577 Å and 5525 Å. The signal from the filters was directed on to a detector by a rotating mirror to measure the oxygen line emission and the background "green continuum". The oxygen emission from the region between 90 and 110 km was about 120 R on the upleg and about 110 R on the downleg. The preliminary profiles shown in Fig. 3 were determined by assuming that the emission is due to the Chapman three-body collision mechanism. To obtain these profiles the densities for N_2 and O_2 were used from the CIRA 1965 model together with the following constants: $k_1 = 1.5 \times 10^{-33}$ cm^6 s^{-1} for the Chapman mechanism, $q_1 = 10^{-17}$ cm^3 s^{-1} for quenching due to N_2, $q_2 = 2.1 \times 10^{-13}$ cm^3 s^{-1} for quenching due to O_2 and $q_3 = 1.3 \times 10^{-13}$ cm^3 s^{-1} for quenching due to O. The peak concentration determined for these preliminary profiles is about 5×10^{11} cm^{-3} near 97 km. This is one of the few times when atomic oxygen measurements have been obtained using more than one technique (see 17]).

A comparison of the three measurements does not yield as much agreement as might have been hoped. Above 110 km, the NO release and corrected mass spectrometer profiles agree much better than the accuracy claimed for either technique. The two profiles show a similar slope change near 120 km which is related to the density bulge already discussed. The ascent and descent profiles of the mass spectrometer agree well above 97 km. Below 97 km, the descent profile is probably more accurate. The green-line photometer and mass spectrometer profiles do not agree very well in shape or maximum density. Work is presently underway to assign error bars to each measurement so that the significance of these differences can be determined. It can be concluded, however, that all three techniques indicate an atomic oxygen density about a factor of two less than the Jacchia 1971 model [18].

4. Dynamic Model Calculations

The neutral species considered are O_1, O_2, O_3, O^1D, $O_2{}^1\Delta_g$, OH, H, HO_2, H_2O, H_2O_2, H_2, and Ar. The diurnally varying number densities from 50 to 150 km are obtained through a finite difference solution (at 0.1 km intervals) of a system of mass and momentum conservation equations [19—21]. The numerical approach is essentially that introduced by Shimazaki [21] but modified at the boundaries and in the volume integrations, following George et al. [19]. The falling sphere temperature and density profiles are used up to 108 km for the 2300 LT flight [22]. Using these data, the initial species distributions are calculated assuming mixing to the turbopause and diffusive equilibrium above. The total number density is obtained by integrating the hydrostatic equation where the sea level mass density and the mean molecular weight profiles are taken from the U.S. Standard Atmosphere (1962). The turbulent diffusion coefficients used are based upon observations of turbulence in chemical trails.

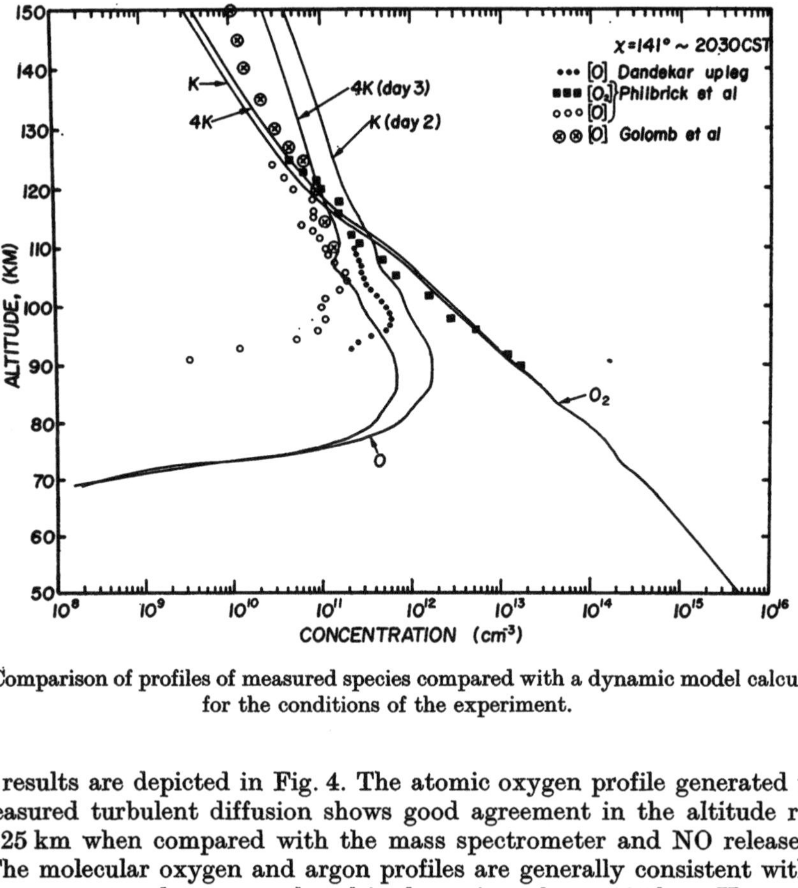

Fig. 4. Comparison of profiles of measured species compared with a dynamic model calculation for the conditions of the experiment.

The results are depicted in Fig. 4. The atomic oxygen profile generated using the measured turbulent diffusion shows good agreement in the altitude region 105–125 km when compared with the mass spectrometer and NO release profiles. The molecular oxygen and argon profiles are generally consistent with the mass spectrometer data over the altitude region above 105 km. However, it should be emphasized that this difference between the theory and measurements below 105 km could be due in large measure to the lack of turbulence measurements below 90 km. Above 125 km, the differences between the atomic oxygen measurements and the theory are related to the difference in the total mass density.

References

[1] S. P. ZIMMERMAN, N. W. ROSENBERG, A. C. FAIRE, D. GOLOMB, E. A. MURPHY, W. K. VICKERY, C. A. TROWBRIDGE and D. REES, Space Research XIV, 81 (1974).
[2] B. W. REINISCH and M. A. MACLEOD, EOS Trans. Am. Geophys. Un. 54, 379 (1973).

[3] M. A. MacLeod, T. J. Keneshea and B. W. Reinisch, EOS Trans. Am. Geophys. Un. **54**, 381 (1973).
[4] B. S. Dandekar, EOS Trans. Am. Geophys. Un. **54**, 380 (1973).
[5] D. Golomb and R. E. Good, EOS Trans. Am. Geophys. Un. **54**, 380 (1973).
[6] C. R. Philbrick, G. A. Faucher and E. Trzcinski, EOS Trans. Am. Geophys. Un. **54**, 380 (1973).
[7] T. J. Keneshea and S. P. Zimmerman, Trans. Am. Geophys. Un. **54**, 381 (1973).
[8] K. Bibl and B. W. Reinisch, Final Rep. AFCRL-72-0064 (1972).
[9] M. A. MacLeod, J. Atmos. Sci. **23**, 96 (1966).
[10] R. S. Narcisi, C. R. Philbrick, M. A. MacLeod and N. W. Rosenberg, EOS Trans. Am. Geophys. Un. **53**, 462 (1972).
[11] T. J. Keneshea and M. A. MacLeod, J. Atmos. Sci. **27**, 081 (1070).
[12] C. R. Philbrick, G. A. Faucher and R. A. Wlodyka, AFCRL Rep. 71-0602 (1971).
[13] C. R. Philbrick, G. A. Faucher and E. Trzcinski, Space Research XIII, 255 (1973).
[14] D. Golomb and R. E. Good, Space Research XII, 675 (1972).
[15] R. E. Good and D. Golomb, Space Research XIII, 249 (1973).
[16] D. Golomb and R. E. Good, J. Chem. Phys. **49**, 4176 (1968).
[17] D. Offermann and A. Drescher, Submitted to J. Geophys. Res. (1973).
[18] L. G. Jacchia, Smithson. Astrophys. Obs. Spec. Rep. No. 332 (1971).
[19] J. G. George, S. P. Zimmerman and T. J. Keneshea, Space Research XII, 695 (1972).
[20] T. J. Keneshea and S. P. Zimmerman, J. Atmos. Sci. **27**, 831 (1970).
[21] T. J. Shimazaki, J. Atmos. Terr. Phys. **29**, 723 (1967).
[22] A. C. Faire, E. A. Murphy and R. O. Olsen, EOS Trans. Am. Geophys. Un. **54**, 379 (1973).

Space Research XIV — Akademie-Verlag, Berlin 1974

ATMOSPHERIC DENSITY, TEMPERATURE AND WINDS MEASURED DURING ALADDIN II

A. C. FAIRE[a], E. A. MURPHY[a] and R. O. OLSEN[b]

[a]Air Force Cambridge Research Laboratories, Bedford, Mass., USA
[b]Atmospheric Sciences Laboratory, White Sands, New Mexico, USA

Results are reported and comparisons made of measurements of neutral atmospheric density, temperature and winds obtained during the Aladdin II rocket program at Eglin, Florida on 12 April 1972. Data were acquired from independent measurements provided by a passive inflatable sphere of 1-meter diameter and two 7-inch rigid falling spheres instrumented with omnidirectional accelerometers. The observations are compared in the altitude regions where the measurements overlapped. All three sets of data contain considerable oscillatory perturbations which, below 85 km, consist of relatively small positive and negative departures relative to the US Standard Atmosphere 1962. Preliminary analysis of northern hemisphere pressure maps and wind data from the inflatable sphere suggest the onset of a spring transition during the time of the measurements. Possible effects of the transition are discussed.

1. Introduction

Neutral atmospheric density, temperature and wind data are reported that were acquired from three rocket-borne falling sphere measurements made during the Aladdin II program at Eglin, Florida, on 12 April 1972. A passive inflatable sphere of 1-meter diameter was launched onboard a Super Loki Dart rocket vehicle at 1905 LT, followed by two 7-inch falling spheres launched onboard Nike Iroquois vehicles at 2045 and 2300 LT. Although the primary objective of the three *in situ* measurements was to provide an accurate reference frame for correlated aeronomical computations, other by-products of geophysical interest emerged during analysis of the data. The relatively short time that elapsed between measurements enabled a comparison to be made of two different techniques used to determine density and temperature in overlapping regions of the atmosphere.

2. Experimental Procedure

Inflatable Sphere. The inflatable sphere system for deriving atmospheric density and winds is a passive technique which is dependent upon radar acquisition of the sphere when it is deployed from a Super Loki Dart rocket at a nominal apogee of 115 km. Near apogee, a pyro-delay train ignites an expulsion charge and the folded sphere is ejected from a canister. Rapid inflation is accomplished by the vaporization of an isopentane liquid. The sphere is fabricated from 0.5 mil mylar which is metallized to produce a good reflective target for radar tracking. Density and wind data are determined by processing the radar observations using appropriate aerodynamic and trajectory equations.

Instrumented 7-inch Falling Sphere. The 7-inch rigid falling sphere, instrumented with an omnidirectional accelerometer, is deployed from an appropriate rocket vehicle at a programmed time during ascent. The accelerometer senses atmospheric drag on the sphere during its free flight, thus providing two sets of measurements (during the upleg and downleg portions of the sphere trajectory) with discrete temporal and spatial separation. Details of the technique have been reported previously [1—3].

3. Results

Density. Density results from each of the three measurements are plotted in Fig. 1 as departures from the US Standard Atmosphere 1962 (USSA-62) [4]. Also, plotted for comparison is the mid-latitude spring/fall model from the US Standard Atmosphere Supplements, 1966 (USSAS) [5]. The 7-inch sphere launched at 2045 was flown onboard a chemical release rocket and the upper end of the measurement was limited by the imposed weight constraints of the combined payloads. Unfortunately, there were also no usable data from this measurement below 68 km. Prominent wave structure is observed that increases in amplitude with height up to 96 km where the data terminate, and at this altitude the density departure relative to USSA-62 is 47% and about 25% from the corresponding value of the seasonal model. Oscillations are also evident in the density departure curve of the 2300 7-inch sphere data but there are some noticeable differences in the amplitude and phase of the oscillations in the altitude region where the data can

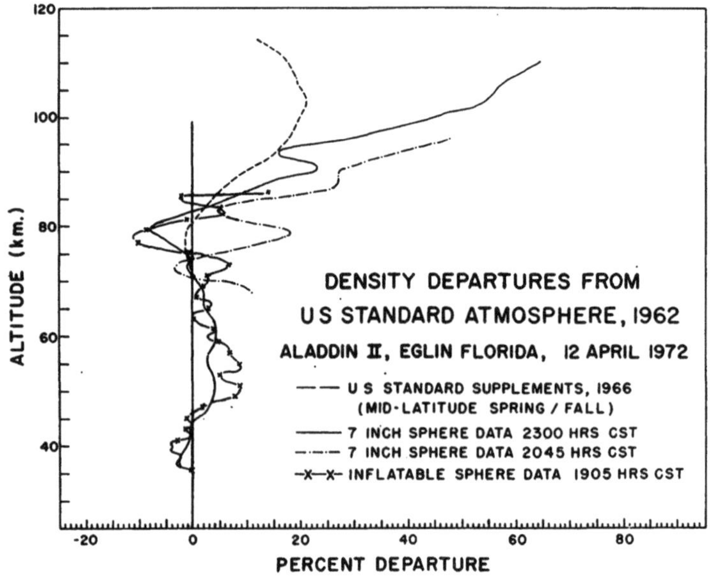

Fig. 1. Density departures from USSA-62, 1905, 2045 and 2300 LT, 12 April 1972.

be compared with that of the 2045 measurement. For example, in the vicinity of 80 km, the 2300 data show negative departures of about 8% from USSA-62, while the 2045 data show positive departures of about 18%. Above 82 km both measurements show departures that are positive with respect to USSA-62. While the smoothed data from the inflatable sphere launched at 1905 exhibit an oscillatory pattern, the mean value of the curve is in quite satisfactory agreement with that obtained from the 7-inch sphere at 2300. The observed oscillatory pattern may be due partly to the effects of winds; however, these density ratios are not typical of previous measurements, nor can they be explained in terms of errors

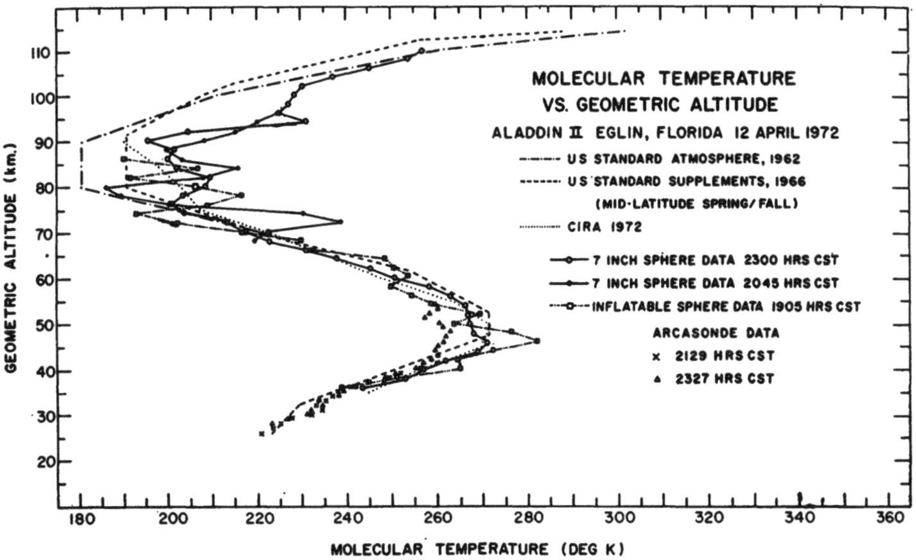

Fig. 2. Molecular-scale temperature, inflatable sphere 1905 LT, 7-inch spheres 2045 and 2300 LT, 12 April 1972.

associated with the inflatable sphere technique. The inflatable sphere descends with relatively small velocities thus making it more sensitive to winds and small amplitude density fluctuations.

It is apparent that the agreement is good between the inflatable sphere and the 7-inch sphere results obtained at 2300. In fact, the two measurements lie within 0—9% of each other up to 86 km. Even more significant is the fact that these data tend to show resemblance to spring/summer conditions as indicated by the departures relative to USSA-62 which are similar to those of the appropriate models from USSAS-66. The phase and amplitude differences observed in the 2045 data and those obtained from the first and third measurements suggest that in between a small transient disturbance occurred but that the atmosphere returned to original conditions by 2300. Density oscillations due to dynamical changes in the atmosphere appear to have a high probability of occurrence during the transition seasons and may be related to these observations.

Molecular-Scale Temperature. Molecular-scale temperatures calculated from the 7-inch sphere and the inflatable sphere densities are plotted in Fig. 2 with meteorological rocket-sonde data obtained at 2129 and 2327 LT. The 2300 sphere temperatures lie close to model values from USSAS-66 up to approximately 75 km, whereas between 75 and 100 km they are higher than those of the models with variations ranging from about 5 to 33 K. Also, these data are in very close agreement with the CIRA 1972 model for April, 30° N. The 2045 data show considerably more structure with temperatures that vary about 5 K less than model values to about 27 K higher in the altitude range 68—96 km. In the stratosphere very good agreement is observed between the meteorological rocket-sonde and the 2300 sphere data and model values, except near the stratopause. In the stratosphere, temperatures derived from the inflatable sphere densities are in general agreement with those obtained from the 2300 7-inch sphere results. The maximum differences range from 0 to 11 K between 36 and 70 km and the average lapse rate in the lower mesosphere is in good agreement with that of the models. Above 70 km somewhat larger variations from model values are observed that have approximately the same amplitude and wavelength. It is probable that phase differences are attributable to differences in the times of observations.

Winds. Wind data from the inflatable sphere experiment are plotted in Fig. 3. At the lower altitudes (36—60 km), the winds are generally light and variable.

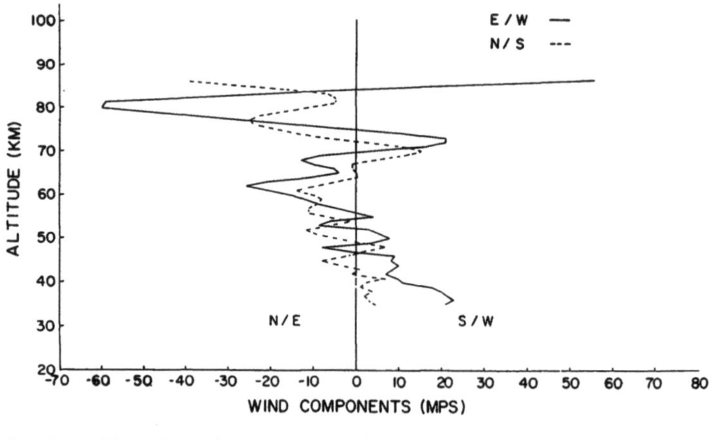

Fig. 3. Zonal and meridional winds components from inflatable sphere, 12 April 1972, 1905 LT.

Above 60 km the horizontal wind components increase in magnitude, with the zonal wind components generally larger than the meridional components above 60 km. Large wind shears are noted in the region 75—86 km.

Pressure maps of the northern hemisphere for the 5, 2 and 0.4 mb levels on 12 April 1972 have been studied. The winter-time circulation pattern in the lower, mid and upper stratosphere commences to break down on 5 April and by 12 April polar vortex splitting is evident. It is possible that large scale disturbances, revealed in the pressure maps for the period of the Aladdin II measurements may be related to the small transient disturbance observed in the data. Further examination of the pressure maps reveal that by 19 April summer conditions are commencing and by 26 April the summer regime is firmly established.

4. Discussion and Conclusions

Density, temperaure and wind data obtained from three falling-sphere measurements made at Eglin, Florida, during Aladdin II suggest that a spring transition was in progress during the period of measurement. This conclusion is supported by an analysis of northern hemisphere pressure maps which appear to have features of a typical seasonal transition. Wilkins [6] describes the seasonal transition as a period of drastic rearrangement of stratospheric structure and motion and assumes the transition line, with opposing wind vectors, to be a region of potentially great instability, wave generation and intense turbulence. The oscillations observed in the data presented could very well be related to lower atmospheric dynamics that were triggered by the spring transition. Within the experimental error the measurements are considered to be self consistent.

Acknowledgments

The authors wish to express their thanks to Dr. K. S. W. Champion of AFCRL for helpful suggestions made while reviewing the manuscript. They also wish to acknowledge the valuable technical assistance provided by Kenneth McGee of AFCRL, Gustave Stromberg and his associates at Wentworth Institute, Boston, Mass. and the personnel of the Atmospheric Sciences Laboratory, White Sands, New Mexico.

References

[1] A. C. Faire, in: 1966 Natl. Telem. Conf. Proc., Boston, Mass., 1966 (p. 70).
[2] A. C. Faire and K. S. W. Champion, Space Research V, 1039 (1965).
[3] A. C. Faire and K. S. W. Champion, Space Research VII, 1046 (1967).
[4] COESA, US Standard Atmosphere, 1962, Government Printer, Washington, D. C.
[5] COESA, US Standard Atmosphere Supplements, 1966, Government Printer, Washington, D. C.
[6] E. M. Wilkins, ATC Rep. No. B-94000/2 TR-18, Advanced Tech. Center, Dallas, Texas, 1972.

Space Research XIV — Akademie-Verlag, Berlin 1974

COMBINED TEMPERATURE, DIFFUSION COEFFICIENT AND DENSITY MEASUREMENTS OF PHOTOLUMINESCENT AlO RELEASES

D. Rees[a], D. F. Kitrosser[b] and D. Golomb[b]

[a]Department of Physics and Astronomy, University College London, London, UK
[b]Air Force Cambridge Research Laboratories/LKC, L. G. Hanscom Field, Bedford, Mass., USA

The objectives of Project Aladdin II, a complex rocket campaign at Eglin Air Force Base, Florida, USA (30° N 87° W) were to make a comprehensive study of the temperature, density, composition, movements and ionization of the atmosphere and their variations through the night of 12—13 April 1972. This paper presents the combined results of two independent series of measurements of neutral temperature and diffusion coefficient obtained by CRL and UCL experimenters during launches at both evening and morning twilight. Agreement between the two sets of temperature results was within the computed experimental error limit (± 30°K), while the diffusion coefficient measurements had a standard difference of 10% above 130 km altitude, compared with a computed standard error of about 15%.

Air density at morning twilight, calculated from the temperature and diffusion coefficient measurements, was about 50% higher than that at evening twilight between 130 and 160 km. The reality of this anomaly is supported by the excellent agreement between both sets of temperature and diffusion coefficient data.

1. Introduction

Ground-based photographic and spectroscopic observations of photoluminous chemical release experiments from rockets have been used for about ten years as a means of determining upper atmosphere temperature and density. The main advantage of the method is that it is independent of the calibration problems associated with mass spectrometer experiments and provides a far higher temporal and spatial resolution than satellite drag methods.

On previous occasions the compatibility of the chemical release results with those of other methods has been reported (rocket mass spectrometers [1], satellite drag and models derived from satellite drag [2—4]). Three research groups, Cambridge Research Laboratories (CRL), University College London (UCL) and Upper Atmosphere Research Group, Weapons Research Establishment, have produced most of the published temperature and density measurements from chemical release experiments. The compatibility of the results of the techniques developed by the latter two groups has already been established and reported [5, 6]; however, the first occasion when it was logistically feasible to check the compatibility of the other two techniques with those utilized by CRL occurred during the Aladdin II rocket campaign at Eglin Air Force Base in April 1972. At the same time, the results of the chemical release techniques could be cross-checked with two alternative methods (18 and 100 cm falling spheres and neutral mass spectrometer). This paper will only discuss the measurements of neutral atmosphere structure derived from CRL and UCL observations of the chemical release experiments, a comparison with falling sphere and mass spectrometer data being the subject of a future paper.

2. Experimental Techniques

2.1. Measurement of Neutral Temperature

Both CRL and UCL temperature measurements were derived from spectroscopic observations of the AlO ($A^2\Sigma - X^2\Sigma$) molecular bands produced by resonance with sunlight* from the release of TMA puffs during the two twilight rocket launches (Clara, 1843 CST 12 April 1972; Hope, 0430 CST 13 April 1972). The altitude range covered by the two sets of measurements was approximately 120 to 170 km.

Before comparing the reduced data from both sets of measurements, it is worthwhile to contrast the methods of obtaining and reducing the AlO spectra. The two spectrometers employed (Fastie-Ebert CRL, Czerny-Turner UCL) each have approximately the same light gathering power for a given resolution. The CRL instrument is used with a 2 Å slit width and a scan period of 5 sec, while the UCL instrument has a 4 Å slit width and a 12 sec scan period. Both instruments use cooled photomultipliers and scan a wavelength range of the order of 8—10 nm about the 484 nm AlO (0-0) band, so that the main contribution to noise in the data is $n^{1/2}$ due to statistical fluctuations, where n is the number of photons counted during the sample period used during the data reduction. The wider slit and longer scan period of the UCL spectrometer essentially produces a less noisy spectrum than the CRL spectrometer. However, the inherent errors due to the error in determining the slit width (slit function) of the UCL instrument and the effects of AlO clouds drifting with time are appreciably greater, off-setting the less noisy spectrum. Averaging of data from several scans of a given cloud is employed to improve the statistics of the results with both techniques, the improvement being somewhat more effective with the greater number of scans per cloud obtainable with the CRL instrument.

Both CRL and UCL reduction techniques are based on comparisons of synthetic spectra, constructed for a convenient range of thermospheric temperatures from the same molecular constants and solar spectrum corrected for Fraunhofer line absorption, with each observed spectrum. Both techniques therefore are subject to any systematic error inherent in either the molecular constants or the solar spectrum. We estimate the maximum absolute uncertainty of these combined errors to be ± 30 °K.

The exact methods of analysis of the data differ considerably. The UCL data measurements, counted and recorded on eight-track paper tape during the experiment, are normalized with respect to wavelength and intensity and subjected to a weighted least-squares analysis with respect to the synthetic spectra constructed for a range of temperatures. The computed results of the analysis are

---

* In accordance with the suggestion of Prof. H. P. Broida, the process of photon-induced emission of resonance lines or bands should be called *photoluminescence*.

a temperature and its standard deviation. The CRL analysis is somewhat more rigorous in that a probability function is computed to establish the likelihood that a particular observed profile results from a normal error distribution about each synthetic profile, for a range of temperatures. We have established, however, that, if the only sources of noise are background skylight and the $n^{1/2}$ of the sample of photons, the methods are essentially identical. For other forms of noise, i.e. electrical switching, transient or detected RF interference causing spurious counts, the CRL analysis should be superior.

2.2. Molecular Diffusion Coefficient

Reduction of photographic records of the expansion of the photoluminous AlO clouds enables a measurement of $D_{1,2}$, the diffusion coefficient of contaminant into the ambient atmosphere, to be made. Both CRL and UCL techniques are based on fitting, by a least-squares method, a gaussian model of the type

$$I = I_0 \exp\left(-r^2/r_e^2\right)$$

to two-dimensional intensity vs radial distance r, profiles of the clouds obtained from microdensitometry of calibrated films of the features. A second least-squares fit to the equation

$$r_e^2(t) = r_0^2 + 4D_{1,2}t$$

for a series of values of r_e (the gaussian radius) vs time t, is then solved to obtain $D_{1,2}$; r_0 is a mathematical 'initial radius', a quantity derived from the least-squares analysis. The CRL and UCL photographic, calibration, microdensitometer and numerical reduction techniques have been described previously [6, 7].

2.3. Air Density

From profiles of neutral temperature T and diffusion coefficient $D_{1,2}$, air density ϱ may be computed according to the gas-kinetic relation [8]

$$\varrho = \frac{3(kTm_{\mathrm{H}}M_{1,2})^{1/2}}{8\pi^{1/2}D_{1,2}\sigma_{1,2}^2\Omega(1,1)^*}$$

where m_{H} = mass of hydrogen atom, $M_{1,2}$ = mean molecular weight = $2M_1M_2/(M_1 + M_2)$, $\Omega(1,1)^*$ = quantity indicating the deviation of a real molecule from a rigid sphere. It is a tabulated function of $t^* = kT/\varepsilon$, where ε is the maximum energy of attraction between the molecules (depth of potential well). We adopt a typical value $k/\varepsilon \approx 10^{-2}$ (°K^{-1}).

If we assume $\sigma_2/\sigma_1 \propto (M_2/M_1)^{1/3}$, $M_2 = 43$ (molecular weight of AlO) and $\sigma_1 = 3.6 \times 10^{-8}$ cm (typical for a diatomic molecule with molecular weight 28—32), we obtain a collision diameter of AlO, $\sigma_2 = 4.1 \times 10^{-8}$ cm. For an ambient molecular weight of 22.5 (at 150 km), $\sigma_{1,2} = 3.7 \times 10^{-8}$. We shall assume that the mean collision diameter varies with the molecular weight of the ambient

$$\sigma_{1,2} = \sigma_2[1 + (M_1/M_2)^{1/3}]/2$$

where M_1 is an altitude-dependent molecular weight. This variation amounts to only 5% from 120 to 170 km. The density is derived from

$$\varrho = \frac{7.2 \times 10^{-5}\, T^{1/2}}{[1 + (M_1/M_2)^{1/3}]^2 [1 + M_2/M_1]^{1/2} \Omega(1,1)^* \, D_{1,2}} \ \mathrm{g\ cm^{-3}}$$

where T and $D_{1,2}$ are experimentally determined, $\Omega(1,1)^*$ is taken for the appropriate T, M_1 is taken from CIRA 1972 Model Atmosphere [9]. Thus, the experimental error in ϱ should be largely determined by the error of $D_{1,2}$ and by the square root of T. The overall estimated experimental error of ϱ is $\pm 25\%$; however, an additional systematic error of $\pm 10\%$ is possible due to the uncertainty of $\sigma_{1,2}$.

3. Comparative Experimental Measurements

Temperature profiles obtained during both evening and morning twilight launches are shown in Fig. 1. The average discrepancy between CRL and UCL measurements of a given cloud is 27 °K, while over the whole series, the UCL measurements average 11 °K higher than the CRL measurements. Both these statistical results are consistent with our earlier estimates of random experimental error and indicate that including the effects of a possible common systematic error to the techniques, the standard deviation of measurements of a particular AlO cloud should not exceed 40 °K.

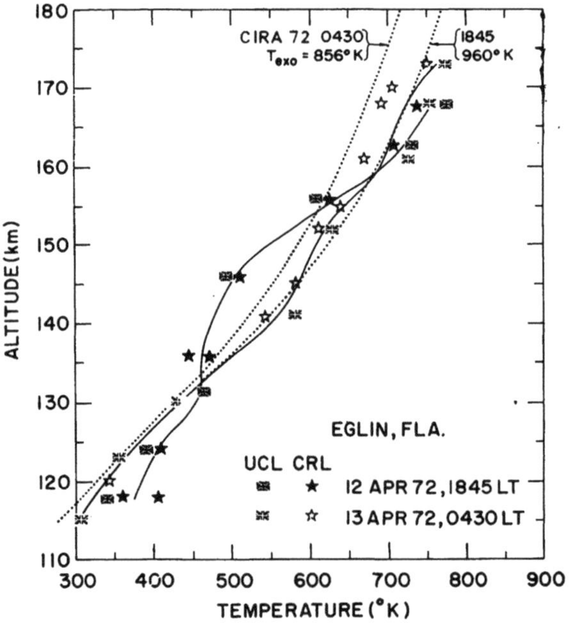

Fig. 1. Neutral temperature, 120—170 km altitude, over Elgin Air Force Base, Florida, at evening and morning twilight, 12/13 April 1972.

Fig. 1 shows several distinctive features on both profiles which, on the basis of the above analysis, are quite significant. Below 130 km and above 160 km the evening temperatures are higher than those in the morning. This could have been anticipated for the greater altitudes, but indicates a strong variation of about 50 °K at the lower level also (which cannot, of course, be called a diurnal variation on the basis of two measurements). Moreover, between 130 and 160 km, there is even stronger evidence that the evening temperatures were substantially (i.e. ~50 °K) lower than the morning temperatures. Reference to two CIRA 1972 [9] profiles indicates that the morning temperature profile was close to a model curve but about 50 °K higher than might have been expected (above 130 km) for the location, local time, season, solar and geomagnetic activity levels, while the evening profile, although closer, on average, to the relevant model, showed anomalous behaviour, being too high below 130 km and too low between 130 and 160 km.

Diffusion coefficient profiles are shown in Fig. 2, but will be discussed only in relation to the derived density profiles (Fig. 3). With the exception of one cloud near 140 km during the dawn experiment where there was a discrepancy of 40% between CRL and UCL measurements, the agreement between the two sets of results was considerably better than expected. The average difference between the two sets of results for a given cloud was about 10%, while, taken over the whole series, there was a systematic bias of less than 2%.

The most noticeable feature of the density profiles is that the morning densities are consistently almost 50% higher than the evening values above about 140 km altitude. Although we would expect the maximum absolute error in density to be

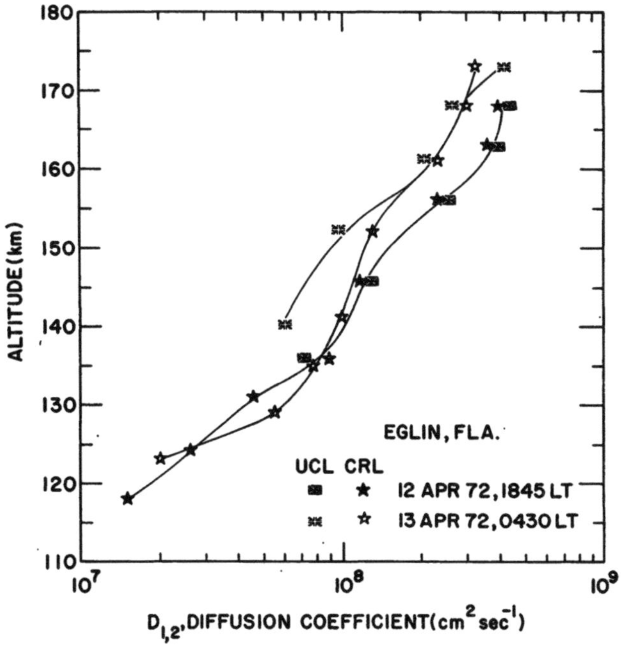

Fig. 2. Diffusion coefficient, 120—170 km altitude, over Elgin Air Force Base, Florida, at evening and morning twilight, 12/13 April 1972.

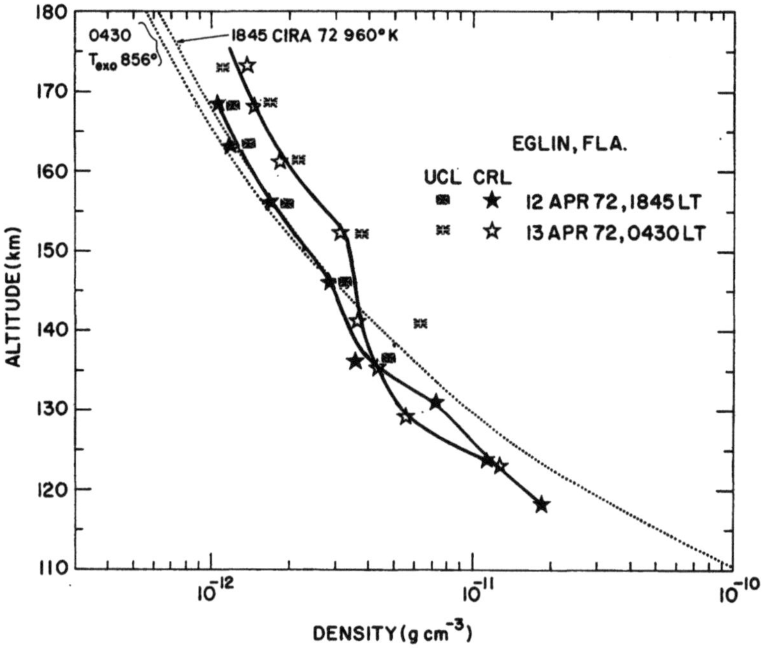

Fig. 3. Air density, 120—170 km altitude, over Elgin Air Force Base, Florida, at evening and morning twilight, 12/13 April 1972.

about 35%, the results during a given experiment and also from two different launches should be self-consistent to about 10—15%. Therefore, the systematic dawn-dusk difference, much larger than the 10% (with reversed phase) model variation, should be highly significant. There is no plausible explanation for the anomalous density results — the effect on air density due to the low evening temperatures between 130 and 160 km should not exceed 10—15% and that should be greatest near 160 km, while the observed evening density reduction extends to below 140 km.

While the evening density profile is consistent with the CIRA 1972 model for $T_{exo} \sim 900\,°K$, the morning profile above 140 km would require $T_{exo} \sim 1100$ to $1200\,°K$, a value substantially in excess of that obtained by extrapolating the observed temperature values ($\sim 950\,°K$), and of the predicted value for the time, place and solar and geomagnetic activity levels. Localized variations of the required magnitude (30—50%) have frequently, however, been recorded by satellite accelerometers and mass spectrometers, and are generally interpreted as evidence of gravity waves.

Below about 130—135 km the density results become systematically and increasingly lower than the CIRA 1972 model values. The discrepancy has been noted previously [5, 6] and appears to be associated with the experimental technique. Partially, this may be due to the difficulty, when computing diffusion coefficients, of correctly assessing and allowing for the distortion of the lower altitude clouds due to wind shears, the effects of which become negligible at greater altitudes (i.e. above 135 km). Additionally, during both experiments photography of the lower altitude releases was hampered by tropospheric cloud. Below about 120 km we suggest that, on occasions, the expansion of clouds and trails is enhanced by turbulence eddy diffusion, even though the appearance of the clouds may not be irregular, characteristic of clouds below the so-called turbopause [5].

4. Summary

The comparative measurements of neutral temperature and molecular diffusion coefficient which were made during the two twilight rocket experiments of the Aladdin II campaign showed that the differing experimental and analytical techniques employed by the CRL and UCL groups produced compatible results. We conclude that the absolute standard error of individual temperature measurements should not exceed 40 °K while that of ambient density computed from the measurements of neutral temperature and molecular diffusion coefficient should not exceed 35%.

Between 130 and 160 km altitude the neutral temperature at dawn was found to be about 50 °K higher than at the previous dusk, while, below 130 km and above 160 km, the dusk temperature was higher than that observed during the dawn experiment. At the lower altitudes (120—130 km) the magnitude of the variation (50 °K) was substantially greater than that allowed in atmospheric models.

The dawn density measurements above 140 km were consistently 50% higher than those during dusk. The magnitude of the variation and its opposite phase to that predicted by models should reflect the variability of the real as opposed to the model atmosphere.

Acknowledgments

The chemical release payloads of the Aladdin II campaign were prepared at the Chemical Physics Branch, Air Force Cambridge Research Laboratories, and launched at Eglin Air Force Base with the full assistance of the range staff. The UCL experimental team was supported by grants from the British Science Research Council and the European Office of Aerospace Research, who assisted greatly in coordinating their participation in the programme.

References

[1] U. von Zahn, J. Geophys. Res. **75**, 5571 (1970).
[2] K. S. W. Champion, Space Research X, 347 (1970).
[3] D. Rees, J. Brit. Interplan. Soc. **24**, 643 (1971).
[4] D. Rees, in: Physics and Chemistry of Upper Atmospheres, D. Reidel Publ. Co., Dordrecht 1973 (p. 11).
[5] D. Rees et al., Phil. Trans. Roy. Soc. London A **271**, 631 (1972).
[6] K. H. Lloyd et al., Planet. Space Sci. **20**, 761 (1972).
[7] D. Golomb et al., Space Research XII, 733 (1972).
[8] J. O. Hirschfelder et al., Molecular Theory of Gases and Liquids, Wiley, New York 1954.
[9] CIRA 1972, COSPAR International Reference Atmosphere, Akademie-Verlag, Berlin 1972.

Space Research XIV — Akademie-Verlag, Berlin 1974

WIND PROFILES OVER HEISS ISLAND

L. A. Andreyeva and L. M. Uvarova

Hydrometeorological Service, Moscow, USSR

Wind measurement results obtained on Heiss Island (80°37′N) in 1970—1971 are given. Wind profiles are obtained in the 80—160 km height region by observing luminous trails and spherical clouds released from rockets. A wind velocity greater 300 m s^{-1} connected with increased geomagnetic activity was observed.

Until now the most northern station where wind profiles have been obtained over a large height interval was Esrange, Kiruna, Sweden [1]. Wind velocity data obtained over Heiss Island are given in [2]; the wind velocity was measured from isolated spherical clouds. The wind profiles from 80 to 160 km were obtained on Heiss Island in February 1970 and March 1971.

Sodium clouds were formed as long trails or by near-simultaneous formation of some luminous spheres. Photography was carried out from two stations by cameras with f 2.5 lenses and with focal lengths of 25 and 10 cm. Experimental details are given in Table 1.

Table 1

Experimental Details

Launch number	Date	Launch time (Moscow)	Cloud form	*Kp*			*K*		
				0	−1	−2	0	−1	−2
1	2	3	4	5			6		
1	26 Feb. 70	16.33	long trail	4	2	2^+	4	3	3
2	23 Feb. 70	16.54	three spher. clouds	3^-	3^-	3	4	4	3
3	14 Mar. 71	01.59	six spher. clouds	5^+	6	5^-	4	6	5
4	17 Mar. 71	19.59	long trail	3^-	1	2^+	3	3	4
5	18 Mar. 71	20.14	long trail	1^+	1^-	1^+	6	3	4

The planetary Kp index and the local K index, characterizing the geomagnetic situation during the experiments on Heiss Island, are given in columns 5 and 6 of Table 1. In these columns the three-hour interval during which the wind data are obtained is denoted by 0; intervals 3 and 6 hours before the experiments are denoted by -1 and -2.

Three wind profiles (1, 4 and 5 according to the launch number) are obtained from 85 to 165 km. The height resolution for profile 1 is somewhat lower than in the two others. The standard error of measurement of the wind velocity at heights up to 120 km was not as a rule greater than 3 m s^{-1} and, at greater heights, 5 m s^{-1}.

Figs. 1, 2 and 3 show values of the full wind velocity and meridional and zonal wind components. Fig. 4 shows the dependence of wind direction on height.

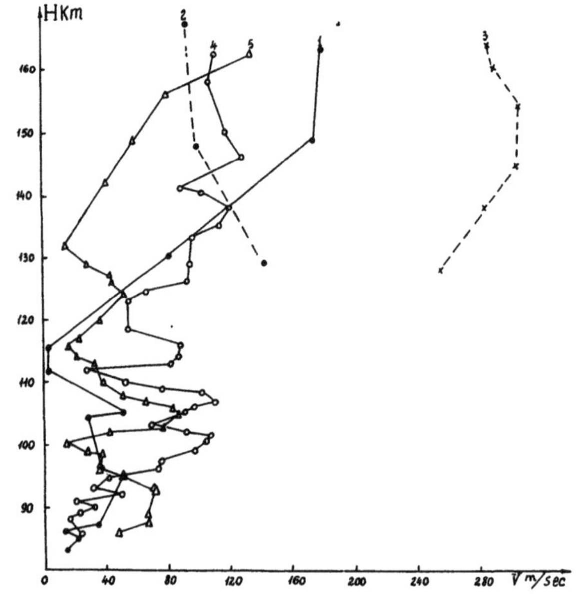

Fig. 1. Height profiles of full wind velocity.

It is clear that the wind profiles in polar latitudes as well as in mid-latitudes are characterized by great irregularity. The greatest velocity variations are observed in the 90—115 km height region. Above 120—130 km the wind profile behaviour is quieter, except for the profile obtained on 17 March, when the velocity variations with amplitudes of the order of 45 m s^{-1} were observed up to 150 km. These variations are caused mainly by zonal component variations. Above 120 km the meridional component gradually increases with height with a vertical gradient of 5—6 m s^{-1} km^{-1}. Analogous wind velocity increase with height is observed in profiles 2 and 5 (Fig. 3). The greatest vertical gradients of the horizontal wind velocity are recorded in the 90—110 km height interval and are equal to about 40 m s^{-1} km^{-1}. Significant wind direction variations were recorded at heights up to 130 km. Above this height the wind changes its direction by 1—1.5 degrees per kilometre on the average.

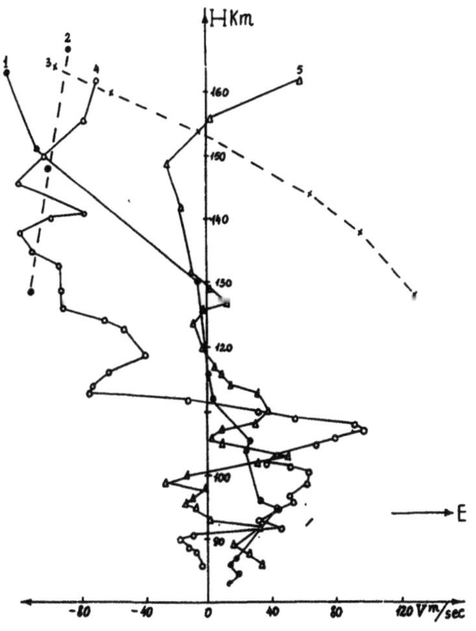

Fig. 2. Height profiles of the zonal wind velocity component.

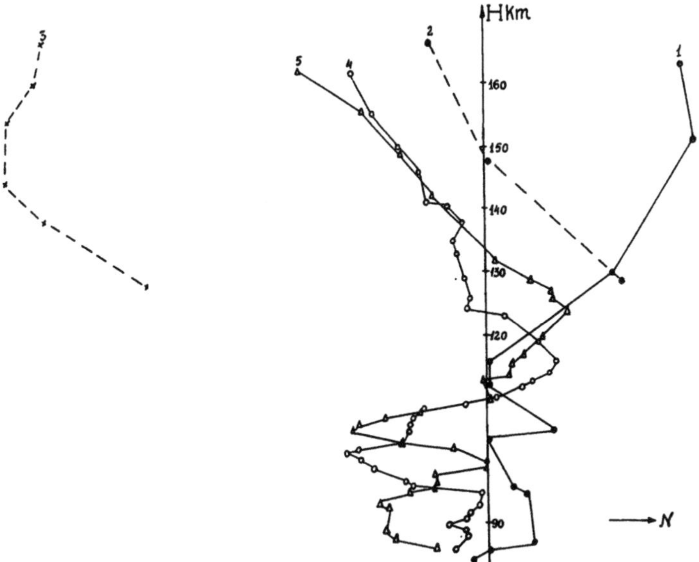

Fig. 3. Height profiles of the meridional wind velocity component.

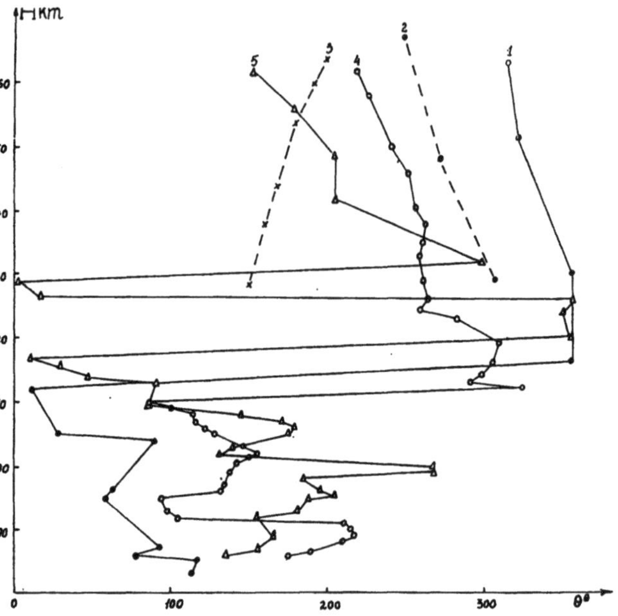

Fig. 4. Wind direction dependence on height.

The characteristic globular trail structure is observed on 17 and 18 March in the 101—107 km and 95—107 km height regions respectively. In the cloud formed on 26 February well developed globulars were not recorded. It is seen from the Figures that significant vertical gradients and great profile variability were also recorded on 17 and 18 March. All three clouds were formed in the evening twilight on the descent.

Turbulence was absent in the cloud of 26 February 1971; the occurrence of turbulence on 17 and 18 March 1971 may be connected with different polar thermospheric stability on those days.

The case of 14 March 1971 when high wind velocity was observed in the 130—165 km height region is of particular interest. The wind velocity from 143 to 155 km was greater than 300 m s^{-1}, the meridional component being by far the larger. From 13 to 17 March a moderate geomagnetic storm and reduced ionospheric electron densities were registered [3]. The experiment on 14 March was carried out at a time when the local K index and the planetary Kp index had values of 5—6. The experiment on 17 March, when the wind velocity greatly decreased, coincided with the period when the magnetic storm stopped.

Rees estimated [1] the W—E wind velocity dependence on variability of the S—N magnetic field component ΔX (in gamma) to be given by $V = 2.5\Delta X$ (in m s^{-1}). Such a distinct dependence for the wind profiles mentioned is not found. This may be explained by the fact that Esrange, Kiruna, and Heiss Island have different positions with respect to the auroral oval.

Acknowledgments

The contributions of V. F. Garbuzenko and A. N. Belova to the success of these experiments are greatly appreciated.

References

[1) D. Rees, J. Brit. Interplan. Soc. **24**, 233 (1971).
[2] L. A. Andreyeva, L. A. Katasyev and D. B. Uvarov, Meteorol. and Hydrolog. 1, 35 (1970).
[3] Cosmic Data, Monthly Review, March 1971, Nauka, Moscow.

Space Research XIV — Akademie-Verlag, Berlin 1974

WIND MEASUREMENTS USING MAGNETOMETER, OPTICAL AND STATIC PRESSURE SENSORS

A. S. Butko

USSR

A brief description of the method of wind measurement by means of large meteorological rockets is presented. The method permits the wind to be measured in the day time and during periods of full moon at night. Some results obtained during the "Sun-Atmosphere 1971" experiment between altitudes of 80 and 120 km are presented.

The instrument assembly consisted of a three-component magnetometer [1] and an optical sensor to give the orientation of the rocket, and an ionization gauge, which was arranged as a static pressure gauge [2].

To calculate the wind velocity and direction it is necessary to determine the orientation of the absolute velocity vector of the rocket. This is done using readings of the static pressure gauge along with the data on rocket orientation.

Then a velocity triangle is constructed consisting of three vectors: the ballistic velocity vector, the total velocity vector and the wind velocity vector. The value and orientation of the ballistic velocity vector can be determined from the trajectory data. Assuming that the vertical component of neutral wind is equal to zero, the value and the direction of the wind velocity vector is obtained.

In Fig. 1, which shows the character of wind variations in the upper and lower mesosphere, the data derived from two MP-12 rocket firings carried out during the "Sun-Atmosphere 1971" experiment are plotted. In Fig. 1a the data of the firing on 24 September 1971 are presented, and in Fig. 1b the data of the firing on 1 October 1971 and those obtained by the radio-drift method at the same time at an altitude of about 112 km [3].

As is shown in [4], 24 September falls during a comparatively quiet period of geomagnetic activity whereas 1 October falls during a period of disturbance.

Comparison of the zonal and meridional wind components shows that on 24 September the changeover of the zonal wind component from westerly to easterly occurred at an altitude of approximately 101 km, and on 1 October at an altitude of 104 km. Also, at an altitude of 113 km the reversal of the zonal flow from easterly to westerly took place, with the velocity of the zonal flow achieving approximately 200 m s^{-1} at an altitude of 115 km. Above 115 km a tendency to the reversal of the zonal component is again seen. A zonal component curve, obtained in the firing on 24 September does not show any such peculiarities.

The reversal of the meridional component in the firing on 24 September from

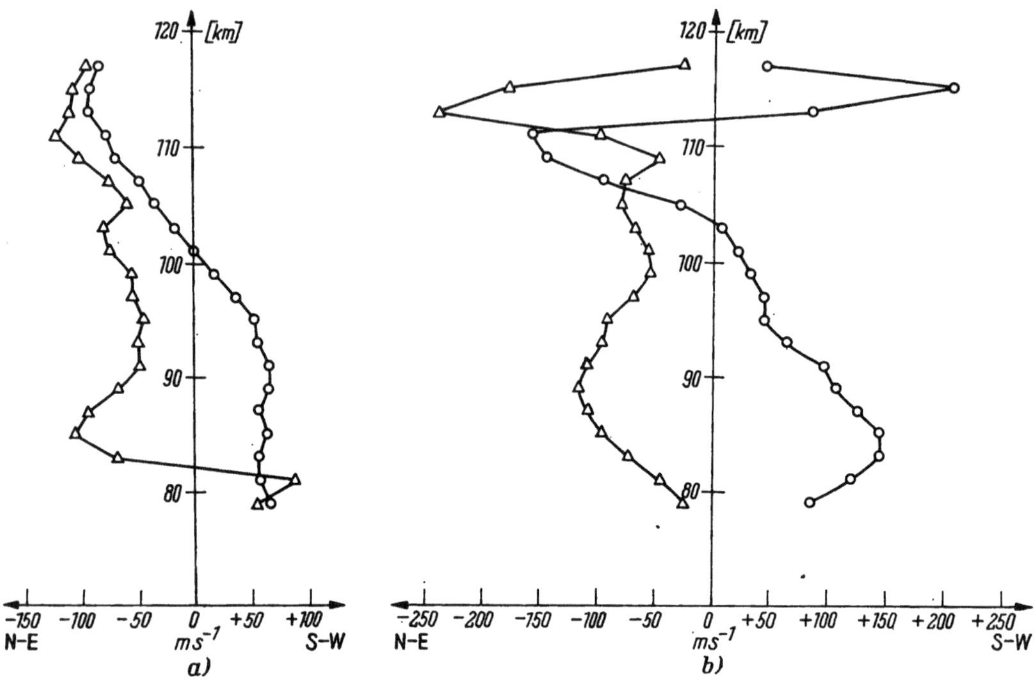

Fig. 1. a) 24 September 1971; b) 1 October 1971. ○ = zonal component, △ = meridional component.

southerly to northerly occurred at an altitude of about 83 km, whereas in the firing on 1 October it was below 80 km.

It should be noted that according to the measurement data of the radio-drift method [3] the magnitude of the wind velocity at an altitude of about 120 km on 24 September was about 60 m s^{-1}, and on 1 October about 200 m s^{-1}, which agrees quite well with our data.

References

[1] Yu. V. Afanasyev, in: Ferrozonds, "Energiya" 1969.

[2] V. M. Bederov, E. A. Besyadovsky, I. N. Ivanova, G. A. Kokin and A. S. Ozerova, Trudy CAO 77 (1967).

[3] The Results of Ionospheric Observations 5, Sib IZMIRAN, Irkutsk 1972.

[4] Z. A. Alexandrov, G. S. Ivanov-Kholodny, T. V. Kazachevskaya, Z. A. Katasyov, G. A. Kokin, O. P. Kolomiytsev, V. V. Mikhnevich, A. P. Popov, U. P. Teslenko and V. F. Tulinov, Meteorol. i Gidrol. No. 9 (1972).

Space Research XIV — Akademie-Verlag, Berlin 1974

MEASUREMENT OF NEUTRAL WINDS AT 360 km FROM A CHEMICAL RELEASE EXPERIMENT CONDUCTED AT SONMIANI

M. S. Ahmad, M. Hanif and P. J. Siddiqui

Pakistan Space and Upper Atmosphere Research Committee, Karachi, Pakistan

The super-rotation of the upper atmosphere derived from changes in orbital inclination of decaying satellites has been explained by several workers in terms of ion motion and neutral air pressure gradients. This theory predicts a wind pattern in which the direction and magnitude of the neutral gas velocity vector varies with local time. A locally fabricated Dragon 2 B rocket (Pak Shahpar 1) carrying a sodium vapour release payload was launched from Sonmiani at 0010 UT (0437 LMT) on 28 April 1972 to obtain measurements of neutral air winds at 360 km. The measurements show that the neutral air winds at this height are of the order of 500 m s^{-1} and are directed west-south-west in general agreement with the theory. The 'visible-radius' data have been used to compute the molecular diffusion coefficient at this height giving a value of 4.4×10^{10} cm^2 s^{-1}.

I. Introduction

The super-rotation of the upper atmosphere derived from changes in orbital inclination of decaying satellites [1] has been explained by several workers in terms of the pressure gradient and ion drag forces and secondary contributions such as viscous and Coriolis forces [2—4]. These investigations predict a wind pattern in which the direction and magnitude of the transport vector varies with local time, and in particular indicates, at certain times of the day, a significant 'east-to-west' component of neutral air winds. Unfortunately observational evidence at heights in the neighbourhood of the F2 region has been rather scanty— only a few luminous vapour releases at heights greater than 250 km and some Doppler shift measurements of airglow emission spectra have been made.

It was therefore considered useful to plan the first Pakistan-made Dragon 2B (Pak Shahpar 1) launching from the Sonmiani range on 28 April 1972 in such a way as to enable neutral air wind measurements to be made. Thus a payload was chosen which would allow sodium vapour releases at two discrete heights (approx. 200 and 400 km). In view of an earlier French Dragon launching from Sonmiani on 29 March 1968 in the evening hours ($\sim$ 1900 LMT) with a similar payload it was decided to launch Pak Shahpar 1 in the morning hours ($\sim$ 0500 LMT). The actual launch time was chosen to correspond to the solar depression angle of 12°. Although this experiment was only a partial success as the lower of the two planned releases of sodium vapours did not materialize, it did succeed in giving very interesting results as described below.

2. Neutral Air Winds at 360 km

A photograph of the quasi-spherical cloud of the released sodium vapours was taken at 265 seconds after lift-off; this photograph corresponds to a diffusion time of 22 seconds. The height of the release was obtained as 360 km, from the observed trajectory of the vehicle and the instant of the release. The cloud centre, c, defined in Fig. 1, moves showing the bulk motion of the neutral atmosphere at the height of the release. The neutral air winds thus determined

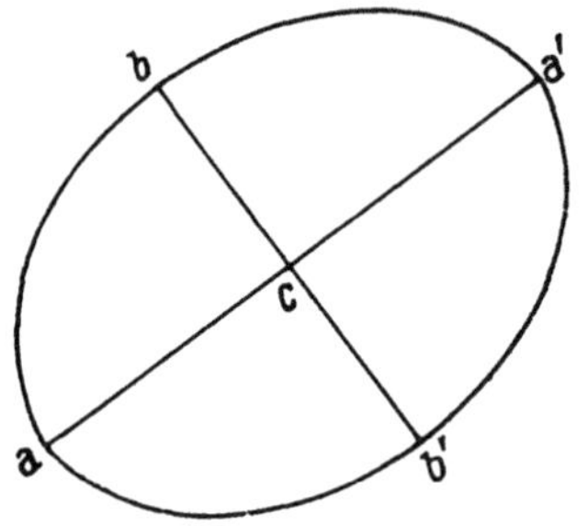

Fig. 1. Features a′, a, b′, b and c in a quasi-spherical cloud.

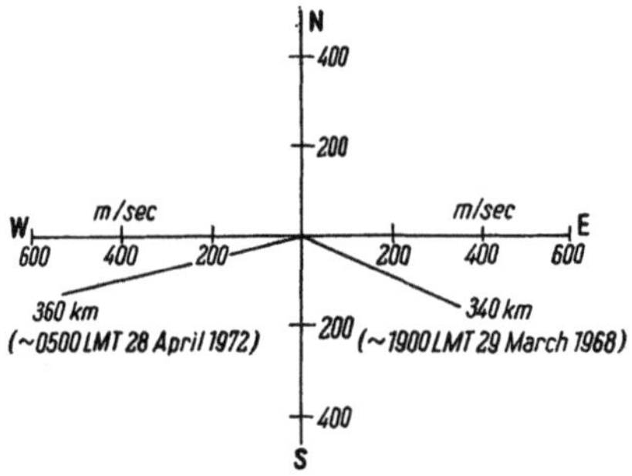

Fig. 2. Comparision of neutral air winds at Sonmiani at 360 km (∼ 0500 LMT on 28 April 1972) and 340 km (∼ 1900 LMT on 29 March 1968).

have a magnitude of ∼ 500 m s^{-1} and are directed to the west-south-west. Fig. 2 presents this result together with the corresponding data obtained from an earlier French Dragon firing (29 March 1968) from Sonmiani at 19 h LMT [5]. On comparing these data with the theoretical pattern given by Kohl for wind velocities at 400 km [4] one notes that the directions of the winds obtained at 05 h LMT as well as at 19 h LMT are in good agreement with the predicted pattern. The magnitudes of the winds are, however, larger on both occasions. Results from Doppler shifts observed in airglow emission spectra, although difficult to interpret, have indicated a west-to-east velocity of about 400 m s^{-1} in the evening [6] while in the morning hours between 03 h and 05 h the winds are directed east-to-west [7] in general agreement with the above data. In view of the fact that the measurements reported here are local atmospheric motions and therefore liable to short-lived disturbances, perfect agreement with a theoretical pattern would not be expected.

3. Diffusion Coefficient at 360 km

Computations of the molecular diffusion coefficient were made using the relationships applicable to the case of a quasi-spherical cloud [8]:

$$\sigma^2 = \frac{1}{2} r_e^2 \left\{ \ln \frac{e \varrho^2}{2} - \ln \sigma^2 \right\}^{-1}$$

$$\sigma^2 = \sigma_0^2 + 4D(t - t_0)$$

where σ^2 = time dependent variance of the Gaussian function of the column density of sodium atoms along the line of sight, σ_0 its initial value, r_e = radius of the sodium cloud just visible against the background sky, ϱ = maximum value of r_e, D is the diffusion coefficient of sodium in the atmosphere and $t - t_0$ = diffusion time.

The 'visible radius' r_e is obtained as an average of the four values corresponding to the features a, a′, b, b′ and the cloud centre c as shown in Fig. 1. The data thus obtained, and plotted in Fig. 3, represent the growth of the sodium cloud

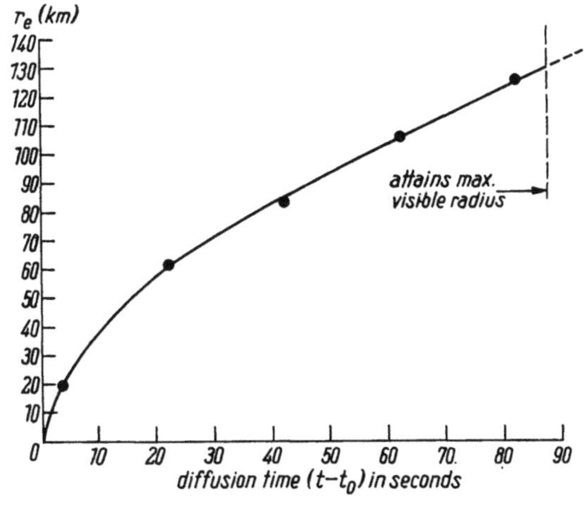

Fig. 3. Growth of quasi-spherical sodium release at 360 km (28 April 1972).

released at 360 km in this experiment. The maximum size ϱ of the visible radius required in the solution of the above transcendental equation was taken as 130 km corresponding to the time the cloud ceased to be distinguished aginst the background sky. It may be mentioned that the solar depression angle at this time was 10.5°. Fig. 4 presents the plot of r_e^2 and σ^2 against diffusion time $t - t_0$. The molecular diffusion coefficient at 360 km thus obtained has a value of 4.4×10^{10} $cm^2\ s^{-1}$. This value is in good agreement with those obtained at Wallops Island [9] under nearly similar solar activity conditions.

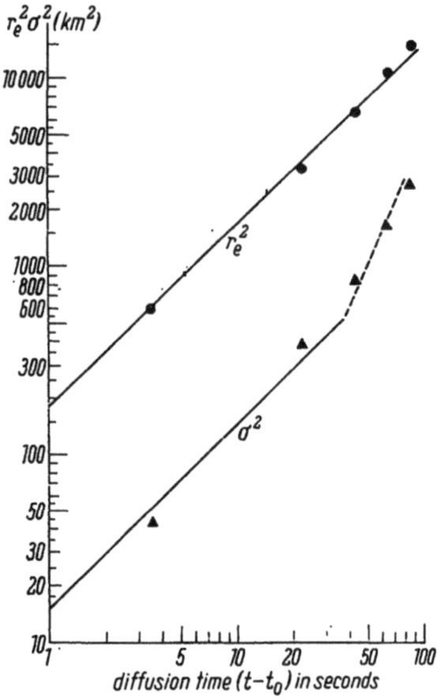

Fig. 4. Plot of r_e^2 and σ^2 against diffusion time.

4. Conclusions

The launching of the first Pakistan made Dragon 2B on 28 April 1972 has enabled observations of neutral air winds at 360 km at 05 h LMT to be made. The measurements indicate that the winds have a magnitude of $\sim$ 500 m s^{-1} and are directed 'west-south-west' at this time of the day. A previous firing with a French Dragon rocket on 29 March 1968 at 19 h LMT on the other hand, gave a value of 364 m s^{-1} directed to the 'east-south-east'. Although on both of these occasions the magnitudes of the observed winds are relatively large they are, nevertheless, in fair agreement with the theoretical pattern given by Kohl [4]. Results from the Doppler shift measurements of airglow emission spectra [6, 7], though not easy to interpret compared with drifts of luminous vapour clouds, seem to agree with these results.

The molecular diffusion coefficient at 360 km computed from the 'visible radius' data has a value of 4.4×10^{10} cm^2 s^{-1}. The result is in agreement with the data obtained at Wallops Island [9] in nearly similar solar activity conditions.

Acknowledgment

The authors wish to record their thanks to Messrs W. Hussain, Nafisuddin, M. Pervez and I. Husain of SUPARCO for their valuable assistance in reading plate-coordinates, preparing input for electronic data processing and diagrams for this paper.

The IMB 360/40 of Habib Bank Limited of Karachi was used in the data processing required in this study. The authors wish to express appreciation and thanks to their Computer Division for the prompt manner in which they have provided processing facilities at short notice.

References

[1] D. G. KING HELE, Planet. Space Sci. **12**, 835 (1964).
[2] R. A. CHALLINOR, Planet. Space Sci. **16**, 557 (1968).
[3] R. A. CHALLINOR, Planet. Space Sci. **17**, 1097 (1969).
[4] H. KOHL, Space Research X, 550 (1970).
[5] M. SHAFI AHMAD, Space Research X, 561 (1970).
[6] M. A. BIONDI and W. A. FEIBELMAN, Planet. Space Sci. **16**, 431 (1968).
[7] E. B. ARMSTRONG, Planet. Space Sci. **17**, 957 (1969).
[8] O. COTE, GCA Techn. Rep. No. 62-12-N (1962).
[9] Geophysics Corporation of America, GCA Tech. Rep. No. 62-13-N (1962).

Space Research XIV — Akademie-Verlag, Berlin 1974

ATMOSPHERIC DENSITIES FROM CANNON BALL 2 AND MUSKET BALL SATELLITE ORBITAL OBSERVATIONS

K. S. W. Champion, D. F. Gillette and I. M. Hussey

Air Force Cambridge Research Laboratories, Bedford, Mass., USA

The orbits of the Cannon Ball 2 (1971-67C) and Musket Ball (1971-67D) satellites, launched at 0011 UT on 7 August 1971, have been accurately observed. Atmospheric densities are obtained for perigees within the altitude range 121 km to 158 km and for latitudes between 88° S and 88° N. The experimental data are compared with the Jacchia 1964, 1970, 1971 and 39 H (1973) models. The average values and the standard deviations of the differences between the measured values and the model predicted values are presented. Although the average values of the differences have become progressively smaller with the newer models, the standard deviation was least for the Jacchia 1964 model in most cases.

1. Introduction

The details of the Cannon Ball 2 and Musket Ball high mass-to-area ratio satellites are described in [1, 2]. Up to 17 days of atmospheric densities derived from observations of the orbits of these satellites (satellite drag) were given in [1], and some of the high resolution accelerometer density data in [2]. In this paper a summary of the results for most of the lifetimes of Cannon Ball 2 (177 days) and Musket Ball (31 days), both launched at 0011 UT on 7 August 1971, is presented. The average density at perigee (more accurately, at half a scale height above perigee) was deduced from analysis of beacon and skin track radar observations of the two satellites. The method used consists of a numerical integration scheme combined with a differential orbit correction procedure where the discrepancies between computed and observed satellite positions are reconciled by adjusting the assumed atmospheric model.

2. Results

Since the perigee altitude must vary smoothly with time the computed values, which have some small short-term fluctuations, are represented by a 14th degree polynomial. No changes are made in the computed density values but only slight changes in the heights to which they are assigned. This yields the so-called "smooth" density ratio curves.

Musket Ball. Using an integration interval of 1.0 day, the atmospheric density ratios (satellite drag compared with the appropriate Jacchia model) are shown in

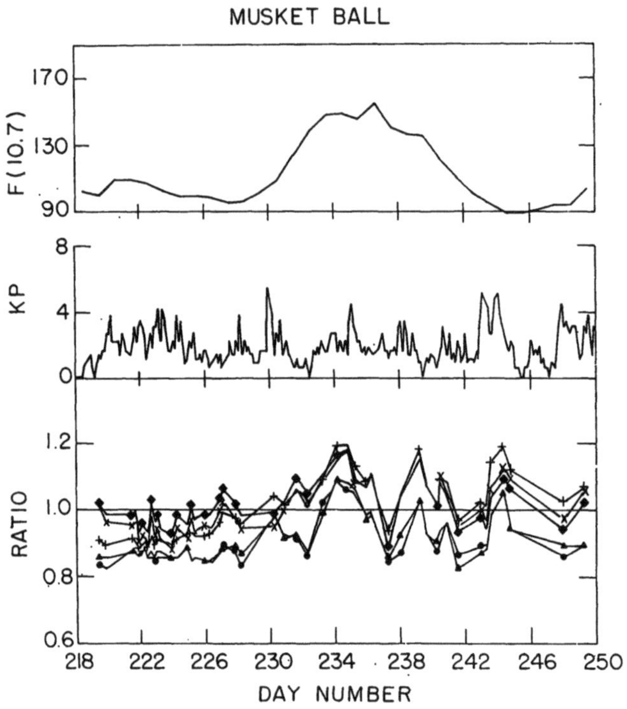

Fig. 1. Ratio of atmospheric densities (derived orbital drag to the Jacchia 1964, 1970, 1971 and 39H (private communication, 1973) models) for Musket Ball together with Kp and $F_{10.7}$ for the period 7 Aug. to 6 Sept. 1971.
● Jacchia 1964 (raw data); ▲ Jacchia 1964 (smoothed data); + Jacchia 1970 (smoothed data); × Jacchia 1971 (smoothed data); ◆ Jacchia 39H (1973) (smoothed data).

Fig. 1. The ratios to the Jacchia 1964 model [3] are given for both the unsmoothed and smoothed perigee altitudes. It can be seen that the differences in the values are small. The smooth values are given for four Jacchia models [4, 5]*. It can be seen that for almost the whole period shown, the ratios to J64 are significantly different from those to the later models. From visual inspection no one of the three later models is obviously better than another. The residual fluctuations appear to have some (but not high) correlation with the values of Kp and $F_{10.7}$, the flux of solar radio emission at wavelength 10.7 cm.

Table 1

Musket Ball

Model	Smooth J 64	Smooth J 70	Smooth J 71	Smooth 39 H
Mean	−0.0931250	−0.0018750	−0.0010417	0.0122917
Standard deviation	0.0697662	0.0955828	0.0768180	0.0656138

*) Jacchia 39H (1973) model by private communication.

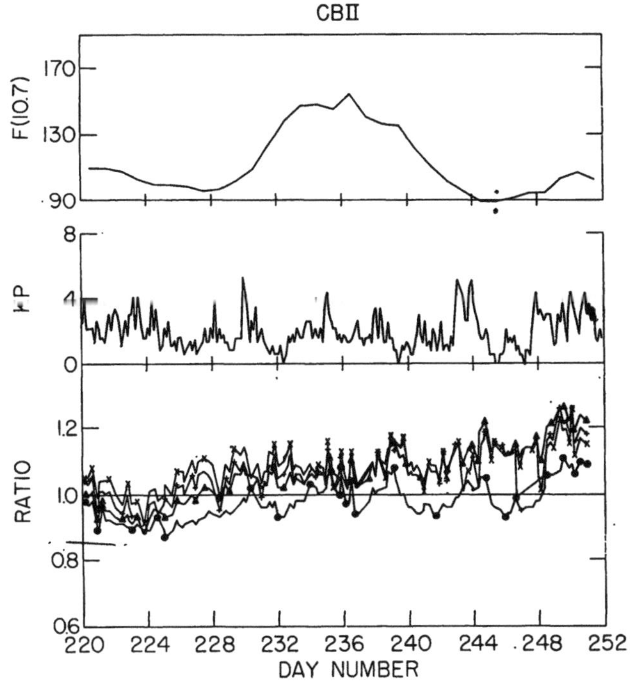

Fig. 2. Ratio of atmospheric densities (derived orbital drag to the Jacchia 1964, 1970, 1971 and 39H (private communication, 1973) models) for Cannon Ball 2, Kp and $F_{10.7}$ for the period 8 Aug. to 7 Sept. 1971.

● Jacchia 1964 (smoothed data); ▲ Jacchia 1970 (smoothed data); + Jacchia 1971 (smoothed data); × Jacchia 39H (1973) (smoothed data).

The statistical analysis of these data is shown in Table 1. For the 21 days of Musket Ball drag density data presented, the mean of the differences between the measured values and the predicted values is extremely small for J71 (−0.10%) and J70 (−0.19%) and large for J64 (−9.3%), but the standard deviation of the differences is least for Jacchia 39H (6.6%) and J64 (6.98%).

Cannon Ball 2. After injection into orbit the perigee altitude of Cannon Ball 2 was 134 km, at local time 1600. It reached a maximum altitude of 142 km 26 days after launch, a minimum of 131 km 52 days after launch, 158 km after 83 days, 122 km after 156 days and 131 km just before re-entry. The initial perigee latitude was 6° N, this precessed to 88° N after 26 days, to 88° S after 83 days, to 88° N after 134 days and to 85° S before re-entry.

Figs. 2—4 show the ratios of satellite drag deduced atmospheric density to the four Jacchia models and the values of Kp and $F_{10.7}$ for comparison. The modelling of the density variation with $F_{10.7}$ seems reasonably good. There is also a great deal of scatter, some of which may correlate with geomagnetic activity, but much of which is probably due to synoptic atmospheric variations. There is a longer term variation with a period of approximately 54 days which appears to be related to the location of perigee. When perigee is near the geographic pole the density is large compared with the models and when perigee is near the equator the density is low compared with the models.

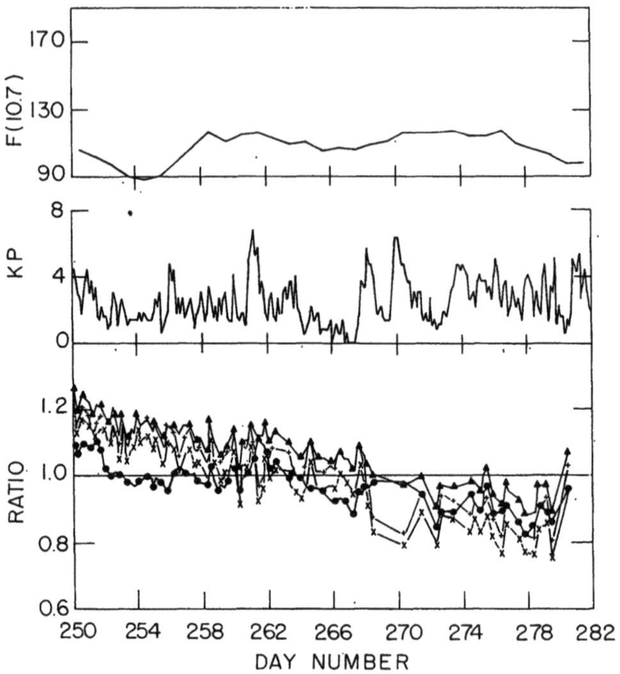

Fig. 3. As Fig. 2 for period from 7 Sept. to 7 Oct. 71.

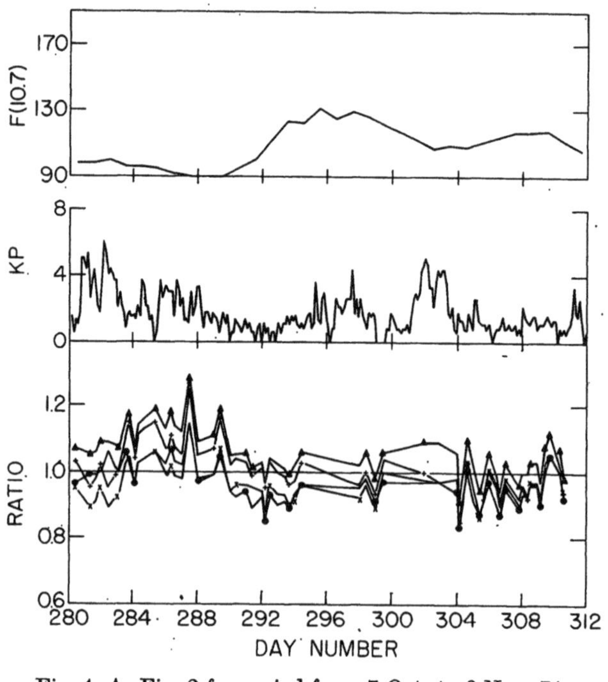

Fig. 4. As Fig. 2 for period from 7 Oct. to 6 Nov. 71.

Comparison of the Musket Ball and Cannon Ball 2 data for the same time period shows the poorer time resolution of the Musket Ball data. This is due to fewer radar observations. This makes detailed comparison of the data more difficult but, in the time period days 222–224 where temporal resolutions are comparable, the time variations of the relative density values are almost identical, although the absolute values of the Musket Ball data are 5% lower than those of the Cannon Ball 2 data. In general, during the period of comparison (days 220–250), although there are differences in the shapes of the relative density curves (probably due primarily to the difference in location of the two perigees) the variability and detailed structure are very similar. For example, the times of relative density maxima and minima correspond almost exactly.

The statistical analysis of 166 days (there is a gap in the processed data for the period 30 November–9 December 1971) of Cannon Ball 2 data is shown in Table 2. The mean of the differences between the measured values and the predicted values is largest for J64 (–4.8%) and least for J71 (–0.39%), but the standard deviation of the differences is least for J64 (7.6%). This suggests that, at least for conditions under which these data were obtained (which includes the period of the Musket Ball data but is much longer), the most accurate of these models would be J64 with a reduction in density between 120 and 150 km of about 4.8%.

Table 2

Cannon Ball 2

Model	Smooth J 64	Smooth J 70	Smooth J 71	Smooth 39 H
Mean	–0.0481718	0.0113377	–0.0038767	–0.0155263
Standard deviation	0.0761054	0.1407007	0.1159787	0.1166047

Acknowledgments

The authors wish to thank Jack Slowey of SAO and Arnold Bramson of IBM for assistance in developing the programs used to reduce the data presented in this paper.

References

[1] K. S. W. Champion, F. A. Marcos and R. A. Schweinfurth, Space Research XIII, 223 (1973).
[2] K. S. W. Champion and F. A. Marcos, Space Research XIII, 229 (1973).
[3] L. G. Jacchia, Smithson. Astrophys. Obs. Spec. Rep. No. 170 (1964).
[4] L. G. Jacchia, Smithson. Astrophys. Obs. Spec. Rep. No. 313 (1970).
[5] L. G. Jacchia, Smithson. Astrophys. Obs. Spec. Rep. No. 332 (1971).

Space Research XIV — Akademie-Verlag, Berlin 1974

COMPARISON OF DENSITIES DEDUCED FROM RADAR DETERMINED ORBITAL DECAY AND ACCELEROMETER MEASUREMENTS

H. R. Rugge

Space Physics Laboratory, The Aerospace Corporation, El Segundo, Calif., USA

A comparison has been made of atmospheric density as deduced from high resolution radar-determined orbital decay data and from data supplied by a uniaxial MESA accelerometer flown onboard two low altitude satellites in June—July and August—September 1970. The radar deduced densities, having an effective six hour time resolution, were determined at altitudes of 143 km and 169 km, very close to one-half scale height above perigee for the two satellites respectively. The accelerometer-deduced densities at these same altitudes were obtained on both the approaching-perigee and leaving-perigee portion of each of 59 orbits for the lower altitude satellite and each of 250 orbits for the higher altitude satellite.

A detailed time comparison of the densities derived from radar-tracking and the accelerometer is presented. The *average* densities obtained on both flights from both types of data are as follows:

Altitude	143 km	169 km
Accelerometer	3.52×10^{-12} g cm^{-3}	0.98×10^{-12} g cm^{-3}
Radar orbital decay	3.43×10^{-12} g cm^{-3}	0.91×10^{-12} g cm^{-3}

While the detailed comparison is not quite so good, agreement is obtained well within the relatively small uncertainties of the measurements. A preliminary comparison of the experimental data with current atmospheric models is made for one of the satellites.

1. Introduction

The purpose of this paper is to compare atmospheric densities deduced from radar-determined orbital decay measurements and from accelerometer measurements on two low altitude satellites. Relatively few comparisons of "*averaged*" orbital decay measurements have been made with data obtained by accelerometers flown onboard the low altitude ($h \leq 200$ km) satellites which measure the "*instantaneous*" orbital decay (i.e. deceleration). Such comparisons are important for two reasons. First, a properly calibrated accelerometer can measure the force resulting from atmospheric drag at all positions along the orbit below 200 km to an accuracy of a few per cent. This accelerometer data can check on the accuracy and consistency of orbital decay data which are taken at relatively few tracking stations. Secondly, the instantaneous density deduced from the accelerometer can be compared with the averaged density deduced from orbital decay in an effort see how well models, which use the average data as input, represent the real changing atmosphere, as measured by the accelerometer.

The data used in this paper were obtained with high accuracy tracking radars and with well-calibrated accelerometers flying on two stably oriented satellites. The single-axis accelerometer, which was pointed along the direction of motion of each satellite, was a Bell MESA (Miniature ElectroStatic Accelerometer) similar to those flown on the LOGACS experiment and the OV1-15, OV1-16, and OV1-20 satellites. The accelerometers were carried onboard the satellites 1970-48A and 1970-61A and produced results in June—July 1970 and August—September 1970, respectively.

2. Radar-Determined Orbital Decay Measurements

The orbital decay measurements used in this study were analyzed by Schusterman and Pearson of The Aerospace Corporation. The density model used was that of Walker-Jacchia [1] as modified by Bruce [2]. The density was evaluated at half a density scale height above perigee to minimize the effect of using a density model with what may be an incorrect scale height. This altitude corresponds closely to 143 km for the satellite 1970-48A and 169 km for 1970-61A for the time periods under consideration in this study.

3. Accelerometer Data

The densities deduced from the accelerometer at 143 km for 1970-48A and 169 km for 1970-61A were obtained for the approaching-perigee and leaving-perigee portion of each orbit for which such data were available. There was considerable variability ($\geqq 20\%$) between these two data points on a fair number of orbits. The two density values obtained for each orbit were averaged to represent a density at the respective altitudes of 143 and 169 km which should be more representative of the value at the geographic latitude of perigee where the radar-deduced density is evaluated. This radar-determined density occurred at a latitude of about 50° N while the accelerometer-deduced densities were obtained at latitudes near 70° N for the approaching-perigee part of the orbit and near 30° N for the leaving-perigee part of the orbit.

4. Comparison of Deduced Densities

Accelerometer-deduced densities were obtained for the satellite 1970-48A at 143 km for a total of 59 orbits in late June and early July 1970. Accelerometer-deduced densities for 1970-61A at 169 km were obtained for $\simeq 250$ orbits in late August and early September 1970 and were the average of the two values obtained for each orbit. These two data sets for 1970-48A are shown in Fig. 1. It should be noted that there is very close agreement between the two curves, except for the fact that the orbital decay curve has an effective response time of 6 hours (4 orbits), while the accelerometer data has an effective response time of 1.5 hrs (1 orbit). Thus, the orbital decay curve cannot follow rapid fluctuations in density as indicated by the accelerometer curve.

Similar comparisons can be made for the data obtained at 169 km from the satellite 1970-61A. A sample of the comparison of some data is shown in Fig. 2.

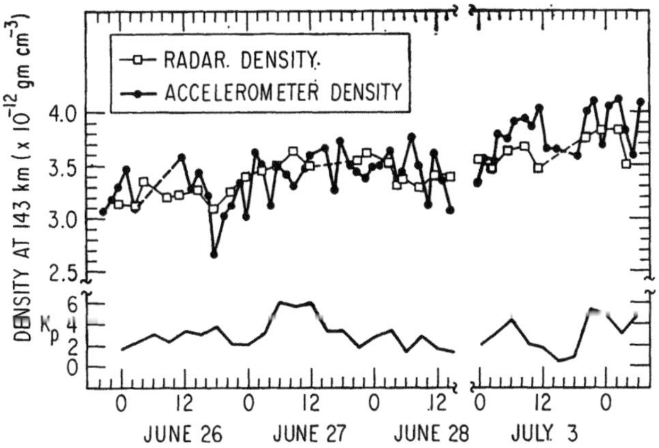

Fig. 1. Comparison of the accelerometer-deduced density and the orbital decay (radar)-deduced density at an altitude of 143 km. Kp for the time period of interest is presented at the bottom of the figure. Satellite 1970-48 A.

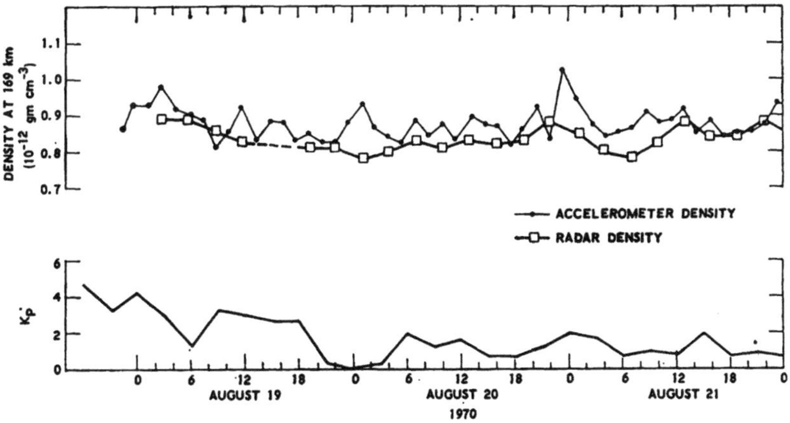

Fig. 2. A sample of the comparison of the accelerometer-deduced density and the orbital decay (radar)-deduced density at an altitude of 169 km during mid-August 1970. Kp for the same time period is presented at the bottom of the figure. Satellite 1970-61 A.

This is representative of all of the data taken for the full 250 orbits. The agreement is good, but there is clearly a systematic offset of the accelerometer data from the orbital decay data, probably indicating that the value chosen for the drag parameter $C_D A/m$ for the satellite was too small by 3 to 5%. Also there are effects of auroral heating in the high latitude accelerometer-deduced densities.

5. Comparisons of 1970-48A Data with Models

Preliminary comparisons of both orbital decay data and the accelerometer data from the satellite 1970-48A have been made with the Jacchia 1970 [3] and 1971 [4] atmospheric models at the altitude of 143 km. The average ratio of the orbital decay-deduced density to the 1970 Jacchia (J70) model density is $\simeq 1.2$. The similar ratio for the 1971 Jacchia model is $\simeq 1.1$. The comparisons are made at a geographic latitude of 52° N. A similar comparison is made in Fig. 3, but with Kp set equal to zero in the J70 and J71 models to emphasize and isolate the response of the density to geomagnetic activity.

A comparison of the two model atmospheres with the accelerometer data is shown in Fig. 4 for the first portion of the accelerometer data which are available for this satellite. The average accelerometer-deduced density to model ratios (including all time periods for which accelometer data are available) are 1.17 for the J70 model and 1.06 for the J71 model. The J71 model appears to give a better fit to both the low altitude orbital decay and accelerometer data at this altitude of 143 km.

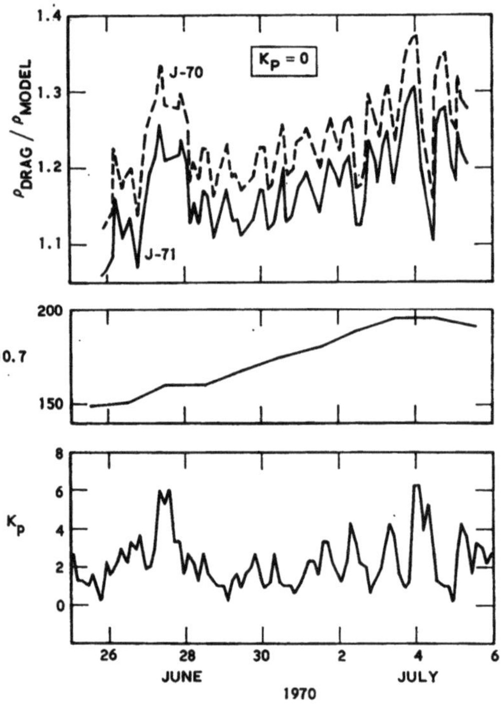

Fig. 3. The ratios of the orbital decay-deduced density to the Jacchia model densities at 143 km with Kp set equal to zero in the model densities. The solar index $F_{10.7}$ and the geomagnetic index Kp are also presented for the same time period. Satellite 1970-48 A.

6. Summary

A comparison between accelerometer-deduced and orbital decay-deduced densities has been made using results obtained from the two low altitude satellites, 1970-48A and 1970-61A. The results indicate, at least for these satellites, that densities deduced at half a scale height above perigee from the orbital decay measurements closely represent average values of density at the same altitudes

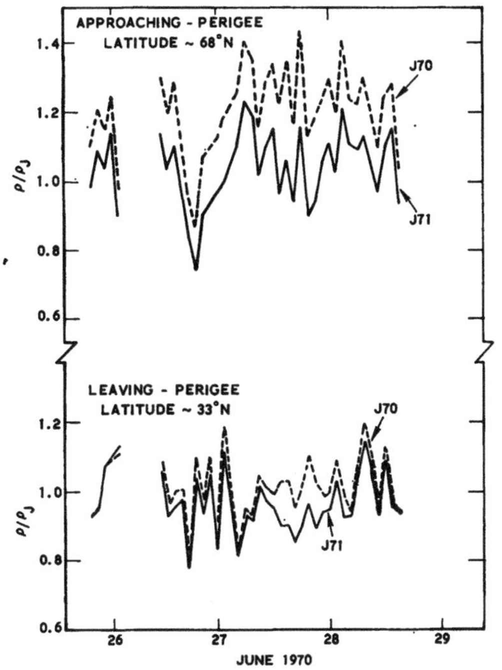

Fig. 4. The ratios of the accelerometer-deduced densities to the Jacchia model atmospheres at 143 km for both the approaching-perigee and leaving-perigee parts of the June 1970 data.

deduced from the accelerometer measurements. An average taken over 59 orbits for 1970-48A and 250 orbits for 1970-61A shows agreement to better than 3% and 8% respectively. This result, if it can be generalized, shows that low altitude atmospheric models based primarily on orbital decay data should be representative of the real atmosphere. Direct comparisons of both the orbital decay and accelerometer deduced densities from 1970-48A with the J70 and J71 models show that densities in both models are less than the observed densities. However, the J71 model gives a better fit to the data. It appears the models may not take the $F_{10.7}$ flux into account properly at this altitude and that mid-latitude densities are fit better than high latitude densities.

References

[1] J. C. G. Walker, J. Atmos. Sci. **22**, 462 (1965).
[2] R. W. Bruce, Rep. TOR-1001 (2110-01)-8, The Aerospace Corporation, El Segundo, California (August 1966).
[3] L. G. Jacchia, Smithson. Astrophys. Obs. Spec. Rep. No. 313 (May 1970).
[4] L. G. Jacchia, Smithson. Astrophys. Obs. Spec. Rep. No. 332 (May 1971).

Space Research XIV — Akademie-Verlag, Berlin 1974

INTERPRETATION OF SHORT PERIOD DENSITY CHANGES SHOWN BY THE DRAG OF SATELLITES

P. BENCZE

Geodetical and Geophysical Research Institute, Hungarian Academy of Sciences, Sopron, Hungary

Short period density changes shown by the drag of satellites during the spring maximum of the semi-annual variation are studied on the basis of neutral temperature and ionospheric parameters. It has been found that there are cases where the only way of matching the various observations is to assume transport processes, which render the reaction of the upper atmosphere to geomagnetic disturbance very variable. Thus subsidiary minima are due partly to the departure of the actual behaviour of the upper atmosphere from the correction for geomagnetic activity based on the general reaction of the upper atmosphere to geomagnetic disturbance.

1. Introduction

For the short-term prediction of the orbit of low altitude satellites the mathematical description and prediction of short-period density changes are needed. Therefore the investigation of the nature of short period density changes is important not only from the point of view of the physics of the upper atmosphere, but also from that of satellite tracking. Such density changes have been revealed by the investigations of King-Hele and Walker [1], Cook and Scott [2] and Ill et al. [3]. These authors have found, on the basis of the drag of satellites, subsidiary density minima occurring in different years during the spring maximum of the semi-annual variation occasionally, but always between 200 and 1100 km. According to Ill et al. [3] these changes cannot be attributed to errors of correction, but they are real density changes corresponding to temperature variations of about 100—200 °K. We shall study here these short-period density changes.

The origin of the subsidiary minima cannot be studied independently of the production of the semi-annual variation. Volland [4] relates the semi-annual variation of density in the thermosphere to the dissipation of the energy of tidal waves from the lower atmosphere. The leakage of the wave energy into the thermosphere is influenced by the semi-annual change of the prevailing wind system and thus it creates a semi-annual variation of density. Below about 400 km composition changes are thought to be of small influence, but above that height the composition changes predominate over the total density and temperature variations. Mayr and Volland [5] have shown that the semi-annual variation in the neutral composition can be explained by a circulation system induced by solar radiation heat input at the equator and auroral heating. Taking into account the conclusions of these investigations it is reasonable to consider the results of studies referring to changes of upper atmosphere parameters during geomagnetic disturbances.

2. Analysis of Subsidiary Minima

Three intervals, studied also in detail by Ill et al. [3], were chosen to demonstrate the origin of subsidiary minima occurring during the spring maxima of the semiannual variation of density. In Fig. 1 the upper curves show the temperature changes during the first two intervals determined from the drag of Echo 2 on the basis of Jacchia's model. Both curves are corrected for the geomagnetic effect and also for the day and night effect. Below the temperature curves the variation of *Ap* is illustrated. As can be seen, there is no strong correlation between the variation of temperature and geomagnetic activity represented by the *Ap* index. It is also obvious that April 1968 is more disturbed than April 1967. In the latter

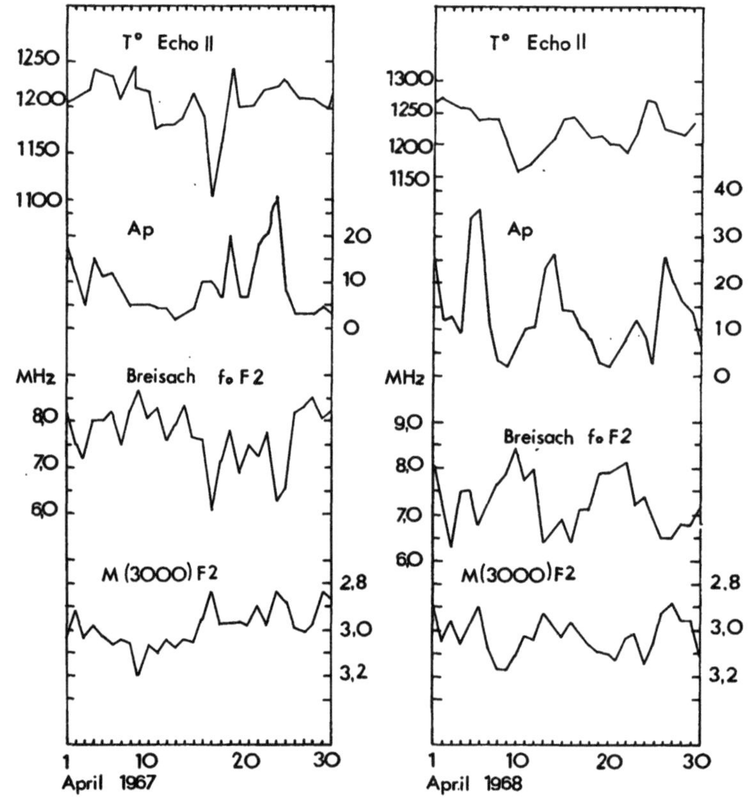

Fig. 1. Variation of the temperature determined from the drag of Echo 2, geomagnetic activity index *Ap*, the daily averages of the F2 layer critical frequency *fo*F2 and the transmission factor M/3000/F2 in April 1967 and April 1968.

case the subsidiary minimum of 17 April occurs during a relatively quiet period, while the density minimum of 11 April 1968 is preceded by a magnetically disturbed period.

For the interpretation of these density changes the ionized component of the upper atmosphere has also been studied. In the third row of curves the variation of the daily averages of the F2 layer critical frequency *f*oF2 is shown. As is known, *f*oF2 indicates the maximum electron density of the F region. On 17 April 1967 the electron density also shows a minimum, simultaneously with the density minimum in the neutral atmosphere. The changes in both quantities, attributable to the small increase of geomagnetic activity, are excessively large compared, for example, with the electron density minimum of 24 April due to the increased geomagnetic activity. In this case the temperature curve, corrected for the geomagnetic effect, does not show significant changes. In April 1968 the density of the neutral atmosphere and electron density show generally opposite variations. It seems that both parameters follow geomagnetic activity with some delay, but the electron density in the opposite sense. In Fig. 1, as in the case of *f*oF2, the daily averages are used to show also the variation of the transmission factor M/3000/F2. As is known, M/3000/F2 is approximately inversely proportional to the height of maximum electron density. In both cases the transmission factor indicates changes opposite to those of the maximum electron density. Ionospheric effects are not restricted to any geographical region, however for illustration only one station has been used.

The curves referring to the third interval are shown in Fig. 2. As in Fig. 1

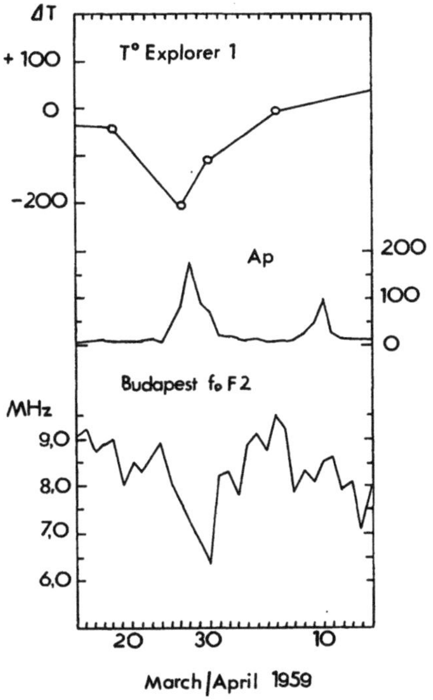

Fig. 2. Variation of the temperature residuals, geomagnetic activity index *Ap* and the daily averages of the F2 layer critical frequency *f*oF2 in March 1959.

the upper curve represents the variation of temperature residuals obtained on the basis of the drag of Explorer 1 by Jacchia et al. [6]. The second curve shows the variation of the geomagnetic index Ap, while the bottom curve depicts the change of the daily averages of foF2. The subsidiary minimum occurs simultaneously with the maximum of geomagnetic activity and the decrease in foF2.

3. Discussion and Conclusions

During the first interval studied here both the total density and the maximum electron density show minima on 17 April 1967. As regards the ionospheric effect, it might be interpreted simply by heating, but the decrease of temperature derived from the drag of satellites on the basis of simplified expressions is not consistent with this interpretation. The decimetric solar flux, $F_{10.7}$, generally used as an index of solar EUV radiation does not show significant changes. Even if it did, a corresponding temperature decrease would reduce the molecular constituents more than atomic oxygen, thus increasing again the electron density decreased by the reduction in the EUV flux. For the detailed study of the subsidiary minimum of 17 April 1967 the ionospheric parameters from 1500 LT were used assuming that steady state conditions were present. The data of the Breisach ionospheric station representing temperate latitude conditions were used. The maximum electron density decreases on 17 April to about 60% of its previous value. According to Mayr and Mahajan [7] a temperature decrease of 100 °K due to the variation in EUV radiation, and corresponding to our case, results in a decrease of the maximum electron density to 85% of its former value. Consequently, as was to be expected, the change of maximum electron density cannot be explained simply by heating due to the EUV flux.

A possible way of matching the various observations is to consider transport effects, which cause a decrease of the concentration of atomic oxygen, the major constituent of the atmosphere in the F region. The decrease of the concentration of atomic oxygen, compared with its former value, could be brought about by the thermospheric wind induced during geomagnetic storms. The meridional component of the storm time wind, as has been demonstrated by Mayr and Volland [8], acts as a sink and decreases the atomic oxygen concentration. The vertical component has also the same effect if its increase is more rapid than the decrease with height of atomic oxygen.

As regards the other two intervals, the subsidiary minimum of 11 April 1968 might be interpreted as the result of the combined effect of the semi-annual meridional wind component and the thermospheric wind induced during geomagnetic disturbances. The delay of the change in temperature derived from satellite drag with respect to the Ap index suggests that in this case the effect of the semi-annual meridional wind prevails. The coincidence of the subsidiary minimum with the maximum in geomagnetic activity during the third interval can be explained again by the effect of the storm time wind. This wind field produces a minimum in atomic oxygen concentration during the time of geomagnetic storms [8].

Thus the conclusion can be drawn that subsidiary minima can be regarded partly as the result of a departure of the actual behaviour of the upper atmosphere from the correction for geomagnetic activity based on the averaged reaction of the upper atmosphere to geomagnetic disturbances.

References

[1] D. G. King-Hele and D. M. C. Walker, Planet. Space Sci. **17**, 197 (1969).
[2] G. E. Cook and D. W. Scott, Planet. Space Sci. **17**, 107 (1969).
[3] M. Ill, F. Barlier and F. Kohler, Space Research X, 485 (1970).
[4] H. Volland, Planet. Space Sci. **17**, 1709 (1969).
[5] H. G. Mayr and H. Volland, Annls Géophys. **27**, 513 (1971).
[6] L. G. Jacchia, J. W. Slowey and I. G. Campbell, Planet. Space Sci. **17**, 49 (1969).
[7] H. G. Mayr and K. K. Mahajan, J. Geophys. Res. **76**, 1017 (1971).
[8] H. G. Mayr and H. Volland, Planet. Space Sci. **20**, 379 (1972).

Space Research XIV — Akademie-Verlag, Berlin 1974

EARTH RADIATION PRESSURE AND THE DETERMINATION OF DENSITY FROM SATELLITE DRAG

J. W. Slowey

Smithsonian Astrophysical Observatory, Cambridge, Mass., USA

The effect of earth radiation pressure relative to that of atmospheric drag increases with height through the lower thermosphere. While it can be entirely neglected as a correction in the determination of density at lower heights, it becomes significant somewhat below 1000 km and rapidly becomes comparable in magnitude with drag above 1000 km. The effects of earth radiation pressure on the orbit of the balloon satellite 1963 30D were calculated during two series of intervals when the orbit was entirely in sunlight. The first of these was when the perigee was very high so that hydrogen was expected to be the dominant atmospheric constituent. The second was when the perigee was lower and helium was expected to be the dominant constituent. Two sets of calculations were performed, one using a constant mean model and the other using a seasonal-latitudinal model, based on reported Tiros 7 data, for the albedo and infrared radiation. The large differences between the two sets of calculations indicate the inadequacy of the mean model. Atmospheric densities deduced from the observed accelerations in the intervals in which hydrogen dominates are also given. Some interesting features appear in these densities as a function of latitude. The diurnal mean of the hydrogen density seems to be about 2.3 times the Kockarts-Nicolet value for solar minimum.

We will refer to the solar radiation reflected by the earth in proportion to the albedo and to the infrared radiation emitted by the earth collectively as earth radiation. The contribution to the total acceleration of an orbiting artificial satellite due to the pressure of this radiation is comparatively small at the heights of most satellites used in atmospheric drag analyses. However, it is significant for some of the higher satellites [1, 2], although it has not usually been taken into account.

We have obtained orbits and accelerations [3] of the balloon satellite Dash 2, 1963 30D, using accurately reduced Baker-Nunn observations in the early portion and radar observations toward the end, when the perigee was lower. Accelerations have been determined by least-squares fitting of the elements and by detailed analysis of mean-anomaly residuals [4]. We decided to use some of the 1963 30D data to assess the effect of earth radiation under conditions where it would constitute an appreciable portion of the total acceleration. The effect was expected to become comparable with that of atmospheric drag slightly above 1000 km and to remain so to very great heights. At such heights, both effects can be completely overwhelmed by the effect of solar radiation pressure; however, the net effect of direct radiation pressure vanishes when the orbit is entirely in sunlight. The study of 1963 30D was limited, therefore, to intervals when this was the case. Two series of intervals were selected: in the first, taken early in the lifetime, the perigee height was between 3500 and 3100 km; in the second, considerably later, the perigee height was between 1300 and 900 km. The atmosphere was expected to consist almost solely of hydrogen during the first series and of between 90 and 95%, by weight, of helium during the second. The effects of earth radiation on the orbital elements were determined by evaluating the Gaussian form of the Lagrangian equations by means of mechanical quadrature, both doubly over the visible cap of the earth and, in order to obtain only the net secular and long-periodic effects, around the satellite's orbit. The earth's albedo and infrared radiation were both included. The former was assumed to be mainly a diffuse reflection following Lambert's law, with a small specular component equal to 5% of the total reflectoin. The area-to-mass ratio of the satellite was taken to be 37.9 $cm^2 g^{-1}$, and the radiation pressure forces were increased by 12.0% relative to what would result from the radiation stopped by the satellite, in order to allow for partially diffuse reflection from the surface. The amount of this increase was determined empirically from an analysis of the effects of direct solar radiation pressure on the orbit. The value of the solar constant was assumed to be 2.00 cal $cm^{-2} min^{-1}$.

Computations were made from two different models of the earth radiation. The first assumed a constant albedo over the whole earth and infrared radiation, also without any variation, as necessary to balance the heat input implied by the albedo and the assumed value of the solar constant. This model was similar to that used by Prior [1, 2], who assumed an albedo of 0.32, constant over the earth. The second model was based on four seasonal means of the albedo and infrared radiation as functions of latitude, obtained from Tiros 7 observations [5]. Because of their contribution to the tangential force, the latitudinal variations were expected to be quite important. The Tiros 7 albedo and infrared measurements seem to be slightly inconsistent. Table 1 gives means over the earth that we computed both for the seasonal albedo models and for the seasonal infrared models. The average of the four means of the albedo is 0.368; this implies that, on the average, a fraction equal to 0.632 of the solar radiation striking the earth and its atmosphere is absorbed in some way. With our adopted value of the solar constant, this would require an average power output over the earth's surface of 0.316 cal^{-2} $cm^{-2} min^{-1}$ in order to maintain heat balance. The average of the four seasonal means of the infrared radiation is 0.339 cal $cm^{-2} min^{-1}$, about 7% greater than the above value. We did not attempt to resolve this discrepancy, which may be related to calibration problems [6, 7] experienced with the instrumentation. The

Table 1

Means over the Earth of the Four Seasonal Models of Earth Albedo and Infrared Radiation reported from Tiros 7 Data

Seasonal model	Albedo	Infrared radiation (cal $cm^{-2} min^{-1}$)
June—August 1963	0.348	0.388
September—November 1963	0.395	0.340
December 1963—February 1964	0.358	0.337
March—May 1964	0.372	0.342

means of the Tiros albedo models are in good agreement with the range 0.35 to 0.40 obtained from observations of earthshine on the moon as quoted by Wyatt [8]. The Tiros albedo models, therefore, were adopted without modification for our second model. We also adopted a value of 0.368 for our first model, that of a constant albedo. It should be mentioned, however, that we would have selected a somewhat smaller value if the satellite considered had had a lower inclination, since the albedo is systematically smaller at lower latitudes. We were primarily interested in determining the results of the variations in albedo that a satellite "sees" around its orbit. Such variations are of much greater consequence than is a relatively small uncertainty in the value of the mean. The possibility of systematic error in the assumed infrared radiation is of very little practical concern since, even with realistic latitudinal variations included, the effects produced were found to be quite small compared with those due to the albedo.

Values of the orbital acceleration (the rate of change of the anomalistic period) obtained with the two different radiation models are plotted in Fig. 1. As can be seen, the differences between the results obtained with the two models are comparable in magnitude with the effect itself. It is clear, therefore, that the constant-albedo model is inadequate for the determination of densities from satellite

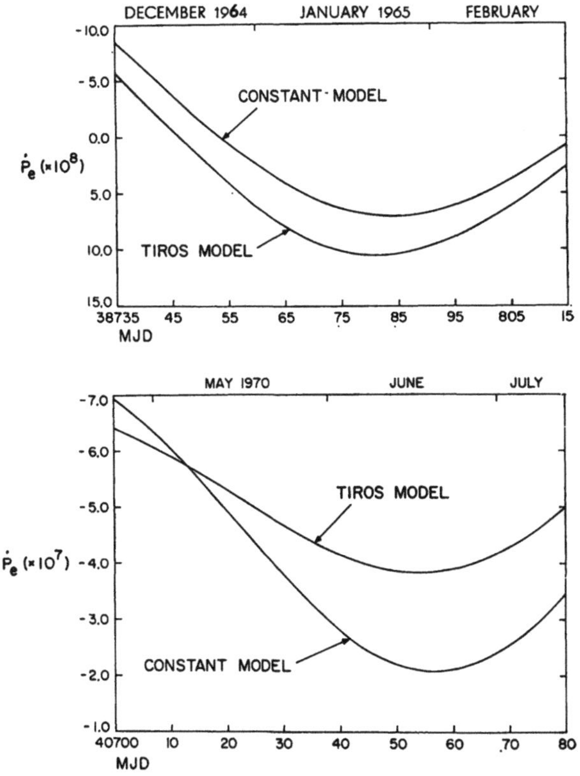

Fig. 1. Orbital acceleration due to earth radiation pressure acting on satellite 1963 30D during two periods computed by using two different models of the radiation. An albedo of 0.368 was used with the constant model; the Tiros model includes latitudinal and seasonal variations in both the albedo and the infrared radiation.

drag—a more detailed model is obviously required to achieve the accuracy we need. Even at somewhat lower heights, the size of the relative error would appear to make computation with the constant model an essentially meaningless exercise. The question remains of how the results from the Tiros 7 model would compare with those from a representation with spatial and temporal variations of the albedo. We are not in a position to make such a comparison, but examination of atmospheric accelerations derived by using the Tiros model indicates that the model is satisfactory for our purposes.

In Table 2 we have tabulated, as a function of the Modified Julian Date (MJD = JD−2400000.5), 10-day means $\dot{P}$ of the observed orbital accelerations of 1963 30D [3] for each of the five intervals over which the earth radiation pressure calculations were made. We have also tabulated the corresponding 10-day means $\dot{P}_e$ of the acceleration due to earth radiation pressure obtained with the Tiros model, the derived atmospheric component $\dot{P}_a$ of the acceleration ($\dot{P}_a = \dot{P} - \dot{P}_e$), the mean computed atmospheric component $\dot{P}_c$ obtained from Jacchia's [9] most recently published model and from a variable model for the drag coefficient [10—12], the ratio $\dot{P}_a/\dot{P}_c$ of the derived to the computed acceleration, the latitude ϕ of the perigee, and the difference in right ascension $\Delta\alpha$ between perigee and the sun.

The results from the Tiros model are most encouraging. There is little doubt that, for the most part, the radiation pressure effect has been successfully removed and what remains, whatever its origin, is largely a drag effect. In the second series the ratios are consistently close to unity and are not systematically related to the radiation pressure accelerations. The average value of the drag coefficient used to obtain the computed accelerations was about 2.9. Where the radiation pressure accelerations are about equal to those of atmospheric drag, the average would be a little low. The situation would be worse if we increased the hydrogen concentration in the atmospheric model, as recent observations would seem to require, since only about 90% of the assumed mass is helium and the remainder is mostly hydrogen. On the other hand, where the ratios are higher they could be reduced by adding more hydrogen, which is about 4% of the mass there, where the radiation pressure accelerations are relatively small. Better overall agreement in the second series would have been obtained if smaller albedos had been used. The argument is not too compelling, however, when the uncertainty that must be associated with the atmospheric model is considered, especially in view of the very large seasonal-latitudinal variations that occur in the helium concentration.

The derived atmospheric accelerations for the first series are quite interesting. The perigee moves through middle and low latitudes in the summer hemisphere, just after the solstice, at local times centered about 6 am. The main difference among the three intervals considered is that the perigee moves south through northern latitudes in the middle one. Daily values of the derived atmospheric accelerations were converted to densities in the usual manner [4], with an assumed drag coefficient of about 3.6. In Fig. 2, these densities are plotted for each interval as a function of the latitude of perigee. The pattern of maxima and minima (especially toward higher latitudes) appears to be rather stable with respect to latitude. Also striking is the apparent decrease in density toward the poles. This may be due in part to inadequacies of the albedo model, which could be invoked to explain the slightly positive accelerations at two of the extremes, but seems largely a real effect in the drag.

The values of $\dot{P}_c$ for the first series in Table 2 were computed by using a modified

Table 2

Ten-day Means of Orbital Accelerations, including Ratios of Derived to Computed Atmospheric Components and Geometric Parameters

First Series

MJD	$10^8\dot{P}$	$10^8\dot{P}_e$	$10^8\dot{P}_a$	$10^8\dot{P}_c$*	$\dot{P}_a/\dot{P}_c$	ϕ	$\Delta\alpha$
38405.0	−0.5	1.4	−1.9	−5.45	0.35	−45°	295°
15.0	−0.6	2.6	−3.2	−5.56	0.58	−39	284
25.0	−2.0	3.1	−5.1	−5.29	0.96	−32	273
35.0	−4.2	3.0	−7.2	−5.00	1.44	−21	262
45.0	−4.2	2.2	−6.4	−5.61	1.14	−10	251
55.0	−5.1	1.1	−6.2	−4.64	1.34	3	240
65.0	−6.6	−0.4	−6.2	−5.20	1.19	19	230
38565.0	−6.4	−0.6	−5.8	−7.07	0.82	59	311
75.0	−3.8	1.8	−5.6	−7.71	0.73	52	300
85.0	−4.6	4.1	−8.7	−6.92	1.26	45	288
95.0	−1.2	5.6	−6.8	−7.22	0.94	37	277
38605.0	−1.8	6.3	−8.1	−8.26	0.98	28	266
15.0	−0.6	5.8	−6.4	−6.80	0.94	18	256
25.0	−0.2	4.6	−4.8	−7.27	0.67	6	245
35.0	−1.7	1.7	−3.4	−7.76	0.44	− 6	235
38745.0	0.1	−1.0	1.1	−5.70	−0.19	−56	304
55.0	1.5	4.0	−2.5	−6.74	0.37	−48	292
65.0	2.4	7.9	−5.5	−6.00	0.92	−39	280
75.0	1.2	10.1	−8.9	−6.26	1.42	−30	268
85.0	1.3	10.3	−9.0	−6.39	1.41	−20	256
95.0	0.6	8.8	−8.2	−5.94	1.38	− 9	245
38805.0	−1.6	6.2	−7.8	−6.26	1.25	3	234

Second Series

MJD	$10^7\dot{P}$	$10^7\dot{P}_e$	$10^7\dot{P}_a$	$10^7\dot{P}_c$	$\dot{P}_a/\dot{P}_c$	ϕ	$\Delta\alpha$
40565.0	−8.41	−4.88	−3.53	−4.47	0.79	60°	310°
75.0	−8.32	−4.30	−4.02	−4.42	0.91	49	295
85.0	−7.12	−3.70	−3.42	−4.02	0.85	37	281
95.0	−5.21	−3.17	−2.04	−2.84	0.72	26	267
40604.5+	−6.06	−2.83	−3.23	−3.27	0.99	15	254
40715.0	−18.5	−5.6	−12.9	−13.1	0.99	−63	316
25.0	−20.1	−5.0	−15.1	−14.5	1.04	−52	302
35.0	−18.7	−4.4	−14.3	−11.1	1.29	−40	288
40761.5++	−10.9	−3.9	− 7.0	− 6.3	1.11	− 8	252

* These accelerations were computed with a modified hydrogen distribution.

+ nine-day means; ++ seven-day means.

Average effective heights to which data apply are, chronologically for the five intervals: 3660, 3590, 3470, 1450, and 1120 km.

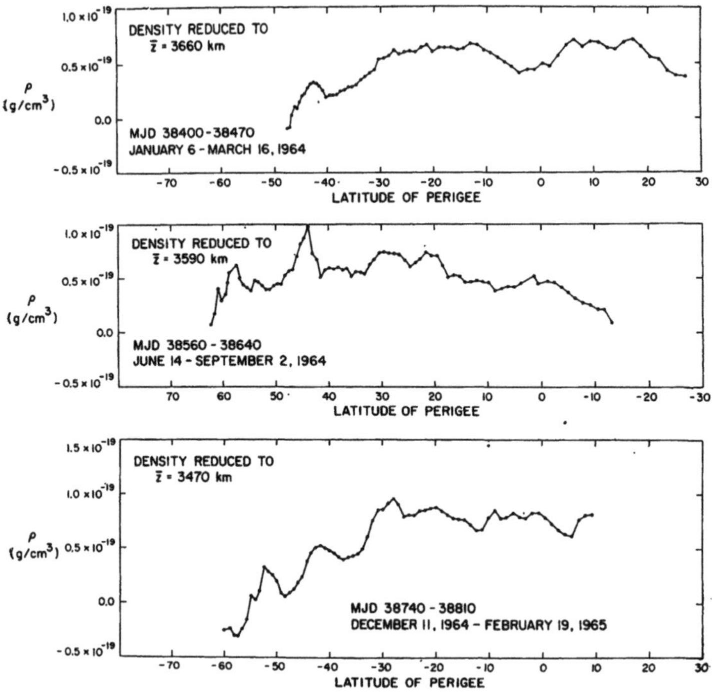

Fig. 2. Atmospheric densities determined from derived orbital accelerations of satellite 1963 30D reduced to the standard heights indicated and plotted as a function of latitude. Densities corresponding to accelerations having positive sign do not, of course, exist but have been plotted as negative values for the sake of completeness. Note that the latitude in the center strip is reversed in order to show latitudinally stable features present in both hemispheres.

version of the hydrogen model of Kockarts and Nicolet [13, 14] as incorporated in Jacchia's atmospheric model. The hydrogen concentrations were increased by a factor of 2.3, and the diurnal temperature variation for hydrogen was adjusted to give a diurnal density ratio of 2. As can be seen from Table 2, the observed accelerations agree almost exactly with the computed ones at lower latitudes in the northern hemisphere (second interval) but are about 1.3 times the computed ones in the southern hemisphere (first and third intervals). The difference is believed to be due to the semiannual variation in the hydrogen concentration [15, 16]. A semiannual variation was not included in the computation of $\dot{P}_c$, since the variation in Jacchia's model becomes essentially zero at the heights involved here. If the density of hydrogen varies with that of the other constituents, the factor for the northern-hemisphere data would be increased to about the value for the southern-hemisphere data. This is because the former data are from a period near a minimum in the usual semiannual variation, while the latter are from times when the semiannual variation goes through zero. If we assume a diurnal ratio of 2 [17, 18], the diurnal mean densities from each of the three intervals would be about 1.3 times the assumed means, or 3.0 times the Kockarts-Nicolet value for solar minimum. If, on the other hand, the semiannual variation for hydrogen were to lag two months behind that for the other constituents [15], the correction would work in the opposite direction. The data from all three intervals would then be in close agreement with the assumed density of 2.3 times the Kockarts-Nicolet value. We prefer this interpretation, which gives a result in excellent agreement with the trend indicated by recent *in situ* measurements [18].

Acknowledgment

This work was supported in part by grant NGR 09-015-002 from the National Aeronautics and Space Administration.

References

[1] E. J. Prior, in: Dynamics of Satellites, Springer-Verlag, New York 1970 (p. 303).
[2] E. J. Prior, in: The Use of Artificial Satellites for Geodesy, American Geophysical Union, Washington 1972 (p. 197).
[3] J. W. Slowey, Smithson. Astrophys. Obs. Spec. Rep. No. 356 (1974).
[4] L. G. Jacchia an J. W. Slowey, Smithson. Astrophys. Obs. Spec. Rep. No. 100 (1962); also in Smithsonian Contr. Astrophys. **8**, 1 (1963).
[5] NASA, Rep. SP-96, National Aeronautics and Space Administration, Washington 1966.
[6] A. Arking and J. S. Levine, J. Atmos. Sci. **24**, 721 (1967).
[7] W. R. Bandeen, M. Halev and I. Strange, NASA Tech. Note D-2534 (1965).
[8] S. P. Wyatt, in: Dynamics of Satellites, Academic Press, New York 1963 (p. 180).
[9] L. G. Jacchia, Smithson. Astrophys. Obs. Spec. Rep. No. 332 (1971).
[10] L. G. Jacchia and J. W. Slowey, Smithson. Astrophys. Obs. Spec. Rep. No. 348 (1972).
[11] G. E. Cook, Planet. Space Sci. **13**, 929 (1965).
[12] G. E. Cook, Annls Géophys. **22**, 53 (1966).
[13] G. Kockarts and M. Nicolet, Annls Géophys. **18**, 269 (1962).
[14] G. Kockarts and M. Nicolet, Annls Géophys. **19**, 370 (1963).
[15] B. A. Tinsley, Space Research X, 582 (1970).
[16] H. C. Brinton and H. G. Mayr, Space Research XII, 751 (1972).
[17] J. A. Quesette, J. Geophys. Res. **77**, 2997 (1972).
[18] A. Vidal-Madjar, J. E. Blamont and B. Phissamay, J. Geophys. Res. **78**, 1115 (1973).

Space Research XIV — Akademie-Verlag, Berlin 1974

SATELLITE MEASUREMENTS OF NEUTRAL ATMOSPHERIC COMPOSITION IN THE ALTITUDE RANGE 150 TO 450 km

C. R. Philbrick

Air Force Cambridge Research Laboratories, Bedford, Mass., USA

Measurements have been obtained from similar pairs of mass spectrometers flown on the OV3-6 and OV1-15 satellites. On each satellite, an instrument using an enclosed ion source and one using a semi-open ion source were used to measure the neutral species. The atmospheric species measured include O, N_2, Ar, N and NO. The results from these experiments have allowed a study of the variations of the atmosphere with local time, latitude, altitude and various geophysical parameters. The $[O/N_2]$ ratio has an average value of unity near 200 km but can vary over a considerable range. The composition measurements exhibit significantly larger variations than presently allowed by atmospheric models and point out the inadequacies of static diffusion models.

1. Introduction

Mass spectrometer measurements from satellites during the past few years have significantly improved our understanding of the composition of the neutral upper atmosphere and variations which occur there. This paper summarizes some of the results obtained from the OV3-6 and OV1-15 satellite mass spectrometer experiments. Prior to these two satellites, most of the data on upper atmospheric composition had been obtained from the Explorer 17 satellite [1] and from several rocket experiments, e.g. [2—4]. More recently, the OGO 6 satellite [5—7] has provided data in the region between 450 and 600 km. Also, during the past few years, incoherent scatter radar experiments [8, 9] have provided important insights into atmospheric variations.

2. Description of the Experiments

The OV3-6 satellite was launched in December 1967 into a circular polar orbit near 440 km. The mass spectrometers were mounted parallel to the spin axis on opposite ends of the satellite. The spin vector was in the plane of the orbit and the instrument axes became parallel with the velocity vector near the equator on either side of the orbit. Mass Spectrometer I utilized an enclosed ion source geometry and complete orbit tape recorded data were obtained for local times between midnight and sunrise during the period of December 1967 through March 1968. The Mass Spectrometer II was of a semi-open ion source design. The instrument recorded data over a complete orbit between noon and sunset for the same time period. Real-time data (10-minute acquisitions) were obtained from both instruments until re-entry in March 1969. Fig. 1 shows the instrument configuration for each of the RF quadrupole mass spectrometers. All of the ion source

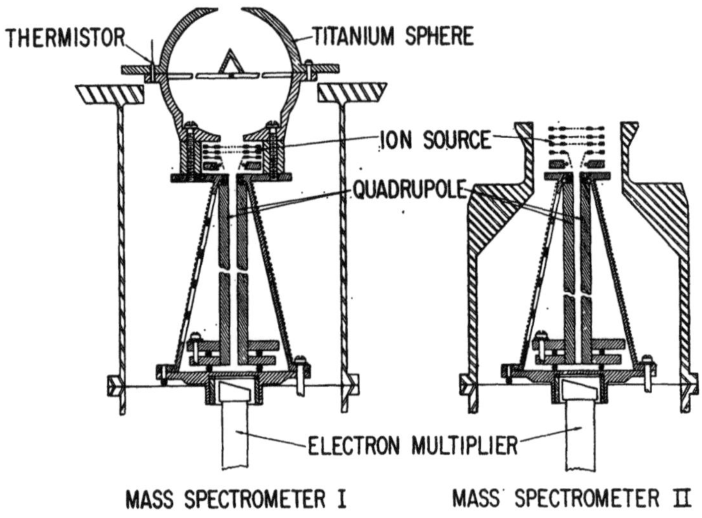

Fig. 1. The instrument configuration of Mass Spectrometers I and II which were flown on the OV3-6 and OV1-15 satellites.

parts and the sampling sphere were made of commercially pure titanium or were plated with titanium in an effort to reduce the recombination of atomic oxygen [10]. A similar ion source was used for each instrument. The ion source was cylindrically symmetric and a three mil diameter tungsten filament was used. The instruments were programmed to sweep the mass range from 1 to 32 amu in a 5 second period.

The OV1-15 satellite was launched into an elliptical orbit at altitudes between 160 and 1800 km on 11 July 1968. Data were obtained by two instruments similar to those on the OV3-6 satellite until re-entry during November 1968. The instruments were mounted on opposite sides of the satellite with their axes perpendicular to the satellite spin axis. The spin axis was actively controlled in an effort to maintain it perpendicular to the orbit plane. In this configuration the instrument samples along the direction of motion once each spin period. The spin period was initially about 6 seconds and became about 12 seconds near the end of satellite life time. The programmed operation of the instrument included measurement of each of four masses (He, O, N_2, Ar) for a period of 12 seconds followed by a mass scan of 1 to 44 amu during 12 seconds.

3. Discussion of the Results

The data from the OV3-6 satellite has led to several interesting conclusions [11—14]. Since the orbit was nearly circular at $\sim$400 km altitude, the latitudinal and temporal variations could be readily studied. The mass spectrometer and density gauge data from the same satellite have been compared and found to be in general agreement. The atomic oxygen was found to exhibit a broad maximum in the mid afternoon between 1400 and 1530 LT. The density maximum appeared to be in the summer hemisphere and poleward of the subsolar point. The molecular nitrogen concentration was found to change considerably with local time and exhibited a larger latitudinal variation than did the atomic oxygen. The N_2 concentration was a few per cent in the late afternoon and was only a few tenths of one per cent in the early afternoon. The N_2 concentration reached a maximum near 1530 to 1600 LT. Comparison of the local time variation of O and N_2 led to the conclusion that the times of the density and temperature maxima are out of phase near 400 km. It was inferred from the data that the N_2 variation may provide a good indication of atmospheric temperature. This result is in agreement with the incoherent scatter results which have shown that the temperature maximum occurs during the late afternoon near 1600 LT [8, 9].

The N_2 concentration was found to have a large variability in the auroral region which is associated with energy sources in the polar atmosphere. A density trough which is more pronounced in the molecular nitrogen profile occurs inside the auroral oval near the magnetic pole in most orbits.

In the equatorial latitudes, relatively large (10—50%) variations occur. These troughs in the neutral atmosphere are correlated with the magnetic equator. The fact that these variations are strongly tied to the magnetic equator suggests that strong coupling exists between the neutral atmosphere and the ionosphere. In most cases, symmetric variations were observed on both sides of the magnetic equator located in the regions 10—15° from the magnetic equator. In a few cases the observed effect was a single trough centered on the magnetic equator. This effect has also been observed recently in the OGO 6 data [6].

Titanium surfaces were used for the ion sources and entrances in an effort to reduce the recombination of atomic oxygen. After a period of a week or so in orbit, the surfaces have encountered a large flux of atomic oxygen and a rather stable surface condition exists. After this period, the adsorption and desorption of atomic oxygen appears to follow a Langmuir characteristic with short and long period components. The recombination to molecular oxygen occurs only during the ram portion of a spin cycle and suggests a recombination coefficient of about 10^{-2} [12]. During the wake portion of a spin cycle O_2 drops below the limit of detectability and O is the major desorption peak.

The elliptical orbit of the OV1-15 satellite allows study of vertical profiles of the atmospheric species [15]. Data for one orbit of the OV1-15 with an equatorial perigee is shown in Fig. 2. The density was determined by comparison of the total measured composition at 200 km for six orbits (between orbits 1110 and 1130) with the mass density determined from an accelerometer on the satellite [16]. This determination of instrument sensitivity was necessary since an offset occurred between laboratory calibration and first turn-on in orbit. In Fig. 2 the measured points are shown with a connecting line to aid in following the profiles. The O and N_2 profiles cross near 180 km for this orbit.

In order to display some of the interesting features in the data, the 200 km crossings for ascent and descent are shown in Fig. 3. The values $[O/N_2]$, $[N_2]$, $[O]$ are shown with the latitude, local time, date and *Ap* index through the satellite lifetime. The $[O/N_2]$ ratio shows an average value slightly less than unity at 200 km. The ratio drops rapidly between orbits 100 and 300 due largely to the change in atomic oxygen density during this period. This large change in atomic

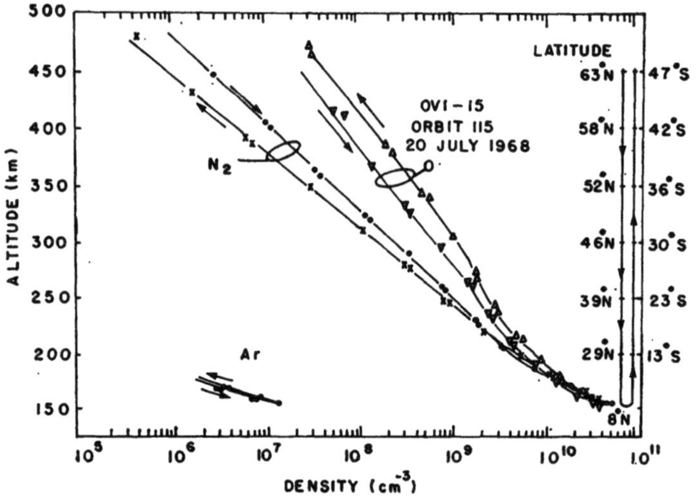

Fig. 2. Altitude profiles for O, N_2 and Ar are shown with the corresponding latitude for descent and ascent of orbit 115 of the OV 1-15 satellite.

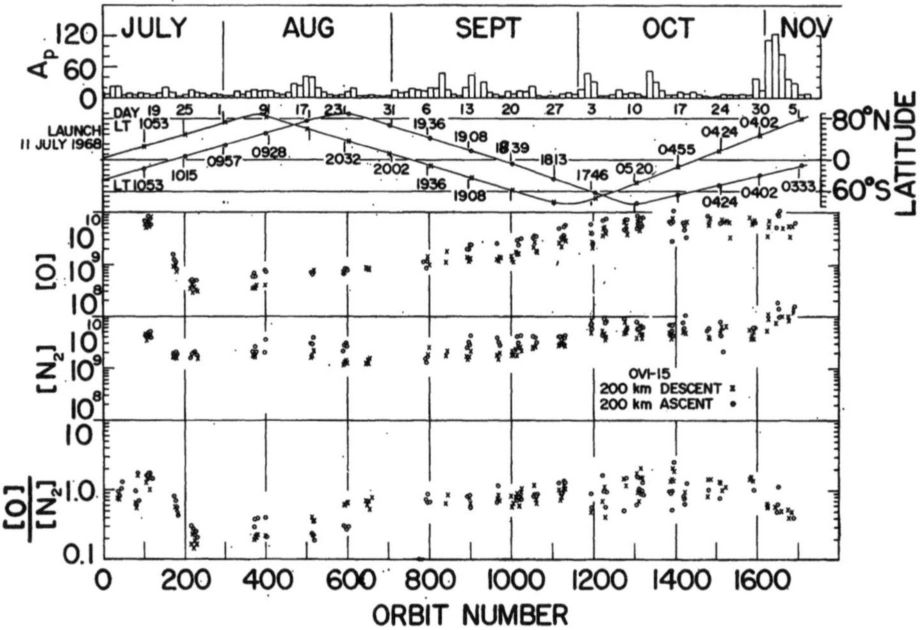

Fig. 3. Measurements of the [O/N_2] ratio, [N_2] and [O] are shown for the 200 km crossings for ascent and descent obtained by Mass Spectrometer I on OV 1-15.

oxygen is difficult to understand. It is interesting that the decrease during this period and the gradual increase in later orbits is in the direction expected for the combined annual and semi-annual variation [17]; however the magnitude observed is much larger than would be allowed in current models. Also the [N_2] variation is in phase with the [O] variation although smaller. The fact that the [N_2] is in phase with the [O] variation is not in agreement with the model of Mayr and Volland [17].

Between orbits 500 and 700 the descent portion of the orbit was shadowed and the ascent was sunlit. From orbit 700 to 900 both the ascent and descent were shadowed and for orbits beyond 900 the data collection period was sunlit. During that period from orbit 500 to 700 the ascent and descent values of [O] are the same but the [N_2] values are significantly higher on the sunlit ascent than on the shadowed descent. This shows the sensitivity of the [N_2] to the atmospheric temperature.

Near the end of the lifetime, a geomagnetic disturbance occurred which caused a large increase in the [N_2]. The [O] was not substantially changed and thus the [O/N_2] ratio decreased. The [N_2] increased by about a factor of two at 200 km and the [Ar] increased about a factor of four at 150 km.

The composition measurements from the OV3-6 and OV1-15 cannot be satisfactorily compared with presently available models. Many of the features which have been used for empirical models of atmosphere density will not apply because of their two-dimensional character. Advances in the dynamical or three-dimensional models should develop rapidly as more composition data become available.

Acknowledgments

The author wishes to express appreciation to many of the individuals who have contributed to the data reduction efforts on these satellites: M. Gardner, R. McInerney, E. Robinson, I. Hussey, E. Cronin, Dr. R. Marcon, P. Pruneau, L. Dalton, D. Williams, R. Desrochers, R. Fioretti and A. Markey. Appreciation also goes to R. A. Wlodyka, G. S. Federico, W. H. Dodson and R. B. E. Moren who performed the electrical and mechanical design of these experiemts and are responsible for the fact that each of these four instruments performed well from launch to re-entry.

References

[1] C. A. Reber and M. Nicolet, Planet. Space Sci. **13**, 617 (1965).
[2] A. O. Nier, J. H. Hoffman, C. V. Johnson and J. C. Holmes, J. Geophys. Res. **69**, 979 (1964).
[3] A. E. Hedin and A. O. Nier, J. Geophys. Res. **71**, 4121 (1966).
[4] A. A. Pokhunkov, Annls Géophys. **22**, 92 (1966).
[5] A. E. Hedin, H. G. Mayr, C. A. Reber, G. R. Carignan and N. W. Spencer, Space Research XIII, 315 (1973).
[6] A. E. Hedin and H. G. Mayr, J. Geophys. Res. **76**, 1688 (1973).
[7] D. R. Taeusch, G. R. Carignan and C. A. Reber, J. Geophys. Res. **76**, 8318 (1971).
[8] J. P. McClure, J. Geophys. Res. **76**, 3106 (1971).
[9] P. Waldteufel and L. Cogger, J. Geophys. Res. **76**, 5322 (1971).
[10] J. E. Morgan and A. J. Schiff, McGill University Final Rep. Contr. AF 19 (628) 2425, AFCRL Rep. 66-64 (1965).

[11] C. R. PHILBRICK and J. P. McISAAC, Space Research XII, 743 (1972).
[12] C. R. PHILBRICK, R. A. WLODYKA and M. E. GARDNER, EOS Trans. Am. Geophys. Un. **51**, 378 (1970).
[13] C. R. PHILBRICK, R. A. WLODYKA and R. S. NARCISI, Trans. Am. Geophys. Un. **49**, 143 (1968).
[14] C. R. PHILBRICK, J. P. McISAAC and M. E. GARDNER, EOS Trans. Am. Geophys. Un. **52**, 294 (1971).
[15] C. R. PHILBRICK, R. S. NARCISI and R. A. WLODYKA, Trans. Am. Geophys. Un. **49**, 722 (1968).
[16] F. A. MARCOS, R. McINERNEY, J. CORBIN, R. FIORETTI and N. GROSSBARD, AFCRL Rep. 72-0608 N. 417 (1972).
[17] H. G. MAYR and H. VOLLAND, J. Geophys. Res. **77**, 6774 (1972).

Space Research XIV — Akademie-Verlag, Berlin 1974

MAGNETIC STORM ASSOCIATED CHANGES IN NEUTRAL COMPOSITION OF THE ATMOSPHERE AT MID-LATITUDES OBSERVED BY THE ESRO 4 GAS ANALYSER

G. W. PRÖLSS and U. VON ZAHN

University of Bonn, Bonn, FRG

Data from the neutral atmospheric composition experiment aboard ESRO 4 are presented for the magnetic storm event on 21/22 February 1973. A large decrease in the O/N_2 ratio which lasts for more than 10 hours is observed at mid-latitudes. The ionosphere responds to this composition change with strong negative storm effects. Possible explanations for this event are briefly discussed.

1. Introduction

It is well known that magnetic storms trigger upper atmospheric disturbances which show a rather complex and variable pattern of behaviour. One important aspect of these upper atmospheric storms—as we may call them in analogy to magnetic storms—are changes in the neutral composition. As early as 1956 Seaton [1] in his attempt to explain ionospheric storm data suggested significant changes in the atomic to molecular oxygen ratio during magnetically disturbed periods; however, it was only in recent years that direct and continuous observations of the different atmospheric constituents have become possible [2, 3]. In this paper we present early data obtained by the gas analyser aboard the ESRO 4 satellite during the magnetic storm of 21/22 February 1973 which show a large composition change in the neutral atmosphere at mid- to low-latitudes.

2. Observations

ESRO 4 is a polar-orbiting satellite (inclination $= 91°$, period $\approx 98^m$). At the time of the storm the perigee was located in the southern hemisphere at about 30° S, 1100 LT, and the perigee height was approximately 253 km. The gas analyser aboard ESRO 4 is a RF monopole spectrometer which sampled the number densities of atomic oxygen, molecular nitrogen, helium, and argon every 9 seconds. Details on the satellite orbit and the gas analyser have been given elsewhere [4].

During the second half of February 1973 a series of moderate and moderately strong magnetic storms occurred. Fig. 1 indicates the magnetic disturbance variation during two days of this period. During the early morning hours of

21 Feburary both the Kp index and the horizontal magnetic component at the southern hemisphere station Toolangi (38° S 145° E) show the peak and recovery phase of a moderate disturbance beginning the day before. After a period of relative quiet between 06 and 16 UT a larger magnetic storm commenced reaching

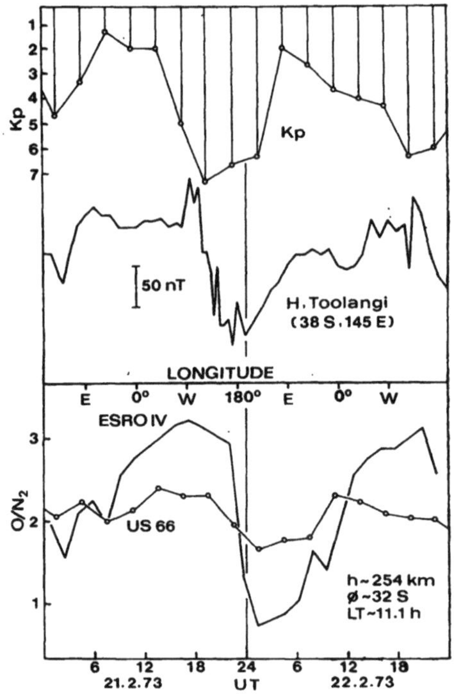

Fig. 1. Time variation of magnetic and atmospheric parameters during the 21/22 February 1973 magnetic storm event. The upper panel shows the Kp index and the Toolangi H component and the lower panel shows the O/N_2 ratio observed by the ESRO 4 gas analyser compared with that predicted by the US Standard Atmosphere 1966.

a peak Kp index of 7_+ and a maximum depression in the Toolangi H component of $-150\,\gamma$. 22 February shows the rapid and slow recovery phase of this storm, which is interrupted at about 16 UT by a new disturbance lasting for the rest of the day.

The lower panel of Fig. 1 shows the disturbance of the neutral upper atmosphere associated with these magnetic storms. The number density ratio O/N_2 serves as a rather sensitive indicator for atmospheric perturbations and is plotted here for a height of 254 km and a latitude of 32° S. We see that apart from a time lag this atmospheric parameter reflects the magnetic disturbance variation very well. Remarkable is the large range of variation (from about 3.2 following the magnetically quiet period down to 0.74 at the peak of the disturbance), indicating that the diffusive equilibrium is not preserved during the atmospheric storms. Further to corroborate this point, we have calculated the composition changes to be expected from the magnetic storm associated heating of the upper atmosphere using the US Standard Atmosphere (1966). Fig. 1 shows that the resulting changes in the O/N_2 ratio are much too small to fit the observations.

To show the spatial distribution of the disturbance we have plotted the atmospheric storm associated changes in atomic oxygen, molecular nitrogen, total mass density and in the N_2/O ratio as a function of (geographic) latitude using the undisturbed conditions around 2027 UT on 21 February as a reference. Fig. 2 shows a "snapshot" of the latitudinal variation in the southern hemisphere at the peak of the disturbance around 0120 UT on 22 February. Note that the density changes have not been normalized to a fixed altitude.

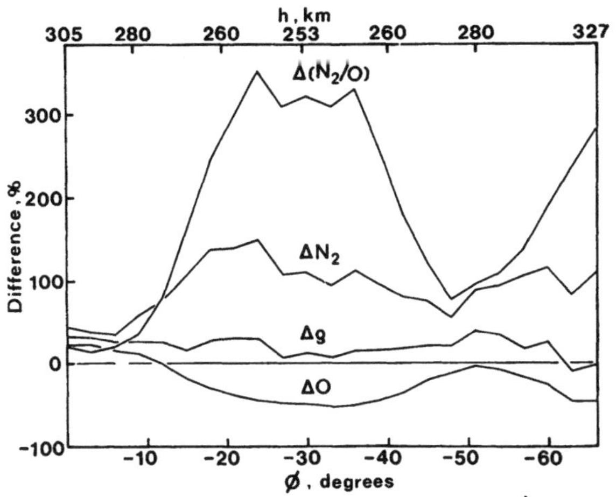

Fig. 2. Latitudinal variation of the atmospheric storm associated changes in atomic oxygen, molecular nitrogen, total mass density, and in the N_2/O ratio during the 21/22 February 1973 event. Plotted is the difference between the highly disturbed orbit 1346 at about 0120 UT on 22 February and the undisturbed orbit 1343 at about 2027 UT on 21 February.

The most striking features of the data presented in Fig. 2 are the large increase in the molecular nitrogen density by up to 150% and the simultaneous decrease in the atomic oxygen density to about one-half of their quiet orbit values in mid- to low-latitudes, indicating that at that time the most disturbed regions are not located in the auroral zones but are relatively close to the equator. Due to the fact that the atomic oxygen density decreases while the nitrogen density increases, the total mass density deviates comparatively little from its quiet orbit value in the disturbed regions. In contrast, the N_2/O ratio proves to be a very sensitive indicator for this kind of disturbance, increasing by more than 300%. This parameter also shows that another centre of disturbance is located in the polar zones. Finally, it should be noted that for the whole duration of the storm the equatorial regions appear only slightly affected by the disturbance except that both the O and N_2 densities show an increase by some 10 to 40%, probably indicating a moderate increase in atmospheric temperature.

A large number of papers have dealt with ionospheric disturbances associated with magnetic storms, and various theories have been proposed to explain the anomalous decrease in ionization observed during these periods. At present the most promising explanation for this phenomenon is based on the assumption that composition changes in the neutral upper atmosphere cause a drastic increase in the ionospheric loss coefficient. It is therefore of some interest to check the ionospheric response to the large decrease in the O/N_2 ratio observed during the magnetic storm period under investigation. Fig. 3 shows the time variation of the critical frequency of the F2 region (which serves as a measure for the peak electron density in this region) for the southern hemisphere station Mundaring (32° S 116° E). For comparison we also plot the average time variation of the

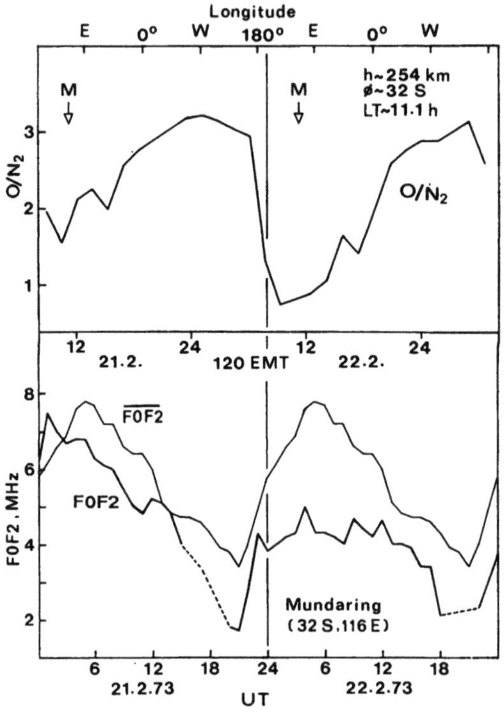

Fig. 3. Ionospheric response to neutral composition changes during the 21/22 February 1973 magnetic storm event. The upper panel shows the O/N_2 ratio observed by the ESRO 4 gas analyser and the lower panel shows the critical frequency of the F2 region (F0F2) during the storm event as compared with the average time variation of this frequency during February 1973 ($\overline{\text{F0F2}}$).

critical frequency for February 1973 and the observed changes in the O/N_2 ratio during the storm period (note that the times at which the satellite probes the atmosphere above Mundaring are indicated by an arrow). Without going into detail, it is at once clear from these data that the ionosphere responds to the observed composition changes with large negative storm effects. Our data therefore strongly support the theory first proposed by Seaton [1] that negative ionospheric storm effects are caused by changes in the neutral composition.

3. Discussion

The purpose of this paper was to describe the large changes in the neutral composition at mid- and low-latitudes observed during a magnetically disturbed period. The data presented indicate that the diffusive equilibrium may be severely disturbed during these times, implying that static atmospheric models are inadequate to deal with such situations. In recent years a number of papers have dealt with the dynamic response of the upper atmosphere to energy deposition in the auroral zones (see, for example [5—7]). Should the observed atmospheric disturbance be caused by heating, some of the results obtained in these studies would be directly applicable to our situation. However, at present no mechanism appears to be known which would deposit sufficient heat energy in the mid- to low-latitudinal regions. Large ionospheric currents and particle precipitation events—often thought to be the principal heat sources in the auroral zones—are rarely observed at lower latitudes. Theoretical studies and direct measurements [8, 9] also indicate that the rise in neutral temperature within SAR arcs is insufficient to explain our observations, especially since most of this energy deposition should occur at heights above 300 km. Prölss et al. [10] investigated the charge-exchange mechanism by which part of the energy contained in the magnetic storm ring current is injected into the low-latitude upper atmosphere. Accepting, however, present ideas about the ring current, this energy input would be too small significantly to heat the lower thermosphere.

Another possible contribution to the observed changes in the neutral composition could arise from a general redistribution of the different atmospheric constituents during magnetically active periods. Reber et al. [11], for example, observed that minima in the N_2 density located at $\pm 30°$ latitude (0700 LT) are filled in and disappear during magnetic storms. If this spatial distribution is partially controlled by ion drag an interesting feedback may occur: A disturbance in the distribution may cause composition changes which in turn cause loss of ionization. This then reduces the ion drag, which may further disturb the distribution, etc. Clearly more information on the dynamics of the neutral upper atmosphere during magnetically active periods (including perhaps the coupling between ionosphere and neutral atmosphere) is needed fully to understand the observations presented in this paper.

Acknowledgments

Thanks are due to K. H. Fricke and H. Ding for their extensive help in data processing. Magnetic records and ionospheric data were obtained through the help of W. Paulishak from World Data Center A. This research is supported through grant WRK 107 of the Bundesministerium für Forschung und Technologie, Bonn.

References

[1] M. J. Seaton, J. Atmos. Terr. Phys. 8, 122 (1956).
[2] D. R. Taeusch et al., J. Geophys. Res. 76, 8318 (1971).
[3] C. R. Philbrick and J. P. McIsaac, Space Research XII, 743 (1972).
[4] R. Jaeschke and J. Tunaley, European Space Research Organization Tech. Rep. No. SN-114 (ESTEC) (1972).
[5] H. G. Mayr and H. Volland, Planet. Space Sci. 20, 379 (1972).
[6] T. Shimazaki, Space Research XII, 1039 (1972).
[7] P. B. Hays et al., Planet. Space Sci. 21, 559 (1973).
[8] R. G. Roble and R. E. Dickinson, Planet. Space Sci. 18, 1489 (1970).
[9] R. G. Roble et al., Planet. Space Sci. 18, 431 (1970).
[10] G. W. Prölss et al., J. Atmos. Terr. Phys. 35, 1889 (1973).
[11] C. A. Reber et al., J. Atmos. Terr. Phys. 35, 1223 (1973).

Space Research XIV — Akademie-Verlag, Berlin 1974

ON THERMOSPHERIC COMPOSITION AND TEMPERATURE VARIATIONS DURING GEOMAGNETIC DISTURBANCES

Y. A. Romanovsky and V. V. Katyushina

USSR

According to mass-spectrometric data from the Cosmos 274 satellite, significant variations of thermospheric composition and temperature during geomagnetic disturbances on 25 March 1969 ($Kp = 4^{+}-5$) have been found. The increase of neutral gas temperature in the sub-auroral zone, determined according to atmospheric composition variations, was about $850° \pm 150$°K. The variations of atmospheric neutral composition during the disturbance are characterized by morning-evening asymmetry: the increase of atmospheric temperature and density during the morning hours is greater in comparison with that in the evening hours at the same geomagnetic latitudes in the northern and southern hemispheres. The effect observed is thought to be due to the precipitation of electrons with energies of 10 to 40 keV, the maximum of which is usually observed during the morning hours.

Recent upper atmosphere investigations have shown that the neutral atmosphere response to a geomagnetic disturbance is characterized by atmospheric temperature and density increase [1, 2], composition changes [3] and the generation of global atmospheric wind systems [4].

Data on some features of thermospheric composition and temperature variations at altitudes of about 200—300 km during a geomagnetic disturbance are given here. These data were obtained during mass-spectrometric experiments aboard the Cosmos 274 satellite. The satellite was launched into an orbit with the following parameters: apogee 320 km, perigee 220 km, inclination 65°. The measurements of ion and neutral composition were carried out by the Bennett radio-frequency mass spectrometer MX-6407 P [5] with an ion source of the semi-open type. In this paper data on variations of the neutral gas composition obtained from four satellite orbits (16—19) over the period 0720—1505 UT on 25 March 1969 during a geomagnetic disturbance ($Kp = 4^{+}-5$) which occurred during the recovery phase of the strong geomagnetic storm ($Kp = 8$) on 24 March 1969 [6] are considered. To characterize the thermospheric neutral composition variations during undisturbed conditions, the data from orbit 32 which covered the undisturbed period from 0720 to 0825 on 26 March 1969 ($Kp < 2^{-}$) are presented.

Neutral constituents of 28 and 32 amu, identified as molecular nitrogen and molecular oxygen respectively, are normally recorded. The magnitude of the atomic oxygen peak was significantly smaller than those of these molecular constituents. The variations of the relative atomic oxygen concentration in the thermosphere were evaluated according to the changes of molecular oxygen peak, assuming that the value of the atomic oxygen recombination coefficient inside the device did not change during the experiment. The reliability of the conclusion is confirmed by data presented in Fig. 1, where the dependence of the values of the ratio O/N_2 as function of altitude, obtained on orbit 32 under undisturbed conditions (dots) and calculated using the Jacchia 1971 [7] neutral atmosphere model

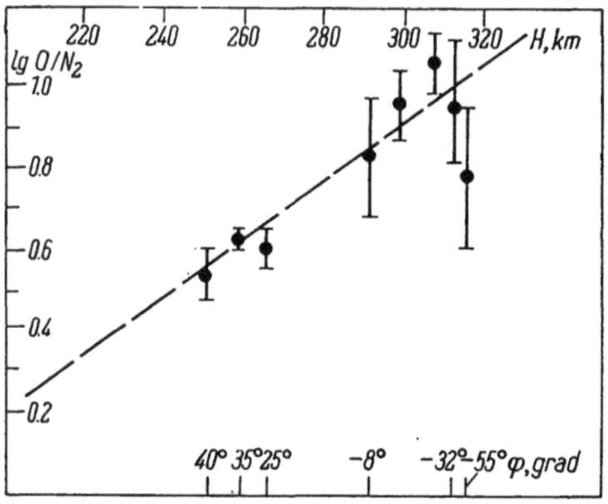

Fig. 1. The dependence of the O/N_2 concentration ratio on altitude under undisturbed conditions, according to experimental data obtained on orbit 32 (dots) and to the Jacchia 1971 model [7].

(dashed line) are compared. The comparison of the data presented shows that under undisturbed conditions good agreement of experimental and model values of O/N_2 is observed except in the subauroral zone where experimental values of this ratio are lower than model ones.

Fig. 2 shows the $[N_2]$ and [O] concentration dependence on altitude according to measurements on orbits 16—19 during the geomagnetic disturbance. Solid lines show values of $[N_2]$ and [O] calculated by the Jacchia model [7] taking into account the latitude, local time and solar and geomagnetic activity levels which correspond to each measurement. The experimental data were fitted to the model. According to Fig. 2 it follows that the Jacchia model gives in general a good description of $[N_2]$ and [O] variations during the magnetic disturbance period; however, some experimental data depart significantly from the model. To characterize the departure of experimental (exp) $[N_2]$ and [O] observations from the model (mod) Fig. 3 shows the values

$$\delta[n] = \log\,[n]_{\text{exp}} - \log\,[n]_{\text{mod}}$$

as functions of geomagnetic latitude Φ and local time LT. The data presented demonstrate the difference of δ variations in northern and southern hemispheres. Thus, at latitude $\Phi = 40°$ N $\delta[N_2] = 0$, while at latitude $\Phi = 40°$ S $\delta[N_2] = 0.55$. The maximum values of δ correspond to measurements carried out in southern subauroral latitudes during morning hours.

Assuming that there are no significant changes in neutral temperature and composition at the turbopause, one may use the exosphere temperature T_∞ of the Jacchia 1971 model. Thus, for $\delta = 0$ $T_\infty = 1000\,°K$. The maximum δ values correspond to a temperature increase up to $\Delta T_{\max} = 850 \pm 150\,°K$. Within the experimental accuracy [O] and $[N_2]$ show the same values of $T_{\max}$.

It is necessary to note, however that there is a tendency for a temperature excess, of that determined by the oxygen over that determined by molecular nitrogen, to exist at low latitudes ($\Phi = 30°$).

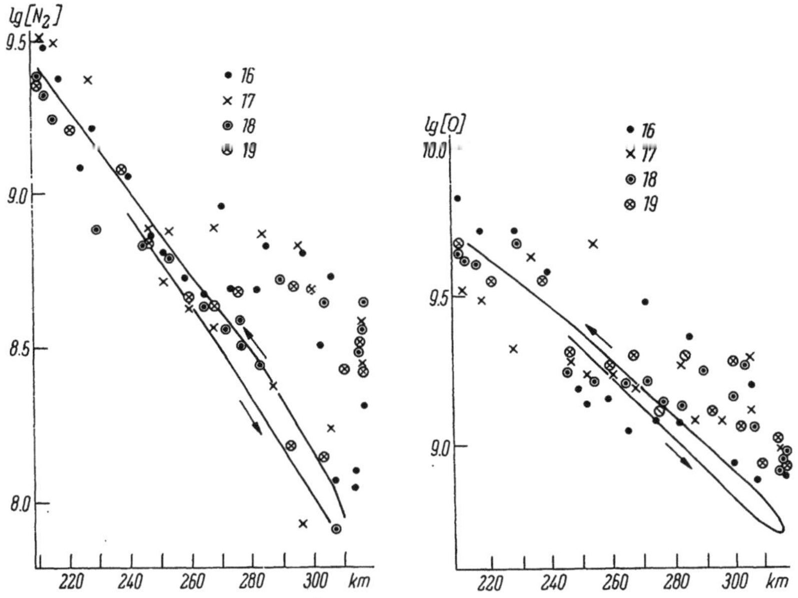

Fig. 2. Comparison of data on $[N_2]$ in cm^{-3} (left) and [O] in cm^{-3} (right) obtained on orbits 16, •; 17, ×; 18, ⊙; and 19, ⊗ during geomagnetic disturbance with Jacchia 1971 model (solid line). Arrows indicate the sequence of data acquisition along the satellite orbit.

The different rates of increase of $[N_2]$ and [O] with latitude during the geomagnetic disturbance result in a significant difference between the O/N_2 ratios at middle and subauroral latitudes at night and morning hours; e.g., at latitude $\Phi = 50°$ S at 4—5 h LT this ratio is an order of magnitude less compared with its value in undisturbed conditions. The analogous effect of thermospheric composition and temperature variations in the morning sector of the southern subauroral zone was observed under quiet conditions (orbit 32). However, in this case $\Delta T_{max} = 400\,°K$ and the zone of the effect is smaller (Fig. 1).

Thus, in this experiment the dramatic thermospheric temperature and composition variations were observed to result from geomagnetic disturbance. During the equinox the occurrence of a latitudinal asymmetry in the response of the neutral atmosphere to geomagnetic activity is hardly probable. Moreover the similarity of phenomena occurring in such geomagnetic conjugate areas may be considered to be determined [8]. Thus it is natural to try to connect the effect detected with phenomena at high latitudes which have their maximum of occurrence in the morning hours and their minimum in the evening hours. Such phenomena which take place during geomagnetic disturbances are variations of the flux of precipitating electrons with energies between 10 and 40 keV [8], i.e. drizzle. We consider that the neutral atmosphere responds to disturbances during the geomagnetic storm not only by Joule heating of the atmosphere by the auroral electrojet and the precipitation of soft auroral electrons [1—4], but also by the precipitation of the more energetic component (electrons with energies 10—40 keV)

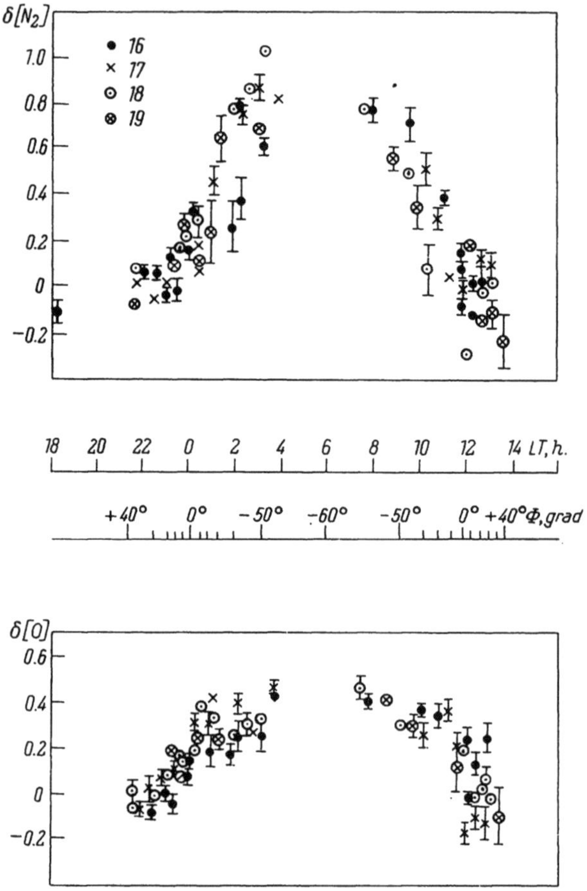

Fig. 3. Departures of the experimental data from the Jacchia model as a function of geomagnetic latitude Φ and local time, LT (symbols as in Fig. 2).

of the auroral and quasitrapped radiations that excite mantle emissions in auroral and subauroral regions. The existence of such a connection would be essential for a theory of neutral atmospheric disturbances [4].

Acknowledgments

The authors wish to express sincere thanks to Dr. V. T. Istomin for valuable discussions.

References

[1] J. E. Blamont and J. M. Luton, paper presented at COSPAR Meeting Leningrad 1970.
[2] M. Roemer, Space Research XI, 965 (1971).
[3] D. R. Taeusch, G. R. Carignan and C. A. Reber, J. Geophys. Res. **76**, 8318 (1971).
[4] H. Volland and H. G. Mayr, J. Geophys. Res. **76**, 3764 (1971).
[5] A. E. Rafal'son and A. M. Schereschevsky, Mass Spectrometricheskiye Priybory, Moscow 1968.
[6] Solar Geophysical Data, No. 297 (1969).
[7] L. G. Jacchia, Smithson. Astrophys. Obs. Spec. Rep. No. 332, May 1971.
[8] S. I. Akasofu, Polar and Magnetospheric Substorms, D. Reidel Publ. Co., Dordrecht, Holland 1968.

Space Research XIV — Akademie-Verlag, Berlin 1974

ATOMIC OXYGEN PROFILES DETERMINED BY EUV ABSORPTION ANALYSIS

G. Schmidtke*, K. Rawer, Th. Fischer and W. Lotze

Institut für physikalische Weltraumforschung, Freiburg, FRG

The EUV experiment aboard the satellite AEROS continuously measures the intensity of solar radiation at wavelengths 58.4 and 30.4 nm. For sunrise and sunset conditions the solar zenith angle at the satellite reaches values greater than 90° so that atmospheric absorption occurs at heights below that of the satellite. Profiles of atomic oxygen are determined for these conditions; two examples are shown for the height range 200 to 450 km.

1. The AEROS EUV Spectrometer

AEROS (AEROnomy Satellite) was launched on 16 December 1972 into a polar orbit, the altitude varying between 200 and 800 km. The concept of AEROS was based on integrated data evaluation of solar EUV photon fluxes from 16 to 106 nm, atmospheric densities (neutrals, ions and electrons) and temperatures (again neutrals, ions and electrons). The neutral densities and temperatures are investigated using a mass spectrometer, the Neutral Atmosphere Temperature Experiment and atmospheric drag analysis. Ion densities and temperatures are measured using a mass spectrometer and retarding potential analyser.

Examples of the EUV spectrometer observations are shown in Fig. 1, in channel 1 from 32 to 107 nm and in channel 2 from 16 to 57 nm. With this instrument the measurement of the major energy input to the upper atmosphere is accomplished.

The alignment of the satellite spin axis (which is parallel to the reference axis of the EUV spectrometer) towards the sun is accurately achieved from time to time only, so that the pointing error may reach 5°. Therefore, the diffraction pattern recorded by the planar grating type spectrometer is shifted depending on β, the misalignment of the sun and the satellite spin axis. With deviations of less than 1.5 degrees from exact solar pointing (mode I), the spectrum is scanned stepwise during one spin period (6 seconds). The influence of this misalignment can be noticed in Fig. 1, where spectral lines are distorted typically. A "defolding" operation is therefore executed by a computer program during data reduction. In mode III the in-flight calibration of the multipliers is accomplished. In addition strong airglow emissions at night will be detected.

---

* This co-author wishes to dedicate his effort to Professor K. Rawer on the occasion of his 60th birthday.

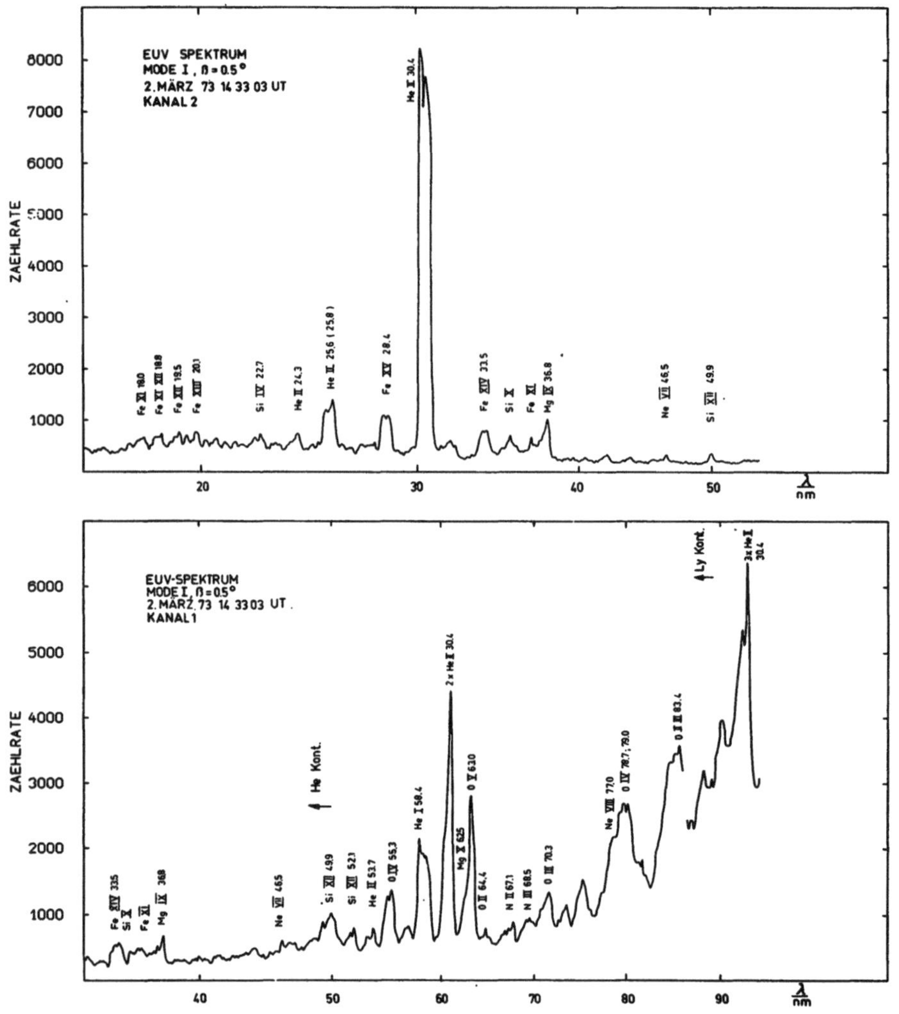

Fig. 1. Typical EUV count rates in mode I (i.e. the pointing error being less than 1.5°), demonstrating the influence of satellite spin.

If the spectrometer is turned on (preferably at sunrise conditions over the southern hemisphere) two selected wavelengths (58.4 nm, 30.4 nm) are measured continuously (mode IV). By telecommand this mode of operation can be continued during the whole orbit (mode V). Measurements in modes IV and V will be of interest here.

2. Determination of Density Profiles for Atomic Oxygen

The two resonance lines of helium at 30.4 and 58.4 nm have been chosen, because (i) their absorption cross sections have been investigated more extensively than those of other solar emission lines within the selected wavelength range;

(ii) their emission from the sun is very intense; (iii) their emission is more evenly distributed over the sun and more steady than radiation from more highly ionized atoms; (iv) no "radiation hardening" [1] has to be taken into account; and (v) they are easy to produce in the laboratory, which helps to align and calibrate the spectrometer. From the measurements at the two wavelengths density profiles for atomic oxygen are derived according to the theory given by Chapman [2] and Swider [3]. The cross sections used are shown in Table 1 [4].

Table 1
Absorption Cross Sections (in 10^{-22} m^2)

$\lambda(nm)$	$\sigma(O)$	$\sigma(N_2)$	$\sigma(O_2)$
30.4	9	12	17
58.4	13	23	23

As a first approximation the nitrogen distribution and the scale heights for atomic oxygen have been taken from the model atmosphere at 700 K as given by Jacchia [5]. Then the atomic oxygen densities have been computed for each channel separately. From these profiles the scale heights have been derived, which were used for the next approximation, and so on. Final results for observations made

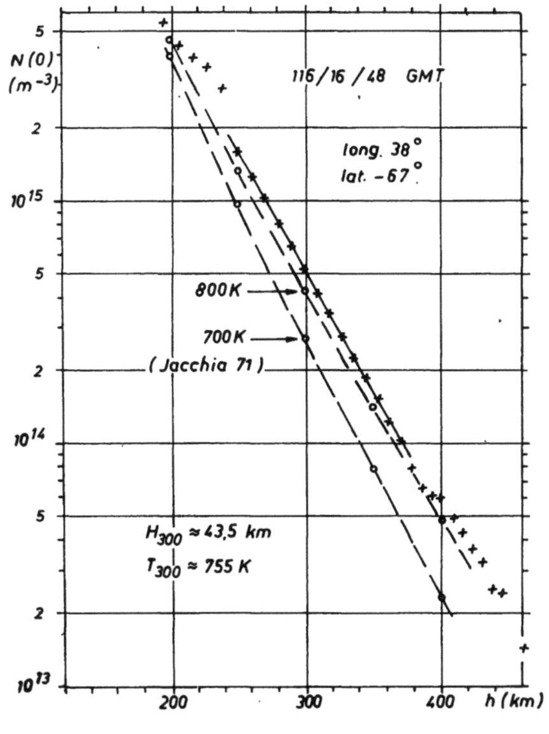

Fig. 2. Atomic oxygen densities (orbit 2002) on 116/16/48 GMT (day/h/min), 1973.

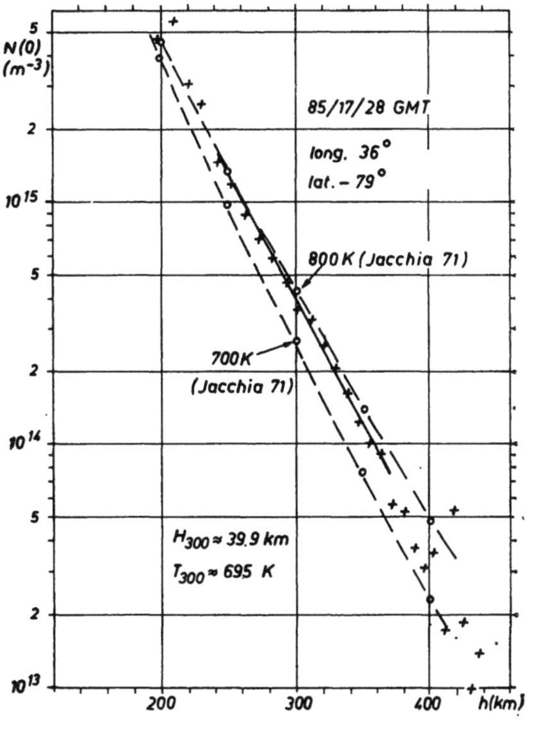

Fig. 3. Atomic oxygen densities (orbit 1524) on 85/17/26 GMT (day/h/min), 1973.

during two orbits are shown in Figs. 2 and 3*. The agreement between the derived densities for both spectral lines is good. The highest deviations between 300 and 360 km are about $\pm 5\%$ for orbit 2002 (Fig. 2) and $\pm 3\%$ for orbit 1524 (Fig. 3). Some points agree to better than 0.5%. The data below 250 km have slightly lower significance due to noise such as stray light from hydrogen Lyman-alpha. The data points above 400 km are affected by statistical fluctuations. The absolute values deduced here depend, of course, on the accuracy of the absorption cross sections of atomic oxygen with density errors inversely proportional to the cross section errors.

These measurements do not produce vertical, but oblique, profiles (the orbit is fixed in local time at about 0300 and 1500 LT at the equator, but there is, of course, a transition at higher latitudes where the measurements have been made). However, it can be assumed that the atomic oxygen densities do not change much with latitude and longitude at the same height within distances of some hundred kilometres.

Acknowledgment

This work was supported by the Bundesministerium für Forschung und Technologie (WRK 192, 226, and 254).

References

[1] H. E. Hinteregger, L. A. Hall and G. Schmidtke, Space Research V, 1175 (1965).
[2] S. Chapman, Proc. Phys. Soc. 43, 26 (1931).
[3] W. Swider, Jr, Planet. Space Sci. 12, 761 (1964).
[4] R. E. Huffman, Canad. J. Chem. 47, 1823 (1969).
[5] L. G. Jacchia, Smithson. Astrophys. Obs. Spec. Rep. No. 332 (1971).

---

* Density scale heights H' have been derived from the profiles at an altitude of $h = 300$ km. Virtual (atomic oxygen) temperature values have simply been deduced from the relation

$$kT_0 = m_0 g H'.$$

These are the values indicated in Figs. 2 and 3.

Space Research XIV — Akademie-Verlag, Berlin 1974

MOLECULAR OXYGEN CONCENTRATIONS AT THE EQUATOR DERIVED FROM SOLAR ULTRAVIOLET ABSORPTION MEASUREMENTS

B. H. Subbaraya, S. Prakash, V. Kumar and P. N. Pareek

Physical Research Laboratory, Ahmedabad, India

Nitric oxide filled ion chambers with lithium fluoride windows were flown from Thumba to measure the noon time absorption profile of solar Lyman alpha radiation. The results were used to compute molecular oxygen concentrations in the height region 65 km to about 95 km. It is found that the measured oxygen concentrations are in general less than the standard atmospheric models of Groves; large differences between the two flights exist in the 80—90 km region.

1. Experiments

Nitric oxide filled ion chambers with lithium fluoride windows were flown from Thumba (8°31′N 76°52′E) on two Nike Apache rockets, 10.37 and 10.43. On both flights the ion chamber was operated in the unity gain mode. Several amplifiers were employed to cover a wide dynamic range and yet to maintain linearity of the amplifier response [1]. Aspect information was obtained by an on-board sun sensor; the ion chamber data were corrected for changing aspect angle.

2. Analysis of Data

On both the flights the ion chambers were used to measure the absorption of solar ultraviolet radiation in the wavelength range 1050—1340 Å from 65 km to about 95 km in the earth's atmosphere. This wavelength band includes the hydrogen Lyman alpha radiation (1216 Å) which contributes nearly 90% of the total flux in the band, at the top of the atmosphere. Thus the measured flux is essentially due to Lyman alpha and the absorption profile represents, to a first approximation, the atmospheric absorption profile of the solar Lyman alpha radiation. At these wavelengths, molecular oxygen is the main absorbing agent in the earth's atmosphere, the net contribution of the other atmospheric gases being nearly two orders of magnitude less. Hence the measured absorption profile can be used to derive molecular oxygen concentrations.

Although the absorption cross section for molecular oxygen varies across the Lyman alpha line [2], it has been found [3] that use of a constant value for the cross section does not lead to appreciable errors in the derived molecular oxygen concentrations. The data of flight 10.37 were analysed earlier assuming a constant value of 1×10^{-20} cm^2 for the cross section [4]. However, Hall [5] has recently shown the presence of errors of the order of 30% in the computed molecular oxygen concentrations near the top of the absorption profile. The errors decrease with decreasing height and become 10—15% near unit optical depth.

In this paper the data of flight 10.37 have been re-analysed, along with the data of flight 10.43, following the method outlined by Hall. Computational and system noise errors are generally negligible in the height range 75—90 km, being of the order of 2%. Below 75 km the errors increase with decrease in height, due to the smallness of the measured ion chamber currents, and reach values of 10—20% around 65 km. Above 90 km the computational error increases rapidly with increase in height. The errors for flight 10.43 are in general less than those for flight 10.37.

3. Results

Fig. 1 shows the results for flight 10.37 along with the atmospheric model of Groves for 10° N [6]. The measured concentrations are everywhere smaller than those published earlier, by 15% to 30%. Fig. 2 shows the data for flight 10.43.

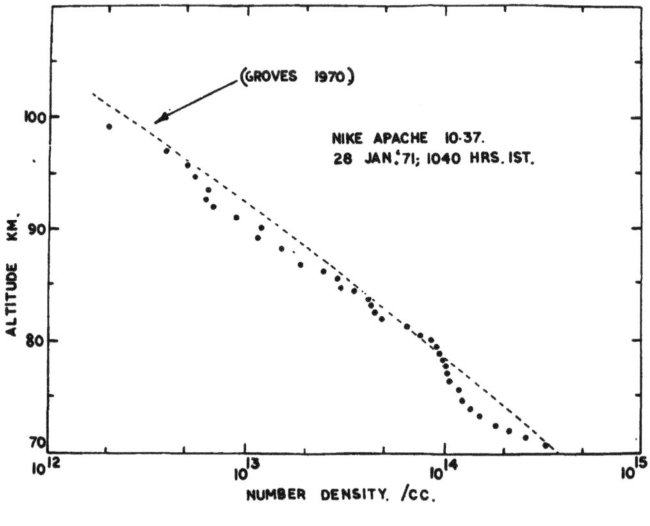

Fig. 1. Molecular oxygen concentrations derived from the Lyman alpha absorption profile of flight 10.37 compared with the standard atmospheric model.

Here again the observed concentrations are smaller than the model values. But the most marked feature of this flight is a sudden depression in the oxygen density profile around 80 km. The densities decrease by a factor of 5 in a height of 2 km. This is followed by a gradual recovery with the measured densities approaching model values around 90 km. The data have been carefully studied to ensure that the observed effect is genuine and cannot be attributed to any instrumental malfunction. Rocket performance was normal and no large changes in rocket aspect occurred in this height region. However, such a large depletion of molecular oxygen is not easily explained on the basis of presently known chemical or dynamical processes in the mesosphere.

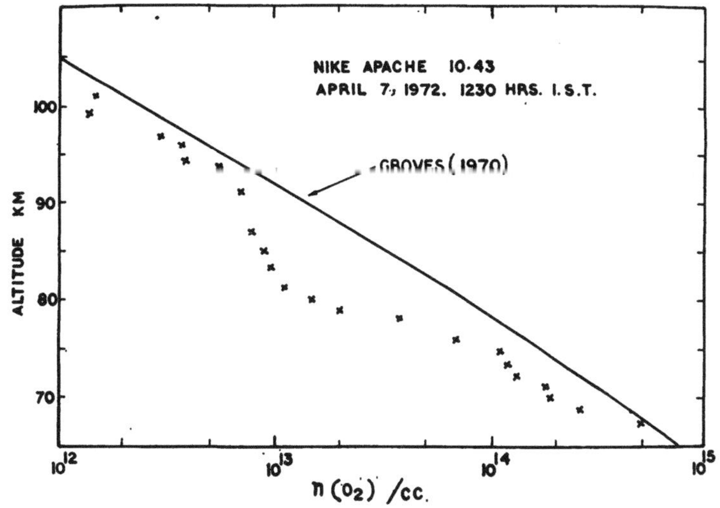

Fig. 2. Molecular oxygen concentrations derived from the Lyman alpha absorption profile of flight 10.43 compared with the standard atmospheric model.

4. Discussion

The results of the two flights show that the molecular oxygen concentrations at Thumba are in general lower than those given by standard atmospheric models, and are quite variable. Order of magnitude variations in the molecular oxygen concentrations at the 100 km level have been reported by Felske et al. [7] from satellite measurements of Lyman α absorption at sunset. Quessette [3] has also reported anomalous observations in the 90—100 km region. But his data show anomalously high concentrations. All these results point to the possibility of large variations in neutral atmospheric densities at mesospheric and lower thermospheric levels. Meteorological rocket soundings at Thumba have revealed that the mesopause at Thumba is, in general, lower than that at middle latitudes, and that there is a high degree of variability in the height of the mesopause and in the pressure there (Pisharoty, private communication). The present observations are consistent with these findings.

Acknowledgments

The authors would like to thank the staff of TERLS for their efforts in making the experiments successful. Mr. Madhavan Nair of TERLS supplied the solar aspect sensor for the two flights. B. K. Dash provided help with the instrumentation and Mr. J. R. Trivedi computational assistance; these are gratefully acknowledged.

References

[1] B. H. SUBBARAYA, B. K. DASH and S. PRAKASH, to be published in J. Inst. Electronics and Telecomm. Engrs, India **19**, 397 (1973).
[2] Y. OGAWA, J. Geophys. Res. **73**, 6759 (1968).
[3] J. A. QUESSETTE, J. Geophys. Res. **75**, 839 (1970).
[4] B. H. SUBBARAYA, S. PRAKASH and P. N. PAREEK, J. Atmos. Terr. Phys. **34**, 1141 (1972).
[5] J. E. HALL, J. Atmos. Terr. Phys. **34**, 1337 (1972).
[6] G. V. GROVES, AFCRL-70-0261, Air Force Surveys in Geophysics No. 218, Office of Aerospace Research US Air Force (1970).
[7] D. FELSKE, L. MARTINI, B. STARK and J. TAUBENHEIM, Space Research XII, 651 (1972).

Space Research XIV — Akademie-Verlag, Berlin 1974

ABSORPTION BY THE UPPER ATMOSPHERE IN THE MIDDLE ULTRAVIOLET REGION

T. TOHMATSU[a], T. OGAWA[a] and T. WATANABE[b]

[a]Geophysics Research Laboratory, The University of Tokyo, Tokyo, Japan
[b]Department of Physics, Tokyo University of Education, Tokyo, Japan

The attenuation of solar middle ultraviolet radiation was measured in the stratosphere and mesosphere by two or three photometers aboard sounding rockets flown from 1965 to 1973 from Uchinoura, Japan. The atmospheric absorption in the altitude range between 20 and 65 km was determined at three spectral regions centred at 2300, 2550 and 2900 Å. The attenuation of this solar radiation is ascribed entirely to absorption by atmospheric ozone. The vertical distribution of ozone density was obtained from the analysis of data from seven rocket experiments.

1. Introduction

The determination of atmospheric absorption in the ultraviolet using sounding rockets provides one with valuable information on the structure of the topside ozonosphere. The vertical distribution of atmospheric ozone density in the stratosphere and mesosphere can be obtained by continuously measuring the solar MUV radiation as a function of altitude. Since 1965 at Uchinoura, Japan (30°15′ N 131°05′ E), vertical profiles of atmospheric ozone density in the altitude range between 20 and 65 km have been obtained using rocket-borne photometers at two spectral bands, 2550 and 2900 Å [1, 2].

Several research workers have discussed whether or not ozone was the only absorbing species in the upper atmosphere [2—4]. In order to clarify this ambiguous point measurement of the absorption at 2300 Å has also been made since 1972. The present paper contains the results of the analysis of seven rocket experiments conducted from 1965 to 1973.

2. Instrumentation

The photometric assembly used before 1970 consisted of three solar photometers which were sensitive to radiations in the spectral bands centred at 2550, 2900 and 4500 Å with half-sensitivity widths of about 150 Å [5]. In the experiments after 1972 a detector working at 2300 Å was added. The four photometers were mouted in the plane normal to the rocket's spin axis looking outward through the windows attached to the rocket's skin. The detectors received solar radiation intermittently as the rocket spun at the rate of 2—3 s^{-1}. The angle between the spin axis and the line of sight of the detector was adjusted before the flight so that the input of solar radiation was a maximum.

3. Procedure of Data Reduction

Fig. 1 is an example of the plot of altitude dependence of the observed solar irradiance at various spectral bands which are normalized with respect to those observed at the top of the atmosphere. The reduction of experimental data can

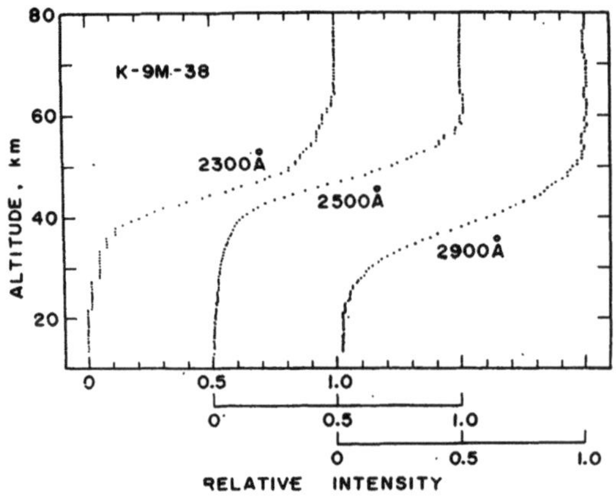

Fig. 1. The vertical distribution of solar MUV irradiance observed with the photometric assembly on the sounding rocket K9M38.

usually be done with the assumption that Beer's law is valid at all wavelengths of interest. A correction is applied to allow for the finite width of the sensitivity band of the photometers. This is done by simulating numerically the relation between the output intensity and the overhead columnar density of ozone, allowing for the variations of filter transmission, spectral sensitivity of the detector and relative solar irradiance [6—12] and using the solar zenith distance and experimental photoabsorption cross section [13—18] as known parameters. For the data of the 2300 Å photometer there is a minor effect of the attenuation by the Herzberg continuum of molecular oxygen.

The second important correction is related to the change of rocket attitude. This is unnecessary when the rocket is well stabilized and has negligible precession. However, the eventual change of rocket attitude due to precession introduces a spurious change of signal because of the finite angular response of the photometers. This kind of geometrical correction is possibly by referring to the change of the signal of the 4500 Å detector. Because the solar radiation at 4500 Å is practically free from attenuation by ozone, the change, if it is present, is entirely due to the angular dependence of the detector's sensitivity.

4. Analytical Result and Discussion

The overhead columnar density of ozone can be obtained as a function of altitude by fitting the experimental irradiance plot as shown in Fig. 1 to the predicted curves given by Fig. 2. The data from the three MUV wavelengths can be

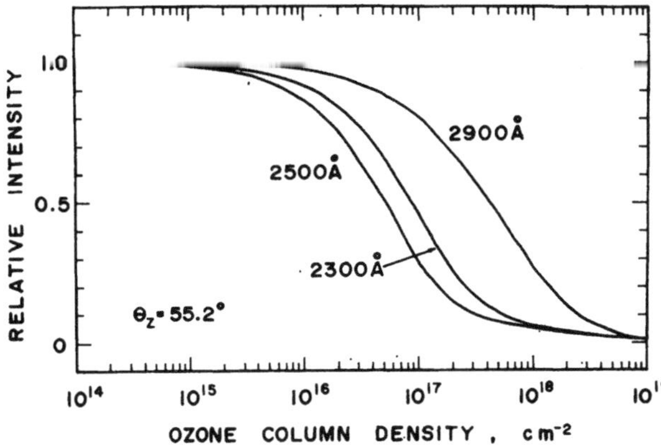

Fig. 2. Predicted attenuation curves of solar MUV radiations measured with the photometric assembly on K9M38.

used to complement each other at different ranges of ozone columnar density, namely at different altitude ranges in the atmosphere. But all data give good results in the common altitude range between 30 and 50 km. Therefore the continuity of the data from the different photometers can be checked using the data from this altitude range.

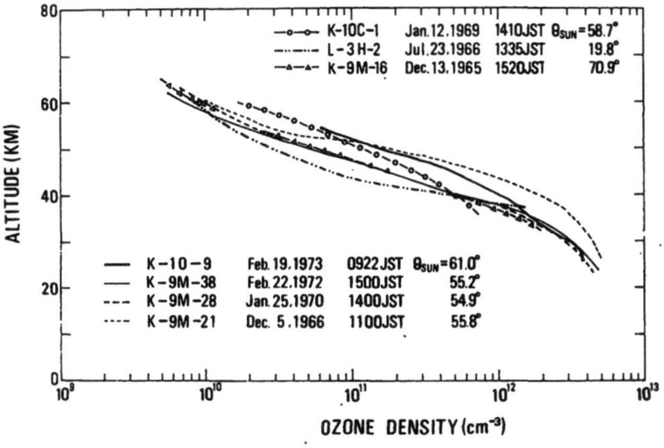

Fig. 3. Vertical distribution of atmospheric ozone density determined in the seven rocket experiments.

The volume density of ozone can be calculated by differentiating numerically the columnar density with respect to altitude. Fig. 3 illustrates the vertical profiles of the atmospheric ozone density obtained for the seven photometric experiments conducted from 1965 to 1973. The observed ozone profile is generally in agreement with the new theory of photochemistry of ozone that takes into consideration the reduction of ozone concentration in a moist atmosphere [19—21]. The second ozone peak as reported in previous papers [1, 22] could not be found by the new analysis. Likewise the new analysis did not allow for the presence of anomalous absorption in the MUV spectral region so far as the data in the altitude range between 30 and 55 km were concerned.

References

[1] T. NAGATA, T. TOHMATSU and H. TSURUTA, Space Research VIII, 639 (1968).
[2] T. NAGATA, T. TOHMATSU and T. OGAWA, Space Research XI, 849 (1971).
[3] A. J. KREUGER, Science **166**, 998 (1969).
[4] J. F. NOXON, Science **168**, 1120 (1970).
[5] T. TOHMATSU, in: Small Rocket Instrumentation Techniques, North-Holland, Amsterdam 1969 (p. 195).
[6] N. L. WILSON, R. TOUSEY, J. D. PURCELL, F. S. JOHNSON and C. E. MOORE, Astrophys. J. **119**, 590 (1954).
[7] H. H. MALISTON, J. D. PURCELL, R. TOUSEY and D. E. MOORE, Astrophys. J. **132**, 746 (1960).
[8] C. R. DETWILER, D. L. GARRET, J. D. PURCELL and R. TOUSEY, Annls Géophys. **17**, 263 (1961).
[9] V. P. KACHALOV and A. V. YAKOVLEVA, Izv. Krimskoy Astron. Obs. **27**, 5 (1962).
[10] R. TOUSEY, Space Sci. Rev. **2**, 3 (1963).
[11] L. DUNKELMAN and R. SCOLNIK, J. Opt. Soc. Am. **49**, 356 (1959).
[12] R. T. BRINKMAN, A. E. S. GREEN and C. A. BARTH, JPL Tech. Rep. No. 32-951 (1966).
[13] NY TZI-ZE and CHOONG SHIN-PIAW, Chinese J. Phys. **1**, 38 (1933).
[14] A. VASSY, Annls Phys. **16**, 145 (1941).
[15] Y. TANAKA, E. C. Y. INN and K. WATANABE, J. Chem. Phys. **21**, 1951 (1953).
[16] E. C. Y. INN and Y. TANAKA, J. Opt. Soc. Am. **43**, 870 (1953).
[17] E. VIGROUX, Annls Phys. **8**, 709 (1953).
[18] E. VIGROUX, Annls Géophys. **25**, 307 (1969).
[19] B. G. HUNT, J. Geophys. Res. **71**, 1385 (1966).
[20] E. HESSTVEDT, Geophys. Norv. **27**, 1 (1968).
[21] T. SHIMAZAKI and A. R. LAIRD, J. Geophys. Res. **75**, 3221 (1970).
[22] E. HILSENRATH, L. SEIDEN and P. GOODMAN, J. Geophys. Res. **74**, 6873 (1969).

Space Research XIV — Akademie-Verlag, Berlin 1974

VERTICAL OZONE PROFILES FROM OBSERVATIONS OF ECLIPSING SATELLITES

R. B. Lee III and D. S. McDougal

NASA Langley Research Center, Hampton, Virginia, USA

An improved method has been developed for remotely determining the vertical ozone profile from ground-based observations. The technique consists of photometrically observing a satellite as it approaches the earth shadow. The resulting attenuation of the apparent satellite intensity due to absorption is measured simultaneously in two narrow bands centered at regions of high and low ozone absorption. The effective ray path altitude is determined by incorporating the effects of atmospheric refraction, the finite size of the sun, and solar limb darkening. The attenuation measurements are corrected for predicted absorption due to air molecules and dust particles. The data are then inverted to yield the vertical number density profile of ozone, assuming a spherically symmetric atmosphere.

The NASA Satellite Photometric Observatory (SPO) was used to observe the PAGEOS and Echo 2 satellites for this experiment. This instrument consists of a mobile 61 cm telescope, a four-channel photometer, and a four-axis tracking mount. Data from nine PAGEOS passes obtained in March 1972 and one Echo 2 pass obtained in February 1969 have been reduced to yield ozone profiles from 15 to 50 km. Comparison is made between the profiles and the ozone model of the US Standard Atmosphere Supplements, 1966. The PAGEOS data show an average peak concentration of about 5×10^{12} molecules cm^{-3} near 20 km and are in excellent agreement with previous measurements.

1. Introduction

Ozone plays an important role in atmospheric dynamics and heating because of its strong absorption of solar ultraviolet radiation. This absorbed energy is the principal source of thermal energy in the stratosphere and is responsible for the temperature increase with altitude in the stratosphere.

Various methods have been used to determine the vertical structure of ozone. These methods include ozonesonde, infrared emission spectroscopy, ultraviolet backscatter, and several stellar and solar occultation techniques. Previous ground-based measurements of eclipsing satellites have been performed using various techniques [1—3]. In this paper, a new method for obtaining stratospheric ozone concentrations from solar occultation measurements is described in which simultaneous observations are obtained by a large telescope of satellite-reflected radiation at 6022 Å and 7029 Å.

2. Instrumentation

Ozone concentrations were obtained from photometric measurements of eclipsing satellites using the NASA Satellite Photometric Observatory (SPO) [4, 5]. The SPO is a mobile observatory with a 61 cm (24 inch) Cassegrain telescope, a four-axis mount specifically designed for horizon to horizon tracking of satellites, and a four-channel detector comprised of four EMI-9558 QA phototubes with S-20 responses. Two narrow band filters are used with peak wavelengths centered at 6022 Å and 7029 Å and with half intensity bandwidths of 14 Å and 18 Å, respectively. The measured signals are recorded on both oscillograph paper and magnetic tape. The satellite intensities are determined in terms of stellar magnitudes and calibrated using standard astronomical techniques [6].

3. Satellites

The **PA**ssive **GEO**detic Earth **S**atellite (PAGEOS), launched in 1966, is a 30.5 meter (100 feet) diameter inflatable sphere constructed of vapor deposited aluminium on a mylar substrate. The exterior surface reflects specularly with a surface reflectivity of approximately 0.89 between 6000 Å and 7000 Å [7, 8].

The Echo 2, launched in 1964, was a 41.2 meter (135 feet) diameter inflatable sphere constructed of rolled aluminum on a mylar substrate and overcoated with an amorphous phosphate substance. The exterior surface reflects essentially diffusely with a reflectance of approximately 0.59 between 6000 Å and 7000 Å during the period of interest in 1969 [7]. The Echo 2 re-entered the earth's atmosphere in 1969.

4. Theory

The technique of obtaining ozone concentrations from ground-based observations of eclipsing satellites can be described as follows. As a satellite approaches the earth's shadow, the satellite is illuminated by solar radiation that has passed through the earth's atmosphere (Fig. 1). The sunlight is attenuated due to absorption, scattering, and refraction. The attenuated solar radiation is reflected from the surface of a satellite towards the observer who is in darkness. The

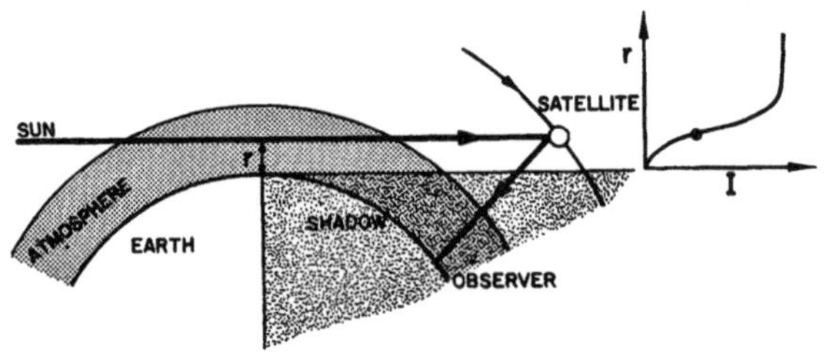

Fig. 1. Schematic diagram of sun-satellite-observer geometry and horizontal attenuation at tangential altitude r.

attenuation of monochromatic radiation from a point source reflected from a perfectly spherical satellite is given by Beer's law:

$$I(r, \lambda)/I(r_\infty, \lambda) = \exp\left[-\sum_i \sigma_i(\lambda) N_i(r)\right], \tag{1}$$

where $I(r, \lambda)$ is the measured intensity, r is the tangential altitude of the ray from the sun to the satellite, λ is the wavelength of the radiation, $I(r_\infty, \lambda)$ is the measured intensity when the solar radiation passes above the atmosphere at tangential altitude r_∞, $\sigma_i(\lambda)$ is the scattering-absorption coefficient for the ith constituent, and $N_i(r)$ is the columnar density between the sun and the satellite of the ith constituent. Eq. (1) can be applied to ground-based observations of satellites because the attenuation from a satellite to the observer is nearly constant for satellites whose apparent position does not appreciably change during the occultation process. This is the case for the experiment as the zenith angle of the PAGEOS did not change more than three degrees during the occultation process.

Two wavelength regions are chosen since there is no suitable region near the visual spectrum where the attenuation due to other species (i.e. air and aerosols) is negligible relative to that of ozone. One narrow band is centered on a Chappius region of strong ozone absorption (6022 Å) and the other on a region of lesser absorption (7029 Å). For these two wavelengths, the relative difference in the scattering coefficient for air and aerosols is much less than the relative difference in the absorption coefficient for ozone [1, 9]. This choice results in the ratio of the 6022 Å and 7029 Å measurements being more sensitive to ozone absorption than measurements in a single wavelength region.

The ratio of the intensities at these wavelengths, normalized outside the atmosphere, is given by:

$$[I(r, \lambda_6)/I(r, \lambda_7)]/[I(r_\infty, \lambda_6)/I(r_\infty, \lambda_7)] = \exp\left[-\sum_i \sigma_i N_i(r)\right], \tag{2}$$

where $\lambda_6 = 6022$ Å, $\lambda_7 = 7029$ Å, and $\sigma_i = \sigma_i(\lambda_6) - \sigma_i(\lambda_7)$. By taking the natural logarithm of each side, the horizontal columnar density of ozone is found in terms of the observed attenuation of the normalized intensity ratio and the assumed attenuation due to air molecules and aerosols.

The horizontal columnar density of a spherically symmetric atmosphere can be expressed by the Abelian integral:

$$N(r') = 2\int_{r'}^{\infty} [n(x)x/(x^2 - r'^2)^{1/2}]\, dx, \tag{3}$$

where $r' = r + R_E$ is the radius from the center of the earth to the tangency point, x is the radius from the center of the earth to the elemental length dx, and R_E is the radius of the earth at the sub-tangency point. The vertical number density $n(x)$ may be found from the inversion of Eq. (3):

$$n(x) = -\frac{1}{\pi}\int_x^{\infty}\left[\frac{dN(r')}{dr'}(r'^2 - x^2)^{-1/2}\right] dr'. \tag{4}$$

There is an additional practical consideration for the use of two-color measurements. The PAGEOS is known to exhibit fluctuations in its brightness which have been attributed to variations in the satellite's local radius of curvature [8, 10, 11]. Such brightness variations can occur over periods comparable with the duration of satellite occultation (approximately one minute for the PAGEOS), hence a significant error could be introduced if the satellite measurements are made at one wavelength. However, since the intensity at each wavelength is similarly effected, the ratio of intensities is essentially unaffected by variations of local radius of curvature of the satellite.

A point source representation of the sun is not realistic for high altitude satellites. The cone of solar radiation that reaches the satellite from the sun is spread over an appreciable altitude range above the earth at the point of tangency. Rays that come from the top of the sun undergo less attenuation than do rays from the bottom of the sun so that the resultant attenuation may be significantly different from the attenuation of rays coming from the solar center [12]. The following technique is used to find an effective altitude within the cone where the mean attenuation occurs. The sun is divided into horizontal strips. The solar flux from each strip is found by taking into account the effect of refraction [13] and solar limb darkening. A mean tangential altitude is determined for each strip, then the attenuation due to an assumed atmosphere is found through Eq. (1). The solar flux for each strip is multiplied by the attenuation and summed, to find the total flux of the sun. This total flux is divided by the unattenuated solar flux to determine the total mean attenuation. Lastly, an effective altitude is found which corresponds to this mean attenuation.

The effective tangential altitude was found to be from -10 km to $+6$ km from the tangential altitude corresponding to the center of the sun, depending on the location in the atmosphere. The corrections become smaller as the satellite approaches the earth's shadow, since the sun appears to shrink vertically due to refraction. For example, at 12 km the vertical dimension of the solar cone is reduced by a factor of 6 compared with the cone's vertical dimension at 80 km. The effective altitude is determined for each wavelength band. Since the 6022 Å band is more attenuated by ozone absorption, the effective altitude for this band is slightly lower than the altitude for the 7029 Å band. In order to determine the ratio of the intensities at a common altitude, the intensities are evaluated at slightly different moments in time. This time difference for the PAGEOS is usually less than one second, but with maximum differences of up to two to three seconds this time difference could produce an error due to variations in satellite shape. However, the errors introduced by this and other effects are estimated to be less than $\pm 10\%$.

5. Data Reduction

The satellite intensity is measured in two bands as a function of time as the satellite approaches the earth shadow. From a knowledge of the satellite orbital elements and the exact universal time of measurements, the sun-earth-satellite geometry is known and the effective tangent altitude is determined as described in § 4. Satellite intensity measurements at each wavelength are evaluated at one second intervals. These data are smoothed using a running arithmetic mean over seven data points to obtain the final data to be processed as described above.

To illustrate the technique, the intensity ratio normalized to outside the atmosphere, from the PAGEOS pass of 15 March 1972, 11 h UT, is plotted against tangential altitude in the top part of Fig. 2. The measurements are compared with the expected attenuation using the USSAS 1966 [14] air molecule and ozone standard model and an aerosol distribution [9] that has been modified to account for the general decrease in aerosol content since 1968 [15].

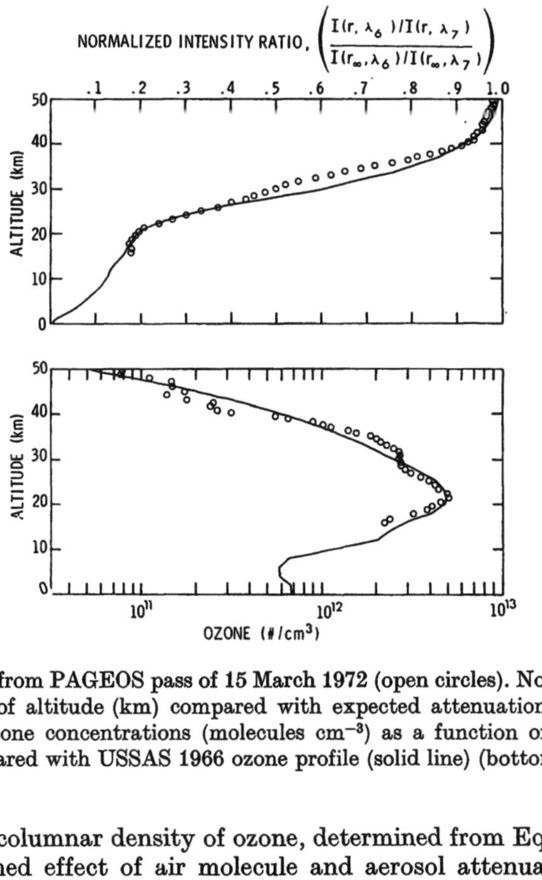

Fig. 2. Observations from PAGEOS pass of 15 March 1972 (open circles). Normalized intensity ratio as a function of altitude (km) compared with expected attenuation (solid line) (top). Corresponding ozone concentrations (molecules cm^{-3}) as a function of altitude (km) compared with USSAS 1966 ozone profile (solid line) (bottom).

The horizontal columnar density of ozone, determined from Eq. (2) and corrected for the assumed effect of air molecule and aerosol attenuation, is used in Eq. (4) to solve for the vertical number density of ozone. This correction is fairly small. For example, at 25 km approximately 92% of the attenuation is due to ozone. The corresponding ozone concentrations for the 15 March 1972, 11 h UT PAGEOS pass are plotted in the lower portion of Fig. 2. The results are compared with the USSAS 1966 ozone standard.

6. Results

A total of nine PAGEOS passes were observed from Yuma, Arizona, from 13 through 19 March 1972. The dates, universal times, and geographic locations of the tangency point are given in Table 1. The locations are grouped about 45° N latitude. Four of the passes are located over the North Atlantic Ocean and five passes are located over the continental United States. All of these measurements

Table 1

Location of Tangency Point

Satellite	Date	Time (UT)	W. Longitude (deg.)	N. Latitude (deg.)
PAGEOS	13 Mar. 1972	1138	82	32
	15 Mar. 1972	0844	39	37
	15 Mar. 1972	1144	84	38
	16 Mar. 1972	0847	40	41
	16 Mar. 1972	1147	85	41
	17 Mar. 1972	1152	86	44
	18 Mar. 1972	0854	41	47
	18 Mar. 1972	1154	87	46
	19 Mar. 1972	0857	42	48
Echo 2	11 Feb. 1969	0410	147	43

occurred at local sunrise. In addition, one Echo 2 pass, observed from Yuma, Arizona, on 11 February 1969, has a tangency point located over the Pacific Ocean at local sunset.

The vertical ozone profiles for the nine PAGEOS and one Echo 2 passes are shown in Fig. 3, with only a representative number of data points from each profile plotted. The profiles as a group are in good agreement with the ozone standard, illustrated by the solid line.

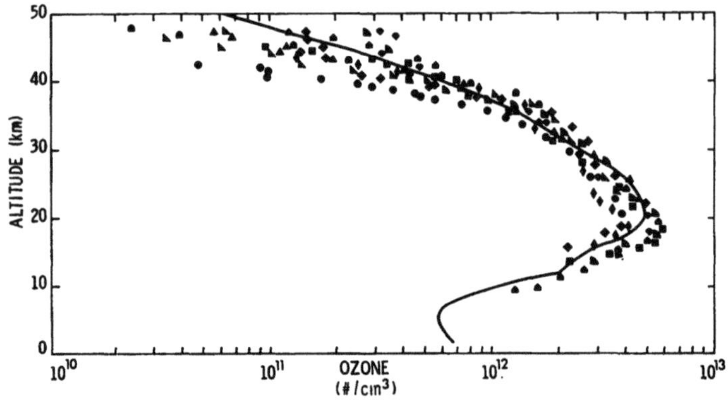

Fig. 3. Ozone concentrations (molecules cm^{-3}) as a function of altitude (km) for nine PAGEOS passes: 13 Mar. 72, 11h UT, ●; 15 Mar. 72, 08h UT, ■; 15 Mar. 72, 11h UT, ◆; 16 Mar. 72, 08h UT, ▲; 16 Mar. 72, 11h UT, ◣; 17 Mar. 72, 11h UT, ▪; 18 Mar. 72, 08h UT, ◓; 18 Mar. 72, 11 h UT, ●; 19 Mar 72, 08 h UT, ♦; and for one Echo 2 pass, 11 Feb. 69, 04 h UT, ▲. Comparison is made with the USSAS 1966 profile (solid line).

The PAGEOS measurements were averaged over 2 km intervals to produce a mean ozone profile. This profile is compared with the ozone standard in Fig. 4. The error bars show the standard deviation of the individual measurements and the dashed lines correspond to the standard deviation of the ozone standard [16].

Within the error limits of both profiles, the measured and standard profiles are in excellent agreement over the altitude range from 48 to 14 km. The measured profile shows a maximum of 5×10^{12} molecules cm^{-3} at 20 km. The measurement error bars increase above 40 km due to limited attenuation.

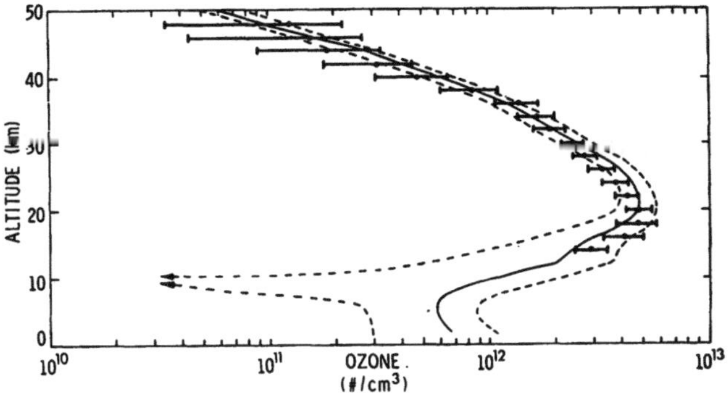

Fig. 4. Ozone concentrations (molecules cm^{-3}) as a function of altitude (km) for the mean profile (solid circles) and standard deviation (error bars) of the PAGEOS measurements of March 1972 compared with the USSAS 1966 profile (solid line) and standard deviation (dashed lines) obtained from [16].

7. Conclusions

The technique of ground-based observations of eclipsing satellites has been shown to be an inexpensive method for determining the ozone concentrations in the stratosphere. Results to date are in excellent agreement with other measurements. Using this technique, latitudinal, longitudinal, and long-term temporal variations of the vertical structure of ozone may be determined; consequently, the technique has the potential to add significantly to the state of knowledge concerning the distribution of ozone.

Acknowledgments

The authors gratefully acknowledge Mr. F. Doll and Mr. C. L. Rogers, Goodyear Aerospace Corporation, and Mr. B. Currier and Mr. F. Diehl, Wyle Corporation, for their contribution during the observation expedition in March 1972, and Mrs. L. Quinn, Langley Research Center, for her aid in data reduction. Particular thanks are due to Mr. G. M. Keating, Langley Research Center, for his comments and assistance in preparing this paper, and to the Marine Corps Air Station, Yuma, Arizona, for its support during the expedition.

References

[1] S. V. Venkateswaran, J. G. Moore and A. J. Kreuger, J. Geophys. Res. **66**, 1751 (1961).
[2] V. G. Fesenkov, Soviet Astronomy-AJ, **11**, 1 (1967).
[3] F. Link, Eclipse Phenomena in Astronomy, Springer-Verlag, New York 1969.

[4] D. GRANT, NASA Contract Rep. 66304 (1966).
[5] R. L. HOSTETLER et al., NASA Contract Rep. 791 (1967).
[6] R. H. HARDIE, Astronomical Techniques, Univ. of Chicago Press, Chicago 1962.
[7] D. C. ROMICK, R. H. EMMONS and R. J. PRESKI, NASA Contract Rep. 66899 (1969).
[8] D. S. McDOUGAL, R. B. LEE III and D. C. ROMICK. The Use of Artificial Satellites for Geodesy, Amer. Geophys. Un. Monogr. 15, Washington, D.C. (1972).
[9] L. ELTERMAN, AFCRL Rep. 68-0153 (1968).
[10] K. E. KISSELL, Space Research IX, 53 (1969).
[11] K. E. KISSELL and D. E. SMITH, Space Research XII, 1523 (1972).
[12] R. G. ROBLE and R. B. NORTON, J. Geophys. Res. **77**, 3524 (1972).
[13] P. B. HAYS and R. G. ROBLE, Planet. Space Sci. **16**, 1197 (1968).
[14] U. S. Committee on Extension to the Standard Atmosphere, United States Standard Atmosphere Supplements, 1966, GPO, Washington, D.C. 1966.
[15] J. ROSEN, private communication, Department of Physics and Astronomy, Univ. of Wyoming (1973).
[16] A. J. KREUGER and R. A. MINZNER, private communications, NASA Goddard Space Flight Center (1973).

Space Research XIV — Akademie-Verlag, Berlin 1974

A STUDY OF THE TIME LAG BETWEEN THE 27-DAY VARIATIONS OF THERMOSPHERIC DENSITY AND 10.7 cm SOLAR RADIATION

G. PAUL, H. VOLLAND and M. ROEMER

Astronomical Institutes of the University, Bonn, FRG

Drag data of ten satellites operating between 1958 and 1967 and covering a height range between 250 and 750 km have been analysed in order to determine the time lag of the quasi-periodic 27-day variation of the thermospheric density behind the corresponding 10.7 cm solar radio emission. It is shown that the average time lag at low latitudes is about 1.7 days. In addition, a significant local time dependence of the lag has been found with maximum amplitude of 2.0 days near local midnight and minimum amplitude of 1.4 days near local noon. This behaviour is interpreted as a linear modulation of two atmospheric waves with periods of 24 hours and 27 days respectively, which are generated within the thermosphere.

Satellite drag data of ten satellite orbits have been analysed in order to determine the time lag between the thermospheric density variations and the corresponding 27-day rotational effect of the 10.7 cm solar radio emission [1]. The drag data have been taken from the publications of the Smithsonian Astrophysical Observatory [2—6]. Table 1 gives the names of the satellites, their perigee heights and epochs. The density values have been transferred into exospheric temperatures by using the Jacchia 71 model [7]. The data were reduced to a geomagnetic activity of $Kp = 3$.

Table 1

Satellite Data

Symbol	Satellite	$\bar{z}_\pi$ (km)	Epoch (MJD)
1958 α	Explorer 1	350	36243.0—39600.0
1958 β1	rocket	655	36710.0—37415.0
1958 β2	Vanguard 1	658	36342.5—37580.0
1959 α1	Vanguard 2	563	36625.0—39600.0
1959 α2	rocket	563	36645.0—37387.5
1959 η	Vanguard 3	516	36836.0—37928.0
1960 ξ1	Explorer 8	426	37247.0—39600.0
1961 δ1	Explorer 9	*	37355.0—38488.4
1962 $\beta\tau$2	Injun 3	250	38013.0—39595.0
1963 ι1	Explorer 17	270	38127.0—38311.0

* Different standard heights between 300 and 760 km.

The association between the 27-day variations of these exospheric temperatures and 10.7 cm solar fluxes is investigated by applying three methods of analysis, namely regression, cross-correlation and spectral analysis. In particular, the method of linear regression has been used with data taken over a range of time intervals, termed regression intervals. The lengths of the regression intervals have been varied in order to find the optimum length of the intervals. It comes out that the time lag depends on the length of the intervals. The correlation between exospheric temperature, and hence atmospheric density, and 10.7 cm emission decreases with increasing regression interval as Fig. 1 shows. Here the average

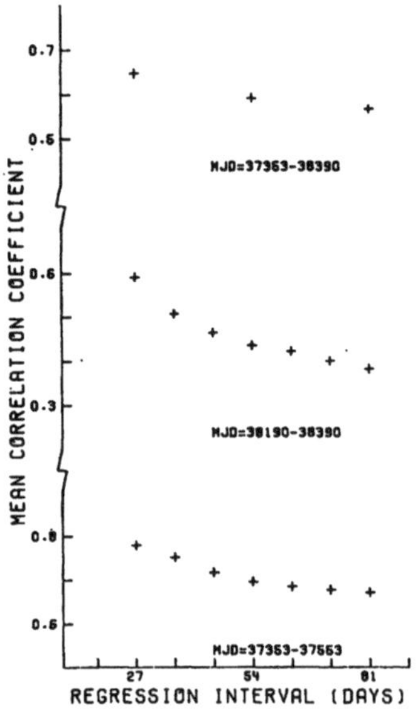

Fig. 1. Average correlation coefficient plotted versus length of the regression interval for three different time sectors of the data of satellite 1961 δ1.

correlation coefficient is plotted versus length of the regression interval for three different time sectors of the satellite 1961 δ1. From this it is apparent that the optimum interval is 27 days. Adopting this time interval we obtained an average time lag of

$$\Delta t = 1.71 \pm 0.04 \text{ days}$$

at low latitudes ($|\phi| < 40°$). A somewhat smaller value of the time lag has been obtained from the cross-correlation method. Here we found an average value of

$$\Delta t = 1.37 \pm 0.19 \text{ days.}$$

However in comparing this value with the result of the regression method one should take into account that the time lag obtained from the regression method depends on the length of the regression interval. The cross-correlation method can be considered as a regression over a large regression interval, and indeed the analysis of the data of the satellite 1961 δ1 showed that the results of both methods agree if we select an appropriate interval length for the regression method. The most important advantage of the regression method is the possibility of eliminating time sectors of the data which do not show significant 27-day variations.

The spectral analysis method gave an average value of the time lag of

$$\Delta t = 1.93 \pm 0.76$$

and a period of

$$P = 27.3 \pm 0.3 \text{ days.}$$

Here the magnitude of the rms error indicates the difficulties involved in this method. Because of the limited temporal length of the data only a small resolution of the spectral analysis is expected, and a small error in the period gives rise to a great error in the time lag. Nevertheless, we notice that the results of the two other methods are within the error bar of the result of the spectral analysis.

The regression method has been used in order to find out whether the time lag depends on local time, as has already been suggested by Roemer [8]. In Fig. 2, grouped means of the time lag taken at different local times are plotted versus

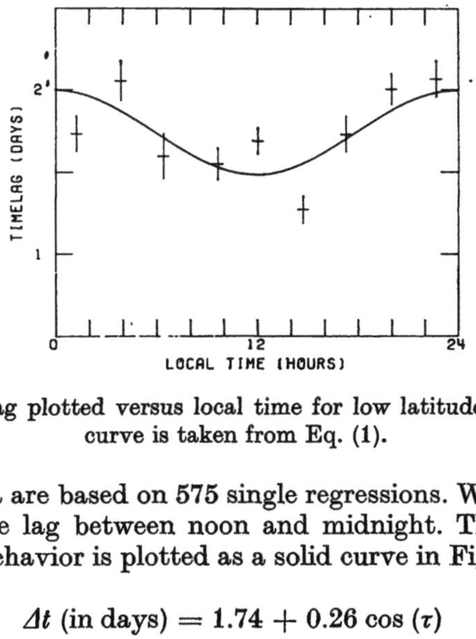

Fig. 2. Weighted time lag plotted versus local time for low latitudes ($|\phi| < 40°$). The solid curve is taken from Eq. (1).

local time. These data are based on 575 single regressions. We notice a significant difference in the time lag between noon and midnight. The best fitting curve approximating this behavior is plotted as a solid curve in Fig. 2. It is represented by the formula:

$$\Delta t \text{ (in days)} = 1.74 + 0.26 \cos(\tau) \tag{1}$$

for low latitudes ($|\phi| < 40°$); τ is the local time. This curve has been obtained from a correlation analysis with a correlation coefficient of $r = 0.74$ and a statistical significance of 97%.

A theoretical value for the average time lag of the order of two days within the height range between 300 and 400 km has already been determined by using a simple one-dimensional model [9]. In order to interpret the local time dependence of the time lag as indicated in Eq. (1) we start from the generally accepted idea that long lived activity regions on the sun with excessive EUV radiation near coronal condensations are responsible for the quasi-harmonic 27-day period. Then the total EUV heat input into the thermosphere is to a first order approximation

$$Q = (Q_0 + \Delta Q \cos \omega_0 t)\,[1 - 2P_1^1 (\cos\theta) \cos(\omega_1 t + \lambda) \ldots]. \tag{2}$$

Here Q_0 is the global averaged heat input, ΔQ is the amplitude of the excessive heat input during the 27-day period of frequency $\omega_0 = 2\pi/27$ days; $2P_1^1(\cos\theta) = 2\sin\theta$, with θ the colatitude, describes the meridional structure of the diurnal variation due to the earth's rotation with frequency $\omega_1 = 2\pi/1$ day. λ is the longitude and t the time. The density variation due to the 27-day effect is, in complex representation,

$$\Delta\varrho = \Delta Q[G_0(\omega_0)\exp(j\omega_0 t) - \sin\theta\{G_1(\omega_1')\exp\{j(\omega_1' t + \lambda)\} + G_1(\omega_1'')\exp\{j(\omega_1'' t + \lambda)\}\}\ldots] \tag{3}$$

with $G_n(\omega)$ a system transfer function and $\omega_1' = \omega_1 + \omega_0$ and $\omega_1'' = \omega_1 - \omega_0$. Because $\omega_1 \gg \omega_0$, we have $\omega_1' \sim \omega_1'' \sim \omega_1$ and $G_1(\omega_1') \sim G_1(\omega_1'') \sim G_1(\omega_1)$. Thus

$$\Delta\varrho \sim \Delta Q\,|G_0(\omega_0)|\left\{\cos\omega_0(t - t_0) - 2\sin\theta\left|\frac{G_1(\omega_1)}{G_0(\omega_0)}\right|\cos\omega_0 t\cos[\omega_1(t - t_1) + \lambda]\right\}$$
$$= A\cos\omega_0(t - \Delta t) \tag{4}$$

with

$$\tan(\omega_0\Delta t) = \frac{\tan(\omega_0 t_0)}{1 - 2\left|\dfrac{G_1(\omega_1)}{G_0(\omega_0)}\right|\dfrac{\sin\theta}{\cos(\omega_0 t_0)}\cos(\tau - \omega_1 t_1)}. \tag{5}$$

Here we used the symbols

$$G_n(\omega) = |G_n(\omega)|\exp\{-j\omega t_n(\omega)\}$$
$$\tau = \omega_1 t + \lambda \quad \text{local time (in hours)} \tag{6}$$

and changed to the real representation of $\Delta\varrho$ in (4) and (5).

It has been shown [10] that at thermospheric heights above 300 km the transfer function G_n can be approximated by

$$G_n(\omega) = \frac{1}{1/R_i + n(n+1)/R^* + j\omega C} \quad \text{for } \omega \lesssim \omega_1,\ n \leqq 2 \tag{7}$$

where R_i and R^* are measures of the wave dissipation due to vertical heat conduction and horizontal ion drag, respectively, and C represents the adiabiatic increase of the internal energy of the neutral gas. From the result of Eq. (1) we take the value of $t_0 = 1.74$ days. Moreover, the time delay between the maximum of the diurnal variation of the solar heat input and the corresponding maximum of the diurnal component of the density is $t_1 = 3$ hours.

Because of the relations (see Eq. (7))

$$\tan(\omega_0 t_0) = \omega_0 C R_i \qquad \tan(\omega_1 t_1) = \frac{\omega_1 C}{1/R_i + 2/R^*} \tag{8}$$

we find from (7)

$$|G_1(\omega_1)/G_0(\omega_0)| = \frac{\omega_0 \sin \omega_1 t_1}{\omega_1 \sin \omega_0 t_0} = 0.0665,$$

and (5) becomes at the equator ($\theta = 90°$)

$$\Delta t = \frac{1}{\omega_0} \arctan \frac{0.428}{1 - 0.145 \cos (\tau - 3^h)} \sim 1.74 + 0.23 \cos (\tau - 3^h) \quad (9)$$

(Δt in days).

Δt of Eq. (9) has maximum (minimum) values of $\Delta t = 2.00$ (1.54) days, which agrees quite well with the empirical formula Eq. (2) apart from the 3-hour difference in the local time behaviour. At 40° latitude the maximum (minimum) amplitudes reduce to 1.93 (1.58) days, which is still in the range of the accuracy of the data.

We conclude that the diurnal variation of the time delay of thermospheric density behind solar 10.7 cm flux for the 27-day solar rotational effect with a minimum near noon is consistent with our model of a linear superposition of atmospheric waves with different periods (24 hours and 27 days), different response times and amplitudes.

References

[1] G. Paul, Bundesmin. f. Forschung u. Techn., FBW 73-10 (1973).
[2] L. G. Jacchia and J. Slowey, Smithson. Astrophys. Obs. Spec. Rep. No. 100 (1962).
[3] L. G. Jacchia and J. Slowey, Smithson. Astrophys. Obs. Spec. Rep. No. 171 (1965).
[4] L. G. Jacchia and J. Slowey, Smithson. Astrophys. Obs. Spec. Rep. No. 326 (1970).
[5] M. Roemer, Smithson. Astrophys. Obs. Spec. Rep. No. 199 (1966).
[6] J. Slowey, Smithson. Astrophys. Obs. Spec. Rep. No. 157 (1964).
[7] L. G. Jacchia, Smithson. Astrophys. Obs. Spec. Rep. No. 332 (1971).
[8] M. Roemer, Space Research VII, 1091 (1967).
[9] H. Volland, Planet. Space Sci. **17**, 1709 (1969).
[10] H. Volland and H. G. Mayr, J. Geophys. Res. **76**, 3764 (1971).

Space Research XIV — Akademie-Verlag, Berlin 1974

DYNAMIC PROCESSES AS DERIVED FROM THE MEAN CIRCULATION IN THE UPPER MESOSPHERE AND LOWER THERMOSPHERE

A. Ebel

Institute of Geophysics and Meteorology, University of Cologne, Cologne, FRG

Seasonal and latitudinal variations of the mean vertical motion and of momentum and heat sources and sinks, as derived from models of the mean atmospheric structure in the region from 70 to 100 km, are discussed. The results suggest a close relationship between the mean meridional circulation and the seasonal and latitudinal changes of various phenomena near the mesopause, one of which is ionospheric absorption. Near the equator and beyond 30° latitude the absorption is enhanced when either downward motions or heat sinks are present.

1. Introduction

The average structure of the atmosphere as described by empirical models of wind, temperature, pressure and composition can be interpreted as a response to dynamic processes which are maintained by external (e.g. solar radiation) and internal (e.g. dissipation) sources and sinks of heat and momentum. In the region from about 70 to 100 km altitude, measurements obtained by rocket and ground-based methods lead to the conclusion that very effective internal sources (and sinks) must exist which vary with height, latitude and season. Since the number of measurements has substantially increased in recent years and improved atmospheric models are now available a numerical investigation of the sources is carried out applying monthly mean meridional cross sections of horizontal wind, temperature and pressure published by Groves [1, 2]. It is assumed that the cross sections represent time and zonal averages. Then the steady-state equations of the mean meridional circulation as given by Gilman [3] can be used to evaluate $\omega = dp/dt$ (p, pressure; t, time) or the velocity component normal to the isobaric surfaces, the heat sources (or sinks) and the zonal momentum sources (or sinks). These sources are said to drive the mean meridional wind. Thus the assumption that the models really represent the mean N-S wind is the most serious one of this analysis, and it should be mentioned that it has still to be proven—empirically by a better longitudinal resolution of observation sites, theoretically by investigations similar to this study. The results shown in § 2 are felt to support the assumption mentioned above, because they roughly correspond to other upper atmospheric phenomena depending on the mean circulation (§ 3). It is evident from the theoretical concept that the sources represent combined effects of different physical processes (heat: radiation, conduction, chemical reactions, transport by eddies and waves; zonal momentum: mainly transport by eddies and waves).

2. Results

For convenience pressure instead of height has been used as the vertical coordinate in the calculations; therefore, the results shown refer to isobaric levels. The original data permit the construction of numerous representations showing the spatial and seasonal behaviour of the momentum and heat sources and the vertical velocity component. The necessary selection is carried out with the aim of trying an application of the results to the winter anomaly of ionospheric absorption.

The mean vertical motion or ω is obtained by integrating the continuity equation. The error resulting from the assumptions on the magnitude of ω at the upper boundary of the atmospheric layer depends on height and latitude. In Figs. 1a, b

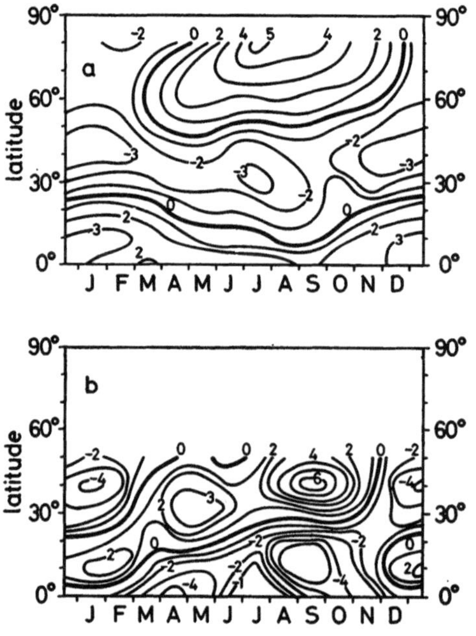

Fig. 1. Seasonal and latitudinal variation of the average velocity component (cm s^{-1}) perpendicular to the isobaric surfaces: a, $p = 2.8$ N m^{-2}, $h \sim 74$ km; b, $p = 0.21$ N m^{-2}, $h \sim 89$ km.

it decreases from about 30% at 70° and 50° latitude respectively, to considerably lower values at the equator. The difference in the latitudinal extent of the figures comes from the wind model [1] which enables computations between 0° and 40° latitude at the upper levels and, applying extrapolation methods, between 0° and 80° latitude at lower heights. The seasonal variation of the vertical wind at the level $p = 2.8$ N m^{-2} ($\sim$74 km) and $p = 0.21$ N m^{-2} ($\sim$89 km) differs markedly. In the lower layer major temporal changes only occur beyond 50° latitude with a downward motion during the winter and an upward motion during the summer. The situation is considerably altered at the upper level where the direction of the vertical component reverses between 20—30° latitude near the solstices, having an upward motion near the equator during the winter and downward motion during the summer. It is evident from Fig. 1b that strong annual and semiannual variations exist with phase changes of three months for the latter period between latitudes 0° and 40°. As a consequence, the vertical motion near 40° goes in the opposite direction during the summer and downward during the winter at both levels. The heat sources behave quite similarly to the vertical motion and their contours are nearly parallel to the contours of the vertical velocity. Sources and sinks correspond to upward and downward motion respectively. This means that cooling or warming by vertical motion must be compensated by other processes, for instance radiation, heat conduction, and transport by fluctuation effects. Horizontal transport by the mean meridional circulation is less effective.

In contrast to the heat sources, the momentum sources derived from the zonal momentum equation are mainly dependent on the mean horizontal motion. A meridional cross section for 1 January/1 July is shown in Fig. 2. The seasonal

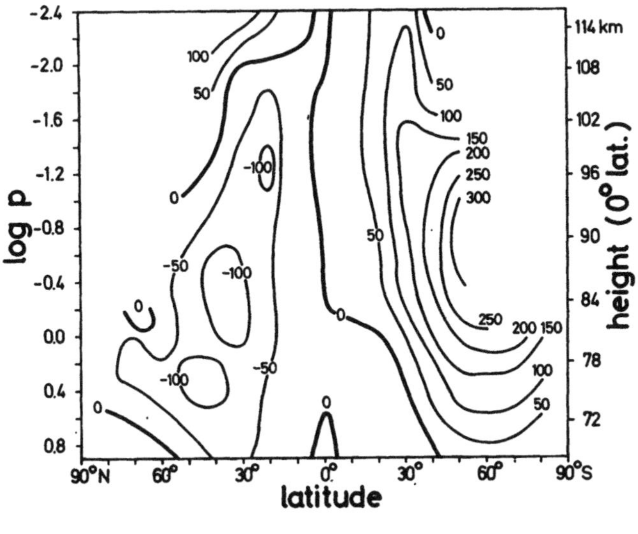

Fig. 2. Zonal momentum sources (positive) and sinks (negative) in units of 10^{-5} m s^{-2}. 1 January (N lat.) and 1 July (S lat.). p in N m^{-2}.

changes of different levels are well correlated at all latitudes except near the equator where the values are less significant. Between latitudes 40° and 50° maxima are obtained during the summer and minima during the winter. The reversal from negative to positive values and vice versa takes place during February and October/November. The zonal momentum source is mainly built up by fluctuation of the W-E wind and the vertical velocity component around their zonal and time averages [3]. Thus it can be regarded as a fluctuation parameter containing information on transport effects caused by different physical processes such as tides, gravity waves and small- and large-scale eddies.

3. Concluding Remarks

When the temporal and spatial variations of ω and the momentum and heat sources are compared with other phenomena near the mesopause indicating circulation effects, qualitative correspondence can easily be found in many cases. Thus, the period of noctilucent clouds roughly corresponds to the period of upward motion at 50° latitude in the lower thermosphere (Fig. 1b). On the other hand, the winter anomaly of ionospheric absorption is observed during the period of downward motion at latitudes greater than about 30°. It now seems generally accepted that ionospheric absorption strongly depends on atmospheric transport in the lower thermosphere [4], and influences even of stratospheric layers have been suggested [5].

The analysis of the mean atmospheric structure between heights of 70 and 100 km enables a somewhat more detailed investigation of the relationship between mean circulation and ionospheric absorption as derived by George [6]. As a result, the winter anomaly takes place when strong sinks of momentum and heat prevail in the lower thermosphere. During the summer, strong sources are found. If momentum and heat sinks are regarded as a result of processes of atmospheric stabilization one can say that the winter anomaly is accompanied by stabilizing processes above the mesopause. Considering that momentum production is very small near the equator, this statement also holds for the equatorial zone where maximum ionospheric absorption and cooling rates are found during the equinoctial months. A simple way to show the different behaviour of the layers above and below the mesopause with respect to absorption (given by the total A-figure in [6]) is to compare it with the seasonal and latitudinal variation of the heat sources (or ω) applying statistical methods. A sufficient correlation is found for the upper layer only. There exist latitudinal differences which are demonstrated in Fig. 3 by means of relative linear regression coefficients a_1/a_0 for the total A-figure and the heat sources H, related by the equation $A = a_0 + a_1 H$. It follows that the total ionospheric absorption is generally increased with increasing cooling rates or increasing downward motion. This effect is as important near

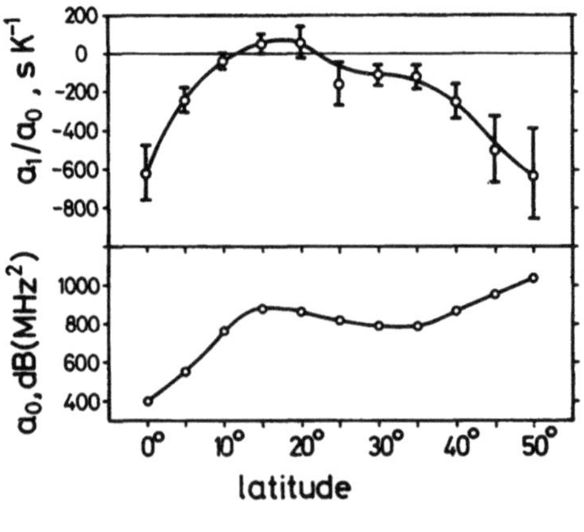

Fig. 3. Linear regression for total absorption $A = a_0 + a_1 H$ (H, heat source in K s^{-1} at the level $p = 0.21$ N m^{-2}).

the equator (equinoctial maximum of total absorption) as at 50° latitude (winter maximum). In the latitude range from 10° to 30° geomagnetic effects are probably more important than circulation effects and thus lead to small regression coefficients. It is hoped that a more comprehensive analysis of ionospheric absorption with respect to circulation phenomena will improve the understanding of lower thermosphere dynamics.

Acknowledgment

Parts of this work have been supported by the Deutsche Forschungsgemeinschaft under grant Eb 56/1.

References

[1] G. V. Groves, J. Brit. Interplan. Soc. **22**, 285 (1969).
[2] G. V. Groves, AFCRL-71-0410, Environmental Research Papers No. 368 (1971).
[3] P. A. Gilman, Tellus **16**, 160 (1964).
[4] C. F. Sechrist, Jr, in: Thermospheric Circulation, MIT Press, Cambridge, Mass., and London 1972 (p. 261).
[5] K. Labitzke and H. Schwentek, Z. Geophys. **34**, 555 (1968).
[6] P. L. George, J. Atmos. Terr. Phys. **33**, 1893 (1971).

Space Research XIV — Akademie-Verlag, Berlin 1974

PHOTOMETRY OF ARTIFICIAL EARTH SATELLITES

V. N. Lebedinets, A. A. Gen, L. A. Kapkov, Yu. G. Kaufman, V. P. Prorok

Hydrometeorological Service, Moscow, USSR

The photoelectric apparatus for four-colour photometry of artificial earth satellites, made at the Institute of Experimental Meteorology (Obninsk), is discussed. With the apparatus it is possible to observe satellites brighter than $+5^m$. A four-axis mount is used for the tracking. Experimental results are recorded on paper tape in a form suitable for direct input to a computer. The upper atmospheric parameters which can be determined with the help pf photometric observation of artificial earth satellites are discussed. The method of mathematical simulation of the experimental data is suggested. The spectral analysis of the satellite Explorer-19 light curve is performed and the spectral density of this curve is determined.

1. Introduction

On the basis of multicolour photometry of artificial earth satellites it is possible to solve some scientific problems [1—6], for example, to investigate variations of atmospheric transparency at 15—60 km altitude in the near-ultraviolet, visible and near-infrared spectral regions, to measure the ozone concentration at 20 to 50 km, to investigate variations of the upper atmospheric density, to investigate absolute and relative variations of satellite reflectivity in different spectral regions, and to identify satellites.

At present some photoelectric photometers for such observations are known [6—10]. Some defects of these instruments include: a) manual tracking [7—9]; b) large field of view [7, 9]; c) successive measurements in different spectral regions and therefore an unjustifiably small time constant [6, 9, 10]; d) use of dc amplifiers, requiring control over changes of the photomultiplier dark current; e) chart recording, requiring continuous manual data processing. The last two defects are common to all the photometers described.

2. Instrumentation

The electrophotometer FEF-1, making it possible to observe satellites brighter than $+5^m$, was developed and made at the Institute of Experimental Meteorology. The measurements are carried out simultaneously in four parts of the spectral region 390—830 nm at a rate of 20 measurements per second. The photometer is mounted on a photographic camera AFU-75, provided with a semi-automatic tracking system. A photometer consists of three blocks (Fig. 1).

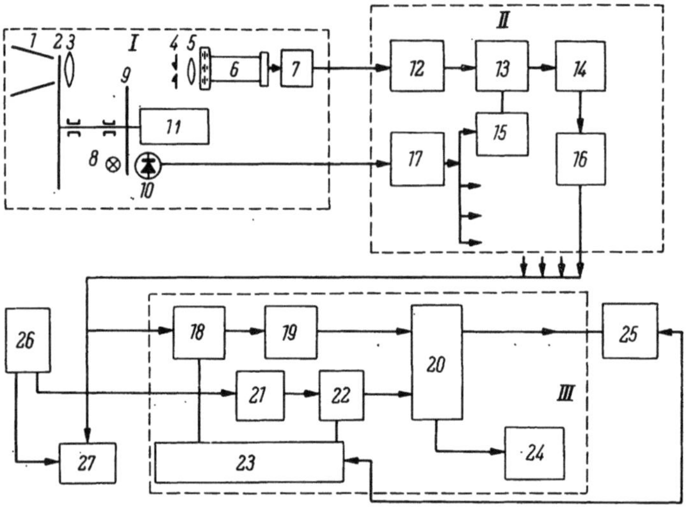

Fig. 1. Instrumentation system block diagram.

I. Optical-mechanical block: 1, blind; 2, main modulator; 3, objective, 56 mm diameter; 4, beam-limiting aperture; 5, Fabry lens; 6, photomultiplier with a filter; 7, preamplifier; 8, illuminator; 9, auxiliary modulator; 10, photodiode; 11, motor.
II. Amplifier block: 12, narrow-band amplifier; 13, phase detector; 14, integrator; 15, phase converter; 16, direct-current amplifier; 17, reference signal.
III. Recording block: 18, commutator; 19, digital voltmeter; 20, power amplifier; 21, control time relay; 22, time counter; 23, command block; 24, computer word formation block; 25, perforator; 26, quartz clock; 27, five-channel pen recorder.

2.1. Optical-Mechanical Block (I)

The optical-mechanical block I (Fig. 1) consists of four identical optical channels assembled in one case which is fixed to the tracking part of the camera. The light beam is modulated by a rotating segmented disk. Photomultiplier signals are amplified by preamplifiers and transmitted to the main amplifier block by a cable, which is also used for telephone communications between the operators.

2.2. Amplifier Block (II)

The amplifier block II has four identical channels, each consisting of a narrow-band amplifier, a phase-locked receiver, an integrator, an autoranging system and a dc amplifier. The reference signal is formed by an individual channel.

2.3. Recording Block (III)

For operative visual control the signals and time marks from the quartz clock are registered by a five-channel recorder H-327-5. At the same time for data processing and the analysis of data by the computer, the signal is translated into a digital code using the digital voltmeter VK-7-10A and is punched on to paper using a tape punch PL-80 or PL-150. The time counter has a control time relay, which makes it possible to keep it set to the quartz clock with an accuracy to 10^{-2} sec.

3. Mathematical Processing of the Photometric Curve

The observations made as a satellite enters eclipse are analysed mathematically. It is shown [11], that if the stationary random process $\xi(t)$ is given in the form of a time-series at discrete intervals, then the forecast $\hat{\xi}(t + n\Delta t)$ is expressed by

$$\hat{\xi}(t + n\Delta t) = \sum_{k=0} C_k \xi(t - k\Delta t) \tag{1}$$

The C_k coefficients of this series can be calculated if the spectral density $G(\omega)$ of the process $\xi(t)$ is known.

Analysing the photometric curve of the eclipse of a satellite, it is necessary to forecast this curve for a period equal to the eclipse duration, e.g. for 10—20 seconds. For a reliable forecast for such a period, at $\Delta t \sim 0.3$ sec the observation period should not be less than 5 min.

In the case of significant trend of the photometric curve the application of Eq. (1) requires the preliminary data to be filtered and the low-frequency trend removed [12]. If there is some information on the satellite shape and form, on its orbit and on atmospheric transparency, then, instead of formal methods of trend removal, the photometric curve can be used taking into account all additional data.

The satellite light curve is expressed by

$$x(t) = AE_0 P^{m(t)} r^{-2}(t) \Phi_F(S, \Psi, \alpha_1, \alpha_2, \alpha_3) \tag{2}$$

where A is the proportionality coefficient, depending on the satellite size, E_0 is the illumination in the absence of the atmosphere, P is the atmospheric transparency coefficient, m is the air mass, corresponding to the zenith distance z of the satellite (at $z \leqq 80°$ $m \approx \sec z$), r is the distance from an observer to the satellite, Φ_F is the phase function of a body with form F, S is the reflection coefficient of the satellite surface, Ψ is the phase angle and α_i are the angles defining the satellite orientation with respect to an observer.

Let us suppose that the satellite is a deformed sphere and $\dot{\alpha}_1 \gg \dot{\alpha}_2$ and $\dot{\alpha}_3$. Then during one satellite pass it is possible as a first approximation to take into account the rotation of the satellite about one axis only. Finally, taking $\dot{\alpha}_1 = \text{const.} = \omega_0$, we write (2) in the form of

$$x(t) = AE_0 P^{m(t)} r^{-2}(t) \Phi_0(S, \Psi) f(\alpha_2, \alpha_3) \sum_{k=-N}^{N} a_k \exp(ik\omega_0 t) \tag{3}$$

where $\Phi_0(S, \Psi)$ is the phase function of a sphere [9], and $f(\alpha_2, \alpha_3)$ is a function describing the orientation of the satellite spin axis. Eq. (3) may be written as

$$x(t) = x_1(t) x_2(t) x_3(t) \quad \text{where} \quad x_1 = AE_0 P^m r^{-2} \Phi_0;$$

$$x_2 = f(\alpha_2, \alpha_3); \quad x_3 = \sum_{k=N}^{N} a_k \exp(ik\omega_0 t)$$

On the basis of such approximation we shall deal with the influence of various factors on the satellite light curve spectrum. To estimate the Fourier transformation $X(\omega)$ [13] of the photometric curve $x(t)$ we shall use Duhamel's theorem. The convolution of X_2 with X_3 represents the superposition of the X_2 and X_3 broadened spectral lines. Neglecting the second-order terms, we have

$$X(\omega) = 2^{-1} \int_{-\infty}^{\infty} X_1(\omega - v) \sum_{k=0}^{N} (a_k + a_{-k}) X_2(k\omega_0 - v) \, dv. \tag{4}$$

Using (4) we can evaluate the contribution made by the two factors $x_1(t)$ and $x_2(t)$ to the resultant spectrum $X(\omega)$.

We have calculated the spectral densities of the Explorer 19 light curve (Fig. 2) recorded as discussed in § 2, for which the assumptions used in (3) are fulfilled

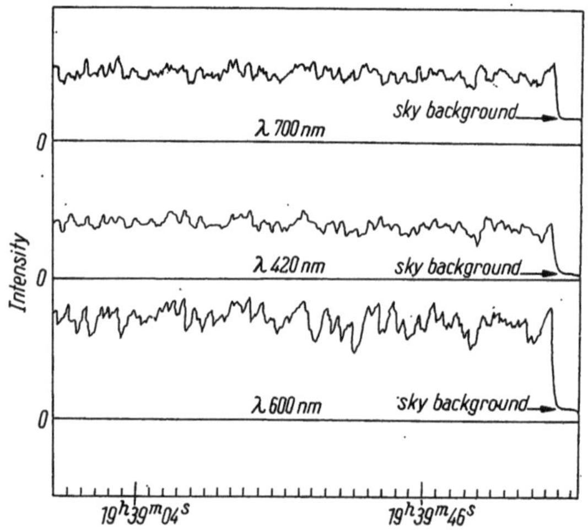

Fig. 2. Part of the light curve of Explorer 19 (6 Oct. 1972).

satisfactorily. Fig. 3(a) shows the spectral density $G(\omega)$ of the fluctuations of the Explorer 19 light curve, calculated with no allowance for deformations caused by the factor $x_1(t)$. It is seen that there are some well-defined peaks, which can be interpreted as the frequency f_0 (0.1 Hz) and its harmonics. The low-frequency part (<0.05 Hz) of the spectrum, however, has much greater intensity. In Fig. 3(a), it is difficult to know how to interpret the peaks at frequencies of 0.02 and 0.04 Hz. Consideration of the factor $x_1 = AE_0P^{m_\gamma-2}\Phi_0$ and the removal of its effect from the spectrum makes the spectrum interpretation much easier (Fig. 3(b)). Peaks at frequencies 0.1, 0.2 and 0.3 become narrower and higher, and peaks at frequencies 0.02 and 0.04 Hz disappear. The low-frequency part (<0.04 Hz) of the spectrum still present can be explained by the fact that, for Explorer 19, approximation (3) is a rather poor one; precession and other effects cannot be ignored.

Spectral characteristics of the satellite light curve can also be used for some other problems, for example, when determining the satellite rotation period, when studying variations of the rotation period under the action of atmospheric and geomagnetic activity variations. It is seen from (4) that when studying the satellite rotation frequency variations it is necessary to take into account the distortions caused by the factors x_1 and x_2. The influence of $x_1 = AE_0P^{m}r^{-2}\Phi_0$ can be

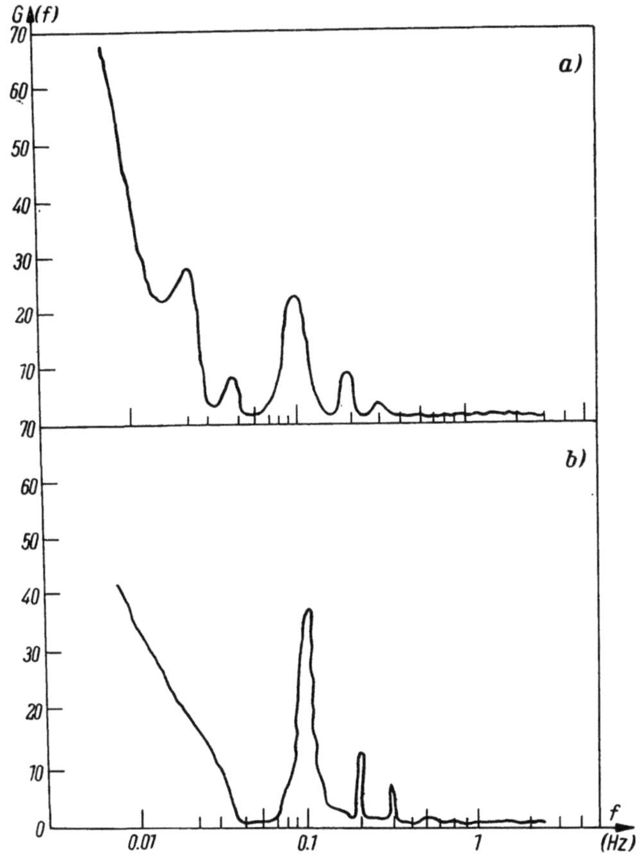

Fig. 3. Spectrum of Explorer 19 photometric curve (6 Oct. 1972). (a) without taking into account the low-frequency deformations, (b) taking into account the low-frequency deformations.

removed since this function can be calculated if the orbital elements of the satellite, atmospheric transparency and reflection coefficients of the satellite S are known. To calculate $x_2(t)$ additional information on satellite orientation is needed.

4. Summary

The apparatus and the method of mathematical processing of the data described allow one to carry out the spectral analysis of the satellite photometric curves, to determine the characteristic rotation periods of satellites and to calculate the attenuation of light reflected by the satellite as it enters into the earth's shadow.

References

[1] F. LINK, Space Research XI, 70 (1961).
[2] S. V. VENKATESWARAN, J. G. MOORE and A. J. KRUEGER, J. Geophys. Res. **66**, 1751 (1961).
[3] V. G. FESENKOV, Observation of Artificial Satellites **5**, 97 (1967).
[4] V. M. GRIGOREVSKY, Bull. st. optc. obs. AES, **54**, 9 (1969).
[5] D. E. SMITH et al., COSPAR Inform. Bull. No. 56, 70 (1970).
[6] R. H. EMMONS et al., Astron. J. **72**, 939 (1967).
[7] ST. AMAND, Annls Géophys. **11**, 435 (1955).
[8] J. G. MOORE and S. D. ELLIOTT, Appl. Optics **2**, 625 (1963).
[9] F. LINK and J. ZACHAROV, Bull. Astron. Inst. Czech. **17**, 151 (1966).
[10] P. N. BOIKO and V. S. MATYAGIN, Bull. st. opt. obs. AES, **54**, 19 (1969).
[11] I. I. GICHMAN and A. V. SKOROCHOD, Introduction to the Theory of Random Processes, Nauka, Moscow 1965.
[12] J. S. BENDAT and A. G. PIERSOL, Measurement and Analysis of Random Data, J. Wiley and Sons, Inc., New York 1967.
[13] B. R. LEVIN, Theoretical Foundations of Statistical Radio-Engineering, Soviet Radio, Moscow 1969.

The Ionosphere

Space Research XIV — Akademie-Verlag, Berlin 1974

RECENT RESULTS FROM SATELLITE BEACON MEASUREMENTS

A. V. DA ROSA

Radioscience Laboratory, Stanford University, Stanford, Calif., USA

A review is made of the techniques for measuring ionospheric electron content, the most important parameter in the study of transionospheric propagation. Data collected since 1964 have yielded a synoptic description of the behavior of the electron content in mid-latitudes. Empirical relationships between the level of solar activity and the electron content have been developed permitting the prognostication of the electron content values. Construction of such prognostication schemes has been stimulated by current efforts to create accurate satellite-borne navigation systems. Marked discrepancies between prognostication and observation which occur during ionospheric storms are being studied to identify their causes. Electron content bite-outs during solar eclipses fall off with distance from totality more rapidly than simple theory predicts suggesting the action of eclipse induced neutral winds. Gravity waves propagating in the thermosphere leave a signature in the electron content records. Studies of such records have allowed the identification of the position of the gravity wave source, and its radiation pattern. A one-to-one relationship between these waves and polar substorms has been revealed. Electron content measurements have been used to monitor the protonosphere with good time resolution. Protonospheric storms have been observed with this technique. Slab thickness data obtained from content measurements have been used to determine the neutral air temperature in the thermosphere.

The earliest uses of radio beacons aboard satellites date back to the Sputnik era. Thus, a classical paper on the determination of the ionospheric electron content was written by Garriott [1] based on data collected from Sputnik 3 in 1958/59. The first beacons designed for ionospheric propagation studies were placed on low orbiting satellites and yielded a nearly instantaneous picture of the electron content along a given track. Typical satellites of this type were BE-B and BE-C, the former in a high inclination orbit essentially scanning the ionosphere along a N—S track and the latter in a low inclination orbit scanning in an E—W direction. Scientists from all over the world made good use of the availability of these beacons and considerable amounts of data were collected. Although the interest in low orbiting beacons has waned, it has not disappeared. For example, Spain is about to launch its first satellite, the INTASAT, which has several of the features of BE-B; in addition, DNA 002 to be placed in a circular orbit at 440 km altitude with an inclination of 98.5° in 1973 will contain a beacon to be used mainly for scintillation studies. Beacons placed on geostationary satellites have the distinct advantage of permitting the continuous surveillance of a fixed volume of the plasmasphere and, for this reason, are much more popular at present. This paper is confined to results obtained from such beacons.

Use of geostationary satellites for propagation studies began in 1964 when the signals from the VHF telemetry transmitter aboard Syncom 3 became available. This satellite was parked over the international date line so that only stations in the Pacific area could collect data, a fact that put the University of Hawaii and Stanford University in a privileged position. These organizations have observed geostationary satellites in an almost uninterrupted manner since 1964 up to the present. Additional impetus to such work was provided by the launching of ATS 1, in 1966. This satellite was also parked over the international date line and substituted Syncom 3 which had drifted out of range of the West Coast of the US. Again, no special equipment was used aboard: the VHF telemetry transmitter served as a beacon. The same occurred with the next two satellites of the ATS series: ATS 3 and ATS 5. These, however, incorporated a simple modification that caused a small amount of the third harmonic of the VHF signal to be radiated, thus effectively acting as a dual frequency beacon useful in protonospheric observations as described later. The first (and possibly, last) geostationary satellite to be equipped with a specially designed beacon, is ATS-F to be launched in mid-1974. The beacon will emit signals at 40, 140, and 360 MHz, modulated by a 1 MHz tone. The 40 and 360 MHz carriers will also be modulated by 100 kHz. Such a combination of frequencies (all coherent) will allow a wide range of propagation experiments.

Signals from both low orbit and geostationary satellite beacons are used mainly in one of two different propagation experiments: scintillation studies and measurement of electron content. One simple but significant result of scintillation studies is the discovery that, in certain regions of the world, gigahertz signals transmitted through the ionosphere may suffer strong amplitude and phase disturbances. This is particularly true in the equatorial and the auroral regions and may have a profound influence on the cost of satellite-to-ground communication systems: enough additional effective radiated power must be provided to avoid signal loss during deep fading periods. Such behavior came as somewhat of a surprise to engineers and scientists comfortably convinced that the scintillation index was proportional to $f^{-\eta}$, where η was supposed to be about 2. It turned out that the exponent is itself a function of the depth of scintillation and may become a small number or even a negative one.

Electron content measurements are based either on the birefringence of the ionosphere or on its dispersivity. The latter can manifest itself by the difference in propagation times of signals of different frequencies or by the fact that the refraction of such signals is not the same. Thus, by observing the different angles of arrival of signals of different frequencies coming from a satellite it is possible to determine the electron content. This technique, at present, does not present advantages capable of balancing the cost of the necessary radio-interferometers. When two signals of different frequencies are transmitted from a satellite through the ionosphere, they experience different retardations which can be measured by comparing the phases received on the ground. With the usual combination of frequencies (say 140 and 360 MHz, as on ATS-F) absolute values of electron content cannot be directly measured because the difference in retardation is such that the observed phase differences can be several hundred times 360°, and the resulting ambiguity cannot be simply resolved. On the other hand, this technique is extremely sensitive to changes in electron content. For instance, the 137/412 MHz pair on ATS 3/5 allows the resolution of better than 10^{14} el m^{-2}. Since on the middle of an equinoctial day, during sunspot maximum, the slant electron content can be some 10^{18} el m^{-2}, a resolution of better than 1:10^4 can be obtained. The question of determining the absolute value of electron content from this so-called phase-path difference technique was resolved a long time ago for the case of low orbiting satellites [2] but only recently for the case of geostationary ones [3]. The method employed with geostationary satellites is a hybrid one and uses information from Faraday rotation measurements. An uncertainty of less than 10^{16} el m^{-2} can be achieved.

The equipment used in the phase-path difference method consists of phase-locked receivers and is, therefore, expensive and not very appropriate to unattended operation. These disadvantages, together with the difficulties in the determination of the absolute value of electron content, led workers in this field to the idea of the "group delay experiment". In this case, the phases of the modulation envelope of two carriers at different frequencies are compared. In the case of ATS-F, for instance, the phases of the 1 MHz envelopes of either the 40 or the 140 MHz carrier can be compared with that of the 1 MHz modulation envelope on the 360 MHz carrier. Since the wavelengths involved are much larger than those used in the phase-path difference method, the ambiguity now is also much larger and can be resolved easily. Table 1 shows the ambiguities and uncertainties of the group delay experiment on ATS-F. Uncertainties are based on a conservative estimate of $\pm 10°$ uncertainty in the phase angle measurement. One basic assumption in this experiment is that the relative phase of the modulation envelopes of the signals as they leave the satellite is known.

Table 1

ATS F Group Delay Experiment

f_m (MHz) \ f (MHz)	40	140		
0.1	12	170	← I_A	m$^{-2} \times 10^{-16}$
	± 0.3	± 5	← I_U	m$^{-2} \times 10^{-16}$
1.0	1.2	17	← I_A	m$^{-2} \times 10^{-16}$
	± 0.03	± 0.5	← I_U	m$^{-2} \times 10^{-16}$

Ambiguities (I_A) and uncertainties (I_U) in the measurement of electron content using either 40 or 140 MHz signals in conjunction with the 360 MHz reference signal. Two modulation frequencies are considered: $f_m = 0.1$ MHz and 1 MHz.

By far the most popular method of determining the ionospheric electron content is that based on the measurement of the Faraday rotation of a linearly polarized signal, in general, at VHF. The great advantage of this method is the simplicity of the equipment both on the ground and on the satellite. The latter may simply be an existing VHF telemetry transmitter, one or two watts of effective radiated power being sufficient. On the ground, the most commonly used equipment is one employing a mechanically rotating Yagi antenna, developed by Professor John Titheridge of New Zealand. Recently more sophisticated VHF polarimeters, using fixed antennas, have appeared on the market. The main drawbacks of the Faraday rotation method are the substantial uncertainties associated with the conversion of Faraday rotation angle into electron content values and the fact that the method becomes progressively less sensitive as the observation site moves closer to the geomagnetic equator.

Adequate electron content information has been accumulated over the last few years by stations in or near New Zealand and by those in the US (including Hawaii and Puerto Rico). Other areas of the world are not as well covered. Figs. 1, 2 and 3 show typical results. The collected data have been put to good use in both science and engineering.

The increasing use of transionospheric transmissions between earth and artificial satellites or other astronomical bodies is generating a growing demand for better knowledge and more accurate predictions of the behavior of the electron content of the plasmasphere. With the proliferation of geostationary satellites, communications engineers require knowledge of the maximum bandwidth that can be transmitted through the ionosphere with a tolerable phase distortion. The dispersion due to the medium may cause the bandwidth to be limited by the electron content; the maximum bandwidth is then a function of the time of day, the season of year, and the phase of the solar cycle. A related problem, also dependent on the value of the electron content, is the distortion suffered by short pulses transmitted through the ionosphere.

Efficient launch operations of some satellites as well as accurate position keeping of satellites already in orbit require the determination of the spin-axis orientation with a precision of about 0.1°. This can be done by measuring the polarization angle of a wave radiated from a satellite antenna mounted along the spin-axis. To obtain the desired accuracy, it is imperative that the Faraday rotation in the plasmasphere be taken into account. The transmission of time signals via satellites offers definite advantages over the use of HF but again, for the highest accuracy, transmission delays due to the electron content must be taken into account. The capacity of communications channels between a satellite repeater and ground can be doubled by using polarization separation of two transmissions employing the same frequency. To accomplish this, orthogonal, linearly polarized antennas are used, and it is necessary to compensate for the Faraday rotation in the ionosphere when orienting the ground antennas. Predictions of the expected Faraday rotation are very useful in designing the system.

Perhaps the most acute need for good ionospheric data is related to the development of precision navigation systems. The one system that seems to be most advanced in its development is the DNSS (Defense Navigation Satellite System) of the Department of Defense of the US. This system employs a constellation of four geo-synchronous satellites, one of which is geostationary while the other three are placed in such an orbit that their ground track is a circle of some 30 geographical degrees in diameter centered on the subsatellite point of the geostationary spacecraft. The three coordinates of the user (latitude, longitude and altitude) are determined by ranging measurements to the satellites. Four, instead of three, satellite-to-user distances have to be measured because additional information is needed for the synchronization of the user's clock. Thus, with four measurements, the four unknowns, x, y, z and vt, are determined. To obtain t, the mean propagation velocity v has to be known from independent information and, since this depends on the electron content in the intervening medium, the latter must be known. One should point out here that there is a requirement that the user be passive, otherwise the capacity of the system would be severely restricted. If the user were able to interrogate the satellite then there would be no problem of user clock synchronization.

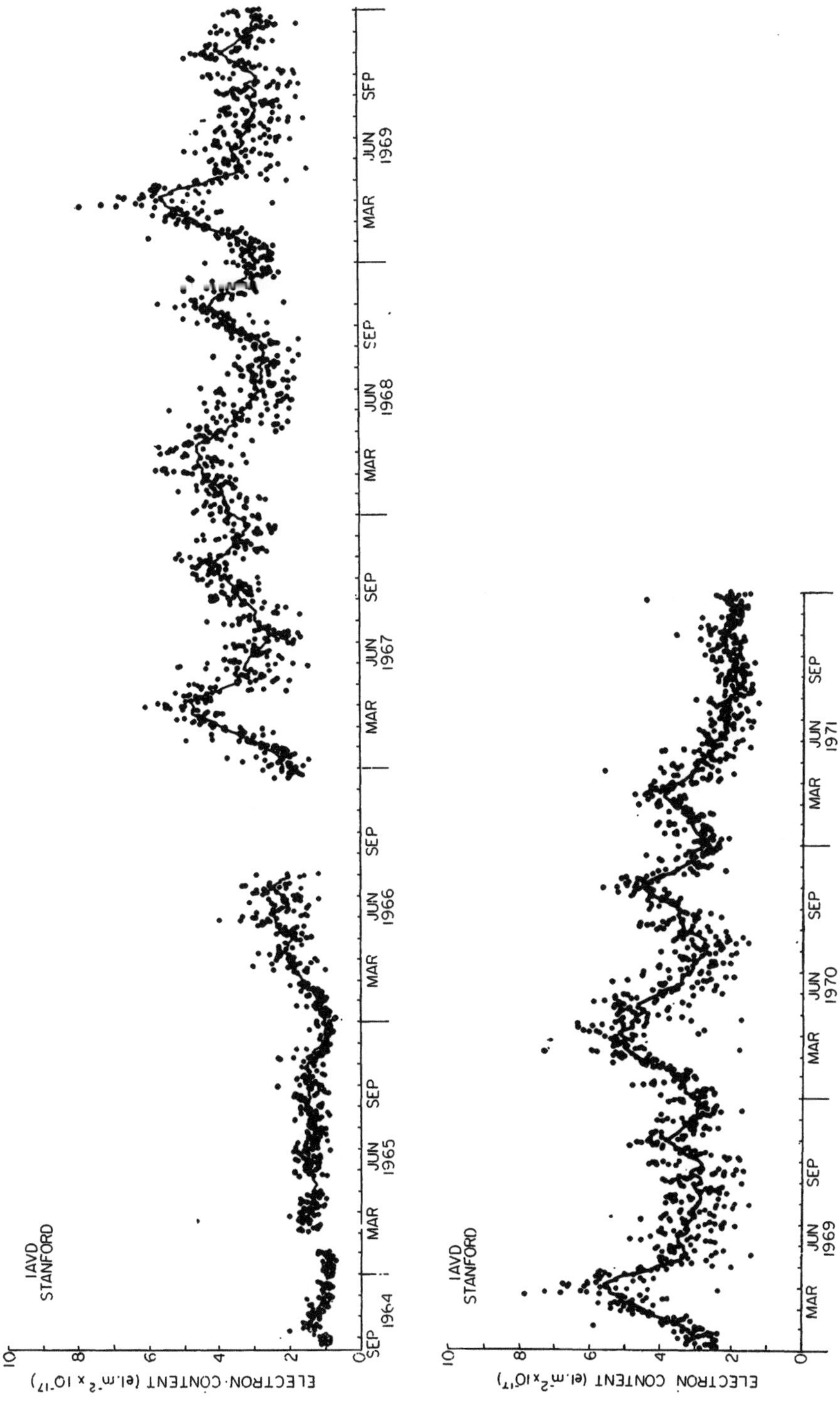

Fig. 1. The daytime average value of electron content (0900—1500), IAVD, for Stanford. The solid line is a 31-day running mean through the data. Syncom 3 observations were used up to July 1966 and AST 1 observations began on the last days of that year.

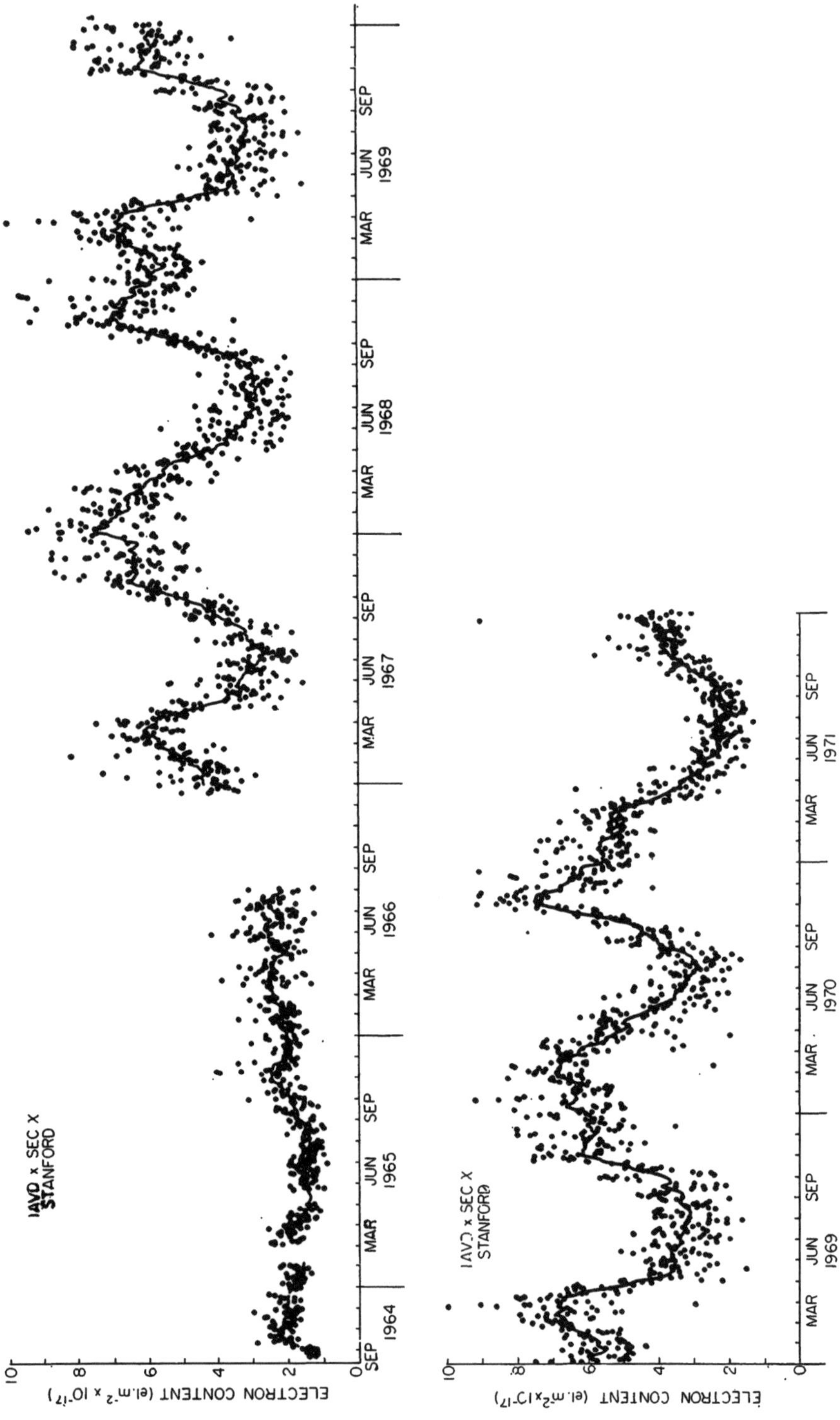

Fig. 2. The data of Fig. 1 are plotted with a correction for an overhead sun.

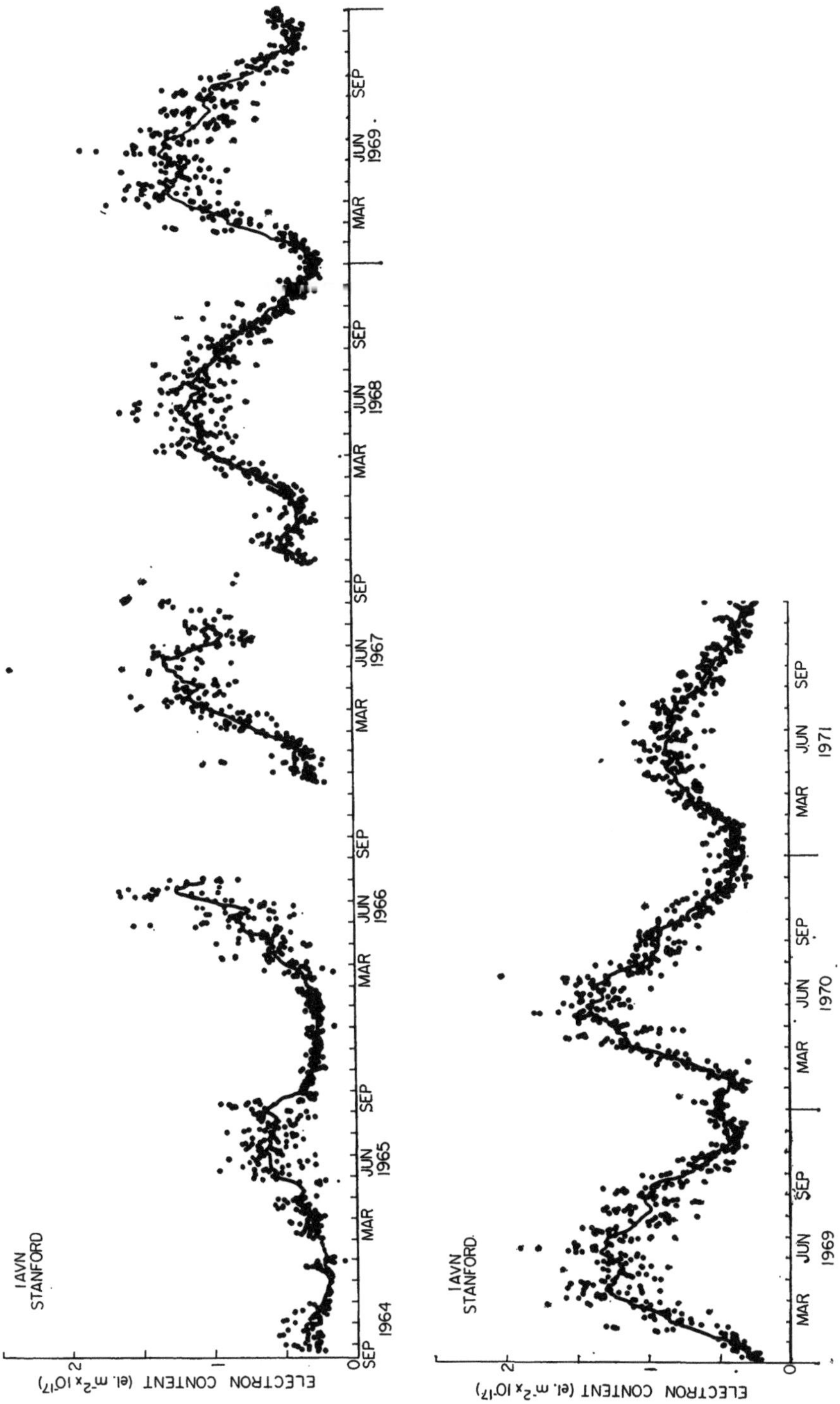

Fig. 3. The night-time average value of electron content (2100—0300).

As the system is to have a precision of a few meters, uncertainties in ionospheric delay become important even at the L-band frequencies used. For this reason considerable effort has been dedicated to the analysis of the ionospheric effects on the accuracy of the system and on finding methods to reduce the propagation uncertainties. Several organizations have constructed electron content prognostication schemes. Since CCIR has developed tables of coefficients that permit the estimation, on a worldwide basis, of values of $f0F2$ (and consequently of N_{max}) given time of day, day of the year, degree of solar activity, and geographic position, and since there is a relationship between the electron content I and N_{max} some of the schemes use the CCIR tables as a starting point. This requires knowledge of the behavior of the slab thickness τ:

$$\tau \equiv \frac{I}{N_{max}}.$$

An auxiliary prognostication scheme has to be created to estimate τ. This has been done by the Air Force Cambridge Research Laboratories [4] using observed values of I and N_{max}. The University of Illinois [5] used real time electron content values and predicted N_{max} to obtain an estimated slab thickness. This value of τ is then used at a second location together with predicted N_{max} to estimate the electron content. The scheme developed by the Applied Physics Laboratory used the predicted values of N_{max} in conjunction with theoretically derived slab thickness values to obtain I. DBA Systems, Inc., derived its prognostication of τ from worldwide top and bottom side soundings [6]. It is the only scheme, at present, that can be applied worldwide.

To avoid the accumulation of errors resulting from the need to prognosticate two quantities (τ and N_{max}), Stanford University [7] developed a scheme in which the electron content is directly prognosticated from its historical behavior in a manner similar to that used in deriving the CCIR coefficients for $f0F2$. The disadvantage of such a method is the present dearth of data that limits the coverage to North America and Hawaii.

Since the rms accuracy of all proposed schemes is about the same at present, (except for the one of Illinois which appears to be more precise), we will look at some of the Stanford results. Figs. 4, 5, and 6 compare the observed diurnal curves of electron content with those predicted by the prognostication scheme and with the 31-day mean curve. In Fig. 4 one can observe a good agreement between observation and prognostication. The 30-day mean is somewhat too high owing to the fact that the 30-day mean solar activity was higher than the activity on the day under consideration. Fig. 5 again shows good agreement between observation and prognostication. In this case, however, the mean is too low because the solar activity on the day in question was much higher than the mean. Fig. 6 shows a case in which prognosticated values depart drastically from observation owing to an (unpredictable) ionospheric storm.

From the above it can be seen that, although statistically the prognostications may be good, on any given day a serious error may be made. Hoping to ameliorate the above situation, the idea of supplementing the prognostication by some updating method was introduced. This requires near real time observation of electron content followed by geographical and temporal extrapolation. Clearly such a scheme can only succeed if there is substantial correlation between the behaviors of the electron content at the user and at the observer. In Fig. 7, σ_1/σ represents

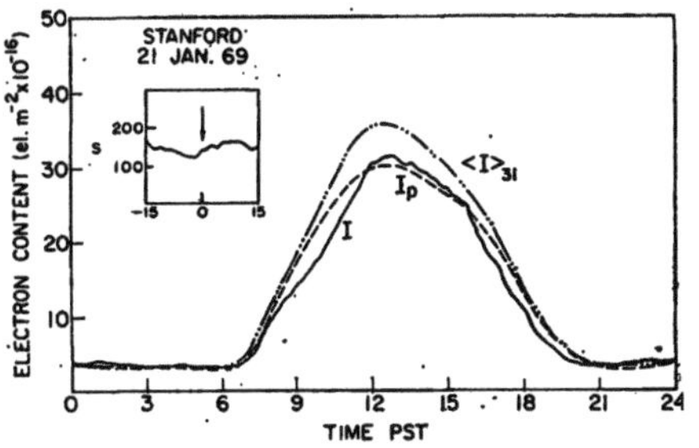

4

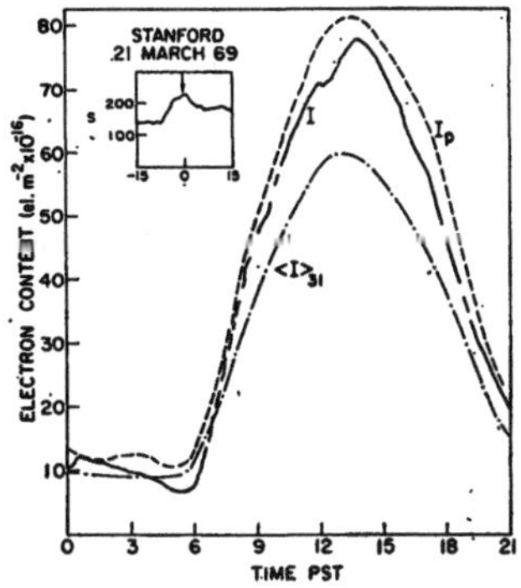

5

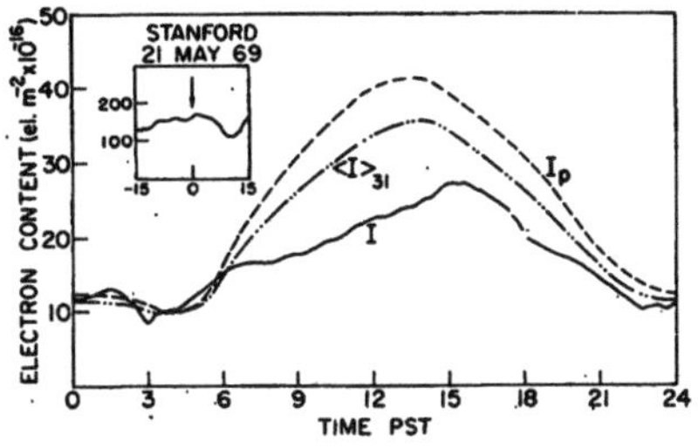

6

Fig. 4, 5 and 6. Comparison between the observed electron content I, the prognosticated value I_p and the 31-day mean value $\langle \hat{I}_{31} \rangle$. Note that the ionospheric storm on 21 May 1969 caused a severe failure of the prognostication scheme. The small graph shows a plot of the solar radio flux S (2.8 GHz) versus time, covering a period of 30 days centered on the day of the main plot.

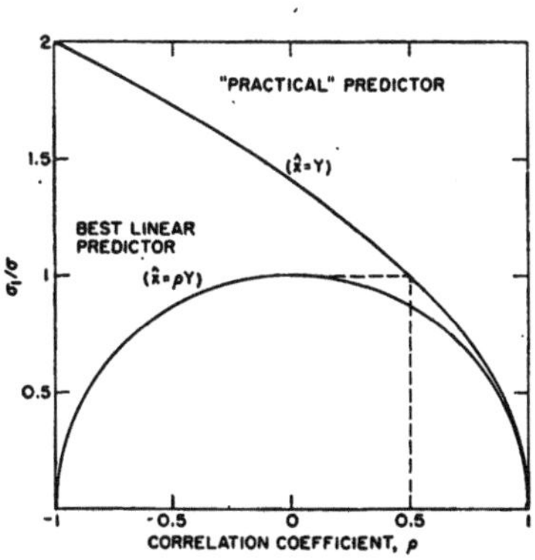

Fig. 7. If X and Y are two random variables with zero mean and the same standard deviation σ, and if ϱ is the correlation coefficient between these variables, then an estimate $\hat{X}$ of X can be obtained from an observation of Y. The error reduction (or amplification) σ_1/σ in the estimate is given as a function of the correlation coefficient. The best linear predictor yields absolutely correct values for $\varrho = \pm 1$ but requires the *a priori* knowledge of ϱ, which is impracticable. The "practical" predictor should not be used if ϱ is expected to be less than 0.5 otherwise there is a worsening of the estimate.

the error reduction (or amplification) factor resulting from updating, for various values of ϱ, the correlation coefficient between the electron contents at these two points. Clearly ϱ is unknown to the operators of the system. It can be seen that

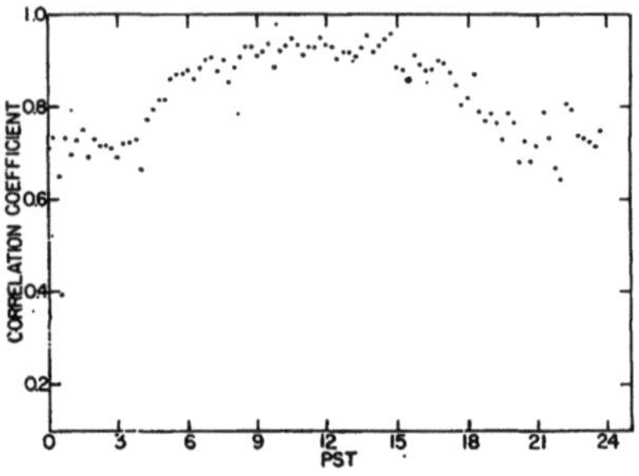

Fig. 8. Correlation between electron content values at Ely, Nevada, and Stanford, California, a distance of 700 km. A 30-day interval beginning on day 40 of 1967. Same *local* time at the two stations.

there is an improvement *only if* $\varrho > 0.5$; otherwise updating will increase the prognosticated error. For this reason, it becomes of interest to study the degree of correlation of the ionospheric electron contents at two points. Fig. 8 shows that for stations separated some 700 km and at roughly the same latitude the correlation coefficient during most of the day can be very high, thus permitting effective updating. On the other hand, in Fig. 9, the correlation coefficient between Hawaii and Stanford is seen to be always insufficient to permit improvement by updating.

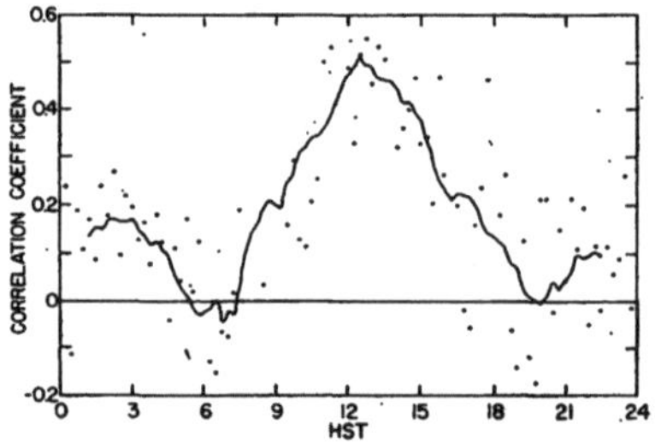

Fig. 9. Correlation between electron content values at Honolulu, Hawaii, and Stanford, California, a distance of 4000 km. A 30-day interval beginning on day 40 of 1967. Same *local* time at the two stations.

Table 2 summarizes the results obtained with the Stanford prognostication scheme. Four different ranging uncertainties (at 1.6 GHz) are shown. These are rms uncertainties over the whole period indicated in the table. ε_1 is the error in the prediction algorithm being tested, i.e., the residual uncertainties after prognostication and updating. ε_2 is the error resulting from the simple assumption that the electron content at the prediction location is the same as that at the observing location. ε_3 is the error when prognosticated values are not updated. ε_4 is the error incurred when ionospheric effects are completely ignored. ε_4 is equal to several meters and is considered unacceptable for many applications. When the user is not too far from the observing site, ε_1 and ε_2 are comparable with one another and are smaller than ε_3. This means that for short distances, especially if the two sites are at nearly the same latitude, it makes no difference whether one uses the proposed algorithm or simply assumes that the electron content at the user is the same as at the observer. When the user is far from the

Table 2

Summary of rms Residual Ranging Uncertainties (at 1.6 GHz) when different electron content estimation schemes are employed

A Prediction at	B Predictor from	Dist. (km)	Period	ε_1	ε_2	ε_3	ε_4
Stanford (1)	Stanford (2)	670	1 Jan. 68 / 31 Dec. 68	0.34	0.35	0.46	3.75
Stanford (2)	Stanford (1)	670	1 Jan. 68 / 31 Dec. 68	0.34	0.35	0.57	3.94
Edmonton	Ft. Collins	1420	1 Dec. 68 / 13 Nov. 69	0.30	0.43	0.31	3.00
Ft. Collins	Edmonton	1420	1 Dec. 68 / 13 Nov. 69	0.34	0.44	0.30	3.36
Stanford (1)	Hawaii	3830	1 Jan. 65 / 31 Dec. 65	0.23	0.96	0.19	1.26
Hawaii	Stanford (1)	3830	1 Jan. 65 / 31 Dec. 65	0.39	1.04	0.25	2.27
Arecibo	Stanford (1)	5910	8 Dec. 67 / 18 Apr. 68	0.57	0.85	0.23	4.13

See text for explanation of $\varepsilon_1 \ldots \varepsilon_4$ (ε in meters).
(1) ATS-1; (2) ATS-3

observer, updating tends to degrade the results: it is best to use prognostication with no correction. At intermediate distances, ε_1 and ε_3 are comparable and are the smallest errors. In such cases it makes little difference whether the algorithm is used with or without updating.

The fact that present prognostication schemes fail drastically on some occasions indicates that our knowledge of the different processes that influence the ionosphere is still incomplete. Thus, we do not understand the reasons for the large observed day-to-day variations in the ionospheric electron content, the mechanisms that cause ionospheric storms are not clearly grasped, and even the seasonal behavior of the electron content is not satisfactorily explained although several theories have been advanced. Clearly, the solar ionizing radiation is one of the driving forces that control our plasmasphere and one that is reasonably well understood. It would therefore be instructive to remove from the observed behavior of the ionospheric electron content, that component that is due to changes in solar EUV: the residual variations in electron content will then contain the effect of all other agencies (or would if all processes were linear). One difficulty with this approach is the fact that we do not have good continuous observations of the EUV spectrum and must instead rely on an indirect index of solar ionizing radiation: the solar radio flux at 2.8 GHz.

Scatter diagrams of electron content over Stanford versus solar radio flux were prepared covering the period 1964—1971 and a linear regression was determined for every third day (Fig. 10). Under the assumption that the short term response of the ionosphere to fluctuations in solar radiation is the same as the long term response, it is possible to write an empirical expression for the daily average value of electron content I_{mean}, in terms of the flux of solar radio noise S:

$$I_{mean} = I_0 + \beta S$$

where I_0 and β depend on the day of the year but not on the level of solar radiation. Given daily values of S, it is then possible to estimate I_{mean}. Fig. 11 shows the predicted (solid line) and the observed (crosses) values of electron content for

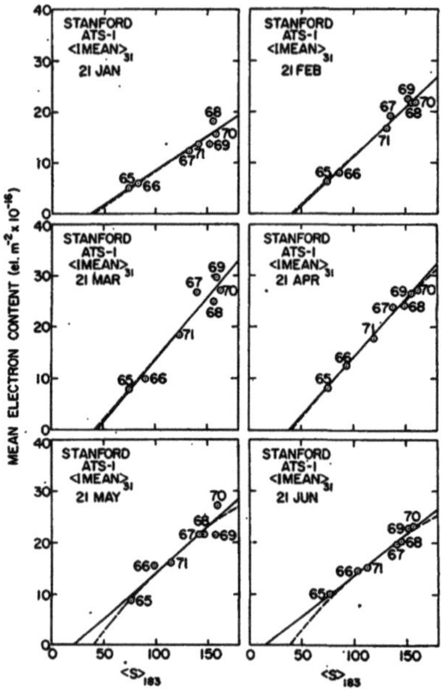

Fig. 10. Examples of regression between solar radio flux S and daily mean electron content I_{mean} (IMEAN on graph).

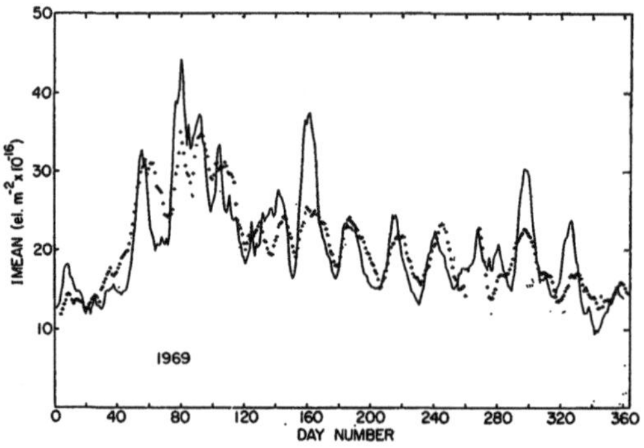

Fig. 11. Comparison between predicted (solid trace) and observed values of daily mean electron content (crosses). Data for Stanford ATS 1.

1969. It can be seen that the general trend of the prediction is correct but that there are many marked discrepancies whose cause is not known.

It is also instructive to examine what happens when one assumes that throughout a year the sun's activity is constant: this will bring out the seasonal behavior shown in Fig. 12. At low solar activity ($S = 60$) there is more ionization in the summer than in the winter, by just the amount one would expect based on the smaller zenith angle and longer days in the summer. As the solar activity increases, equinoctial humps begin to appear, the vernal one being more pronounced than the autumnal. Waldman [8] has proposed an ingenious explanation for this phenomenon based on the variation of neutral hydrogen content in our atmosphere.

The influence of neutral winds on the degree of ionization in the plasmasphere has been recognized by theoreticians (cf. King and Kohl [9]) and more recently has been used as explanation for a number of features in the behavior of the ionosphere. The most obvious effect caused by the wind is the action of its meridional component in lifting (equatorward wind, at night) or lowering (poleward wind, in daytime) the ionization and thus altering the average recombination rate. This effect is similar to that caused by electric fields; which of these two causes is the dominant one depends on magnetic activity. Stubbe [10] showed

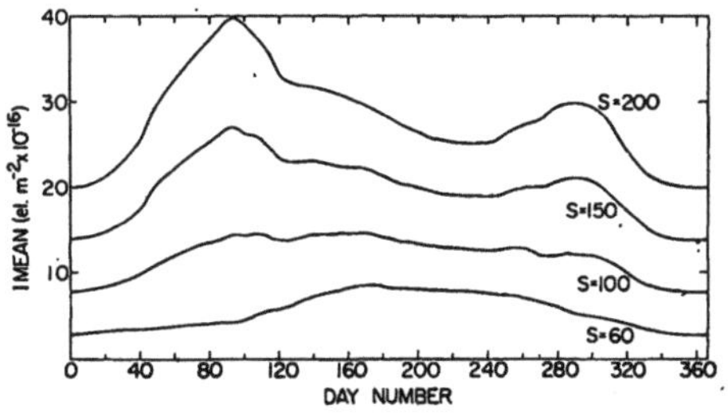

Fig. 12. Seasonal behavior of the daily mean electron content.

that on a quiet day winds are clearly dominant. On a disturbed day, electric fields are greatly enhanced and override the wind effect. These facts are important when one wishes to examine the wind effect on observed electron content: one must select quiet days.

A recent study of wind effects was made by Bendito [11] who analyzed the interesting feedback mechanism that exists in the plasmasphere: Let us fix our attention on the conditions prevailing during midday when the neutral winds are essentially poleward and tend to depress the ionization. Consider a step increase in solar radiation. The production of ionization will increase almost simultaneously and will quickly result in an increase in electron content. Since in the F region (which is the main contributor to the electron content) the main restraining force for neutral winds is the ion drag, the increase in ionization will cause a decrease in the average recombination coefficient. This causes the ionization to grow further in a positive feedback loop that searches a new equilibrium point in which h_{max} is substantially higher. After some time delay [12] the neutral atmosphere reacts to the ineteased solar radiation by an increase in temperature. This causes a reduction in the pressure gradients per unit mass that drive the wind so that there is another increase in the electron content. The conclusion of Bendito's analysis is that the wind effect will cause a relative increase in electron content larger than the relative increase in solar radiation that triggered the changes. In addition, a time delay in the ionospheric response to solar changes is predicted. Experimental observations made during quiet periods, in which the solar energy had sufficiently well marked changes, confirms the theoretical predictions. Fig. 13 shows the cross correlation function between electron content at Stanford and solar radio noise. It can be seen that for all years, except 1965, the electron content changes lag those of the solar activity. In 1965, the fluctuations in solar activity were too small to permit any significance to be attached to the cross correlation variations.

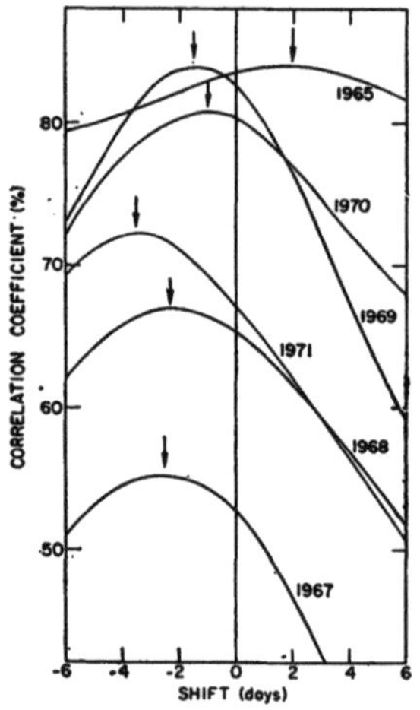

Fig. 13. Cross correlation between electron content and solar radio flux showing that the content lags by about 2 days (1965 results have no significance, see text).

Wind effects on the ionosphere can also be detected by observing the response of the electron content to solar eclipses. Fig. 14 shows the 25% bite-out in electron content observed at Eastville, Va, USA, during the 7 March 1970 solar eclipse. The volume of the plasmasphere under scrutiny was in the path of totality. Simultaneous observations of electron content made at stations in which maximum obscuration of the sun was 45%, showed non-detectable bite-outs. This lack of reaction can be explained by postulating a neutral wind flowing towards the volume of the atmosphere cooled by the sun's shadow [13].

An interesting use of electron content data is made by workers dealing with gravity waves. Such waves leave signatures in electron content records in the

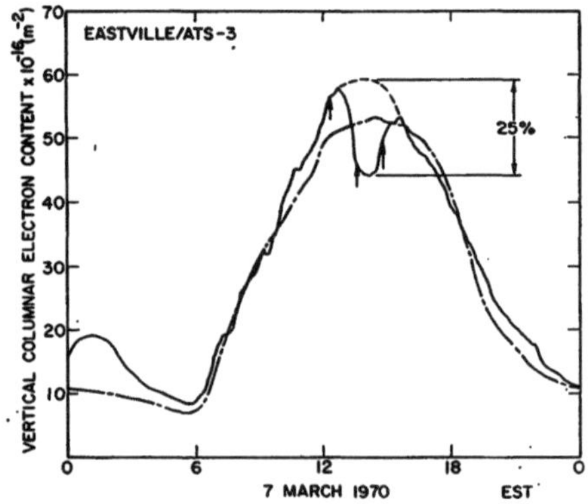

Fig. 14. Electron content bite-out during the solar eclipse of 7 March 1970, observed in the path of totality.

form of bumps. Davis [14] using a four-station network measured the vector velocity of these gravity waves and observed a systematic diurnal oscillation in the travel direction. Average direction was equatorward along the magnetic meridian but a diurnal swing of $\pm 20°$ was clearly detectable. This suggested a source in the auroral oval at a fixed geomagnetic time. As the earth spins, the source moves eastward causing the change in direction. Davis was able to pin down the position of the source in the evening sector of the oval and to establish a one-to-one relationship between polar substorms and individual bumps in electron content. Substantial contributions to the understanding of the gravity wave generation and propagation mechanism were made.

Titheridge, in work not yet published, has been able to use a combination of electron content and foF2 data to determine the temperature of the neutral atmosphere in roughly the 400 km altitude region. Based on theoretical considerations, a linear relationship between slab thickness and neutral temperature was derived:

$$\tau = \tau_0 + \alpha T_n$$

in which α hovers between 0.220 and 0.225 and τ_0 depends on season and time of day and has values between 2 and 22 km. Calculated night-time values of T_n agree with satellite drag values to within $\pm 5\%$. Behavior of daytime values resembles more that of values derived from incoherent backscatter. If further investigation along the lines described above consolidates our confidence in this method of temperature measurement, then a useful tool has been developed for the study of atmospheric processes.

As pointed out in the beginning of this report, electron content measurements of the protonosphere have been carried out by Almeida. This is done by using a combination of phase-path difference and Faraday rotation measurements; the former yields the *total* electron content I up to the satellite while the latter measures essentially the integral:

$$I_F = \frac{1}{B_{LI}} \int_0^S N B_L \, ds$$

where the electron concentration N is weighted by the longitudinal component of the geomagnetic field. Thus the difference between I and I_F can be written:

$$I_W \equiv I - I_F = \int_0^S N\left(1 - \frac{B_L}{B_{LI}}\right) ds \equiv \int_0^S NW \, ds .$$

Clearly, if W has the property of being equal to zero up to a given height and unity above that height, then I_W is a measure of the electron content above that height. It turns out that for certain geometries of observation, the weighting function W behaves in a manner acceptably close to the ideal one mentioned above, as can be seen in Fig. 15. Under such conditions it is possible to observe, on a continuous basis and with good time resolution, the protonospheric content. Results are shown in Fig. 16 which display the depletion of the protonosphere following a storm and its eventual build up to normal values. The sharp bump on 14 May 1969 has been interpreted as the signature of a "blob" of plasma peeled off from the protonosphere by the storm-enhanced dawn-to-dusk electric fields and ejected into the interplanetary space.

As pointed out when prognostication schemes were discussed, there are a number of days during the year when the electron content behavior departs markedly from the expected. These are, by definition, ionospheric storms, whose

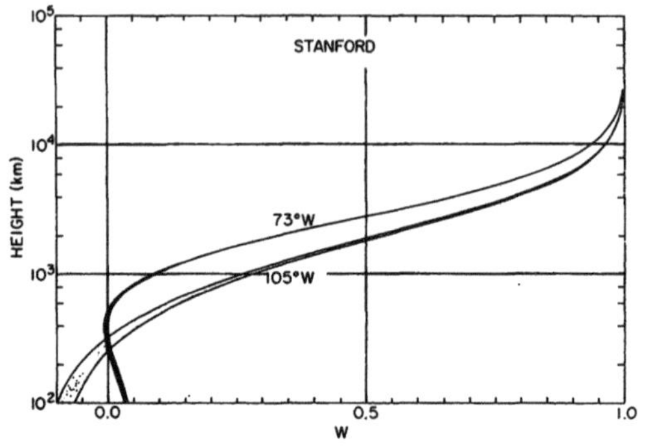

Fig. 15. Plots of the weight function W versus height for an observer at Stanford. Two families of W are shown corresponding to geostationary satellites positioned at 73° W and 105° W.

close association with geomagnetic storms was recognized early. Although ionospheric storms can easily be recognized from the examination of ionosonde or incoherent scatter records, the electron content is by far the simplest and most direct way of observing this phenomenon. This is specially true when modern polarimeters are used, yielding real time electron content plots. Fig. 17 is a plot of observed electron content (solid line) compared with the 7-day average (dotted line, taken as reference) and shows a typical ionospheric storm. The numbers that appear in the figure are a "disturbance index D" defined as the ratio (as percentage) of the mean quadratic deviation of electron content from its average to the average value. Day 81 is seen to be essentially undisturbed ($D = 6\%$).

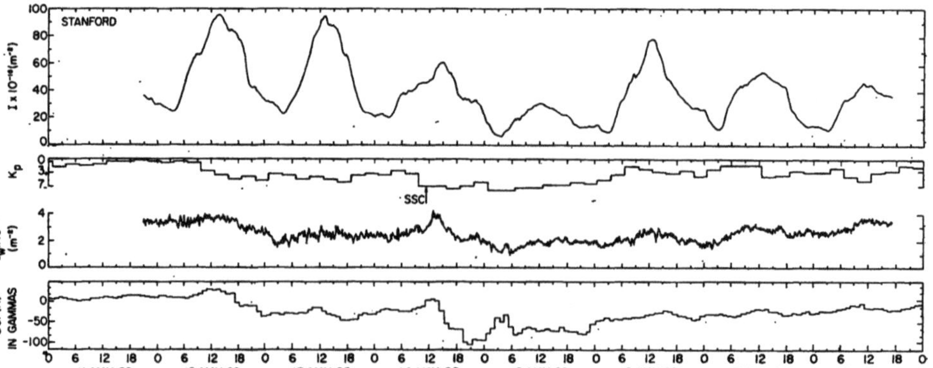

Fig. 16. Protonospheric electron content observation, during an uninterrupted period of 7 days in May 1969, from Stanford. The geostationary satellite was parked at 73.4° W.

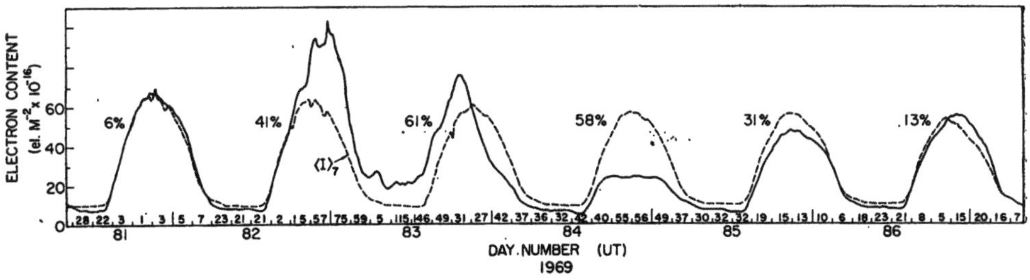

Fig. 17. Comparison between observed electron content (solid line) and the seven-day mean (dashed line).

The storm starts on day 82 and is characterized by an enhancement of electron content followed eventually by a depression and a subsequent recovery. Examination of electron content records shows that the behavior of this quantity is strongly dependent on the local time of storm onset, i.e. a given storm will cause effects that are dependent on the longitude of the observing station. Thus, it would be of great interest to borrow a technique from the geomagneticians and, by using data from a number of equal latitude stations more or less evenly distributed in longitude, expand the storm induced deviation of electron content in a Fourier series as a function of longitude:

$$R(\tau) = R_0(\tau) + \sum_n R_n(\tau) \sin [nl + \Phi_n(\tau)]$$

where $R = (I - \langle I \rangle)/\langle I \rangle$, $\langle I \rangle$ is the reference value of electron content, τ is the time elapsed since the storm onset, l is the longitude and R_n, Φ_n are Fourier coefficients; R_0 is independent of longitude and corresponds to *Dst* of geomagnetism.

The behavior of R_0, R_n, and Φ is of great interest to the investigator attempting to uncover the storm producing mechanisms. Unfortunately, at present, there is an insufficient number of electron content observing stations to allow a good determination of these quantities. If a collection of similar storms, differing only

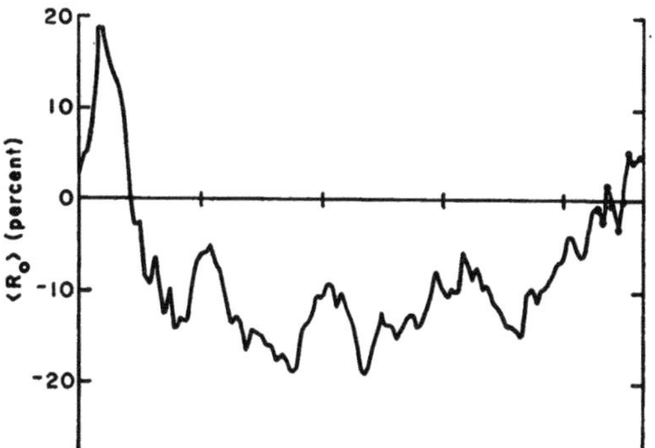

Fig. 18. Storm time variation of ionospheric disturbance at midlatitude. From Mendillo [4].

in the universal onset time were available, then one would have an ergodic process and R_0 could be calculated from a large number of storms observed at one location rather than from one storm observed from many locations. As individual storms differ greatly from one another such a calculation of R_0 yields only their average behavior $\langle R_0 \rangle$. Mendillo [14] calculated R_0 for a total of 28 storms (all with $Ap > 30$) and used the time of the sudden commencement as the time origin ($\tau = 0$). Fig. 18 displays his results and shows clearly the positive, the negative and the recovery phases lasting some 12, 72 and 24 hours respectively.

Although the morphology of storms has been under study for some time, there is room for considerably more investigation based on vast masses of data collected all over the world. A complete study of the dependence of this morphology on various factors such as seasons and phases of the solar cycle will without doubt be helpful in building a theory of storms.

Thus, I would like to encourage colleagues all over the world to take part in a comprehensive program of observation of the ionospheric electron content.

Acknowledgments

The present work was supported by NASA Grant NGR-05-020-001.

References

[1] O. K. Garriott, J. Geophys. Res. **65**, 1139 (1960).
[2] F. de Mendonça, J. Geophys. Res. **67**, 2315 (1962).
[3] O. G. Almeida, Tech. Rep. SEL-72-019, Stanford University 1972.
[4] J. A. Klobuchar and R. S. Allen, Air Force Surveys in Geophysics **222**, AFCRL 1970.
[5] N. N. Rao, M. Y. Youakim and K. C. Yeh, SAMSO TR-71-163 (1971).
[6] R. B. Bent, S. K. Llewellyn and M. K. Walloch, SAMSO TR-72-239 (1972).
[7] H. Waldman and A. V. da Rosa, Tech. Rep. SEL-71-047, Stanford University 1971.
[8] H. Waldman, Tech. Rep. SEL-71-048, Stanford University 1971.
[9] J. W. King and H. Kohl, Nature **211**, 699 (1965).
[10] P. Stubbe, J. Atmos. Terr. Phys. **32**, 865 (1970).
[11] J. L. Bendito, Tech. Rep. SEL-73-023, Stanford University 1973.
[12] J. R. Herman and S. Chandra, Planet. Space Sci. **17**, 815 (1969).
[13] O. G. Almeida, H. Waldman and A. V. da Rosa, J. Atmos. Terr. Phys. **34**, 713 (1972).
[14] M. Mendillo, Nature Phys. Sci. **234**, 23 (1971).

Space Research XIV — Akademie-Verlag, Berlin 1974

IONOSPHERIC ELECTRON CONTENT AND ITS HORIZONTAL GRADIENTS AT MIDDLE AND HIGH LATITUDES

V. A. Misyura, I. I. Kapanin, V. A. Podnos, O. F. Tyrnov, A. M. Tsymbal, G. N. Zintchenko, A. K. Surkov and V. M. Trubitsyn

Radiophysical Faculty, State University, Kharkov, USSR

Diurnal, latitudinal and seasonal dependences of the ionospheric electron content and its horizontal gradients obtained from Doppler and Faraday records of coherent satellite signals (Cosmos 321, 356, 381, Intercosmos 2 and Transit) received at middle and high latitudes from 1970 to 1972 are analysed.

The electron content N_0 was determined using the first-order approximation of geometrical optics [1, 2]. Cases in which errors were significant were excluded [3], or corresponding corrections were made where necessary for higher approximations (using available Doppler and Faraday records at several frequencies).

The Cosmos 381 (satellite height 1000 km) signals received in the winter of 1970—1971 at several stations spaced in geographic latitude φ (Kharkov, 50.0° N; Belomorsk, 64.5° N; Arkhangelsk, 64.6° N; Murmansk 69.0° N) allowed the latitudinal dependence of $N_0(\varphi)$ to be studied at different local times. The diurnal dependence of $N_0(t)$ averaged over many satellite passes for different latitudes from 40° N to 72° N was also investigated. Typical variations of $N_0(t)$ are presented in Fig. 1. The smallest N_0 values are observed near Belomorsk, i.e. in the region of the main trough. Near Murmansk from 0200 to 0700 LMT, N_0 values are somewhat higher than those in the vicinity of Kharkov. This is related to the presence of significant ionization sources due to charged particles at high latitudes even at night; they are practically absent at middle latitudes. For $\varphi = 68°$ N to 70° N, the diurnal change of energetic particle intensities seems to be insignificant since $N_{0\,\max(\text{day})}/N_{0\,\min(\text{night})} \approx 4$ in this region. Note that the values of $N_{0\,\max}/N_{0\,\min} \approx 2-3$ obtained from high latitude Doppler records ($\varphi = 69°$ N) of Transit signals in the summer of 1972 are smaller compared with those in winter. In winter, as φ decreases, the ratio $N_{0\,\max}/N_{0\,\min}$ increases (up to $\sim$15 at middle latitudes). However, the monotonic latitudinal variation of the ratio is distorted in the vicinity of the main trough (Fig. 1), due to the diurnal development and motion of the trough. Moreover, the sharpest N_0 increase is observed in this region after sunrise.

A characteristic feature of the winter diurnal variation of N_0 (Fig. 1) is the night-time maximum between $\varphi = 46°$ N and 54° N. It is most pronounced at $\sim$50° N and disappears at 44° N and 56° N. A similar night-time N_0 increase has been observed by others [4, 5]. In the winter the dependence of $N_0(t)$ derived from Faraday records of Cosmos 321 (at altitude 350 km) signals received near

Irkutsk (52° N), showed a maximum at about midnight. For satellite altitudes less than 300 km, this night-time maximum does not always appear, and is less pronounced than for the higher altitudes (~1000 km).

The middle latitude Faraday records of satellite signals (Cosmos 321, Intercosmos 2) received during February—March 1970 allowed the $N_0(t)$ dependences to be derived and compared with vertical sounding and incoherent scatter data

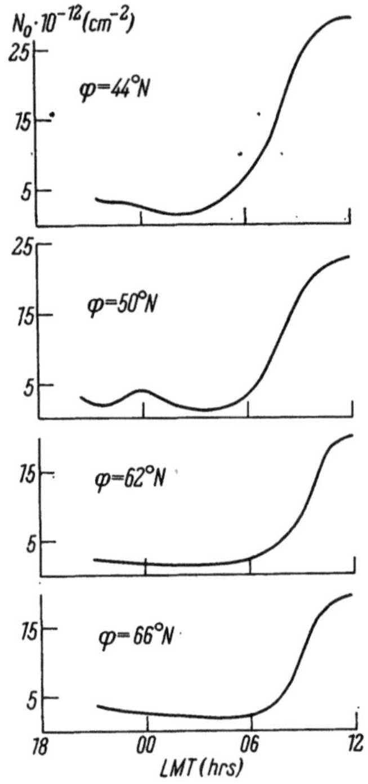

Fig. 1. Diurnal dependences of the total electron content N_0 to altitude 1000 km for different latitudes, Cosmos 381, winter1970—1971.

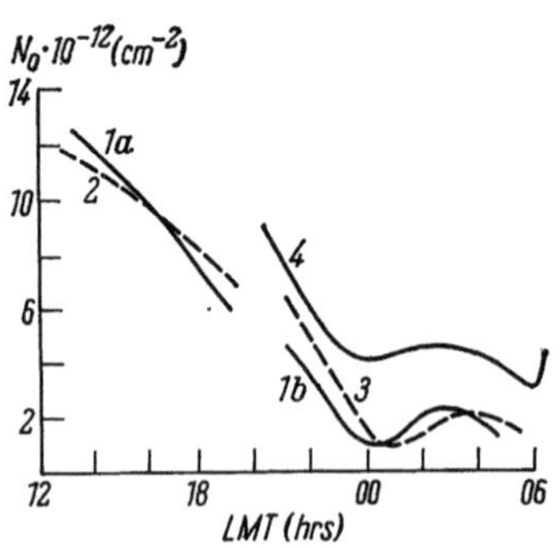

Fig. 2. Diurnal dependences of the total electron content N_0 for middle latitudes: 1a, Cosmos 321, average altitude $\bar{z} = 275$ km; 1b, Cosmos 321, $\bar{z} = 370$ km; 2, vertical sounding; 3, incoherent scatter, $z \approx 350$ km; 4, Intercosmos 2, $z \approx 800$—1000 km. 1a, 1b, 2, 3 refer to February—March 1970 and 4 to March 1969.

(Fig. 2). It is seen that the night-time maximum $N_0(t)$ occurs at about 0300 LMT, which agrees with [6—9].

For spring, autumn and winter between 40° N and 70° N, it is seen that at middle latitudes N_0 values usually decrease with latitude for any solar activity level. For separate satellite passes over 42° N—47° N, a trough is observed in the N_0 curve which may be smoothed after N_0 values for many satellite passes are averaged. At middle latitudes, the winter N_0 dependence on solar zenith angle is defined by the relation $N_0 \propto (\cos \chi)^n$ where n changes from $^1/_3$ to 1. In winter from 55° N to 60° N, N_0 decreases rapidly (Fig. 3) with the largest changes in the

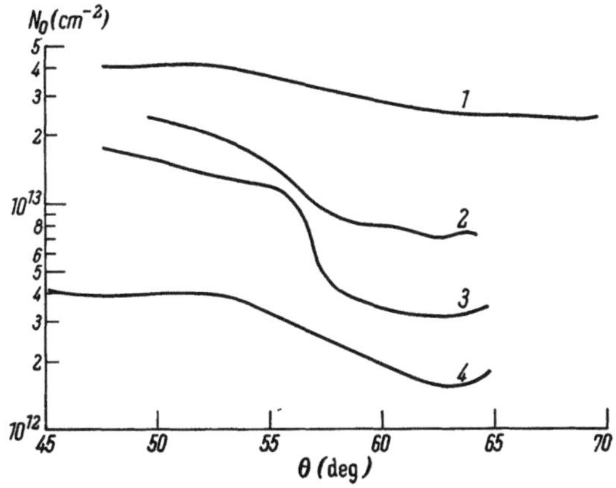

Fig. 3. Latitudinal dependences of the total electron content N_0, Cosmos 381, $z \approx 1000$ km: 1, 29 December 1970, 1100 LMT; 2, 3 January 1971, 1000 LMT; 3, 14 January 1971, 0900 LMT; 4, 9 January 1971, 2300 LMT.

morning. This latitude range may be apparently considered as a boundary region between middle and high latitudes. Also, in the vicinity of 62° N, a N_0 trough was observed which, unlike the mid-latitude one, was narrower; the trough occurred for both high (~1000 km) and low (~350 km) satellite altitudes. Its latitude changed with time of day from 59° N to 65° N: at night the trough shifted to lower latitudes, by day it returned to higher latitudes. The winter trough was observed for evening and morning hours, being more pronounced in the morning. Its occurrence was of irregular character in reference to day and time of day. Some negative correlation seems to exist between trough occurrence and Wolf numbers. In middle of autumn the trough was observed even at night. Between 70° N and 71° N, another trough was sometimes observed. Comparison of the seasonal and diurnal trough behaviour with space-time distribution of aurorae indicates their close connection. It is not unlikely that auroral phenomena and trough formation have a common cause. The behaviour of N_0 at the latitudes discussed is not always observed: in summer, during the last phase of spring and initial phase of autumn, the value of N_0 usually decreases from middle to high latitudes [3, 10].

Satellite signal records made simultaneously at three spaced stations enabled us to derive γ_h, the total relative horizontal gradient of N_0, and its components, the latitudinal and longitudinal gradients, γ_θ and γ_φ [3]. ($\gamma_h = N_0^{-1} \nabla_r N_0$, where

∇_r is the horizontal component of ∇.) Characteristic measurements are illustrated in Fig. 4 and in [3, 11—13]. The observations analysed for spring and winter periods showed that the magnitudes of the gradients and their dependence on time of day are different at middle and high latitudes. The greatest γ_φ-values were usually characteristic of sunrise and sunset, being $(3-5) \times 10^{-3}$ km^{-1} and $(5-8) \times 10^{-3}$ km^{-1} at middle and high latitudes respectively.

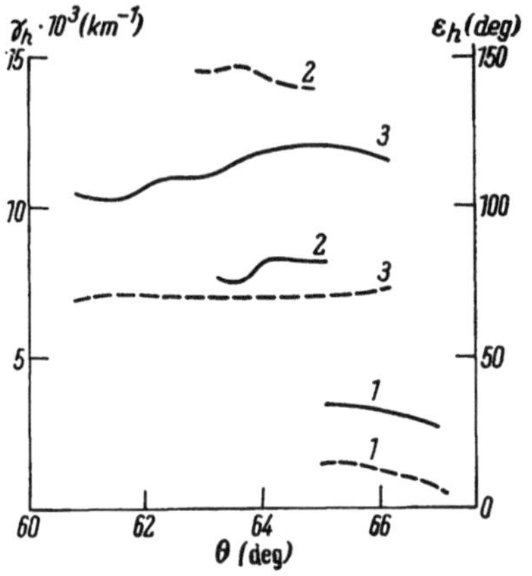

Fig. 4. Latitudinal dependences of the magnitude of the total relative horizontal gradient γ. (solid line) and its orientations (ε_h is the azimuth, dashed line), Cosmos 381, $z \approx 1000$ km winter 1971: 1, 14 January, 0500 LMT; 2, 15 January, 0700 LMT; 3, 6 January 1000 LMT

The greatest γ_h values at middle latitudes are also observed in the morning and evening, by day γ_h is somewhat smaller and by night it is smallest ($\sim(2-8) \times 10^{-4}$ km^{-1}). Around midday, γ_h is directed approximately along a magnetic meridian. The mid-latitudinal γ_θ value changes from $(1-2) \times 10^{-4}$ km^{-1} to $(5-7) \times 10^{-4}$ km^{-1} and is usually smaller than γ_φ. The presence of γ_θ is well seen in the latitudinal curves of N_0. The estimates made showed that in the intervals of 38° N—55° N and 66° N—70° N, γ_θ values were about $(1-3) \times 10^{-4}$ km^{-1} and $(3-5) \times 10^{-4}$ km^{-1} respectively. Between 55° N and 66° N, γ_θ may be greater than γ_φ, being $(3-6) \times 10^{-3}$ km^{-1}.

At high latitudes, unlike middle latitudes, for the winter, 2300—1000 LMT, observations, γ_h at night often exceeded its morning value. The night-time gradient is south-east or north-east at high latitudes, being dependent on the presence of the trough and on the location of the receiving station with respect to the trough; at middle latitudes it may have different orientations [14]. These differences of orientation seem to be caused by the high latitude N_0-trough and energetic charged particle flux which to some extent—and by night completely—determines the N_0-behaviour at these latitudes. Fig. 4 shows values of the gradient γ_h and its azimuthal bearing ε_h at high latitudes. The gradient orientation at 1000 LMT (curve 3) is approximately north-east: this is connected with the fact that our measurements were conducted northward of the trough where N_0 was increasing with latitude. The trough latitude for this satellite pass was about 60° N. At 0700 LMT (curve 2) the gradient orientation was south-east: no trough was observed, and N_0 increased southward. The gradient was the smallest at 0500 LMT (curve 1). In this case the longitudinal gradients were insignificant and the trough was located southward; therefore the gradient had a nearly northward orientation.

Further statistical data appear to be required for more accurate definition of the features discussed and for their reliable physical interpretation.

References

[1] V. A. Misyura, in: Ionosph. Issled., Nauka, Moscow 1969 (p. 120).
[2] V. A. Misyura et al., in: Issled. Kosm. Prostranstva, Nauka, Moscow 1965 (p. 138).
[3] V. A. Misyura et al., paper presented at XV General Assembly of IGGU, Moscow 1971.
[4] G. K. Hartmann et al., Mitteilungen aus dem Max-Planck-Institut für Aeronomie, No. 48 (I), Springer-Verlag, Berlin—Heidelberg—New York 1972.
[5] F. Bertin and J. Papet-Lepine, Radio Sci. **5**, 899 (1970).
[6] J. E. Titheridge, J. Geophys. Res. **73**, 2985 (1968).
[7] J. V. Evans, J. Geophys. Res. **70**, 1175 (1965).
[8] J. V. Evans, J. Geophys. Res. **70**, 4331 (1965).
[9] A. F. Nagy, F. G. Fontheim et al., J. Geophys. Res. **75**, 4665 (1969).
[10] L. Liszka, J. Atmos. Terr. Phys. **29**, 1243 (1967).
[11] V. A. Misyura et al., Space Research XI, 1027 (1971).
[12] V. A. Misyura et al., Proc. 10th All-Union Conf. on Propagation of Radio Waves, Section 1, Nauka, Moscow 1972 (p. 29).
[13] V. A. Misyura et al., Proc. 10th All-Union Conf. on Proapagtion of Radio Waves, Section 1, Nauka, Moscow 1972 (p. 39).
[14] Narayana Rao, IEEE Trans. Anten. Propag. **17**, 111 (1969).

Space Research XIV — Akademie-Verlag, Berlin 1974

EXPERIMENTS WITH EUROPIUM VAPOUR CLOUDS IN THE UPPER ATMOSPHERE

S. DRAPATZ, H. FÖPPL, G. HAERENDEL, L. HASER, K. W. MICHEL and A. VALENZUELA*

Max-Planck-Institut für extraterrestrische Physik, Garching, FRG

The feasibility of europium-vapour clouds for magnetospheric studies was examined in three rocket experiments carried out in Chamical, Argentina, on 2, 4 and 11 November 1972. Two canisters filled with a mixture of Eu/Ba/CuO and Eu/CuO respectively were released at an altitude of about 160 km from two rockets. The Ba:CuO ratio in the first mixture was nearly stoichiometric, so that with approximately 20% of the Eu only a small quantity of Ba was vaporized. In the third rocket a 2.5 kg shaped charge with a europium liner was ignited at an altitude of 240 km. The drift of the neutral and the ionized clouds was measured. By means of spectroscopic observations from the ground, the mechanism of ionization by solar radiation of these europium clouds was studied, and theoretical considerations on the visibility were confirmed.

1. Introduction

The method of measuring electric fields in the upper atmosphere in magnitude and direction with ionized barium clouds has been studied extensively [1]. This method, however, has some important drawbacks. Of these the most significant two are that the clouds can only be seen during twilight and with clear skies, and that the barium content of the photosphere of the sun leads to strong Fraunhofer absorption lines (for instance, at the resonance line of Ba II at 4554 Å, the central intensity is only 5% of the surrounding continuum). To overcome the latter disadvantage, experiments were carried out to determine whether the few other elements which have strong resonance lines in the visible region for the ionized state could be used for ion cloud experiments. Three rocket experiments were launched from Chamical, Argentina, on 2, 4, and 11 November respectively, using Eu/Ba/CuO and Eu/CuO mixture in two rockets and a 2.5 kg shaped charge with a Eu liner in the third experiment. In the following, details of the experimental method and the results of the observations are presented.

2. Theoretical Considerations

Theoretical calculations indicate that the elements most suitable for magnetospheric studies are Ca (B. P. 1760 °K), Sr (B. P. 1648 °K), Ba (B. P. 1895 °K), Eu (B. P. 1764 °K), Yb (B. P. 1700 °K) [2]. Of these Eu and Yb are especially

* Of the National Commission on Space Research, Argentina.

appropriate, because they are very rare in the sun [3]. The ionization potential for the five elements is lowest for Ba (5.21 eV), followed by Eu (5.66 eV), while for Yb it is relatively high (6.2 eV). Serious consideration, however, should be given to the time constant for ionization, the strength of the resonance lines, and the visibility. The time constant should be reasonably short in the region of 10 to 100 sec.

In Fig. 1 are presented the main features of the Eu I energy level diagram. As in the case of Ba, the Eu I spectrum is essentially a two-electron spectrum with relatively few intense lines. The inner 4f shell of Eu I has a stable half-filled

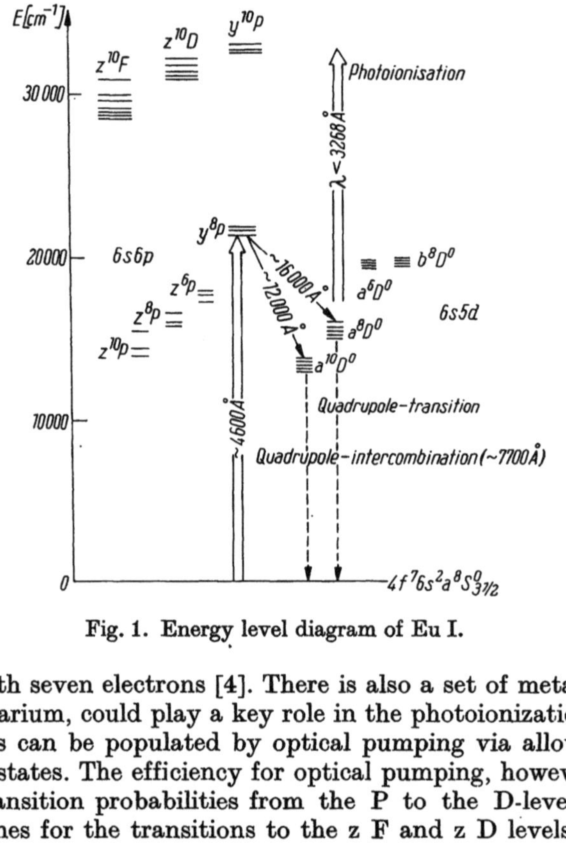

Fig. 1. Energy level diagram of Eu I.

configuration with seven electrons [4]. There is also a set of metastable D-levels which, as with barium, could play a key role in the photoionization process. The metastable levels can be populated by optical pumping via allowed transitions involving the P-states. The efficiency for optical pumping, however, depends on the unknown transition probabilities from the P to the D-levels. From these D-levels some lines for the transitions to the z F and z D levels in the spectra should also be observed. With some assumption for the f-values and life times for the D-levels based upon the similarity of the Eu and Ba spectra, the ionization time constant was estimated by Drapatz [5] to be in the range of 50 to 100 sec. The emission rate of the resonance lines of Eu II should be 3 to 4 times stronger than for Ba II.

3. Experimental Method

The experiments were launched at the Argentine Space Research Organization CNIE launch site in Chamical, Argentina, using three Argentine Rigel rockets. Groups of the University of Mendoza and the University of Cuyo participated as experimental observers. Each payload had two separate cloud experiments and one electron-density measurement (capacity probe [6]) on board besides the europium canisters. The shaped charge was a 2.5 kg TNT/RDX explosive.

For the preparation of the canisters for generation of europium clouds, the same procedure was chosen as for the barium clouds [7]. A thermite mixture with an excess of the metal to be vaporized was prepared. The europium metal itself

Table 1

List of Containers in the Three Rigel Rockets

Exp. No.	Launch Date	Altitude (km)	Mixtures in the Containers (gm) Components		Moles	Filling Weight	Metal Exess (moles)	
63	2 Nov 1972	141	Ba	162	1.18	522		
			CuO	136	1.71		Eu	1.31
	1910 LT		Eu	252	1.66			
	2328 UT		NaN_3	1				
		190	Li	208	30.0	762	Li	16.0
			CuO	554	7.0			
		190	Li	11	1.59	58 + 2500	Li	1.59
			Ba	47	0.34	explosive	Ba	0.34
64	4 Nov. 1972	158	Eu	392	2.58	541	Eu	1.24
			CuO	148	1.86			
	1909 LT	184	Ba	1200	8.73	1506	Ba	4.91
	2327 UT		CuO	304	3.82			
			NaN_3	2				
		178	Ba	87	0.63	98 + 2500	Ba	0.63
			Li	11	1.59	explosive	Li	1.59
65	11 Nov. 1972	180	Ba	414	3.01	531	Ba	1.55
			CuO	116	1.46			
	0408 LT		NaN_3	1				
	0826 UT							
		225	Ba	1200	8.73	1506	Ba	4.91
			CuO	304	3.82			
			NaN_3	2				
		248	Eu	212	1.40	212 + 2500	Eu	1.40
						explosive		

was obtained by reduction of the oxide with aluminium powder between 1000 °C and 1100 °C in a vacuum oven, followed by distillation of europium. The very reactive metal was then melted to a lump and machined under argon or oil. For the shaped charge a special liner with a 20° cone angle was machined. In Table 1 the weights and mixing ratios for the different containers are listed. In the first experiment, in order to save the precious europium, the part of the metal which is burned was replaced by barium, which has a lower vapour pressure. Though the enthalpy of formation of the two oxides is about the same, europium evaporates more readily; in fact barium was hardly seen in the spectrum of the cloud.

4. Experimental Observations

Ground observations were carried out with black and white and colour photography and with a six-channel photomultiplier. The intensities in six different lines, which were selected by filters, were recorded on magnetic tape. The observed spectra are shown in Fig. 2. The lower diagram shows the spectrum of a europium cloud 50 sec after release for an evening launch; the upper two graphs are the spectra from a morning launch. The blue lines of Eu I in the 4600 Å region are the most prominent for all three spectra. The three strong lines at 4129 Å, 4205 Å and 4436 Å are from Eu II and are clearly visible; however, they are always weaker than the Eu I lines, contrary to barium where after a few minutes the intensity of the neutral cloud is negligible compared with the ionized cloud.

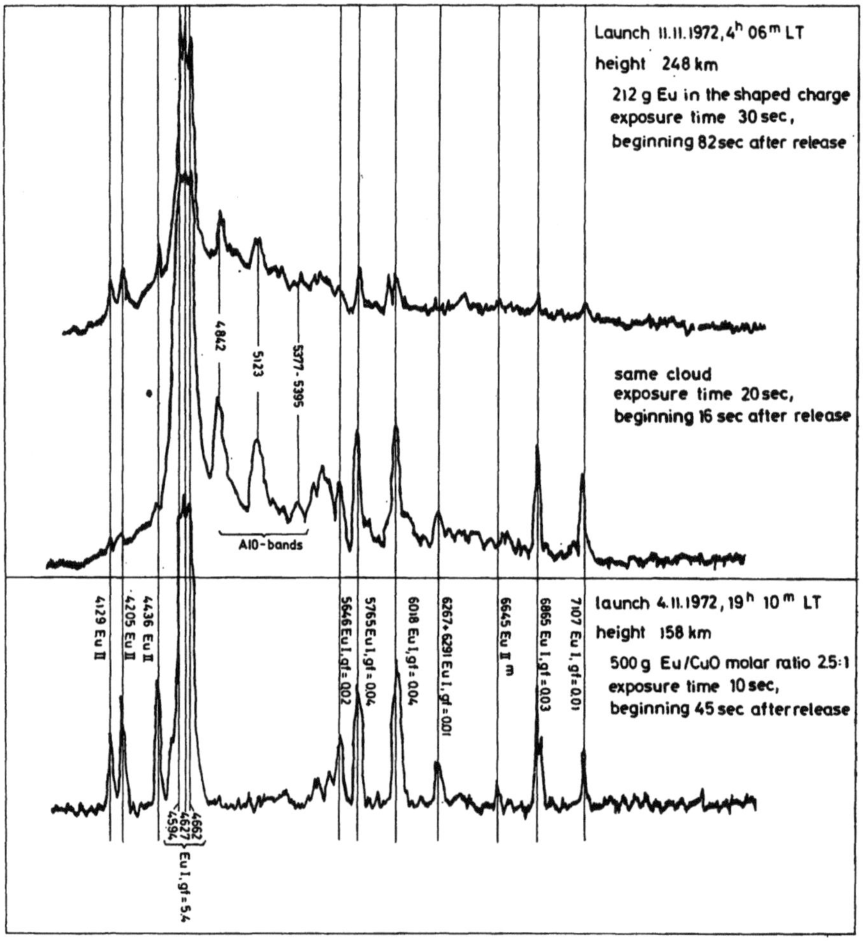

Fig. 2. Photometer curves of Eu evaporations over Chamical, Argentina. Grating spectrograph with image intensifier and smear-film slit 0.5 mm.

Just at the limit of detectability one line can be seen at 6645 Å for the transition from the metastable $^9D_6$ to the z^9P_5 level. All observed lines on the spectra were identified. The three bands, which were seen in a cloud produced by a shaped charge are caused by AlO. The two most striking findings of this experiment are the relatively weak transitions to the metastable levels and the considerably lower rate of ionization than that observed for Ba.

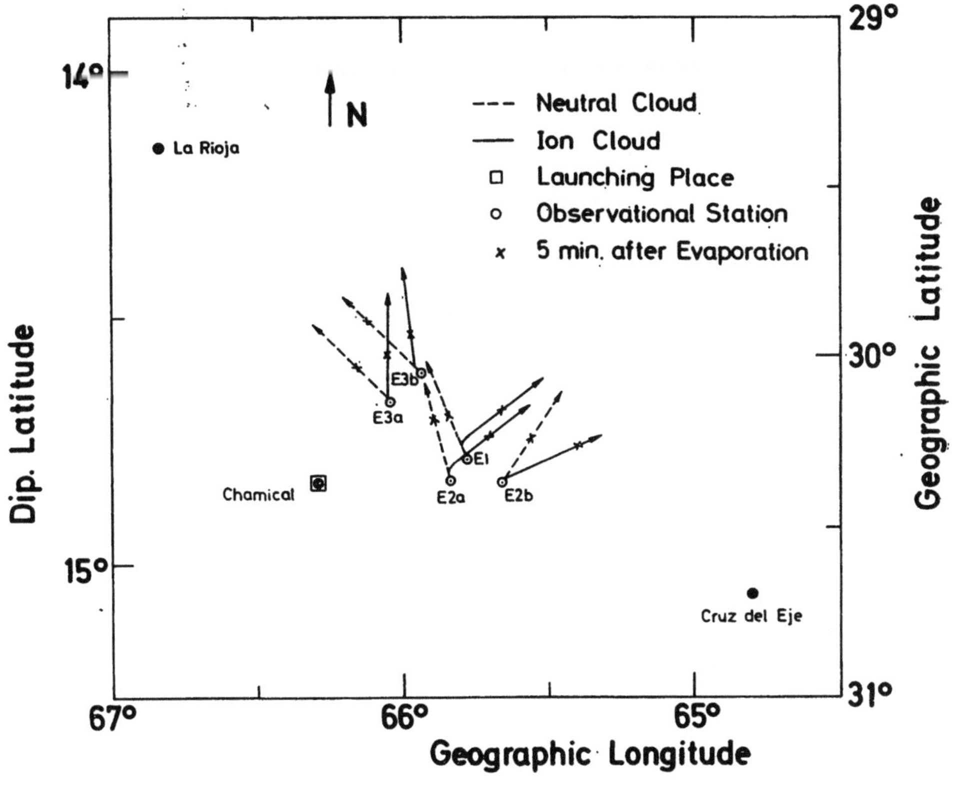

Fig. 3. Geographical coordinates of the observation points and the observed cloud drifts.

In Table 1 the first three columns give the launch times and the altitude at which the clouds appeared. The excess of metal which should theoretically be vaporized when all CuO is reduced is listed in the last column. In the case of barium about 20% of the excess is seen as vapour in the cloud. The large quantity of lithium atoms in two of the charges are due to the low atomic weight of Li, and give rise to very striking red clouds.

The geographical coordinates of the launching site and the observation points are given in Fig. 3 along with the ground projection of the motion of the neutral and the ionized clouds which were generated. The motion of the clouds is discussed in detail in [8]. Figs. 4 and 5 represent photographs taken after the two evening launches on 2 and 4 November 1972. Note in Fig. 4 the very intense lithium cloud

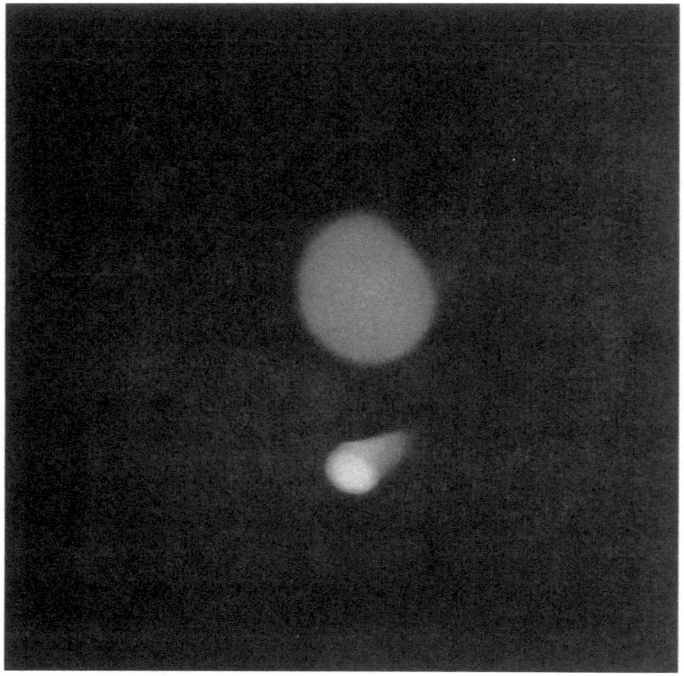

Fig. 4. Evening launch on 2 November 1972, observed from Cruz del Eje, 314 sec after release, exposure time 6 sec.

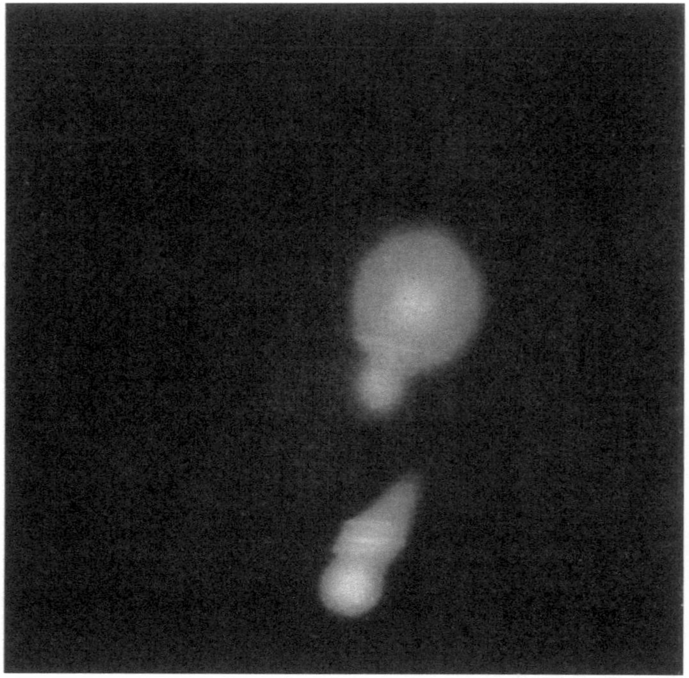

Fig. 5. Evening launch on 4 November 1972 observed from Cruz del Eje, 340 sec after release, exposure time 18 sec.

at 190 km altitude and the neutral and ionized europium clouds at 141 km altitude. The colours of Eu I and II are very similar, unlike those of Ba. The ionized cloud, however, can easily be distinguished by its field-aligned structure. In addition to these two clouds, note in Fig. 5 also a barium cloud at 184 km

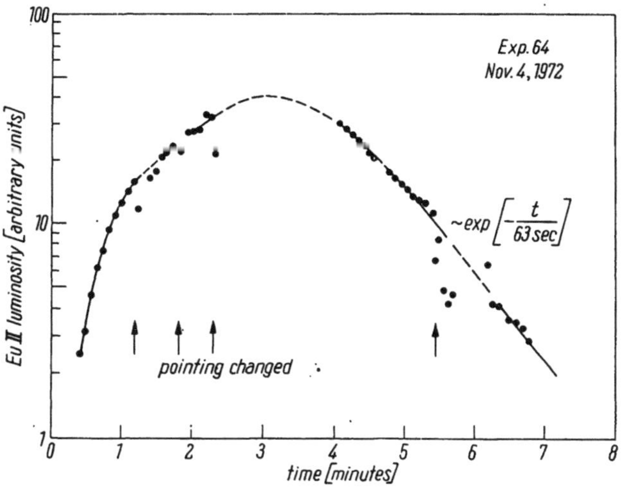

Fig. 6. Luminosity of Eu II versus time after an evening launch on 4 November 1972.

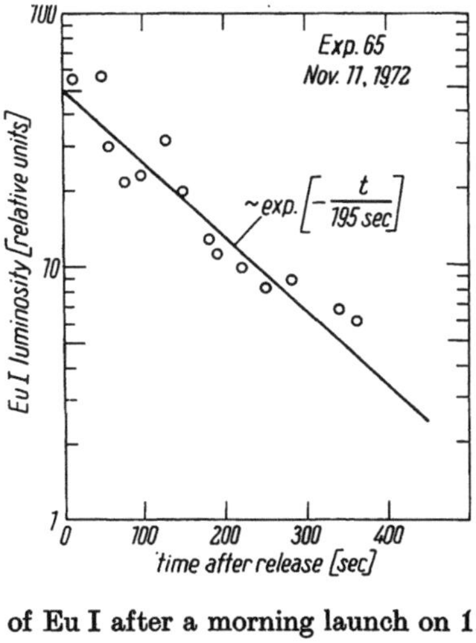

Fig. 7. Luminosity of Eu I after a morning launch on 11 November 1972.

altitude. The lithium cloud seen is due to 0.3 g of lithium, corresponding to an 11 g Li liner in the shaped charge with 3% vaporization, compared with about 190 g of excess europium that was used for the generation of the blue cloud. The two ionized clouds at 158 km and 184 km altitude respectively show a slightly different direction of the electric field.

The variation of the Eu II luminosity versus time is shown in Fig. 6 during the evening event. If the exciting radiation of the sun, especially for ionization, remains constant and the cloud as a whole remains in the field of view of the photometer ($\sim 8° \times 8°$), a monotonic increase of intensity with time is expected. However, these conditions are not valid here, since the decrease of the ionizing UV radiation of the sun makes an evaluation of the data in terms of ionization rather difficult. It should be noted that the cloud reached an altitude which was 30 km lower than was expected; thus illumination by the sun was appreciably reduced.

A more reliable time constant for ionization was obtained from the evaluation of the morning launch. In Fig. 7 is plotted the total intensity of the neutral europium cloud with time, as obtained from a black and white film. From this an ionization time of 195 ± 25 sec seems to be realistic.

5. Summary and Conclusion

Experimental observations substantiate the feasibility of europium as another possible element for ion cloud experiments. The fact that the rate of ionization is only about one tenth of that of barium is not necessarily a disadvantage. Whenever large volumes in space are to be filled with visible ions, particularly at high altitudes, europium is a good choice. Also, Eu II is intensely illuminated by the sun, and thus its resonance lines are strongly visible.

Acknowledgments

The authors are grateful for the very effective support of the National Commission on Space Research of Argentina. For their help in delivering the three Rigel rockets and for launching and tracking the experiments we are greatly indebted. A. Fritsch prepared with great skill the Eu-liner for the shaped charge.

References

[1] G. Haerendel, Solar-Terrestrial Physics **4**, 87 (1970).
[2] L. Biermann, R. Lüst, Rh. Lüst and H. H. Schmidt, Z. f. Astrophys. **53**, 226 (1961).
[3] E. A. Müller, Origin and Distribution of the Elements, Pergamon Press, Oxford 1968 (p. 155).
[4] H. N. Russell and A. S. King, Astrophys. J. **90**, 155 (1939).
[5] S. W. Drapatz, Planet. Space Sci. **20**, 663 (1972) and S. W. Drapatz, L. Haser and K. W. Michel, Z. Naturf. **29a**, March 1974.
[6] F. Melzner and H. H. Rabben, Plasma Waves in Space and in the Laboratory **2**, Edinburgh University Press, Edinburgh 1970 (p. 77).
[7] H. Föppl, G. Haerendel, L. Haser, J. Loidl, P. Lütjens, R. Lüst, F. Melzner, B. Meyer, H. Neuss and E. Rieger, Planet. Space Sci. **15**, 357 (1967).
[8] A. Valenzuela, H. Kappler, G. Haerendel, H. Föppl, H. Hippman and R. F. Woodman, Proc. 1st L. V. Berkner Symp. on Irregularities in the Equatorial Ionosphere, Dallas, Texas, May 1973.

Space Research XIV — Akademie-Verlag, Berlin 1974

NEUTRAL WIND VELOCITIES CALCULATED FROM TEMPERATURE MEASUREMENTS DURING A MAGNETIC STORM AND THE OBSERVED IONOSPHERIC EFFECTS

O. P. Saxena and K. K. Mahajan

Radio Science Division, National Physical Laboratory, New Delhi, India

Neutral temperature measurements made from the OGO 6 satellite, reported by Blamont and Luton, are used to calculate the neutral wind velocities during the magnetic storm of 27 September 1969. The neutral densities required for these calculations are taken from the Jacchia 1971 model and the electron density profiles from the *f*oF2 data, with appropriate modelling above and below the F2 peak. These velocities when compared with those calculated for magnetically quiet conditions indicate little change during the magnetic storm at low latitudes, but a significant change at mid- and high latitudes. The observed ionospheric effects during the storm, which show a large decrease in the F2 region electron density at mid- and high latitudes and practically no change at the low latitudes, are discussed in relation to the calculated wind velocities.

1. Introduction

The atmospheric pressure gradients produced by solar heating are large enough to cause horizontal neutral winds [1]. King and Kohl [1] were the first to estimate the neutral wind velocity by using the average values of the density and temperature gradients reported by Jacchia [2] and Anderson [3] and were able to conclude that these winds can have significant effects on the ionospheric behaviour. Since then, based upon the model atmospheres [4, 5] and in particular the Jacchia 1965 model [5], several authors have calculated the wind systems including their diurnal, latitudinal and seasonal variations (e.g. [6], and see [7] for the latest review). As the model atmospheres are based upon satellite drag measurements, these models essentially give the density gradients and not the pressure gradients. Thus the wind systems derived from these models could be grossly in error, if the pressure gradients used in the calculations are different from those really existing in the neutral atmosphere.

Unfortunately, at the time the various wind systems were calculated, no direct measurements of latitudinal variations of neutral temperature were available. It is only recently that latitudinal profiles of neutral temperature have been measured on the OGO 6 satellite. Blamont and Luton [8] have reported measurements between altitudes of 200 and 320 km at local times from 1200 to 1700 for the period 24 September to 4 October 1969, which also includes the severe magnetic storms of 28 and 30 September 1969. We have used temperatures from these measurements, the neutral densities from the Jacchia 1971 model [9] and the method of Geisler [6]. In this paper we shall present results on the wind velocities calculated during the quiet and the magnetic storm period. We shall also discuss the results of these calculations with reference to the observed magnetic storm effects on the ionosphere.

2. Calculations of Wind Velocity

In our calculations we have considered only the steady state equation of neutral motion. We have omitted unimportant terms like the Coriolis force, centrifugal force and the non-linear acceleration terms [7]. For simplicity, we have ignored the effects of electric fields and assumed the ion motion to be along the field line alone, with the velocities of the ions and the neutral particles being equal. Thus in the equation of motion we are left with the following major terms: (i) neutral pressure gradient, (ii) ion drag and (iii) viscous forces. The simplified equation of neutral motion is

$$\frac{d^2U_y}{dh^2} = \left(\frac{\nu_{\text{in}}}{N_{\text{n}}} N_{\text{i}} U_y \sin^2 I - F_y\right)\Big/\mu \qquad (1)$$

where U_y = meridional wind velocity, h = altitude, N_{i} = ion concentration, N_{n} = neutral concentration, $\mu = \eta/\varrho$ = kinematic viscosity coefficient, η = viscosity coefficient, ϱ = neutral mass density, I = magnetic dip angle, ν_{in} = ion neutral collision frequency, $F_y = -(1/\varrho)\,(\partial p/\partial y)$ is the meridional pressure gradient force, reckoned positive from south to north, and p = atmospheric pressure.

Eq. (1) has been solved numerically by the Runge-Kutta method for heights between 110 and 500 km. The wind velocity is assumed to be zero at 110 km and that sets the lower boundary condition. The altitude gradient of the velocity dU_y/dh is taken to be zero at 500 km and this provides the upper boundary condition.

The assumptions are reasonably correct, because at lower heights the pressure gradients are expected to be zero or negligible; Lindzen [10] has found that, even if non-zero values for velocity are assumed at the lower boundary, the calculated values of velocity above about 150 km remain virtually unaffected.

As μ increases exponentially with increasing height, d^2U_y/dh^2 tends to zero. Thus dU_y/dh = constant for $h \to \infty$. This constant must also be zero, as otherwise one needs a shearing force to maintain a gradient and this cannot be provided by any of the other forces like pressure gradient, Coriolis force and ion drag (see also [7]).

3. Input Parameters

We have calculated the latitudinal profiles of the meridional wind velocity for the period 24 September to 4 October 1969. The meridional pressure gradients have been calculated by using the method adopted by Geisler [6]. These calculations, as already mentioned, are based upon the OGO 6 neutral temperature measurements of Blamont and Luton [8] and the neutral densities of the Jacchia 1971 model atmosphere [9] appropriate to the measured temperatures. Blamont et al. [11], however, in their recent work find the measured temperatures to be not consistent with the densities from the Jacchia 1971 model atmosphere. Since no atmospheric model based upon simultaneous measurements of neutral density and temperature is yet available; the procedure adopted by us seems to be probably the best choice.

The ion (electron) density profiles required are obtained by using the ionosonde observed values from 1000 to 1400 LT of the peak electron density by appropriate modelling. These models are based upon an analysis of a large number of incoherent-scatter measurements of electron density profiles at several stations and in particular at Arecibo, Puerto Rico (18.4° N 67° W).

The F2 peak was assumed to be around 300 km and below the F2 peak electron density was assumed to decrease exponentially with a scale height of 70 km. Above the peak the distribution was assumed to be of the α-Chapman type and of the form

$$N_i = N_m \exp \frac{1}{2}\left[1 - \frac{h - h_m}{H} - \exp\left(-\frac{h - h_m}{H}\right)\right] \tag{2}$$

where N_m is the peak electron concentration and H is the neutral scale height. At heights about 4 scale heights above the F2 peak, the electron density was assumed to decrease exponentially, with the plasma scale height H_p given as

$$H_p = \frac{k(T_e + T_i)}{m_i g} \approx \frac{3kT_n}{m_i g} \tag{3}$$

where k = Boltzmann's constant, g = acceleration due to gravity, and m_i = mass of the atomic oxygen ion which is the main ionic constituent in the topside ionosphere. Eq. (3) assumes the ion and the neutral temperatures T_i and T_n to be equal and the electron temperature T_e to be twice the ion temperature. The ion drag coefficient is taken from Stubbe [12] and is of the form

$$\frac{\nu_{in}}{N_n} = 9.3 \times 10^{-10}(T_n/1000)^{0.37}\ (\mathrm{cm^3\,s^{-1}}). \tag{4}$$

Eq. (4) assumes collisions of O^+ mainly with oxygen atoms, which is reasonable for heights around the F2 peak.

The kinematic viscosity coefficient (μ) profile up to 400 km is taken from Matuura and Nagata [13]. These values are extrapolated to greater heights by assuming that μ increases exponentially above 400 km with a scale height of 70 km.

4. Results and Discussion

Fig. 1a shows the latitudinal behaviour of neutral temperature as obtained from the measurements on OGO 6 reported by Blamont and Luton [8]. These temperatures are plotted for a quiet day (25 September 1969) and for a storm day (30 September 1969). It can be noted from the figure that the neutral temperature and thus the pressure is generally maximum at the poles. On the quiet day there is some evidence of another temperature maximum around the sub-solar point which is near the geographic equator. During the magnetic storm the temperature becomes higher at the higher latitudes (beyond 40°) in both hemispheres, especially in the auroral and polar zones. At latitudes below 40° there is very little change in the temperatures. It should be remarked that, since the pressure bulges are at the poles during the magnetic storm, the neutral winds will be directed towards the equator. On the quiet day, the winds will be polewards below about 50° latitude and equatorwards at the higher latitudes.

The latitudinal behaviour of the noontime foF2 values on the quiet and the storm days is shown in Fig. 1b. The noon-time values are the averages of hourly

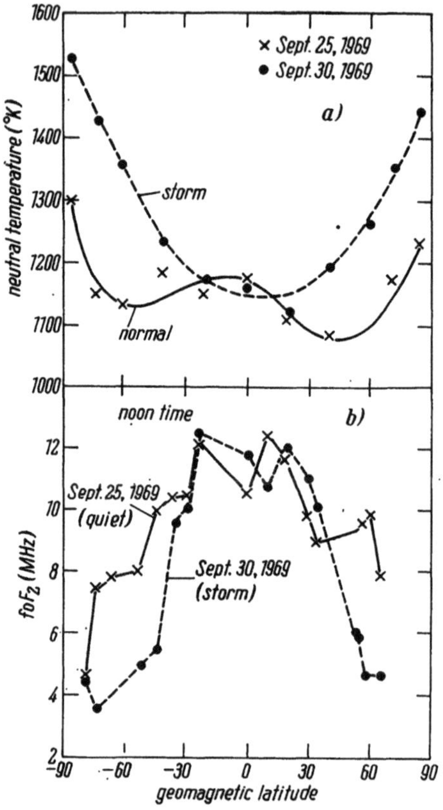

Fig. 1. Neutral exospheric temperatures vs. geomagnetic latitude on a quiet (solid line) and a storm (dashed line) day (after Blamont and Luton [8]). The latitudinal behaviour of noontime foF2 on the quiet and the storm days is also shown.

values between 1000 and 1400 local time. Most of the data used are from the stations in the Asian zone. One can easily note the significant decrease in the peak electron density during a magnetic storm at high latitudes (above 40°) in both the hemispheres, whereas below 40° there is no change.

The latitudinal behaviour of the vertical ion drift velocity on the quiet and the storm days is shown in Fig. 2. Instead of the horizontal velocity, we have plotted the vertical component of the ion drift velocity, because it is then easy to understand the effects of these winds on the ionosphere. The vertical component is calculated from the equation

$$W_i = -U_y \sin I \cos I . \qquad (5)$$

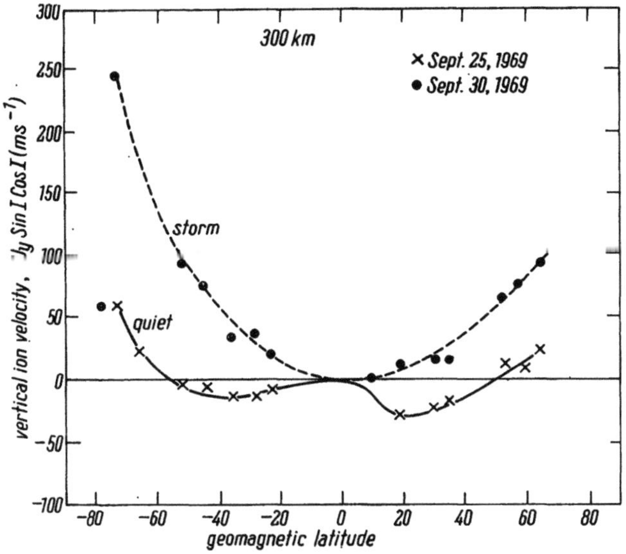

Fig. 2. Vertical ion drift velocity vs. geomagnetic latitude on the quiet (solid line) and the storm (dashed line) days.

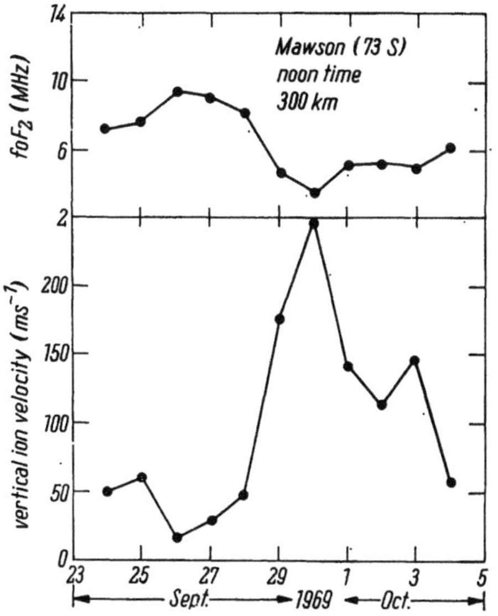

Fig. 3. Behaviour of the vertical ion drift due to neutral winds for a high latitude station Mawson (geom. lat. 73° S, geom. long. 103° E, magnetic dip 69° S) during the storm period.

In this figure the upward ion drift velocity is taken as positive. It can be noted that the vertical ion drift during the magnetic storm is higher at high latitudes (reaching up to about 250 m s^{-1} at 70° S). It is zero at the equator (because $I = 0$) and less than 90 m s^{-1} at low latitudes on a quiet day as well as on a storm day. Comparing with Fig. 1a, one can notice that the vertical ion drift follows very closely the variation of the temperature gradient.

Fig. 3 shows the calculated vertical ion drift for a high latitude ionospheric station like Mawson (geom. lat. 73° S, geom. long. 103° E, magnetic dip 69° S) for the complete period of temperature measurements by Blamont and Luton [8]. The velocities show a significant increase during the storm period. The *f*oF2 values measured by ionosonde are also plotted. The high velocities correspond to days of low electron concentration.

The above results indicate that the wind velocity during magnetic storms is directed towards the equator. Thus the wind would move the ionization up along the magnetic field lines to regions of low electron loss. This should, in fact, result in an increase of electron concentration at the F2 peak levels during magnetic storms. The observed effects, on the other hand, indicate a strong decrease in the F region electron densities and thus it appears that the neutral winds cannot explain the ionospheric effects caused during the magnetic storms. However it is likely that the neutral composition changes significantly with the result that there is a general decrease of the ratio of the atomic to molecular constituents (i.e. O/O_2 and O/N_2) during magnetic storms at high and mid-latitudes.

Acknowledgment

One of the authors (O. P. S.) is grateful to the CSIR for the grant of a Junior Research Fellowship during the course of this work. The authors also thank Prof. J. E. Blamont for supplying the data in a form readily usable for these calculations.

References

[1] J. W. King and H. Kohl, Nature **206**, 699 (1965).
[2] L. G. Jacchia, Space Research V, 1152 (1965).
[3] A. D. Anderson, Nature **209**, 656 (1966).
[4] CIRA, COSPAR International Reference Atmosphere 1965, North-Holland Publ. Co., Amsterdam 1965.
[5] L. G. Jacchia, Smithson. Contrib. Astrophys. 8, 215 (1965).
[6] J. E. Geisler, J. Atmos. Terr. Phys. **28**, 703 (1966).
[7] H. Rishbeth, J. Atmos. Terr. Phys. **34**, 1 (1972).
[8] J. E. Blamont and J. M. Luton, J. Geophys. Res. **77**, 3534 (1972).
[9] L. G. Jacchia, Smithson. Astrophys. Obs. Spec. Rep. No. 332 (1971).
[10] R. S. Lindzen, J. Geophys. Res. **72**, 1591 (1967).
[11] J. E. Blamont, J. M. Luton and J. S. Nisbet, paper presented at XVIth COSPAR Plenary Meeting, Constance 1973.
[12] P. Stubbe, J. Atmos. Terr. Phys. **30**, 1965 (1968).
[13] N. Matuura and T. Nagata, Rep. Ionosph. Space Res. **16**, 185 (1962).

Space Research XIV — Akademie-Verlag, Berlin 1974

A ROCKETBORNE RIOMETER FOR THE STUDY OF THE LOWER IONOSPHERE

Y. V. Somayajulu and M. B. Avadhanulu

Space Research Section, Radio Science Division, National Physical Laboratory, New Delhi, India

A rocketborne riometer was flown on a Nike-Apache rocket (ISRO 30.01) on 2 January 1970 at 1212 LT to derive the electron density profile by measuring the changes in cosmic noise absorption in the lower ionosphere. Two additional experiments were included in the same payload to measure electron density: a propagation experiment and a Langmuir probe.

In this paper the flight performance of the riometer is described. It is found that the changes in cosmic-noise absorption with altitude are consistent with the absorption changes measured by the propagation experiment in the 60—80 km altitude region. Above 85 km altitude, the large electron densities cause impedance changes of the riometer antenna. The antenna detuning is utilized to estimate the electron densities in this region. The measurement accuracy is expected to be better when the D region absorption is high, as during solar flare or PCA event.

1. Introduction

It was proposed by Somayajulu and Mitra [1] that a rocketborne riometer could be used to measure the ionospheric absorption profile during its flight through the D region to derive the electron densities in this region. The technical feasibility of a rocketborne riometer was tested on a Nike-Apache rocket, 14.256 II, launched from Wallops Is. on 17 March 1967 [2]. Based on the results of this flight another payload was designed and successfully flown aboard a Nike-Apache rocket (ISRO 30.01) from Thumba, India, on the geomagnetic equator on 2 January 1970 at 1212 LT ($\chi = 30°$). It reached a peak altitude of about 125 km. This payload also carried two back-up experiments to measure electron densities: one using a standard absorption experiment with a ground-based transmitter and the other using a Langmuir probe. The details of the payload and of the experiments are described in [3] and the results in [4]. This paper presents the description of, and the results obtained from, the rocketborne riometer experiment.

2. Principle of the Technique

The riometer uses the galactic cosmic noise as its signal; during its passage through the ionosphere this suffers absorption. The rocketborne riometer on its passage up through the D region receives increasing cosmic-noise power with height owing to the reduction in the total D region absorption suffered by the cosmic noise. The total absorption undergone by a radio wave may be expressed

$$A(\text{dB}) = \int_{\text{ground}}^{h} NK_0\,(\omega, \omega_{\text{H}}, \nu)\, dh \tag{1}$$

where N is the electron density which is height-dependent, K_0 is the specific absorption, i.e. absorption in dB per unit path length, and is a function of the exploring wave frequency ω, ω_{H} is the gyrofrequency and ν is the collision frequency which is also a function of height. Eq. (1) may then be used to derive the electron density profile from a knowledge of ω, ω_{H} and the altitude profile of collision frequency.

3. Description of the Rocketborne Riometer

The riometer employs the standard principle. The operating frequency is chosen to be 15.115 MHz based on the following considerations: (i) the availability of a clear channel; (ii) the operating frequency should be greater than the F2-layer critical frequency foF2 so that deviative absorption is small; (iii) the operating frequency should be low enough for non-deviative absorption in the D region to be large enough to give adequate measurement accuracy and height resolution. The 3 dB (power) bandwidth of the riometer is chosen to be 6 kHz as a compromise between the large bandwidth required for receiving as large a cosmic noise signal as possible and the susceptibility to interference when using large bandwidths. The integration time constant of the output is half a second.

To provide an adequate signal, a 20 ft dipole antenna was used. This antenna was deployed in flight at an altitude of about 50 km. The Nike-Apache rocket was initially spun at about 9 rev/sec. After deployment of the antenna, the spin rate of the rocket dropped to about 3 rev/sec which is just above the critical level for rocket stability.

4. Data Analysis

Using Eq. (1) the absorption ΔA in a given height range may be expressed as

$$\frac{\Delta A}{\Delta h} = K_0(\omega, \omega_{\text{H}}, \nu)N. \tag{2}$$

Eq. (2) requires the evaluation of K_0 which is given by

$$K_0 = \frac{5\pi e^2}{mc\nu}\, C_{5/2}\left(\frac{\omega}{\nu}\right) \tag{3}$$

for the ordinary mode under transverse propagation conditions. In Eq. (3) e and m are electron charge and mass and c is the velocity of light. Eq. (3) is valid for the non-deviative case ($\mu \approx 1$). The equation for the extraordinary mode is more complicated and has been given by Sen and Wyller [5]. These equations require the evaluation of the Dingle integrals $C_p(x)$; the Burke-Hara polynomial expansion [6] is used. The collision frequency model is derived assuming the atmospheric model based on CIRA 1965 and using the relationship given by Phelps [7],

$$\nu = 9 \times 10^7\, p\ (\text{s}^{-1}) \tag{4}$$

where p is the total atmospheric pressure in mmHg.

The height resolution of electron density values that could be obtained is dependent on the absorption change in the altitude range, which should be greater than the minimum detectable capability of the instrument, 0.2 dB. Therefore altitude slabs are chosen so that the absorption change in a slab exceeds 0.2 dB. Typically the slab thickness is about 5 km. Since the variation of collision frequency in such a slab is considerable, this is taken into account when determining the electron density in each slab.

Since the differential absorption at 15 MHz is very small, treating the received signal entirely in terms of ordinary wave absorption leads to a maximum error of 6%. The error due to this reduction procedure is expected to be less than 5%. Therefore, for the analysis the observed absorption profile is taken as representative of the ordinary wave absorption.

5. Results

5.1. Electron Densities from Riometer Absorption Measurements

The telemetry record of the riometer output shows two modulations, with different time periods: one has a large period and superimposed on this is the other having a period corresponding to the spin of the rocket. It is observed that the period as well as the amplitude of the large period modulation gradually decreases

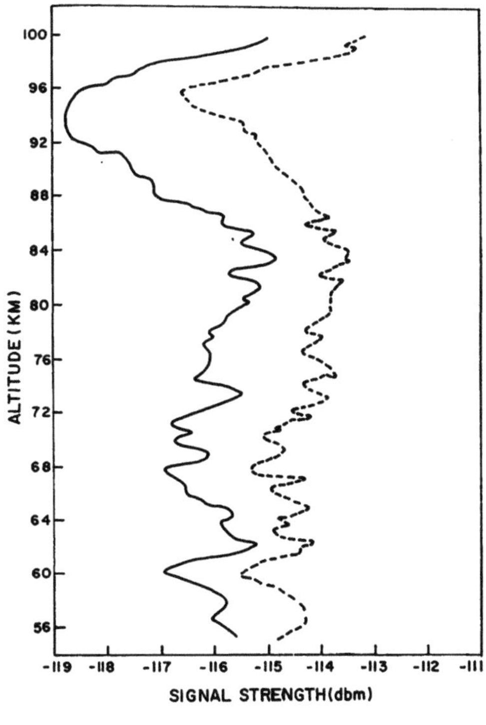

Fig. 1. Signal strength observed by the rocketborne riometer on ISRO 30.01 vs altitude. --- peak; —— valley.

with increasing time of flight. It is possible that this modulation is due to the precession of the rocket. The origin of the second modulation, i.e. one corresponding to the spin period of the rocket, is not clear. However, it is interesting to note that Stone et al. [8], while measuring the antenna reactance at 4.44 MHz from a sounding rocket, also observed a variation in the output at twice the spin period. These aspect-dependent variations are attributed to several possible phenomena. One source of variation is the inhomogeneous plasma distribution about the vehicle. The modulation at the spin frequency of the rocket in the present case

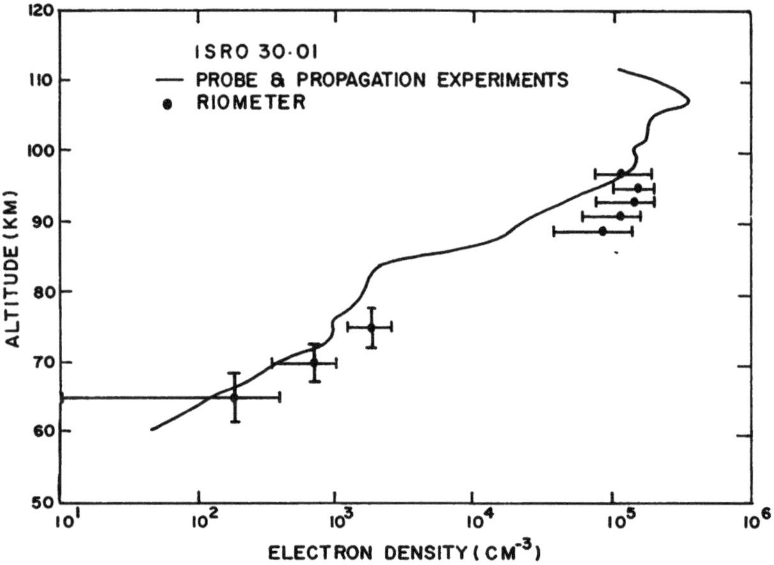

Fig. 2. Electron densities derived from the riometer experiment (60—80 km from riometer absorption, above 85 km from riometer antenna detuning) and electron density profile obtained from probe and propagation experiments.

may be due to such inhomogeneities; these modulations are smoothed out to derive the cosmic-noise signal power. The signal intensity of the riometer plotted as a function of altitude after such smoothing is shown in Fig. 1. The two profiles correspond to the peaks and valleys of the large-period modulation. It can be seen from Fig. 1 that there is a general increase in signal intensity from 60 to 80 km. The change in signal strength observed during the flight gives the altitude variation of cumulative absorption. In Fig. 2 are shown the electron density values obtained from the absorption changes, based on the procedure described in the previous section, and the electron density profile obtained from the probe and propagation experiment included in the same payload.

5.2. Electron Density Determination from Antenna Detuning

Above 85 km the electron density increases rapidly and merges with the E region. In this region significant detuning of the antenna occurs [9]. Referring to Fig. 1 we see that the riometer output decreases smoothly and rapidly until it reaches a deep minimum at an altitude of about 95 km. Thereafter, the riometer signal recovers to its original value.

From the power attenuation observed, the change in the reactive part of the antenna impedance is calculated [10]. Electron density is then derived from the change in the capacitance [9]:

$$\frac{C_a'}{C_a} = K = 1 - \frac{80.6N}{f^2} \tag{5}$$

where f is the exploring wave frequency in kHz, C_a' is the antenna capacitance in the ionosphere and C_a is the antenna capacitance on the ground. The electron density values thus derived in the altitude region above 85 km using the antenna detuning are shown in Fig. 2.

6. Discussion

It may be seen from Fig. 2 that the electron density values obtained from the riometer absorption experiment agree well with those derived from the propagation experiment. The values are somewhat on the higher side, presumably due to the smoothing used. The agreement might be improved by a better knowledge of the origin of the modulation.

The electron densities derived from the antenna detuning are again in good agreement with the values derived from the propagation and probe experiments as far as the general altitude variation is concerned. There is, however, a systematic difference in the absolute values, the values derived from antenna detuning being higher than those obtained from the other experiments by a factor of 2—5. As pointed out by Kane et al. [11], this systematic difference is probably due to the capacitance of the sheath around the antenna.

7. Conclusion

A new technique, using a rocketborne riometer proposed for determining the electron densities in the D region, has been tried out successfully. This technique has the advantage that it requires only a receiver, and a separate transmitter at the ground and the ground support required for a conventional propagation experiment are not necessary. However, the accuracy is rather limited because of the small amount of the total D region absorption during normal conditions. On the other hand, this technique would yield better measurements when there is large ionospheric absorption as, for example, during a period of high solar activity, a solar flare or a PCA event.

References

[1] Y. V. Somayajulu and A. P. Mitra, RSD Sci. Rep. No. 29, National Physical Laboratory, Delhi, India (1966).

[2] Y. V. Somayajulu, Report submitted to NASA and INCOSPAR (1968).

[3] Y. V. Somayajulu, M. B. Avadhanulu, D. S. Kushwah, S. C. Garg and K. S. Zalpuri, RSD Sci. Rep., National Physical Laboratory, Delhi, India, to be published, 1974.

[4] Y. V. SOMAYAJULU, M. B. AVADHANULU, K. S. ZALPURI and S. C. GARG, Space Research XI, 1131 (1971).
[5] H. K. SEN and A. A. WYLLER, J. Geophys. Res. **65**, 3931 (1960).
[6] E. H. HARA, J. Geophys. Res. **68**, 4388 (1963).
[7] A. V. PHELPS, J. Appl. Phys. **31**, 1723 (1960).
[8] R. G. STONE, J. FAINBERG and J. K. ALEXANDER, Planet. Space Sci. **17**, 1437 (1969).
[9] J. E. JACKSON and J. A. KANE, NRL Rep. 5345 (1959).
[10] S. S. SANDLER, Planet. Space Sci. **11**, 817 (1963).
[11] J. A. KANE, J. E. JACKSON and H. A. WHALE, J. Res. Nat. Bur. Std. **66D**, 641 (1962)

Space Research XIV — Akademie-Verlag, Berlin 1974

JOULE HEATING DUE TO THE EQUATORIAL ELECTROJET AS OBSERVED BY ROCKETBORNE PROBES

S. SAMPATH[a], T. S. G. SASTRY[a], K. OYAMA[b] and K. HIRAO[b]

[a] Physical Research Laboratory, Ahmedabad, India
[b] Institute of Space and Aeronautical Science, University of Tokyo, Tokyo, Japan

A Nike-Apache rocket carrying a proton precession magnetometer and an electron temperature probe was launched on a magnetically quiet day, near noon, from Thumba, India. The purpose of the experiment was to study Joule heating in the E region of the ionosphere, due to the equatorial electrojet current. The current density (J) profile derived from the magnetometer data, the electron temperature (T_e) profile obtained from the electron temperature probe and the neutral temperature (T_g) profile from the CIRA model are used to study the Joule heating effect. The excess of electron temperature over that of the neutrals $T_e - T_g$, which is a consequence of Joule heating, correlates well with the current density J. To relate J and $T_e - T_g$ quantitatively the Cowling conductivity (σ_3) profile and the calculated profile of mean fractional energy loss per collision (λ) of electrons with the neutrals are used. The energy needed to give the observed increase in temperature, $T_e - T_g$, has also been calculated theoretically and compared with the energy delivered by the electrojet.

1. Introduction

The electron temperature distribution under equilibrium conditions is determined by the rate of heat input to the ambient electron gas and the rate of energy loss by various loss processes. A general heating is produced by the absorption of solar radiation. Much of the absorbed energy appears as kinetic energy of photoelectrons. The energetic photoelectrons lose their energy to the ambient electrons, ions and neutrals. It is assumed that the ambient electrons suffer frequent elastic collisions among themselves and attain a Maxwellian distribution, characterized by the electron temperature T_e. The ions and neutrals, except in the upper F region, attain equilibrium amongst themselves and are considered to have a common temperature T_g. The electron gas cools by collisions with neutrals and ions. The heat loss to positive ions is important only above 200 km, while below this height the heat loss to neutrals is more important.

Theoretical electron temperature profiles based on the above mentioned assumptions have been derived by various workers [1—4]. It is found that the electron temperature increases with height throughout the ionosphere. T_e is equal to T_g below 150 km, above which T_e exceeds T_g. The difference $T_e - T_g$ increases above 150 km and reaches a maximum near 250 km. According to these findings there is no thermal non-equilibrium below 150 km. These conclusions are based upon the assumption that radiation from the sun is the only source of heat energy and that local heating effects such as Joule heating are negligible. However, for calculating the temperature distribution in the equatorial and auroral ionospheres, Joule heating effects have to be taken into account.

Joule heating in the equatorial E region due to the electrojet has been studied theoretically by Kato [5]. Joule heating affects the thermal equilibrium, by raising T_e above T_g in a way similar to the heating produced by the absorption of solar radiation. In the present discussion, the energy produced by Joule heating is related to the observed increase in T_e over T_g under the assumption that the energy loss of electrons by collisions with neutrals is the only loss process.

2. Flight Data

A Nike-Apache rocket carrying a proton precession magnetometer, an electron temperature probe, a capacitance probe and a gyro-plasma probe, was launched on a magnetically quiet day, 25 April 1971 ($Kp = 2$) at noon from Thumba, near the magnetic dip equator. The proton magnetometer and the electron temperature probe gave satisfactory measurements during the ascent up to 160 km. The electron temperature probe used in this experiment is the same as that described in [6].

3. Flight Results

The proton magnetometer measured a difference field ΔH of $\pm 110\,\gamma$ as it traversed the electrojet. The vertical distribution of electric current density is derived from the slope of the difference field by the usual method [7]. The electron temperatures measured by the temperature probe and the gas temperature given by the CIRA 1965 model have been used to derive the temperature difference

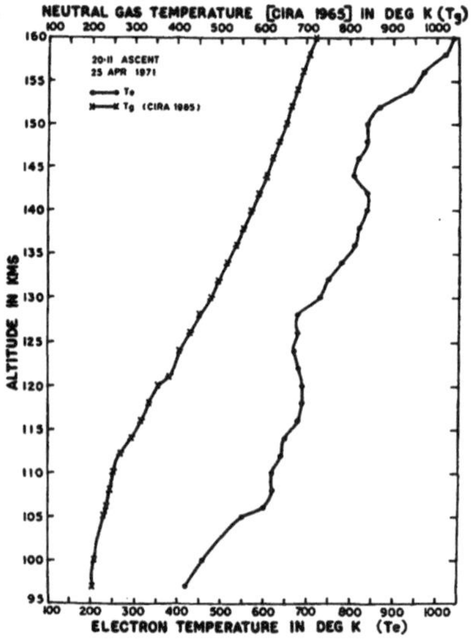

Fig. 1. Vertical profiles of measured electron temperature T_e and the CIRA 1965 model neutral gas temperature T_g.

$T_e - T_g$. Since the electron density measurements from the same flight were not available, vertical profile of electron density obtained by a rocketborne probe launched from the same range under similar magnetic conditions, near local noon, has been used here [8]. The Cowling conductivity (σ_3) profile is calculated, taking

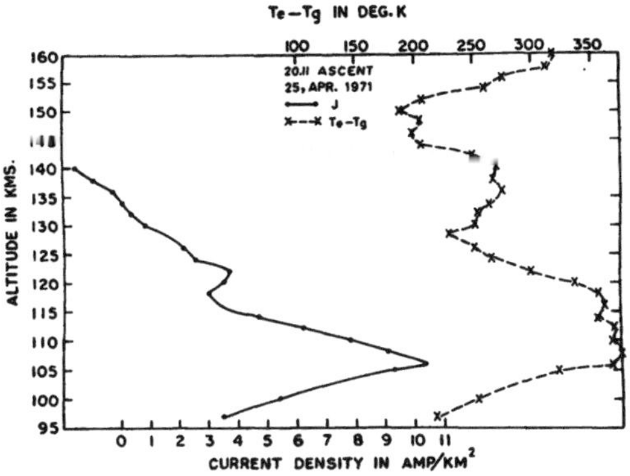

Fig. 2. The current density J derived from the magnetometer data and the difference temperature $T_e - T_g$ plotted as functions of altitude.

the σ_1 and σ_2 profiles of Maeda and Matsumoto [9]. The vertical profile of the electron temperatures obtained from the measurements is shown in Fig. 1, together with the CIRA 1965 model gas temperature profile. The current density J derived from the magnetometer is shown in Fig. 2.

4. Calculations

The power generated by the electrojet is calculated and correlated with the observed difference $T_e - T_g$. It is assumed that this difference $T_e - T_g$ is due to the electrojet currents only. The power Q generated by Joule heating can be written as

$$Q = J^2/\sigma_3. \tag{1}$$

The heat input by the electrojet Q_E is obtained as a function of altitude, and is shown in Fig. 3.

If we consider that the loss of energy due to collisions alone is important, under equilibrium conditions, the balance between the heat input by Joule heating and power dissipation by collision with neutral particles is soon established. Following Kato [5] the energy input can be written

$$Q = \frac{3}{2} k n_e \nu_e \lambda (T_e - T_g) \tag{2}$$

where k is Boltzmann's constant, n_e is the electron density, ν_e is the collision frequency of electrons with neutrals and λ is the mean fractional energy loss per collision. From (1) and (2)

$$\lambda = \frac{2}{3} \frac{J^2}{\sigma_3} \frac{1}{k n_e \nu_e (T_e - T_g)}. \tag{3}$$

By putting in the known values of J, σ_3, $T_e - T_g$, n_e and ν_e, the parameter λ is calculated as function of altitude. This profile is shown in Fig. 3.

In order to obtain an estimate of Joule heating, the heat input needed to give the observed increase in temperature is calculated. Following Dalgarno et al. [2], if Q_T is the energy transferred to the electron gas in unit time, at equilibrium,

$$Q_T = -\frac{3}{2} k n_e \left\{ \left(\frac{dT_e}{dt}\right)_O + \left(\frac{dT_e}{dt}\right)_{N_2} + \left(\frac{dT_e}{dt}\right)_{O_2} + \left(\frac{dT_e}{dt}\right)_+ \right\} \tag{4}$$

where dT_e/dt is the rate of temperature change by the electrons in collision with the neutrals, the neutral species being denoted by subscripts. In the present case the last term, which denotes the loss due to collisions with positive ions, is negected. The expressions adopted for these calculations are

$$\left(\frac{dT_e}{dt}\right)_O = -10^{-14} n(O) T_e^{1/2} (T_e - T_g) \tag{5}$$

$$\left(\frac{dT_e}{dt}\right)_{N_2} = -\left\{7.6 \times 10^{-16} T_e + \left(\frac{G\bar{\nu}}{n(N_2)}\right)_{rot}\right\} n(N_2)\,(T_e - T_g) \tag{6}$$

$$\left(\frac{dT_e}{dt}\right)_{O_2} = -4.7 \times 10^{-16} n(O_2) T_e (T_e - T_g). \tag{7}$$

n is number of particles cm^{-3}, T in °K and dT/dt in deg s^{-1}.

To evaluate Eqs. (5), (6) and (7), the number densities $n(O)$, $n(N_2)$ and $n(O_2)$ are taken from the CIRA 1965 model. The rotational loss parameter $\left(G\bar{\nu}/n(N_2)\right)$ in Eq. (6) is calculated according to Dalgarno and Moffett [10]. The energy Q_T needed to maintain the observed temperature difference is then calculated, and is shown in Fig. 3.

5. Discussion

It is evident from Fig. 1 that the measured electron temperatures (T_e) are more than the CIRA model neutral gas temperatures throughout the E region. In Fig. 2, the $T_e - T_g$ and the current density, J, profiles are shown for comparison.

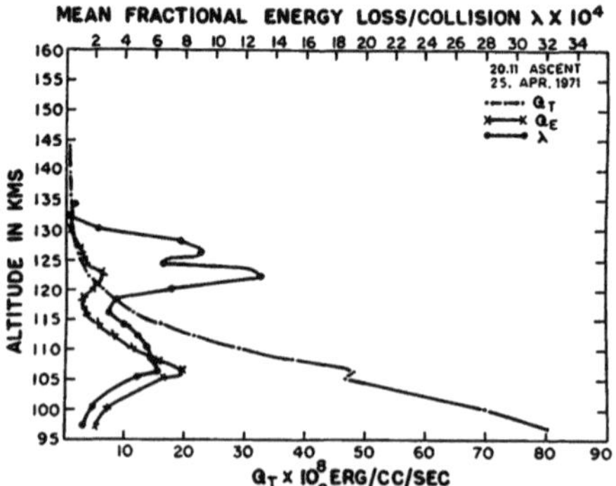

Fig. 3. Experimentally obtained values of Q_E the power generated by the electrojet, theoretical estimates of Q_T the power needed to maintain the observed temperature difference $T_e - T_g$, and the mean fractional energy loss λ per collision of electrons with neutrals plotted as functions of altitude.

It is seen that there is good correspondence between $T_e - T_g$ and J throughout the main electrojet layer around 106 km. The $T_e - T_g$ profile shows a rather broad maximum compared with the current density maximum, implying that the heating is not localized, but extends slightly above the electrojet layer.

The λ values shown in Fig. 3, are slightly higher than those used by Kato [5]. The calculated λ values show that $\lambda > 2m_e/M$, where m_e is the electronic mass (9.1×10^{-28} g) and M is the mass of the heavy particle in collision (5×10^{-23} g). This implies that the collisions are inelastic in the region considered.

The vertical profile of λ shows a large peak between 120 and 130 km. Part of this increase in λ is due to the small increase in current J in this altitude range. This increase in J is due to a second current layer often seen in equatorial J measurements [12]. The broad peak in $T_e - T_g$ around 110 km may partly be due to the effect of this second current layer. Even after removing the effects of this possible second current layer, λ shows a peak around 125 km. The magnitude of this peak depends largely on the rate of fall of electrojet current with altitude above the peak of the electrojet at 107 km. Examining Eq. (3), ν_e falls rapidly with altitude and $[kn_e\nu_e(T_e - T_g)]^{-1}$ has large values beyond 115 km. Since J^2/σ_3 has significant values between 115 and 130 km, λ shows large values in this region. This means that the electrons in this region attain large energies between collisions. Therefore, one can expect the energy transfer per collision of electrons with neutrals to be large in this region.

The Q_T profile shows a slight deviation from its normal trend at 106 km. The Q_E value at 106 km is nearly one third of the Q_T value at the same altitude. This implies that the measured strength of the electrojet current J can account for only one third of the observed difference in temperature $T_e - T_g$. The difference between Q_T and Q_E decreases above 106 km.

The discrepancy between the theory and the experiment can be attributed to the following factors: (i) The experimentally measured electron temperatures could be slightly higher than the actual temperatures in the region. (ii) The CIRA model gas temperatures T_g used in the calculation may not represent the actual as temperatures in this region. The measurements by Knudsen and Sharp [11] at mid-latitudes suggests that $T_i \cong T_g$ and that the measured T_g values agree with the model gas temperatures up to 120 km. However, there are no direct gas temperature measurements over the equator. (iii) The dependence of ν_e on T_e can increase the ν_e values used here by a factor $(T_e/T_g)^{1/2}$ which is of the order 1.6 around 106 km. This increase in ν_e will reduce the λ values by a factor 0.6 near 106 km. The new λ value thus obtained for the region around 106 km agree with the λ value used by Kato [5] for a difference in temperature of 440 °K.

6. Conclusions

1. The pronounced peak of $T_e - T_g$ around 106 km is due to the Joule heating effect of the electrojet. 2. The electron temperatures measured are higher than the CIRA model gas temperatures all over the E region. 3. The observed electrojet strength J can account for only one third of the observed temperature difference $T_e - T_g$. 4. The λ values obtained show that in the region of the electrojet the collisions between electrons and neutrals are inelastic.

To obtain a good understanding of the phenomenon of Joule heating it is necessary to measure T_e, T_g, J and n_e in simultaneous or near simultaneous flights. It is also necessary to confirm the above findings by further experiments over the equator.

References

[1] W. B. HANSON and F. S. JOHNSON, Memoires Soc. R. Liège, Series 5, 300 (1961).
[2] A. DALGARNO, M. B. MCELROY and R. J. MOFFETT, Planet. Space Sci. **11**, 463 (1963).
[3] A. P. WILLMORE, Proc. Roy. Soc. **281**, 140 (1964).
[4] J. E. GEISLER and S. A. BOWHILL, J. Atmos. Terr. Phys. **27**, 457 (1965).
[5] S. KATO, Planet. Space Sci. **11**, 1297 (1963).
[6] K. HIRAO and K. OYAMA, J. Geomag. Geolect. **22**, 303 (1970).
[7] T. S. G. SASTRY, Space Research X, 778 (1970).
[8] S. PRAKASH, S. P. GUPTA, B. H. SUBBARAYA and C. L. JAIN, Nature **233**, 56 (1971).
[9] K. MAEDA and H. MATSUMOTO, Rep. Ionosph. Space Res. Japan **16**, 1 (1962).
[10] A. DALGARNO and R. J. MOFFETT, Planet. Space Sci. **9**, 439 (1962).
[11] W. C. KNUDSEN and G. W. SHARP, J. Geophys. Res. **70**, 143 (1965).
[12] T. S. G. SASTRY, J. Geophys. Res. **73**, 1789 (1968).

Space Research XIV — Akademie-Verlag, Berlin 1974

CORRELATION BETWEEN DAYTIME ELECTRON TEMPERATURE AND DENSITY VARIATIONS AT LOW LATITUDES*

K. Spenner, A. Dumbs, W. Lotze and H. Wolf

Institut für Physikalische Weltraumforschung, Freiburg, FRG

**Dedicated to Professor K. Rawer on the occasion of his 60th birthday*

Electron temperature and density measured on board the satellite AEROS are presented. Good inverse correlation between temperature and density in three altitude regions (220—400, 400—600, 600—800 km) is obtained at low and middle latitudes. By day (1445 LT) the electron temperature has a minimum in the equatorial belt which is about 1000 K lower than north and south of this belt. The electron density shows a maximum in the same belt, the importance of which depends on the altitude. During night-time (0245 LT) the electron temperature shows a nearly constant value of about 850 K independent of latitude and altitude.

1. Introduction

During the last few years several satellite and rocket measurements of electron temperatures and densities have been made [1—3]. The purpose of this paper is to present low altitude electron density and temperature data from the satellite AEROS for given local times as functions of geographic latitude. The data have been collected over a period of 20 days at the beginning of 1973. Since apogee and perigee change slowly, height profiles could be obtained on a statistical basis.

1.1. Description of the Experiment

The aim of the experiment Retarding Potential Analyser (RPA) on board the satellite AEROS is to measure the energy distribution of electrons in the range 0—30 eV, and also the ion temperature and density. The sensor is of planar geometry and has a diameter of 8 cm. It consists of a double retarding grid and several additional grids which cover almost the whole diameter of the sensor. The limiting surface is a collector with an area of 2 cm^2 surrounded by a guard ring. Each grid (30 lines per cm) has a transparency of 83%. All parts of the sensor are gold plated.

1.2. Spacecraft

The spacecraft is the Aeronomy satellite AEROS launched on 16 December 1972, perigee 218 km, apogee 865 km, orbital inclination 96.9°; the motion of the perigee was —3.49° per day.

2. Technique

The electron temperature is derived from the slope of the current-voltage curve and the electron density from the knee of this characteristic. Electron densities achieved in this way may be taken as relative values; for absolute values they need to be corrected. The temperature values have an accuracy of better than 5% with respect to experimental and data processing errors. The absolute accuracy depends on the degree of confidence of the probe theory [4]. A few comparisons of the derived electron temperature with incoherent scatter data from Malvern show coincidence within 100 K in the temperature range between 2200 K and 2600 K.

As the satellite has a sun-synchronous orbit, the data refer to the same local time during day (1445 LT) and night (0245 LT). In order to eliminate seasonal effects, only orbits of a 20-day period have been selected. As the perigee changes 3.5° per day, at constant latitude different altitudes are sampled a few days later. Other influences, such as variable solar fluxes, could not fully be eliminated, but data with extreme solar fluxes were not used for constructing averages of electron densities and temperatures. In order to separate height effects from latitude effects to a certain extent, all data are classified in three height ranges: 220–400 km, 400–600 km and 600–860 km, so that during the day the lowest range always covers the F2 layer [5].

Under the above conditions statistical altitude and latitude profiles can be derived; data from all longitudes are included.

3. Results

All measured daytime electron temperatures are plotted in Fig. 1. Data points of a given orbit and altitude range are connected, and the different ranges are distinguished by open circles, large and small points. The average temperature value is computed for each altitude range and given in the figure. At all height ranges, the temperature decreases strongly in the equatorial region. The change of the electron temperature from orbit to orbit can be large, but each orbit shows the same behaviour. The altitude range near 500 km has the strongest decrease, by more than 1000 K. At middle latitudes for all altitude ranges, the electron temperature is about 350 K higher in the winter hemisphere than in the summer hemisphere. Electron temperatures measured at night are given in Fig. 2. Up to $\pm 40°$ latitude all these temperatures are near 850 K, with no temperature gradient with height.

Depending on the geographic latitude the electron density shows the expected behaviour (Fig. 3). We find the strongest equatorial increase of density in the region 401–600 km where the electron temperature decreases. From orbit to orbit the variation of electron density is smallest for large altitudes. Due to the strong effect of the altitude the electron density profiles still exhibit a modulation which could be eliminated by applying a narrower height grid. Therefore, we feel that only the averaged values may be significant. This becomes very clear in Fig. 4, where the satellite, unfortunately, changed its altitude by an important amount, so that latitude and height effects are mixed.

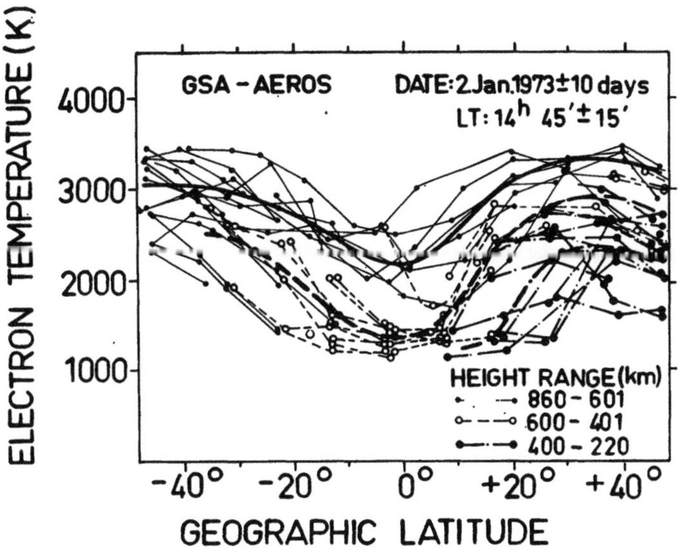

Fig. 1. Electron temperatures versus geographic latitude at 1445 LT for three altitude regions described by circles, small and large points joined by fine lines. The heavy lines give the averaged temperatures in the various regions: see key on figure.

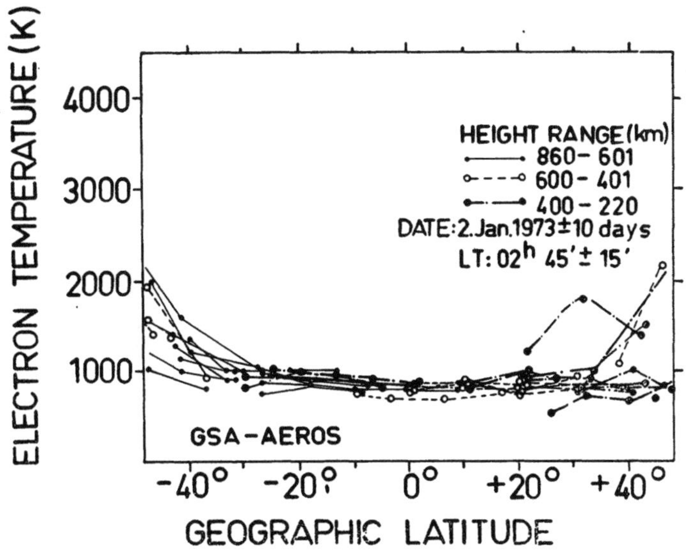

Fig. 2. Electron temperatures versus geographic latitude at 0245 LT; symbols and lines as in Fig. 1.

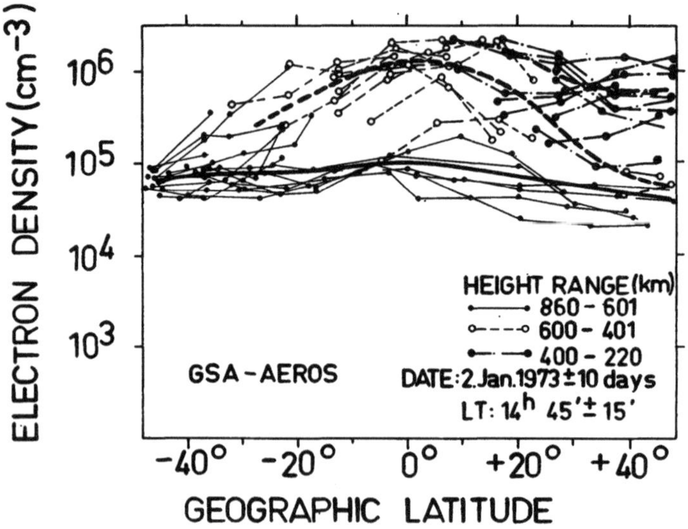

Fig. 3. Electron density versus geographic latitude at 1445 LT; symbols and lines as in Fig. 1.

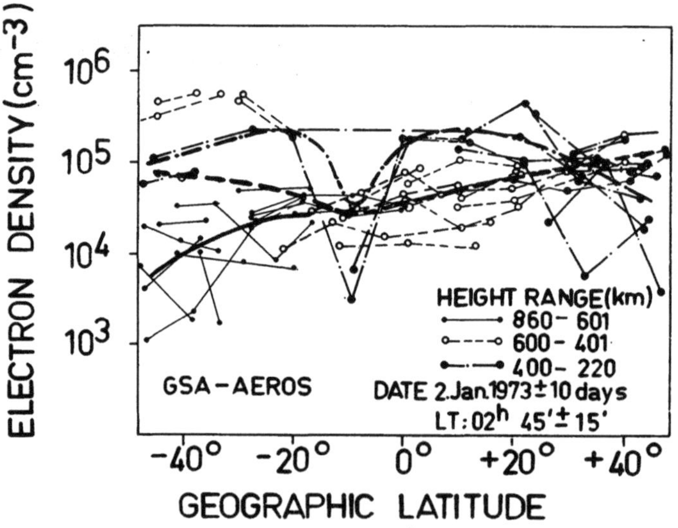

Fig. 4. Electron density versus geographic latitude at 0245 LT; symbols and lines as in Fig. 1.

4. Conclusions

The daytime thermal electron measurements show good inverse correlation between density and temperature in all height ranges within $\pm 40°$ latitude. The most important temperature decrease and density increase is found in the altitude range of 400—600 km. The electron density in the 600—860 km range is lower in the winter hemisphere where the temperature is higher. During night-time the electron temperature is independent of height and latitude.

Acknowledgments

The authors wish to thank all those who have helped to make possible the AEROS project which was financially sponsored by the Ministry of Education and Sciences of the FRG. We are grateful to Professor Rawer for delegation of this work and to Dr. Taylor for preparing incoherent scatter data.

References

[1] L. H. Brace, Space Research X, 633 (1970).
[2] E. S. R. Maier, J. Geophys. Res. **74**, 815 (1969).
[3] W. B. Hanson et al., J. Geophys. Res. **74**, 2229 (1969).
[4] W. Pfister, Space Research XII, 1261 (1972).
[5] W. Becker, Space Res. XII, 1241 (1972).

Space Research XIV — Akademie-Verlag, Berlin 1974

PROTON WHISTLERS IN THE EQUATORIAL F REGION

Ya. I. Likhter[a], Ya. P. Sobolev[a], F. Jiříček[b] and P. Tříska[b]

[a] Institute of Geomagnetism, Ionosphere and Radio Propagation (IZMIRAN), Academy of Sciences, Moscow, USSR

[b] Geophysical Institute of the Czechoslovak Academy of Sciences, Prague, Czechoslovakia

At low heights (200—330 km) in the vicinity of the geomagnetic equator, signals analogous to proton whistlers were recorded on board of the Intercosmos 5 satellite. These proton whistlers are peculiar in that the asymptotic frequency, which is presumed to be the proton gyrofrequency where the proton whistlers are formed, decreases rapidly from 340 Hz to 260 Hz as the satellite moves only 8° to higher latitudes. Thus the proton whistlers are believed to be formed close to the geomagnetic equator. This phenomenon can be explained by an ion-cyclotron wave propagating down the line of force and, at and below the "cross-over" level, propagating as an electron (right-hand polarized) wave.

Analysis of ion whistlers, received on board of earth satellites, makes it possible to determine the ion composition [1] and yields useful data for ELF-wave propagation studies. The minimum height at which ion whistlers have hitherto been received was 442 km for night-time and 640 km for daytime conditions [2]. Hybrid proton whistlers, detected by the Intercosmos 5 satellite over the equatorial area, have propagated to the satellite both from below and from above from the opposite hemisphere after crossing the equator [3].

The Intercosmos 5 satellite carried apparatus for VLF measurements together with a magnetic tape recorder which stored broadband analogue data in the frequency band of 0.05—15 kHz [4]. During revolutions No. 532 (7 January 1972) and 915 (2 February 1972) of Intercosmos 5, proton whistlers were received at very low heights, close to the electron density maximum of the F2 layer (i.e. at 280—330 km for rev. No. 532 and 200 km for rev. No. 915). These signals consisted of fractional-hop whistlers, denoted 0_+ according to [5], followed by whistlers from the opposite hemisphere that crossed the equator (1_- whistlers). Unlike the hybrid proton whistlers described in [3], during revolutions 532 and 915 of Intercosmos 5, the proton trace was observed only together with a 1_- whistler. Fig. 1 shows whistler traces from rev. No. 532, and Table 1 gives the essential data referring to them.

The 0_+ whistlers were usually of lower amplitude and their spectrum mainly contained frequencies below 1 kHz, whereas the spectrum of individual 1_- whistlers showed traces up to 10 kHz. The time delay of the 1_- whistlers with respect to 0_+ whistlers increased with the distance from the geomagnetic equator, this being associated with a longer propagation path. Since the signals propagated at low altitudes where $f \ll f_{0e}$ and $f \ll f_{He}$ (f = wave frequency, f_{0e} = plasma frequency, f_{He} = electron gyrofrequency), the whistler dispersion D could be

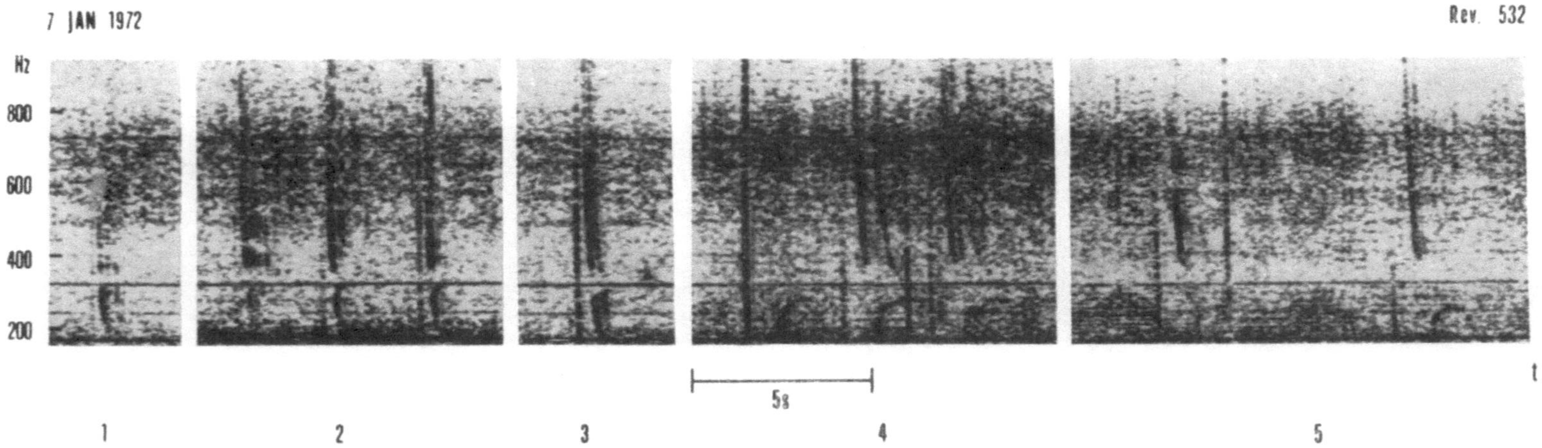

Fig. 1. 0_+ and 1_- fractional hop whistlers recorded by the Intercosmos 5 satellite over the South American equatorial area on 7 January 1972. Essential data referring to proton whistlers Nos. 1—5 are given in Table 1.

Table 1

No.	UT	LT	Satellite position °S	°W	Satellite height (km)	Trajectory max. height (km)	Dispersion ($s^{1/2}$) D_1	D_2	D_3	f_{Hi} (Hz)	f_{12} (Hz)	α
						Signal groups with proton whistlers (see Fig. 1)						
1	092950	0510	22.4	65.0	335	430	2.54	5.9	—	340	320	0.118
2	093029	0518	24.4	63.0	322	500	2.37	8.7	—	330	300	0.180
3	093048	0522	25.2	62.0	317	575	2.74	10.0	—	320	270	0.298
4	093220	0543	29.2	57.2	292	800	2.65	15.1	33.4	290	220	0.439
5	093250	0550	30.6	55.5	284	1100	—	17.3	—	260	180	0.539
						Signal groups without proton whistlers						
6	093359	0608	33.6	51.4	268	—	2.72	21.0	—	—	—	—
7	093710	0700	40.8	38.4	236	—	—	28.5	—	—	—	—

determined according to the formula $D = tf^{1/2}$. From Table 1 it follows that the dispersion of 0_+ whistlers displayed very slight changes due only to slight changes of the satellite altitude. The disperion of the electron trace of 1_- whistlers varied between 5.9 $s^{1/2}$ and 28.5 $s^{1/2}$. In one event (0932 UT, Fig. 1), a 2_- whistler occurred, i.e. a signal which had crossed the equator twice. Assuming the propagation of all three signal types 0_+, 1_-, 2_- along the same path, the following relation must be satisfied as to the dispersion: $D_3 = D_1 + 2D_2$ (D_1, D_2, D_3 = dispersion of 0_+, 1_- and 2_- whistlers, respectively). Calculations yielded $D_1 + 2D_2 = 32.85\ s^{1/2}$, $D_3 = 33.4\ s^{1/2}$. The discrepancy of $<2\%$ is explained by an error in determining D. One can thus conclude that all signals propagated along the same path. Since the signals crossed the area of the Brazilian magnetic anomaly, the dipole model of the geomagnetic field could not be used. Therefore, the calculated data on the geomagnetic field structure were taken from [6]. Fig. 2 shows geomagnetic

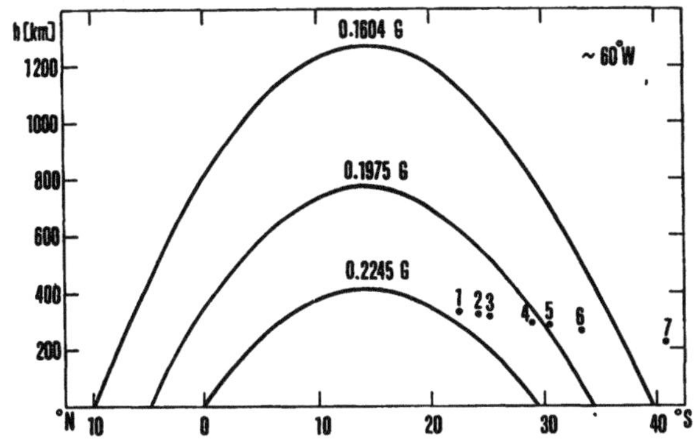

Fig. 2. Geomagnetic field lines at 60° W [6] and the geographic latitude and altitude h; the field intensity over the geomagnetic equator is given in gauss. Points 1, 2, 3, ... 7 denote the IK-5 satellite's positions, referring to the data in Table 1 and Fig. 1.

field lines beginning at points of different geographic latitude on the earth's surface in the northern hemisphere, the longitude being 60° W. At each line, the geomagnetic field intensity over the geomagnetic equator is given in gauss. The figures 1, 2, 3 etc., denote the satellite's position when whistlers were recorded, as shown in Fig. 1 and Table 1.

Obviously, one can assume that proton whistlers still retain their equatorial values of the gyrofrequency and cross-over frequency f_{12} after crossing the equator. Thus, based on the gyrofrequency f_{Hi} of the proton whistlers, the value of the magnetic field, as well as the corresponding height over the equator where this field value is equal to that calculated as in [6], can be found. In this way it has been found that the maximum height over the equator for signals with a proton trace varied between 430 and 1100 km.

From the proton gyrofrequency f_{Hi} and the cross-over frequency f_{12} of ion whistlers, the fractional concentration of hydrogen ions has been calculated according to [1]:

$$\alpha = (264/255)\,[1 - (f_{12}/f_{Hi})^2].$$

The α-height profile above the geomagnetic equator is given in Fig. 3. The area over the geomagnetic equator through which the signals propagated was already sunlit (about one hour after sunrise), so that values of α refer to daytime conditions.

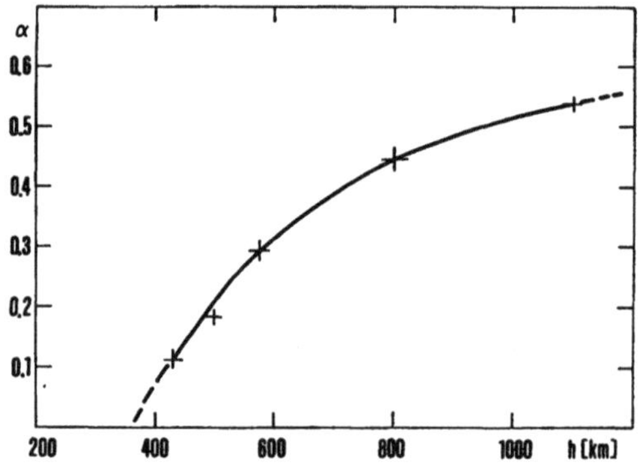

Fig. 3. Variation of the fractional concentration of hydrogen ions α with altitude h above the geomagnetic equator, derived from proton whistlers shown in Fig. 1.

From the values of α obtained, the altitudinal variation of f_{12} has been determined ($D = 0$), as well as the frequency for which the refractive index of the left-hand polarized wave n_L is equal to zero ($L = 0$) (7). The corresponding results are given in Fig. 4. Since the relative contents of helium (β) and oxygen (γ) ions were not known, calculations of $D = 0$ and $L = 0$ were made for two extreme conditions:

$$\beta = 1 - \alpha; \qquad \gamma = 0 \tag{1}$$

$$\beta = 0; \qquad \gamma = 1 - \alpha \tag{2}$$

For other possible combinations of β and γ, the curves $D = 0$ and $L = 0$ will be located between D_1 and D_2 and between L_1 and L_2 respectively. It can be seen from Fig. 4 that no proton whistlers could be produced at satellite heights

< 335 km because of a very low proton concentration ($f_{12} \simeq f_{\mathrm{Hi}}$). One can actually see in Fig. 1 that 0_+ whistlers do not show any proton trace. Since proton traces can be found in all 1_- signals, shown in Fig. 1 and listed in Table 1, one can believe, in accordance with Fig. 4, that they originated at a height above 335 km.

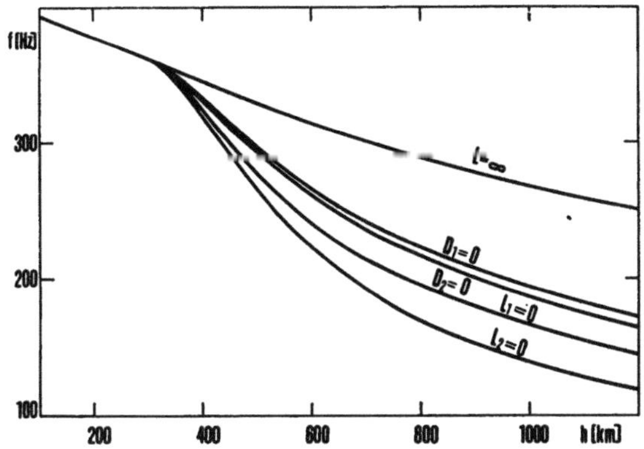

Fig. 4. Variation with altitude h of the frequencies where $L = 0$, $D = 0$ and $L = \infty$ for a model ionosphere with α values given in Fig. 3. Subscripts 1 and 2 refer to Eqs. (1) and (2).

The electron trace of 1_- whistlers shows an absence of signals between f_{12} and a frequency f_c higher than f_{12} (see Fig. 1), this being obvious evidence that transition from the electron to the ion cyclotron wave proceeds without coupling of these waves. This is in good agreement with the situation near the equator, where the angle between the wave normal and the geomagnetic field is large and there is little probability of mode coupling.

Let us consider the spectral changes of a signal propagating from the northern hemisphere to the satellite at a height of 292 km over the southern hemisphere (spectrogram No. 4 in Fig. 1 and corresponding data in Table 1). The maximum height up to which the signal rises over the equator is about 800 km. According to Fig. 4 the possibility of ion-cyclotron wave generation exists at a height slightly above 335 km. But, along the field line crossing the equatorial plane at about 800 km, left-hand polarized waves can only propagate in the frequency band from 220 to 290 Hz. Farther from the equator the signal will descend to the satellite, and two possibilities occur here:

(a) The wave will not change its polarization after reaching the level $D = 0$, and it will penetrate to lower levels, still remaining an ion-cyclotron wave. Therefore it is not able to cross the level $L = 0$, because $n_L{}^2 < 0$ below this level.

(b) The wave polarization will change after crossing the level $D = 0$, and a right-hand wave (electron mode) will propagate downward showing no peculiarities at the level $L = 0$. The wave can only reach the satellite in this latter case. Then the path from the level $D = 0$ down to the satellite will introduce an additional time delay of the signal which does not cause a significant change to the dynamic spectrum; the described signals retain such characteristic properties of proton whistlers as the cross-over frequency and the proton gyrofrequency where they were formed, near the equator. Since the whole spectrum of these signals, received on board the satellite consists in this case of right-hand polarized waves.

one can hope that they may be able to penetrate the lower ionosphere down to the ground, so that it may be possible to receive them at ground stations at equatorial latitudes.

It is suggested that explanation (b) is the more probable.

References

[1] S. D. SHAWHAN and D. A. GURNETT, J. Geophys. Res. **71**, 47 (1966).
[2] S. D. SHAWHAN, J. Geophys. Res. **71**, 29 (1966).
[3] YA. I. LIKHTER et al., Space Research XIII, 695 (1973).
[4] F. JIŘÍČEK et al., Studia Geoph. et Geodet. **17**, 49 (1973).
[5] R. L. SMITH and J. J. ANGERAMI, J. Geophys. Res. **73**, 1 (1968).
[6] M. A. KUTIMSKAYA and V. A. KUZMIN, Geom., Aeronomy and Solar Phys., Rep. No. 14 SIBIZMIRAN, Irkutsk 1970.
[7] D. A. GURNETT et al., J. Geophys. Res. **70**, 1665 (1965).

Space Research XIV — Akademie-Verlag, Berlin 1974

SIMULTANEOUS MEASUREMENTS OF SOME IONOSPHERIC PARAMETERS AT ALTITUDES 100–170 km

A. D. Danilov, T. V. Kazatchevskaya, V. V. Selantiev, V. K. Semenov, V. G. Hryukin and U. K. Chasovitin

Hydrometeorological Service of the USSR, Moscow, USSR

The results of simultaneous measurements of several ionospheric parameters are presented. The rocket MR-12 was launched on 1 December 1972 at 1539 LT. The electron temperature was measured by a cylindrical Langmuir probe, the electron concentration by a high-frequency impedance probe, the ion composition by an RF mass spectrometer; these and the intensity of the ionizing radiation were measured at altitudes between 100 and 170 km. Comparison of the data obtained gives some conclusions on both the effective recombination coefficient and the source of E region ionization at twilight.

The geophysical rocket MR-12 was launched on 1 December 1972 at 1539 LT at Volgograd. The solar zenith angle was 86° and the solar radio-emission flux at 10.7 cm was 94×10^{-22} W m^{-2} Hz^{-1}. The magnetic conditions were quiet: the measurements made directly at Volgograd gave a K index equal to 0.

Electron concentration [e] was measured by a high-frequency impedance probe and electron temperature by a cylindrical Langmuir probe. The error of the measurements was about 15% for [e] and ±100 °K for T_e. The methods and devices used are described in [1, 2]. The electron density distribution is shown in Fig. 1. This profile is typical for pre-sunset conditions and is very like daytime profiles. It is characterized by a sharp gradient of [e] at altitudes 80–90 km and by the monotonic increase of [e] above that level and by the slightly defined maximum in the E region. At 119 km a weak maximum of [e] corresponding to an Es layer, registered on the ionograms, is observed. The experimental values of the electron temperature are presented in Table 1.

Table 1

Experimental Values of Electron Temperature

h (km)	100	110	120	130	140	150	160	170	180
T_e (°K)	400	450	500	500	550	550	600	650	800

Measurements of the ion composition at 120–180 km were made by an RF mass spectrometer during both ascent and descent. The method of measurement, the reduction of the data and the relative ion concentrations obtained, are described in [3]. With the help of the electron concentration measured in this flight we were able to give the absolute ion concentrations shown in Fig. 1.

It is worth mentioning that in this flight anomalous behaviour of O^+ ions was detected at 170—180 km. At these heights the ion peak 16^+ began rapidly to decrease and simultaneously ion peaks 13^+ and 14^+ appeared. No changes in NO^+ and O_2^+ behaviour were recognized at that moment. We explain the observed

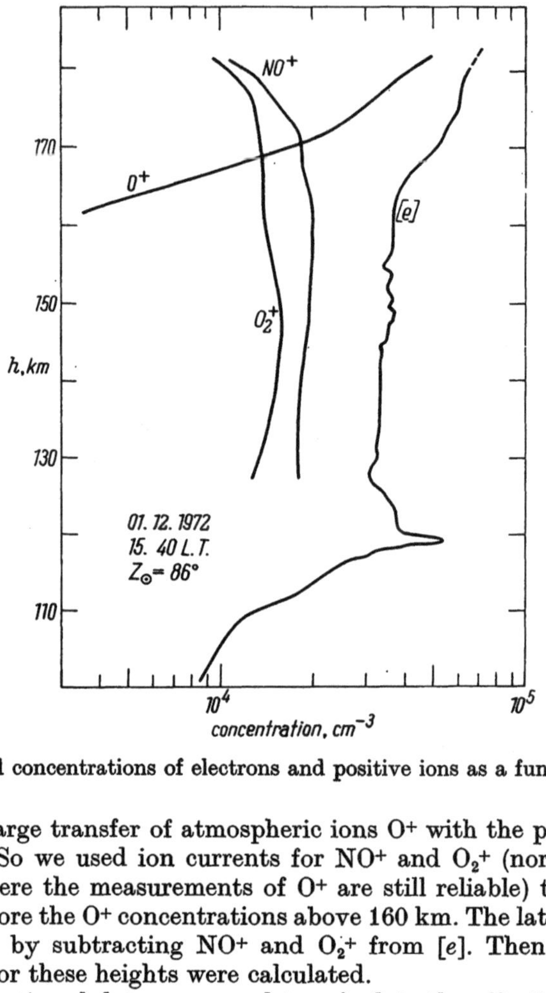

Fig. 1. Measured concentrations of electrons and positive ions as a function of height *h*.

effect by the charge transfer of atmospheric ions O^+ with the products of rocket contamination. So we used ion currents for NO^+ and O_2^+ (normalized to $[e]$ at $h = 160$ km, where the measurements of O^+ are still reliable) together with the $[e]$ profile to restore the O^+ concentrations above 160 km. The latter were obtained at 170—180 km by subtracting NO^+ and O_2^+ from $[e]$. Then the relative ion concentrations for these heights were calculated.

The above mentioned data were used to calculate the effective recombination coefficient α' [4]:

$$\alpha' = \frac{[NO^+]}{[e]} \alpha^*_{NO^+} + \frac{[O_2^+]}{[e]} \alpha^*_{O_2^+}, \tag{1}$$

$\alpha^*_{NO^+}$ and $\alpha^*_{O_2^+}$ being the values of the dissociative recombination coefficient of NO^+ and O_2^+ ions and $[NO^+]/[e]$ and $[O_2^+]/[e]$ the relative ion concentrations measured by a mass spectrometer. The values of α^* used were [5]:

$$\alpha^*_{NO^+} = 4.5 \times 10^{-7} \frac{300}{T_e} \text{ cm}^3 \text{ s}^{-1} \tag{2}$$

and

$$\alpha^*_{O_2^+} = 2.2 \times 10^{-7} \frac{300}{T_e} \text{ cm}^3 \text{ s}^{-1} \tag{3}$$

the T_e values being taken from the experiment.

The values of α' obtained are shown in Fig. 2 by the solid line; the broken lines show the values of α' obtained from Mitra and Banerjee model [6] for low solar activity. One can see from the figure that the solid line based on simultaneous

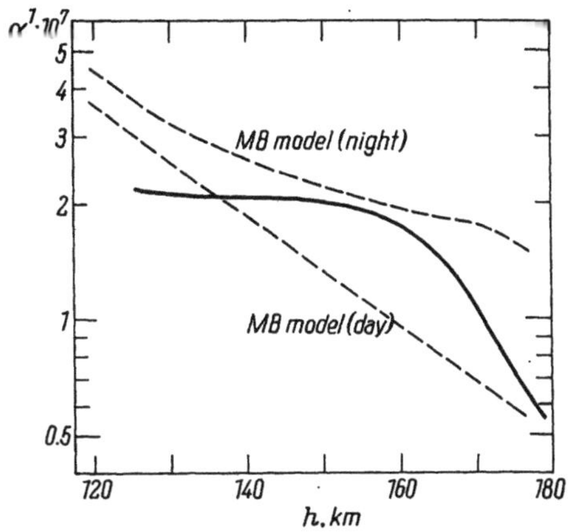

Fig. 2. Variation with altitude h of effective recombination coefficient α' (cm^{-3} s^{-1}): —— from observations, ---- two models of Mitra and Banerjee [6].

measurements of ion composition and T_e in one pre-sunset flight lie between the model values for day and night. We do not intend to discuss here the complicated question of α' determination in the ionosphere (done in detail in [7, 8]), but only to mention that the α' values obtained in this work are higher than those given by most of the ground-based methods, but are in good agreement with a photochemical model [6].

Some discrepancy may be seen between observed and model values of α' below 140 km. We think that this difference reflects different ideas about the T_e value in the E region. If T_e is taken equal to T_n as found from the results of incoherent scatter experiments then α' is higher at $h < 140$ km than we obtained with $T_e > T_n$ (as the probe results usually give).

The experimental data described were used for the calculation of the recombination rate $\alpha'[e]^2$. The profile of this value so obtained is shown in Fig. 3 by a solid line. At altitudes below 125 km where there were no measurements of the ion composition in this flight we used the relative ion composition taken from the empirical model [9]. These results are shown in Fig. 3 by points and the $\alpha'[e]^2$ profile is presented below 125 km by a chain line.

To measure the ionizing radiation two devices with $CaSO_4$—Mn thermophosphor surfaces were installed on the rocket. As described earlier [10] the measurements of EUV radiation with $\lambda < 1300$ Å as well as of the soft electron flux with $E_e \geqq 1$ keV may thus be carried out. These devices were installed so that only one of them viewed the sun. Measurements with different filters and without a filter were made during the flight. Aluminium ($\varrho = 1.5 \times 10^{-3}$ gram cm^{-2}), LiF and film ($C_6H_{10}O_2$, $\varrho = 1 \times 10^{-4}$ gram cm^{-2}) filters were used. The device which viewed the sun gave the data on solar EUV radiation. In the measurements with the LiF filter the intensity of the Lyman α line ($\lambda = 1216$ Å) was found to be equal to 5 erg cm^{-2} s^{-1} at $h = 100-110$ km.

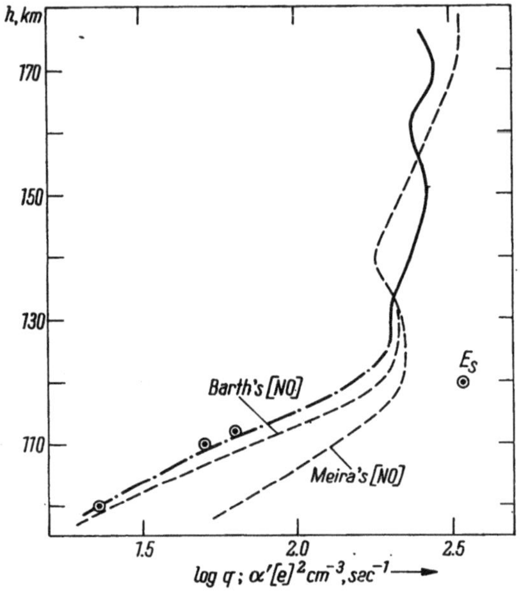

Fig. 3. Variation with altitude h of loss rate by recombination $\alpha'[e]^2$ (cm^{-3} s^{-1}):
—— observed, ⊙ and —·— observed but with ion concentrations from the model [9].
---- Variation of electron production rate q by solar EUV radiation and by solar Lyman α radiation for two models for NO distributions [11, 12].

The integral intensities W in the spectral regions $\lambda < 100-150$ Å and $\lambda < 25$ Å were obtained in the measurements with the film and Al filters respectively. These values were compared with the corresponding values for the solar spectra during solar minimum activity [7] $W^{\odot}_{\min}$ and the following results were obtained at $h = 150-180$ km:

$$\lambda < 100-150 \text{ Å} \qquad W/W^{\odot}_{\min} = 2.5$$

$$\lambda < 25 \text{ Å} \qquad W/W^{\odot}_{\min} = 80$$

The scattered solar radiation in the Lyman α line was measured with a LiF filter by a device which did not view the sun directly. The intensity was found to be 1.7×10^{-2} erg cm^{-2} s^{-1} sr^{-1} ($\sim$13 KR) at $h = 182-183$ km.

The intensity of the corpuscular flux ($E \gtrsim 5$ keV) W_c was obtained by the same device without filters and with a tube filter. It was found to be 6×10^{-4} erg cm^{-2} s^{-1} sr^{-1} at $h \geqq 100$ km. It is necessary to note that the value W_c was measured with an error of a factor 2, the errors for the Lα measurements being 30—50%.

The above-mentioned data on solar EUV radiation were used for the calculation of the total electron production rate q at altitudes 100—180 km during the rocket flight. The results of the calculations [7] for $Z_{\odot} = 86°$ were used and corrected with the help of our measurements for $\lambda < 25$ Å and $\lambda < 100-150$ Å. The ionization of NO molecules by the direct Lα radiation at $h = 100-120$ km was taken into account. Our measurements of Lα intensity and two NO distributions [11, 12] were used. Corresponding q-values are shown in Fig. 3 by the broken lines. The contribution to the q-values from scattered Lα radiation and from the corpuscular ionization was too small to be taken into account.

As one can see from Fig. 3 good agreement is observed between the experimental data on $\alpha'[e]^2$ and values obtained for the electron production rate q. This means, first, that the different experimental data obtained in the same experiment are consistent, and second, that at these heights the distribution of charged particles is governed by photochemical processes. The ionization of the E region during the experiment was produced by the direct Lα radiation. Our values of $\alpha'[e]^2$ agree better with the q values based on Barth's [NO] model than on Meira's [NO] model.

As is seen from Fig. 3, there is strong disagreement between $\alpha'[e]^2$ and q at 120 km. At this altitude the weak Es layer was observed (see Fig. 1). It is thought now that an Es layer is mainly formed by the wind shear mechanism [13], which collects ions (for example meteoric ions) with a low recombination coefficient. For this case the real α' coefficient at 120 km has to be smaller than that given by Eq. (1).

Summarizing, we may say that the results of the simultaneous measurements of several parameters in one rocket flight agree within the limits of modern photochemical theory. The data obtained allow the altitude profile of the effective recombination coefficient to be drawn.

References

[1] G. P. Komrakov et al., Kosm. Issled. **8**, 278 (1970).
[2] U. B. Burakov et al., Trudy IEM **16**, 47 (1970).
[3] V. K. Semenov and A. D. Danilov, paper presented as COSPAR-IAGA-URSI Symposium on Methods of Measurements and Results of the Lower Ionosphere.
[4] A. D. Danilov and M. N. Vlasov, Photochemistry of Ionized and Excited Particles in the Lower Ionosphere, Hydrometeoizdat, Moscow 1973.
[5] M. Biondi, Can. J. Chem. **47**, 1711 (1969).
[6] A. P. Mitra and P. Banerjee, Space Research XI, 1019 (1971).
[7] G. S. Ivanov-Kholodny and G. M. Nicolsky, Solntse i Ionosphera, Nauka, Moscow 1969.
[8] K. B. Serafimov, Ionosph. Issled. **19**, 99 (1969).
[9] A. D. Danilov and M. N. Vlasov, paper presented at IRI Working Group, URSI Assembly, Warsaw 1972.
[10] T. V. Kazatchevskaya, Kosm. Issled. **5**, 637 (1967).
[11] L. G. Meira, J. Geophys. Res. **76**, 202 (1971).
[12] C. A. Barth, Annls Géophys. **22**, 198 (1966).
[13] J. D. Whitehead, J. Geophys. Res. **76**, 3127 (1971).

Space Research XIV — Akademie-Verlag, Berlin 1974

METEOR IONS IN THE POLAR IONOSPHERE: ROCKET MASS-SPECTROMETRIC MEASUREMENTS AND THEORETICAL CALCULATIONS

A. D. Zhlood'ko, V. N. Lebedinets and V. B. Shushkova

Hydrometeorological Service of the USSR, Moscow, USSR

Results are presented of mass-spectrometric measurements of Mg^+ ion concentration at altitudes 100—190 km, according to data obtained during four rocket launchings from Heiss Island in 1969—1972. Maximum concentrations vary from 3×10^2 cm^{-3} to 3×10^4 cm^{-3} and are observed in two altitude ranges, 100—110 and 160—170 km. Minimum estimates of equilibrium concentrations of meteor ions formed as a result of active meteor ionization at values of the effective recombination coefficient of 10^{-8}—10^{-10} cm^3 s^{-1} are obtained theoretically. Comparison with the experimental data shows that this source can explain the observed Mg^+ concentrations in the E region. Above 140 km it is necessary to invoke other sources of ion production. Such possible sources are discussed.

1. Introduction

Aerosols, atoms, molecules and ions of meteor origin are generated as a result of sputtering, evaporation and fragmentation of meteoroids in the upper atmosphere. The most probable mean composition of meteor matter is (in %): SiO_2 38.4; MgO 23.84; FeO 12.45; Al_2O_3 2.50; CaO 1.95; Na_2O 0.98; K_2O 0.17; Fe 11.76; Ni 1.34; FeS 5.73. The presence of the typical meteor ions Mg^+, Fe^+, Si^+, Ca^+, Na^+ etc. has been established in the upper atmosphere by many authors from rocket mass-spectrometric measurements [3—16]. For example, Istomin [3] found ions Mg, Ca, and Fe at altitude 104 ± 5 km. The abundance of magnesium in the layer is approximately 7×10^9 cm^{-2}, the abundance of iron is almost the same and of calcium slightly less. According to [16] the abundance of Mg^+ is almost an order of magnitude less and equal to 7.9×10^8 cm^{-2}; according to [13] the abundance of Mg^+ is 7×10^9 cm^{-2}, Fe^+ 10^{10} cm^{-2} and Ca^+ 3×10^8 cm^{-2}. Smith et al. [7] discovered a layer consisting of magnesium at altitude 93 km with abundance 5×10^8 cm^{-2}, and another layer at altitude 113 km with a silicon abundance about 10^9 cm^{-2}. Narcisi and Bailey [4] discovered two layers with meteor ions at altitudes of 98 km and 111 km.

2. Experiment

Measurements of ion concentrations in the polar ionosphere were obtained using radio-frequency mass spectrometers M6407 aboard MP-12 rockets. Concentrations of ions 24+ ($^{24}Mg^+$) observed during four rocket launchings on Heiss Island are presented in Fig. 1. These results and earlier observations [12, 14] show that the concentration of such ions in the polar atmosphere changes by 2—3 orders of magnitude; the threshold of sensitivity of the mass spectrometer is 10^2 cm^{-3}.

In Fig. 1 only the data of ions 24^+ are presented, but during the measurements heavy magnesium isotopes 25^+, 26^+ and sodium ions 23^+ were recorded as well.

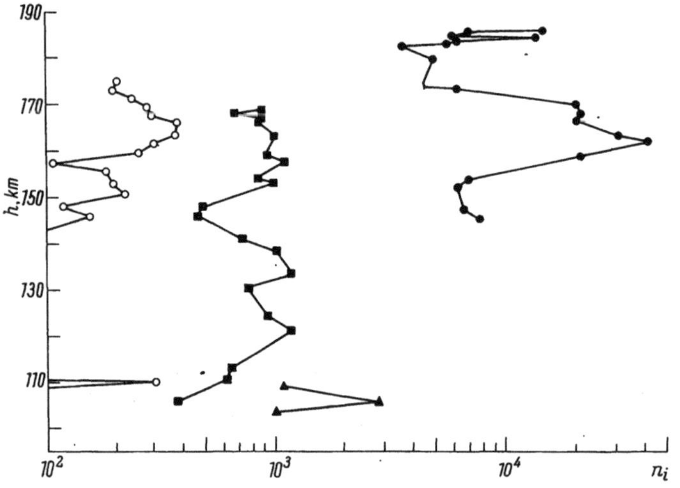

Fig. 1. Variations of ions 24^+ (Mg^+) concentrations (n_i in cm^{-3}) in the polar ionosphere according to massspectrometric data: ▲, 18 March 1969, 2358 LT; ⊙, 28 November 1970, 0007 LT; ■, 18 February 1972, 0625 LT; ●, 21 February 1972, 1727 LT.

Their concentrations where usually less than the concentration of ions 24^+. During the experiment on 18 February 1972 the following mean ratio values were obtained: $24^+/25^+ \approx 2.9 \pm 0.9$; $24^+/26^+ \approx 5.2 \pm 3.3$; $26^+/25^+ \approx 0.3 \pm 0.1$. During the flight on 21 February 1972 the ratios of concentrations were: $24^+/25^+ \approx 4.9 \pm 1.4$; $24^+/26^+ \approx 5.4 \pm 1.7$; $26^+/25^+ \approx 0.9 \pm 0.1$. There is close agreement between the relative concentrations of these ions in mid-latitude and high-latitude regions.

3. Theoretical Calculations

Different mechanisms of meteor ion formation are possible: (i) active meteor ionization of meteor atoms by the collision of sputtered and evaporated atoms with atmospheric molecules, (ii) photoionization of meteor atoms, (iii) charge exchange between meteor atoms and atmospheric ions, (iv) ionization by solar particles which may be significant in the polar ionosphere associated with solar disturbances. To calculate the equilibrium concentration of meteor ions it is necessary to know the intensity of all these processes of ionization and also the effective recombination coefficient of meteor ions α_{eff}. Since different processes of recombination of meteor ions are possible besides radiative recombination, the effective recombination coefficient of meteor ions will be significantly higher than the coefficient of radiative recombination; according to approximate estimates [17, 18] $\alpha_{eff} = 10^{-8}$—10^{-10} cm^3 s^{-1}.

To obtain the minimum equilibrium concentration of meteor ions at different values of α_{eff} we consider only one type of ionization—active meteor ionization. Three mechanisms of active meteor ionization are possible during the motion of a meteoroid in the atmosphere: (1) ionization of meteor atoms at the collision of sputtered meteor atoms with atmospheric molecules; this mechanism is effective for very small particles, e.g. micrometeorites; (2) ionization of evaporated atoms which is effective for meteoroids producing meteors; these meteoroids when decelerated are heated so much that they lose all their mass as a result of evaporation; (3) ionization of atmospheric molecules which collide with the meteoroid.

The most detailed calculation of the rate of active meteor ionization caused by molecules reflected from a meteoroid was made by Fiocco [19]. Nevertheless Fiocco assumed that the accommodation coefficient was equal to zero, so that

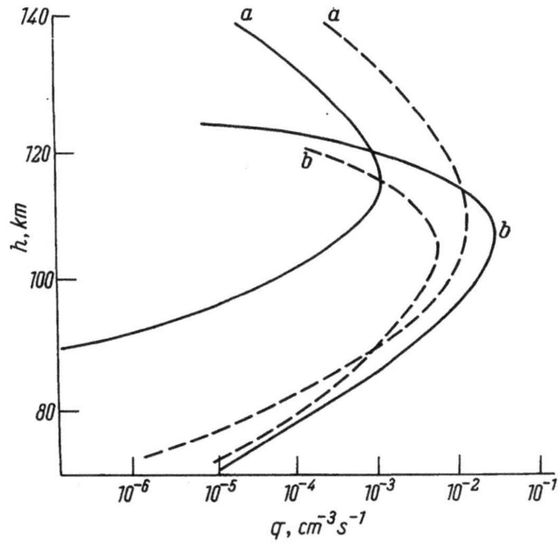

Fig. 2. The rate of active meteor ionization due to meteor (solid lines) and atmospheric (dotted lines) ions of micrometeorites (a) and meteors (b).

the velocity of the reflected molecules is overestimated by a factor of approximately two; also he did not take into account the decrease of meteoroid mass due to sputtering and evaporation, and therefore the ionization at low altitudes was overestimated: a contribution to the ionization at an altitude not reached by the meteoroid was calculated. Also the ionization of meteor atoms was not considered.

The interaction of meteoroids of different sizes with the atmosphere has been considered by us in [20—24]. The rate of active meteor ionization due to meteoric (solid lines) and atmospheric (dotted lines) ions of micrometeorites (a) and meteors (b) is presented in Fig. 2.

Taking into account that in the real atmosphere atmospheric molecules may be ionized not only by meteoroids entering the atmosphere but by other sources of ionization, we calculate the equilibrium concentrations of atmospheric N_a^+ and meteor N_m^+ ions to estimate the efficiency of active meteor ionization. The results of calculations are presented in Fig. 3. The effective recombination coefficient of atmospheric molecular ions in the E region was assumed equal to 1×10^{-7} cm^3 s^{-1} [27, 28].

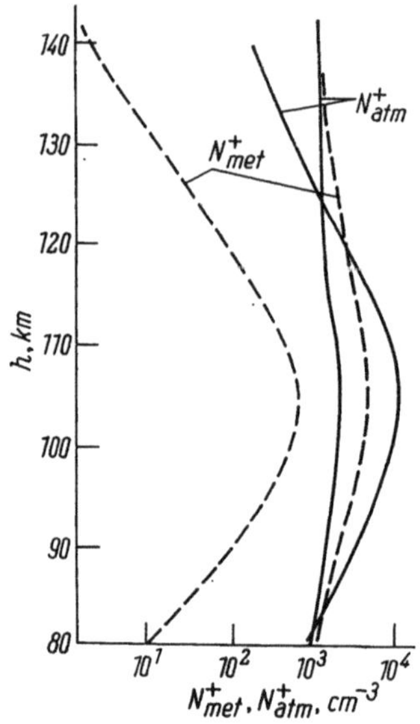

Fig. 3. Equilibrium concentrations of atmospheric N^+_{atm} and meteor N^+_{met} ions at $\alpha_\mu' = 10^{-10}$ cm^3 s^{-1} (solid lines) and $\alpha_\mu' = 10^{-8}$ cm^3 s^{-1} (dotted lines).

4. Discussion

Comparison of the results of theoretical calculations with the experimental data shows that Mg^+ concentrations observed in the E region may be explained by the mechanism of active meteor ionization considered. Thin Es layers with high concentration of meteor ions (10^3—10^5 cm^{-3}) which are often observed, may apparently be explained by redistribution of the ions according to the wind-shear theory. Very great variations of Mg^+ concentration from 3×10^2 to 3×10^4 cm^{-3} are observed at altitudes of 150—170 km. Minimum values (10^2—10^3 cm^{-3}) may be explained by active meteor ionization, but to explain maximum concentration values it is necessary to invoke other sources of ionization. In the polar atmosphere other sources are possible: (i) the entry of solar particles; (ii) the entry of streams of easily melted meteoroids (like ice from a cometary core) and (iii) a change of atmospheric conditions which causes an increase in the rate of charge-exchange aeronomic reactions. Ions observed in the upper atmosphere are not only brought there by meteors, but also from the oceans [29].

References

[1] H. C. Urey and H. Craig, Geochim. Cosmochim. Acta **4**, 36 (1953).
[2] V. N. Lebedinets, Doctor's Thesis, Obninsk, USSR (1966).
[3] V. G. Istomin, Dokl. Akad. Nauk USSR, **137**, 1102 (1961).

[4] R. S. Narcisi and A. D. Bailey, J. Geophys. Res. **70**, 3686 (1965).
[5] R. C. Sagalyn, M. Smiddy and W. P. Sullivan, Space Research VII, 449 (1967).
[6] J. M. Young, C. I. Johnson and J. C. Holmes, J. Geophys. Res. **72**, 1473 (1967).
[7] C. R. Smith, H. C. Brinton, M. W. Pharo III and H. A. Taylor, Jr., J. Geophys. Res. **72**, 2357 (1967).
[8] R. S. Narcisi, A. D. Bailey and L. Della Lucca, Space Research VII, 446 (1967).
[9] R. S. Narcisi, Space Research VII, 186 (1967).
[10] F. Arnold, W. Berthold, B. Betz, P. Lämmerzahl and J. Zahringer, Space Research IX, 256 (1969).
[11] R. A. Goldberg and Z. Y. Blumle, J. Geophys. Res. **75**, 133 (1970).
[12] A. D. Zhlood'ko and N. M. Klyueva, Space Research X, 746 (1970).
[13] T. D. Whitehead, Rev. Geophys. **22**, 119 (1966).
[14] A. D. Zhlood'ko, Space Research XI, 1051 (1971).
[15] A. Johannesen and D. Krankowsky, J. Geophys. Res. **77**, 2888 (1972).
[16] C. R. Philbrick, R. C. Narcisi, R. E. Good, H. S. Hoffman, T. J. Keneshea, M. A. MacLeod, S. P. Zimmerman and B. W. Reinisch, paper presented at XV COSPAR Plenary Meeting, Madrid 1972.
[17] R. S. Narcisi, Annls Géophys. **22**, 224 (1966).
[18] E. E. Ferguson and F. C. Fehsenfeld, J. Geophys. Res. **73**, 6215 (1968).
[19] G. Fiocco, J. Geophys. Res. **72**, 3497 (1967).
[20] V. N. Lebedinets and V. B. Shushkova, in: Physics and Dynamics of Meteors, D. Reidel Publ. Co., Dordrecht, Holland 1968.
[21] V. N. Lebedinets and V. B. Shushkova, Astron. Vestnik, USSR **2**, 37 (1968).
[22] V. N. Lebedinets and V. B. Shushkova, Planet. Space Sci. **18**, 1653 (1970).
[23] V. N. Lebedinets and V. B. Shushkova, Planet. Space Sci. **18**, 1659 (1970).
[24] V. N. Lebedinets, A. V. Manochina and V. B. Shushkova, Planet. Space Sci. **21**, (1973).
[25] V. N. Lebedinets, Space Research XI, 307 (1971).
[26] K. Fujutaka and T. Ogawa, J. Atmos. Terr. Phys. **33**, 687 (1971).
[27] G. S. Ivanov-Kholodny and G. M. Nikolsky, The Sun and the Ionosphere, Nauka, Moscow 1969.
[28] A. R. Serafimov, Ionosph. Issled. No. 19, 99 (1970).
[29] C. R. Burnett, W. E. Lammer, W. T. Novak and V. L. Sides, J. Geophys. Res. **77**, 2934 (1972).

Space Research XIV — Akademie-Verlag, Berlin 1974

ION COMPOSITION DURING THE FORMATION OF A MID-LATITUDE Es LAYER

A. C. AIKIN[a], R. A. GOLDBERG[a] and A. AZCÁRRAGA[b]

[a]Laboratory for Planetary Atmospheres, NASA/Goddard Space Flight Center, Greenbelt, Md, USA

[b]Instituto Nacional de Tecnica, Departamento de Quipo y Armamento, Torrejon de Ardoz, Spain

The positive ion composition within a mid-latitude sporadic E layer has been measured with the aid of a rocketborne ion mass spectrometer launched from El Arenosillo, Spain on 3 July 1972 at 0743 LMT. Ionograms taken before and during the rocket flight showed a developing sporadic E layer near 114 km. Rocket data showed peaks in electron density and metallic ions at this same height. Both the maximum and total content of the metals are observed to be greater on the downleg than on the upleg measurement.

1. Introduction

Mid-latitude sporadic E (Es) is commonly observed on ionograms as a low-lying horizontal trace. It is the result of an electron density enhancement in a narrow altitude range in the vicinity of the E region maximum. This enhancement is thought to be caused by horizontal wind shears acting in the presence of the vertical component of the terrestrial magnetic field [1—3]. Long-lived ions (e.g. metallic ions) must be a part of the ion composition of the Es layer to support the wind shear theory. Metallic ions have in fact been observed within Es layers by rocketborne ion mass spectrometers [4, 5].

In this work we report the results of a recent rocketborne ion mass spectrometer, which was fired into a small Es layer apparently during the period of formation. The launching took place at El Arenosillo, Spain (37.1° N 6.7° W) on 3 July 1972 at 0743 LMT for a solar zenith angle χ of 57.1°. The rocket reached an apogee of 122.8 km and passed thorugh an Es layer on both upleg and downleg 90 seconds later. Metal ions were detected within the Es layer.

2. Instrumentation

The Nike-Apache payload consisted of several instruments to measure the properties of the E region. Ion composition was determined by a pumped quadrupole ion mass spectrometer whose characteristics are listed in Table 1 [6, 7].

Electron density was measured by means of the Faraday rotation technique which utilized linearly polarized radio waves at 3.030 and 6.985 MHz propagated to rocketborne receivers with linearly polarized antennas. Comparison of the absolute period of the rocket spin ($\sim$5 c/s) with the apparent period observed

Table 1
Spectrometer Characteristics

Characteristic	Value
Entrance aperture diameter (mm)	0.742
Rod length (cm)	12.7
Sampling interval (s)	1.99
Sweep range (amu)	$4 \rightarrow 96$
Time/unit mass (s)	0.0185
RF (MHz)	2.518
Attractive potential (V)	−10

by the airborne receivers was used to obtain the electron density height profiles [8]. The overall electron density measurement probably has an accuracy of at least $\pm 5\%$ but the downleg data have less height resolution.

Total positive ion density was measured by means of a Gerdien condenser. In order to obtain the upleg ion density profile the Faraday rotation technique was utilized for absolute density and fine structure was resolved by use of the Gerdien condenser. The downleg profile was determined exclusively from the radio propagation experiment. A set of Geiger-Müller tubes, sensitive to electrons of energy greater than 40 and 100 keV, was also included as part of the payload.

The ion composition data were reduced to absolute values by normalizing the distribution to the total density profile, taking free molecular flow mass weighting effects into account [9]. This is extremely important in the reduction of the downleg data, where results are much more sensitive to mass weighting effects. The mean rocket coning angle was used for aspect correction of the data. Coning angle variations ($\pm 2.5°$) could introduce fluctuations from the plotted values of as much as 10% for the heavy mass ions such as Fe^+. Above 105 km a background current occurred on each spectrum and this is attributed to soft energetic particles below 15 keV. This current was subtracted from the ion current for each mass prior to normalization.

Care was taken in the vicinity of the Es peak to produce a more accurate value for each constituent by adjusting it to the extrapolated value of the ion density N_i determined for the precise height at which the specified constituent was measured. However, the height resolution of the mass spectrometer was 0.8 km near 114 km altitude. This led to inaccuracies for the density maximum determination of each constituent within the layer. A comparison of the relative ionic abundances within the layer with those near 100 km showed similar values [10], suggesting that the height resolution limit did not produce severe errors in the representation of the distribution within the layer.

3. Results and Discussions

Fig. 1 shows three ionograms taken before, during, and after the rocket firing. The sporadic E trace indicated a true height near 114 km. This is in agreement with the ion density enhancement at 114 km determined from the rocket experiments. As time progresses the sporadic E layer is observed to be less patchy and more intense. The sporadic E critical frequencies for the time of the ionograms

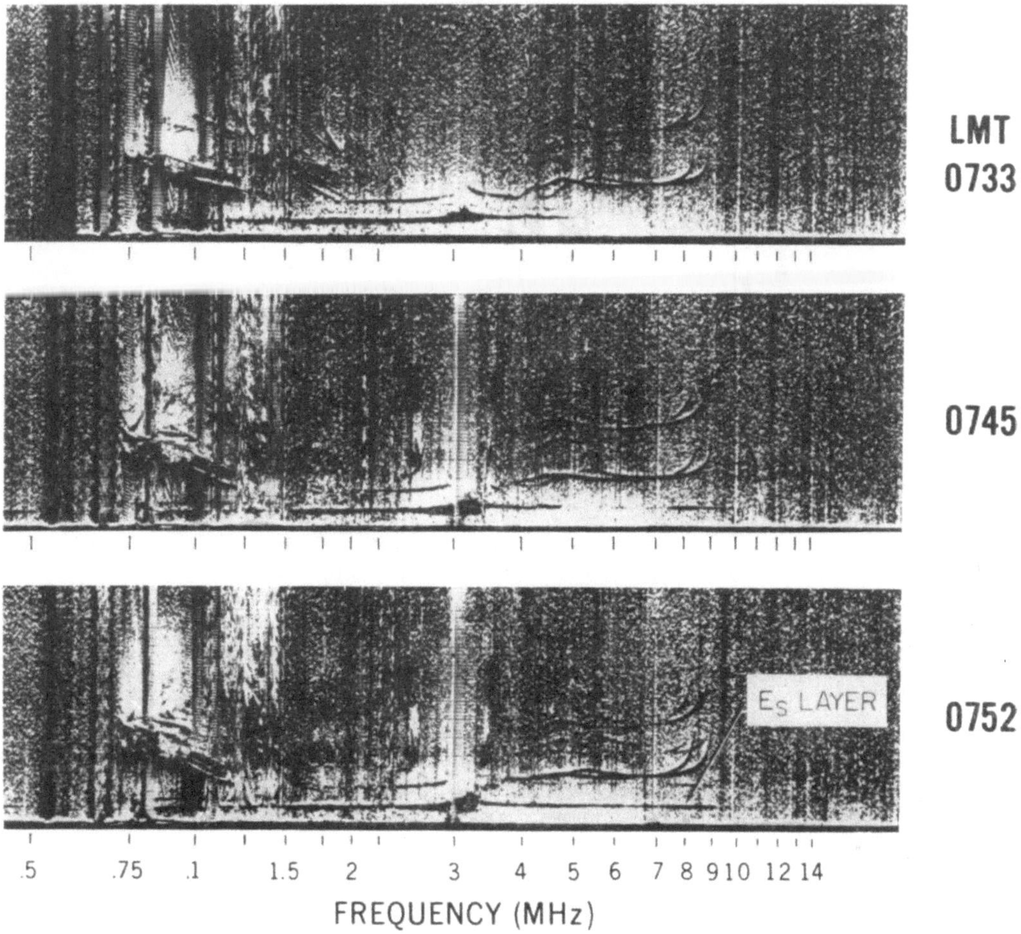

Fig. 1. Ionograms recorded before, during, and after the rocket flight period, illustrating the Es layer formation. 3 July 1972. El Arenosillo.

were 5.2 MHz at 0733 LMT, 10 minutes before the rocket firing, and 10 MHz for the ionograms taken at 0745 and 0752. The primary difference between the last two ionograms shown in Fig. 1 is the lack of patchiness in the 0752 trace.

The history of the sporadic E layer can be determined by plotting the critical frequency of the layer (*f*oEs) as a function of time. This is shown in Fig. 2. Also indicated on this figure are the times of rocket transit through the sporadic E layer and the total flight period. There were no ionograms between 0758 and 0815 at which time *f*oEs had dropped to 4.7 MHz. This time period is indicated by the dashed line. A survey of morning-time ionograms for the two-week period prior to launch showed a regular low-level occurrence of Es with nominal peak values of *f*oEs between 4 and 6 MHz. However, these ionograms were spaced at one hour intervals, making it diffcult to determine whether the short-term enhancement of *f*oEs during the rocket flight was unique in that two-week period.

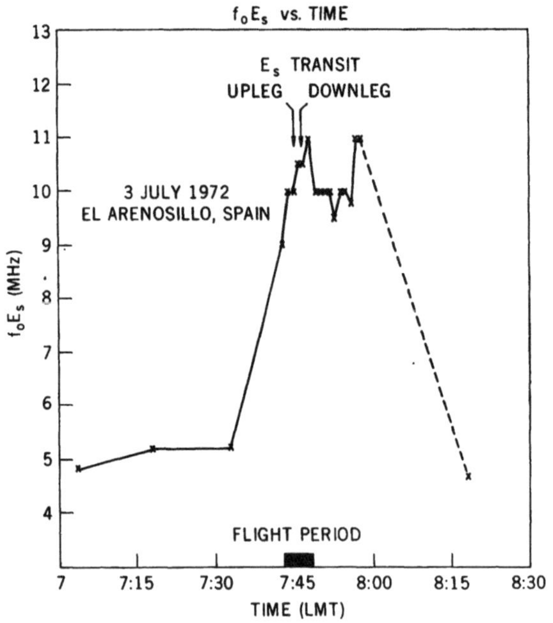

Fig. 2. Variation of *f*oEs during the time period including the rocket measurement. The times of payload transit through the Es layer are also indicated.

The observed ion composition is given in Fig. 3. It is seen that while NO^+ and O_2^+ are the major ions, there are many species of metallic ions present: included are 23^+, Na^+; 24^+, Mg^+; 28^+, Si^+; 39^+, K^+; 40^+, Ca^+; 45^+, Sc^+; 52^+, Cr^+; 56^+, Fe^+; and 58^+, Ni^+. Other isotopes of these constituents were also seen but are not shown. There is a well defined enhancement of the metals in the vicinity of 114 km which produces a 50% increase in the total electron or ion density N_i. This corresponds to the sporadic E trace observed on the ionogram.

The width of the metal layer at 10^4 cm^{-3} ion density is less than 1 km on the upleg; however, on the downleg portion of the flight the width is nearly 2.5 km. This thickening is symmetrical between top and bottomside. In addition the downleg maximum of the metallic ion layer was encountered one kilometer below the height for upleg. The downleg data below 110 km are not displayed because of the difficulty in evaluating the necessary mass weighting corrections in the transition region below this height.

The maximum concentration of metallic ions within the sporadic E layer is approximately a factor of two larger on the downleg than on the upleg. The major contribution to this enhancement appears to be 28^+, which is identified as Si^+. Above the Es region 28^+ behaves differently from the metal ions. Based on this, one can determine the contribution of N_2^+ to the observed 28^+ ion density. Metallic constituents other than Si^+ also show an increase in the downleg case. A notable exception is Fe^+, which may be enhanced but not observed because the spectrometer height resolution prevents an accurate peak value determination. The total content of metal ions on downleg is also significantly larger, as can be seen by both the increased thickness and magnitude of the layer. Within the sporadic E layer there is less O_2^+ and NO^+ for the downleg than for the upleg.

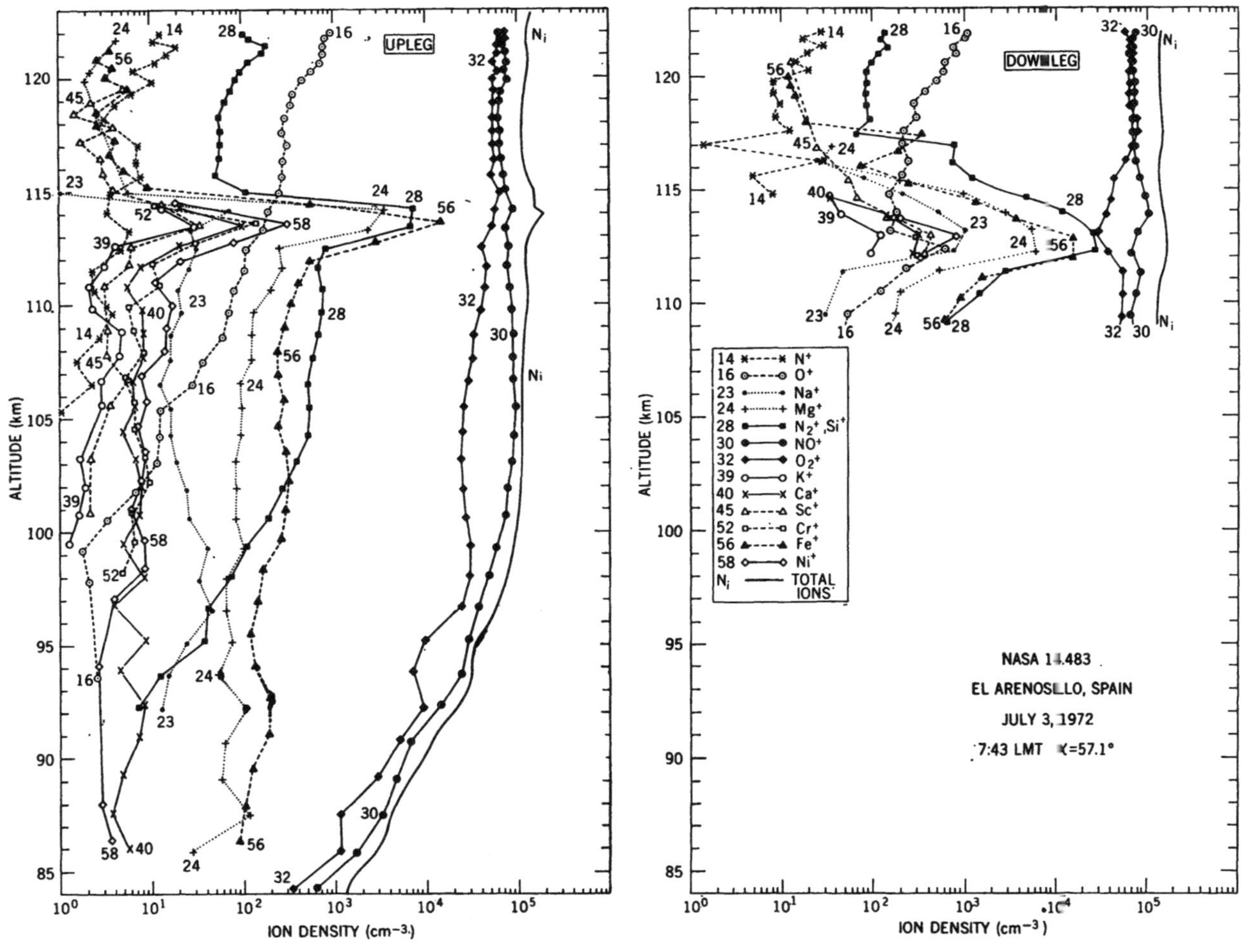

Fig. 3. Observed distribution of positive ion species in the daytime E region on 3 July 1972 over El Arenosillo. Upleg and downleg portions of the flight are indicated. The total electron (or ion) density is N_i.

The upleg and downleg passages of the rocket are separated by 90 seconds in time and 25.7 kilometers in horizontal range. It is not possible to separate temporal and spatial effects and the observed changes in ion composition may conceivably be due entirely to spatial irregularities. If one assumes that the variations in metal content are temporal in nature and that the beam of the ionosonde antenna contains both the upleg and downleg crossings of the sporadic E layer, then additional information can be derived on the formation of the sporadic E layer.

The negligible change in electron density is consistent with the foEs observed by the ionosonde which indicates no measurable change between the upleg and downleg sporadic E layer during rocket crossing times. However it should be remembered that sporadic E critical frequencies are sensitive functions of the electron density gradient as well as the actual electron density concentration.

At later times Fig. 2 shows an foEs increase to 11 MHz. This is consistent with further enhancement of the metal ions within the sporadic E region. It is not known whether the layer shown on the downleg was simply compressed or if new ions were transported from above and below the sporadic E height level.

Fresh meteoric material may have been added. 3 July was during the time period of the predicted β Taurids meteor shower. Cuchet [11] has shown that there is a greater probability of mid-latitude sporadic E having a critical frequency greater than 8 MHz during periods of enhanced meteoric activity. The present data confirm that Es layers are very sensitive functions of the metal ion content.

4. Conclusions

A measurement is presented of the daytime E region ion composition during the period of formation of a mid-latitude sporadic E region. Nine species of metallic ions were detected between 85 and 122 km. There was significant enhancement of these metallic ions in the vicinity of 114 km at the location of a sporadic E layer as observed from ionosonde data. Comparison of upleg with downleg information showed that the maximum value of metals within the sporadic E layer had increased by a factor of two and the total content by a somewhat larger factor. Shortly thereafter the critical frequency of the Es layer attained its maximum value of 11 MHz. The data substantiate the role of metallic ions in the formation of mid-latitude sporadic E layers.

References

[1] J. D. Whitehead, J. Atmos. Terr. Phys. **20**, 49 (1961).
[2] J. D. Whitehead, in: Ionospheric Sporadic E, Pergamon Press, Oxford, 1962 (p. 276).
[3] G. Chimonas and I. W. Axford, J. Geophys. Res. **73**, 111 (1968).
[4] J. M. Young, C. Y. Johnson and J. C. Holmes, J. Geophys. Res. **72**, 1473 (1967).
[5] R. S. Narcisi, Space Research VIII, 360 (1968).
[6] R. A. Goldberg and L. J. Blumle, J. Geophys. Res. **75**, 133 (1970).
[7] R. A. Goldberg and A. C. Aikin, J. Geophys. Res. **76**, 8352 (1971).
[8] A. C. Aikin, J. A. Kane and J. Troim, J. Geophys. Res. **69**, 4621 (1964).
[9] Hsue-shen Tsien, J. Aeronaut. Sci. **13**, 653 (1946).
[10] R. A. Goldberg and A. C. Aikin, Science **180**, 294 (1973).
[11] M. L. Cuchet, C. R. Acad. Sci. Paris **264**, 952 (1967).

Space Research XIV — Akademie-Verlag, Berlin 1974

THE REFRACTION OF VLF RADIO SIGNALS (CHORUS) BY THE LOWER IONOSPHERE

R. A. Usher[a], M. J. Rycroft[a] and W. Gibbons[b]

[a]Department of Physics, University of Southampton, UK
[b]Department of Physics, University of Sheffield, UK

A Petrel sounding rocket (P47H) was launched during a VLF chorus event at 0515 UT on 19 November 1970 from South Uist, Outer Hebrides, Scotland ($L \simeq 3.4$, close to the plasmapause). The rocket carried a uni-axial VLF radio receiver with seven narrow band channels (Southampton) and a Langmuir probe (Sheffield). The outputs of the channels in the chorus band, for example at 2.8 kHz, were modulated with a period of half the spin period of the rocket. From the variation of the angle between the wave normal and the axis of precession it is concluded that the downleg observations are not inconsistent with propagation of a downcoming whistler mode wave closely along the geomagnetic field (at 71.5° to the horizontal). This is not so on the upleg, when the rocket passed through an electron density enhancement, which is considered to be a well-defined sporadic E layer, at an altitude of 107.5 km. There, assuming propagation in the magnetic meridian, the angle between the wave normal and the geomagnetic field became 30° and the angle the wave normal made with the vertical was increased to 48°. The refraction of VLF signals by the lower ionosphere is studied in detail using refractive index surfaces determined from the electron density data provided by the Langmuir probe and a model of the neutral atmosphere. Between altitudes of 126 km and 108 km the ray direction deviates up to 10° to the north of the geomagnetic field line, but below the sporadic E layer the ray is closely field-aligned. Thus it is shown that the surfaces of constant electron density in the sporadic E layer make an angle of up to $64° \pm 7°$ with the horizontal.

1. Introduction

The Petrel rocket P47H was the first payload to be launched, at 0515 UT on 19 November 1970 from South Uist, Outer Hebrides, Scotland ($L \simeq 3.4$), as part of a joint research programme of the Universities of Southampton and Sheffield. The purpose of the programme is to study the characteristics of propagation of very low frequency (VLF) radio waves in the lower ionosphere. The observations to be discussed here were obtained during the successful flight of P47H whilst chorus signals were simultaneously recorded on the ground. Such chorus signals are believed to be triggered close to the equatorial plane of the magnetosphere by the transverse resonance, or gyroresonance, plasma instability. They are amplified by phase coherent cyclotron resonance with energetic electrons, and propagate through the magnetospheric plasma in the whistler mode. These topics have been discussed in a recent review [1].

2. Observations

The payload, which has already been described in detail [2], consists of an electrostatically screened loop aerial of 60 turns of copper wire. This is matched to a low-noise, broad-band preamplifier, from which the signal is fed to seven narrow-band (~100 Hz bandwidth) amplifiers, with AGC (automatic gain control) compression, in parallel. The design considerations, specification and performance of these selective channels have been given by Vávra [3]. The chorus band on South Uist ($L \simeq 3.4$, in the vicinity of the plasmapause) is observed by the five channels at 2.2, 2.8, 3.5, 4.4 and 5.4 kHz; the two remaining channels are at 9.6 kHz and, in order to study the wave field in the ionosphere of the signal transmitted by GBR Rugby, 16 kHz. The rectified and integrated outputs of these channels are sampled and the analogue outputs are telemetered back to the ground where they are stored on magnetic tape for subsequent digitization. For the observations at 2.8 kHz to be discussed here the sampling rate is 240/second.

3. Spin Modulation Ratio

The period of spin of the rocket about its longitudinal axis was found to be 125 milliseconds by studying the modulation of the telemetry signal strength. The output of the VLF channels showed a periodicity with a period equal to one half the rocketpin speriod; maixmum ($\mathscr{E}_{\max}$) and minimum ($\mathscr{E}_{\min}$) induced VLF signals are observed for loop orientations 90° apart. The depth of the spin modulation was calculated from this sinusoidal output variation (see [2], Fig. 4) using a least squares method. If the spin modulation ratio R is defined by the relation

$$R = \frac{\mathscr{E}_{\max} - \mathscr{E}_{\min}}{\mathscr{E}_{\max} + \mathscr{E}_{\min}}, \tag{1}$$

then the angle θ between the wave normal of a circularly polarized single plane wave and the rocket spin axis is given by

$$R = \tan^2 {}^1/_2\theta. \tag{2}$$

The variation of the angle θ is presented in Fig. 1; θ reaches large values, $\lesssim 30°$, on the upleg above an altitude of 106 km, and up to 125 km. At flight times beyond 160 seconds, θ exhibits a jagged, quasi-periodic variation. This is interpreted as being due to the precession of the rocket. A specially developed least squares sine fit computer program was used to find the best values of period, phase angle and amplitude of the oscillation. Considering observations in the region of quasi-constant electron density over the last half of the flight, the period of p ecession was found to be 13 seconds, with a precession cone half angle of 4.5° centred on a mean value of $\theta = 22°$. The variation of θ due to precession was then removed from these values by subtraction, both for the last half of the data and for the first half by extrapolating the best fit sinusoid back in time. An estimate of the typical standard deviation of θ is $\pm 2.0°$, calculated from the spread of the differences between the computed best fit values and the observed values in the last half of the data.

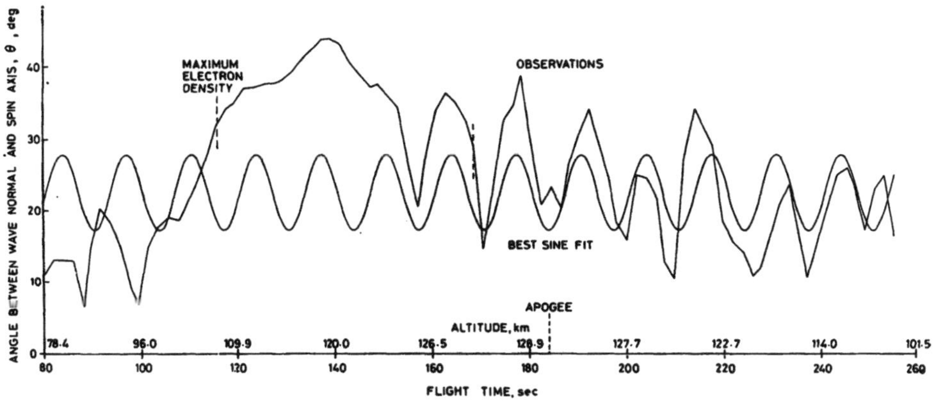

Fig. 1. Angle between wave normal of 2.8 kHz chorus signals and rocket spin axis θ derived from the observations as a function of flight time for P47H. Also shown is the best sine fit to the observations for flight times > 168 sec. This sinusoidal variation, due to precession of the rocket, is extrapolated back in time.

The angle θ is related to the angle of elevation of the rocket spin axis above the horizontal, $\alpha = 74.5°$, and the angle that the wave normal makes with the horizontal β, by

$$\cos\theta = \sin\alpha\sin\beta + \cos\alpha\cos\beta\cos\phi, \tag{3}$$

where ϕ is the geomagnetic bearing of the rocket, 288.5° E. It is here assumed that the wave is propagating with its wave normal in the magnetic meridian, or close to it. Then Eq. (3) can be solved for β, the angle between the wave normal and the horizontal in the magnetic meridian. The variation of β with flight time is shown in Fig. 2.

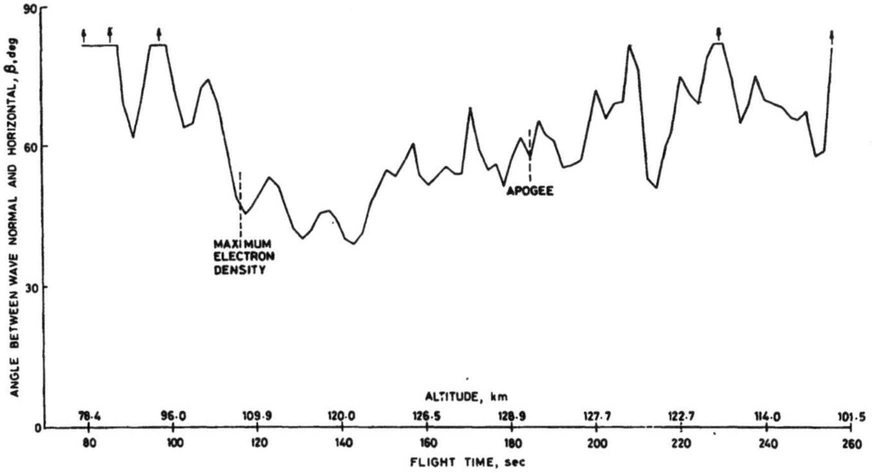

Fig. 2. Angle β between the wave normal of 2.8 kHz chorus signals, assumed to be propagating in the whistler mode as a single plane wave with its wave normal in the magnetic meridian, and the horizontal as a function of flight time for P47H. The arrows show where β approaches 90°.

In the latter part of the flight β varies irregularly about a mean value which is close to 71.5°, the angle the geomagnetic field makes with the horizontal. These results are not inconsistent with the belief that the chorus signal, observed in the lower ionosphere on the downleg portion of the flight, can be regarded as a single plane wave propagating in the whistler mode with its wave normal aligned along the geomagnetic field.

In the earlier part of the flight β is reduced to $\sim 42°$ near 120 km altitude. This is interpreted as showing that the wave is being significantly refracted on the upleg portion of the flight, between 106 and 125 km.

4. Electron Density Variation

The reduction of the Langmuir probe observations to derive the ionospheric electron density N_e profile has already been discussed [2]. Briefly, the current I to the probe, when biased to a potential V above space potential, is given as

$$I = N_e e S \left(\frac{kT}{2\pi m}\right)^{1/2} \left(1 + \frac{eV}{kT}\right)^n, \tag{4}$$

where the other symbols have their usual meanings. By plotting $\log I$ against $\log (1 + eV/kT)$ for the accelerating region of each sweep, made at 2-second intervals, n is found. An average value of n, calculated to be 0.56, is used in Eq. (4) to determine successive values of N_e, again using the accelerating region of each sweep.

The upleg profile, which has been published (see [2], Fig. 8), shows that the rocket passed through a well-defined electron density enhancement in the E region with the maximum electron density occurring at an altitude of 107.5 km. Fig. 3 shows the variation of electron density at the rocket as a function of flight

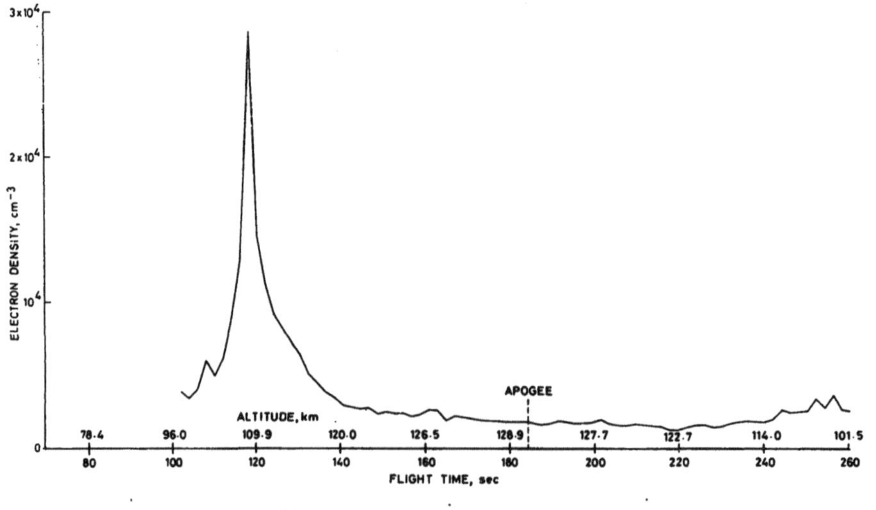

Fig. 3. The electron density at the rocket as a function of flight time for P47H. The rocket passed through a marked sporadic E layer at 107.5 km on the upleg, which was not observed on the downleg.

time. The downleg results, similarly derived, are included here. It is clear that no marked enhancement is observed at 107.5 km on the downleg, at a horizontal distance of only 45 km from the upleg crossing. This enhancement can probably be ascribed to a sporadic E layer of limited spatial extent. However it could be ascribed to a short lived ($\sim$20 sec) Es event of the type observed in auroral regions and reported by McNamara [4]. The reasons for believing the enhancement to be a spatial, rather than a temporal, variation include the following: the South Uist ionosonde indicated a sporadic E layer, sometimes blanketing, which persisted for several hours; the downleg profile is in good agreement with the upleg one in every respect except the enhancement; the Langmuir probe observations give no evidence of the fine structure which might be expected in an auroral Es layer. Although none of these points provides conclusive evidence, when considered together they suggest a stable sporadic E layer structure with little temporal variation on a time scale of the order of two minutes. It is possible that the electron density enhancement could be due to secondary ionization produced by the precipitating energetic electrons in gyroresonance with the whistler mode chorus signals [1]. A somewhat similar electron density enhancement was observed [4] as a Langmuir probe, ejected from a rocket launched from Churchill, Canada, passed through an active visual auroral form, also at 107 km altitude. Such ionization could then be redistributed to produce a well-defined Es layer.

The electron density observations are displayed as a function of time in order to facilitate their comparison with the results on wave normal geometry presented in Figs. 1 and 2. It is clear that, on the upleg, the downcoming chorus signals are being significantly refracted by the higher part of the sporadic E layer.

5. The Refraction of Chorus Signals by the Lower Ionosphere

For a certain electron density value and a model of the neutral atmosphere, the whistler mode refractive index surface is calculated at suitable frequencies using a standard computer program. The roots of the Booker quartic are found for wave normal angles of up to $\pm 90°$ from the geomagnetic field direction. The real part of the root corresponding to a downgoing propagating wave is chosen. The imaginary part of the root and thus of the refractive index is negligible at these heights. The surfaces have been plotted at 2 km height intervals above the sporadic E layer, from 108 km to 128 km, and half of these are shown for 2.8 kHz in Fig. 4. Also shown by dashed lines in Fig. 4 are the refractive index surfaces below the sporadic E layer, at altitudes of 96, 100, 104 and 107 km. The direction of the geomagnetic field is shown by $\boldsymbol{B}_{\mathrm{E}}$.

If the refracting medium varies slowly in space a wave is continuously refracted, that is, as it propagates the direction of the wave normal changes. Snell's law states in general that if a plane wave is incident obliquely on a boundary between media of refractive indices $\mathfrak{n}_{\mathrm{I}}$ and $\mathfrak{n}_{\mathrm{T}}$, at an angle θ_{I} to the normal to the boundary, then

$$\mathfrak{n}_{\mathrm{I}} \sin \theta_{\mathrm{I}} = \mathfrak{n}_{\mathrm{T}} \sin \theta_{\mathrm{T}}, \tag{5}$$

where θ_{T} is the angle between the transmitted wave normal and the normal to the boundary. Usually the wave normal direction of a wave propagating through the lower ionosphere is found assuming a horizontally stratified medium. In this case, however, it is the wave normal angles that are known at different heights in

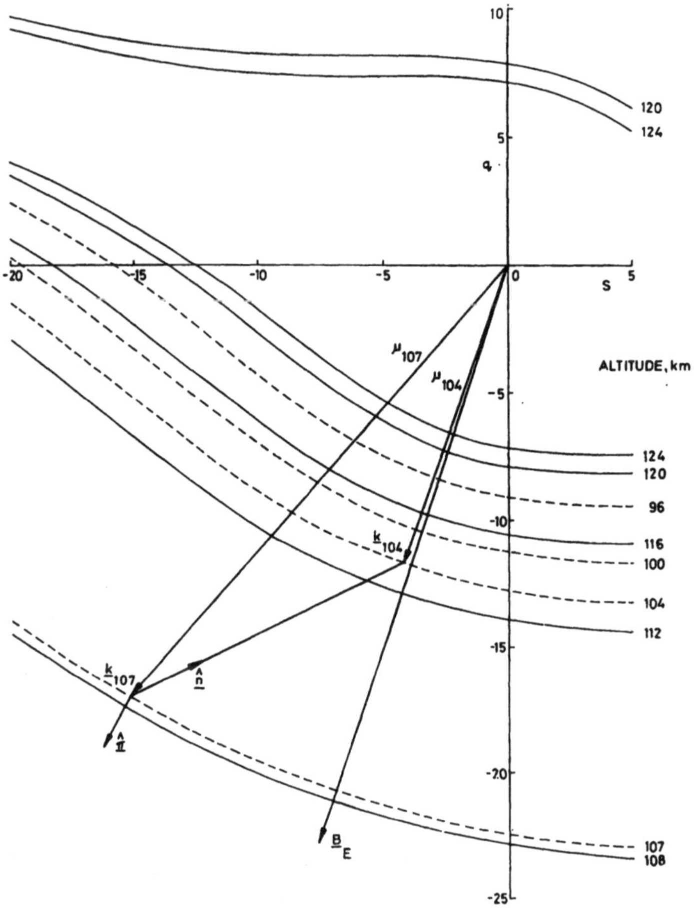

Fig. 4. Refractive index surfaces for 2.8 kHz whistler mode signals at different heights in the lower ionosphere on the upleg (solid lines) and downleg (dashed lines) of P47H. Wave normals, $\boldsymbol{k}$, are shown at altitudes of 104 and 107 km: the normal to the boundary between layers at these altitudes is shown by $\hat{\boldsymbol{n}}$. The ray direction at 107 km altitude is shown by $\hat{\boldsymbol{\pi}}$.

the ionosphere (see Fig. 2). Thus information can be obtained here on the structuring of the medium.

Using Snell's law and a simplified ray tracing technique [5], the refraction of the wave by the ionosphere can be studied in some detail. The ionosphere is divided into layers 2 km thick, corresponding to the refractive index surface spacing, and the wave normals at these specific heights are drawn at the angles to the horizontal obtained from Fig. 2 to intersect the respective refractive index surfaces shown in Fig. 4. Then a line drawn normal to the boundary between any two of these layers passes through the respective refractive index vector terminations [5] (see, for example, $\hat{\boldsymbol{n}}$ in Fig. 4). The orientation of the boundaries of these layers at 120, 116, 112, 108, 107, 104, 100 and 96 km altitude is indicated on the right-hand side of Fig. 5, with the corresponding electron densities.

The ray direction at each height (see, for example, $\hat{\boldsymbol{\pi}}$ in Fig. 4) is found by constructing the normal to the refractive index surface where the refractive index vector, in the direction of the wave normal, terminates. The ray direction of the downcoming chorus signal is thus obtained at altitude intervals of 2 km, and these directions are plotted relative to the geomagnetic field direction on the left-hand side of Fig. 5.

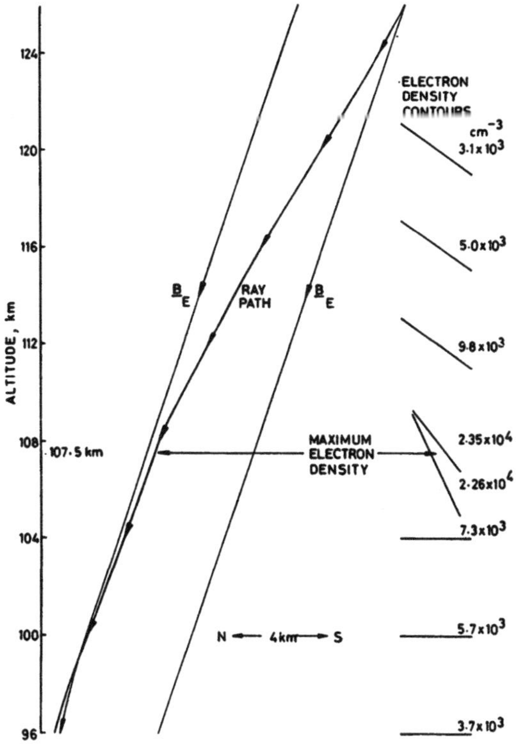

Fig. 5. Diagram in the geomagnetic meridian plane illustrating how the ray direction of 2.8 kHz chorus signals varies with altitude in the lower ionosphere, for P47H, in relation to the geomagnetic field direction B_E. The inclination of the electron density contours, i.e. the boundaries between the refracting layers, are also shown at eight different altitudes.

The method outlined above is only valid if the medium is slowly varying, as determined by the condition [6] that

$$\left| \frac{3}{4} \left(\frac{1}{\mu^2} \frac{d\mu}{dh} \right)^2 - \frac{1}{2} \frac{1}{\mu^3} \frac{d^2\mu}{dh^2} \right| \ll k^2, \tag{6}$$

where μ is the real part of the refractive index, k the propagation constant in free space, and h the height. Having determined the differentials to sufficient accuracy by graphical methods, it is concluded that the graphical method of deriving ray paths [5] is valid from 128 km to 108 km and from 107 km downwards. A full wave treatment for a downcoming whistler mode wave is only necessary for the ionosphere in the sporadic E layer between 107 km and 108 km. The effects of not using a full wave treatment could occur outside this region; for example, waves partially reflected from the two sides of the sporadic E layer could interfere as discussed by Pitteway and Jespersen [7] using a theoretical model.

6. Conclusions

The electron density enhancement in the E region is observed only on the upleg at 107.5 km altitude, and not on the downleg at a horizontal distance of only 45 km (at 288.5° E geomagnetic) and 136 seconds later. This enhancement can probably be ascribed to a sporadic E layer of limited spatial extent, or possibly to a temporally short lived auroral-type Es event. With the observations reported here, it is not possible to distinguish between spatial and temporal variations.

In this analysis, a non time-varying electron density profile is assumed. The observations have been used to estimate the inclination of the sporadic E layer to the horizontal on the assumption that the layer is tilted in the local magnetic meridian. The wave normal direction of the downcoming chorus signals, during the upleg of the flight, shows strong refraction throughout the region studied in detail from 128 km down to 96 km. The wave normal becomes more horizontal from 128 km down to 118 km, where it is at 42 ± 2° to the horizontal. From 118 km to 108 km the wave is gradually refracted, the wave normal moving a few degrees towards the vertical. At 107.5 km, corresponding to the position of the maximum electron density, the wave is strongly refracted, the angle between the wave normal and the horizontal increasing to 71.5° at 103.5 km altitude. Below 96 km the wave normal becomes almost vertical (Fig. 2).

The boundaries of the constant refractive index layers are shown by the ray tracing technique to be inclined at an angle of typically 35 ± 5° to the horizontal from 124 km to 112 km (Fig. 5). At 108 km the boundary is inclined at 51 ± 7° to the horizontal, and at 107 km, the region of the strongest refraction, the boundary is inclined at 64 ± 7° to the horizontal. Below 104 km, down to 96 km, the boundaries are horizontal; this is the region where the wave normal direction is within 20° of the vertical. On the downleg the wave is not strongly refracted (Fig. 2).

The ray direction for the downcoming chorus signals is obtained from altitudes of 126 km to 96 km, and is shown in Fig. 5. The ray propagates closely along the geomagnetic field. Between 126 km and 108 km the ray propagates at up to only 10° off the field line, to the north. The ray changes direction appreciably at 108 km, being within ±2° of the geomagnetic field between 108 km and 96 km altitude.

Acknowledgments

The authors are grateful to the Science Research Council for financing this project, to D. Hickinson and N. Jenkinson of Dunford Hadfield (Sheffield) Limited for building the electronic equipment, to members of the launch team at South Uist, to colleagues at our Universities for useful discussions, and to S. Powell for making significant mathematical contributions and for computer programming.

References

[1] M. J. RYCROFT, Radio Sci. 7, 811 (1972).
[2] M. J. RYCROFT, R. A. USHER et al., Space Research XII, 1499 (1972).
[3] J. VÁVRA, A rocket-borne very low frequency receiver, University of Southampton Report (April 1970).
[4] A. G. MCNAMARA, Can. J. Phys. 47, 1913 (1969).
[5] R. A. HELLIWELL, Whistlers and Related Ionospheric Phenomena, Stanford University Press, Stanford 1965.
[6] K. G. BUDDEN, Radio Waves in the Ionosphere, Cambridge University Press, Cambridge 1961.
[7] M. L. V. PITTEWAY and J. L. JESPERSEN, J. Atmos. Terr. Phys. 28, 17 (1966).

The Magnetosphere

Space Research XIV — Akademie-Verlag, Berlin 1974

PARTICLE ENTRY INTO THE RING CURRENT REGION OF THE EQUATORIAL MAGNETOSPHERE

T. A. Fritz[a*], D. J. Williams[a], P. H. Smith[b], R. A. Hoffman[b], J. N. Barfield[c] and A. Konradi[d]

[a]Space Environment Laboratory, NOAA/ERL, Boulder, Col., USA
[b]NASA/Goddard Space Flight Center, Greenbelt, Md, USA
[c]Environmental Data Services, NOAA, Boulder, Col., USA
[d]NASA/Johnson Space Center, Houston, Texas, USA

We present a review of initial results obtained from the NASA S^3-A satellite (Explorer 45) concerning the behavior and dynamics of protons associated with the storm-time and quiet-time extraterrestrial ring current at the equatorial plane out to radial distances of 5.5 earth radii. These initial results all pertain to the evening sector where the position of the plasmapause during quiet times is well above the S^3 apogee. During quiet times the proton energy spectrum exhibits a peak in the interval from 100 to 200 keV with protons of energies > 100 keV representing most of the particle energy density around $L \sim 4$. During storm conditions, the intensities of the higher energy protons decrease while the intensities of protons from 10 to 100 keV are greatly enhanced making them the dominant contributor to the storm-time particle energy density. Values of $\beta > 1$ are observed during magnetic storms. The plasmasphere is apparently responsible for the decay of the storm-time ring current into the quiet-time particle distribution. Initial results for injection of electrons are also reviewed briefly.

1. Introduction

We present a short review of the initial results of various studies being conducted to determine the behavior of magnetospheric particle populations near the equatorial plane at altitudes below the geostationary orbit. These studies are based on data obtained by the NASA Small Scientific Satellite, S^3-A (Explorer 45), which is providing the first total proton energy density measurements in the storm-time ring current region. Prior to the present studies, studies in the geostationary orbit have shown effects of particle injection and acceleration [1, 2] along with effects of drift in the geomagnetic and geoelectric field [3—6]. The protons associated with the ring current have been directly observed only in one case [7] but the magnetic signature of the ring current has been observed on several occasions [8—11].

2. Description of Satellite, Instrumentation, and Program Concept

S^3-A was launched on 15 November 1971 into an elliptical orbit having an apogee of 5.24 earth radii, a perigee of 220 km, a period of 7.82 hours, and an inclination of 3.5° [12]. The local time features of the early orbits are shown in

* Work done while spending a year in residence at the Max-Planck-Institut für Aeronomie, Institut für Stratosphärenphysik, Lindau/Harz, FRG.

Fig. 1: apogee moves westward in this diagram about 12° per month. Proton intensities are measured from 0.73 keV to 2800 keV in 31 differential channels [13] and electron intensities are measured from 0.73 keV to 560 keV in 20 differential channels. Three-axis magnetic field measurements [14] and single-axis dc electric field measurements [15] are made as well as electric and magnetic noise measurements. The location of the plasmapause can be identified by the dc

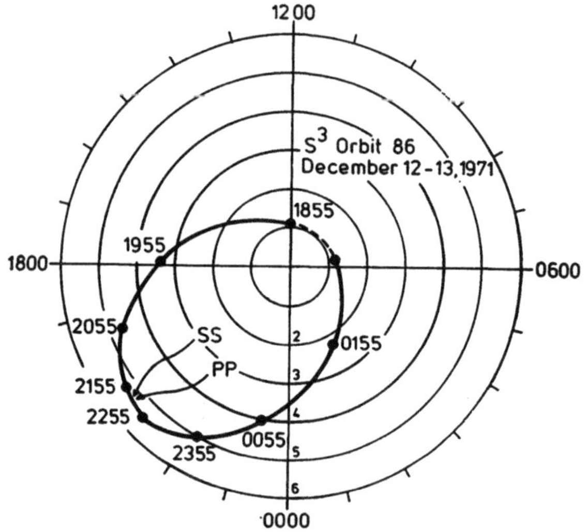

Fig. 1. S^3 orbit as a function of L value in earth radii and local time [UT times shown on orbit along with PP (plasmapause position) and SS (substorm start) are connected with substorm shown in Fig. 4] [17].

electric field detector [15]. The satellite contains an operational data processing system which permits complete format changes in flight in order to concentrate on specific problems. This feature has been used extensively.

The S^3 concept is unique. A single data tape is distributed to all experimenters in the program. The satellite is considered as a single experiment with each experimenter free to use any and all of the data regardless of his particular connection with the instrument making the measurements. The experimenter team consists of eight scientists representing five separate institutions. There are also a number of associate members in this team. With this operational situation, magnetospheric problems can be viewed and approached with a complete set of simultaneous data bearing on the problem.

3. Dynamics of Particle Entry

3.1. Quiet-time Ring Curren

Since the orbit of S^3 confines the satellite to the local evening quadrant during the time period discussed all measurements taken during quiet conditions are within the plasmapause. The quiet-time ring current is observed as a peak in the proton energy spectrum between 100 and 200 keV and is located roughly around $L \sim 4$. The spectrum shows a decrease from <1 keV to 10 keV, a relative minimum from 40 keV to 80 keV and a relative maximum from 80 keV to 300 keV, above which it steadily decreases with increasing energy [16]. An example of this distribution is given in Fig. 2 as Orbit 97 which was recorded during the quiet period preceding the large 16—17 December 1971 magnetic storm (and in Fig. 5 of Williams et al. [17] before the onset of a substorm).

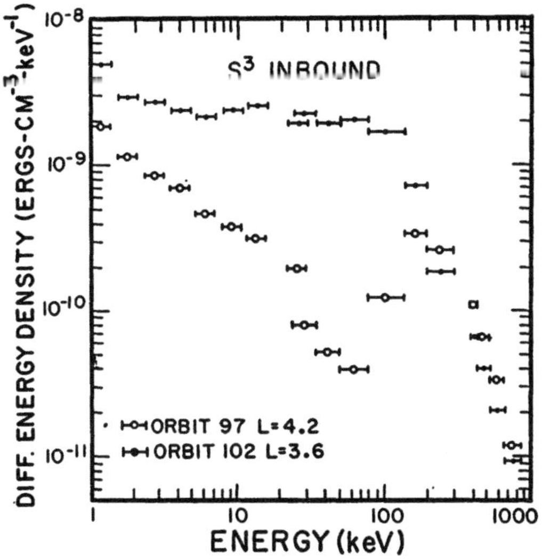

Fig. 2. Proton energy density spectra taken on the inbound portion of the S^3 orbit during a quiet period (Orbit 97 at $L = 4.2$ preceding and during the symmetric storm-time ring current in the second magnetic storm (Orbit 102 at $L = 3.6$) of the 16—18 December 1971 activity [16].

3.2. Magnetic Storms

We have shown that the storm-time ring current is produced by the injection of protons with energies from 1 keV to 140 keV with particles from 20 keV to 100 keV producing the largest increase from the quiet-time distribution [16]. An example of this distribution is presented in Fig. 2 (Orbit 102) which was recorded during the second of the large storms of 16—17 December 1971. It is noted that the minimum in the quiet-time spectrum is filled in and the higher energies are decreased, while protons with energies from 1 to 138 keV contribute only 20% or less to the energy density out to $L \sim 4$ during magnetically quiet periods; their enhancement during magnetic storms combined with a depletion of protons with energies greater than about 170 keV make them the dominant (>90%) contributors to the storm-time energy densities [16]. We have found that the accepted phenomenological fact of $\beta_{max} \lesssim 1$ where β is the ratio of the particle energy density to the magnetic field energy density is not true. Values of β as large as 6 have been observed using measurements of both the magnetic field and the particle distributions [18].

The inner edge of the storm-time ring current has been observed as close as $L = 2.5$ and is usually associated with the plasmapause [15, 16] with the inner edge appearing a maximum of $\sim$1 earth radii inside the position of the S^3 definition of the plasmapause [15]. We have shown that conditions suitable for the initiation of the ion-cyclotron instability do exist in this interaction region between the plasmapause and the inner edge of the ring current but that the subsequent evolution of the data does not follow the predictions of this instability based on linear theory [19]. We have argued that a dynamic mechanism such as a wave-particle instability like the ion-cyclotron resonance interaction must be the process governing the "loading" of magnetic field lines and the trapping capabilities of the magnetosphere as opposed to a maximum value of β [18]. This instability is apparently responsible for the decay of the storm-time ring current distribution of Fig. 2 back into the quiet-time distribution of Fig. 2 as the plasma density in the plasmasphere recovers by filling from the underlying ionosphere and expands to its pre-storm position beyond the quiet-time ring current position. This instability is probably also involved in the build-up of the ring current during main phase magnetic storms. For strong pitch angle diffusion the lifetime of a proton approaches a minimum value of $\tau = 2T_B(\alpha_0)^{-2}$, where T_B is the bounce period and α_0 the size of the loss cone [20]. τ is at least an hour at $L = 4$. In contrast Davis [21] has provided evidence that the injection of particle energy during a main phase storm is burst-like with individual bursts having durations of several tens of minutes in coincidence with peaks in substorm activity 16, 18]. Therefore if substorm activity is rapid and prolonged the protons will be enhanced at a rate faster than the operating loss mechanism can get rid of them.

3.3. Detail Features of Particle Entry (i.e. Substorms)

In studying the details of particle entry into the equatorial region covered by S^3, it is most convenient to consider isolated substorms. It is difficult to locate an isolated substorm during a period of low magnetic activity and also find the position of the plasmapause below the satellite at apogee ($L \sim 5.2$). The plasmapause is an important boundary apparently governing the earthward extension of the substorm and particle injection [22, 23]. We have found that the initial enhancements of protons associated with the first substorm of a magnetic storm do not exhibit the clear dispersion of arrival times at the satellite as a function of energy and pitch-angle which is exhibited in proton injection events that are associated with later substorms throughout the development of a magnetic storm and which have been reported at the geostationary orbit [23]. The satellite is usually within the plasmasphere for these initial substorms.

The protons appear to be injected at essentially all pitch angles. We have used this fact to report the first observation of pitch angle dispersion in the protons arriving at the satellite [24]. In Fig. 3 we show three of nine complete pitch angle scans at 8.5 second intervals. The pitch angle distributions clearly show the evolution with time of a $\sin^n\alpha$ distribution into a triple hump distribution. According to Lew [25] the relationship between the drift time of a proton and its energy and equatorial pitch angle α_{eq} is

$$t = t_0 + \frac{44}{LE} \frac{\varphi - \varphi_0}{360} \frac{G}{F} (\alpha_{eq})$$

where t_0 and t are the injection and observation times in minutes, φ_0 and φ are the injection and observation azimuths in degrees, L is McIlwain's parameter,

E is the proton energy in MeV and G/F is a function of the equatorial pitch angle and is of the order of unity. In Fig. 3 we have plotted the time of observation against $G(F)^{-1}$; this gives a straight line, the slope of which indicates the drift in azimuth and the intercept the arrival time of 90° pitch angle protons. Then

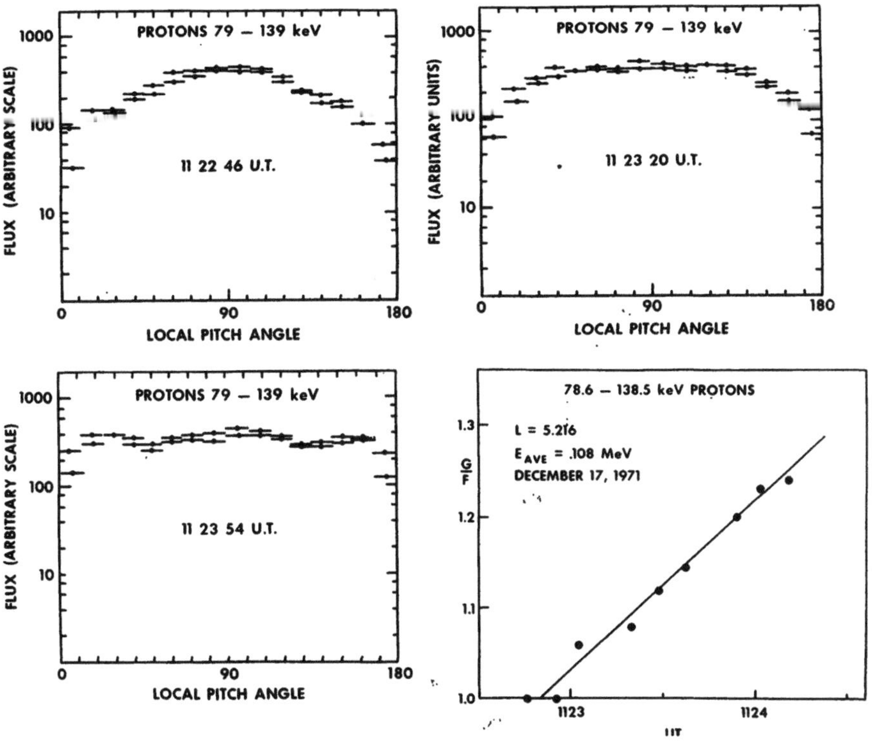

Fig. 3. Time development of the pitch angle distributions (in degrees) of 79—139 keV protons at 34 second intervals observed at $L = 5.22$ (2100 LT) on 17 December 1971 [24]. The function $G(F)^{-1}$ (see text) versus time for all data in this series is also plotted.

from the above equation we can determine that these protons were injected into the magnetosphere 20° W of midnight at ~1117 UT. Two substorms were registered on the College magnetometer starting about 1104 and 1118 UT. We therefore believe that we see pitch angle dispersion resulting from drifting protons injected into the magnetosphere near midnight [24]. The plasmapause was located at $L \sim 4.3$ on the outbound portion of this orbit and at $L \sim 4.0$ on the inbound portion. The energy of the enhanced protons is usually from 20 keV to 80 keV as in Fig. 2 with the minimum in the quiet-time distribution "filling in". For an example associated with a single substorm see Fig. 5 of Williams et al. [17]. There appears to be a decrease in the higher energy protons in the individual substorm injections as was noted in § 3.2 for magnetic storms.

The injected protons are able to penetrate into the plasmasphere but are attenuated [17, 22] and the majority of their energy density may stay outside the plasmapause [17]. The behavior of electrons has been studied for only two substorms thus far. In one case the satellite remained within the plasmapause during the entire substorm and only proton enhancements were observed [22]. In the other case the satellite moved outside the plasmasphere approximately 20 minutes after the initiation of the substorm and observed dispersion effects in the arrival of the electrons. These dispersions are presented in Fig. 4 for the substorm depicted in Fig. 1. The vertical bars are the width of the energy channels and the horizontal bars are an estimate of timing errors. A dispersion curve due to gradient drift in a dipole field based on injection at midnight is shown for comparison.

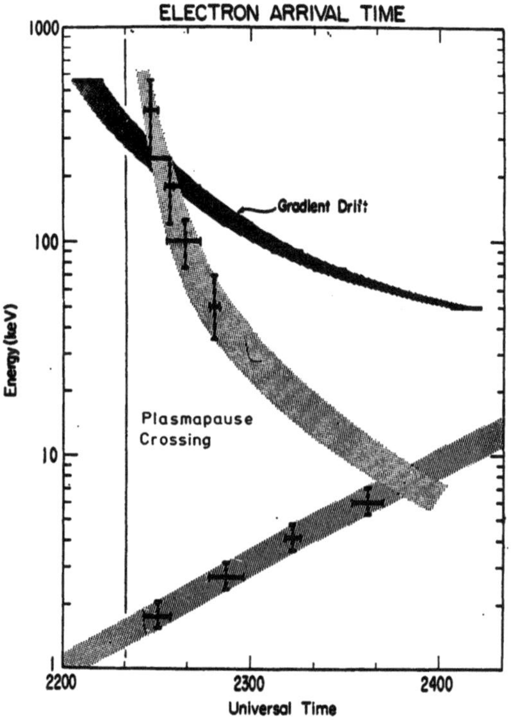

Fig. 4. Electron energy dispersion characteristic observed for a substorm occurring at ~2150 UT on 12 December 1971 [17].

We have compared these results with flow patterns expected from a model geomagnetic-geoelectric field constructed specifically to explain particle signatures observed at ATS-5 in geostationary orbit [5, 17]. Electron trajectories in this McIlwain model were calculated for all values of the magnetic moment of interest. *No* electrons originating in the geomagnetic tail could reach the S^3 position in this model via a continuous flow process. The low energy electrons had all of the characteristics associated with the electron plasma sheet, but cannot be associated with a simple inward convection from the geomagnetic tail in a steady electric field. A discontinuous change at the time of the substorm initiation is apparently required to explain the sudden appearance of these particles at altitudes well within the geostationary orbit. This agrees basically with the plasma clouds concepts advanced by De Forrest and McIlwain [4]. After the plasma cloud has been injected to low altitudes, drift in the static electric field model does seem to explain subsequent particle observations [17].

4. Conclusion

We have attempted to review the results from initial studies of particle dynamics with the S^0 satellite. We have shown the first observations of the complete quiet-time and storm-time ring current proton distributions and have shown that values of $\beta > 1$ do occur during magnetic storms. We have also shown that the plasmasphere exhibits a strong influence over the characteristics of particle injection and apparently governs the decay of the storm-time ring current back to the quiet-time distribution via a wave-particle instability similar to (if not) the ion-cyclotron instability. If the position of the plasmapause is below the satellite, proton injection appears to be consistent with the patterns determined at the geostationary orbit of injection over a narrow region near midnight and subsequent drift in the geoelectric and geomagnetic fields.

A discontinuous change in the geoelectric field at the time of substorm initiation may be required to explain the sudden appearance of particles at the S^3 orbit.

Acknowledgments

The authors acknowledge the participation of Dr. L. J. Cahill of the Univ. of Minnesota, Dr. D. A. Gurnett of the Univ. of Iowa, and Dr. N. Maynard of NASA/GSFC as members of the S^3 experimenter team. One of us (TAF) wishes to thank Professor G. Pfotzer and Dr. D. J. Williams for the special arrangements which have made the year of residence at MPI/Lindau and this review work a possibility.

References

[1] G. K. Parks and J. R. Winckler, J. Geophys. Res. **73**, 5786 (1968).
[2] T. W. Lesniak and J. R. Winckler, J. Geophys. Res. **75**, 7075 (1970).
[3] R. L. Arnoldy and K.-W. Chan, J. Geophys. Res. **74**, 5019 (1969).
[4] S. E. De Forrest and C. E. McIlwain, J. Geophys. Res. **76**, 3587 (1971).
[5] C. E. McIlwain, in: Earth's Magnetospheric Processes, D. Reidel Publ. Co., Dordrecht, Holland 1972 (p. 268).
[6] E. G. Shelly, R. G. Johnson and R. D. Sharp, Radio Sci. **6**, 305 (1971).
[7] L. A. Frank, J. Geophys. Res. **72**, 3753 (1967).
[8] E. J. Smith, D. J. Coleman, Jr, D. L. Judge and C. P. Sonett, J. Geophys. Res. **65**, 1858 (1960).
[9] L. J. Cahill, Jr, J. Geophys. Res. **71**, 4505 (1966).
[10] P. J. Coleman, Jr, and W. D. Cummings, J. Geophys. Res. **76**, 51 (1971).
[11] M. Sugiura, B. G. Ledley, T. L. Skillman and J. P. Heppner, J. Geophys. Res. **76**, 7552 (1971).
[12] G. W. Longanecker and R. A. Hoffman, J. Geophys. Res. **78**, 4711 (1973).

[13] T. A. FRITZ, NOAA Tech. Rep. (1973).
[14] L. J. CAHILL, JR, J. Geophys. Res. **78**, 4724 (1973).
[15] N. C. MAYNARD and D. P. CAUFFMAN, J. Geophys. Res. **78**, 4745 (1973).
[16] P. S. SMITH and R. A. HOFFMAN, J. Geophys. Res. **78**, 4731 (1973).
[17] D. J. WILLIAMS, J. N. BARFIELD and T. A. FRITZ, J. Geophys. Res. **79**, 554 (1974).
[18] R. A. HOFFMAN, J. Geophys. Res. **78**, 4771 (1973).
[19] D. J. WILLIAMS, T. A. FRITZ and A. KONRADI, J. Geophys. Res. **78**, 4751 (1973).
[20] C. F. KENNEL, Rev. Geophys. **7**, 379 (1969).
[21] T. N. DAVIS, J. Geophys. Res. **74**, 6266 (1969).
[22] J. N. BARFIELD, D. J. WILLIAMS and N. MAYNARD, Trans. Am. Geophys. Un. **54**, 415 (1973).
[23] P. H. SMITH, R. A. HOFFMAN and T. A. FRITZ, Trans. Am. Geophys. Un. **54**, 415 (1973).
[24] A. KONRADI, D. J. WILLIAMS and T. A. FRITZ, J. Geophys. Res. **78**, 4739 (1973).
[25] J. S. LEW, J. Geophys. Res. **66**, 2681 (1961).

Space Research XIV — Akademie-Verlag, Berlin 1974

THE SPATIAL DISTRIBUTION OF QUASI-TRAPPED ENERGETIC ELECTRONS OBSERVED ABOARD THE SATELLITE "SHINSEI"

H. Takeuchi, T. Imai, S. Kumagaya, M. Wada and Y. Miyazaki

Cosmic Ray Laboratory, Institute of Physical and Chemical Research, Itabashi, Tokyo, Japan

Preliminary analyses using the data of a plastic scintillator are carried out. A contour of the counting rate in the $B-L$ diagram at the fringe of the inner radiation belt is checked in the electron energy range greater than 120 keV. The counting rate specifying the contour is taken as roughly corresponding to the intensity of 100 electrons $cm^{-2} s^{-1} sr^{-1}$. The results agree with those obtained by the previous investigators concerning position and slope in the $B-L$ diagram. Electron intensity enhancements are observed several times in the lower energy channels in the limited regions of $L = 1.5-1.7$ over the Caribbean Sea. The results are near but a little different from those of $L = 1.2$ and $L = 2.0$ enhancements reported by Paulikas and Freden.

1. Introduction

In this report, preliminary analyses are presented on the quasi-trapped electrons observed with the detectors aboard the first Japanese scientific satellite (1971-080A, Shinsei) launched on 28 September 1971. Experimental procedures are briefly explained in § 2. Preliminary analyses on several events which are interpreted as precipitation of electrons are presented in § 3. The comparison with one of the previous investigations is briefly remarked in § 4.

2. Experiment

The satellite is spin-stabilized at the rate of about two spins per second. The orbital parameters of the satellite in the early stage are as follows: epoch, 2 Oct. 1971, 0000 UT; semimajor axis, 7750.31 km; eccentricity, 0.06430; inclination, 32.05°; perigee height, 874 km; apogee height, 1870 km; anomalistic period, 113 minutes.

The measurements are carried out with a plastic scintillator and three Geiger counters including two mica window counters. Only the results obtained with the scintillator are reported here. The scintillation counter system consists of a plastic scintillator of cylindrical shape, 14.5 mm diameter and 5 mm height, and five pulse height discriminators which provide electron threshold energies of approximately 70, 120, 400, 1000 and 2500 keV respectively, together with the use of the entrance window of aluminium 4 mg cm^{-2} thick. A collimator made of aluminium cylinder is used with its axis placed perpendicular to the spin axis of the vehicle.

The geometric factor and the mean aperture angle of the collimator are $G = 0.20\ \text{cm}^2\ \text{sr}$ and $2\bar{\theta} = 24°$ respectively.

Counts of eight kinds of signals in the time interval of 3.88 seconds (nearly equal to the frame period of the PCM telemetry used in the experiment) are stored in the eight-word counters, each of which consists of eight bits, and are stored by the on-board tape recorder. The tape recorded data were acquired on command at Kagoshima Space Center (31° N 131° E) from late September 1971 to early February 1972. Real-time telemetry data were taken till the middle of July 1972.

3. Results

In order to check the validity of the data a counting rate contour in the $B-L$ diagram* is examined, which corresponds to 100 counts per 4 seconds (corresponding approximately to 100 counts $\text{cm}^{-2}\ \text{s}^{-1}\ \text{sr}^{-1}$), for the electron energies greater than 120 keV. Its shape and the slope in the diagram are similar to those obtained by previous investigators, e.g., by Temny [2] for electrons of energies greater than 50 keV in the lower part of the inner radiation belt, or by Paulikas, Blake and Freden [3] for electrons of energy greater than 300 keV at the lower fringe of the inner radiation belt. The relation of the contour of such quasi-trapped electrons obtained by the scintillator to those of the trapped proton belt agrees approximately with those shown by previous investigators. Detailed analyses are now being carried out.

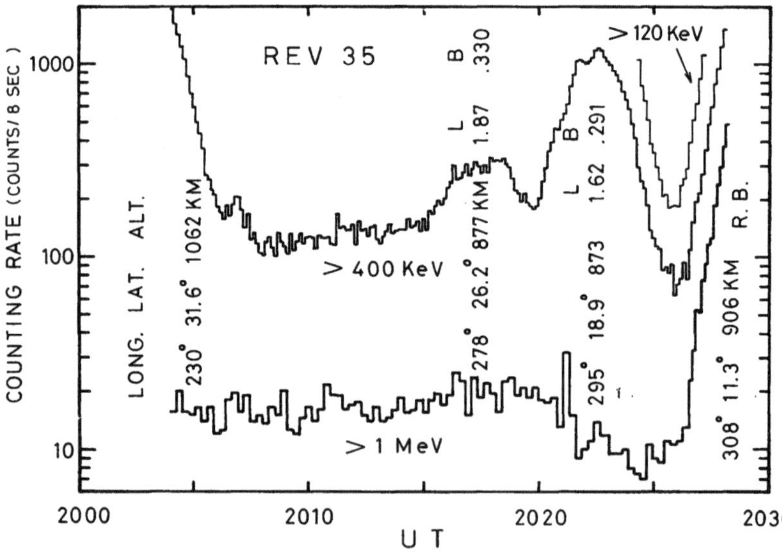

Fig. 1. Counting rate vs universal time (UT), Rev. 35. R. B. indicates the inner radiation belt.

---

* A table of B and L prepared by B. Venkatesan using the programme by C. E. McIlwain [1] is used.

Several events which are to be interpreted as electron precipitation were observed on almost every day from 28 September to 12 October 1971 at the lower fringe of the inner radiation belt over the Caribbean Sea. Typical counting rate curves which show one or two peaks among the background counting rates are shown in Fig. 1 and 2. They are seen in the results corresponding to electron

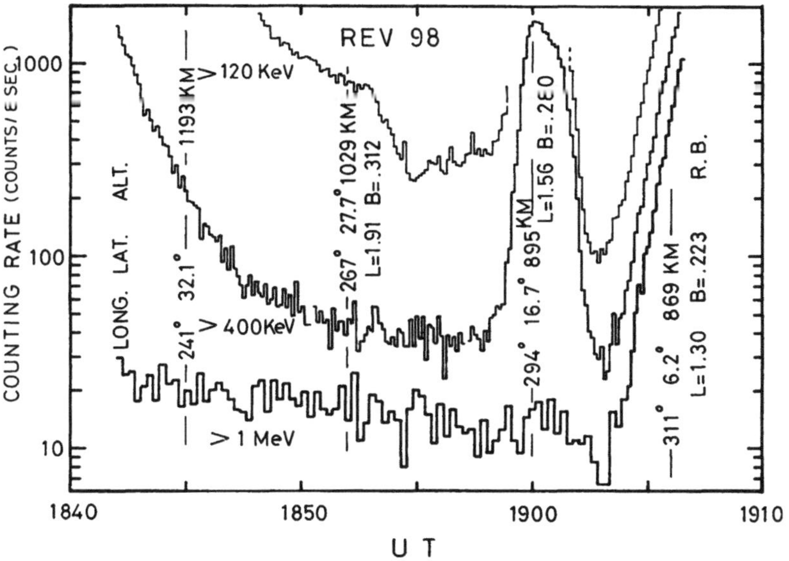

Fig. 2. Counting rate vs universal time (UT), Rev. 98.

energies greater than 400 keV, and to other threshold energies. There are some exceptions in the higher altitude observations after 11 October. In these cases a small peak is observed in the data of the 1 MeV channel in each observation.

After the last record which showed such events, the vehicle rose to higher altitudes in the region concerned, and consequently penetrated into the inner radiation belt.

The times, geographic locations, and the other characteristics of the peak events are summarized in Table 1. The positions of the peaks in the $B-L$ diagram are plotted in Fig. 3. They seem to lie almost in the interval of L between 1.5 and 1.7. Two points lie at nearly $L = 2$, corresponding to the first peaks in the two peak events (Rev. 35, and Rev. 60), see Table 1. In order to show the position of the lower edge of the inner radiation belt, a line described in the paper by McIlwain [1] is included in Fig. 3.

4. Remarks

Similar events were reported previously by Paulikas and Freden [4] for electrons of $\sim$0.9 MeV at fairly low altitudes. They observed sharp peaks at $L = 2.0$ and at $L = 1.2$ at various places. There are indications of events at $L = 2.0$ in our results, too, but no evidence is obtained for $L = 1.2$ in our case.

Table 1

Summary of the Peak Events observed in the Electron Energy greater than 400 keV

Rev	Date 1971	UT (h m)	Long. (°E)	Lat. (°N)	Alt. (km)	LT (h)	Approx. peak c.r. (counts/4 sec)	L	B (gauss)
10	28 Sep.	2119.9	296.9	16.0	918	17.1	450	1.54	0.272
23	29 Sep.	2114.8	271.3	22.5	872	15.8	280	1.65	0.313
35-1	30 Sep.	2017.5	277.8	26.2	877	14.8	150	1.87	0.330
35-2	30 Sep.	2022.6	295.0	18.9	873	16.0	600	1.62	0.291
48	1 Oct.	2048.2	273.2	23.3	875	15.0	120	1.70	0.317
60-1	2 Oct.	1916.4	263.1	30.8	981	12.8	90	2.03	0.329
60-2	2 Oct.	1924.8	293.1	21.6	877	15.0	110	1.72	0.303
61	2 Oct.	2117.2	262.5	22.5	882	14.8	85	1.59	0.307
73	3 Oct.	1956.7	291.6	16.0	871	15.4	500	1.53	0.282
98	5 Oct.	1900.2	293.9	16.7	895	14.6	800	1.56	0.280
99	5 Oct.	2050.6	257.2	20.8	927	14.0	450	1.51	0.290
123	7 Oct.	1803.1	294.6	18.3	955	13.7	500	1.62	0.279
148	9 Oct.	(1707.7)[a]	300.6	17.1	1008	13.2	(2000)[a]	1.59	0.262
161	10 Oct.	(1735)[a]	295.0	18.7	1072	13.2	(6000)[a]	1.66	0.266
174	11 Oct.	(1805)[a]	277.4	16.0	1079	12.6	(10000)[a]	1.52	0.262
186	12 Oct.	(1639)[a]	289.6	18.2	1158	12.0	(>10000)[a]	1.66	0.258

[a]) See the explanation of crosses in the caption of Fig. 3.

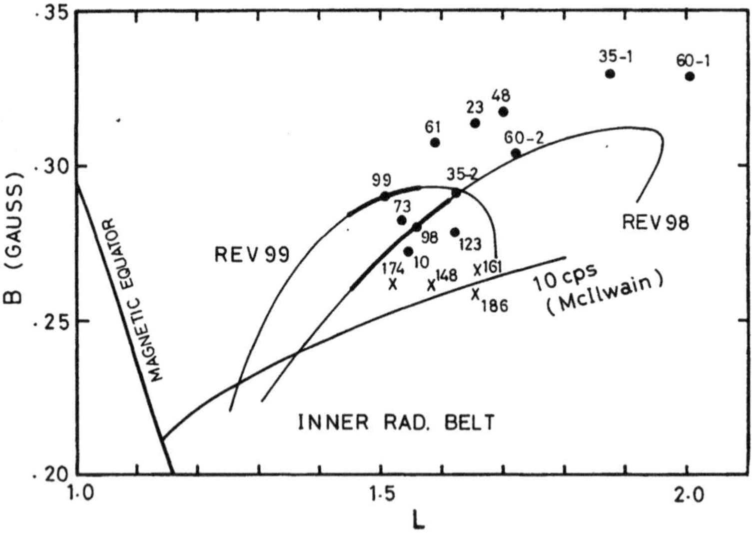

Fig. 3. The position of the peaks in the $B-L$ diagram. The thick portions in the lines showing the trajectories of Revs. 98 and 99 show the regions of appreciable intensity enhancement. Full circles show the peak positions observed in the revolutions denoted by the numbers attached to them. Crosses correspond to the cases in which the peaks cannot be determined due to overflow in the memory counter and are only estimated from the counting rates before and after the supposed peak positions.

Acknowledgment

This work was performed in co-operation with the Institute of Space and Aeronautical Science, University of Tokyo. International support of the project as well as the collaboration of the colleagues are gratefully acknowledged. Thanks are due to Dr. Y. Hakura who kindly lent the authors the table of *B* and *L* by Venkatesan.

References

[1] C. E. McIlwain, J. Geophys. Res. **66**, 3681 (1961).
[2] V. V. Temny, Space Research V, 489 (1965).
[3] G. A. Paulikas, J. B. Blake and S. C. Freden, J. Geophys. Res. **71**, 3165 (1966).
[4] G. A. Paulikas and S. C. Freden, J. Geophys. Res. **69**, 1239 (1964).

Space Research XIV — Akademie-Verlag, Berlin 1974

SPATIAL DISTRIBUTION OF CHARGED PARTICLES BENEATH THE RADIATION BELTS AS MEASURED ON BOARD THE SATELLITE INTERCOSMOS 3

M. BÁNÓ[d], J. DUBINSKY[d], S. FISCHER[c], T. GOMBOSI[e], N. L. GRIGOROV[b], Á. HOLBA[e], S. N. KUZNETSOV[b], V. A. KUZNETSOVA[b], V. N. LUTSENKO[a], S. PINTÉR[f], N. F. PISARENKO[a], I. A. SAVENKO[b], A. SOMOGYI[e], P. V. VAKULOV[b], L. VANICSEK[e], S. N. VERNOV[b] and A. V. ZAKHAROV[a]

[a]Space Research Institute of the USSR Academy of Sciences, Moscow, USSR
[b]Research Institute of Nuclear Physics of the Moscow State University, Moscow, USSR
[c]Astronomical Institute of the Czechoslovak Academy of Sciences, Ondřejov, Czechoslovakia
[d]Institute of Experimental Physics of the Slovak Academy of Sciences, Kosice, Czechoslovakia
[e]Central Research Institute of Physics of the Hungarian Academy of Sciences, Budapest, Hungary
[f]Geophysical Institute of the Slovak Academy of Sciences, Hurbanovo, Czechoslovakia

Some properties of charged particle fluxes observed under the radiation belts at $L \leqq 3$ are described. In these regions, for electrons with energies $E_e > 40$ keV there is a characteristic longitude dependence of the electron intensity. It is shown that, as the electrons drift round the earth, most of them are lost in the region of the Brazilian anomaly, where their trajectories move down into the atmosphere. Summarized and averaged data measured during the violent geomagnetic storm of 17 October 1970 and immediately after it, during a time of moderate geomagnetic disturbance (19 October 1970) are given.

The intensities of very energetic particles, i.e. cosmic rays and their secondaries, are given as a function of L. The values obtained are in good agreement with those measured during the solar activity cycle 19, in the years 1959—1960. Attention is drawn to a significant increase of such penetrating particle intensities in equatorial regions at longitudes 100—150°.

1. Introduction

The intensities of electrons with energies $E_e > 40$ keV and protons with energies $E_p > 1$ MeV were measured on board the satellite Intercosmos 3 by means of several semiconductor detectors and end-window GM counters. Furthermore, the intensities of more energetic particles were measured by means of two shielded GM counters of types SBT-9 and STS-5. A full description of the detectors is given in [1].

The satellite Intercosmos 3 was launched on 7 August 1970 into an orbit with apogee 1300 km, perigee 210 km and equatorial inclination 48.5°.

This paper deals mainly with the data measured in regions with $h_{min} < 0$, where h_{min} is the minimum altitude of the mirror point, i.e. in regions below the radiation belts.

2. Electron Loss

Increased intensities of >40 keV electrons with trajectories $h_{min} < 0$ were observed fairly frequently. Fig. 1 shows a typical pitch angle distribution of the dissipated electrons as measured in the northern hemisphere. Quantitative analysis of this distribution is rather difficult because of the large opening angle, $\pm 30°$, of the detector. However, it can be seen that intensities of particles moving away from the earth ($\alpha < 90°$) are somewhat smaller than those of particles travelling towards the earth ($\alpha > 90°$). This indicates that part of the particles registered had pitch angles $<70°$. Arriving from the southern hemisphere, they plunge immediately into the atmosphere. It may thus be concluded that the loss of electrons is connected with a continuous decrease of the altitude of the mirror points of trapped particles. Most of the observed particles have pitch angles of

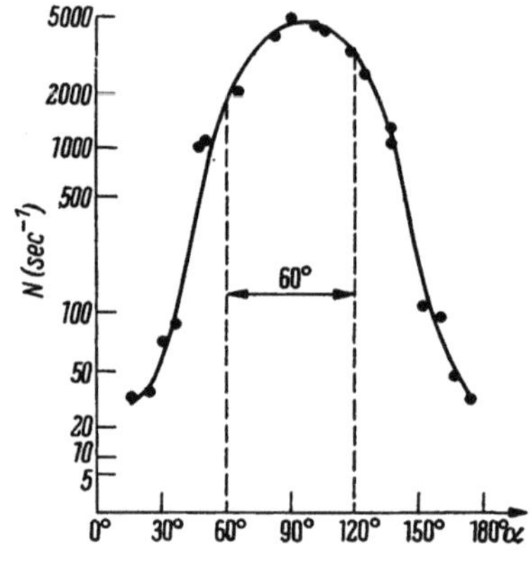

Fig. 1. Angular distribution of electrons with energies > 40 keV in the northern hemisphere. 21 August 1970, 1509—1515 UT (orbit 211). The angle α is that between the detector axis and the direction of the magnetic field. The pitch angle of a particle travelling along the detector axis is $180° - \alpha$. $L = 1.744—1.838$; $B = 0.407—0.406$ gauss; $H_{\min} = -989$ to -927; $h = 672—482$ km; $\lambda = 118—150°$ E.

about 90° at the point of observation and drift continuously eastwards while bouncing from one hemisphere to another.

The mirror points of these particles go down into the dense parts of the atmosphere in the region of the Brazilian anomaly. Thus the particles drifting round the earth move into the atmosphere and are lost. This type of loss is characteristic for shells up to $L = 3—4$. If the reduction of the mirror point altitude takes place continuously at all geographic longitudes, then an increase of the electron intensity should be observed when the region of magnetic anomaly is approached from the west. Fig. 2 shows the average of the intensities of dissipated particles measured at the shells $L = 1.7$, $L = 2.5$ and $L = 3$ in the northern hemisphere within the intervals of geographic longitudes $200° \leqq \lambda \leqq 220°$, $270° \leqq \lambda \leqq 290°$, $300° \leqq \lambda \leqq 310°$, on the 17 October 1970 and 19 October 1970, i.e. during and immediately after the violent magnetic storm which started on the 16 October 1970. Data measured on the 17 October 1970 were obtained half a day after the sudden commencement of the storm, and those of the 19 October 1970, one and a half days after the storm, which took place on 17—18 October. It may be seen from Fig. 2 that the intensities of electrons are, at all L shells, higher on 17 October than those observed on 19 October by up to an order of magnitude. These latter values do not differ much from values obtained during magnetically quiet periods. This means that the time to build up the stationary distributions of particles after a strong magnetic storm is about 0.5 to 1.5 days. Maximum values of the intensity of electrons are reached in the interval $270° \leqq \lambda \leqq 290°$ at all L shells and during both time periods under consideration, whereas minimum values, which are one or two orders of magnitude less than the maximum values, are observed in the interval $300° \leqq \lambda \leqq 310°$. This indicates that electrons are lost at the longitudes of the Brazilian anomaly.

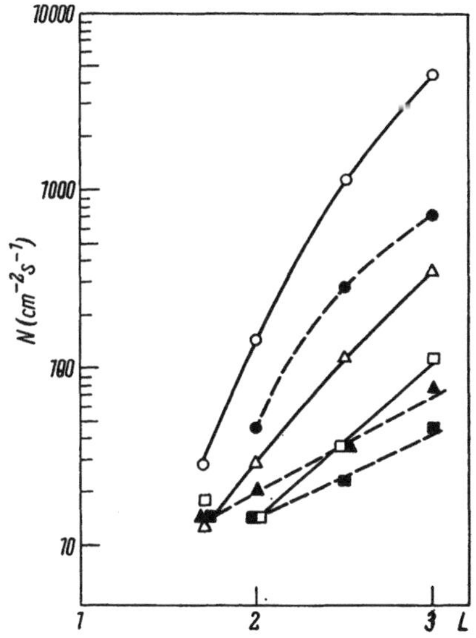

Fig. 2. Intensity of electrons with energies > 40 keV as a function of L in the longitude intervals △, ▲, $200° \leqq \lambda \leqq 220°$; ○, ●, $270° \leqq \lambda \leqq 290°$; □, ■, $300° \leqq \lambda \leqq 310°$. Open points, 17 October 1970; solid points, 19 October 1970.

3. Penetrating Radiation

A few interesting results obtained by means of the GM counters SBT-9 and STS-5 will be considered. The counter SBT-9 shielded on all sides by copper ~3 g cm^{-2} thick was placed outside the satellite; the counter STS-5 inside the satellite was shielded by ~0.5 g cm^{-2} Al on one side and ~5—10 g cm^{-2} of material on the other. In regions with $h_{min} < -300$ km these counters registered mainly cosmic rays (i.e. particles arriving directly from outside the magnetosphere) and particles generically related to them.

In agreement with results published in [2] and [3], the flux of penetrating particles was observed to be ~0.5 particle cm^{-2} s^{-1} near the geomagnetic equator and to increase up to ~2 particles cm^{-2} s^{-1} at higher latitudes. Values of these penetrating particle fluxes through a few dozen orbits are shown in Fig. 3 as a function of L. The data were averaged in intervals of 0.1 L. Plotted in the same figure are analogous data obtained on Explorer 7 in 1959—1960 (curve 1) and on Cosmos 17 in May 1963 (curve 2) [2]. It may be seen that the data of Intercosmos 3 agree with those indicated in [3]. In [2] it was pointed out that the discrepancy of the curves may be explained in terms of the change of the flux of galactic cosmic radiation due to the 11-year cycle of solar activity. Curve 1 was obtained during the period near the maximum of the solar cycle number 19, whereas curve 2 belongs to the period near minimum solar activity.

The data of Intercosmos 3 were measured 11 years after the flight of Explorer 7, i.e. during the maximum of solar cycle number 20. Agreement of the data obtained after 11 years, i.e. the length of a solar activity cycle, shows that the bulk of the particles observed consists of galactic cosmic rays and penetrating particles related to them.

Detailed analysis of the latitude variation of penetrating particle fluxes has shown that the intensities at the geomagnetic equator are different during different orbits. In order to make this point clear, the intensity measured at the geomagnetic equator, i.e. $L \leqq 1.1$, was plotted as a function of longitude (see Fig. 4). It may be seen that the intensity varies as the longitude changes. At longitudes $100° \leqq \lambda \leqq 150°$, an increase significantly exceeding the measuring errors of measurement may be seen. This longitude region nearly coincides with the region of maximum strength of the geomagnetic field. Field strengths of the geomagnetic fields along the equator are represented by the upper curve in Fig. 4.

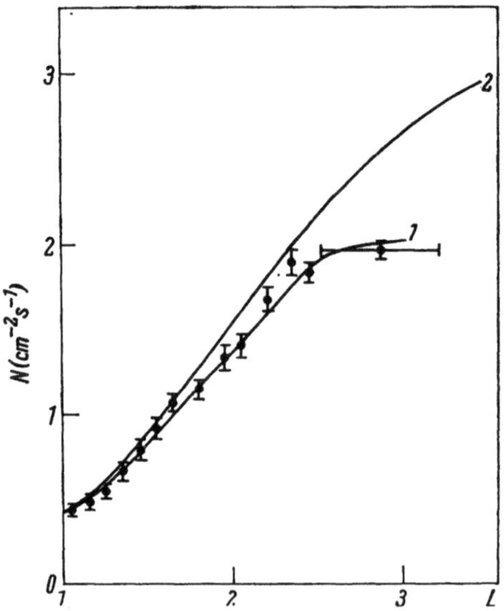

Fig. 3. Intensity of penetrating radiation as a function of L. Curve 1, Explorer 7 [3]; curve 2, Cosmos 17 [2]; solid points, Intercosmos 3 (present paper).

At places of maximum geomagnetic field strenght the intensity of cosmic rays and secondaries produced directly by them would be expected to have minimum values. The increase of intensity observed at the place of maximum geomagnetic field strength can only be explained in terms of particle storage in the geomagnetic field due to some unknown mechanism. Such a mechanism should, however, be effective at all longitudes. It seems, however, that the conditions of storage of the particles, or reduced loss into the atmosphere, are more favourable in the longitude interval mentioned above.

Fluxes of high energy electrons as observed in equatorial regions are given in [4]. Assuming an integral energy spectrum of electrons $N(> E) \sim E^{-1}$ and extrapolating the fluxes given in [4] up to energies as high as 5—10 MeV we may conclude that the bulk of the "surplus radiation" observed in equatorial regions consists of electrons with energies 5—10 MeV. It is, however, difficult to explain why the increase observed at longitudes $100° \leqq \lambda \leqq 150°$ by means of the counter STS-5 is much less than that observed by means of the SBT-9.

Further experimental studies, of an investigation of electron fluxes with energies 1—10 MeV, are needed to find the correct explanation of the increase* of the penetrating particle fluxes in equatorial regions at longitudes $100° \leq \lambda \leq 150°$

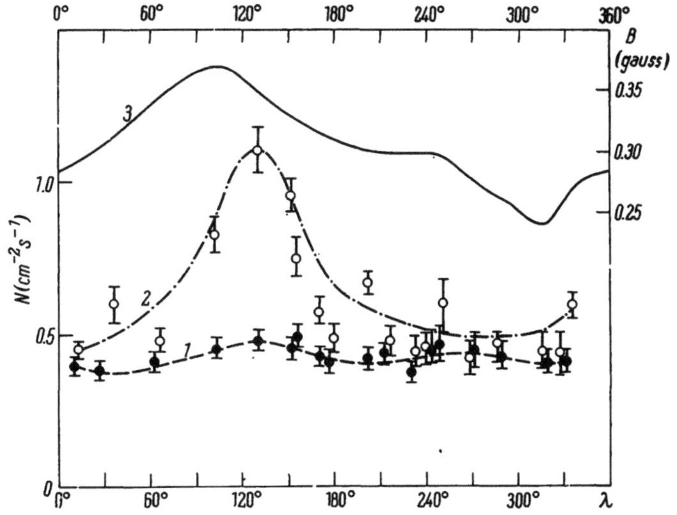

Fig. 4. Longitude dependence of penetrating radiation near the geomagnetic equator ($L \leqq 1.1$) on orbits 188, 209, 217 and 224. 1, STS-5 ($\sim$0.5 g cm^{-2} Al); 2, SBT-9 ($\sim$3 g cm^{-2} Cu); 3, B (gauss).

References

[1] P. V. Vakulov, V. Vesely, J. Votta, N. L. Grigorov et al., Kosm. Issled. **10**, 382 (1972).
[2] S. N. Vernov, A. E. Chudakov, P. V. Vakulov et al., Izv. Akad. Nauk SSSR **28**, 2058 (1964).
[3] W. C. Lin, D. Venkatesan and J. A. Van Allen, J. Geophys. Res. **68**, 4885 (1963).
[4] N. L. Grigorov, R. N. Basilova et al., Kosm. Issled. **11**, 160 (1973).

* *Note added in proof:* The increase can be accounted for in terms of activation of the counter walls when passing through the radiation belts. See paper published in *Kosmicheskie Issledovaniya.*

Space Research XIV — Akademie-Verlag, Berlin 1974

OBSERVATIONS OF 5—45 MeV PROTONS AT $L \lesssim 3$ AND $L \gtrsim 7$

R. C. FILZ[a], L. KATZ[a], P. L. ROTHWELL[a], B. SELLERS[b], F. A. HANSER[b] and E. HOLEMAN[c]

[a]Air Force Cambridge Research Laboratories, Bedford, Mass., USA
[b]Panametrics, Inc., Waltham, Mass., USA
[c]Franklin Pierce College, Rindge, New Hampshire, USA

This paper reports new results on trapped proton and solar proton fluxes obtained in late 1972. The data were obtained aboard satellite 1972-076B launched on 2 October 1972 by an instrument which detects protons from 5 to 45 MeV in five energy channels. The high inclination of the satellite orbit permits the study of both magnetospherically trapped proton fluxes in the South Atlantic anomaly and solar proton fluxes over the polar regions. The L-dependence and pitch-angle dependence of trapped protons with $H_{\min}$ between 700 and 760 km are presented and the data are compared with earlier data taken at low altitude. The time-dependence, energy spectra and pitch-angle distribution of solar protons measured during the solar event of 30 October—2 November 1972 are also presented and discussed.

1. Introduction

The particle instrument flown on Air Force satellite 1972 076-B, apogee 750 km, perigee 729 km, inclination 98.4°, performs a simultaneous six-channel energy analysis and identification of protons, deuterons, tritons, helium three and alpha particles. The detector contains four solid state detectors arranged as a telescope and a photomultiplier-plastic scintillator anticoincidence shield. The telescope has a 6° half angle view cone and a geometric factor of 0.0257 cm² sr. The total energy analysis is performed by two 300 mm², 5 mm thick lithium drifted silicon detectors acting as a single 10 mm detector. The plastic scintillator anticoincidence shield strongly reduces false counts from side or rear entering particles. The signals from the various components of the telescope are combined to obtain, a particle identifier signal in five bins and a total energy signal in six bins yielding a matrix of 30 channels which are accumulated and telemetered. For protons the energy channels are 5.0—7.0 MeV, 7.0—12.2 MeV, 12.2—18.2 MeV, 18.2—28 MeV and 28—45 MeV, with an additional channel to indicate protons of an energy not normally capable of depositing sufficient energy to trigger the counter without causing an anticoincidence.

All aspects of the system appear to be operating normally as evidenced, for example, by the operation of the detectors while spinning through the trapped proton mirror plane in the South Atlantic. The lack of noise in the data as received is exemplified in the 28—45 MeV trapped proton pitch angle distribution shown in Fig. 1.

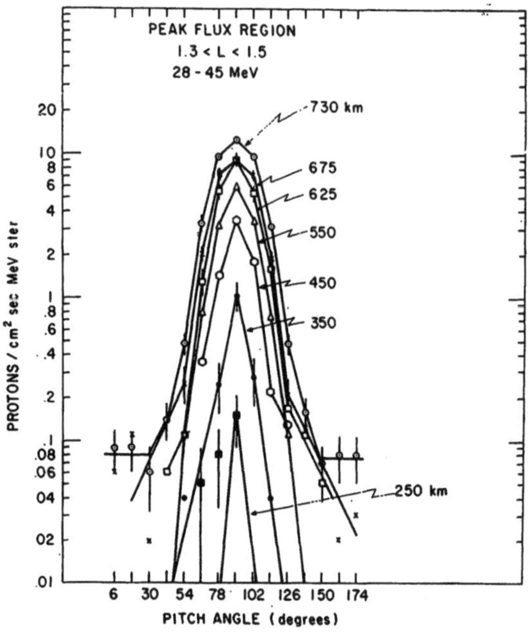

Fig. 1. The 28—45 MeV proton pitch angle distribution measured for H_{min} 250—730 km and $1.3 \leqq L \leqq 1.5$.

2. Trapped Proton Observations

In order to analyze the large amount of data received, it was first necessary to divide the measurements into intervals of McIlwain's L parameter [1] and B, the magnetic field strength. In this case H_{min}, the minimum altitude reached by the trapped proton guiding center in its longitudinal drift, was used in place of B. B was determined to be a linear function of H_{min} for constant L. Fig. 1 shows the pitch angle distribution of 28—45 MeV protons measured in the peak flux region for different intervals of H_{min}. The distributions are narrower at lower H_{min} due to losses below ~200 km.

Since the background is negligible, we have obtained omnidirectional trapped proton fluxes from the data by simply dividing the total counts into 0.1 earth radii intervals of L and 10 km intervals of H_{min}. For the period 7 October to 6 November 1972, the L distribution of the omnidirectional proton fluxes is shown in Fig. 2 for H_{min} between 700 and 760 km. Wide differences are exhibited here in the location of maximum flux from the lowest energy of 6 MeV with a peak between L 2.0 and 2.4, and from 28—45 MeV with a sharp peak at $L = 1.4$.

Utilizing the 700—765 km H_{min} data for the 7 October—6 November 1972 period, we have also derived energy spectra which exhibit three general shapes: (i) at low L values 1.1—1.4 very flat; (ii) at middle L values 1.4—2.2 fairly steep with fairly characteristic breaks at 9 and 15 MeV; (iii) at high L values greater than 2.2 very steep spectra with slope increasing with L. These spectra are very similar to observations made earlier by other experimenters [2]. The altitude dependence of the omnidirectional peak fluxes for appropriate L intervals is smoothly varying with altitude; however, they cannot be fit by a single power law with altitude.

In order to compare our 1972 measurement with nuclear emulsion measurements made with recoverable satellites in 1961—1962, we constructed a model to fit the 1972 data. The model was extrapolated to 55 MeV at each individual

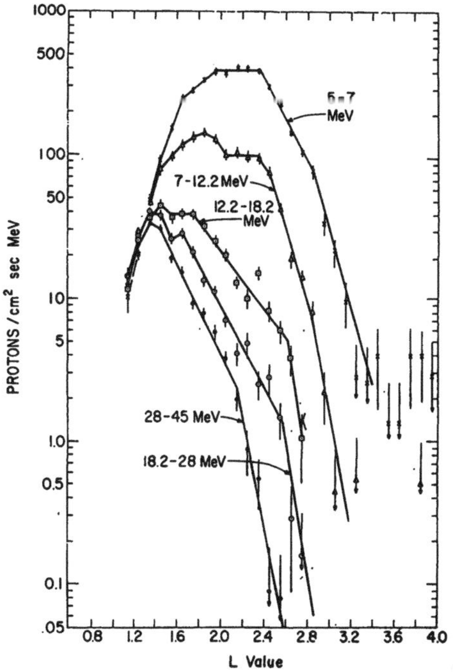

Fig. 2. The L dependence of the trapped proton fluxes for the five energy bins for the period 7 October—6 November 1972 and the H_{min} interval 700—765 km. The errors are purely statistical and the solid lines are drawn to indicate the data trend.

H_{min}, L point and integrated over the satellite orbits used in the nuclear emulsion 1961—1962 study by Filz et al. [3]. A correlation of the 1961—1962 55 MeV flux data shows excellent agreement which indicates that over the 11-year solar cycle the fluxes have returned to their original value. The 1972 data are also compared with data obtained by Achtermann et al. [4] for two different L shells. The fluxes recorded by Achtermann et al. appear a factor of ~ 2 higher at 20 to 50 MeV and as much as a factor of ~ 10 higher at the lower energies. While the 20—50 MeV difference can be expected from solar cycle changes, it is not clear that such a drastic change in the 6 MeV protons is due to this cause.

3. Solar Particles

Solar protons were observed at high latitudes during the 29 October—1 November 1972 time period. In this paper we would like to concentrate on the fluxes, energy spectra and pitch-angle distributions of these particles. The 5—7 MeV channel first detected solar protons on 29 October 1972 at ~2300 h UT. The flux recorded in this channel peaked at ~1200 h UT on 31 October at 70 protons/(cm^2-MeV-sec). By 0300 h UT on 1 November the flux had decreased to 3 protons/(cm^2-MeV-sec). The 28.0—45.0 MeV proton channel, on the other hand, recorded a flux maximum of 0.08 protons/(cm^2-MeV-sec) at ~0100 h UT on 31 October, or approximately 11 hours before the 5—7 MeV proton flux

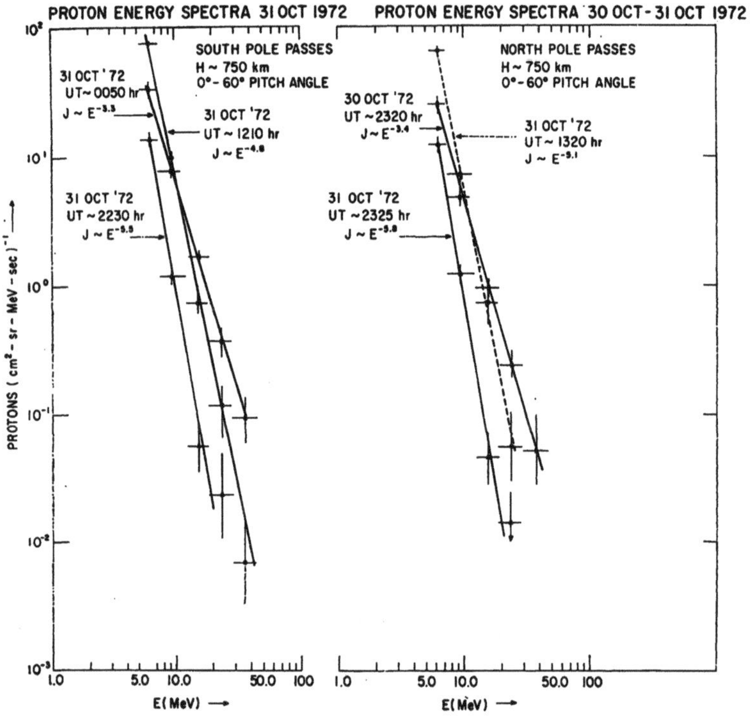

Fig. 3. Solar proton energy spectra measured over the North and South poles at similar times.

peaked. The time-histories show excellent agreement with those measured outside the magnetosphere by satellite OV5—6 [5]. The South Pole passes showed a slightly higher flux than the North Pole passes, an asymmetry pointed out by Evans and Stone [6]. The three other proton channels whose energy limits are between those of the other two channels show similar behavior.

Proton energy spectra are shown in Fig. 3. A single power law function fits the proton spectra over the energy region of interest. These spectra are similar to those reported for the large 28 May 1967 solar event by Lanzerotti [7]. North and South polar passes taken within less than one hour of one another show no significant difference in spectral shape. Over larger periods of time spectra become characteristically softer.

Fig. 4 shows some pitch-angle distributions of 5—7 MeV solar protons. The flux decreases rapidly beyond 90° due to the atmospheric loss cone. The data were separated into north-south, day-night passes. The symbol $\bar{\Lambda}$ denotes the average invariant latitude over which the distribution was taken. The two curves for each pass correspond to data taken on the polar plateau and between the high-latitude cutoff and the flux peaks. The data indicate that the solar particle flux is isotropic in the upper hemisphere, except perhaps over the North Pole at local night. The seemingly higher fluxes along the field line for this particular pass could, however, be due to statistical fluctuations.

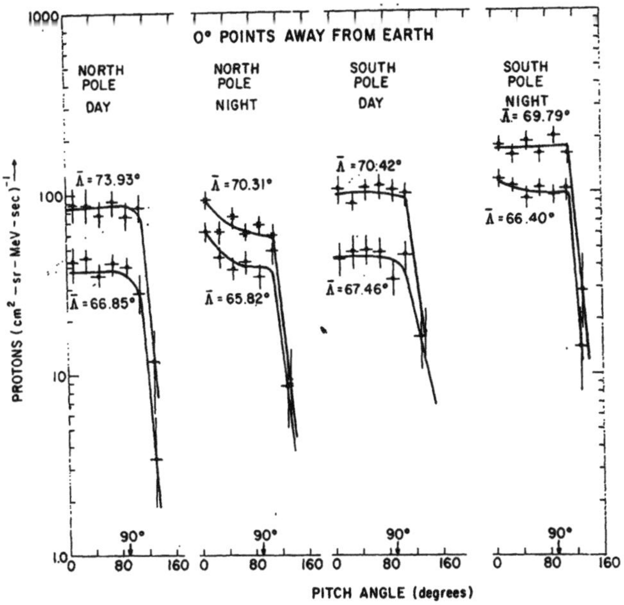

Fig. 4. Pitch-angle distributions of 5—7 MeV solar protons for north and south, day and night passes 31 October 1972. The vertical error bars denote the statistical uncertainty in the flux and the horizontal error bars the 20° angular interval over which the data were taken.

4. Conclusions

A preliminary analysis of proton data obtained from the particle instrument on Air Force satellite 72-1 has yielded important information on trapped protons in the South Atlantic at this phase of the solar cycle. When extrapolated to 55 MeV energy and integrated over the same orbits, the 1972 data are in agreement with nuclear emulsion data taken in 1961—1962 at the same phase of the solar cycle. Solar protons during the 29 October—1 November period were also measured on the S72-1 satellite. The 28.0—45.0 MeV proton flux reached its maximum value approximately 11 hours before the 5—7 MeV proton flux. The proton energy spectra progressively softened with time and were consistent with a single power law fit. The 5—7 MeV proton flux was isotropic in the upper hemisphere.

References

[1] C. E. McIlwain, J. Geophys. Res. **66**, 3681 (1961).
[2] See e.g. W. H. Hess, The Radiation Belt and Magnetosphere, Blaisdell Publ. Co., Waltham, Mass. 1968.
[3] R. C. Filz and E. Holeman, J. Geophys. Res. **70**, 5807 (1965).
[4] E. Achtermann, S. C. Freden and D. Hovestadt, Report, Max-Planck-Institut für Physik und Astrophysik, Extraterr. 6/67 (May 1967).
[5] L. Katz, P. L. Rothwell, G. K. Yates, B. Sellers and F. A. Hanser, Proc. 7th ESLAB Symp. on Correlated Interplanetary and Magnetospheric Observations, D. Reidel Publ. Co., Dordrecht, Holland 1974 (p. 413).
[6] L. C. Evans and E. C. Stone, J. Geophys. Res. **74**, 5127 (1969).
[7] L. J. Lanzerotti, J. Geophys. Res. **74**, 2851 (1969).

Note added in proof. The omnidirectional trapped proton fluxes presented in Fig. 2 should be multiplied by $\pi/2 = 1.57$. This allows slightly better agreement with the Achtermann data (see text) but has little effect on the correlation with the emulsion results of Filz and Holeman [3]. This is a correction related to the spinning of the detector through the trapped mirror plane distribution and does not change either the directional fluxes or the solar particle fluxes.

Space Research XIV — Akademie-Verlag, Berlin 1974

OBSERVATIONS OF THE ENTRY OF SOLAR PROTONS INTO THE MAGNETOSPHERE BY USE OF RIOMETERS

J. J. Berthelier and M. Pirre*

Groupe de Recherches Ionosphériques, Saint-Maur-des-Fossés, France

During the solar proton event of 28 January 1967 we associated two sharp variations of the ionospheric absorption measured by riometers at a few stations in Antarctica with the passage of discontinuities of the interplanetary magnetic field, convected by the solar wind past regions where field lines from these stations are connected to interplanetary field lines through the magnetopause. For very low and moderate reconnection rates the ordered magneto-tail length and the corresponding cross tail electric field are found to be 1050 R_e and 500 R_e, 0.2×10^{-4} V m^{-1} and 1.5×10^{-4} V m^{-1} respectively.

1. Introduction

The entry of solar protons into the magnetosphere has been extensively studied during the last few years both experimentally [1—3] and theoretically [4—6]. One of the most interesting aspects of this problem is that it leads to a better understanding of the gross features of the magnetosphere. We present in this paper a number of observations made during the solar proton event of 28 January 1967 both at high latitude stations (riometer measurements of ionospheric absorption) listed in Table 1 and in the interplanetary medium (interplanetary magnetic field and particle measurements), which enabled us to deduce good estimates of the ordered magneto-tail length and of the cross tail electric field.

Table 1

Place	Longitude (+, east)	Latitude (+, north)	Invariant latitude
Vostok	106.9°	−78.5°	84.5°
Dumont d'Urville	140.0°	−66.7°	80.5°
Wilkes	110.3°	−66.2°	80.2°
McMurdo	166.4°	−77.5°	80.0°
South Pole	—	−90.0°	74.8°
Byrd	−120.5°	−80.0°	68.7°

* Assistant à l'U.E.R. Sciences d'Orleans, Orleans, France.

2. Observations

On 28 January 1967 a solar flare probably occurred at 0730 UT some 60° beyond the western solar limb [7]; this emitted intense fluxes of energetic protons the arrival of which at the orbit of the earth was first detected near 0830 UT by satellite particle detectors (Fig. 1) and high latitude riometers (Fig. 2). We shall restrict our study to two interesting periods of this PCA event, the first one from

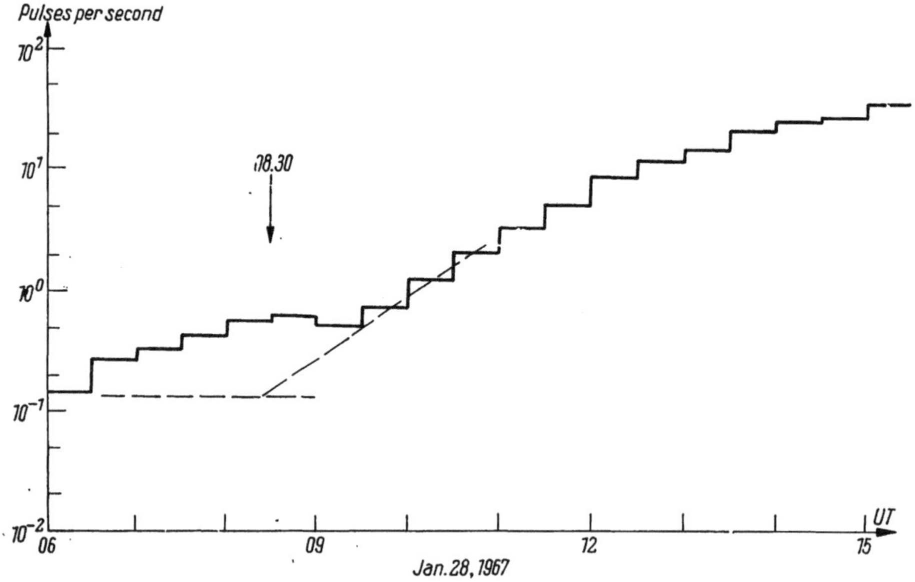

Fig. 1. Time history of solar protons $0.3 \leqq E_p \leqq 10$ MeV measured on board Explorer 33 from 0600 to 1500 UT, 28 January 1967.

the onset to ~1500 UT on 28 January, the second between 2100 UT 28 January and 0600 UT 29 January.

Fig. 1 shows the time history of solar proton $0.3 \leqq E_p \leqq 10$ MeV as seen on Explorer 33 with a uniform rise from 0830 UT to 1500; similarly Fig. 2 displays the variations of ionospheric absorption measured on 30.1 MHz in five southern and one northern high latitude stations during the same period. Due to the date of the event, the antarctic stations are in a very favourable situation since the D region is continuously illuminated by the sun at Byrd, South Pole and McMurdo and there is only about six hours of darkness at Dumont d'Urville and Wilkes. In this case it is possible to observe the variations of proton fluxes bombarding the upper atmosphere over the stations almost continuously and with a very good time resolution. Superimposed on the normal increase of absorption an irregular variation indicated by arrows appears on the records from Dumont d'Urville and Wilkes. The absorption increases by 1 dB with a very similar time variation in both cases, starting at 1000 ± 0010 and 1045 ±0010 UT respectively and lasting approximately $1^1/_2$ hours at each station. No such features are seen at the other stations. The sudden increase of 1 dB at Byrd near 0900 UT is in fact due to auroral absorption: there was substorm activity at this time at Byrd and the record from the conjugate station of Great Whale River shows a similar variation at the same time; since the upper atmosphere is in complete darkness the similarity between the two increases in absorption is proof of the auroral origin of this supplementary absorption.

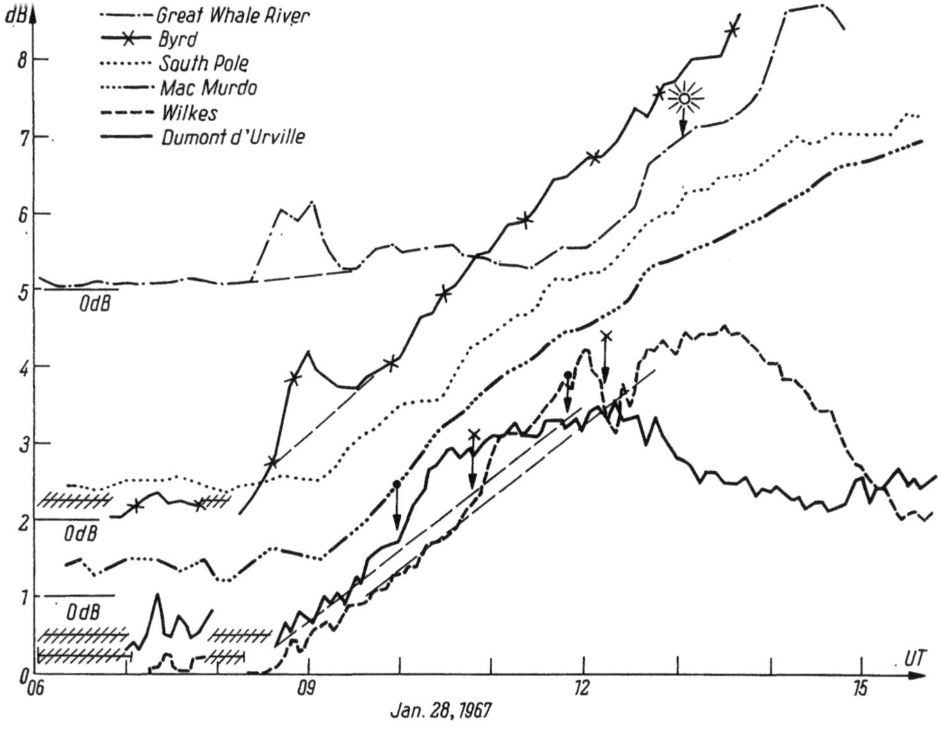

Fig. 2. Ionospheric absorption variations at five southern and one northern high latitude stations from 0600 to 1500 UT, 28 January 1967. 30 MHz.

Similarly to Fig. 1, Fig. 3 shows the history of solar protons from 28 January, 2100 UT to 29 January, 1200 UT. The most striking feature is a sharp decrease in the fluxes between 0030 and 0730 UT. The effect of this flux decrease is readily observed at all six antarctic stations, as shown in Fig. 4, except in the case of Byrd where auroral absorption due to substorm activity from 0300—0500 and from 0700—0800 UT makes the phenomenon less discernible. In Table 2 are listed the times of onset of the decreases in absorption for the different stations and the time lags with respect to 0030 UT, the time of onset of the decrease in solar proton fluxes.

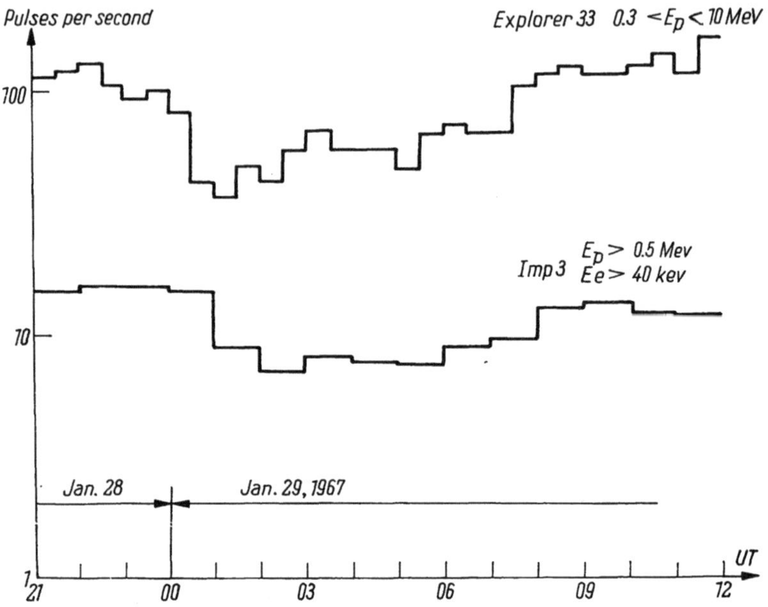

Fig. 3. Time history of solar protons from 2100, 28 January to 0900 UT, 29 January 1967.

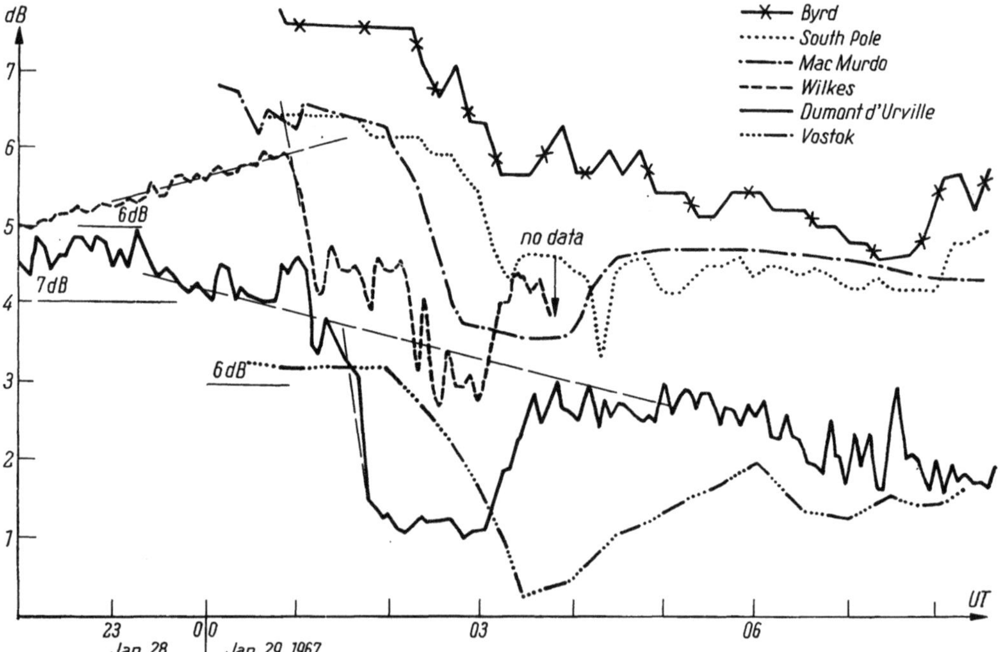

Fig. 4. Ionospheric absorption variations at six southern high latitude stations from 2200, 28 January to 0900 UT, 29 January 1967. 30 MHz.

Table 2

Place	Time of onset (UT) 29 Jan. 1967	Time lag (min)	Electric field (V m^{-1})
Vostok	0230 $\pm$ 20′	120 $\pm$ 20	$(1.7 \pm 0.3) \times 10^{-4}$
Dumont d'Urville	0145 $\pm$ 10′	75 $\pm$ 10	—
Wilkes	0100 $\pm$ 10′	30 $\pm$ 10	$(1.75 \pm 0.5) \times 10^{-4}$
McMurdo	0215 $\pm$ 10′	100 $\pm$ 10	$(1.65 \pm 0.2) \times 10^{-4}$
South Pole	0300 $\pm$ 10′	150 $\pm$ 10	$(1.25 \pm 0.15) \times 10^{-4}$

3. Interpretation

In these case of the first anomalous variation of the absorption during the early phase of the PCA event, no sudden time variation of solar proton fluxes may be advocated in view of the Explorer 33 data; moreover such a change would immediately be seen at auroral zone stations such as Byrd [5]. Fig. 5 shows the

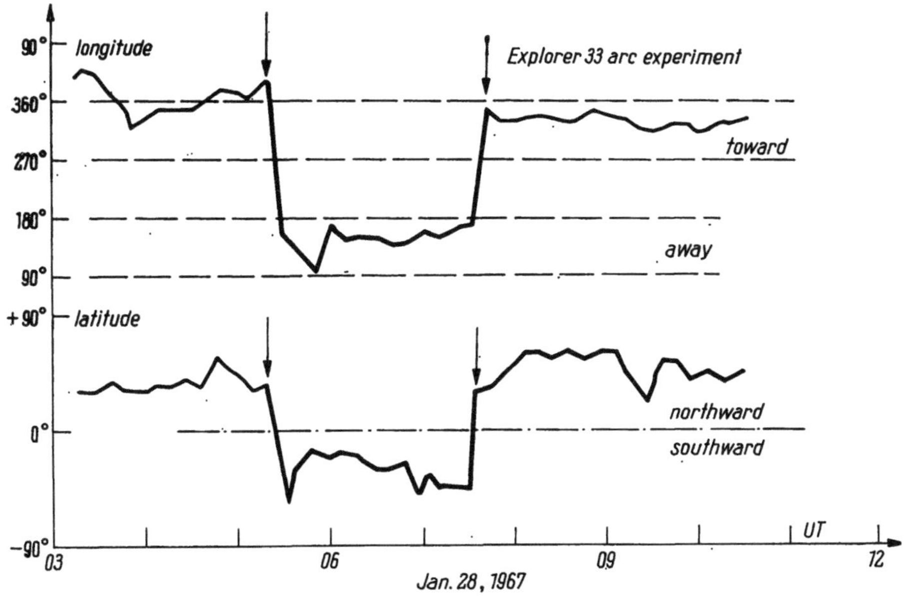

Fig. 5. Interplanetary magnetic field components from 0300 to 1200 UT, 28 January 1967.

5 min average longitude and latitude angles of the interplanetary magnetic field (IMF) between 0300 and 1200 UT. A very sharp and drastic change in the orientation of the IMF occurs between 0530 and 0730 UT, the field changing from northward and toward the sun to southward and away from the sun. Such a filamentary structure is not infrequent in the interplanetary medium and has

been observed to be generally associated with rapid variations of the proton fluxes ([8] and references cited). We think that, in the case under study, such a filament might conceivably have contained a higher intensity proton population which would be conveyed along the magnetopause at the solar wind velocity. The time at which it sweeps past Explorer 33 is some two hours before the flare; consequently no solar protons are seen. When the solar protons reach the orbit of the earth the higher intensity inside the filament is immediatly "sensed" by stations whose field lines are connected through the magnetopause to the interplanetary filament. The rapid rise time of the variation in absorption ($\sim$10 min) precludes any diffusion process such as advocated by Michel and Dessler [9] and favours the reconnection model developed by Morfill and Quenby [5]. There is, however, no possibility of distinguishing between the "kink" and adiabatic models. If we simply assume direct reconnection we find an ordered tail length of $\sim 1050 \pm 50\, R_e$ using Wilkes riometer data and $850 \pm 50\, R_e$ using Dumont d'Urville riometer data. The magnetic depth D of a station as defined by Morfill [10] is an increasing function of the smallest distance between the magnetopause and the intersection of the trajectory of protons precipitated over that station with a cross section of the tail at 25 R_e. In our case, the magnetic depth of Dumont d'Urville $D = {}^1/_2(D_{3.4\,\mathrm{MeV}} + D_{20\,\mathrm{MeV}})$ is $\sim$0.46 GV and that of Wilkes only $D \sim 0.11$ GV. Thus, if both field lines reconnected in the same way across the magnetopause, reconnection should occur at a much larger distance in the case of Dumont d'Urville. Our conclusion is that at the time of the event the field line from Dumont d'Urville was being convected sunward; such a sunward convection at an invariant latitude of 81° and magnetic local time of $\sim$2100 is in agreement with OGO 6 electric field measurements. During the whole period 0300–1500 UT, except during the passage of the filament, the IMF was strongly northward and with a weak eastward component; in such a situation the sunward-antisunward convection boundary is shifted to higher latitudes during the evening quadrant [11]. Calculation of the convection electric field by means of Morfill's D parameter yields a value $\sim 0.2 \times 10^{-4}$ V m^{-1} when using Wilkes data; according to our hypothesis discussed above Dumont d'Urville data are meaningless for this calculation. The value of 0.2×10^{-4} V m^{-1} is quite low, but it must be emphasized again that the IMF direction was such as to give a very low reconnection rate.

The decrease in absorption at all high latitude stations on 29 January is clearly due to the lowering of proton fluxes with a sharp decrease near 0030 UT. As in the preceding case we have analysed the time variations of absorption in order to determine the time lag between the onset of the event at the Explorer 33 orbit and the beginning of the decrease of absorption at each station. In order to have an estimate of the ordered tail length we can safely take data from South Pole and Vostok which are deep enough in the tail, with D parameters ranging from 0.3 to 0.42 GV. These measurements yield a length of the order of $500 \pm 50\, R_e$ which is significantly shorter than the value found for 28 January. We think that this is an indication of an increased reconnection rate due to a more favourable position of the IMF which is less northward in the interval 2300–0600 UT than the day before. Electric field values derived for McMurdo, South Pole and Vostok data give about $(1.5 \pm 0.2) \times 10^{-4}$ V m^{-1}, which is in reasonable agreement with values given by Morfill and Scholer [12]. It must be pointed out that the theoretical magnetospheric model used by Morfill [10] would indicate direct entry of protons $E_p \geqq 5$ MeV through the magnetopause at distances $\leqq 15\, R_e$ for Wilkes and

Dumont d'Urville. Although we are well aware of the fact that the model used by Morfill is not exactly adapted to our situation, we feel that the main refinement needed concerns the high latitude dayside magnetosphere.

4. Conclusion

Using riometer absorption data during the 28 January 1967 PCA event we have been able to derive values of ordered tail lengths of $\sim(1050 \pm 50)\, R_e$ and $\sim(500 \sim 50)\, R_e$ corresponding to cases with respectively very low and moderate reconnection rates due to different directions of the interplanetary magnetic field. The corresponding cross tail electric fields are $\sim 0.2 \times 10^{-4}\ \mathrm{V\,m^{-1}}$ and $(\sim 1.5 \pm 0.2) \times 10^{-4}\ \mathrm{V\,m^{-1}}$.

Acknowledgments

We are greatly indebted to Dr. S. M. Krimigis and Dr. D. S. Colburn for providing respectively the particle and interplanetary magnetic field data from Explorer 33. IMP 3 data are due to the generosity of Prof. K. A. Anderson and solar wind parameters, from the MIT experiment on Explorer 33, were kindly supplied to us by Dr. Howe. The riometer at McMurdo is operated by McDonnel Douglas and data were made available to us by Dr. A. J. Masley. Riometer data from South Pole, Byrd, Vostok and Great Whale River were generously supplied by Dr. G. C. Reid and those from Wilkes by Dr. Sulzberger from the Australian Antarctic Division. Expéditions Polaires Françaises and TAAF are to be thanked for providing maintenance of the geophysical laboratory at Dumont d'Urville and we acknowledge the work of our colleagues in this station under very severe conditions. This research was conducted in part under the auspices of a CNES grant/CNES—66-035.

References

[1] L. C. Evans and E. C. Stone, J. Geophys. Res. **74**, 5127 (1969).
[2] J. A. Van Allen, J. F. Fannel and N. F. Ness, J. Geophys. Res. **76**, 4262 (1971).
[3] A. C. Durney, G. Morfill and J. J. Quenby, J. Geophys. Res. **77**, 3345 (1972).
[4] R. Gall, S. Bravo, A. Orozco, J. Geophys. Res. **77**, 5360 (1972).
[5] G. Morfill and J. J. Quenby, Planet. Space Sci. **19**, 1541 (1971).
[6] M. Scholer and G. Morfill, Planet. Space Sci. **20**, 1051 (1972).
[7] R. P. Bukata, P. T. Gronstal, R. A. R. Palmeira, K. G. McCracken and U. R. Rao, Solar Phys. **10**, 198 (1969).
[8] M. Scholer, J. Geophys. Res. **77**, 2762 (1972).
[9] F. C. Michel and A. J. Dessler, J. Geophys. Res. **75**, 6061 (1970).
[10] G. Morfill, J. Geophys. Res. **77**, 4010 (1972).
[11] J. P. Heppner, in: Critical Problems of Magnetospheric Physics, Proc. Joint COSPAR/IAGA/URSI Symp., Madrid 1972, published by IUCSTP Secretariat.
[12] G. Morfill and M. Scholer, J. Geophys. Res. **77**, 4021 (1972).

Space Research XIV — Akademie-Verlag, Berlin 1974

OGO 5 OBSERVATIONS OF THE PHYSICAL PROCESSES OCCURRING IN THE DISTURBED POLAR CUSP AND THE CUSP-MAGNETOSHEATH INTERFACE

C. T. RUSSELL[a], R. W. FREDRICKS[b], M. G. KIVELSON[a], M. NEUGEBAUER[c] and F. L. SCARF[b]

[a]Institute of Geophysics and Planetary Physics, University of California, Los Angeles, Calif., USA
[b]Space Sciences Department, TRW System Group, Redondo Beach, Calif., USA
[c]Jet Propulsion Laboratory, California Institute of Technology, Pasadena, Calif., USA

OGO 5 observations of the polar cusp on 1 November 1968 show that the north-south component of the interplanetary field exhibits control over, not only the location of, but also the physical processes occurring in, the polar cusp. When the interplanetary field was southward, the electron temperature in the polar cusp was lower and the currents were stronger than when the interplanetary field was northward. In addition, regions of apparently rapidly varying currents were encountered within the cusp, together with enhanced VLF electric fields. When the interplanetary field was northward, quasi-monochromatic Pc 1 waves close to but below the proton gyrofrequency were observed which were inconsistent with cyclotron resonant generation. Estimates of anomalous resistivity along field lines due to the electrostatic waves lead to estimates of field-aligned potential drops between OGO 5 (at $\sim 3\ R_e$ geocentric) and the ionosphere of the order of 2 kilovolts. When OGO 5 entered the magnetosheath near the cusp-magnetosheath boundary the fluxgate magnetometer recorded the highest level of turbulence that it had ever encountered in any region of space. One possible interpretation of these data is that the attached shock postulated by Walters was present at this time.

1. Introduction

Plasma measurements on a variety of spacecraft have shown the magnetosheath plasma penetrates from the magnetopause to the ionosphere in a thin arc above the dayside auroral oval [1—3]. This region, called the polar cusp by some and the cleft by others, is a permanent feature of most modern models of the magnetosphere and has been probed many times by both low altitude polar and highly eccentric orbiting spacecraft. Nevertheless, many of the properties of the cusp, many of the processes occurring within the cusp and the role of the cusp in magnetospheric dynamics remain unstudied or are subjects of controversy.

In this paper, we discuss measurements made by OGO 5, a three-axis stabilized spacecraft, during one set of multiple encounters with the polar cusp on 1 November 1968. While OGO 5 encounters with the polar cusp are few in number, the high temporal and amplitude resolution provided by the OGO 5 instruments, and the complement of instruments carried by OGO 5 provide information on the polar cusp not presently available from other spacecraft. We first review briefly the solar wind and magnetospheric conditions during these encounters and summarize previous OGO 5 findings; then we examine three separate topics: field-aligned currents in the polar cusp, turbulence in the cusp, and the cusp-magnetosheath interface.

2. Geomagnetic Conditions and OGO 5 Trajectory during the Cusp Encounters

As discussed by Russell et al. [3], a large sudden commencement geomagnetic storm began early on 31 October 1968. During the recovery phase of this storm, on 1 November, another sudden commencement occurred and a new main phase developed. The OGO 5 encounters with the cusp took place as the second main phase developed. The Kp index was 8^+ at this time. Solar wind plasma data were

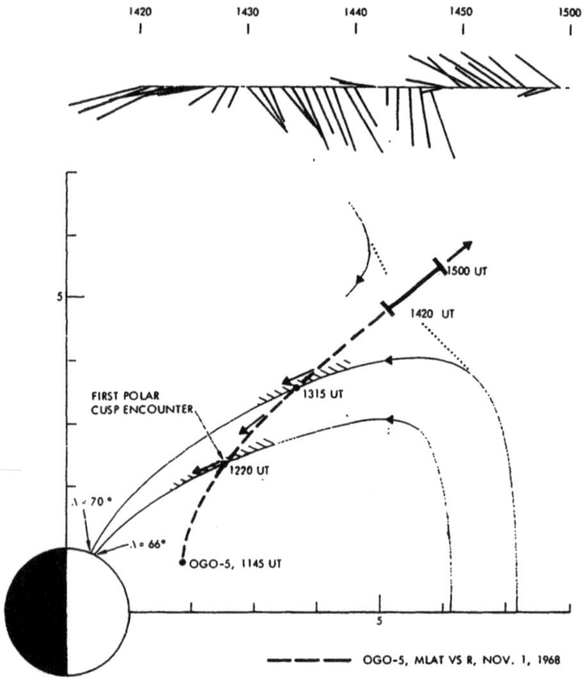

Fig. 1. The trajectory of OGO 5 in the local dipole meridian showing probable magnetic field configuration. Projections of one minute averages of the field in this plane for the period from 1420 to 1500 UT are shown at the top of the figure.

not available for most of the OGO 5 outbound pass. However, magnetosheath measurements were available from Vela 4B late in the pass, at which time the magnetosheath flow velocity ranged from 700 to 1000 km s^{-1} and the density from 9 to 33 cm^{-3}. As will be discussed below the interplanetary magnetic field was alternately strongly southward and strongly northward. Fig. 1 shows the trajectory of OGO 5 in the dipole meridian as well as the expected field line geometry and the locations of two of the cusp encounters. At 1427 UT OGO 5 entered the magnetosheath.

3. Previous OGO 5 Findings

The similarity of the fluxes and energy spectra of the suprathermal electrons measured by the JPL solar wind experiment to those of magnetosheath electrons has been used to identify the regions of suprathermal electron enhancements as encounters with the polar cusp. During these encounters, decreases in the magnetic energy density consistent with the presence of equal densities of magnetosheath protons were observed [3]. There were four encounters with the polar cusp spanning a range of local times from 1100 to 1400 LT. OGO 5 wave measurements have revealed that the polar cusp was quite turbulent with intense ULF magnetic and VLF electrostatic fluctuations [3, 4]. The ULF wave power was more intense transverse to the field than parallel to the field. The VLF electrostatic emissions were identified as being, in part, lower hybrid resonance noise, and possibly, in part, Buneman current driven instabilities. As noted in the initial survey of these data [3], the cusp moved equatorward when the interplanetary field was southward and poleward when it was northward. In later work [5], further dependences of the cusp on the north-south component have been examined. The average energy of the suprathermal electrons decreased for southward fields, and energetic electron fluxes ($E > 50$ keV) were present only for northward interplanetary fields. This latter observation suggests that cusp plasma was present on closed field lines. Further, the currents in the polar cusp weakened when the interplanetary field turned northward.

4. Polar Cusp Currents

There are two types of currents observed in the polar cusp: a broad diffuse current flowing essentially throughout the cusp [6] and narrow intense current filaments [7]. Low altitude measurements [8] appear to include contributions from both current types. Fig. 2 shows one-minute averages of the magnetic field during the first two cusp encounters in a field-aligned coordinate system (Z along the average measured field and Y azimuthal eastward). The internal geomagnetic field of approximately 1000γ has been subtracted from these data. The field-aligned component B_Z has been plotted (solid line) on the same scale as the deviation of the field $\Delta|B|$ from the predicted field strength (dashed line) for that position. There is close agreement between these two variables and the region of magnetic field depression is taken to be the region of enhanced proton flux. It does not show a one-to-one correspondence with the presence of suprathermal electrons as measured by the electron energy density shown on the bottom panel.

Sheets of field-aligned currents flowing on shells of constant L-value will cause magnetic perturbations in the azimuthal direction. The signature of two parallel but opposing sheets of current is a large perturbation of this type, a peak. In both cusp encounters 1 and 2A there are two such peaks. In encounter 2A, during which the interplanetary field was northward, the perturbation and hence the current was much weaker. We expect two current systems with field-aligned segments in the polar cusp region. We will loosely refer to these systems by the type of current closing them in the ionosphere: Pedersen currents and Hall currents. Merging causes a poleward flow of both closed and open field lines. Open field lines are "dragged" across the polar cap and closed field lines flow into this region to replace them and eventually merge with the interplanetary field themselves. The ionosphere resists this motion. Pedersen currents flowing parallel to the electric field in the ionosphere exert a poleward force on the plasma to overcome the ionospheric drag. This current flows wherever field lines are moving and the ionosphere is resisting this motion. We would expect these currents to close in part along field lines if the resistance of the magnetospheric path is

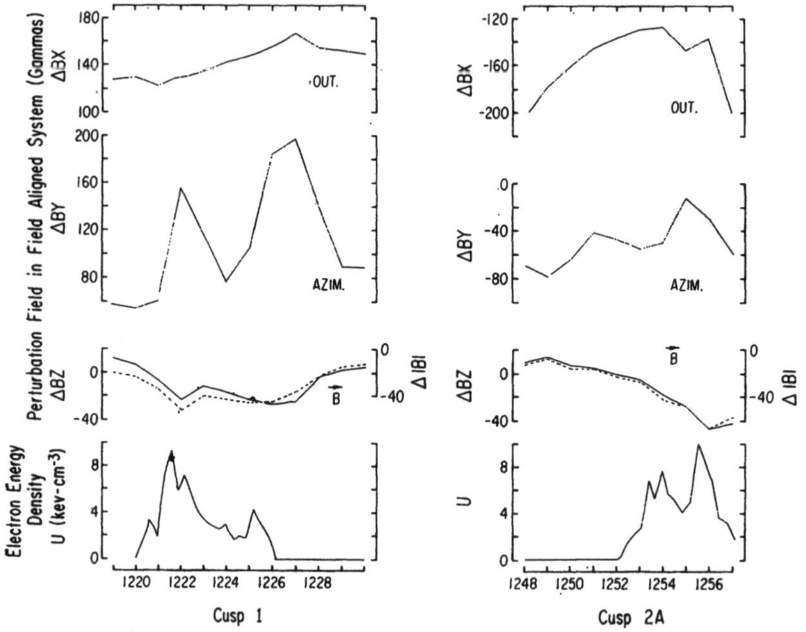

Fig. 2. One minute averages of the magnetic field during the first two cusp encounters. The data are in a field-aligned coordinate system with Y azimuthal eastward and the contribution from the earth's main field has been removed. The lower panel shows the energy density of suprathermal electrons every 19 seconds.

comparable with that of possible ionospheric paths. The perturbation field arising from this current system opposes the earth's main field, effectively reducing the earth's dipole field and allowing the magnetopause to move inward without a change in the solar wind dynamic pressure [9].

Hall currents will flow perpendicular to the applied electric field whenever there is a significant Hall conductivity, and can close along field lines when there is a gradient in the Hall conductivity. Such a gradient is expected in the polar cusp [10] and can lead to field-aligned sheets of current. We interpret our observations and those of other workers [6, 8] to be the closure of these Hall currents. A decrease in either the applied electric field, the Hall conductivity, or the Hall conductivity gradient could cause a weakening of these currents. The rapid motion of the polar cusp in response to changes in the north-south component of the interplanetary field suggests that the merging rate and hence the applied electric field decreased when the field turned northward. Finally, we note that, given a single Hall conductivity enhancement centered on the polar cusp, we expect a westward field perturbation. This agrees with previous work [6, 11]. Thus, perhaps the double twin current sheet structure should be viewed as the primary Hall closure system in the center bounded by two auxiliary currents poleward and equatorward.

Fig. 3 shows the second type of currents observed in the cusp. They are either filamentary or impulsive and last only seconds. Their amplitudes are not much

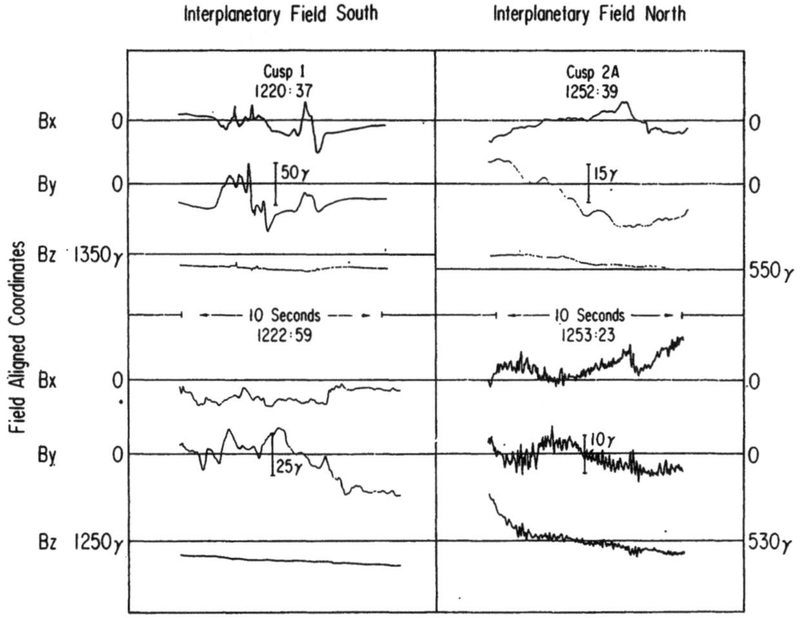

Fig. 3. High resolution magnetic field data (18 msec per sample) showing impulsive currents in the cusp, and ion cyclotron waves (lower right panel).

less than those in the large scale twin current sheets. However, they apparently are much narrower suggesting their current density is much more intense. Features such as these require observations from closely spaced satellites such as the planned mother/daughter mission (IME) before an unambiguous measure of thicknesses and hence their current density can be made.

5. Turbulence

The bottom right-hand panel of Fig. 3 shows ion cyclotron waves which were observed in the cusp only when the interplanetary field was northward and when the cusp plasma was observed on apparently closed field lines [5]. These waves are transverse to the field but have a compressional part. They do not have the characteristics of the pressure anisotropy-driven ion cyclotron instability, but instead may be generated by a current driven instability [12]. D'Angelo [13] has sought to explain the ULF fluctuations observed by OGO 5. He points out that the various strong gradients in the cusp region would naturally lead to drift wave generation, but that, since the polar cusp plasma is expected to be streaming counter to the magnetospheric plasma, the Kelvin-Helmholtz instability should also occur. He considers the latter to be most consistent with the observed characteristics of the waves. However, as shown in the first three events of Fig. 3, much of this "turbulence" can be explained by "impulsive" field-aligned currents rather than waves. These "impulsive" currents are also accompanied by VLF electrostatic turbulence (see [7, Fig. 3]). The anomalous resistivity associated with this noise together with the observed currents could lead to a potential drop of the order of 2 kilovolts along the field line [7]. Such potential drops would significantly energize cusp plasma. Since such field-aligned currents are either narrow filaments or impulsive bursts, the associated electric fields would be filamentary or impulsive. In either case this would lead to a burst-like nature of precipitation in the polar cusp. An association of impulsive field-aligned currents and impulsive precipitation at low altitudes has been observed [14].

6. The Cusp-Magnetosheath Interface

At the top of Fig. 1 are shown the projections in the dipole meridian plane of one minute averages of the geomagnetic field in the region of the cusp-magnetosheath interface. The corresponding part of the orbit trace is shaded with a heavy

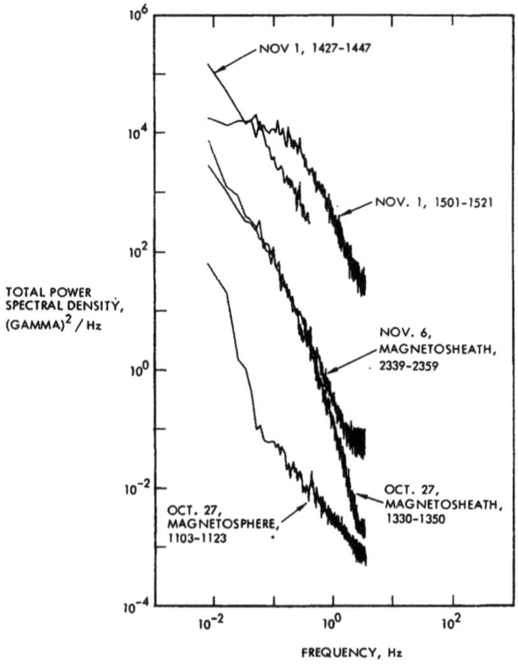

Fig. 4. Power spectra, summed over the three orthogonal directions, at the cusp-magnetosheath interface on 1 November 1968 and in the magnetosheath on neighboring orbits. A magnetospheric power spectrum is shown for reference.

line. High resolution data show that OGO 5 left the magnetosphere at 1427 UT and returned briefly from 1440 to 1442. The top panel of Fig. 1 shows that the magnetospheric field lines were oriented as expected during this interval, i.e. towards the earth and horizontally or slightly downwards. At other times, i.e. 1428—1439 and 1443—1500, the field assumed a variety of orientations. Examination of Explorer 33 interplanetary field data shows that these changes occurred in response to similar changes in the interplanetary magnetic field.

While the orientation of the magnetospheric and magnetosheath fields were rather as expected, the level of turbulence was not. Fig. 4 shows two power spectra obtained on this pass compared with power spectra in similar regions of the magnetosheath on other passes, and with a magnetospheric power spectrum at a similar radial distance. The wave power was two orders of magnitude higher on 1 November 1968 than on the control days. While 1 November was a disturbed day, this is a considerable enhancement of the power spectrum. Such enhanced turbulence is consistent with the suggestion made by Walters [15] of an attached shock at the cusp. On the other hand, Spreiter and Summers [16] have argued against the existence of such a region and instead postulate a contact discontinuity separating flowing magnetosheath plasma from stationary cusp plasma. Such a region should be turbulent also because of the Kelvin-Helmholtz instability but perhaps not as turbulent as behind a shock. We note that both Walters' and the Spreiter-Summers' models were proposed for closed magnetospheres, whereas merging is known often to play a significant role in the behavior of the magnetopause. Thus, while our data are consistent with the occurrence of an attached shock in this region, much more evidence is required for conclusive proof of its existence than we have on this one orbit.

7. Summary

The polar cusp plays a fundamental role in magnetospheric processes. Not only is the cusp an entry region of magnetosheath plasma and the source of daytime aurora, but field-aligned currents flow in the polar cusp communicating the stresses between the magnetopause and the ionosphere. During the very active period studied here broad Hall closure current sheets flow in the cusp, and there are filamentary or impulsive current structures in which the anomalous resistivity associated with VLF electrostatic emissions is capable of generating large potential drops along the field line. The interface region between the cusp and the magnetosheath is a complex, dynamic and turbulent region in which Walters' attached shock may be present. In short, the polar cusp is a veritable plasma physics laboratory containing a broad spectrum of plasma physics phenomena.

Acknowledgments

This work was supported by the National Aeronautics and Space Administration under research grant NGL 05-007-004. Beneficial discussions of this work with F. V. Coroniti, T. E. Holzer and R. L. McPherron are gratefully acknowledged.

References

[1] L. A. Frank, J. Geophys. Res. **76**, 5202 (1971).
[2] W. J. Heikkila and J. D. Winningham, J. Geophys. Res. **76**, 883 (1971).
[3] C. T. Russell et al., J. Geophys. Res. **76**, 6743 (1971).
[4] F. L. Scarf et al., J. Geophys. Res. **77**, 2274 (1972).
[5] M. G. Kivelson et al., J. Geophys. Res. **78**, 3761 (1973).
[6] D. H. Fairfield and N. F. Ness, J. Geophys. Res. **77**, 611 (1972).
[7] R. W. Fredricks et al., J. Geophys. Res. **78**, 2133 (1973).
[8] A. J. Zmuda et al., J. Geophys. Res. **71**, 5033 (1966).
[9] F. V. Coroniti and C. F. Kennel, J. Geophys. Res. **78**, 2837 (1973).
[10] C. F. Kennel and M. H. Rees, J. Geophys. Res. **77**, 2294 (1972).
[11] J. C. Armstrong and A. J. Zmuda, J. Geophys. Res. **75**, 7122 (1970).
[12] R. W. Fredricks and C. T. Russell, J. Geophys. Res. **78**, 2917 (1973).
[13] N. D'Angelo, J. Geophys. Res. **78**, 1208 (1973).
[14] C. T. Russell and R. E. Holzer, in: Particles and Fields in the Magnetosphere, D. Reidel Publ. Co., Dordrecht, Holland 1970 (p. 195).
[15] G. K. Walters, J. Geophys. Res. **71**, 1341 (1966).
[16] J. R. Spreiter and A. L. Summers, Planet. Space Sci. **15**, 787 (1967).

Space Research XIV — Akademie-Verlag, Berlin 1974

ROCKET INVESTIGATIONS OF THE INTENSITY AND COMPOSITION OF CORPUSCULAR RADIATION AT ALTITUDES UP TO 180 km IN THE POLAR REGION

V. F. Tulinov, V. M. Feigin, V. A. Lipovetsky, S. L. Novikov, G. F. Tulinov and Yu. M. Zhuchenko

Central Aerological Observatory, Hydrometeorological Service of the USSR, Moscow, USSR

The corpuscular energy source affecting the polar atmosphere at an altitude range 100 to 180 km is studied. The energy spectra of precipitating electrons measured in the energy range of $0.1-10^2$ keV are reported; the scale of variations of the electron intensity is defined: the precipitating electron intensities vary with geomagnetic activity. The measured intensities of low-energy protons (1—10 MeV) are also discussed. Simultaneous investigations of the neutral polar thermosphere show that the increase in precipitating electron intensity is correlated with atmospheric heating at about 165 km altitude. The contribution of the corpuscular energy source to the heating of the polar atmosphere is typically 100°K during a day for the altitude range considered. However, this value is much less than that of the observed temperature variations.

In order to investigate the corpuscular energy source further and to find out how this source affects the upper atmosphere of the polar region, a number of MR-12 meteorological rockets have been launched from Heiss Island (Franz Jozef Land, $\varphi = 80°37'$ N, $\lambda = 58°03'$ E) in spring 1972 with apogees up to 180 km. Earlier corpuscular radiation measurements (17 MR-12 rockets were launched from Heiss Island during 1969—1971 [1—3]) have been made by Geiger counters. The effective energy threshold of these counters was approximately 40 keV.

These investigations showed, besides other results, a positive correlation between the degree of local geomagnetic disturbance and the precipitating particle intensity at high altitudes over the polar cap [4]. This correlation is depicted in Fig. 1, where vertical lines and arrows show data for 1969—71. The data define the electron intensity at heights 110—180 km, i.e. in the altitude region where the absorption of electrons with energies above 40 keV is negligibly small, and where electrons precipitate from the upper hemisphere, the radiation being isotropic. The scatter of values is associated with variations in the average electron intensity at 110—180 km on a time scale $\geqq 5$ sec, i.e. the period over which the data have been averaged. The upward and downward arrows show the upper and lower limits of the intensity of radiation which are defined, on the one hand by the telemetry system, and on the other hand by the sensitivity of the recording equipment used to measure the soft corpuscular radiation.

The temperature of the polar thermosphere was obtained at the same time by observing luminescent clouds during twilight. The basic results are that in spring the temperature of the polar atmosphere at an altitude of 165 km is usually

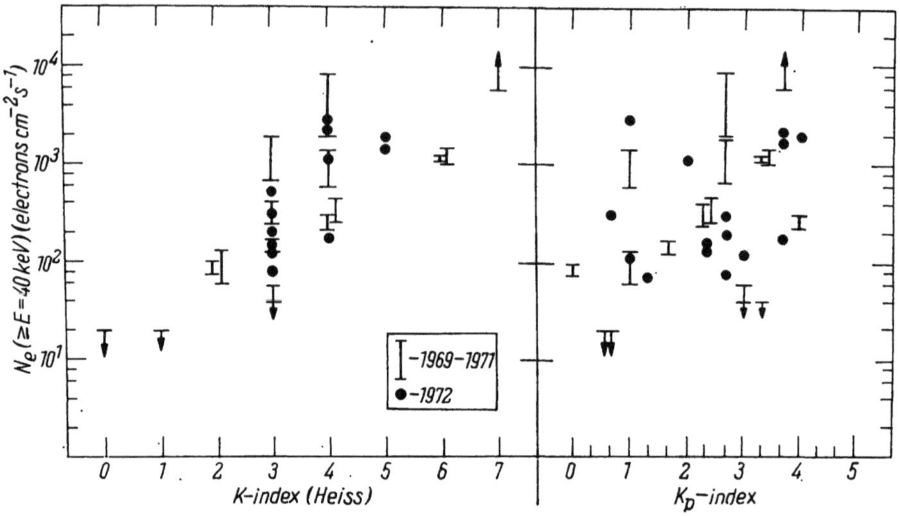

Fig. 1. The intensity of precipitating electrons with energies $\geq$ 40 keV at altitudes 110—180 km versus local (Heiss Island) K index of geomagnetic field.

500 $\pm$ 100 °K; this is considered to be the normal temperature in this period. On a number of nights the temperature exceeded its mean value by several hundred degrees. A strong correlation has been established between the local perturbation of the geomagnetic field and such enhancements in atmospheric temperature, with a time delay of 4 hours. Temperature enhancements were not observed to be associated directly with the increase of Kp index. A comparison of simultaneous measurements of the intensity of precipitating electrons with energies $\geq$40 keV and the polar thermospheric temperature at 165 km altitude shows a good correlation between these variables [4]. The data from a series of rocket experiments carried out in the period from 23 February to 7 March 1970 are good examples of this (see Fig. 2). An increase in the $\geq$40 keV electron intensity from its minimum value, of the order of 10^{-1} cm^{-2} s^{-1} on 23 February to approximately 10^3 cm^{-2} s^{-1} on 2 March was observed to be followed by a 300 °K enhancement of atmospheric temperature at 165 km altitude. On subsequent days decrease in both electron intensity and atmospheric temperature was observed; and on 6—7 March these parameters reverted to their original values.

The correlation between the increases in corpuscular fluxes and in thermospheric temperature indicates a contribution made by precipitating particles to the heating of the upper atmosphere of the polar region. To evaluate this contribution one should have data on the energy spectrum of electrons of low energies ($\sim$1 keV), in order to know the intensity of electrons effectively absorbed in the altitude range under consideration. From measurements of the intensity of electrons with energies $\geq$40 keV the spectrum can be determined only above 40 keV. Variations of electron intensity with altitude, for heights of 70—100 km, where the electrons are absorbed are investigated. The layer of residual atmosphere serves as a filter of variable thickness. In particular the energy spectra of electrons registered on 2 March 1972 (see Fig. 2), were satisfactorily approximated by the exponential function $N\,(>E) \propto E^{-\gamma}$, where $\gamma \simeq 3$. By extrapolating the spectrum to smaller energies one can obtain the value of electron intensity for the range of energies of interest. The calculations made on the basis of the data obtained during the two flights of 2 March 1970 showed that in the altitude range 160—170 km the source of corpuscular radiation is capable of heating the atmosphere at a rate given by $\partial T/\partial t \simeq 10^{-2}$ deg s^{-1}. This allows a reasonable interpretation of the

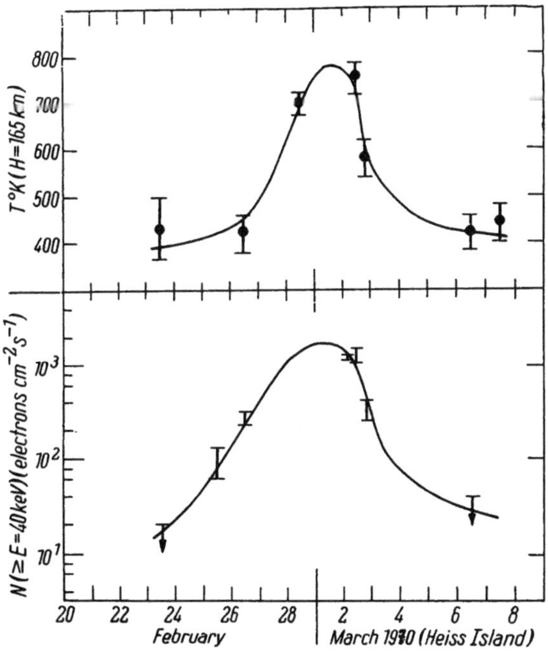

Fig. 2. The intensity of precipitating electrons with energies $\geqq$ 40 keV at altitudes 110—180 km and the temperature of the polar thermosphere at 165 km for February—March 1970 ($Z_{\odot} = 95°$).

effect of the observed temperature enhancement at the altitudes under consideration.

Four MR-12 rockets were launched in 1972 with instruments installed [6] that allowed simultaneous observation of the energy spectra of precipitating electrons with energies of 0.1—10^2 keV, as well as of the proton component of corpuscular radiation (proton energy is 1—9 MeV). The experiment, along with Geiger counters, used spectrometers with channel multipliers [7] and semiconductor detectors for recording protons [8]. First of all, it should be emphasized that the results of 1972 measurements of precipitating electrons with energies $\geqq$40 keV are in agreement with those of 1969—1971. This can be seen in Fig. 1, where the points represent the mean values of such electron intensity at 110—180 km altitude. The results of simultaneous measurements of electron and proton components of corpuscular radiation in the upper atmosphere of a polar region show that the proportion of protons with energies more than 1 MeV in the total corpuscular flux is small (with the exception of rare cases, of intense fluxes of solar cosmic rays generated in flare regions).

The measurements of precipitated electrons in the energy range $\leqq$40 keV are

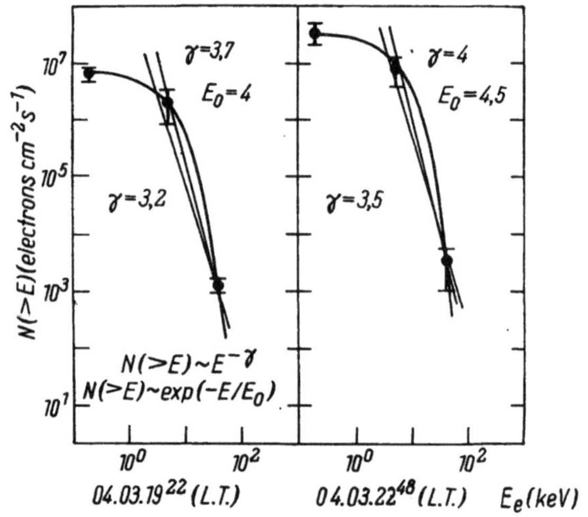

Fig. 3. Integral energy spectrum of precipitating electrons in the polar region at 0403 UT on 4 March 1972.

discussed below. It can be seen from the data in Fig. 3 that the downward extraspolation of the energy spectrum using data on the atmospheric absorption of electrons $\geqq 40$ keV seems not to be correct. Actually, the electron spectrum becomes softer, $\gamma = 3.5-4.5$ in the range of the energies less than 40 keV compared with $\gamma = 3$ for energies above 40 keV. However, below 1 keV the measured energy spectrum of precipitating particles becomes exponential and may be expressed as follows:

$$N(>E) \propto \exp(-E/E_0),$$

where $E_0 = 3.5-4.5$ keV.

These precipitated electron fluxes contribute to the heating of the polar thermosphere at a rate $\partial T/\partial t \simeq 10^{-3}$ deg s^{-1} for the altitude range 160—170 km. These values are an order of magnitude lower than those of preliminary evaluations made on the basis of the experimental data above 40 keV and lead to a diurnal variation of only 100 °K.

References

[1] L. S. NOVIKOV, V. F. TULINOV, G. F. TULINOV and V. M. FEIGIN, Trudy IPG, No. 17, 1973.

[2] V. F. TULINOV, V. M. FEIGIN, V. A. LIPOVETSKI, L. S. NOVIKOV and G. F. TULINOV, Trudy IPG, No. 24, 1974.

[3] V. F. TULINOV, V. M. FEIGIN, YU. M. ZUCHENKO, V. A. LIPOVETSKI and G. F. TULINOV, Geomagn. i Aeron. **13**, 513 (1973).

[4] V. F. TULINOV, V. M. FEIGIN, G. F. TULINOV, L. S. NOVIKOV and L. S. LIPOVETSKI, Kosm. Issled. **11**, 704 (1973).

[5] S. M. POLOSKOV, G. F. TULINOV, J. BLAMONT, M. L. CHANIN, M. MAILLARD, Space Research, X, 519 (1970).

[6] V. M. FEIGIN, Trudy CAO, No. 99, 58 (1971).

[7] V. F. TULINOV, V. M. FEIGIN, V. A. LIPOVETSKI and YU. M. ZUCHENKO, Trudy CAO, No. 122, 1974.

[8] V. F. TULINOV, V. A. LIPOVETSKI and V. M. FEIGIN, Trudy CAO, No. 107, 97, 1973.

Space Research XIV — Akademie-Verlag, Berlin 1974

ESTIMATION OF H^+ FLUXES IN THE POLAR REGIONS

K. K. Mahajan and V. K. Pandey

Radio Science Division, National Physical Laboratory, New Delhi, India

Alouette 2 topside electron density profiles to heights of 3000 km are analysed for low, mid and polar latitudes. The electron scale heights at the low and mid latitudes indicate H^+ to be the dominant ionic constituent for heights above 1000 km. At the polar regions the scale heights are low, indicating the dominance of O^+ ions, even above 1000 km. Thus this lends support to the theories of H^+ escape from the polar regions. The electron distributions at the mid and high latitudes are used to estimate the H^+ flow velocity at the polar regions. The calculated velocity is in reasonable agreement with that expected from the 'polar wind' theory.

1. Introduction

The ion composition measurements on various satellites have shown a large depletion of lighter ions at polar latitudes (see e.g. [1—7]). This depletion has also been seen in the electron concentration at heights around 1000 km and above, especially for the night-time [8—10]. This H^+ decrease has been explained as being due to upward flow of plasma at the poles, because the high latitude field lines open into regions of very low plasma pressure [11—15]. In view of the supersonic nature of the escape, the name polar wind (in analogy to solar wind) has been suggested by Axford [16] and a complete theory of this escape has been worked out, in a series of papers, by Banks and Holzer [13—15]. According to Banks and Holzer, the H^+ flow velocity becomes supersonic in the altitude range 2000—4000 km and the plasma is further accelerated with increasing altitude. While an exact calculation of the flow velocity needs accurate knowledge of temperature and composition of the ions and neutrals, it is, however, possible to deduce approximate values of the flow velocity from topside electron density profiles. The Alouette 2 topside sounder gives electron densities up to heights as great as 3000 km and a large number of these profiles have been published by the Department of Communications, Communications Research Centre, Ottawa, Canada [17]. We have used these profiles in our analysis and in this paper we shall report the values of the H^+ flow velocity estimated from these profiles at polar latitudes.

2. Method

The equation of motion for H^+ ions, parallel to the magnetic field can be given as [18]:

$$m_{H^+}n(H^+)\frac{\partial V_{H^+}}{\partial t} + m_{H^+}n(H^+)V_{H^+}\frac{\partial V_{H^+}}{\partial s} + \frac{kTn(H^+)n(O^+)}{[n(H^+)+n(O^+)]D_{H^+O^+}}(V_{H^+} - V_{O^+})$$
$$+\frac{kTn(H^+)n(N)}{[n(H^+)+n(N)]D_{H^+N}}(V_{H^+} - V_N) = -kT\frac{\partial n(H^+)}{\partial s} - \frac{kTn(H^+)}{N_e}\frac{\partial N_e}{\partial s}$$
$$- kn(H^+)\frac{\partial T}{\partial s} - m_{H^+}gn(H^+) \tag{1}$$

where m_{H^+} = mass of H^+ ion, $n(H^+)$, $n(O^+)$ = number densities of H^+, O^+, $n(N)$ = number density of neutral constituent, N_e = number density of electrons, V_{H^+}, V_{O^+} = velocity of H^+, O^+ ions, T = plasma temperature = $(T_e + T_i)/2$, $D_{H^+O^+}$, D_{H^+N} = diffusion coefficient of H^+ in O^+ and in the neutral constituent respectively, s = coordinate along the magnetic field.

If one considers steady state conditions, neglects height gradients of V_{H^+} and T, and ignores the effects of H^+ diffusion in the neutral atmosphere, Eq. (1) reduces to

$$\frac{n(O^+)(V_{H^+} - V_{O^+})}{[n(H^+)+n(O^+)]D_{H^+O^+}} = -\frac{1}{n(H^+)}\frac{\partial n(H^+)}{\partial s} - \frac{1}{N_e}\frac{\partial N_e}{\partial s} - \frac{m_{H^+}g}{kT}. \tag{2}$$

At polar latitudes for heights around 2000 km, $n(O^+) \gg n(H^+)$ and thus Eq. (2) further reduces to

$$V_{H^+} - V_{O^+} = D_{H^+O^+}\left[-\frac{1}{n(H^+)}\frac{\partial n(H^+)}{\partial s} - \frac{1}{N_e}\frac{\partial N_e}{\partial s} - \frac{m_{H^+}g}{kT}\right]. \tag{3}$$

Eq. (3) is our basic relation used for the calculation of H^+ flow velocity at polar latitudes. As the field lines are nearly vertical at the poles, the altitude gradients of N_e, taken from the N_e-h profiles, can give the parameter $(1/N_e)(\partial N_e/\partial s)$, the reciprocal of the scale height, directly. While no height profiles of H^+ are available, we have estimated the term $(1/n(H^+))(\partial n(H^+)/\partial s)$ from the ion composition measurements on the Explorer 32 satellite, reported by Brinton et al. [19] in the trough region. These measurements correspond to the height range from about 800 km to 2600 km and are for daytime winter conditions. From these measurements we have obtained a value of 8.3×10^{-9} cm^{-1} for $(1/n(H^+))(\partial n(H^+)/\partial s)$ for a height of 2000 km. The quantity $m_{H^+}g/kT$ has been estimated to be 2.5×10^{-9} cm^{-1} assuming a plasma temperature of 2500 °K.

3. Results

Fig. 1 shows the electron density profiles for the low, mid and high latitudes for summer daytime conditions. It can be noted that for heights below about 1000 km there is no relative change in the profile shape as one moves from low to high latitudes. For heights above 1000 km, there is a significant latitudinal variation in the profile shape. At the low and the mid latitudes the electron scale heights are large, thus indicating the dominance of the H^+ ions. At the high latitudes, the scale heights are small, thus showing the dominance of O^+ ions. Similar profiles for night-time conditions are shown in Fig. 2. One again sees a significant change in the electron scale height from the mid to the high latitudes at heights above about 1000 km.

The latitudinal variation of the electron scale height $N_e/(dN_e/dh)$ for heights of 2000 km and 2500 km are shown in Fig. 3. The sudden decrease in the scale

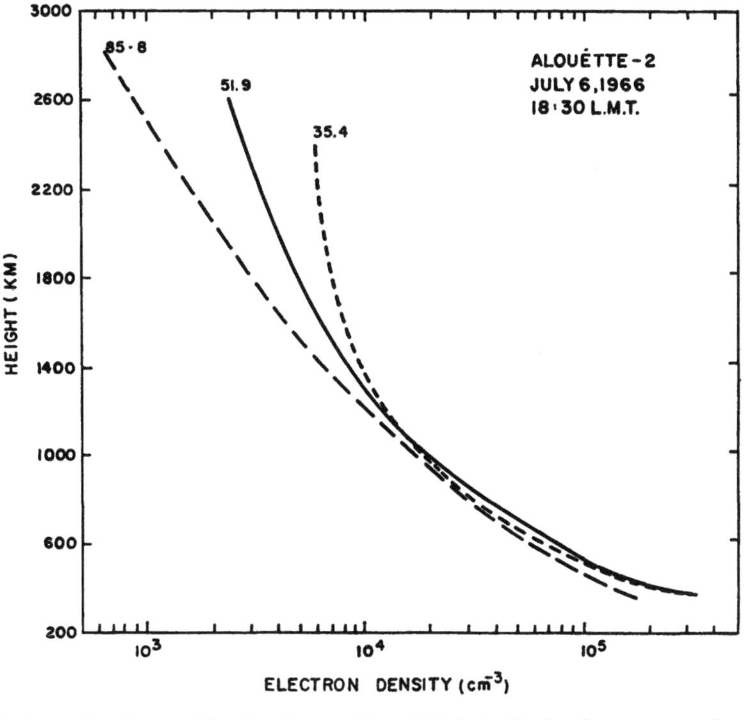

Fig. 1. Electron density profiles for low, mid and high latitudes for summer daytime conditions. The figures on the profiles indicate the geomagnetic latitude.

height at polar latitudes can be clearly noted. These results again indicate that the polar latitudes are dominated by O^+ ions, while the low and mid latitudes have H^+ as the dominant ion. Under conditions of diffusive equilibrium, one would expect H^+ to be a dominant ion at the polar latitudes also and thus their absence at the polar regions indicates an upward flow of H^+ ions. As mentioned in the Introduction, this flow can occur at the poles because there the field lines are not closed, but open into regions of low plasma pressure.

We have calculated the H^+ flow velocity by using Eq. (3). An analysis of a large number of high latitude N_e-h profiles gives an average value of 1.6×10^{-8} cm^{-1} for the parameter $(1/N_e)(dN_e/dh)$ at 2000 km. Substituting for the diffusion

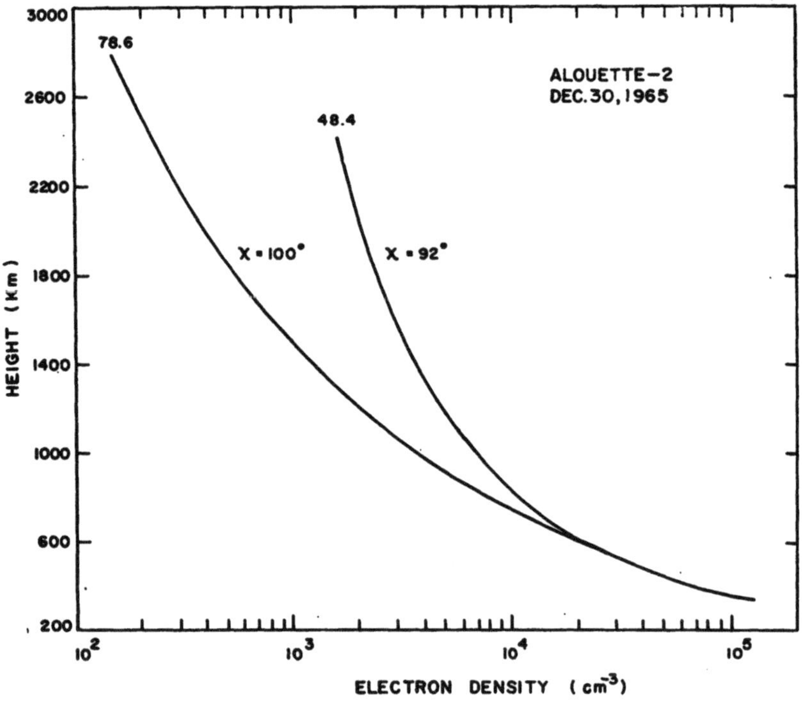

Fig. 2. Electron density profiles for the mid and the high latitudes for winter night-time conditions. The figures on the profiles indicate the geomagnetic latitude.

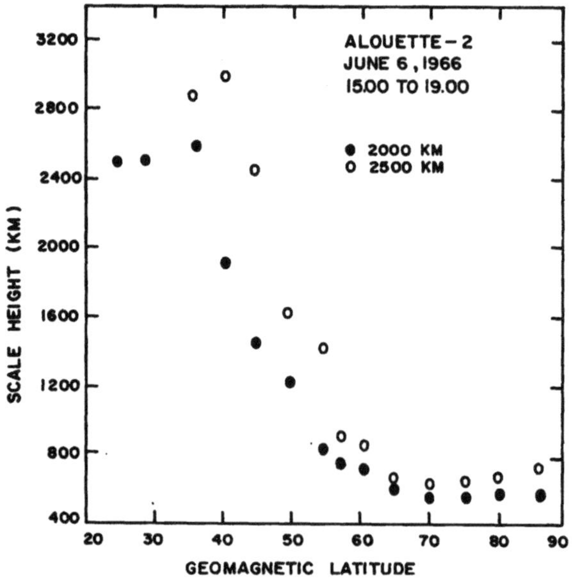

Fig. 3. Latitudinal variation of electron scale height $(= N_e/(dN_e/dh))$ at 2000 km and 2500 km. Large decreases in this parameter at the polar latitudes can be clearly noted.

coefficient value, which can be given by [20]:

$$D_{\mathrm{H^+O^+}} = \frac{AT^{5/2}}{n(\mathrm{O^+})\,[\ln \Lambda - 0.961]}\ \mathrm{cm^2\,s^{-1}} \tag{4}$$

where

$$A = 2.1 \times 10^9\ \mathrm{cm^2\,s^{-1}}$$

$$\Lambda = 5.84 \times 10^3 \left[\frac{T^3}{n(\mathrm{O^+})}\right]^{1/2} \mathrm{deg\ cm^3} \tag{5}$$

one gets an average value of 22 km s^{-1} for $V_{\mathrm{H^+}} - V_{\mathrm{O^+}}$. This value is in reasonable agreement with the supersonic velocities of the order of 20 km s^{-1} at 2000 km expected from the polar wind theory for low solar activity conditions [15]. A flow velocity of 22 km s^{-1} will result in an H^+ flux of 4.4×10^8 cm^{-2} s^{-1} at 2000 km, if one uses a value of 2×10^2 cm^{-3} for $n(\mathrm{H^+})$ from ion composition measurements on Explorer 32 [19]. This flux is in reasonable agreement with values in the range of $(2-7) \times 10^8$ cm^{-2} s^{-1} calculated by Banks and Holzer [15].

4. Conclusions

Alouette 2 topside sounder electron density profiles were used to estimate the H^+ flow velocity and H^+ fluxes at the polar latitudes. These results give a value of 22 km s^{-1} for the H^+ velocity and 4.4×10^8 cm^{-2} s^{-1} for the H^+ flux. These values are in reasonable agreement with those expected from the polar wind theory.

Acknowledgments

The authors express their thanks to the Department of Communications, Communications Research Centre for making available the Alouette 2 ionospheric data. One of the authors (V.K. Pandey) has been a Junior Research Fellow of the Council of Scientific and Industrial Research during the progress of this work.

Note Added in Proof

Further analysis indicates that the H^+ flow velocity is lower for summer conditions. For the case of 6 July 1966 (Fig. 1), the velocity comes to about 5 km s^{-1}, while it is 24 km s^{-1} for the case of 30 December 1965 (Fig. 2).

References

[1] H. A. Taylor, Jr, H. C. Brinton, M. W. Pharo III and N. K. Rahman, J. Geophys. Res. **73**, 5521 (1968).
[2] J. H. Hoffman, Proc. IEEE **57**, 1063 (1969).
[3] H. C. Brinton, R. A. Pickett and H. A. Taylor, Jr, J. Geophys. Res. **74**, 4064 (1969).
[4] H. A. Taylor, Jr, H. G. Mayr and H. C. Brinton, Space Research X, 663 (1970).
[5] S. Chandra, B. E. Troy, Jr, J. L. Donley and R. E. Bourdeau, J. Geophys. Res. **75**, 3867 (1970).
[6] H. A. Taylor, Jr, Planet. Space Sci. **20**, 1593 (1972).

[7] H. A. TAYLOR, JR, and W. J. WALSH, J. Geophys. Res. **77**, 6716 (1972).
[8] L. H. BRACE and B. M. REDDY, J. Geophys. Res. **70**, 5783 (1965).
[9] N. J. MILLER, J. Geophys. Res. **75**, 7175 (1970).
[10] M. J. RYCROFT and J. O. THOMAS, Planet. Space Sci. **18**, 65 (1970).
[11] A. J. DESSLER and F. C. MICHEL, J. Geophys. Res. **71**, 1421 (1966).
[12] A. NISHIDA, J. Geophys. Res. **71**, 5669 (1966).
[13] P. M. BANKS and T. E. HOLZER, J. Geophys. Res. **73**, 6846 (1968).
[14] P. M. BANKS and T. E. HOLZER, J. Geophys. Res. **74**, 6304 (1969).
[15] P. M. BANKS and T. E. HOLZER, J. Geophys. Res. **74**, 6317 (1969).
[16] W. I. AXFORD, J. Geophys. Res. **73**, 6855 (1968).
[17] Alouette II Ionospheric Data: interpolated $N(h)$, Vol. 1 Department of Communications, Communications Research Centre, Ottawa, Canada.
[18] H. G. MAYR, J. M. GREBOWSKY and H. A. TAYLOR, JR, Planet. Space Sci. **18**, 1123 (1970).
[19] H. C. BRINTON, J. M. GREBOWSKY and H. G. MAYR, J. Geophys. Res. **76**, 3738 (1971).
[20] R. W. SCHUNK and J. C. G. WALKER, Planet. Space Sci. **18**, 535 (1970).

Space Research XIV — Akademie-Verlag, Berlin 1974

X-RAYS FROM AN AURORAL SUBSTORM AT ALTITUDES OF 60—80 km AND 15—20 km

W. R. Sheldon, J. W. Kern and J. R. Benbrook

University of Houston, Houston, Texas, USA

A technique for the measurement of high altitude X-rays associated with auroral events is described. The instrument package is boosted to the lower D region by a rocket and descends slowly by parachute. This arrangement facilitates co-ordination with the launching of other rocket probes during events of interest. A flight on 2 February 1973 from the Churchill Research Range provided 17 minutes of X-ray data above 28 km. The flight was made into an intense break-up event and X-ray fluxes were observed during two episodes of the explosive phase. Data from three integral energy channels (> 5 keV, >10 keV and >40 keV) are presented with emphasis on the time when the detector was at atmospheric depths greater than 50 g cm^{-2}. The observed X-ray spectra are relatively soft, particularly during several intense X-ray bursts. Existing treatments of X-ray propagation in the atmosphere do not predict the observed spectra. The reduction of auroral absorption during the later phases of the event suggests precipitation of energetic electrons unaccompanied by the lower energy precipitation responsible for the absorption earlier in the substorm.

A system for measuring the temporal variation in the high altitude X-ray flux occurring in conjunction with other phenomena of geophysical interest has been developed by the University of Houston Cosmic Ray Group [1]. A NaI(Tl) scintillation crystal is employed as the X-ray detector; a Superarcas rocket is used to boost the payload rapidly to high altitude. The payload is deployed on a parachute at apogee and descends at a relatively slow rate. The flight described here had an apogee of 82 km and provided 17 minutes of data above 28 km, a typical float altitude for balloons.

The present results were obtained on a flight made at the Churchill Research Range (CRR) on 2 February 1973 into an active break-up aurora. The Superarcas launching was co-ordinated with launchings of a Black Brant 3 from Gillam (150 miles south of CRR) and a Black Brant 5 from CRR, conducted by colleagues of the National Research Council of Canada. The flight was made into a large break-up event which began south of Gillam. The break-up was preceded for several hours by quiet, slow-moving east-west arcs in the region between CRR and Gillam, and events near the eastern horizon which appeared to be more active and to have north-south components in their movement. Large X-ray fluxes were observed for about one minute immediately after deployment at 0625:40 UT, between altitudes of 82 km and 63 km. The maximum count rate recorded in the >5 keV channel during this portion of the event was 11000 counts/sec. A report on the simultaneous observations of charged particle fluxes above and the X-ray flux below the optical aurora will be presented later. The present report is concerned primarily with the X-ray data at lower altitudes.

The data discussed here are shown in Fig. 1, for times 0643 to 0704 UT. Magnetometer and polar riometer records are shown for this period as well as count-rates for three integral X-ray channels $E > 5$, 10, and 40 keV. The atmospheric depth of the detector is shown at the top of the figure. Features in the data that are of

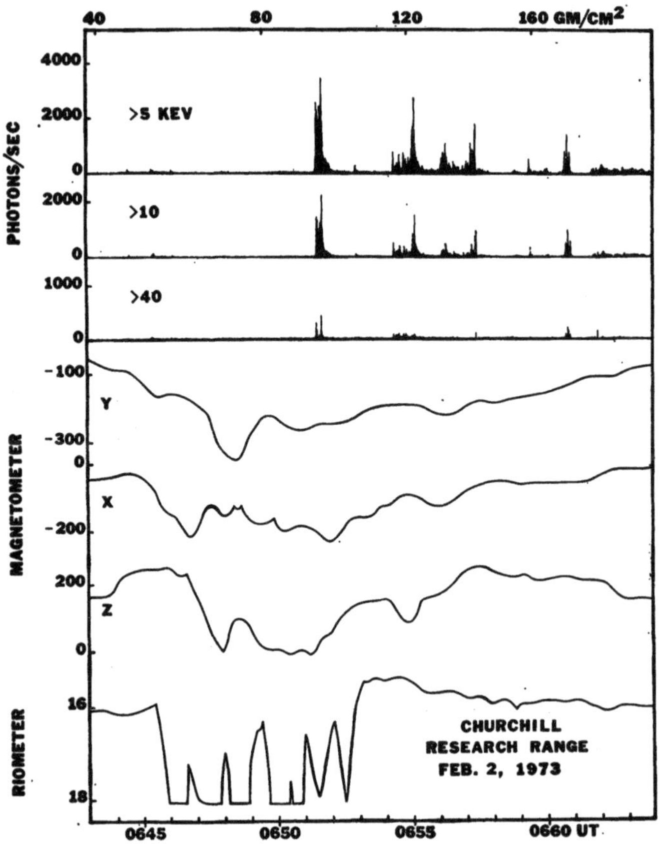

Fig. 1. X-ray count rates from descending parachute-borne detector and ground-based observations made during auroral substorm, 0643—0704 UT, 2 February 1973 at Churchill Research Range. Magnetometer deflections in γ; riometer deflections in dB.

interest are the enhancement in count-rates associated with the magnetometer and riometer activity beginning at 0643 UT, and the series of intense bursts beginning at 0651:30 UT at large atmospheric depths.

The enhancement in the X-ray flux beginning at 0643 UT was accompanied by an increase in the intensity of magnetic activity and a rapid increase in cosmic radio noise absorption. The X-ray fluxes, Y-magnetometer, and polar riometer data are shown in Fig. 2 for 0643—0654. The atmospheric depth of the payload at 0643 was about 50 g cm^{-2}. The integral X-ray spectrum in the more intense bursts during the early portions of this event is relatively hard, and the average fluxes are about $1^1/_2$ times the background flux at this altitude.

At 0651:30 UT an extremely intense burst of X-rays is observed at an atmospheric depth of 100 g cm^{-2}. Count rates of 3600 s^{-1} (>5 keV), 2400 s^{-1} (>10 keV), and 400 s^{-1} (>40 keV) indicate that the observed spectrum is relatively soft, with an e-folding energy of between 10 and 15 keV. At this depth, photoelectric

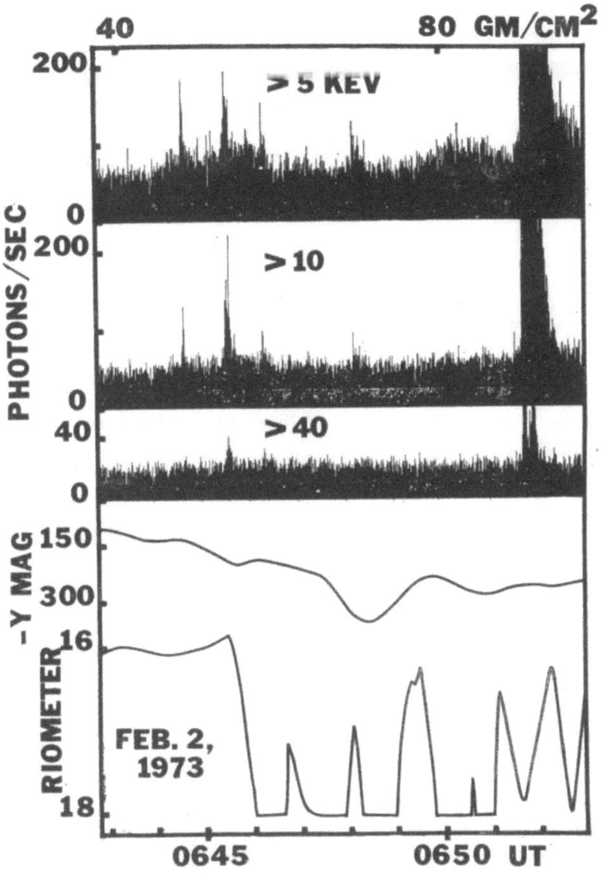

Fig. 2. Detail of X-ray count rate, 0643—0653 UT, 2 February 1973, Churchill Research Range. Magnetometer deflections in γ; riometer deflections in dB.

absorption removes photons with energies <40 keV in short distances, hence the intense fluxes of photons at low energies must be locally produced by Compton scattering of more energetic X-rays. The series of bursts is shown in Fig. 3. Note that the last burst at 0701:20 UT, 170 g cm^{-2}, has nearly the same spectral character as the first.

The energy flux at the top of the atmosphere associated with these bursts can be estimated from the observed integral X-ray flux $J(>E)$ and the total energy absorption coefficients for air given by Evans [2]. Assuming the primary

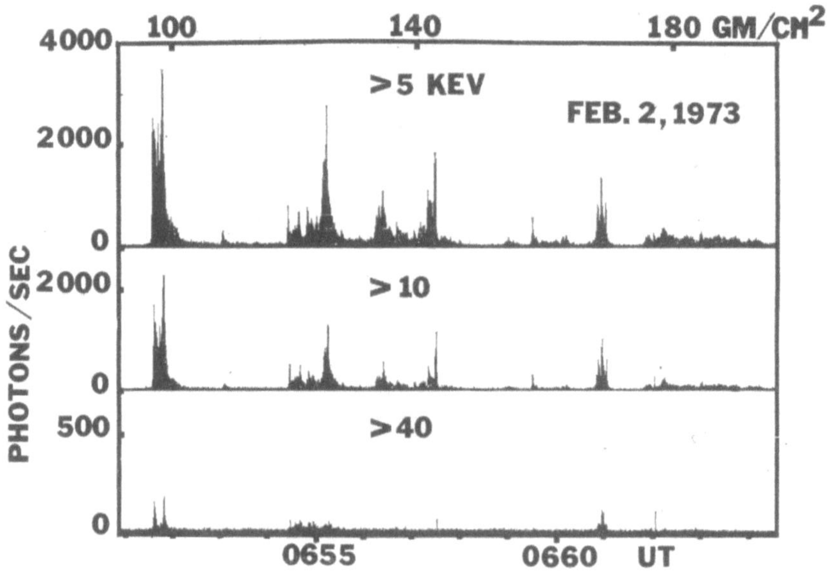

Fig. 3. X-ray bursts at 90—200 g cm^{-2} atmospheric depth (approx. 18—12 km altitude), 0650—0704 UT, 2 February 1973, Churchill Research Range.

X-rays have $E > 50$ keV, we can approximate the absorption coefficient by 2.5×10^{-2} cm^2 g^{-1}. At a depth of 100 g, this gives an attenuation of $e^{-2.5} \sim 10^{-1}$. If the data are fitted to an exponential spectrum, the energy flux is about 10^4 keV cm^{-2} s^{-1} for energies above 5 keV. Thus the energy flux of bremsstrahlung photons ($E > 50$ keV) must be about 10^5 keV cm^{-2} s^{-1} at the top of the atmosphere. This flux can be related to the flux of electrons with $E > 50$ keV required to produce the X-rays. A conversion coefficient of the order of 2×10^{-5} keV^{-1} has been found to relate a differential electron spectrum to the spectrum of bremsstrahlung X-rays at the top of the atmosphere [2, 3]. Using this factor we estimate an integral electron flux ($E > 50$ keV) of about 5×10^6 cm^{-2} s^{-1}. Similar considerations apply to the other bursts observed at even greater depths. The calculation of the incident electron spectrum, using X-ray data from depths >50 g cm^{-2}, deserves further investigation. It appears that present treatments of X-ray propagation in the atmosphere [4—6] do not predict the soft X-ray spectra observed here.

It is of interest to note that proper operation of the detector can be verified through measurements of the Pfotzer maximum in the background X-ray flux during both the ascending and descending portions of the flight. The Pfotzer maximum during descent can be seen as the envelopes of the lowest values in Fig. 2. These values, as well as those measured in the Pfotzer maximum during ascent and measurements made later in the flight at lower altitudes were compared with other measurements of the X-ray background at the CRR. The background values obtained in the present flight were almost exactly the same as those measured with an identical instrument flown on 2 November 1972 beneath a quiet arc and are in close agreement with the results of other workers. These comparisons, taken with the observation that the X-ray bursts were correlated with very intense and active auroral break-up activity, lead us to conclude that the auroral observations reported here, although unusual, were made with a properly functioning instrument.

One feature of interest is the recovery of the polar riometer during the succession of very intense X-ray bursts at about 0653 UT. The radio absorption is enhanced during the burst at 0651:30, but recovers to normal levels before the next burst. This suggests that the electron flux at lower energies is not enhanced during the later stages of the break-up. The energetic electrons associated with the penetrating X-ray burst seen here apparently occur independently of the particle precipitation that produces enhanced auroral absorption.

Acknowledgments

This research was supported by the National Aeronautics and Space Administration under Grant NGR-44-005-133.

References

[1] W. R. Sheldon, J. W. Kern and B. Whalen, paper presented at the XV IUGG General Assembly (IAGA), Moscow 1971.
[2] R. D. Evans, The Atomic Nucleus, McGraw-Hill, New York 1955.
[3] K. K. Vij, D. Venkatesan, W. R. Sheldon, J. W. Kern, J. R. Benbrook and B. A. Whalen, University of Calgary Physics Dept. Preprint (1973).
[4] M. H. Rees, Planet. Space Sci. **12**, 1093 (1964).
[5] G. R. Pilkington and C. D. Anger, Planet. Space Sci. **19**, 1069 (1971).
[6] M. J. Berger and S. M. Seltzer, J. Atmos. Terr. Phys. **34**, 85 (1972).

Space Research XIV — Akademie-Verlag, Berlin 1974

AN ATTEMPT TO EXPLAIN SATELLITE OBSERVATIONS OF HIGH LATITUDE VLF HISS IN TERMS OF GENERATION BY INCOHERENT CERENKOV RADIATION

H. W. English and A. R. W. Hughes

Department of Physics, University of Sheffield, Sheffield, UK

Abstract

A study of high latitude ($\Lambda > 60°$) hiss at 3.2 kHz, 9.6 kHz and 16 kHz using data from the Ariel 3 satellite reveals a diurnal variation in the spectrum and intensity of the emission. During the day the spectrum is relatively flat and the most intense events are of the order of 5×10^{-14} W m^{-2} Hz^{-1}. At night (1600—0400 MLT) the most intense emissions at 9.6 kHz are 10—20 dB greater than those observed during the day. The spectral intensity at 3.2 kHz shows an additional enhancement of 8—10 dB above the night-time intensity at 9.6 kHz.

Calculations of the Cerenkov radiation from low energy (< 1 keV) electrons were made for different plasma frequency models by summing the incoherent contributions from individual particles. The spectrum obtained is dependent on the value of the plasma frequency relative to the gyrofrequency and also on the altitude range over which the emitted radiation is summed. Previous calculations by Jørgensen and by Lim and Laaspere showed a peak in the spectrum which was primarily determined by the altitude to which the calculation was performed.

Our calculations show that it is difficult to explain either the night-time enhancement at 3.2 kHz or the maximum intensities ($\sim 5 \times 10^{-11}$ W m^{-2} Hz^{-1}) using realistic plasma models and recently obtained measurements of low energy electron fluxes.

1. Introduction

A broad band VLF emission known as high latitude or auroral hiss occurs in a circum-polar oval band [1, 2] similar to the auroral oval. Ellis [3] originally suggested that the emission might be Cerenkov radiation from auroral electrons. Investigation of this theory by McKenzie [4, 5], Liemohn [6] and Mansfield [7] showed that the emission intensity predicted by the theory was several orders of magnitude less than that observed by ground and satellite receivers. Jørgensen [8] using new particle data and later plasma density models showed that a spectral intensity of 10^{-14} W m^{-2} Hz^{-1} could be obtained which was still two orders less than the intensity observed on Injun 3 [9, 10]. Lim and Laaspere [11] showed that intensities of 10^{-13} W m^{-2} Hz^{-1} could be explained, but they obtained a maximum in the spectrum between 50 and 70 kHz which does not agree with the observations [8].

We present here new data from the Ariel 3 satellite which show that the maximum intensities are of the order of 10^{-12}—10^{-11} W m^{-2} Hz^{-1} in agreement with the

Injun 5 results [12, 13]. In addition, however, we find that the intensity spectrum has both a shape and a local time variation which cannot be explained by the incoherent Cerenkov theory using realistic models for the region of generation. We have performed the Cerenkov calculation for different ranges of electron energies and different plasma models.

2. The Ariel 3 Experiment

The Ariel 3 experiment has been fully described [14, 15] and is summarized below. The Ariel 3 satellite was launched on 5 May 1967 into a circular low altitude ($\sim$550 km) orbit with an inclination of 80°. The receiver employed a loop aerial and included channels at 3.2, 9.6 and 16 kHz each with a bandwidth of 1000 Hz and a logarithmic response of 75 dB range. The sensitivities of the channels were better than 10^{-11}, 10^{-12} and $10^{-13}\,\gamma^2\,\mathrm{Hz}^{-1}$ at 3.2, 9.6 and 16 kHz respectively. The peak, mean and minimum outputs, having time constants 0.01, 30 and 0.1 seconds respectively, were sampled every 28 seconds and recorded on magnetic tape in the satellite.

3. Observations

We have compared the intensities at three frequencies of individual events using the outputs of the mean reading ($\tau = 30$ s) circuits at 3.2, 9.6 and 16 kHz. The data are initially selected by computer by taking all observations at invariant latitudes of greater than 60° for which the signal intensity is 5 dB greater than noise level on any one channel and greater than noise level on each of the others. The events are then further selected by inspection. We want to consider only broad band noise extending over the frequency range 3.2—16 kHz so events are chosen which have a similar intensity profile with latitude and have a maximum intensity at the same location. Further details of the observations are given in [16].

A total of 1180 hiss events were studied in the periods 5 May—14 June and 17 July—3 August 1967 which gave full local time coverage. The averaged intensity ratios for these events are shown in Fig. 1. The left-hand frames show the variation in the ratio N_1 of the intensity at 3.2 kHz to that at 9.6 kHz with magnetic local time (MLT). The right-hand frames are for the ratio N_2 of the intensity at 16 kHz to that at 9.6 kHz. It may be seen that the ratio N_2 is relatively steady at all MLT whereas N_1 shows a pronounced diurnal variation. A characteristic of this variation is an 8—10 dB intensity enhancement at 3.2 kHz compared with that at 9.6 and 16 kHz between 20 h and 24 h MLT in the northern hemisphere. A similar variation is observed in the southern hemisphere. At noon, the spectrum is relatively flat with a maximum intensity of the order of $5 \times 10^{-14}\,\mathrm{W\,m^{-2}\,Hz^{-1}}$ at 9.6 kHz which is 10—20 dB less than the night-time values.

4. The Cerenkov Calculation

The calculation of the spectral intensity given by the Cerenkov mechanism using the cold plasma theory follows that due to Jørgensen [8]. In this calculation the power spectral density $(dp/df)_e$ produced by a single electron (given by Mans-

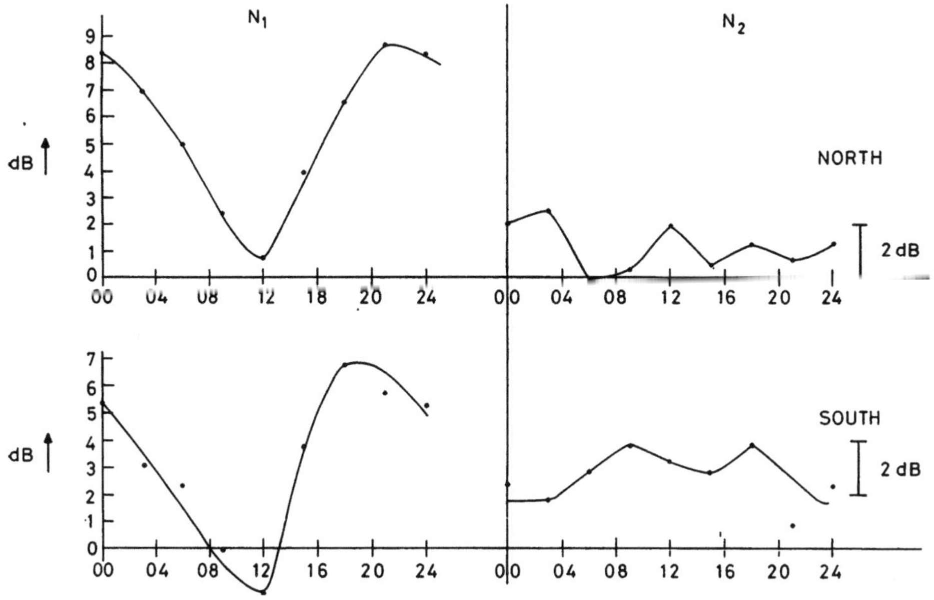

Fig. 1. The diurnal variation in the ratios of the intensities at 3.2 kHz to 9.6 kHz, N_1, and 16 kHz to 9.6 kHz, N_2, for the northern and southern hemispheres.

field [7], McKenzie [5] and discussed by Trulsen and Fejer [17]) was summed over a magnetic flux tube of electrons. The power at the base of the tube due to these particles was then calculated by placing an altitude limit on the summation of 26000 km, as doneby Jørgensen [8]. A similar calculation using this same height limit was performed by Lim and Laaspere [11] who obtained a different spectral shape from that of Jørgensen. Neither of these spectra agree with the night-time spectra observed on Ariel 3; for these calculations $(dp/df)_e$ was given by

$$\left(\frac{dp}{df}\right)_e = \sum_{j=1}^{2} \frac{q^2\pi f}{\varepsilon_0 v_{\parallel}} \frac{\beta_\perp{}^2 T_{11} J_1{}^2(L_0) + \beta_\parallel{}^2 T_{33} J_0{}^2(L_0) - 2\beta_\perp\beta_\parallel T_{13} J_0(L_0) J_1(L_0)}{(B_n{}^2 - 4C_n\varepsilon_1)^{1/2}}\Bigg|_{n=n_j} \quad (1)$$

Here $\beta_\perp = v_\perp/c$, $\beta_\parallel = v_\parallel/c$, J_0 and J_1 are zero and first order Bessel functions of the first kind, $n_{1,2}$ are the two values of the refractive index for which Cerenkov radiation is generated, q is the charge on the electron and T_{11}, T_{33}, ε_1, B_n, C_n, L_0 and $n_{1,2}$ are given in [7, 8, 11]. The total power spectral density $(dp/df)_T$ was then determined for each frequency by dividing the tube into volume intervals ΔV_i and the flux data into energy intervals ΔE_j and summing using the scheme

$$\left(\frac{dp}{df}\right)_T = \sum_{j=1}^{m} \sum_{i=1}^{n} N_{ij}\, \Delta V_i \left(\frac{dp}{df}\right)_{ij} \Delta E_j \quad (2)$$

where N_{ij} is the number density of electrons in the energy interval ΔE_j and volume interval ΔV_i.

In the calculation a dipole magnetic field model was used and the power incident on 1 m² at the satellite altitude (500 km) was determined for 70° latitude.

Little difference was observed when (1) and (2) were recalculated at 80° latitude. Flux measurements of electrons with energies greater than 50 eV used in our calculations were taken from an event described by Ackerson and Frank [18] and are shown together with Evans' [19] results (>1 keV), used originally by Jørgensen [8], in Fig. 2.

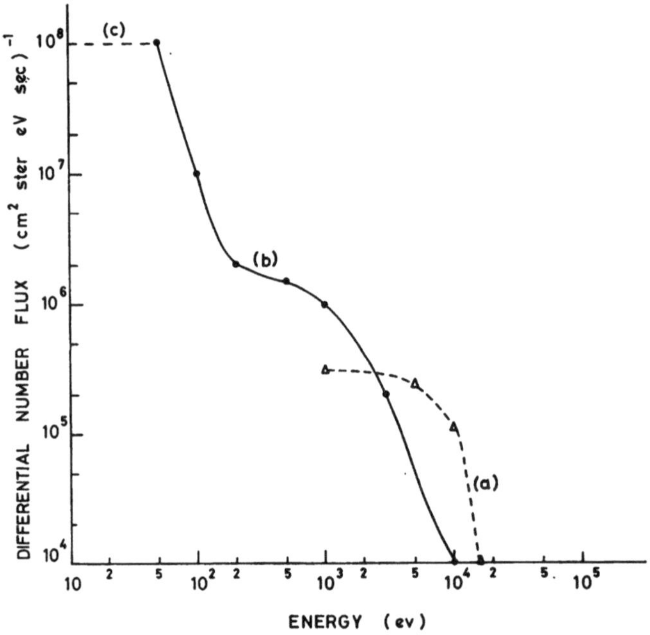

Fig. 2. Electron fluxes used in the Cerenkov calculations measured by (a) Evans [19] and (b) Ackerson and Frank [18]; (c) is an extension of model (b) to 10 eV.

5. The Intensity and Shape of the Spectrum

It can be seen from (1) and (2) that a singularity exists for zero energy particles. However, Stix [20] has pointed out that collisionless damping occurs when the mean thermal speed of the ambient plasma approaches the phase velocity of the wave. Lim and Laaspere [11] took account of this effect by neglecting radiation for which the phase velocity of the wave was less than twice the thermal velocity of the background electrons (temperature taken to be 3000 °K). This was done by restricting the refractive index to values of less than 400. We have adopted the same criterion in some of our calculations.

The shape of the spectrum obtained from the Cerenkov calculation is influenced by a number of factors. Important among these for a given plasma density model (the different plasma models used are shown in Fig. 3) are the altitude to which the integration is performed and the distribution of energetic electrons with altitude. Jørgensen [8] assumed, on the basis of experimental evidence, that the density of energetic electrons is constant with altitude. The assumption of constant energetic electron density means that in a given flux tube there are many

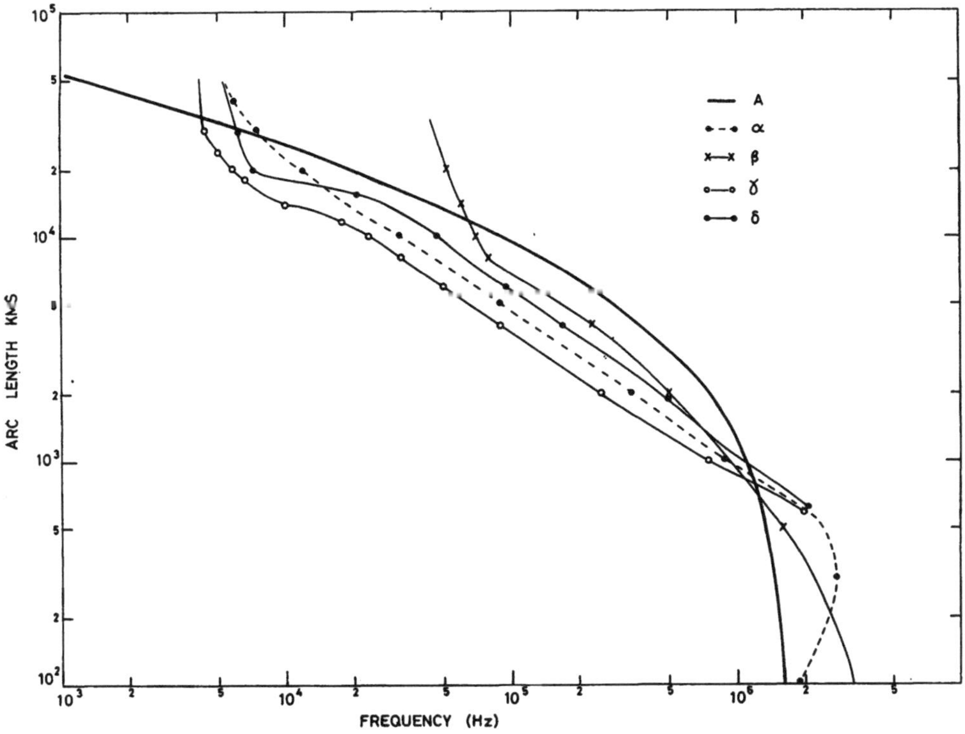

Fig. 3. Gyrofrequency and plasma frequency models for an invariant latitude of 70°: A, centred dipole gyrofrequencies (solid line) and plasma models: α, Jørgensen [8] after Evans [19]; β, Banks and Holzer [21]; γ and δ artificial models.

more electrons at great altitudes (since the cross section area of the tube is $\propto 1/B$) and since whistler mode emissions are limited to frequencies less than both the plasma frequency f_p and the gyrofrequency f_H this gives an increase in the intensity at low frequencies despite the fact that $(dp/df)_e$ decreases with decreasing frequency. The effect of terminating the integration at 26000 km where the gyrofrequency is $\sim$10 kHz ($f_H < f_p$) is to reduce the contribution to the spectrum at all frequencies less than 10 kHz and to produce a peak in the spectrum at this frequency (Fig. 4).

In computing the spectra shown in Fig. 4 we have used Jørgensen's plasma model α and summed the power over all pitch angles (the distribution being taken as isotropic). The upper four curves are for part of the energetic electron flux and given by the curve (b) in Fig. 2. To illustrate the importance of low energy electrons we have used only the flux for electron energies between 50 eV and 1 keV. The solid curves are calculated without applying the damping criterion discussed above and the dashed curves have damping included. As one might expect, the effect of the damping criterion is to attenuate the lower frequencies for which the phase velocities in the whistler mode are smaller. The curves (ε, ε') and (ϱ, ϱ') are for integrations performed to altitudes of 36000 km and 26000 km respectively. The effect of the increase of maximum altitude may be seen to be to shift the peak in the spectrum to slightly lower frequencies. It is not, however, possible with this plasma model to obtain an intensity at 3.2 kHz greater than that at 9.6 kHz. The lower curves in Fig. 4 are for electron fluxes (a) used by Jørgensen [8]. The energy range is from 1 to 12 keV and as expected the calculated ntensity is considerably reduced and the effect of damping is negligible.

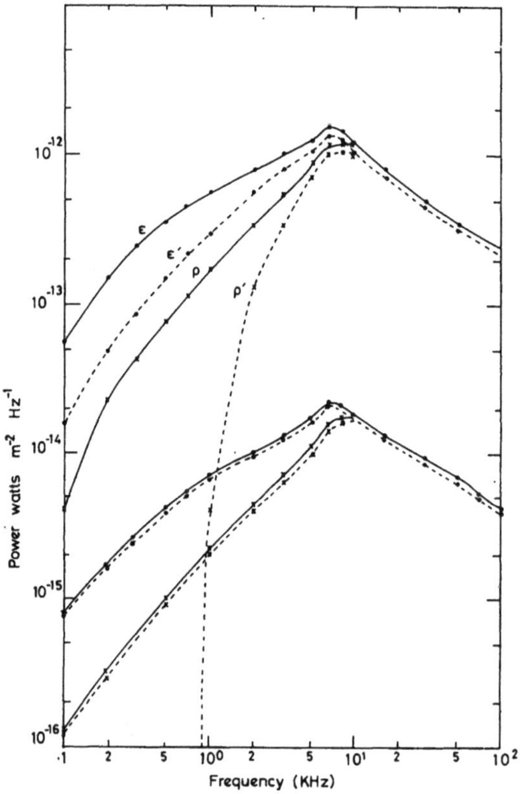

Fig. 4. Spectra obtained from the Cerenkov calculation with (dashed curves) and without (solid curves) damping. The upper four curves are obtained using electron fluxes (b) of Fig. 2 with energies between 50 eV and 1 keV and the lower four curves are for electron fluxes (a) of Fig. 2. The pair of curves (ε, ε'), with and without damping, are for summations carried out to 36000 km and (ϱ, ϱ') are for 26000 km. The same integration altitudes were used in the lower curves.

The plasma frequency model affects the intensity and the shape of the emission spectrum. For $f_p < f_H$, f_p defines the altitude limit to the region of generation for a given frequency and that primarily determines the intensity at that frequency. In addition, $(dp/df)_e$ decreases as f_p increases. This may be seen by simplifying Eq. (1) by assuming zero pitch angle, $v_{||} \ll c$ and neglecting ions. This gives

$$\left(\frac{dp}{df}\right)_e = \pi q^2 \left(\frac{f_H}{f_p}\right)^2 \left(\frac{f}{\varepsilon_0}\right) \frac{1}{v_{||}}. \tag{3}$$

The spectra obtained for the four different plasma frequency models (Fig. 3) are shown in Fig. 5. These were obtained using electron flux data from 50 eV to 1 keV given by (b) in Fig. 2. The spectra are designated by a Greek letter which relates to the plasma model in Fig. 3. Model α is due to Jørgensen [8] and β to Banks and Holzer [21] which was used by Lim and Laaspere [11]. The most recent model by Lemaire [22] can be obtained from β by decreasing it by a factor of 2 and is similar to the lower portion of model γ (<20000 km). When continued

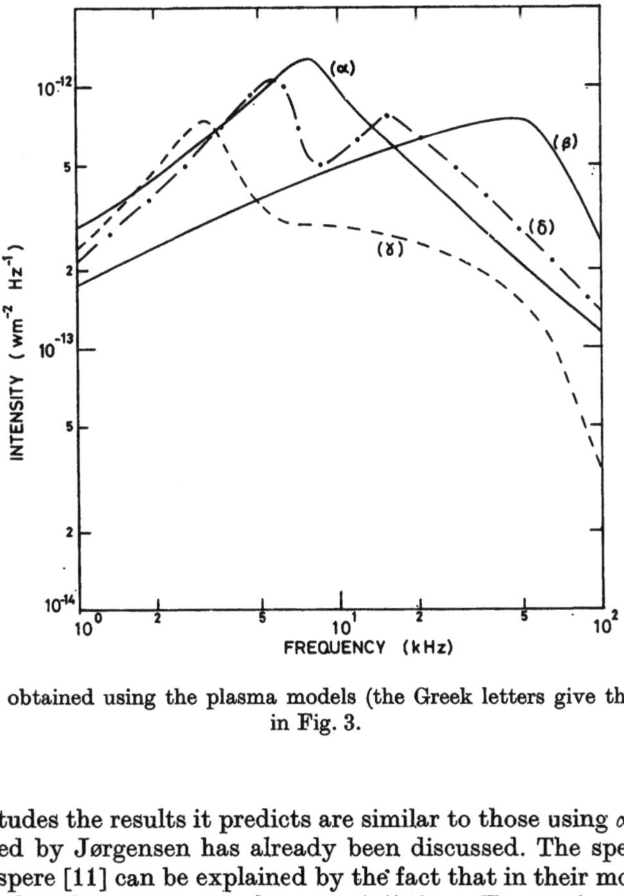

Fig. 5. Spectra obtained using the plasma models (the Greek letters give the model) shown in Fig. 3.

to higher altitudes the results it predicts are similar to those using α. The spectral shape obtained by Jørgensen has already been discussed. The spectra given by Lim and Laaspere [11] can be explained by the fact that in their model β between 1000 and 8000 km the variation in the ratio $(f_H/f_p)^2$ (see Eq. (3)) is small and hence for frequencies greater than about 70 kHz the intensity is primarily determined by the altitude limit to the integration. Above 8000 km $(f_H/f_p)^2$ decreases rapidly which leads to a decrease in the intensity below this frequency and a peak is formed at approximately $f = f_p = f_H$.

In Fig. 5 spectra are also given for models δ and γ from which it may be seen how different spectral shapes may be produced by varying the plasma model, particularly at great altitudes. The spectra of Fig. 5 are calculated without including damping, but is may be seen by referring to Fig. 4 that damping does not appreciably affect the intensity of frequencies greater than about 3 kHz.

It may be noted that the spectrum γ shows a 3 dB enhancement at 3.2 kHz over that at 9.6 kHz. This is very much less than many of the enhancements observed at night on Ariel 3 and it is only obtained from the incoherent Cerenkov calculation by using an artificial plasma density model which differs markedly from the theoretically derived models.

6. The Effect of Energetic Electron Energy on the Cerenkov Spectrum

It is interesting to consider which energetic electrons contribute most to the calculated intensity of VLF hiss and how the shape of the spectrum depends on the particle energy. To study this we have taken the electron spectrum (b) in Fig. 2 and extended it arbitrarily to 10 eV as shown by (c). The results are shown in Fig. 6 which gives the intensity at the three frequencies of interest as a function of energy. The intensity increases with decreasing particle energy for energies down to about 10 eV and below this energy damping becomes important. It may be seen that electrons with energies between about 10 and 70 eV contribute most to the calculated spectrum. In the upper panel of Fig. 6 the shape of the spectrum is given for the corresponding energies below, assuming that the emission is entirely produced by electrons in that energy range. It is noteworthy that the calculated spectrum does vary with particle energy though in no case is the

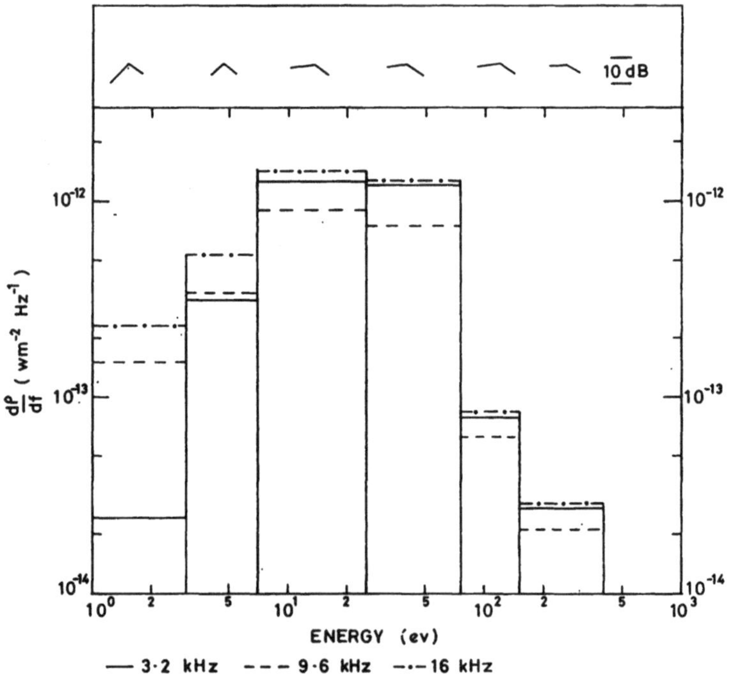

Fig. 6. The intensity at 3.2 kHz, 9.6 kHz and 16 kHz as a function of energy for the electron flux (b) between 50 eV and 1 keV combined with (c). The top panel shows qualitatively the spectral shape of a VLF emission produced by electrons in the energy interval below.

intensity greater at 3.2 kHz than at 9.6 kHz. It is suggested, therefore, that the observed spectral variation in VLF hiss cannot be explained in terms of a variation in the energy of the electron flux assuming that the emission is incoherent Cerenkov radiation.

7. Conclusion

We have shown that a diurnal variation exists in the spectrum of VLF hiss which it is not possible to explain in terms of incoherent Cerenkov radiation.

We have demonstrated how the theoretical spectrum varies as a function of (i) the plasma frequency model assumed, (ii) the electron energy spectrum and (iii) the altitude to which the numerical integration is performed. These three factors are important in determining the theoretical spectrum and are largely responsible for differences in the spectra and intensity obtained by different workers. In addition, our calculations show, using known electron flux data, that electrons with energies between 10 and 70 eV should contribute most to the emission intensity.

Acknowledgments

The authors wish to thank Dr. T. S. Jørgensen for kindly giving them a copy of his Cerenkov computer programme which was used as a basis for calculating their results. They would also like to thank Dr. K. Bullough and Professor T. R. Kaiser for their interest in the work and for many stimulating discussions, and J. Lemaire for helpful private communications.

References

[1] A. R. W. Hughes, K. Bullough and T. R. Kaiser, Space Research XI, 1323 (1971).
[2] A. R. W. Hughes and T. R. Kaiser, in: The Radiating Atmosphere, D. Reidel Publ. Co., Dordrecht, Holland 1971 (p. 336).
[3] G. R. A. Ellis, J. Atmos. Terr. Phys. **10**, 302 (1957).
[4] J. F. McKenzie, Phil. Trans. Roy. Soc. A **255**, 585 (1963).
[5] J. F. McKenzie, Phys. Fluids **10**, 2680 (1967).
[6] H. B. Liemohn, Radio Sci. **69 D**, 741 (1965).
[7] V. N. Mansfield, Radio Physics Lab. Sci. Rep., Dartmouth College, Hanover, N. H. 1964.
[8] T. S. Jørgensen, J. Geophys. Res. **73**, 1055 (1968).
[9] D. A. Gurnett and B. J. O'Brien, J. Geophys. Res. **69**, 65 (1964).
[10] D. A. Gurnett, J. Geophys. Res. **71**, 5599 (1966).
[11] T. L. Lim and T. Laaspere, J. Geophys. Res. **77**, 4145 (1972).
[12] D. A. Gurnett and L. A. Frank, Tech. Rep. 71/19, Univ. of Iowa, Iowa City 1971.
[13] S. R. Mozier and D. A. Gurnett, J. Geophys. Res. **76**, 972 (1971).
[14] K. Bullough, A. R. W. Hughes and T. R. Kaiser, Proc. Roy. Soc. A **311**, 477 (1969).
[15] K. Bullough, A. R. W. Hughes, T. Hudson, D. Hickinson, P. Broomhead and P. Tomlinson, J. Sci. Instrum. (J. Phys. E) **1**, 77 (1968).
[16] A. R. W. Hughes and H. W. English, to be published.
[17] J. Trulsen and J. A. Fejer, J. Plasma Phys. **4**, 825 (1970).
[18] K. L. Ackerson and L. A. Frank, J. Geophys. Res. **77**, 1128 (1972).
[19] D. S. Evans, NASA Rep. X-611-66-376, NASA/GSFC, Greenbelt, Md. 1966.
[20] T. H. Stix, The Theory of Plasma Waves, McGraw-Hill, New York 1962.
[21] P. M. Banks and T. E. Holzer, J. Geophys. Res. **74**, 3317 (1969).
[22] J. Lemaire, private communication 1972.

Space Research XIV — Akademie-Verlag, Berlin 1974

ROCKET EXPERIMENTS ON NON-LINEAR WAVE-WAVE INTERACTION IN THE IONOSPHERIC PLASMA

S. Miyatake, H. Matsumoto and I. Kimura

Department of Electrical Engineering, Kyoto University, Kyoto, Japan

Controlled experiments on non-linear wave-wave interactions in the ionospheric plasma were performed by two recent sounding rockets. The objective of the experiments is to investigate the feasibility of realization of a three-wave coupling process between two electron plasma waves and one ion acoustic wave in the ionospheric plasma. A high-frequency electric field was applied on grids exposed to the plasma in order to transmit the primary electron plasma wave. Detection of the second electron plasma wave as well as the ion acoustic wave was made by long antennas. The frequency (or energy) relation of the three-wave decay process is investigated using phase sensitive detectors. Wide-band reception of the ion acoustic waves was also made which yielded very interesting wave spectra similar to those of VLF emissions.

1. Introduction

It has been stressed that non-linear interactions between various plasma waves should play an important role in the turbulent state of space plasmas. There have been, however, only few quantitative experiments in space until a quantitative study of a non-linear three-wave coupling process between Bernstein waves was recently performed for the frequency spectra of the Alouette topside sounder [1].

In an earlier experiment made by using the K-9M-29 rocket (launched in 1970) [2], an excitation of VLF noises was found, which was closely related with the RF electric field that was applied to the antenna for measuring the electron density in the ionosphere. The frequency of the RF electric field was swept from 0.5 to 10 MHz covering the electron plasma frequency, so that it is suggested that the VLF noises may have been excited through the non-linear three-wave decay process between electron waves and an ion acoustic wave.

The non-linear three-wave process should satisfy the well known relations, that is, $\omega_0 = \omega_h + \omega_l$ for their frequencies and $\boldsymbol{k}_0 = \boldsymbol{k}_h + \boldsymbol{k}_l$ for their wave normals where $(\omega_0, \boldsymbol{k}_0)$ are the frequency and wave normal of a primary wave and $(\omega_l, \boldsymbol{k}_l)$ and $(\omega_h, \boldsymbol{k}_h)$ are those of an excited second and third waves. Therefore the above relations have to be confirmed for the VLF noises to be excited through the non-linear three-wave coupling process.

2. Experimental Equipment

Because of a difficulty in measuring the wave numbers in the experiment with the K-9M-35 rocket, only the frequency relation between an excited VLF wave and an RF electric field applied to a plasma was confirmed. The RF electric field was applied on two separated grids to generate a primary electron plasma wave. Responses of the plasma were received by two pairs of spherical antennas installed between the two grids. VLF waves with frequencies from 0.3 to 30 kHz were received by the one pair of antennas to which a dc bias (0—10 V) was imposed in order to improve the antenna efficiency. The receiver had a sensitivity of 10 μV (at S/N = 10 dB). The RF receiver connected to the other pair of antennas

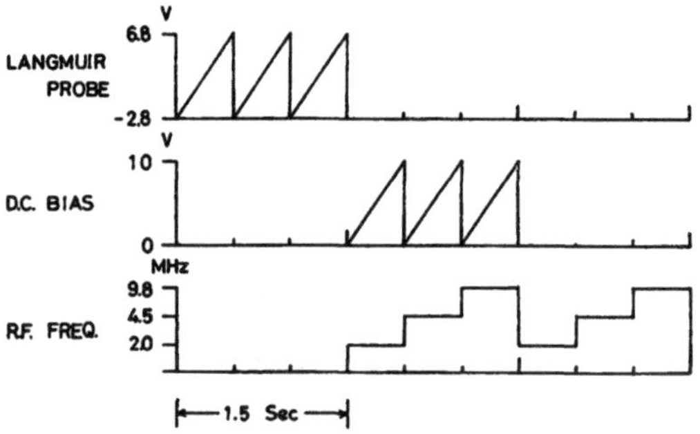

Fig. 1. Timing chart of the experiment by the K-9M-35 rocket.

was tuned to a frequency synchronized with that of the RF electric field applied to the grids. The sensitivity was suppressed to 10 mV in order to keep the equipment stable. The frequency relation was investigated by making use of a phase sensitive detector (PSD). The received VLF signal $A_l \cos(\omega_l t + \theta_l)$ was fed to the PSD where it was compared with the beat between the primary RF signal $A_0 \cos(\omega_0 t)$ and the secondary RF wave $A_h \cos(\omega_h t + \theta_h)$. Unless the relation of $\omega_0 = \omega_h + \omega_l$ is satisfied, the output of PSD, i.e.

$$^1/_2\, A_0 A_h A_l \cos\{(\omega_0 - \omega_h - \omega_l)t + (\theta_h + \theta_l)\}$$

will not appear because of the low pass characteristics of the PSD. The frequency of the RF electric field applied to a pair of the grids immersed in the plasma was chosen to be 2.0 MHz, 4.5 MHz and 9.8 MHz; its amplitude was about 25 V_{p-p}. Though this was much larger than in laboratory experiments, the excited turbulence was not too intense because the potential of the grids varied from space potential due to the difference in mobility between electrons and ions. A timing chart of the experiment is shown in Fig. 1. It took 4.5 sec to achieve one experimental sequence which was divided into three modes of operation. Natural noises in the ionosphere were observed in the first mode. In the second and third modes, an RF electric field of 2.0 MHz, 4.5 MHz and 9.8 MHz was applied to the grids for 0.5 sec and the response of the plasma was observed at the same time. The dc bias applied to the antennas was swept from 0 to 10 V only in the second mode.

This time sequence made it possible not only to distinguish natural noises from waves artificially excited through the non-linear process but also to investigate the effect of the dc bias applied to the antenna immersed in the plasma.

3. Detection of Non-linear Three-wave Interaction

The experimental results of the K-9M-35 rocket (launched in 1971) have shown that VLF waves are excited when an RF electric field is applied to a pair of grids. When the VLF waves are excited, the output of the PSD becomes non-zero showing that the VLF waves and the applied RF field satisfy the frequency relation for the three-wave coupling. Though the sensitivity of the RF receiver

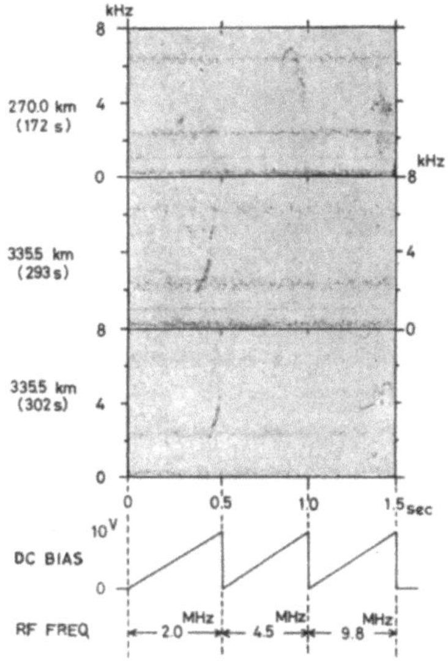

Fig. 2. Frequency spectra of the excited VLF noises due to non-linear three-wave coupling in the ionosphere observed by the K-9M-35 rocket.

was too poor to measure even the intensity of the primary RF wave, it is suggested that two RF waves existed because they were detected by the highly sensitive PSD. Frequency spectra analysed by a Sonagraph as shown in Fig. 2 indicate that the excited VLF waves have frequency spectra like hiss, riser and inverted hook and these isolated noise spectra were clearly correlated with the output of the PSD. The appearance of the VLF noise depends both on the ambient plasma frequency and on the frequency of the RF electric field applied to the plasma. In the case that the applied frequency was equal to 9.8 MHz, the VLF noises were excited at five different regions in the ionosphere. These excitation regions are numbered 1 to 5 in a schematic illustration of the electron density profile shown in Fig. 3. The electron density profiles up to 220 sec after launching and after 370 sec are those measured by the VLF Doppler technique, but from 220 to 370 sec near apogee the profile is drawn schematically. This inferred profile agrees well qualitatively with that observed by an impedance probe also on board. The plasma frequency f_p at the excitation regions is about 7.8 MHz. The ratio of the frequency of the applied RF field f_0 to the plasma frequency f_p is, therefore, 1.2—1.3. For f_0 equal to 2.0 MHz and 4.5 MHz, excitations of VLF waves were also observed when the plasma frequency at the excitation region was around 7 MHz. Judging from the fact that the applied RF electric field was not a pure sinusoidal wave, it is probable that higher harmonics of the original fre-

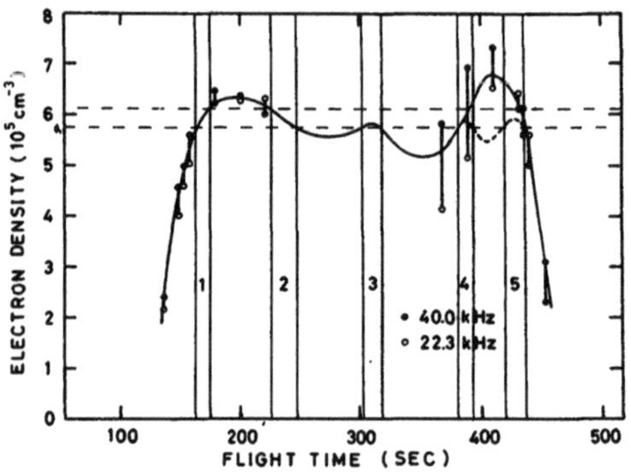

Fig. 3. Dependence of the excitation region on the ambient plasma density which was measured by making use of the VLF waves with frequencies of 22.3 kHz and 40.0 kHz transmitted on the ground. The regions numbered 1 to 5 are those of the excitation of VLF noises due to the RF electric field with a frequency 9.8 MHz.

quency might have played an important role. In a region where the ambient plasma frequency was equal to 2.0 MHz or 4.5 MHz, the collision frequency was not so low and the plasma parameter itself changed so rapidly that excitation did not occur. The experimental results suggest that the VLF noises were excited by the non-linear three-wave coupling process between the electron plasma waves and the ion acoustic wave. The fact that the excitation was observed when the dc bias was more than 5 V is partly because the dc bias "peeled off" an ion sheath around the receiving antenna, resulting in an improvement of the antenna pick up factor. It is also probable that the dc bias accelerates the plasma so that it becomes unstable in the RF electric field.

4. Artificially Excited VLF Noises

A similar experiment was conducted by the K-9M-41 rocket (launched in 1973), with an exciter of the primary electron plasma wave on a separable daughter rocket in order to measure the wavelength exactly. Though this experiment did not check the frequency and wave number relations of the non-linear three-wave coupling because of a problem in the mother-daughter communication, very interesting spectra of artificially excited VLF noises were obtained. Fig. 4 shows a part of the spectra. As is seen, many artificial noises were found in addition to natural whistlers. Both discrete (narrow band) and hiss-like (broad band) noises, very similar to the natural VLF emissions, were simultaneously excited in the surrounding plasma. There is a periodicity in the appearance of noises synchronized with

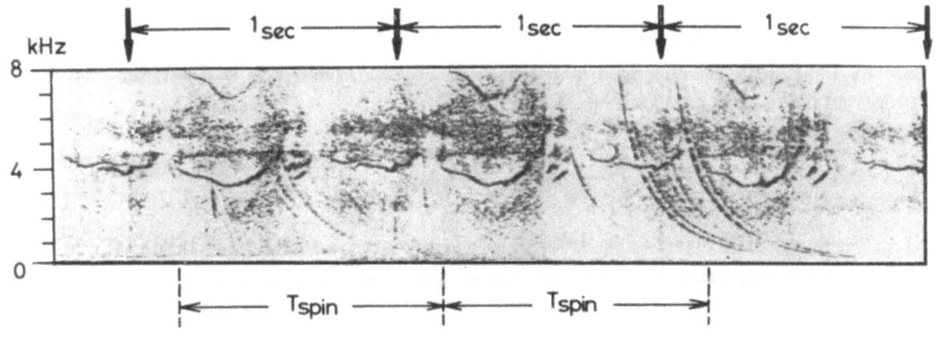

Fig. 4. VLF spectra of artificially excited noises in the ionosphere observed by the K-9M-41 rocket.

the rocket spin period T_{spin}. The discrete hook-like noises were so intense that higher harmonics were also excited up to the fourth (not shown in the figure). Judging from the similarity of the spectra for K-9M-35 and K-9M-41, the discrete noises are not whistler mode waves but ion acoustic waves.

The frequency change of these discrete noises as a function of rocket altitude is given in Fig. 5 as well as the O^+ ion plasma frequency and the lower hybrid resonance frequency. As is seen, the frequency of the discrete noises is correlated

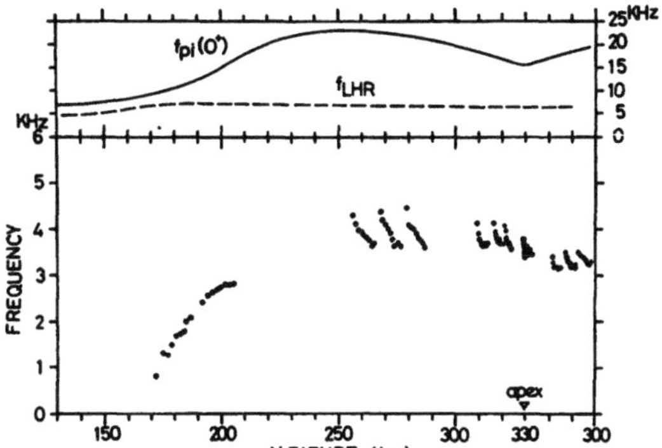

Fig. 5. Frequency change of the artificially excited discrete noises versus altitude.

with the ion plasma frequency, strongly indicating that these noises are ion acoustic waves. The excitation mechanisms of these noises are as yet not clear although there are the following two possibilities. The first is the non-linear three-wave coupling mechanism through which ion acoustic waves are produced by the transmitted electron plasma wave. In this rocket experiment, the RF voltage applied to the exciter of the electron plasma wave was about 30 V_{p-p}. The second is a beam-wave instability. An electron beam emitter was installed on the rocket periphery in the vicinity of the receiving antennas on which the dc bias varying from 0 to 10 V in steps of 1 V was imposed. Therefore, an electron beam current (of the order of 100 μA) was formed in a closed circuit composed of the rocket body, the surrounding plasma, the receiving antennas and the electron emitter. The arrows in Fig. 4 indicate the timing of these steps of the dc bias. It is noticed that the frequency of the discrete noises falls just after this voltage step as seen, around altitude 280 km, in Fig. 5.

5. Conclusion

In this paper, we have experimentally demonstrated the possibility of exciting VLF noises artificially by the non-linear three-wave coupling between electron plasma waves and an ion acoustic wave. This strongly suggests that some natural VLF emissions may be caused by such a non-linear process associated with strong electron plasma waves which are easily excited in the ionospheric plasma. These VLF noises may possibly be excited artificially by an electron beam.

Acknowledgments

The authors would like to express their sincere thanks to Professors K. Maeda, T. Obayashi and K. Hirao for their encouragement and support to this work. They also wish to express their thanks to Messrs. T. Matsuo and A. Nagao for their assistance in data analysis.

References

[1] H. OYA, Phys. Fluids **14**, 2487 (1971).
[2] H. MATSUMOTO, S. MIYATAKE and I. KIMURA, Rep. Ionos. Space Res. Japan **25**, 40 (1971).

Space Research XIV — Akademie-Verlag, Berlin 1974

THEORY OF THE EXCITATION OF THE LOWER OBLIQUE RESONANCE IN THE MAGNETOSPHERIC PLASMA

R. Pottelette and G. Cissoko

Groupe de Recherches Ionosphériques, Saint-Maur-des-Fossés, France

We study the possibility of determining the characteristics of the lower oblique resonance of the magnetospheric plasma by means of a dipole probe of a special type, consisting of two independent spherical electrodes, one of which is used for transmitting and the other for receiving. This lower oblique resonance occurs between the lower hybrid frequency and the electron gyrofrequency. We show that as compared with the predictions of simple theory for a stationary cold plasma, the characteristics of this oblique resonance are affected by the collective motion of the electrons along the magnetic field lines and also by their random thermal motion.

If the collective motion is the dominant effect, the contribution of thermal electrons to the electric currents flowing along the lines of force of the earth's magnetic field can be measured. On the other hand, if the thermal effect is dominant, the electron temperature can be measured.

1. Introduction

The interest in observing the oblique resonances in an anisotropic plasma, as a means of plasma diagnostics, has been increasing considerably during recent years. The first experimental proof of the existence of these resonances was obtained in a laboratory experiment [1]. Since then, other experiments have been done, both in the laboratory [2] and in space [3], which have confirmed their existence. In this paper, we study the possibility of determining the characteristics of the lower oblique resonance of the magnetospheric plasma by means of a dipole (or double monopole) probe consisting of one transmitting sphere E and one receiving sphere R. This lower oblique resonance occurs between the lower hybrid frequency and the electron gyrofrequency. As compared with the predictions of the simple theory for a cold stationary plasma, the characteristics of this lower oblique resonance are greatly affected by bulk motion of the electrons along the magnetic field lines and also by their random thermal motion. Consequently, we study the influence of these two factors in turn. For this purpose, in the frequency range of the lower oblique resonance, we can neglect collisions between particles of the magnetospheric plasma. Moreover, at frequencies much greater than the lower hybrid frequency, we can consider the ions as having infinite mass. We adopt the quasi-static hypothesis.

2. Influence of the Collective Motion of the Electrons on the Characteristics of the Lower Oblique Resonance

The two spheres constituting the dipole probe, are separated by a distance $r_0 = \mathrm{ER}$ that we suppose to be large compared with their radii (Fig. 1); hence they can be considered as points. The emitting sphere E is excited by a source of alternating current $I = I_0 \exp(j\omega t)$ at the angular frequency ω. According to cold plasma theory and in the absence of any dissipative processes, the lower

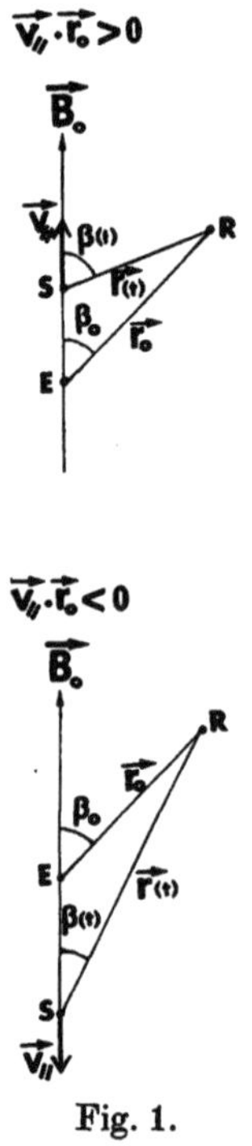

Fig. 1.

oblique resonance is characterized by the fact that the potential created by an alternating point source becomes infinite along a cone whose axis is parallel to the magnetic field $\boldsymbol{B}_0$ and whose half angle β_0 is determined by the plasma density, by the magnetic field strength, and by the working frequency. This conical form of the region most strongly excited by the resonance makes the dipole probe sensitive to field aligned currents. Since our aim is simply to show that the probe is sensitive to the component $\boldsymbol{v}_{\parallel}$ of the electron velocity parallel to the magnetic field, we shall suppose that the other components are zero.

The effect of movement is easy to understand in terms of the impulse response. Suppose that, instead of a sinusoidal current, the sphere E emits a current pulse $I(t) = Q_0\delta(t)$, where Q_0 is the total charge emitted and $\delta(t)$ is the Dirac distribution. This pulse excites a volume of electrons which subsequently is carried along the magnetic field lines with the velocity $\boldsymbol{v}_{||}$. In a frame of reference fixed with respect to the spheres, the motion gives rise to a variation, both in magnitude and in direction of the vector distance $\boldsymbol{r}$ between the apparent point of emission S, which coincides with E at the initial instant $t = 0$ and thereafter moves with the electrons, and the point of reception R which remains fixed (Fig. 1). Consequently the resonance cones corresponding to the different frequencies move successively over the point R. As we can see in Fig. 1 we have to distinguish two cases, according to whether $\boldsymbol{v}_{||} \cdot \boldsymbol{r}_0$ is positive or negative.

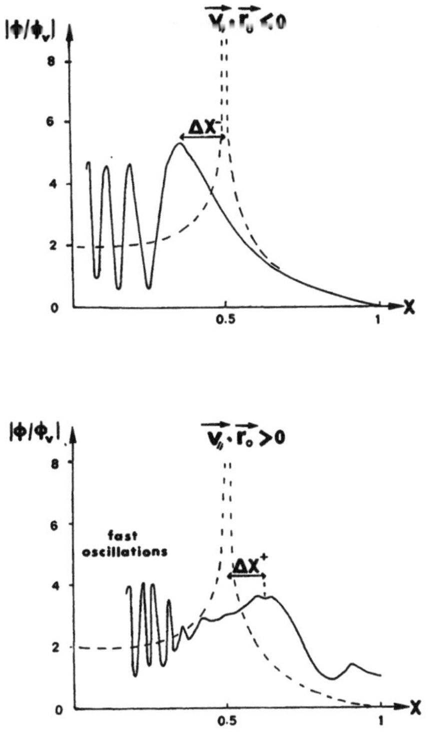

Fig. 2.

The potential induced on the receiving sphere, as a function of time, has been calculated [4]. It takes the form

$$\Phi(\boldsymbol{r}, t) = \frac{C\omega_b}{r} \left\{ J_0(\omega_b t \sin \beta) + \delta'(\omega_b t) + \frac{\sin^2 \beta}{2} [J_2(\omega_b t \sin \beta) - J_0(\omega_b t \sin \beta)] \right\} \quad (1)$$

where C is a constant, J_0 and J_2 are zero-order and second-order Bessel functions, δ' is the first derivative of the Dirac distribution, and ω_b is the electron gyro-frequency. Moreover, we have

$$\boldsymbol{r}(t) = \boldsymbol{r}_0 + \boldsymbol{v}_{||} t \quad \text{and} \quad \sin \beta(t) = r_0 \sin \beta_0 / r(t). \quad (2)$$

We now return to the frequency domain by taking the Fourier transform of $\Phi(\boldsymbol{r}, t)$. Fig. 2 shows, for $\beta_0 = 30°$, the amplitude of the potential $\Phi(\omega)$ normalized with respect to the potential $\Phi_v(\omega)$ that would exist in vacuum as a function of the normalized frequency $\chi = \omega/\omega_b$ and for the two opposite signs of $\boldsymbol{v}_{\parallel} \cdot \boldsymbol{r}_0$. The dotted curves represent the case with zero velocity. The solid curves were constructed for $\boldsymbol{v}_{\parallel} = 63$ km s^{-1}, which is distinctly less than the thermal velocities of the electrons of the magnetospheric plasma. In both cases, the resonance is damped and shifted in frequency, and a series of secondary peaks appears. The frequency shift $\Delta\chi$ has the same sign as $\boldsymbol{v}_{\parallel} \cdot \boldsymbol{r}_0$, and we distinguish its two values, for the same amplitude of $\boldsymbol{v}_{\parallel}$ but with $\boldsymbol{v}_{\parallel} \cdot \boldsymbol{r}_0$ first positive and then negative, by the symbols $\Delta\chi^+$ and $\Delta\chi^-$ respectively. The amplitude of the shift depends on that of the velocity: in particular, $\Delta\chi^-$ is proportional to the square root of $\boldsymbol{v}_{\parallel}$ for small values.

If we interchange the roles of the transmitting and receiving spheres, this is equivalent to the inversion of the direction of $\boldsymbol{v}_{\parallel}$. By so doing, we can measure $\Delta\chi^+$ and $\Delta\chi^-$ and therefore $\boldsymbol{v}_{\parallel}$. This measurement would be particularly interesting in the magnetosphere, because one could deduce from it the contribution of the thermal electrons to the electric currents flowing along the lines of force. Although this contribution is probably very important, no method for measuring t exists at the present time. It seems that the effects of the orbital motion on the ower oblique resonance have been observed aboard the satellite OV1-20S, which is in a polar orbit [3].

3. Influence of the Thermal Motion of the Electrons on the Characteristics of the Lower Oblique Resonance

Until now the influence of the thermal movement on the characteristics of the lower oblique resonance has been studied only in the limiting case where the magnetic field is infinite [5]. In this paper we eliminate this rather restrictive case and consider a finite magnetic field. The potential created by an alternating point source is calculated by means of the full hydrodynamic approximation, in which the electrons are considered as a fluid of the most general type characterized by a tensor pressure and adiabatic transformations. Since we adopt the quasi-static hypothesis, the electric field can be derived entirely from a scalar potential, which is obtained by using the Poisson equation alone. Accordingly, the basic problem is to obtain the scalar potential $\Phi(\boldsymbol{r})$ created by an alternating point source. This involves evaluating a three-dimensional Fourier transform. In view of the cylindrical symmetry of the problem around the magnetic field line passing through the point source, the three-dimensional Fourier transform can be replaced by a Hankel transform for the direction ϱ perpendicular to the magnetic field, and by a one-dimensional Fourier transform for the direction z parallel to the field. Then the potential $\Phi(\boldsymbol{r})$ is given by the formula

$$\Phi(\boldsymbol{r}) = \frac{I}{(2\pi)^2 j\omega\varepsilon_0} \int_0^{\infty} \mathrm{J}_0(k_\perp \varrho) k_\perp \, dk_\perp \int_{-\infty}^{+\infty} \frac{\exp(jk_\parallel z)\, dk_\parallel}{\varepsilon_{\mathrm{L}}(k_\parallel, k_\perp)} \tag{3}$$

where ε_0 is the electric permittivity of free space, $k_\parallel$ and $k_\perp$ are the components of the wave vector $\boldsymbol{k}$, respectively parallel and perpendicular to the magnetic field, and $\varepsilon_{\mathrm{L}}(k_\parallel, k_\perp)$ is the longitudinal dielectric constant.

The singularities of the integrand are solutions of the dispersion equation $\varepsilon_L = 0$. The calculation of the roots of this equation becomes much easier if the electron temperature is assumed to be low, in which case we can proceed by successive approximation. One or other of the two integrals can be performed analytically, but then the remaining integration has to be performed numerically. The final result of the numerical integration is presented in Fig. 3, in which the amplitude of the potential Φ, normalized with respect to the potential Φ_v that would exist in a vacuum, is plotted as a function of the angle β between the magnetic field and the direction of observation, which is at a distance of 200 Debye lengths from the source. As the angle β increases from zero the potential exhibits a sequence of maxima and minima before reaching a large but finite value on a certain cone, whose half angle is slightly different from that given by cold plasma theory (in dotted lines in Fig. 3).

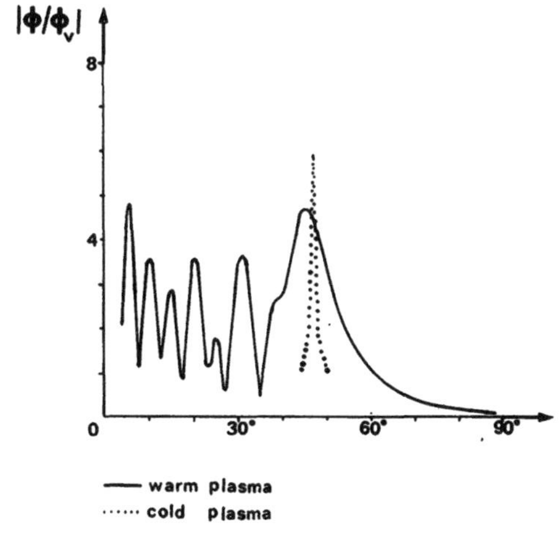

Fig. 3.

Clearly the dipole probe, used in the frequency band of the lower oblique resonance, is likely to be a useful diagnostic tool. The determination of the angle of the resonance cone at a fixed frequency, or of the resonance frequency at a fixed angle, allows us to calculate the plasma frequency and thence the electron density, when the electron gyrofrequency is known. In addition, the fine structure that appears inside the resonance cone enables us to measure the electron temperature.

4. Conclusion

We have seen that, in the magnetosphere, the characteristics of the lower oblique resonance are strongly dependent on both the collective drift movement of the electrons and their random thermal movement. Thus, if we want to study one of these phenomena quantitatively we must first make sure that its effects are dominant under the conditions of the experiment envisaged.

Acknowledgments

The authors are indebted to Dr. L. R. O. Storey for helpful criticism.

References

[1] R. K. FISHER and R. W. GOULD, Phys. Rev. Lett. **21**, 1013 (1969).
[2] A. GONFALONE, J. de Phys., **33**, 521 (1972).
[3] D. A. MCPHERSON and H. C. KOONS, private communication 1973.
[4] R. POTTELETTE, Annls Géophys. **29**, 201 (1973).
[5] N. SINGH and R. W. GOULD, Phys. Fluids, **16**, 1 (1973).

Space Research XIV — Akademie-Verlag, Berlin 1974

THEORY FOR THE INTERPRETATION OF MEASUREMENTS OF THE SIX COMPONENTS OF A RANDOM ELECTROMAGNETIC WAVE FIELD IN SPACE

L. R. O. STOREY and F. LEFEUVRE

Groupe de Recherches Ionosphériques du C.N.R.S., Saint-Maur-des-Fossés, France

We seek the best method for interpreting a set of simultaneous measurements of the three electric and three magnetic components of a wide-band random electromagnetic wave field in a magnetoplasma; the point of observation is assumed to be stationary with respect to the medium. The work has involved three stages: definition, in the framework of classical electrodynamics, of the distribution function that characterizes the statistics of a random electromagnetic wave field in a lossless medium; derivation of the spectral matrix for the electric and magnetic components of a field of this type, in a cold collisionless magnetoplasma; interpretation of a set of experimental measurements so as to obtain the most likely estimates of the parameters of a model of the distribution function. At the same time we obtain a check on the validity of the model, the choice of which should be guided by physical considerations.

1. Introduction

This paper seeks to answer the question as to how one should analyse data on the six components of a random electromagnetic wave field in space, for the purpose of studying natural wave-particle interactions in the magnetosphere. The wave phenomena considered are those that appear to be continuous and structureless, such as VLF and ELF hiss, in contrast to discrete ephenomena such as whistlers or VLF emissions which have clear-cut structures in the frequency-time plane. Their interactions comprise the plasma instabilities that produce them, and also their eventual perturbation of energetic particles not involved in their production.

Now the distinctive feature of VLF and ELF hiss is their randomness: they look like random noise. The causative instabilities act over frequency bands of width comparable with the mean frequency, and doubtless act also over a large volume of space. The field observed at any point in the magnetosphere is of widespread origin, and has arrived at that point by many different paths. Hence it must be quite unlike that of a single plane wave propagated in a definite direction. A more realistic model is a continuum of superposed plane waves of different frequencies, propagated in different directions, with no mutual phase coherence. A random field like this can only be described statistically. In fact, the description that is required is similar to that used for the energetic particles. Just as one asks how the particles of a given type are distributed in energy and in pitch angle, so one should ask how the waves in a given mode are distributed in frequency and in direction of propagation. The waves also must be described by a distribution function. This is the information that the theoretician needs, both to test the output from instability theory and as input to the theory of the perturbation of the particle population by the waves, and this is what the experimenter should be giving him. Hence the first problem that arises is how to define a distribution function for waves. This is dealt with in § 2 for the restricted case of electromagnetic waves in lossless media: the extension of the definition, to make it cover electrostatic waves and absorbing media, poses problems that will not be considered here.

The subject of the present paper is how to determine a wave distribution function experimentally. Actually there are two distinct questions to be answered: firstly, what data should be measured; secondly, how should they be interpreted? Now for VLF and ELF waves in the magnetosphere, there is particular difficulty because their wavelengths are usually much larger than the dimensions of any antenna that could be deployed from a spacecraft, and this fact fundamentally limits the directional resolution. In these circumstances the most that can be done is firstly to collect as many data as possible with antennas that are small compared with a wavelength, and secondly to interpret these data in the best possible way, using whatever knowledge one has of propagation theory and of the properties of the medium. The first point means that all six components of the field—three electric and three magnetic—should be measured: there remains the problem of interpretation. As a preliminary, § 3 examines the simpler direct problem of the statistics of random electromagnetic wave fields: namely, given a field characterized by a known distribution function, what are the statistical properties of its six components? They are found to be described conveniently by a 6×6 spectral matrix, and a general expression is obtained for the elements of this matrix.

Section 4 tackles the inverse problem, which is the true problem in hand: given a set of measurements of the six field components, what can be found out about the distribution function? The need is established to adopt some mathematical model of this function, and maximum-likelihood theory is used to make the best possible estimates for the parameters of the model. The validity of the chosen model can also be tested.

2. The Wave Distribution Function

In the kinetic theory of gases, the distribution function $F_i(t, \boldsymbol{r}, \boldsymbol{v})$ for the particles of type i is defined in such a way that

$$F_i(t, \boldsymbol{r}, \boldsymbol{v})\, d^3r\, d^3v$$

is the number of these particles, at the time t, in the volume element $d^3\boldsymbol{r}$ at the point $\boldsymbol{r}$ in coordinate space, whose velocity vectors lie in the element $d^3\boldsymbol{v}$ at the point $\boldsymbol{v}$ in velocity space.

By analogy, the distribution function $F_i(t, \boldsymbol{r}, \boldsymbol{k})$ for the wave mode i may be defined in such a way that

$$F_i(t, \boldsymbol{r}, \boldsymbol{k})\, d^3r\, d^3k$$

is the amount of wave energy in this mode, at the time t and in the volume element d^3r, due to waves whose wave-number vectors lie in the element d^3k at the point $\boldsymbol{k}$ in wave-number space.

From here on, to simplify the discussion, it will be supposed that the wave field is stationary in time and in space, so the variables t and $\boldsymbol{r}$ can be omitted. Equally it will be supposed that only one wave mode is present, and the subscript i will be dropped accordingly. Then the wave distribution function can be written simply as $F(\boldsymbol{k})$.

When the medium is optically uniaxial, as is the case for a stationary magnetoplasma, it is more convenient to express the wave distribution function in terms of the frequency ω and of the direction of propagation (i.e. the direction of the $\boldsymbol{k}$ vector), as specified by the angle θ between this vector and the optic axis, and by an azimuthal angle ϕ. In this representation, the distribution function will be written $F(\omega, \theta, \phi)$, and the element of wave energy density is $F(\omega, \theta, \phi)\, d\omega\, d\theta\, d\phi$. The relation between this and the previous form of the distribution function is

$$F(\omega, \theta, \phi) = |V_g|^{-1} k^2 \sin \theta F(\boldsymbol{k})$$

where $k = |\boldsymbol{k}|$, and $|V_g| = |(\partial\omega/\partial k)_\theta|$ is the modulus of the group velocity [1].

Another reason why the representation in terms of ω, θ and ϕ is convenient is that the experimenter, by means of narrow-band filters, can select for observation waves whose frequencies ω lie between limits of his choosing. Of course, the observed frequencies are equal to the true wave frequencies only if the point of observation is stationary with respect to the medium, because any motion gives rise to Doppler shifts. For practical purposes, the point of observation is effectively stationary if all Doppler shifts are less than the smallest interval of frequency over which $F(\omega, \theta, \phi)$ varies appreciably. Here this will be assumed to be the case.

3. Statistics of a Random Wave Field with a known Distribution Function

When one seeks to predict the statistics of the six components of a random electromagnetic wave field, given the distribution function, the first question to be answered is how these statistics should be described. Now the field is to be pictured as being made up of a continuum of elementary plane waves of different frequencies and with random phases, propagated through the medium in different directions. Any axial component of any one of these waves is a sinusoid, and thus each component of the total field is the sum of an infinite number of elementary sinusoids, with infinitesimal amplitudes, different frequencies and random phases. It is well known that a signal of this type is a stationary Gaussian random process of zero mean value, and that as such, its statistical properties are described completely by its auto-covariance function or by its mean power spectrum. If, now, the six axial field components are considered collectively, then besides their individual properties, it is necessary to know what statistical relationships exist between them. These may be described either by the cross-covariance functions or by the cross-power spectra, of which there is one for each of the fifteen possible pairs of field components. In sum, the statistics of the six field components are described by their six auto-covariance and fifteen cross-covariance functions, or by the corresponding mean power and cross-power spectra.

For present purposes, it is preferable to work with the power spectra and cross-spectra, rather than to use the covariance functions, because the spectra are related more directly to the wave distribution function.

As a first step towards deriving the relationship, it is convenient to group all the statistical data together in a 6×6 spectral matrix. Take Cartesian coordinate axes $Oxyz$, with Oz parallel to the steady magnetic field. Let E_x, E_y, E_z be the axial components of the electric field of the wave at the point of observation, and H_x, H_y, H_z the components of the wave magnetic field. From these variables, a generalized electric field vector $\vec{\mathscr{E}}$, with 6 components, is defined as follows:

$$\mathscr{E}_{1,2,3} = E_{x,y,z} \qquad \mathscr{E}_{4,5,6} = Z_0 H_{x,y,z}$$

where Z_0 is the wave impedance of free space. Let $\mathscr{E}_i$ be any component of this vector, the subscript i running from 1 to 6. Then the spectral matrix $[S_{ij}(\omega)]$, which has 36 elements, is defined in such a way that its element $S_{ij}(\omega)$ is either the mean power spectrum of the field component $\mathscr{E}_i$ (if $i = j$), or the mean cross-power spectrum of the components $\mathscr{E}_i$ and E_j (if $i \neq j$).

The relationship between the distribution function and the spectral matrix is derived by considering the field to be made up of a continuum of elementary plane waves, as explained previously. For any one such wave, the i-component of its generalized electric field, which will be called $e_i(t, \boldsymbol{r})$, varies sinusoidally with time and in space:

$$e_i(t, \boldsymbol{r}) = \mathrm{Re}\,\{e_i \exp\,[j(\omega t - \boldsymbol{k}\cdot\boldsymbol{r})]\}\,.$$

On the right-hand side of this equation, the symbol Re signifies the extraction of the real part, e_i is the complex amplitude, the angular frequency ω is real and positive, and the wave-number vector $\boldsymbol{k}$ is real because the medium is assumed to be lossless; let θ and ϕ be the angles that specify the direction of this vector. Moreover, let ϱ be the electromagnetic energy density of the wave in coordinate space. Now set

$$a_{ij} = (e_i e_j^*)/\varrho\,. \tag{1}$$

Other things being equal, the coefficient a_{ij} is independent of the amplitude of the wave. With all possible pairs of values of the indices i and j, Eq. (1) defines 36 such coefficients. They are real when $i = j$, and generally complex when $i \neq j$; clearly $a_{ji} = a_{ij}^*$. Their values, as functions of ω, θ, ϕ and of the plasma parameters, can be derived from the magneto-ionic theory. The derivation is straightforward, and will not be given here; it is at this point that propagation theory and the properties of the medium enter the argument. The next and final step is to add together an infinite number of elementary plane waves, of infinitesimal amplitude and mutually independent, to form the total field $\vec{\mathscr{E}}$. On so doing, the elements of the spectral matrix of this random field are found to be given, in terms of the coefficients a_{ij} and of the distribution function F, by the expression

$$S_{ij}(\omega) = \frac{\pi}{2} \int_0^{\pi}\int_0^{2\pi} a_{ij}(\omega, \theta, \phi) F(\omega, \theta, \phi)\, d\theta\, d\phi \tag{2}$$

for ω positive. This is the required result.

4. Interpretation of Experimental Data to determine the Distribution Function

The problem to be considered now is the inverse of that treated in the previous section: given a set of experimental data, comprising the six components $\mathscr{E}_i(t)$ of the wave field measured over the full frequency range of interest and over a time interval $0 \leqq t \leqq T$, to what extent can the wave distribution function $F(\omega, \theta, \phi)$ be determined, and how should this be done? Actually, since the data relating to waves of different frequencies can be separated easily by the use of filters, the real problem is to determine how F varies with θ and ϕ at a given frequency, ω_0 say.

Now since the statistical properties of the field from which the data come are characterized fully by the spectral matrix $[S_{ij}(\omega)]$, and since this is related quite simply to the distribution function, clearly the first step in determining $F(\omega_0, \theta, \phi)$ must be to estimate the $S_{ij}(\omega_0)$ from the data. The frequency components in the raw data that are useful for this purpose are contained in the band centred on ω_0, within which $F(\omega, \theta, \phi)$ is substantially independent of ω and equal to $F(\omega_0, \theta, \phi)$. They can be isolated by means of band-pass filters, then used to estimate the quantities $S_{ij}(\omega_0)$ following standard procedures [2].

With these quantities put on the left-hand side of Eq. (2), the problem is how to invert this equation in order to determine $F(\omega_0, \theta, \phi)$. Now, from the nature of the equation, it is clear that no formal inversion exists. $F(\omega_0, \theta, \phi)$ is a function of two continuous variables, while $S_{ij}(\omega_0)$ is simply a number. The reduced data, however, consist of 36 such numbers, for all possible combinations of the subscripts i and j. In these circumstances, a solution for $F(\omega_0, \theta, \phi)$ can be obtained only by assuming some mathematical model for this function, governed by a finite number of parameters. In principle, one could use a model with as many as 36 free parameters, which obviously would be able to represent a very wide class of functions. In practice, however, the choice of model would be guided by whatever knowledge one has about the origin of the wave field from which the data came, and this would make it possible to use a model characterized by relatively few parameters. Thus the inverse problem is reduced to that of estimating the parameters of the model.

In the present work, the estimation has been based on the principle of maximum likelihood. Starting from Bayes' theorem, an expression has been derived for the statistical likelihood L that the observed data came from a random wave field with a distribution function described by an assumed model characterized by n parameters with given values. Call these parameters $\psi_1, \psi_2, \ldots \psi_n$. Then the best estimate of their values is considered to be the set ψ_i which is most likely, i.e. that for which the likelihood function $L(\psi_1, \psi_2, \ldots \psi_n)$ attains its highest maximum. The condition for stationary L, namely that $\partial L / \partial \psi_i = 0$ for all i, yields a set of n simultaneous equations which has to be solved for the parameter values. These equations are complicated and highly non-linear, but nevertheless they can be solved by iterative numerical methods, such as the Newton-Raphson method; it is advisable to check that the extremum of L to which the iteration converges is indeed the highest maximum. If the assumed model is correct, the result is the best possible estimate—in the above statistical sense—of the set of characteristic parameters.

If the model is incorrect, this fact can be detected by examining the value of the likelihood function at the maximum. The iteration still converges to a parameter set that appears the most likely, but the corresponding likelihood is not as high as it should be. The nominal maximum value of L, for a given model, parameter set, and data set, can be determined theoretically. If the observed value is significantly less than this, it follows that the model is incorrect; unfortunately this test yields no guidance as to how the model should be improved.

The present state of this research is that the interpretation procedure described above has been tested on artificial data and works satisfactorily. Assuming a very simple model for the distribution function, characterized by only three parameters, the 36 elements of the spectral matrix were calculated for a random VLF electromagnetic wave field in the whistler mode. When these "data" were put into the inversion routine, the iteration converged as it should to the assumed values of the parameters. The next step will be to use the procedure to interpret five-component data on high-latitude VLF hiss, recorded by the FR-1 satellite [3]. Future extensions of the theory will include treatment of the case where the point of observation is moving with respect to the medium, one application being to the field of Alfvén waves in the solar wind, and also the case of random wave fields of artificial origin, such as those set up in the magnetosphere by VLF transmitters on the ground.

Acknowledgments

The authors are most grateful to Dr. D. A. Gurnett for suggesting this problem, in a private communication to L. R. O. Storey in 1965. The work was begun while one of us (L. R. O. S.) was a guest worker at the Institute for Extraterrestrial Physics of the Max-Planck Institute for Physics and Astrophysics, Garching (German Federal Republic) at the kind invitation of Dr. G. Haerendel. It forms part of the FR-1 satellite programme, which was carried out jointly by the NASA in the USA, and by the CNES and the CNET in France.

References

[1] G. Bekefi, Radiation Processes in Plasmas, Wiley, New York 1966 (p. 32).
[2] J. D. Means, J. Geophys. Res. 77, 5551 (1972).
[3] L. R. O. Storey, Space Research VII, 588 (1967).

Space Research XIV — Akademie-Verlag, Berlin 1974

ON THE STUDY OF SIMULTANEOUS MAGNETIC RECORDS FROM GROUND STATIONS AND SYNCHRONOUS ORBIT

C. T. Russell[a], R. L. McPherron[a,b] and P. J. Coleman, Jr[a,b]

[a]Institute of Geophysics and Planetary Physics, University of California, Los Angeles, USA
[b]Department of Planetary and Space Science, University of California, Los Angeles, USA

The magnetic field at the synchronous orbit exhibits a wide spectrum of variations, from diurnal changes, through substorm effects to various ULF wave phenomena. While the properties of these waves at synchronous orbit are now moderately well defined, the modification of these properties during their propagation through the magnetosphere and ionosphere to the ground is not well understood, despite the establishment of a ground station at the expected ATS-1 conjugate point. This is because relatively few simultaneous ground-satellite events have been detected.

Data from synchronous orbit have proven to be quite useful in reducing the number of possible interpretations of ground magnetometer data. For example, the current system causing the mid-latitude signature of the expansion phase of substorms is found to close beyond synchronous orbit. Likewise ground magnetometer data have proven to form a vital link in the interpretation of the magnetic records obtained in synchronous orbit. For example, midnight mid-latitude magnetograms have been used to determine the onset times of substorms and thus order substorm associated field changes at synchronous orbit as a function of local time and substorm phase. Recommendations for future studies include using comparable equipment on the ground and in the satellite, and the use of network of stations near the conjugate point.

1. Introduction

The magnetic field variations at synchronous orbit, while consisting of a wide variety of types, are usually clearly separated from one another in time, space and frequency band and are amenable to rather simple and straightforward analyses. At the longest periods, there is the diurnal variation of the magnetic field and the changes in this diurnal pattern in the course of a year [1]. During disturbed periods, the magnetic field at synchronous orbit changes in response to the increased strength of the ring current, magnetopause current and tail current systems [2], with occasional but rare traversals of the magnetopause. At shorter periods, the field at synchronous orbit responds to the various phases of substorm activity [3], sudden impulses [4] and a variety of wave phenomena [5].

Many of the corresponding magnetic field variations at the surface of the earth are moderately well understood. The diurnal variation of the geomagnetic field, the response to geomagnetic storms and the propagation of sudden impulses fall in this category. However, the source and nature of substorm magnetic variations and the source and nature of the various wave phenomena are not well understood despite years of intensive study. Furthermore, the study of the relationship between the ground-based variations and the simultaneous observations of substorm and wave phenomena in synchronous orbit is only just beginning. For reasons we discuss below, some of these studies, while simple in concept, prove to be very difficult in practice.

Since it is our present purpose to prepare for future studies, we will not review in this paper that which we presently understand well, but rather those areas in which much more work remains to be done: substorms and satellite-ground wave studies. Substorms are important because they constitute the process by which most changes in the state of the magnetosphere occur. Satellite-ground studies of waves are important if we ever hope to use ground-based wave data as a diagnostic probe of the state of the magnetosphere. In the following sections, we first review wave observations at synchronous orbit, and then several examples of ground-satellite wave correlations. Next we examine the use of mid-latitude ground station together with synchronous orbit magnetometer data to model substorm current system variations, and finally, we provide a list of technical suggestions for future studies of this type.

2. ULF Wave Observations in Synchronous Orbit

2.1. Substorm Associated Waves

Four main types of ULF waves have been observed at synchronous orbit during substorms. These are: mixed mode Pc 4-5 [6, 7], Pc 1 [5], broadband Pi2 [8]; and bandlimited Pi1 (or Pc3) [3, 9]. Fig. 1 shows power spectra of these phenomena in the principal axis coordinate system. In this system, X is the direction of maximum power spectral density at each frequency, and Z is in the direction of minimum power. The trace labelled T is the sum of the power in the three orthogonal directions.

The mixed mode (MM) Pc 4-5 event is linearly polarized, since the event occurs only on the principal axis (X). More detailed analysis shows that this direction is in the magnetic meridian tilted 32° away from the field. The transverse Pc1 event is nearly circularly polarized since both X and Y powers are roughly equal through the spectral peak. (Note the change in both the frequency and power scales here.) As expected, this and every other Pc1 event analyzed at synchronous orbit has been left circularly polarized with respect to the geomagnetic field. The next three spectra show examples of broadband (BB) Pi2 events. As is illustrated by these three examples, these events range from being purely compressional to moderately transverse and from having significant spectral peaks to having none. The final example shown is a bandlimited (BL) Pi1 event. This event was nearly linearly polarized with its major axis roughly azimuthal.

Fig. 2 shows the occurrence rate of these four classes of micropulsations as a function of local time. These events have been identified and classified solely on the basis of their appearance on high resolution microfilm plots of the time series. In brackets is listed the number of hours per year in which some activity of each type was noted. For Pc 4-5, some activity occurred in each of 171 hours during two years' data. For Pi2, 360 hours with events were found in six months data. We see that the occurrence of both Pc 1 and Pc 4-5 waves peaks in the late

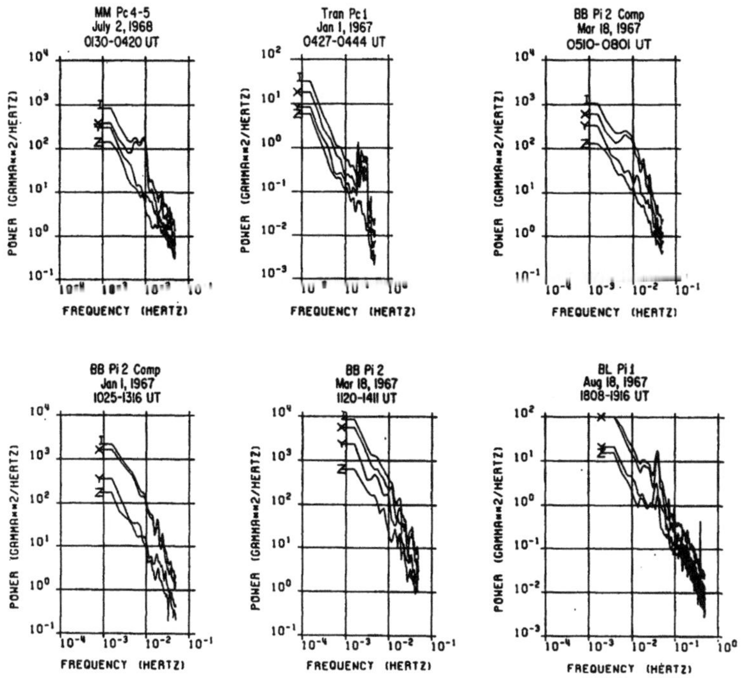

Fig. 1. Power spectra of typical substorm associated micropulsation phenomena (MM, mixed mode; Tran, transverse; BB, broad band; Comp, compressional; BL, band-limited).

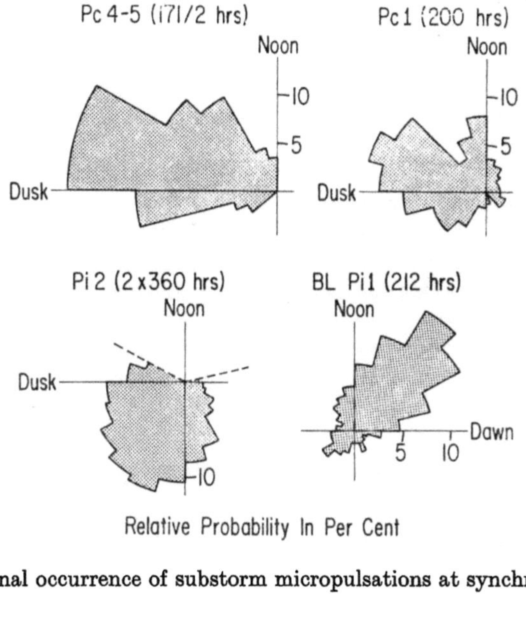

Fig. 2. Diurnal occurrence of substorm micropulsations at synchronous orbit.

afternoon, whereas Pi2 is a pre-midnight phenomenon and bandlimited Pi1 is a mid-morning phenomenon. We note that the majority of Pc1, Pc2 and BL Pi1 events begin within one hour of a substorm expansion onset. Pc 4-5 occur only during very disturbed periods when it is difficult to assign individual substorm onsets.

2.2. Quiet-Time Waves

There are only two wave phenomena at or near synchronous orbit that have been reported as being common during quiet times. These are quiet time Pc 4-5 waves [10] and Pc3 waves [11]. The quiet time Pc 4-5 waves are predominantly transverse and occur predominantly in the early afternoon hours. The average duration of an event is 2.5 hours and the average maximum amplitude during an event is 2.8γ [12]. Quiet time Pc3 waves observed with the Dodge satellite have a most probable period of 30 seconds, peak to peak amplitudes of 2γ and were linearly polarized transverse to the geomagnetic field in the azimuthal direction [11]. These characteristics resemble those reported for the substorm associated bandlimited Pi1 events.

3. Simultaneous Ground-Satellite Wave Observations

3.1. The Compressional Pc4 Event of 14 February 1967

Only one purely compressional Pc4 event has been reported at synchronous orbit [13]. It had a period of 107 sec and a peak to peak amplitude of 10γ. Coherent oscillations of energetic electrons were recorded simultaneously on the same satellite. Despite its uniqueness, this is an important event because it was clearly observed in rapid-run magnetograms from College, Alaska, approximately 600 miles ENE of the ATS-1 conjugate point [14]. Fig. 3 shows simultaneous power spectra of the satellite and ground events. The power at the ground station is some 50 times smaller than at ATS-1. While at the satellite the event was purely compressional, at College the event was left-hand elliptically polarized with an ellipticity of 0.7 and the major axis of the event was oriented at 45° to true north (i.e. NE—SW). Essentially no Z component fluctuations were observed on the ground during this event.

Such an excellent correspondence between the satellite and a ground station 1000 km from the satellite conjugate point would lead one to be very optimistic about the possibility of success of ground-satellite correlations using measurements at the conjugate point. Such a station was set up at Tungsten in the Canadian North West Territories for this purpose. (The conjugacy of ATS and the ground station and the diurnal variation of this conjugacy will be discussed in a later section.) The planned correlations turned out not to be straightforward and were somewhat disappointing. We will examine correlations of two types of wave events to illustrate the difficulties involved.

3.2. Pc1 Observations

Fig. 4 shows the dynamic spectrum of a Pc1 event on 10 October 1969 at the conjugate point of ATS-1 [5]. There is initially a small burst at about 0.2 Hz, then a couple of stronger bursts at 0.4 Hz and then a broadband event centered

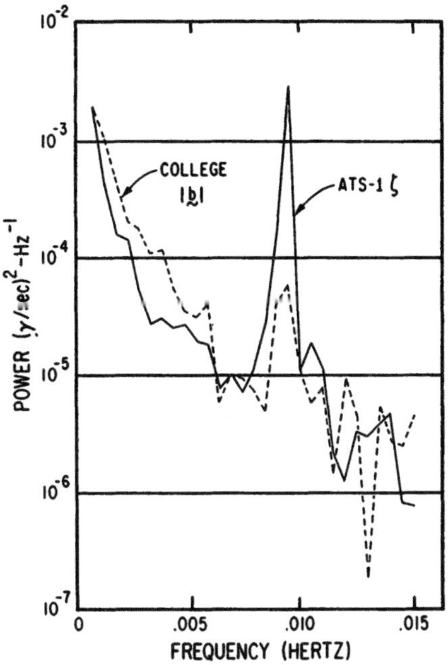

Fig. 3. Simultaneous power spectra of College, Alaska, magnetometer data and ATS magnetometer data during a purely compressional Pc4 event at ATS.

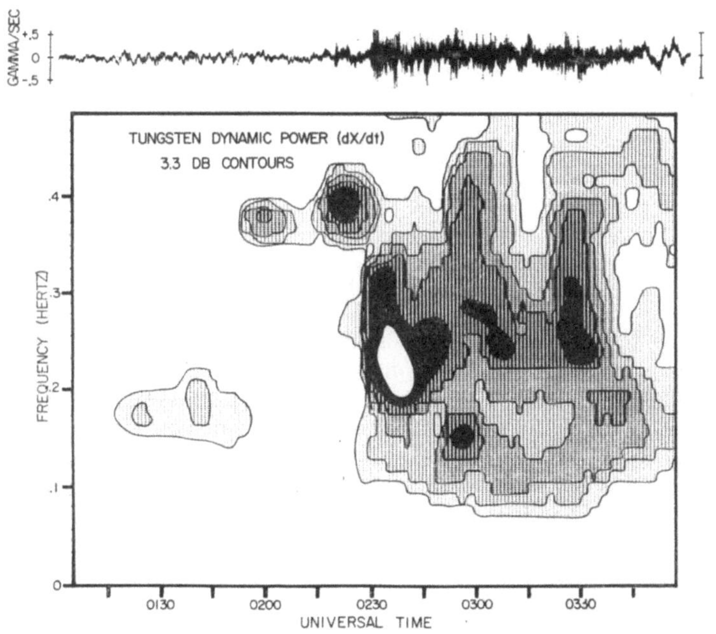

Fig. 4. Dynamic spectrum of Pc1 event at Tungsten, NWT, simultaneous with Fig. 5.

at about 0.25 Hz. Fig. 5 shows the dynamic spectrum generated from simultaneous ATS-1 data [5]. A spacecraft generated interference is responsible for the tones at 0.195 and 0.390 Hz. There is only one period when there is correspondence between the ground and satellite, i.e. 0230—0300. We note that both Tungsten and ATS at times observe signals that the other does not observe. Thus, we conclude that Tungsten and ATS are sensitive to magnetospheric fluctuations in two separate volumes of space which, at least on occasion, have some overlap.

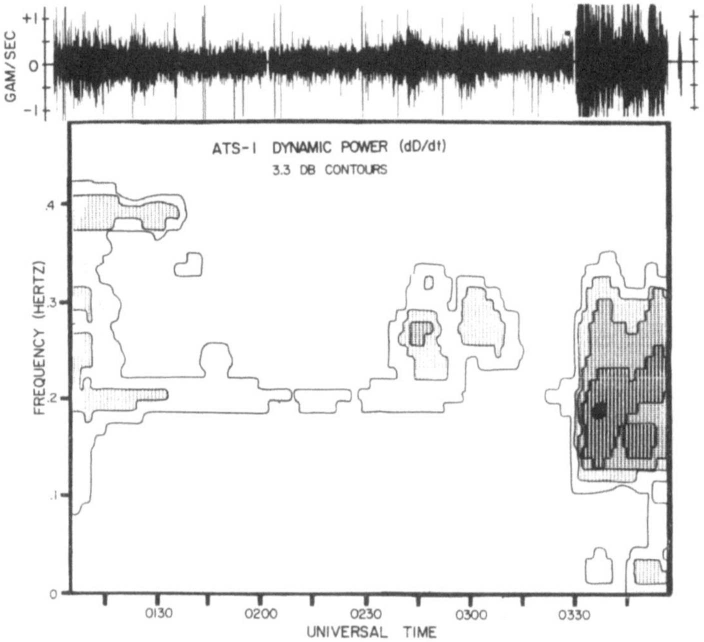

Fig. 5. Dynamic spectrum of Pc1 event at ATS-1.

3.3. Pi1 Observations

To illustrate the degree of correspondence between waves at synchronous orbit and at its "conjugate" ground station, we will examine the results of a systematic search for simultaneous Pi1 events at ATS and Tungsten [15]. Pi1 events are moderately easy to identify by examination of regularly produced time series of the ATS and Tungsten magnetometers. Nine months' data from Tungsten produced 135 possible Pi1 events. Of these 59 were classified as strong and 76 were classified as weak. Thus far, only the strong events have been examined further. Of these 59, only 15 events had complete simultaneous coverage at ATS, 5 had partial coverage and 39 had no coverage. The poor coverage at ATS was due to frequent special operations of ATS at the same time each day which happened to coincide with the peak in the diurnal occurrence pattern of Pi1 events. Of the 15 events with full ATS coverage only one event was seen simultaneously at ATS-1 and the ground. Part of this event is shown in Fig. 6. As noted for Pc1 waves, the ground station records waves not observed at ATS-1 and vice versa. Fig. 7 shows the power spectral density at 0.041 Hz as a function of time for three periods during this event at ATS and at Tungsten. It should be noted that these are the best examples of correspondence in the only one of 12 ground events that showed any correspondence.

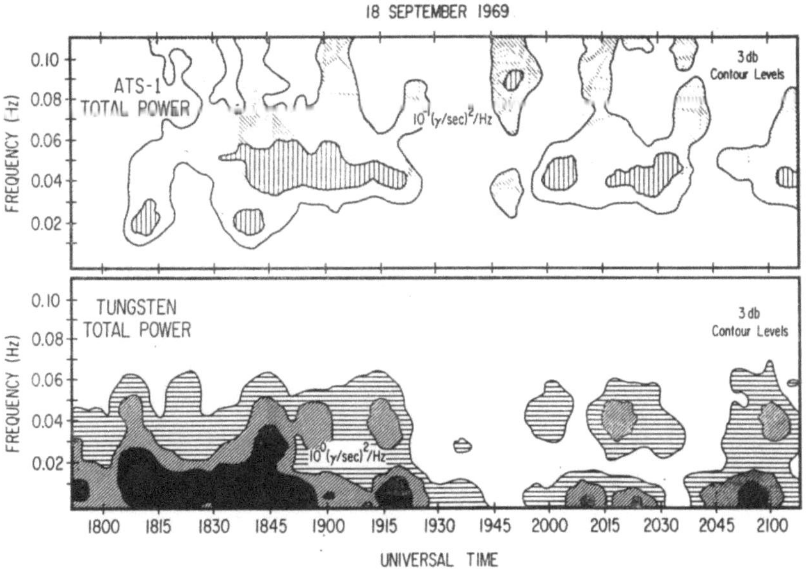

Fig. 6. Dynamic spectra of ATS-1 and Tungsten magnetometer data showing part of a Pi 1 event observed simultaneously at the two sites.

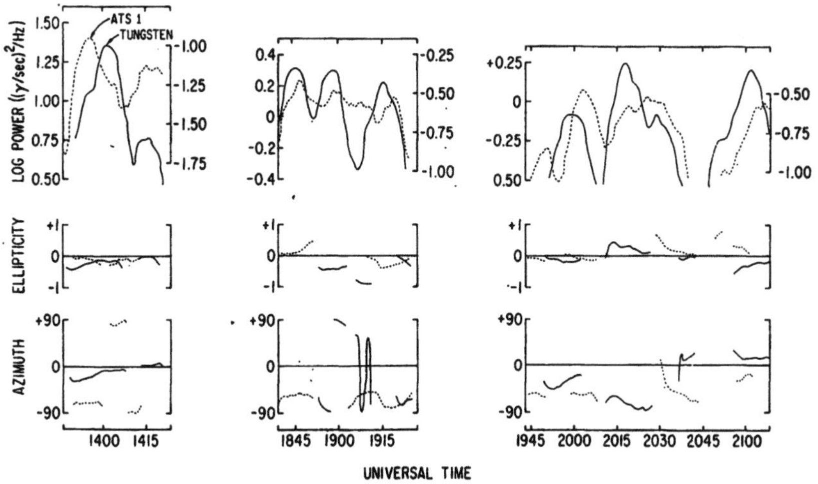

Fig. 7. Power spectral density at 0.041 Hz as a function of time at ATS-1 and Tungsten during Pi 1 event. Also shown are simultaneous measures of ellipticity and the azimuth of the major axis.

4. Three Dimensional Equivalent Current Systems

Equivalent current systems derived solely from ground station data are consistent with many current systems including a two-dimensional current system flowing entirely in the ionosphere. Three-dimensional current systems are expected to occur in the magnetosphere as the result of various physical processes.

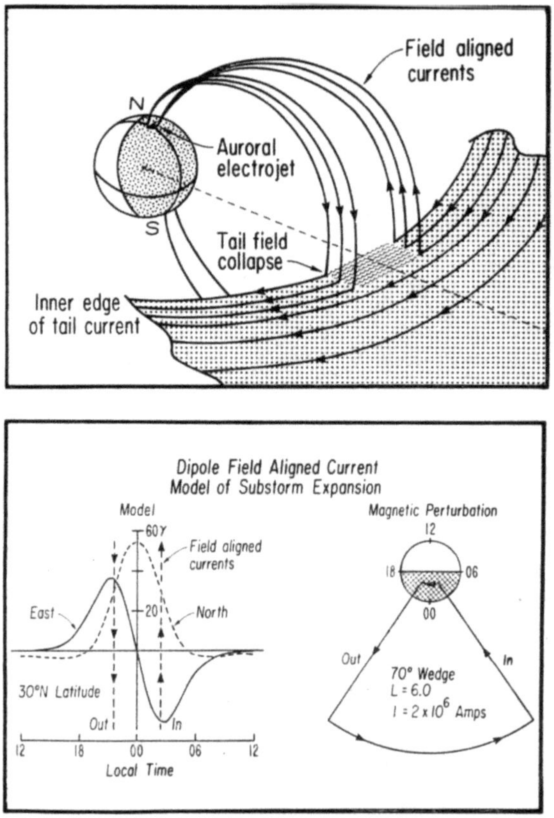

Fig. 8. Model for the substorm expansion phase signature seen in mid-latitude magnetograms.

Magnetic field data from synchronous orbit can be used to study the three-dimensional characteristics of these systems. Fig. 8 shows a model current system proposed to explain the mid-latitude signature of the expansion phase of substorms [16]. A disruption of the tail current diverts current into the auroral oval along field lines. Within the arms of the current wedge, there is an increase in the northward field at mid-latitudes, and outside the arms there is a decrease. There is an eastward magnetic perturbation on the pre-midnight side of the central meridian and a westward field on the post-midnight side. The model has four basic parameters: the central meridian of the substorm, the angular separation of the two field-aligned segments of the current, the strength of the current and the equatorial closure distance. Independent features of the mid-latitude ground signature are sensitive to the central meridian of the current system and its angular extent. However, the amplitude of the perturbation is more sensitive to a function of the current strength and the equatorial crossing distance than either of these parameters separately. Data at synchronous orbit can help to determine both the current and the closure distance. At the very minimum, one can determine whether ATS is inside or outside the current loop. (In all examples examined to date ATS is inside the loop.) This puts a lower limit on the equatorial closing distance and the current. Unfortunately ATS-1 is not only in the geographic equator but also in the magnetic equator and is not sensitive to effects of the field-aligned currents. However, in principle, in other geostationary orbits more than a lower limit could be derived.

Finally, not only are the magnetic records from synchronous orbit a useful supplement to ground records but ground records are often essential to the analysis of the data in synchronous orbit. An example of this is the study of the response of the magnetic field at synchronous orbit to substorms as a function of local time. In this study, the mid-latitude records from the station nearest midnight were used to time the substorm onsets [3], ordering an otherwise bewildering variety of signatures.

5. Conjugacy of Synchronous Orbit and Ground Stations

The diurnal variation in the positions on the earth's surface which are conjugate to a satellite in synchronous orbit is quite large, because at the equator the geomagnetic field is weak and is very sensitive to the various magnetospheric current systems. Sugiura and coworkers [17] have mapped the equatorial magnetospheric field strength with the OGO 1, 3 and 5 satellites and have used this to calculate the conjugate points of ATS-1. Under quiet conditions $Kp = 0-1$, the conjugate point varies from 66.1° geomagnetic latitude at noon to 65.3° at midnight. During disturbed periods, the conjugate latitude moves from 66.5° at noon to 64.8° at midnight. (The dipole field line through synchronous orbit intersects the earth at 67° magnetic latitude.) The ground station that we have used to study ATS conjugate phenomena is at 66° N geomagnetic latitude. Thus, we expect that ATS-1 is outside its conjugate field line at noon and inside it at midnight. This is illustrated in Fig. 9. During the growth phase of substorms when the midnight field lines can become quite tail-like, the Tungsten field line might pass many earth radii tailward of ATS-1.

Thus far we have considered only distortions in latitude. The noon-midnight meridian is close to being a plane of magnetic symmetry. Thus, we would expect field lines to stay close to the dipole magnetic meridian at noon and at midnight. However, near dawn and dusk, we would not expect magnetic fields necessarily to remain in the magnetic meridian. Such distortions have been observed [18]. Again the amount of longitudinal deviation of conjugacy should be a function of both local time and geomagnetic conditions.

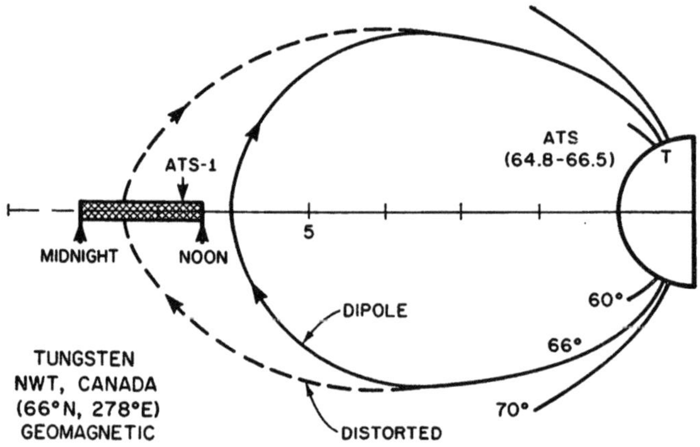

Fig. 9. Changes in equatorial crossing distance of the Tungsten field line (using observations of Sugiura et al. [17]).

6. Recommendations for Future Work

From our experience in establishing the Tungsten ground station and attempting conjugate point studies, we recommend that the following points be considered in future studies.

(i) *Comparable equipment at satellite and ground station.* Correlative studies require the treatment of both sets of data in the same way, and usually require digital data for quantitative results. Similar if not identical data rates are also highly desirable.

(ii) *Prelaunch check out of equipment.* While ground station equipment can be repaired after launch, the system should be checked out before launch. It often requires a lengthy period of time even to determine that there is a problem, let alone identify its cause and repair it at a site that may be quite remote. Since for a variety of reasons data obtained during the first few months after launch are often the most actively studied, successful ground station operation during this period is essential.

(iii) *Use of more than one site.* Since the geographic location of a point conjugate to a geostationary satellite is quite variable in both latitude and longitude, a network of stations in latitude and longitude about the conjugate point is highly desirable. Alternatively, but not equivalently, a movable ground station might be used.

Finally, we note that data in synchronous orbit provide constraints on possible magnetospheric current systems and hence can be used in such studies involving large chains of stations. Such chains might be longitudinal, as in the case of the mid-latitude stations we have discussed, or possibly latitudinal chains at constant geomagnetic longitude.

In conclusion, while many of our conjugate studies have been only minimally successful, we are not convinced that our ultimate goal of being able to use ground observations of wave phenomena to diagnose the state of the magnetosphere is unattainable. We have observed many of the same phenomena in space and on the ground. Our only failure, and it is not a complete failure, is to observe the identical features in space and on the ground simultaneously. We interpret this merely to mean that we have most often been looking in the wrong place.

Acknowledgments

Many of the studies of the ATS data reported here would not have been possible without the able assistance of a number of UCLA graduate students both past and present. We especially wish to thank C. W. Arthur, J. N. Barfield, M. Bossen, C. R. Clauer and B. L. Horning. We are also deeply indebted to R. C. Snare and F. R. George who set up and maintained the Tungsten ground station. This work was supported by the National Aeronautics and Space Administration under research grant NGL 05-007-004 and by the National Science Foundation under research grant GA-28907.

References

[1] W. P. Olson and W. D. Cummings, J. Geophys. Res. **75**, 7117 (1970).
[2] P. J. Coleman, Jr, and W. D. Cummings, Radio Sci. **3**, 762 (1968).
[3] P. J. Coleman, Jr, and R. L. McPherron, in: Particles and Fields in the Magnetosphere, D. Reidel Publ. Co., Dordrecht, Holland 1970 (p. 171).
[4] V. L. Patel and P. J. Coleman, Jr, J. Geophys. Res. **75**, 7255 (1970).
[5] R. L. McPherron et al., Space Sci. Rev. **11**, 111 (1972).
[6] J. N. Barfield et al., J. Geophys. Res. **77**, 143 (1972).
[7] J. N. Barfield and R. L. McPherron, J. Geophys. Res. **77**, 4720 (1972).
[8] R. L. McPherron and P. J. Coleman, Jr, J. Geophys. Res. **75**, 3927 (1970).
[9] R. L. McPherron and P. J. Coleman, Jr, J. Geophys. Res. **76**, 3010 (1971).
[10] W. D. Cummings et al., J. Geophys. Res. **74**, 778 (1969).
[11] M. L. Dwarkin et al., J. Geophys. Res. **76**, 3688 (1971).
[12] W. D. Cummings et al., J. Geophys. Res. **77**, 748 (1972).
[13] J. N. Barfield et al., J. Geophys. Res. **76**, 5252 (1971).
[14] L. J. Lanzerotti and N. A. Tartaglia, J. Geophys. Res. **77**, 1934 (1972).
[15] C. W. Arthur et al., EOS, Trans. Am. Geophys. Un. **53**, 492 (1972) (abstract).
[16] C. R. Clauer and R. L. McPherron, submitted J. Geophys. Res.
[17] M. Sugiura et al., J. Geophys. Res. **76**, 7552 (1971).
[18] D. H. Fairfield, J. Geophys. Res. **73**, 7329 (1968).

Space Research XIV — Akademie-Verlag, Berlin 1974

COMPARISON OF THE TWO-DIPOLE AND EMPIRICAL MAGNETOSPHERIC MODELS

A. E. Antonova and V. P. Shabansky

Institute of Nuclear Physics, Moscow State University, Moscow, USSR

Theoretical values of magnetic fields at high latitudes on the day-side magnetosphere obtained in terms of the two-dipole model are compared with those from the Mead-Fairfield model based on satellite data. It can be seen from the comparison that the field strength near the magnetospheric boundary agrees well for the two models. An insignificant discrepancy which may be qualitatively removed by including the ring current in the two-dipole model is observed in the inner magnetosphere.

Measurements of the day-side magnetospheric field at high latitudes have revealed a region where the field is weaker than the dipolar field. This decrease is connected with the solar wind pressure anisotropy at the magnetospheric boundary, which results in high-latitude minima of the field along distant field lines on the day side. The effect of the shift of the field minimum from the equator to high latitudes in these regions can be seen from the corresponding displacement of the stable trapping zone from equator to high latitudes. In this case the magnetic drift shells become doubly connected, and the second invariant of particle motion is violated in a certain way at the points of shell branching [1]. Analysis [2] has shown that the stable trapping zone boundary calculated in terms of a two-dipole model including the above effects [1] does not contradict the experimental evidence.

The high-latitude field minima are involved in all the known mathematical magnetospheric models. Besides that, all models contain two high-latitude neutral points in the field separating the field lines going to the day side and to the magnetospheric tail. Characteristic values of the coordinates of the neutral points in terms of the above models for quiet conditions, when the distance to the noon magnetospheric boundary L_1 is 10.5 R_e, are: (1) $r_N = 0.93\, L_1$, $\varphi_N \simeq 70°$ (latitude) in Mead's model [3], (2) $r_N = 1.2\, L_1$, $\varphi_N \simeq 45°$ in the two-dipole model [4]*, (3) $r_N = 1.1\, L_1$, $\varphi_N \simeq 63°$ in the paraboloid model [5]. The neutral point should not

* The two-dipole model of the geomagnetic field consists of one dipole located at the earth's centre (with magnetic moment $M = 8.06 \times 10^{25}$ gauss cm^3) and a second dipole (with magnetic moment $M' = j^3 M$) displaced a distance a (in earth radii) along the earth-sun line toward the sun from the first dipole. In the simplest model of [4] (used in Fig. 1) the second dipole is parallel to the z axis of solar magnetospheric coordinates with $a = 33.0$ and $j^3 = 12.7$. Models with the second dipole oriented at an angle ψ relative to the z direction are also considered in [4].

exist in magnetospheric models more realistic than models (1) and (3). Interplanetary plasma in the vicinity of the neutral points forms a cusp or cleft in the geomagnetic field. This cusp is naturally represented in terms of the two-dipole model where the very character of construction of this most simplified model implies that the magnetospheric boundary is determined only by the field lines closed to the earth. (All field lines including the line drawn through the neutral point which are closed to the image dipole should be discarded.)

Recently the results of magnetic field measurements from Explorer 33 and IMP 4 and 5 satellites have made it possible to construct an empirical field model [6, 7] which includes a similar cusp. The best agreement with the experimental data for magnetically quiet periods has been obtained by fitting the appropriate coefficients of the field expansion into a power series in Cartesian solar-magnetospheric coordinates for an arbitrary inclination angle of the geomagnetic dipole

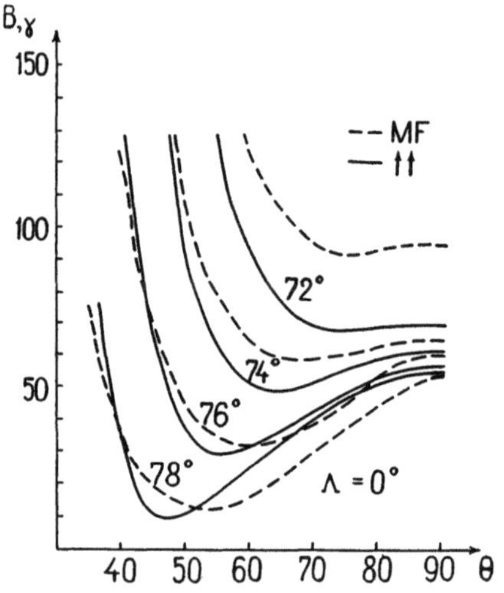

Fig. 1. Magnetic field strength versus co-latitude $\theta = {}^1/_2\pi - \varphi$ in terms of the two-dipole model (solid lines) and Mead-Fairfield's empirical model (dashed lines) in the noon meridian plane. The numerals denote the latitude of the foot of the field line in terms of model (2).

axis to the ecliptic plane. Preliminary results of this model presented in the form of a computer sub-routine [7] have been compared with the two-dipole model results. The solid curves in Fig. 1 show the dependence of magnetic field strength along field lines in the noon meridian plane in terms of the two-dipole model. The dashed line shows the field dependence along the field lines in terms of the empirical model. The comparison shows a good agreement between the empirical and two-dipole models near the magnetospheric boundary both at the equator and in the region of the high-latitude field minimum. A small disagreement is observed when moving deeper into the magnetosphere, which can be most clearly seen at the equator: the field value in terms of the empirical model is somewhat greater than that in terms of the two-dipole model at the given parameters of the latter. This difference can be traced up to the noon meridian and may be eliminated by including a ring current in the two-dipole model and by decreasing the distance to the morning-evening magnetospheric boundary, which corresponds to the pattern actually observed.

The cusp location may be derived from the coordinates of the minimum field point on the extreme field line (the line labelled 78° in Fig. 1). In the two-dipole model the point of minimum is shifted polewards by an angle $\delta\varphi \sim 7-10°$ relative to the corresponding point for the empirical model. As can be seen from comparison of coordinates of the neutral points, this shift is much greater for models (1) and (3). A comparatively low latitude of the model (2) neutral point follows from indirect inclusion of isotropic solar wind pressure which is absent from models (1) and (3). If the empirical model is realistic it follows that the isotropic component of interplanetary plasma pressure exhibits a more considerable effect than that included indirectly in the fairly artificial model (2).

References

[1] V. P. Shabansky and A. E. Antonova, Geomagn. i Aeron 8, 993 (1968).
[2] T. Murayama, J. Geomag. Geoelect. **23**, 205 (1971).
[3] G. D. Mead, J. Geophys. Res. **69**, 1181 (1964).
[4] A. E. Antonova and V. P. Shabansky, Geomagn. i Aeron. **8**, 801 (1968).
[5] I. I. Alekseev and V. P. Shabansky, Planet. Space Sci. **20**, 117 (1972).
[6] G. D. Mead and D. H. Fairfield, paper presented at XV General Assembly of IUGG, Moscow 1971.
[7] D. H. Fairfield, private communication 1971.

Space Research XIV — Akademie-Verlag, Berlin 1974

FAR MAGNETOSPHERIC FIELD OBSERVATIONS BY PIONEER 8: THE DISTANT BOW SHOCK AND THE EXTENDED TAIL

B. Bavassano[a], F. Mariani[a,b] and U. Villante[b]

[a]Laboratorio Plasma nello Spazio, CNR, Frascati, Italy
[b]Istituto Fisico, Università, L'Aquila, Italy

Pioneer 8 magnetic field observations of the bow shock and of the geomagnetic tail at large geocentric distances are investigated. The existence of the bow shock is established as far as 230 earth radii (R_e) from the earth. Magnetic observations during the time of the expected tail encounters at geocentric distances as large as 500 R_e are compatible with the existence of a tail of well defined shape, although with filamentary structures. A reasonable agreement with identifications of the tail based on plasma measurements is found. Some considerations about the observed field strength in the tail and the ratio of thermal to magnetic pressure in the distant magnetosheath are presented.

1. Introduction

In recent years the interaction between the solar wind and the geomagnetic field has been extensively studied in the near-earth region and many aspects of the associated physical phenomena are reasonably understood. [1, 2] are recent reviews. However, not very much is known experimentally about this interaction downstream at distances beyond 80 R_e. Recently, using magnetic data Bavassano et al. [3] and Scarf [4] identified bow shock crossings at distances as large as 200 R_e. Tail observations far downstream were made by Pioneer 8 at about 500 R_e [5—7] and by Pioneer 7 at about 1000 R_e [8—11]. At higher distances, at 3300 R_e, Mariner 4 failed to reveal tail-associated phenomena [12, 13].

In this paper we consider the magnetic field observations of the bow shock and the tail made by Pioneer 8 during the first portion of its flight at $\sim 200\ R_e$. The projection of the trajectory on the ecliptic plane for the period of interest is shown in Fig. 1a; the distance from the ecliptic was never greater than 15 R_e; more information on the trajectory is given by Mariani et al. [14]. We reconsider the tail observations by Pioneer 8 to investigate whether the tail is closed within 500 R_e and the observations are those of a turbulent wake, or whether a well-defined tail is flapping past the spacecraft, or whether the tail has a filamentary structure.

2. The Distant Bow Shock

In the period 19—25 December 1967 Bavassano et al. [3], based on Pioneer 8 magnetic field data, identified more than 30 bow shock crossings, at geocentric distances between 120 and 200 R_e (Fig. 1b). The authors suggested also the prob-

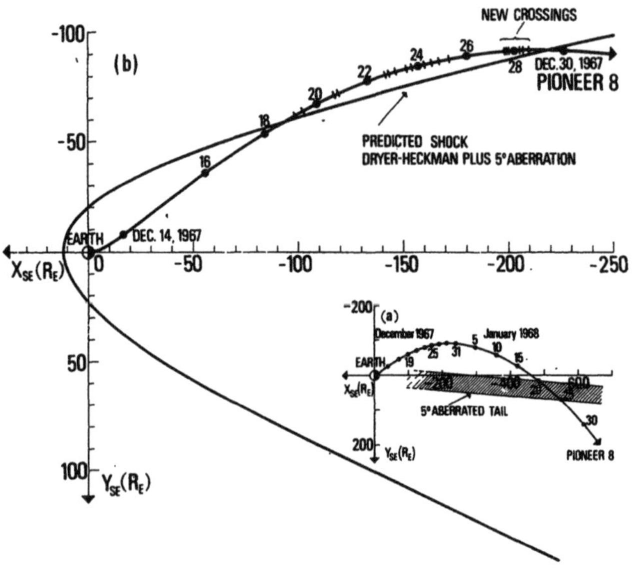

Fig. 1. (a) Ecliptic projection of the Pioneer 8 orbit (solar ecliptic coordinates). A 5° aberrated tail with a 25 R_e radius is also shown. (b) More detailed portion of the orbit showing the locations of the complete set of observed bow shock crossings.

able existence of bow shock crossings at larger distances. Scarf [4] confirmed the whole set of observations and put the final re-entry of Pioneer 8 into the magnetosheath at 28 December.

Here the magnetic field observations after 25 December are reconsidered using better time resolution. From this investigation it has been possible to identify seven more bow shock crossings on 27 and 28 December. Their positions on the ecliptic plane are given in Fig. 1b. The geocentric distance of the farthest crossing is about 230 R_e. As for the shock crossings previously identified, the field magnitude at Pioneer 8 is distinctly higher than that at Explorer 35 in the sunward region of interplanetary space. The increased field magnitude is simply interpreted as a *local* field compression.

The new crossings have been analysed in the same way as described in [3]. Table 1 gives geometric parameters of these furthest bow shock crossings, the

Table 1
Geometric Parameters of the Farthest Bow Shock Crossings

Date	UT	α_N	β_N	ε	$\varrho(R_e)$
27 Dec.	16^h48^m	−29°	280°	16°	219
27 Dec.	17^h56^m	− 1°	295°	4°	220
27 Dec.	19^h1^m	30°	289°	6°	221
28 Dec.	3^h45^m	−24°	289°	8°	225
28 Dec.	9^h25^m	31°	296°	3°	227
28 Dec.	14^h5^m	24°	281°	14°	229
28 Dec.	14^h14^m	28°	289°	12°	229

crossing times and the orientation of the shock normals in solar ecliptic coordinates where the longitude β_N is 0° toward the sun and measured counterclockwise, the latitude α_N is positive above the ecliptic, ε is the 95% error cone angle obtained by the Lepping and Argentiero [15] technique and ϱ is the geocentric distance. Fig. 2 shows the observed magnetic field directional changes and

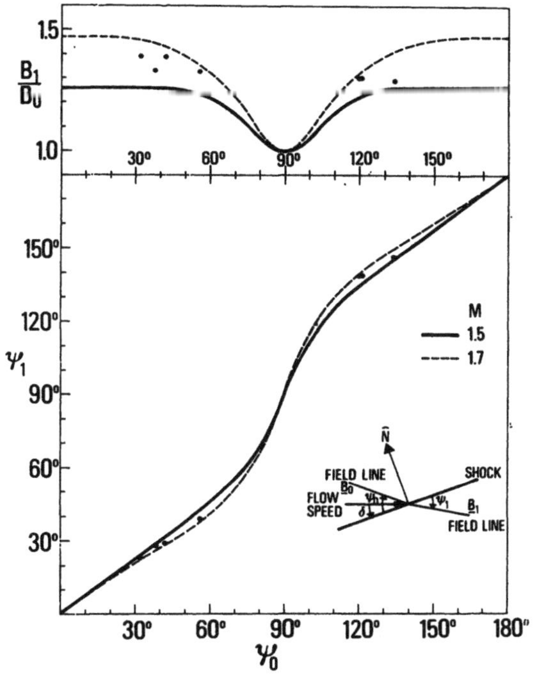

Fig. 2. The ratio B_1/B_0 of the field strengths and the angle ψ_1 versus the angle ψ_0 (ψ_1 and ψ_0 defined in the lower part of the figure). The curves are computed taking the angle δ of the solar wind flow with the shock as 20°, the ratio of thermal to magnetic pressure $\beta = 2$, the specific heat ratio $\gamma = 5/3$. The Mach numbers are those computed using the normal velocity. The dots are from the newly observed crossings.

magnitude amplifications across the shock compared with the theoretical predictions.

The complete set of observations suggests that during the time interval considered the bow shock was moving back and forth across the Pioneer trajectory. Such a variability could be related either to large scale coherent shock motions [16] following variations in the interplanetary medium parameters, or to local oscillations of the bow shock. One needs multiple satellite observations to distinguish clearly between the possible kinds of shock motions.

Unambiguous identification of the final re-entry into the magnetosheath was impossible from our data, due to the simultaneity of an interplanetary perturbation reaching Pioneer 8 on 29 December at $\simeq$ 23 h UT. Two days later, when the perturbation was over, the spacecraft was already well imbedded in the magnetosheath.

3. The Extended Tail

For about two weeks Pioneer 8 was imbedded in the tail region. In this time interval the tail was frequently, even though intermittently, observed [5—7]. Several interpretations of the available magnetic and plasma observations are possible: (i) the tail is closed within 500 R_e so that a turbulent wake is observed; (ii) the solar wind has not diffused into the tail which is flapping past the spacecraft; (iii) the tail has a filamentary structure; (iv) the tail has broken up into bundles disconnected from the earth.

Tail identifications based on magnetic data have been compared with those based on preliminary plasma observations (J. Wolfe, private communication).

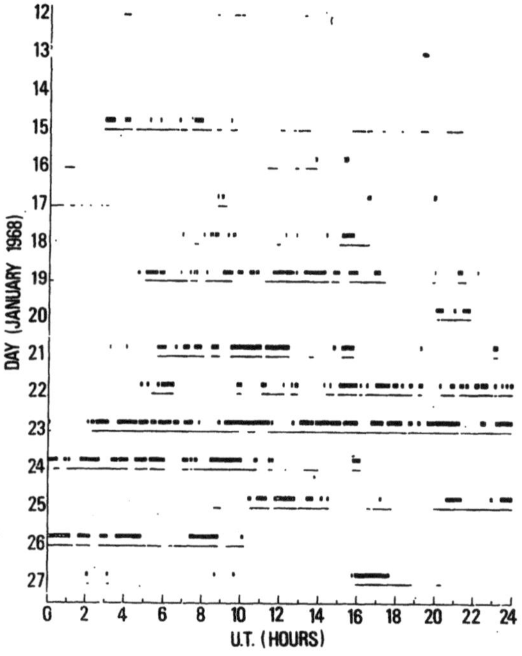

Fig. 3. The time intervals of in-tail observations according to only magnetic data (thick line) or only plasma data (thin line).

A reasonable agreement is achieved only when the criteria used by Mariani and Ness [5] in their early analysis are given quantitatively. In this paper we require that the in-tail magnetic field must have the azimuth ϕ equal to 0° or 180° within $\pm 20°$ and the inclination θ equal to 0° within $\pm 15°$. In the time interval 19—27 January only a few cases occur in which a tail behaviour shows up in the magnetic field and not in the plasma, or vice versa (Fig. 3). Before 19 January magnetic data show some tail-like features only in very few cases, although tail characteristics are suggested by the plasma observations. Thus, a considerable degree of variability of the tail field direction about the predicted one is revealed by the cross comparison of magnetic and plasma data at distances of 400—500 R_e. Our conclusions can be compared with those by Intriligator et al. [17] that, at 1000 R_e downstream, the tentative identification of tail-associated structure based only on either magnetic or plasma data may actually be ambiguous.

Some information on the geometry of the tail at 500 R_e can be obtained from the orientation of the normal to the plane of the discontinuity separating in-tail and out-of-tail fields derived on the reasonable assumption that this discontinuity is tangential. In most circumstances the near-earth magnetopause was indeed interpreted as a tangential discontinuity (see, e.g. [18]). Only sometimes, and only for short time intervals and in connection with certain types of geomagnetic disturbances, a different structure was observed [19, 20]. At greater distances, a tangential structure for the extended discontinuity between the in-tail and the out-of-tail field is compatible with the observations [21, 22].

The set of magnetic data for the whole time interval of tail observations was analysed to detect those discontinuities for which the field on one side has in-tail characteristics. A visual check of the 136 automatically selected cases on the magnetic field plots shows that they well represent all the recognizable crossings.

The normal $\boldsymbol{n}$ to the discontinuity plane, defined by $\boldsymbol{n} = \pm \boldsymbol{B}_1 \times \boldsymbol{B}_2 / |\boldsymbol{B}_1 \times \boldsymbol{B}_2|$ where $\boldsymbol{B}_1$ and $\boldsymbol{B}_2$ are the fields on the two sides of the discontinuity, is identified by two angles: ϕ_n which is its azimuth in the ecliptic plane, and θ_n, which is its inclination with respect to this plane. Due to our tail definition ϕ_n has to be near to 90° or 270°; this is what is found. The inclination is more significant, since information on the cross section of either the tail or the tail filaments can be derived from the angular distribution of θ_n. There is indeed an implicit ambiguity of sign in $\boldsymbol{n}$ which cannot always be resolved *a priori*. Even so, the choice of the determination of $\boldsymbol{n}$ pointing above the ecliptic still leads to some information independent of the sign.

Fig. 4a shows the inclination θ_n for each individual crossing. The error cone angle corresponding to the 80% confidence level is generally 5° to 10°. Different regions labeled A, B and C can be identified in the complicated behaviour of θ_n. In region A, the angle θ_n takes values in its entire range of variability and both tail field orientations are observed. In region B, θ_n is generally large and only sunward fields are observed. In region C, θ_n is on the average smaller and both orientations are observed. A suggested interpretation is that A-type data correspond to a filamentary structure, since very different values of θ_n and both tail polarities are present. However, for B- and C-type data it is possible that a defined tail flapping past the spacecraft is being observed. Indeed, a remarkably non-uniform θ_n distribution should not be typical of thin filamentary structures; rather, it is that expected from a model of a well shaped tail fluctuating with an amplitude small compared with its transverse direction. In particular, in the case of B-type data, large θ_n values and magnetic fields of only one polarity (towards the sun) are seen; this could mean that the spacecraft is crossing the top boundary of the tail upper lobe. For C-type data small θ_n and both polarities are seen; probably the spacecraft looks at the side boundary of both tail lobes. These results do not change when the criteria to identify the crossing are varied.

There are other arguments favouring the idea of a well shaped tail. Fig. 4b shows the time variation of the Pioneer 8 distance from the predicted tail axis, taken parallel to the aberrated direction of the solar wind (with the effect of the dipole tilt included). The temporal profile of the vector velocity of the solar wind is used as observed on board Explorers 33 and 35 (J. Binsack, private communication). We should bear in mind that at 500 R_e a one-degree fluctuation of the solar wind direction corresponds to a fluctuation of about $10 R_e$ in the tail axis intersection with a plane perpendicular to the sun-earth line. According to the Spreiter and Alksne [23] computations, at geocentric distances beyond approximately $130 R_e$ the magnetopause radius should be nearly constant at $25 R_e$. Thus, the magnetopause crossings should be mainly observed at distances from the tail axis of about $25 \pm 10 R_e$. Fig. 4c shows for each hour the total time spent by Pioneer inside the tail, with magnetic field orientation respectively towards and

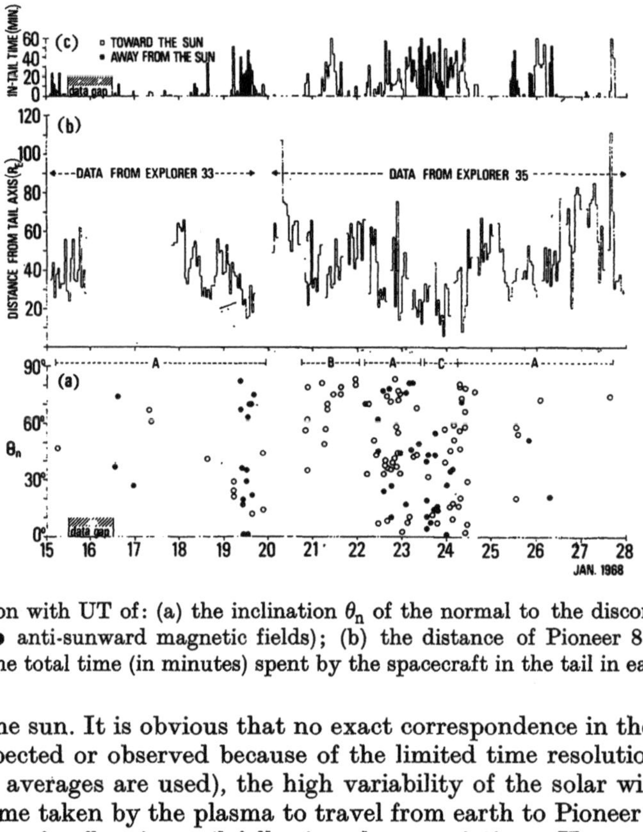

Fig. 4. Variation with UT of: (a) the inclination θ_n of the normal to the discontinuity plane (○ sunward, ● anti-sunward magnetic fields); (b) the distance of Pioneer 8 from the tail axis; (c) the total time (in minutes) spent by the spacecraft in the tail in each hour.

away from the sun. It is obvious that no exact correspondence in the details can be either expected or observed because of the limited time resolution of plasma data (hourly averages are used), the high variability of the solar wind direction within the time taken by the plasma to travel from earth to Pioneer and the unknown inertia of a flapping tail following those variations. However, the gross scale characteristics show a reasonably good correlation. The time spent inside the tail is zero or small when the distances from the tail axis are greater than $40 R_e$; it increases strongly at distances around $30 R_e$, clearly reaching a maximum about 23 January, when a minimum average distance less than $25 R_e$ is predicted. On the afternoon hours of 24 January and the morning hours of 25 January the tail is not observed. Scarf et al. [7] noticed that the plasma conditions were quite variable during this time. Indeed, the Explorer 35 data show a large and systematic trend of the solar wind to flow more from north than usual and this causes a motion of the average position of the tail southward from the spacecraft location.

The frequency distribution of the jumps of the magnetic field intensity across the discontinuity between the in-tail and the out-of-tail fields is shown in Fig. 5a against $r = |B_2 - B_1|/B_2$ where the subscripts 1 and 2 refer to the magnetosheath and the tail respectively. The average value is $\langle r \rangle = 0.41$ with a standard deviation of 0.14. The r-distribution gives information on the ratio β of the thermal to the magnetic pressure on the two sides of the discontinuity. Since the total pressure does not change across a tangential discontinuity, we obtain $\beta_2 = \beta_1 a^2 + a^2 - 1$ with $a = B_1/B_2 = 1 - r$. In the case of negligible thermal pressure in the tail ($\beta_2 = 0$) we can derive from the distribution of Fig. 5a that β_1 is in the range 0.6—5.3 in about 90% of cases with an average value of 2.7. For interplanetary space Ness et al. [24] estimated $\beta_1 = 3$. This coincidence of the values

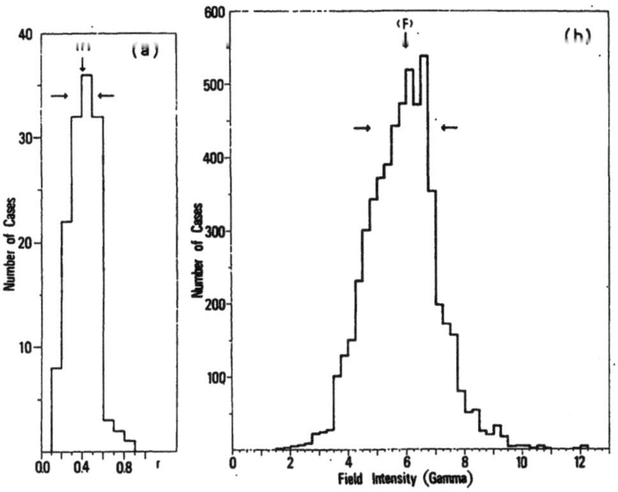

Fig. 5. (a) Histogram of the frequency distribution of the ratio $r = |B_2 - B_1|/B_2$. (b) Histogram of the frequency distribution of the 30-sec averages of the in-tail magnetic field.

of β_1 is interesting, since our results are obtained in the wake region where a reduced density or an increased temperature or both have been observed compared with upstream interplanetary values [25].

The frequency distribution of the in-tail field strength, derived from 30 sec averages, is shown in Fig. 5b. The average magnetic field measured is 6γ, which is remarkably higher than that predicted by extrapolating to $500 R_e$ the radial variation suggested by Behannon [26] and Mihalov et al. [27]. Actually our observed values correspond to their values extrapolated to distances only a little above $100 R_e$; this could mean that the field gradient beyond $\sim 100 R_e$ would be zero or very near to zero. The zero gradient is just the assumption on which Spreiter and Alksne [23] based their model of the magnetospheric tail.

4. Concluding Remarks

The magnetic field measurements on board Pioneer 8 during its passage in the regions of the expected bow shock and the geomagnetic tail have been reconsidered. The existence of the extended bow shock is established as far as $230 R_e$. As regards the tail, a reasonable agreement between identifications based either on magnetic or on plasma data is found. The observed field strength in the tail has an average value of about 6γ. This value is noticeably higher than that predicted by extrapolating the radial gradients found in the near-earth region. The orientation of the surface between the in-tail and out-of-tail magnetic fields has been determined for 136 crossings. Our results can be interpreted as giving support to the idea of a well shaped tail with a transverse radius around $25 R_c$, although with some filamentary structure.

Acknowledgments

We gratefully acknowledge fruitful discussions with Dr. N. F. Ness, who was the main investigator of the magnetic experiment on Pioneer 8. Thanks are also due to Mr. C. S. Scearce of the NASA-Goddard Space Flight Center and Dr. S. Cantarano of the Laboratorio Plasma nello Spazio, CNR, for their help with the engineering development of the instrument.

References

[1] J. H. WOLFE and D. S. INTRILIGATOR, Space Sci. Rev. **10**, 511 (1970).
[2] J. V. KOVALEVSKY, Space Sci. Rev. **12**, 187 (1971).
[3] B. BAVASSANO, F. MARIANI, U. VILLANTE and N. F. NESS, J. Geophys. Res. **76**, 5970 (1971).
[4] F. L. SCARF, J. Geophys. Res. **76**, 7769 (1971).
[5] F. MARIANI and N. F. NESS, J. Geophys. Res. **74**, 5633 (1969).
[6] D. S. INTRILIGATOR, J. H. WOLFE, D. D. MCKIBBIN and H. R. COLLARD, Planet. Space Sci. **17**, 321 (1969).
[7] F. L. SCARF, I. M. GREEN, G. L. SISCOE, D. S. INTRILIGATOR, D. D. MCKIBBIN and J. H. WOLFE, J. Geophys. Res. **75**, 3167 (1970).
[8] N. F. NESS, C. S. SCEARCE and S. CANTARANO, J. Geophys. Res. **72**, 3769 (1967).
[9] J. H. WOLFE, R. W. SILVA, D. D. MCKIBBIN and R. H. MASON, J. Geophys. Res. **72**, 4577 (1967).
[10] D. H. FAIRFIELD, J. Geophys. Res. **73**, 6179 (1968).
[11] U. VILLANTE, R. C. WALKER and A. J. LAZARUS, paper presented at 54th AGU Meeting, Washington, D.C. 1973.
[12] P. J. COLEMAN, JR., L. DAVIS, JR., D. E. HONES and E. J. SMITH, Trans. Am. Geophys. Un. **46**, 533 (1965).
[13] J. A. VAN ALLEN (1966).
[14] F. MARIANI, N. F. NESS and B. BAVASSANO, CNR Lab. Plasma nello Spazio, Int. Rep. LPS-71-22 (1971).
[15] R. P. LEPPING and P. D. ARGENTIERO, J. Geophys. Res. **76**, 4349 (1971).
[16] E. W. GREENSTADT, P. C. HEDGECOCK and C. T. RUSSELL, J. Geophys. Res. **77**, 1116 (1972).
[17] D. S. INTRILIGATOR, J. H. WOLFE and D. D. MCKIBBIN, J. Geophys. Res. **77**, 4645 (1972).
[18] B. U. Ö. SONNERUP and L. J. CAHILL, JR., J. Geophys. Res. **72**, 171 (1967).
[19] B. U. Ö. SONNERUP, J. Geophys. Res. **76**, 6717 (1971).
[20] M. P. AUBRY, M. G. KIVELSON and C. T. RUSSELL, J. Geophys. Res. **76**, 1673 (1971).
[21] J. D. MIHALOV, D. S. COLBURN and C. P. SONETT, Planet. Space Sci. **18**, 239 (1970).
[22] J. D. MIHALOV, W. C. FELDMAN, C. P. SONETT and J. H. WOLFE, Cosmic Electrodyn. **1**, 389 (1970).
[23] J. R. SPREITER and A. Y. ALKSNE, Planet. Space Sci. **17**, 233 (1969).
[24] N. F. NESS, A. J. HUNDHAUSEN and S. J. BAME, J. Geophys. Res. **76**, 6643 (1971).
[25] G. L. SISCOE, F. L. SCARF, D. S. INTRILIGATOR, J. H. WOLFE, J. H. BINSACK, H. S. BRIDGE and V. M. VASYLIUNAS, J. Geophys. Res. **75**, 5319 (1970).
[26] K. W. BEHANNON, J. Geophys. Res. **73**, 907 (1968).
[27] J. D. MIHALOV, D. S. COLBURN, R. G. CURRIE and C. P. SONETT, J. Geophys. Res. **73**, 943 (1968).

The Sun

Space Research XIV — Akademie-Verlag, Berlin 1974

SOME CRITICAL PROBLEMS OF SOLAR FLARE RESEARCH: ENERGETIC PHENOMENA

L. D. de Feiter

Space Research Laboratory of the Astronomical Institute at Utrecht, The Netherlands

Some problems of current interest in the research of energetic solar flare phenomena are reviewed with the aim of determining the open questions in this field. Possible directions in which the solutions may be found are indicated. In this context the following problems are discussed: the need for a determination of the magnetic field structure in the chromospheric and coronal parts of an active region before, during and after the occurrence of a flare, the "thermal" flares and their differences with respect to the impulsive flares, the energetic relation between the non-thermal and the thermal components of the high-energy flare, the implications of repetitive flare instabilities and continuous acceleration models, the nuclear gamma rays and neutrons as manifestations of the solar effects of accelerated nuclei.

1. Introduction

Flare processes cover a large dynamic range, both in the physical properties of the regions of the solar atmosphere that are affected by them, and in the emitted fluxes of photons and particles, and the spectral distributions of these fluxes. The densities and the temperatures in the source regions of the flare emissions vary by several orders of magnitude, and the same is possibly true for the magnetic field strengths. The distribution of the total flare energy among the various components of the flare emissions varies considerably from one event to another. Also, some flares are accompanied by violent motions in the active region and beyond, whereas others develop in a more or less stationary manner. Because of this impressive variety, it seems unlikely that just one single process can be indicated that controls the whole flare development. Instead we should think of a number of different processes, or at least a repetition of one of these under different circumstances and possibly at different locations. The relation between such elementary flare processes, and the conditions under which they develop, determines the overall development of the flare we observe. The combined effects of these processes, and the physical conditions in the region where they develop, cannot easily be unravelled from the observations, unless we gradually develop a certain way of pattern recognition from a study of as broad a range of observational material as possible for a limited number of carefully selected flares.

At this moment we have available to us, at least in principle, sources of information covering the electromagnetic spectrum from the nuclear gamma rays to the plasma frequency of the solar wind in the neighbourhood of the earth, and particle observations ranging in energy from that of a solar-wind particle to that of the relativistic domain. Apart from the further development of theoretical ideas, it seems that we require further insight into the problem of combining the various kinds of observations in constructing a picture of the flare origin and its development. For the discussion that follows I have selected a few examples to illustrate this general statement.

2. Pre-Flare Activity and Magnetic Field Changes

Most flare phenomena occur in the chromospheric and low coronal parts of an active region. The structural changes in these parts that occur during a flare, and in particular the changes in the coronal magnetic field, are only poorly known. If the flare derives its energy from the magnetic field, and many flare theories attempt to explain how the conversion of magnetic field energy into "visible" forms of energy takes place (see e.g. [45]), a better definition of the structure above the active region before, during and after a flare is clearly needed. Also if this field only serves to provide the proper conditions for the flare to occur, observations of coronal structures and their flare-associated changes are urgently needed.

No methods for the direct measurement of coronal magnetic fields are available at this moment. Some promising attempts to derive coronal magnetic field data from specific radio burst observations have recently been made (e.g. [27, 35]. Such methods, however, refer to disturbances already in progress and therefore cannot be used in an assessment of field changes associated with this disturbance. For quiet conditions the application of an extrapolation procedure of the measured photospheric magnetic field, originally developed by Schmidt [38] and extended by Altschuler and Newkirk [2], has contributed significantly to our understanding of large scale coronal structures (see e.g. [29]), and those in the solar wind [36, 37]. Also, Valdez and Altschuler [52] were able to present some evidence for the opening up of coronal field loops following the occurrence of proton flares in the active region concerned. McCabe and Fisher [26] observed a flare spray preceding a moving type IV burst [34], and the trajectory of the source of this radio burst could be shown to lie along the coronal magnetic field line. In spite of these obvious successes of the field extrapolation procedures, their applicability to the flare problem proper is restricted by the assumptions on which they are based, namely the fields should be current free, or at least force free. This restriction implies that the configurations determined by the extrapolation procedure represent those of minimum energy and consequently they will not show where the excess magnetic energy that might be used during the flare process is stored.

Approaching the problem from a different angle, we could start from the close relationship between the coronal density structure, e.g. the dense coronal loops, and the magnetic field topology and use observed changes of the first to infer the changes of the latter. To this end sequences of high resolution pictures of the corona in the XUV and soft X-ray spectral region will be extremely useful, since in particular the soft X-ray pictures show such structures [23]. Possibly such sequences of XUV and soft X-ray pictures will be secured during the ATM Skylab mission to be executed later in 1973 [33], and in any case future solar missions should include instrumentation for these observations. In addition new methods that can lead to a direct determination of the coronal magnetic field should be developed.

As is well known some coronal structures can be easily disturbed through the actions of a nearby, or even distant, flare. Such disturbances also seem to occur prior to the occurrence of a flare; Bruzek [6] reports that about 45% of all major flares are preceded by filament activations, and other kinds of coronal pre-flare activity have been observed by Bruzek and Demastus [7]. All these observations indicate that prior to the occurrence of a flare, changes in the coronal magnetic field occur that might be important in connection with a determination of the flare triggering process. The selection effect that is inherent in the observations obtained thus far, because of the reference made to the existence of suitable filaments or because of limb observations only, will be much reduced for the XUV and X-ray observations mentioned before.

3. Pre-flare Heating and the Thermal Phase

The high temperature flare is best studied in the soft X-ray region. The flare associated soft X-ray burst starts, on the average, two minutes before the Hα flare; it reaches its peak intensity about three minutes after the end of the Hα flare [49]. Even if due allowance is made for the near coincidence in time of background fluctuations, the occurrence of an X-ray precursor is real in a statistical sense. Thomas and Teske [49] also found that this precursor is most conspicuous at about 8—12 Å; Kreplin et al. [22] have found longer lead times for the 0.5—3 Å band as compared with the 1—8 and 8—16 Å bands. The source of this pre-flare increase in the soft X-ray region is presumably a restricted coronal region where the temperature has increased to some 5×10^6 K. The location of this source with respect to the structure elements of the Hα flare, and also with respect to the source of the hard X-ray emission, will be an important datum for theories on the initial flare phases. Although the impulsive phase, to which flare theorists have paid much attention [41], is very conspicuous indeed, it may be that the physical conditions for its occurrence are set by the heating process that leads to the formation of the X-ray precursor. Kahler and Kreplin [19] point to the possibility of the occurrence of a thermal instability of a plasma that is heated beyond a few million degrees (see also [10]). The negative slope of the energy loss as a function of temperature above the peak at about 4×10^5 K makes it possible for the temperature of a gas that is intially in a steady state to attain progressively higher values once a small extra heat input has been provided. Inversely, however, we should expect a progressive cooling of the region once the heat input falls short of the radiative losses at the equilibrium temperature. Considerations of the equilibrium conditions for plasmas in the temperature range concerned, roughly between 4×10^5 K and 2×10^7 K, require a detailed analysis of the elementary processes concerned, taking into account their characteristic times. Also the changes in the surroundings of the heated region should be taken into account. Detailed calculations of these processes in a representative magnetohydrodynamic configuration, along lines similar to those applied by Bessy and Kuperus [4], will provide a deeper insight into the physics of the flaring region.

Some flares do not show an impulsive or explosive phase; they are sometimes referred to as thermal flares [3, 11]. The X-ray emission for these flares does not extend beyond 10—20 keV photon energy and slowly developing bursts are observed in the microwave region, and possibly also in the EUV region [57]. The EUV bursts are emitted by a source located in the transition region between the chromosphere and the corona, whereas the gradual microwave and the soft X-ray burst probably originate in sporadic coronal condensations. The relation between the gradual microwave burst and the gradual soft X-ray burst has been studied by Hudson and Ohki [17], by Shimabukuro [39], and also by Ramaty and Petrosian [32]. In future model calculations the whole complex of coronal and chromospheric regions, comprising the sources of X-ray, microwave, EUV and Hα emission, should be considered and the effects of downward heat conduction from the hot coronal regions must be estimated (e.g. Švestka [44]). For such model calculations to be more effective, observations should be made with adequate temporal and spatial resolution, in particular of the microwave and X-ray sources. Also the inhomogeneous structure of the source regions, as has been derived for the chromospheric flare from spectroscopic analysis (see [43] for a review on this subject) and which probably also exists in the X-ray source [1], has to be taken into account. The energy transfer along very thin filaments aligned with the magnetic field will pose quite restrictive boundary conditions, and unexpected results may emerge.

4. The Non-thermal Component and its relation with the Thermal Flare Emissions

In many flares the initial soft phase is followed by an impulsive phase, which is most clearly observed in the microwave and hard X-ray region of the electromagnetic spectrum. These emissions are produced by energetic electrons: the hard X-ray burst as Coulomb bremsstrahlung, and the microwave burst as magnetic bremsstrahlung. The non-thermal hard X-ray burst occurs in combination with a thermal burst at photon energies below about 10 keV [48]; Kane [20] has shown that in many cases two components can be found, the first one having a power-law photon spectrum whereas the second component has a thermal photon spectrum, see Fig. of [20]. The spectrum of the non-thermal component shows a considerable steepening above about 100 keV [21, 53]. In addition to the two-component bursts found by Kane [20] various other types have been reported in the literature; Frost and Dennis [16] have discussed an event which consists of two non-thermal components, Parks and Winckler [30], Frost [15], and van Beek et al. [54] have discussed events that exhibit periodic and very rapid variations in their time profiles. Tindo et al. [50, 51] present evidence for polarization of the hard X-radiation persisting through the whole event. All these observations indicate that the acceleration process that produces the energetic electrons does not always operate impulsively, but that it sometimes remains operative for a prolonged period of time with a rapidly varying efficiency.

The impulsive EUV bursts observed by Wood et al. [57], and the correlations of hard X-ray bursts with brightenings of small features of the Hα flare [14, 55, 56, 58] indicate a connection between the non-thermal flare and the flare regions in the transition region and in the chromosphere. Positional information on the source of the hard X-ray burst is available for only one event: 27 September 1969 [46]; the position line for the source of the microwave emission is derived from radio interferometric observations [47]. Both position lines are superimposed on an Hα picture of the flare. To the present author it does not seem accidental that a bright kernel in the Hα picture can be found near the crossing point of the two position lines. The sizes of the source regions are less than 1 arcminute for the X-ray source and about 0.5 arcminute for the radio region. In another case points with a size of only one arcsecond have been found flashing concurrently with rapid flashes in hard X-rays [54, 58].

The picture that emerges from the above discussion of observational evidence is already far more detailed than that used as an observational input to flare theories. It indicates that the electron acceleration can take place not just for a very brief period during the flash phase, but that it can continue and be repeated several times, possibly at different locations in the active region. Also, the effects of a beam of electrons on the atmospheric layers traversed seem to be important in connection with the relation between the different flare emissions. Well coordinated observations of these emissions may very well lead to the conclusion that many flare properties can be derived from those of the beam of accelerated electrons.

5. The Acceleration of Nuclei and the Production of Gamma Rays

The presence of energetic protons and nuclei in the solar atmosphere cannot be determined through the radiation produced in interactions with the ambient charged particles or with the magnetic field, as is the case for accelerated electrons. The amount of radiation produced in these interactions is lower than that produced by electrons by a factor $(m_e/m)^2$, where m_e is the electron mass and m the mass of the nucleus considered. Until the recent discovery of nuclear gamma rays, for the large flares of August 1972 [9], our only source of information on the acceleration of protons and other nuclei was their direct detection with spaceborne or ground-based detectors. The information obtained from these measurements cannot be related directly to the properties of the accelerated particles in the source region; we have to allow for considerable changes to occur during the release from the acceleration or the storage region and during the transport through interplanetary space. Many investigations have been devoted to the theory of energetic particle transport through the magnetic field irregularities in interplanetary space. A review of the theoretical ideas has been given by Jokipii [18].

As far as the physics of solar flares is concerned, the critical questions in this field are not so much of a theoretical nature. A more decisive problem is our lack of knowledge of the conditions under which the particles are released into interplanetary space. Moreover, the derivation of the diffusion tensor for energetic particle transport from the statistical parameters describing the magnetic field irregularities is necessarily restricted to the field measurements near the earth's orbit. Also, significant differences of the propagation characteristics of electrons and nuclei, and difficulties with the unique interpretation of particle data (e.g. [40]), require careful judgment in deriving the properties of energetic particles in the source region. The recent discovery of nuclear gamma-ray emission for the large flares of August 1972 [9] is an important step towards a study of the effects in the solar atmosphere produced by energetic nuclei. Computations of the production rates of these gamma-ray lines have been made by Lingenfelter and Ramaty [25], Lingenfelter [24], Cheng [8] and de Feiter [12]. These production rates depend very critically upon the model assumed for the source region; the measured fluxes for the 4 August 1972 event can be produced by energetic protons, having an energy spectrum which is exponential in rigidity, with a total kinetic energy of 10^{28} ergs if the gamma rays are produced by protons moving down into the solar atmosphere, whereas about 10^{30} ergs are needed if the production takes place at the site of acceleration [31]. According to de Feiter and Švestka [13] the first alternative seems to be called for by the apparent association with the production of optical continuum radiation in very powerful flares. In that case the protons could contribute also to the heating of the solar flare plasma traversed, and one might expect that such heating also takes place for less powerful flares, in which the particles may be stopped at larger altitudes. Much more work needs to be done in this area, and in particular the application of previous studies (5, 28, 42] to realistic models of the flare region should be carried out.

6. Summary and Conclusions

In the previous section I have discussed a few problems of solar flare research that need further investigation in the coming years. In selecting these problems I have put considerable emphasis upon the combination of different kinds of observations for specific events. It is this kind of approach, which also was at the origin of such international projects as the Proton Flare Project and CINOF, which, apart from the further development of plasma theoretical models, is most needed in order to understand the most challenging solar phenomenon: the solar flare.

References

[1] L. W. ACTON, Astrophys. J. **152**, 305 (1968).
[2] M. D. ALTSCHULER and G. NEWKIRK, Solar Phys. **9**, 131 (1969).
[3] K. A. ANDERSON, S. R. KANE and R. P. LIN, U. C. L. Berkeley, S. S. L. Series 12, Issue 7 (1971).
[4] R. J. BESSY and M. KUPERUS, Solar Phys. **12**, 216 (1969).
[5] S. BISWAS and B. RADHAKRISHNAN, Solar Phys. **28**, 211 (1973).
[6] A. BRUZEK, in: Solar Flares and Space Research, North-Holland Publ. Co., Amsterdam 1969 (p. 69).
[7] A. BRUZEK and H. L. DEMASTUS, Solar Phys. **12**, 447 (1970).
[8] C. C. CHENG, Space Sci. Rev. **13**, 3 (1972).
[9] E. L. CHUPP, D. J. FORREST, P. R. HIGBIE, A. N. SURI, C. TSAI and P. P. DUNPHY, Nature **241**, 333 (1973).
[10] D. P. COX and W. H. TUCKER, Astrophys. J. **157**, 1157 (1969).
[11] L. D. DE FEITER, Space Sci. Rev. **13**, 827 (1972).
[12] L. D. DE FEITER, Solar Phys. **19**, 207 (1972).
[13] L. D. DE FEITER and Z. ŠVESTKA, Space Research XII, 1547 (1972).
[14] C. DE JAGER, Solar Phys. **2**, 327 (1967).
[15] K. J. FROST, Astrophys. J. **158**, L159 (1969).
[16] K. J. FROST and B. R. DENNIS, Astrophys. J. **165**, 655 (1971).
[17] H. S. HUDSON and K. OHKI; 1972, Solar Phys., **23**, 155.
[18] J. R. JOKIPII, Rev. Geophys. Space Phys. **9**, 27 (1971).
[19] S. W. KAHLER and R. W. KREPLIN, Solar Phys. **14**, 372 (1970).
[20] S. R. KANE, Astrophys. J. **157**, L139 (1969).
[21] S. R. KANE and K. A. ANDERSON, Astrophys. J. **162**, 1003 (1970).
[22] R. W. KREPLIN, D. M. HORAN, T. A. CHUBB and H. FRIEDMAN, in: Solar Flares and Space Research, North-Holland Publ. Co., Amsterdam 1969 (p. 121).
[23] A. S. KRIEGER, G. S. VAIANA and L. P. VAN SPEYBROECK, in: Solar Magnetic Fields, IAU Symp. No. 43, 397 (1970).

[24] R. E. LINGENFELTER, Solar Phys. **8**, 341 (1969).
[25] R. E. LINGENFELTER and R. RAMATY, in: High Energy Nuclear Reactions in Astrophysics, W. A. Benjamin, New York 1967 (p. 99).
[26] M. K. McCABE and R. R. FISHER, Solar Phys. **14**, 212 (1970).
[27] D. L. McLEAN and K. V. SHERIDAN, Solar Phys. **26**, 176 (1972).
[28] K. NAJITA and F. Q. ORRALL, Solar Phys. **15**, 176 (1970).
[29] G. NEWKIRK, in: NASA SP-308, 1972 (p. 11).
[30] G. K. PARKS and J. R. WINCKLER, Solar Phys. **16**, 186 (1971).
[31] R. RAMATY and R. E. LINGENFELTER, in: NASA X-693-73-193, 1973 (p. 301).
[32] R. RAMATY and V. PETROSIAN, Stanford Univ. Inst. for Plasma Res. Rep. No. 465 (1972).
[33] E. M. REEVES, R. W. NOYES and G. L. WITHBROE, Solar Phys. **27**, 251 (1972).
[34] A. C. RIDDLE, Solar Phys. **13**, 448 (1970).
[35] H. ROSENBERG, Solar Phys. **25**, 188 (1972).
[36] K. H. SCHATTEN, in: NASA-SP-308, 1972 (p. 65).
[37] K. H. SCHATTEN, J. M. WILCOX and N. F. NESS, Solar Phys. **6**, 442 (1969).
[38] H. V. SCHMIDT, AAS-NASA Symp. on the Physics of Solar Flares, NASA-SP-50, 1964 (p. 107).
[39] F. I. SHIMABUKURO, Solar Phys. **23**, 169 (1972).
[40] D. F. SMART and M. A. SHEA, Space Research XIV, 471 (1974).
[41] P. A. STURROCK and B. A. COPPI, Astrophys. J. **143**, 3 (1966).
[42] Z. ŠVESTKA, Solar Phys. **13**, 471 (1970).
[43] Z. ŠVESTKA, Ann. Rev. Astron. Astrophys. **10**, 1 (1972).
[44] Z. ŠVESTKA, Solar Phys. **31**, 389 (1973).
[45] P. A. SWEET, Ann. Rev. Astron. Astrophys. **7**, 149 (1969).
[46] R. TAKAKURA, K. OHKI, N. SHIBUYA, M. FUJII, M. MATSUOKA, S. MIYAMOTO, J. NISHIMURA, M. ODA, Y. OGAWARA and S. OTA, Solar Phys. **16**, 454 (1971).
[47] H. TANAKA and S. ENOMÉ, Solar Phys. **16**, 431 (1971).
[48] R. G. TESKE, Astron. J. **72**, 832 (1967).
[49] R. J. THOMAS and R. G. TESKE, Solar Phys. **16**, 431 (1971).
[50] I. P. TINDO, V. D. IVANOV, S. L. MANDELSTAM and A. I. SHURYGHIN, Solar Phys. **14**, 204 (1970).
[51] I. P. TINDO, V. D. IVANOV, S. L. MANDELSTAM and A. I. SHURYGHIN, Solar Phys. **24**, 429 (1972).
[52] J. VALDEZ and M. D. ALTSCHULER, Solar Phys. **15**, 446 (1970).
[53] H. F. VAN BEEK, Thesis, Utrecht 1973.
[54] H. F. VAN BEEK, L. D. DE FEITER and C. DE JAGER, Space Research XIV, 447 (1974).
[55] J. VORPAHL, Solar Phys. **26**, 397 (1972).
[56] J. VORPAHL, J. and H. ZIRIN, Solar Phys. **11**, 285 (1970).
[57] A. T. WOOD, R. W. NOYES, A. K. DUPREE, M. C. E. HUBER, W. H. PARKINSON, E. M. REEVES and G. L. WITHBROE, Solar Phys. **24**, 169 and 180 (1972).
[58] H. ZIRIN and K. TANAKA, Collected Data Reports on August 1972 Solar Terrestrial Events, Report UAG-28, Part I, 1973 (p. 121).

Space Research XIV — Akademie-Verlag, Berlin 1974

DISCUSSION OF X-RAY PICTURES OF THE SUN

H. Bräuninger[a], H. Einighammer[a], G. Elwert[b], J. Feitzinger[b], H. H. Fink[a], and G. Krämer[a]

[a]Astronomisches Institut, Tübingen, FRG
[b]Lehrstuhl für Theoretische Astrophysik der Universität Tübingen, FRG

X-ray pictures of the sun at Fe XVII 15.5 Å and O VII 21.6 Å were taken with Fresnel zone plate cameras used in ESRO and NRL rockets. The correlation of the structures of the coronal condensations with the photospheric magnetic fields and the height of these coronal emission regions are discussed. On one of the flights a small flare source was observed at $\lesssim$ 6 Å coinciding with the top of a loop-like structure as seen at 15.5 Å.

In this paper a few preliminary results derived from X-ray pictures of the sun taken with Fresnel zone plates by the Tübingen Solar Space Research Group are reported. The theoretical resolution of the zone plates as given by the width of the central lobe of the diffraction pattern is of the order of 1″ for wavelengths of the order of 10 Å and diameters of the zone plates of 1 mm. Indeed this resolution can almost be obtained for point sources. Around this small lobe, however, a more or less bright disk is observed, the picture of zeroth order, the so-called halo, the diameter of which is equal to the diameter of the zone plate. The relative brightness of this disk decreases with increasing number of zones. An important property of the zone plate is that its focal length depends on the wavelength. This property is an advantage in so far as a non-monochromatic source is represented sharply only in the light of one wavelength. If there exist, for instance, two lines in the passband of the filter, of which one is focused, a chromatic aberration is caused by the other line. This leads to an additional intensity disk around the focused wavelength. The effects of these aberrations, however, can be reduced by a correction procedure. For the cores of the active regions the halo must not be considered as these bright regions are mainly pictured by the central lobe.

First a report on an ESRO experiment which was launched in March 1971 will be given. Copies of the X-ray pictures obtained with this experiment taken especially in the light of the line of Fe XVII at 15.5 Å and of O VII at 21.6 Å are published in [1]. The pictures show different regions of enhanced emission on the east limb, in particular a very bright one, and some regions on the disk, especially three brighter ones. To discuss the structures and intensity distributions, isodensity pictures have been made by using isodensity film for the cores of the sources. From the focal length of the zone plate one can calculate the diameter of the sun's picture. Therefore it is possible to derive the height of the core of the brightest region at the east limb. In this way the height can be estimated to be 15000—20000 km. The diameter of the innermost core is about 7000 km. This extension means a resolution of approximately 10″.

In order to determine the position of the active regions with respect to the magnetic fields in the photosphere Mt Wilson magnetograms were used to draw magnetograms for the time of observation. By superposing the isodensity pictures of the core regions and these magnetograms Fig. 1 was obtained. Considering the region near the disk centre, the strongest X-ray emission region lies just in the narrow zone in which the change of polarity from negative to positive field direction takes place. It is remarkable that details in structure of the magnetic

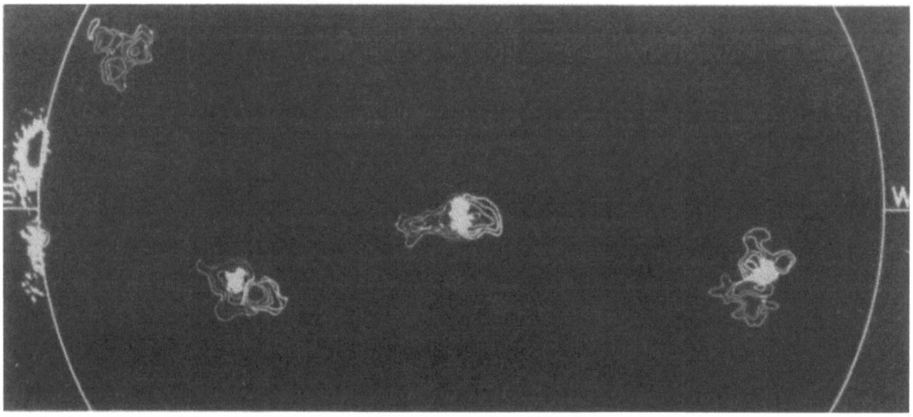

Fig. 1. X-ray picture of 11 March 1972 in isodensity representation of the brightest emission cores superposed on the Mt. Wilson magnetogram of the corresponding activity centres reduced for the time of observation.

field are also found in the X-ray picture. It seems that the magnetic fields holding together the coronal condensations reach the corona practically undistorted. The X-ray source near the west limb shows a relatively small shift relative to the magnetic field pattern. Interpreting this as a projection effect it is necessary to assume a height of 15000–20000 km for the X-ray source also.

On 2 March 1972 further X-ray pictures taken with zone plate cameras were obtained owing to an invitation from Dr. Kreplin to participate in a rocket experiment of the NRL. The picture taken at 15.5 Å (Fig. 2) shows again different long-lived active regions as well as a flare at the east limb which reached its maximum brightness in Hα 12 minutes before the rocket was launched. This flare originated above a bipolar photospheric magnetic field. The X-ray source of this flare has an arch-shaped structure. Fig. 3 shows the superposition of two iso-density pictures. In the same figure the magnetograms of the Mt. Wilson maps are represented as well as the iso-lines of the radio radiation at 9.1 cm according to Stanford observations. The source near the centre of the solar disk lies above one polarity of the bipolar field; the centre of the radio emission coincides with the same polarity of the magnetic field above which the X-ray emitting region is situated. It is therefore to be assumed that the magnetic field strength in the corona attains its greatest value above this polarity in the western direction. This conjecture is confirmed by the calculation of the magnetic field in the corona.

Fig. 4a shows the arch shaped structure of the X-ray source in the flare region, taken in the Fe XVII line. Fig. 4b shows an X-ray picture of the flare derived

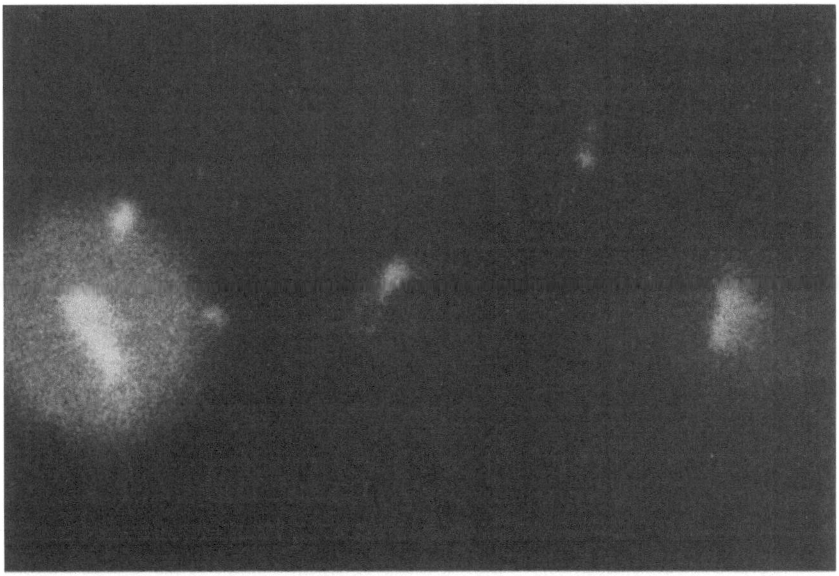

Fig. 2. X-ray picture of 2 March 1972 at a wavelength of 15.5 Å.

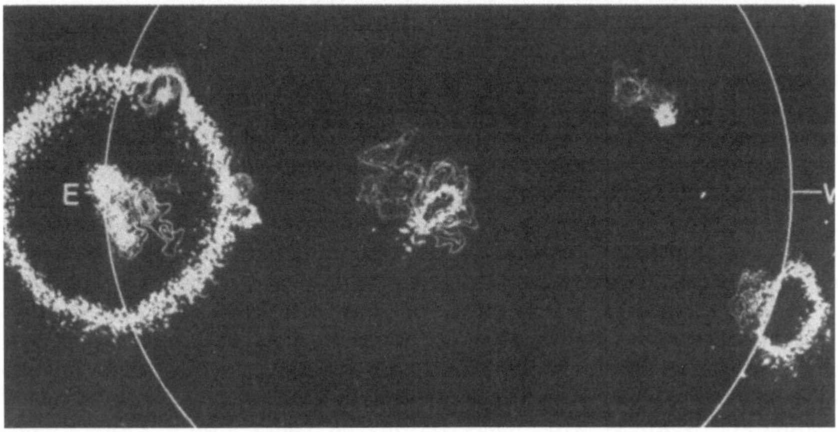

Fig. 3. Isodensity representation of the X-ray picture of Fig. 2 with the Mt. Wilson magnetogram of the corresponding active regions.

for the hard X-radiation below 6 Å on the right side. This last mentioned picture is deduced from a pinhole picture of the flare region; as a result of the large diameter of the pinhole initially a very blurred picture was obtained. The shape of the source determines the intensity decrease of the blurred picture at its limb. Therefore the shape of the source can be reconstructed by a correction of the picture [2]. In this way a relatively small and approximately circular but heterogeneous structured X-ray source results, which is localized in the upper part of the arch. It coincides also with the centre of the "halo" lies in Figs. 2 and 3 which

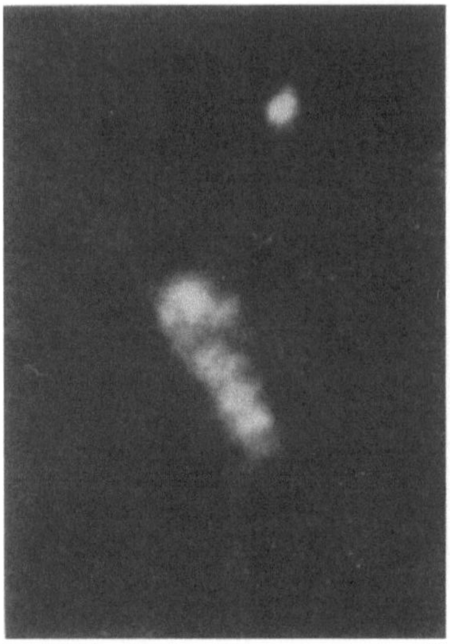

a

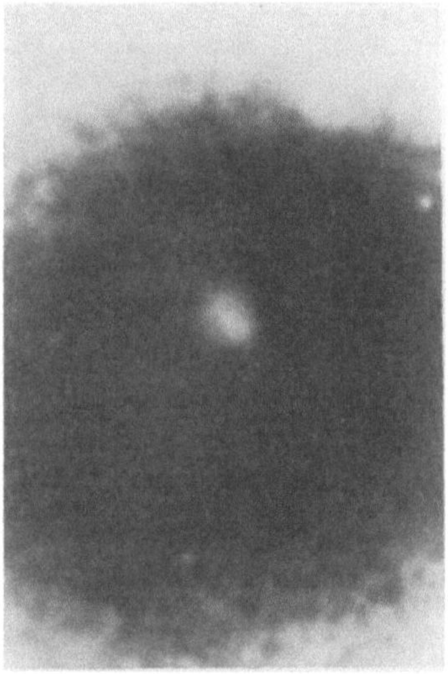

b

Fig. 4. a, Flare region taken at 15.5 Å, on 2 March 1972, 1826 UT (exposure time 200 sec). b, Deblurred pinhole picture of the same flare region (bright spot) at wavelength below 6 Å.

is mainly due to the hard flare X-rays for which the zone plate is completely transparent. As this picture of the small source is produced at shorter wavelengths than that of the arch shaped structure, the temperature of this small source is higher than that of the arch. This statement supports the idea that in the initial phase of a flare particle acceleration takes place in a small volume in the corona, that the fast particles are thermalized by collisions and that in a second stage of the event the hot plasma heats its neighbourhood by heat conduction along the magnetic field lines directed towards the photosphere. Another possibility seems, however, to be that the heating by the primary process occurs not only in the upper parts of the sun's atmosphere but also in the lower ones and that the cooling of the lower parts is produced by radiation loss. The appearance of segments of nearly equal separation in the arched structure indicates, however, that wave phenomena or constrictions of plasma are involved. Concerning wave phenomena the magnetic field guided Alfvén mode of magnetohydrodynamic waves especially comes to mind. The picture might show a residue of the corresponding density fluctuations. The occurrence of hydromagnetic waves with a velocity of approximately 10^3 km s^{-1} which emerge from the flash region was already inferred from disturbances observed in Hα [3—6].

Details will be published in *Solar Physics*.

References

[1] H. Bräuninger, H. J. Einighammer, J. V. Feitzinger, H. H. Fink, D. H. Höhn, H. Koops, G. Krämer, U. Mayer, G. Möllenstedt and M. Mozer, Solar Phys. **20**, 81 (1971).
[2] H. J. Einighammer and R. Hauke, Optik **37**, 557 (1973).
[3] G. E. Moreton, Astron. J. **65**, 494 (1960).
[4] G. E. Moreton, IAU Symp. No. 22, 371 (1965).
[5] F. Meyer, IAU Symp. No. 35, 485 (1967).
[6] Y. Uchida, Solar Phys. **4**, 30 (1968).

Space Research XIV — Akademie-Verlag, Berlin 1974

ANALYSIS OF THE SOLAR X-RAY SPECTRUM OF 20 AUGUST 1971

J. Jakimiec[a], V. V. Krutov[b], S. L. Mandel'stam[b], B. Sylwester[c], J. Sylwester[c] and I. A. Zhitnik[b]

[a]Wrocław University, Wrocław, Poland
[b]Lebedev Physical Institute, Moscow, USSR
[c]Solar-Terrestrial Relations Laboratory, Wrocław, Poland

The analysed spectrum covers the wavelength range 8—13 Å and was obtained during the decaying part of a solar flare. Absolute intensities of the spectral lines and continuum have been determined. A thermal model of the emitting region has been calculated and compared with various models of coronal condensations. It has been verified that an exponential function provides a good approximation for the distribution of the emission measure with temperature.

1. Observations

Solar X-ray radiation was investigated from the geophysical rocket Vertical 2 launched on 28 August 1971 at 0320 UT. The rocket reached an altitude of 463 km and the observations lasted for 542 sec. In the present investigation data received by the following instruments prepared by Lebedev Physical Institute have been used: two Bragg spectrometers with KAP crystals and GM counters recording the X-ray spectrum in the wavelength intervals 7—13 and 11—16 Å, and an X-ray photometer consisting of an aluminium filter and GM counter and recording the radiation in the wavelength interval 8—11 Å. During the flight the spectrometers performed four full scans of the spectral region investigated. A high sensitivity in the region was a valuable feature of the spectrometers; because of this the counting rates were high in the continuous spectrum ($N \gtrsim 50\ \mathrm{s}^{-1}$) and the spectrum is well defined.

There were several active regions on the solar disk on 20 August 1971, but one of them (McMath region No. 11482) predominated greatly over the others. The region was in the phase of its maximum development and produced several flares every day. A large coronal condensation, very intense in X-ray radiation, corresponded to the region (cf. London-Leicester X-ray solar maps in [1]). During the flight a flare shone in the active region 11482 and contributed significantly to the X-ray spectrum. The Hα flare was classified as a subflare SN, nevertheless it lasted for about 24 minutes. It started at 0314, reached its maximum intensity at 0318 and ended at about 0338 UT [1]. Our observations lasted from 0322 to 0331 UT, i.e. they coincided with the descending part of the flare. The flare in the X-ray radiation was an even more intense phenomenon. According to Solrad 9 measurements [1], during our observations the X-ray flux in the range 8—20 Å was enhanced about three times compared with the flux after the flare, and the 1—8 Å flux was enhanced about eight times. Data obtained by other instruments located on the rocket Vertical 2 (a collimated scanning photometer 8—12 Å and a set of pinhole cameras) as well as the London-Leicester X-ray map of the Sun showed consistently that the bulk of the recorded 8—12 Å radiation originated in the intense flare region 11482.

2. Data Analysis

The first aim of the analysis consisted in obtaining reliable absolute intensities of spectral lines and the continuum in order to control the amount of the matter (emission measure) taking part in the emission. The absolute value of the flux in a given line was calculated from the total number of photons recorded in the line (cf. [2, 3]):

$$F = \frac{h\nu \int N \, d\theta}{R_c \eta A}$$

where F is the energy flux (erg cm^{-2} s^{-1}) in the line, falling on the spectrometer, $h\nu$ the energy of a photon, N the counting rate (photons s^{-1}), R_c the integral coefficient of reflectivity of the crystal, η the efficiency of the detector, A the effective area and θ the angle of incidence on the crystal, and the integration is taken over the small interval of the angle containing the line.

It is known that there are some serious difficulties in absolute calibration of the Bragg spectrometer measurements, which cause significant differences between the line intensities obtained by various authors (cf. discussion in [4]). It is especially difficult to obtain reliable absolute values of the crystal reflectivity $R_c(\lambda)$. This is why we have decided to use the measurements of the 8—11 Å photometer for the calibration of the line and continuum intensities, confining our considerations to a spectral interval only a little broader than that controlled by the photometer. The relative distribution of $R_c(\lambda)$ was taken according to the Leicester group measurements [2] and the intensities in the spectrum were normalized according to the condition:

$$A_1 \sum_{\text{lines}} (h\nu)^{-1} F_i \eta_1 + A_1 \int_{8\,\text{Å}}^{11\,\text{Å}} (h\nu)^{-1} \frac{dF}{d\lambda} \eta_1 \, d\lambda = N(8-11)$$

where $N(8-11)$ is the counting rate of the 8—11 Å photometer, A_1, η_1 are the effective area and the efficiency of the photometer and $dF/d\lambda$ the flux in the continuum (erg cm^{-2} s^{-1} Å^{-1}). The X-ray line intensities obtained corresponding to the beginning of our observations are given in Table 1. The intensities are in good agreement with the measurements of Walker and Rugge [3] in the sense that they fall between their values for non-flare sun and for a large 3B flare. The continuum intensities deduced are shown as points in Fig. 1.

Since our observations concerned the soft X-ray range and were made during the decaying phase of the flare, we have good reasons to believe that the radiation observed was of quasi-thermal nature. The flux in a given line or in an unit interval of continuum can then be written in the form:

$$F = \int f(\lambda, T) \, \varphi(T) \, dT$$

Table 1

Intensity of the Spectral Lines

λ(Å)	Ion	Flux (10^{-5} erg cm^{-2} s^{-1})
8.42	Mg XII	4.6
9.17	Mg XI	17.0
9.32	Mg XI	13.0
10.24	Ne X	6.7
11.00	Na X	8.0
11.56	Ne IX	9.9
12.12	Fe XVII }	36.0
12.13	Ne X }	
12.26	Fe XVII	29.0
12.82	Fe XX*	7.2

* This identification of the line is not generally accepted, cf. [12].

where $f(\lambda, T)$ is the energy flux in the line or continuum, observed at 1 AU and produced by a plasma volume of temperature T and of unit emission measure; $\varphi(T)$ defines the thermal model of the emitting region, i.e. the distribution of the differential emission measure with the temperature. In the following computations

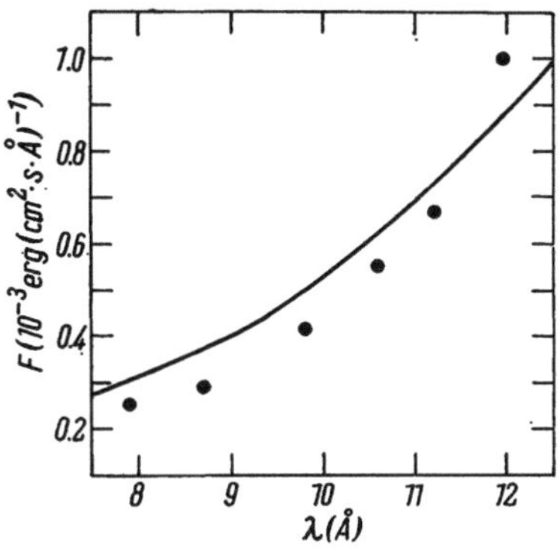

Fig. 1. Comparison between the observed (points) and calculated (line) continuum intensities.

the values of $f(\lambda, T)$ were taken according to the calculations of Mewe [5]. An exponential approximation of $\varphi(T)$ is often used in the literature for the high temperature part of coronal condensations and X-ray flares [6—9]:

$$\varphi(T) = C\, 10^{-\alpha T} \tag{1}$$

where C and α are constant parameters of the model.

In the previous investigations samples of the lines of various elements were taken for adjusting the parameters in the formula (1). This has the disadvantage that the errors of admitted element abundances enter into the analysis and in consequence the validity of the approximation (1) cannot be verified comprehensively. Having at our disposal homogeneous data for a sufficient number of lines, we were able to test approximation (1). The parameters of the distribution (1) have been calculated using a pair of lines of the same element; then the errors of abundances do not influence the obtained value of α. The calculations have been performed for various elements independently and the resulting distributions $\varphi(T)$ are shown in Fig. 2. In the figure each distribution $\varphi(T)$ has been

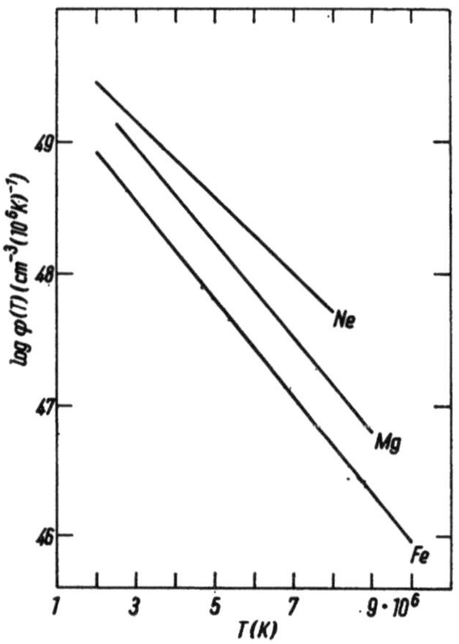

Fig. 2. The distributions of the differential emission measure $\varphi(T)$. Each distribution has been obtained from two lines of the same element: Mg, lines Mg XI 9.17 Å and Mg XII 8.42 Å; Ne, Ne IX 11.56 Å and Ne X 10.24 Å; Fe, Fe XVII 12.26 Å and Fe XX 12.82 Å.

drawn only within the temperature interval which contributes essentially to the formation of the corresponding pair of lines. We see a good agreement between the inclinations α obtained from the lines of various elemants; the differences of the inclination fall within the error limits defined by the accuracy of the line intensity measurements. This indicates that the formula (1) provides a good approximation of the emission measure distribution in the radiating region for the rather wide temperature range $(2-10) \times 10^6$ K.

The relative vertical displacements of the individual distributions $\varphi(T)$ are caused by the errors of the admitted element abundances [5], as well as by systematic errors in the calculation of the spectral line intensities for the individual elements (calculation of the ionization equilibrium and excitation rates). If we assume that the displacements are entirely due to the abundance errors, we obtain the following abundances (relative to hydrogen):

$$A(\text{Mg}) = 3.0 \times 10^{-5}; \quad A(\text{Ne}) = 8.8 \times 10^{-5}; \quad A(\text{Fe}) = 2.3 \times 10^{-5};$$

where A(Mg) has been retained as in [5] and the other abundances have been corrected in order to fit the distributions $\varphi(T)$ in Fig. 2.

A comparison of our thermal model of the X-ray flare region with the models of coronal condensations of other authors, is given in Fig. 3. It must be remembered that our observations, and therefore also our model, concern the radiation of the large coronal condensation greatly enhanced by the flare. In particular, it is seen that the Landini and Monsignori Fossi (L) model [10] may be appropriate for a very large condensation in the region of high temperatures, but it evidently assigns too large an amount of the matter to temperatures below 3×10^6 K.

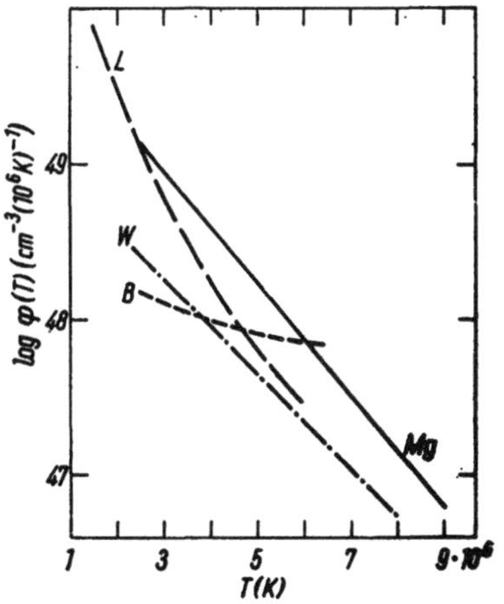

Fig. 3. Comparison of the obtained model of the emitting region (line "Mg") with various models of coronal condensations: L, model of Landini and Monsignori Fossi [10]; W, model No. 1 of Walker [9]; B, model of Batstone et al. [11] (for their condensation No. 1).

Hence also the total emission measure assigned to the condensation by the L model (about 6×10^{49} cm^{-3}) is considerably overestimated (at least by a factor of five).

Fig. 1 shows a good agreement between the observed and calculated continuum intensities. This provides a valuable test of the thermal model of the emitting region as well as of the reliability of the continuum measurements in our spectrum. Fig. 4 shows some examples of the time variations of the spectral lines and continuum observed in the course of our observations. We can see appreciable similarity in the decay of the various lines and continuum. We could not establish any reliable time variation of the parameter α of the distribution $\varphi(T)$ from our observations—the variations fall within the limits of observational errors.

It is worthwhile to note that the total continuum intensity in the range 8—12 Å prevails appreciably over the total intensity of the lines in the same range:

$$\text{lines/continuum} = 0.32,$$

which is in full agreement with the above calculations based on the tables [5].

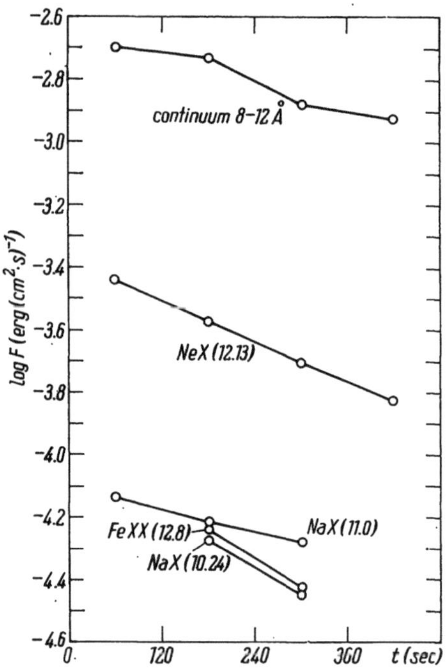

Fig. 4. Examples of the time variation of the spectral lines and continuum.

References

[1] Solar Geophysical Data 1971, Space Environment Services Center, Boulder, Col.
[2] K. EVANS and K. A. POUNDS, Astrophys. J. **152**, 319 (1968).
[3] A. B. C. WALKER, JR, and H. R. RUGGE, in: Solar Flares and Space Research, North-Holland Publ. Co., Amsterdam 1969 (p. 102).
[4] H. V. ARGO et al., Astrophys. J. **160**, 283 (1970).
[5] R. MEWE, Solar Phys. **22**, 459 (1972).
[6] G. CHAMBE, Astron. Astrophys. **12**, 210 (1971).
[7] L. W. ACTON et al., Solar Phys. **26**, 183 (1972).
[8] L. W. ACTON et al., Space Research XIII, 781 (1973).
[9] A. B. C. WALKER, JR, Space Sci. Rev. **13**, 672 (1972).
[10] M. LANDINI and B. C. MONSIGNORI FOSSI, Solar Phys. **17**, 379 (1971).
[11] R. M. BATSTONE et al., Solar Phys. **13**, 389 (1970).
[12] J. P. CONNERADE, Astrophys. J. Lett. **162**, L139 (1970).

Space Research XIV — Akademie-Verlag, Berlin 1974

ANALYSIS OF SOLAR FLARE X-RAY RADIATION WITH BRAGG SPECTROMETERS

R. W. Kreplin and J. F. Meekins

E. O. Hulburt Center for Space Research, Naval Research Laboratory, Washington, D.C., USA

Solar X-ray spectra in the wavelength range 0.5—25 Å have been obtained by several experimenters. Lines of highly ionized elements have been identified and their flux ratios used to calculate crude plasma temperatures. Further advance in the use of Bragg spectrometers in solar flare research, in particular the measurement of flux in lines and continuum and the calculation of emission measures, will depend on absolute calibration. This paper presents the results of measurement of detector efficiency, filter transmission and integrated reflectivity of LiF for the Bragg spectrometers flown aboard OSO 6. The accuracy of the measurements is demonstrated by agreement in the areas of overlap between the reduced spectra from different spectrometers. The continuum radiation from the intense X-ray flare of 12 August 1970, 2015 UT has been analyzed to demonstrate the utility of these calibrations.

1. Introduction

Solar X-ray spectra in the wavelength region 0.5—25 Å have been obtained by a number of experimenters using Bragg crystal spectrometers [1, 2]. In general, the instruments have been adequately calibrated to provide wavelength measurements to about 0.001 Å. Both non-flare and flare X-ray lines have been tabulated and identified, with the result that the presence of H and He like spectra of the ions of Ca, S, Si, Mg, Ni, Cu, K, A, Ne, N, O, and Fe have been clearly identified. Calculations of emission measure, determination of coronal abundance and comparison of the intensities of lines not closely spaced, however, require the conversion of the raw counting rate obtained from the telemetry record into absolute flux units. In the usual arrangement where the Bragg crystal spectrometer is used to obtain solar spectra, three different quantities enter the determination of the overall instrument efficiency. These are the detector quantum efficiency, the transmission of any protective filters and the crystal reflectivity in the first and higher orders.

Of the various crystals used in the 0.5—25 Å region, only KAP (Potassium Acid Phthalate) has received much attention. By examining the work of several authors who have used KAP in Bragg spectrometers, it becomes evident that there are significant problems in the calibration of these instruments for the determination of absolute line intensities.

The Los Alamos group [3] has published data on their measurement of the intensity of the O VIII Lα line at 18.97 Å. They obtained values of from 5×10^{-3}

to 3×10^{-2} erg cm^{-2} s^{-1} from 1964 to 1968 in non-flare conditions using a mica crystal for which they claim an accuracy of $\pm 10\%$ in measurement of integrated reflectivity. It will be instructive to compare their results with those of other authors. Walker and Rugge [4] obtained flux values for the O VIII Lα line of 1.7×10^{-3} erg cm^{-2} s^{-1} during a quiet period in December 1966 and 4.2×10^{-3} erg cm^{-2} s^{-1} during a 3B flare on 13 February 1967. Note that even their flare value is smaller than the measurement of Argo et al. [3] during a quiet period. They admit that the efficiency of their CsI photomultiplier might have changed with time but experience has shown that CsI is quite stable and it is unlikely that changes by a factor of about ten would have occurred. Evans and Pounds [5] found the intensity of the O VIII Lα line to vary from 5.4×10^{-4} to 1.3×10^{-3} erg cm^{-2} s^{-1} in May 1966. These results are about a factor of two less than those of Walker and Rugge. They are as low as our earliest calculations of O VIII Lα intensity, 7×10^{-4} erg cm^{-2} s^{-1} (unpublished), for the NRL flight of 4 October 1966. Recently, Blake and House [6] recalculated the O VIII Lα flux from the same flight [7] and obtained a higher value, 2.1×10^{-3} erg cm^{-2} s^{-1}. They obtained detector efficiencies from NRL but determined their own reflection coefficients. Biegman and Vainshtein [8] have published flux values for our measurements without information on the instrument efficiencies which they used to derive the results. Their value, 1.5×10^{-2} erg cm^{-2} s^{-1}, is about ten times greater than that of Blake and House [6]. Several authors [1, 8, 9] have used published values of line fluxes to make coronal abundance determinations, emission measure calculations, and to compare intensities of lines in different parts of the X-ray spectrum. The agreement of the data in their sources is not particularly good, probably due to differences in the values of integrated reflectivity used to derive flux values. In only a few cases have measurements of $R(n, \lambda)$, the integrated reflectivity at order n and wavelength λ, been published, but we can conclude that differences must exist at least as large as a factor of ten which would indicate that conclusions based on comparisons of intensities published by different authors should be regarded with some caution.

We shall outline our efforts to calibrate the LiF, EDDT and KAP spectrometers employed in the NRL experiment aboard OSO 6.

2. Measurement of Spectrometer Efficiency

Since it is not possible at present to measure the overall efficiency of the spectrometer, we must measure individually the crystal reflectivity, the detector efficiency and the filter transmission.

2.1. Crystal Reflectivity

Meekins [10] has developed a comprehensive technique for the measurement of absolute reflectivity with a double crystal spectrometer. We shall present only a brief discussion here, but emphasize those features of the technique which we feel contribute most significantly to its accuracy.

The response of the detector in a Bragg spectrometer is a function of the source intensity and polarization, the reflectivity and rocking curve of the crystal, the geometry of the system, and the quantum efficiency of the detector itself. In the laboratory the radiation can be confined to a very narrow beam; crystal and detector window areas are sufficiently large that no purely geometrical corrections are necessary over ordinary values of θ. The Henke source produces both characteristic line radiation and polarized continuum. Polarization of the continuum radiation has a significant effect in the determination of absolute crystal reflectivities using a double crystal spectrometer, but it is of no concern in the measurement of line radiation with the single crystal instrument; since the continuum is only slowly varying with angle, its contribution to the spectrum can be easily removed. Since Bragg crystal spectrometers have appreciable reflectivity in orders higher than the first, we make a measurement of relative integrated reflectivity using line radiation by comparing the integrated line intensity in higher orders with that in first order. The measurement is made using the standard Bragg spectrometer arrangement and the Henke source. Since the line radiation from the Henke source is unpolarized, no correction is required for this effect. Once the absolute reflectivity of the crystal is known in first order, the absolute reflectivities (in response to random source polarization) can be calculated for the higher orders.

The measurement of absolute integrated reflectivity in the first order is made using the double crystal spectrometer in the (1, −1) position. A line is selected by setting the first crystal according to the Bragg condition and sweeping the second crystal and detector over the diffracted line. The integrated intensity $E_{\mathrm{L+C}}(1, -1)$ thus obtained contains contributions from both line and continuum. By swinging the second crystal out of the beam, the peak intensity due to line and continuum $I_{\mathrm{L+C}}$ is measured. In the past, it was assumed that the polarization of $I_{\mathrm{L+C}}$ due to a partially polarized continuum contribution was small and could be neglected. Our measurements have shown, on the contrary, that it is important. To observe the continuum, we rotate the first crystal slightly so that the line is no longer observed and only continuum radiation is seen by the second crystal. We make the same series of measurements as before obtaining $E_{\mathrm{C}}(1, -1)$ for the integrated continuum intensity and I_{C} for the peak intensity of the beam incident on the second crystal. To evaluate the polarization effects due to diffraction from the first crystal, the second crystal is rotated 90° about the radiation beam incident on it. We measure in this position the integrated intensity $E_{\mathrm{L+C}}(1, \mathrm{i})$ and the integrated continuum intensity $E_{\mathrm{C}}(1, \mathrm{i})$. In each of the measurements, the value of ω, the angular velocity of the scan, is determined. The continuum contributions in $E_{\mathrm{L+C}}$ and $I_{\mathrm{L+C}}$ are then removed by subtraction taking into account slight variations in ω between different measurements.

Meekins [10] has developed the following equations relating the theoretical expressions for integrated reflectivity to the measured quantities. The integrated reflectivity in response to random source polarization is given by

$$R(n, \lambda) = \{A(n, \lambda)\}^2 \, \frac{1 + f_n^2}{2} \tag{1}$$

where $A(n, \lambda)$ is the diffraction amplitude containing the crystal structure factors and f_n contains the polarization factors. n is the diffraction order. In first order, f is related to the diffraction angle θ by

$$f_1^2 = (\cos 2\theta)^q \tag{2}$$

where $q = 1$ for a perfect crystal and $q = 2$ for an ideally imperfect crystal. f_1^2 is related to our measured quantities by

$$H(\lambda) = \frac{E_L(1, i)}{E_L(1, -1)} = \frac{2f_1^2}{1 + f_1^4}. \tag{3}$$

Experimentally $H(\lambda)$ is obtained for several wavelengths and the resulting values of f_1^2 are used to calculate q in Eq. (2) by a least squares fit. Our results for LiF give values of q of about 1.75 indicating that it is nearly a mosaic or ideally imperfect crystal. Using the fitted values of f_1^2, the diffraction amplitude can be calculated from the experimental relation

$$G(\lambda) = \frac{E_L(1, -1)}{I_L} = \frac{\{A(1, \lambda)\}^2 (1 + f_1^4)}{(1 + f_1^2)}. \tag{4}$$

Now having f_1^2 and $\{A(1, \lambda)\}^2$ we can calculate the integrated reflectivity in first order $R(1, \lambda)$.

Initially, we measured the relative reflectivities of the crystal for different orders:

$$F(n, \lambda) = \frac{\{A(n, \lambda)\}^2 (1 + f_n^2)}{\{A(1, \lambda)\}^2 (1 + f_1^2)}. \tag{5}$$

From this, the absolute value of the integrated reflectivity, (1), in all orders higher than the first can also be calculated. The results for the LiF crystals used

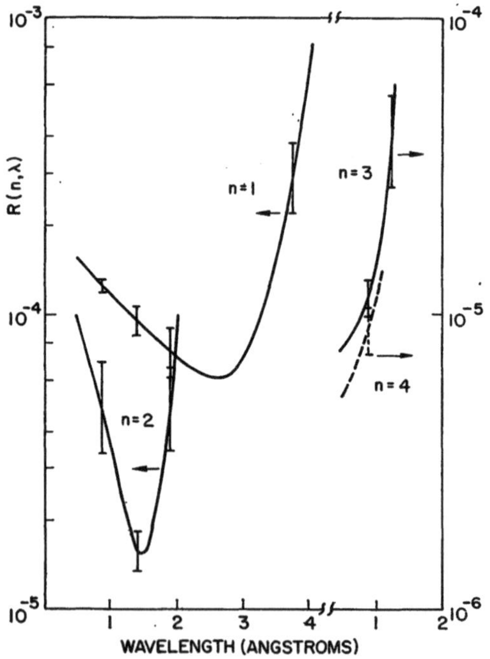

Fig. 1. Experimental measurements of $R(n, \lambda)$ for LiF crystals used in the OSO 6 Bragg spectrometer.

in the OSO 6 instrument are presented in Fig. 1. Corrections for polarization of the continuum contributions have not been made as yet for the EDDT and KAP crystals aboard OSO 6.

2.2. Detector Quantum Efficiency

The detector quantum efficiencies have been calculated from the known mass absorption coefficients of the elements comprising the window and the gas filling using the McMaster et al. [11] mass absorption coefficients and the composition of Indian ruby mica given by Skow [12]. These are shown in Fig. 2 for window

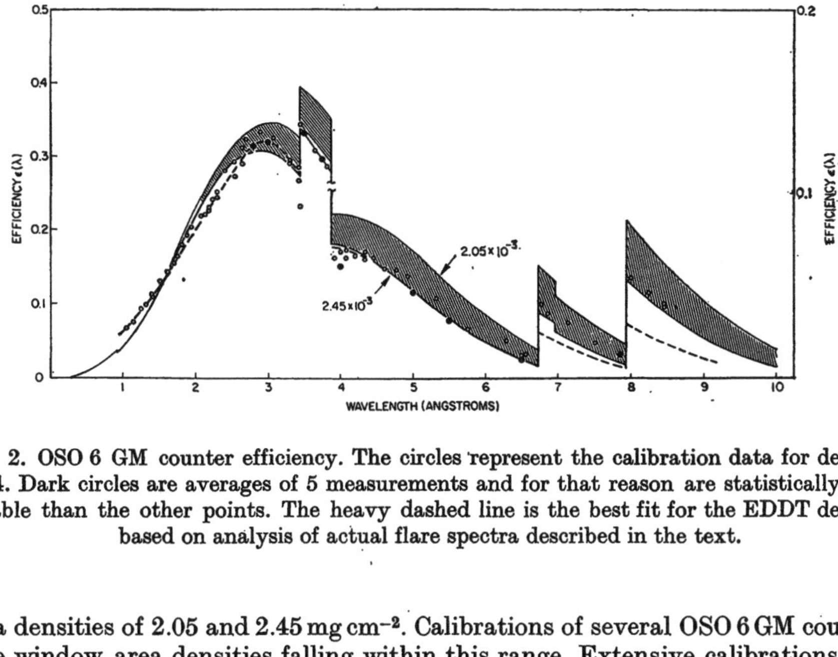

Fig. 2. OSO 6 GM counter efficiency. The circles represent the calibration data for detector 8634. Dark circles are averages of 5 measurements and for that reason are statistically more reliable than the other points. The heavy dashed line is the best fit for the EDDT detector based on analysis of actual flare spectra described in the text.

area densities of 2.05 and 2.45 mg cm^{-2}. Calibrations of several OSO 6 GM counters give window area densities falling within this range. Extensive calibrations were performed on two OSO 6 type GM counters using a double window proportional counter to calibrate the incident X-ray beam. The efficiency is given by

$$\varepsilon(\lambda) = \frac{R_{12}}{R_{41}} \left(\frac{R_{22}}{R_{12}}\right)^{1/2} \left[1 - \frac{R_{32}}{R_{22}}\right] \tag{6}$$

where the R's are the counting rates measured under the conditions described below. The final subscript refers to the counter, i.e. 2 refers to counter 2, the test counter, and 1 refers to the double window calibration counter:

R_{12}, detector 1 removed from beam
R_{22}, detector 1 in beam — no fill gas
R_{32}, detector 1 in beam with fill gas
R_{41}, counting rate measured by double window counter.

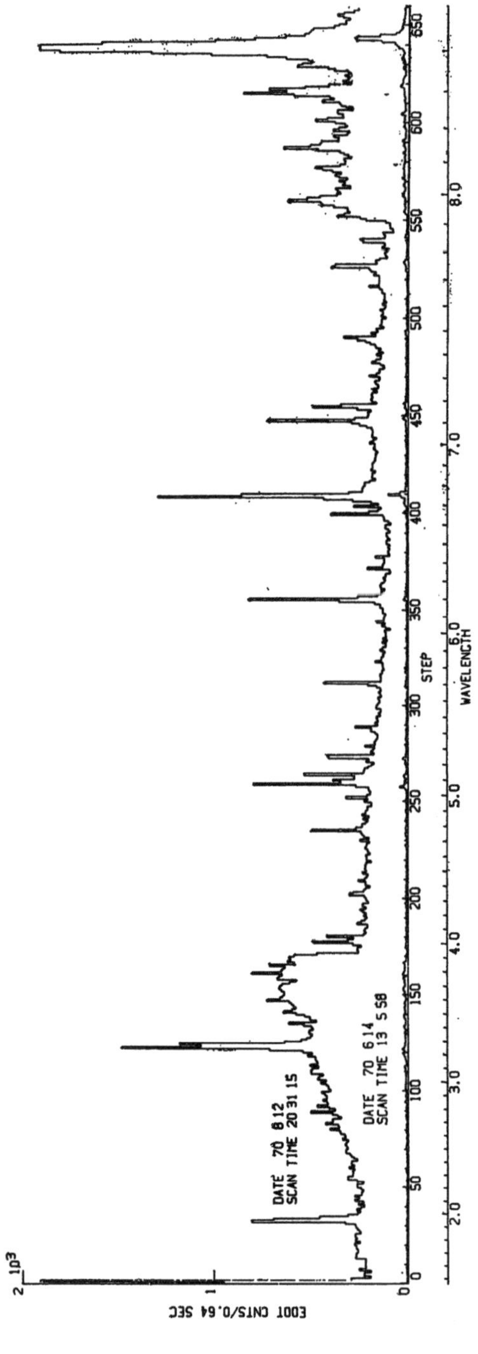

Fig. 3. Uncorrected EDDT spectra from two periods. 12 August 1970 is a flare spectrum and the 14 June 1970 record is from an active period, but at a time when no flares were in progress.

The window areas of counter 1 must be large enough not to restrict the X-ray beam in any way and it is assumed that there is no significant recombination in the double window proportional counter.

The results of the measurements of detector 8634 are shown in Fig. 2 (circles). We observe some significant differences between the calibration of 8634 and the calculated efficiency curves.

Since efficiency calibrations were not performed before the OSO 6 detectors were installed for flight, it was necessary to derive a calibration using the solar continuum flux. Continuum from the solar flare of 12 August 1970 (shown in Fig. 3) was chosen for this purpose. We assumed that the solar continuum spectrum contained no significant discontinuities and that the crystal reflectivity was a continuous function of wavelength over the region of interest. A trial product $R(n, \lambda) \cdot \varepsilon(\lambda) \cdot T(\lambda)$ was formed and divided into the geometrically corrected continuum flux n_t/F; $F = d\lambda/d\theta \cdot$ (instrument cross-sectional area) and $T(\lambda)$ is the filter transmission described in the next section. The shape of $\varepsilon(\lambda)$ was then adjusted to produce an EDDT or LiF spectrum from which the K, A, Si and Al edges were removed. Calculation of an area density, which would accurately describe the fitted experimental curve, required that the assumed composition of Indian ruby mica given by Skow be altered.

2.3. Filter Transmission

Filters consisting of thin plastic films of mylar and paralyne coated with aluminum were used to eliminate response to longer wavelength XUV radiation. Simple transmission measurements at a few selected wavelengths were made and

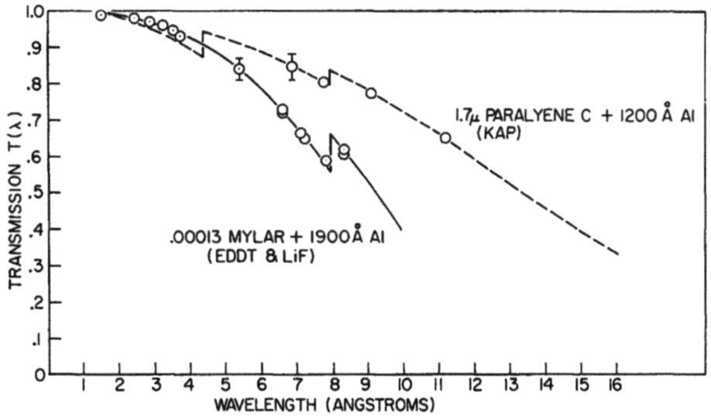

Fig. 4. Transmission of the OSO 6 filters.

fitted with theoretical curves based on the mass absorption coefficients. Fig. 4 shows the transmission curves for the OSO 6 filters.

The effectiveness of these filters plus the mica window of the GM counter in eliminating longer wavelength XUV is demonstrated by the presence of the X-ray absorption edges in the 12 August 1970 flare spectrum (Fig. 4) and by the absence of any sizable signal during the active but non-flare record of 14 June 1970.

3. The Solar Flare of August 1970

Although the corrections for source polarization have not yet been made for the EDDT and KAP spectrometers, it is useful to examine the region of overlap to see if there is agreement in the corrected values of X-ray flux (actually the effects of source polarization should be smallest at the long and short wavelength ends of the scan where comparison is made, and largest in the mid range). In Fig. 5

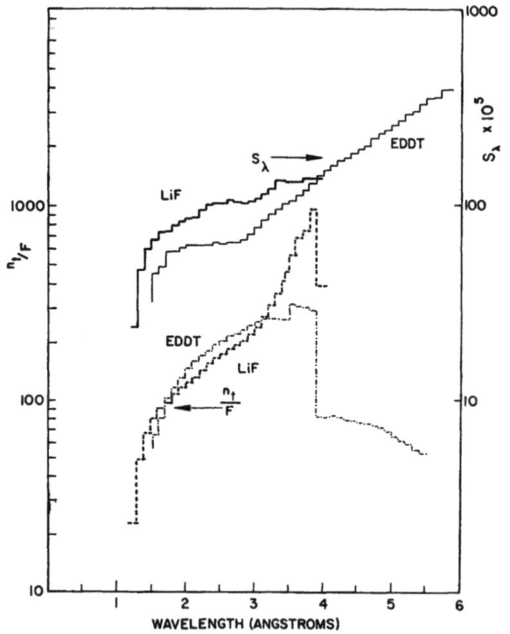

Fig. 5. Reduced continuum spectra for the flare of 12 August 1970 showing the region of overlap of the EDDT and LiF spectrometer data.

we show the reduced data for the flare of 12 August 1970 taken from both the EDDT and LiF spectrometers. n_t/F is the raw counting rate corrected for area factor and S_λ is the flux in photons $cm^{-2}\,s^{-1}$ obtained with the appropriate values of $R(n, \lambda) \cdot \varepsilon(\lambda) \cdot T(\lambda)$. The flux at 2 Å agrees to within about 40%; this is about what could be expected since the spectrum contains not only the wavelength distribution of solar X-ray flux but also its time variation. This particular flare flux rose very rapidly; the flare was of relatively short duration. In Fig. 6 each of the available spectral scans for this flare are plotted. The first scan beginning at 2010:05 UT, decreasing in wavelength, shows the solar flux increasing with time. The second scan begins at 2017:10. At the end of this scan, the X-ray flux at ~4 Å has increased by a factor of 200. The flare maximum occurs just after the beginning of the third scan at 2024:12 and the last two scans show decreasing flux levels. The solar flux is changing so rapidly in this flare that very little can be deduced from the shape of the continuum spectra.

Doschek [13] has used our preliminary calibrations to determine the intensity of the Lα Si XIV line at 6.182 Å from both the EDDT and KAP spectrometers. During the 12 August 1970 flare, he found an intensity of 9.8×10^6 photons cm^{-2} s^{-1} from the KAP spectrum and 7.8×10^6 photon cm^{-2} s^{-1} from the EDDT spectrum. The agreement is within about 26% and must be considered as quite satisfactory considering that the corrections for polarization have not yet been made in the KAP and EDDT integrated reflectivities.

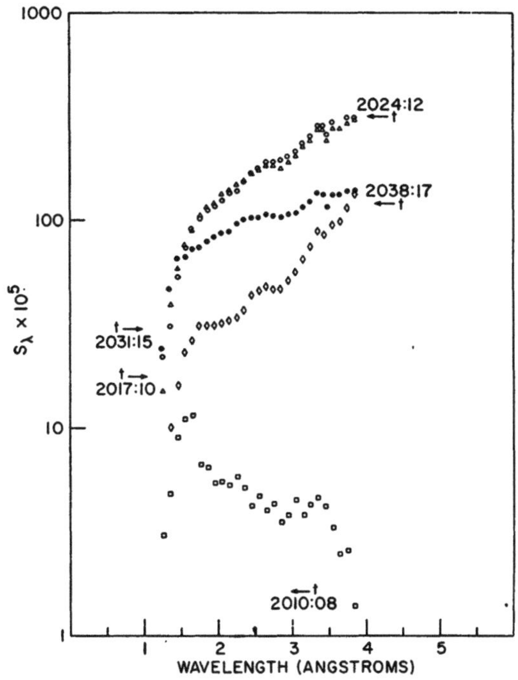

Fig. 6. Reduced continuum spectra from all scans made during the 12 August 1970 flare.

References

[1] A. B. C. Walker, Jr, Space Sci. Rev. **13**, 672 (1972).
[2] G. A. Doschek, Space Sci. Rev. **13**, 765 (1972).
[3] H. V. Argo et al., Astrophys. J. **160**, 273 (1970).
[4] A. B. C. Walker, Jr, and H. R. Rugge, in: Solar Flares and Space Research, North-Holland Publ. Co., Amsterdam 1969 (p. 102).
[5] K. Evans and K. A. Pounds, Astrophys. J. **152**, 319 (1968).
[6] R. L. Blake and L. L. House, Astrophys. J. **166**, 423 (1971).
[7] G. G. Fritz et al., Astrophys. J. **148**, L133 (1967).
[8] I. L. Biegman and L. A. Vainshtein, Soviet Astronomy A. J. **14**, 825 (1971).
[9] G. Chambe, Astron. and Astrophys. **12**, 210 (1971).
[10] J. F. Meekins, Thesis, Catholic University of America, Washington, D.C. (1973).
[11] W. H. McMaster, N. Kerr Del Grande, J. H. Mallet and J. H. Hubbell, University of California, Lawrence Radiation Lab. Rep. UCRL 50174, Sec. II, Rev. I (1969).
[12] M. L. Skow, U. S. Bureau of Mines Information Circular 8125 (1962).
[13] G. A. Doschek, private communication.

Space Research XIV — Akademie-Verlag, Berlin 1974

ELECTRON TEMPERATURE AND EMISSION MEASURE VARIATIONS IN THE SOLAR CORONA

D. M. Horan, K. P. Dere and R. W. Kreplin

E. O. Hulburt Center for Space Research, Naval Research Laboratory, Washington, D.C., USA

Ionization chambers carried by the Solrad 9 and Solrad 10 satellites provide measurements from which the differential emission measure of the multi-thermal coronal plasma can be deduced. In this paper, line emission has been included in the assumed solar emission spectrum permitting the extension of the analysis into the 8—16 Å wavelength region. Earlier work used only the 0.5—3 Å and 1—8 Å bands and limited the assumed solar spectrum to continuum emission. Earlier work was limited to an isothermal coronal plasma.

Analysis of the X-ray data obtained prior to and during the initial phases of the large flare which began about 0300 UT on 2 August 1972 shows that the flare plasma increased in temperature prior to any emission measure increase. During the decaying portion of this large flare, the flare plasma's temperature returned to its pre-flare level more rapidly than its emission measure did. Significant changes in the plasma's lower temperature component, which may be of use in solar activity predictions, were observed as McMath region 11976 rotated on to the visible disk.

For several years, solar X-ray measurements made by ionization chambers sensitive in the 0.5—3 Å and 1—8 Å bands have been used to determine an effective electron temperature and emission measure for the coronal source region. Initially, the technique considered only continuum X-ray emission from thermal bremsstrahlung and radiative recombination [1]. Recently, line emission contributions have been incorporated into the X-ray spectrum [2]. Now the technique has been improved to permit the consideration of a multi-thermal plasma and use data from ionization chambers sensitive in the 0.5—3 Å, 1—5 Å, 1—8 Å, 1—16 Å, and 8—16 Å bands.

The assumed solar emission spectrum is obtained from the equations describing emission due to thermal bremsstrahlung and radiative recombination given by Culhane [3] and the line emission data presented by Tucker and Koren [4]. The spectrum is calculated at 1×10^6 K intervals between 2 and 20×10^6 K inclusive. The multi-thermal plasmas are assumed to be composed of discrete emission measure contributions at 1×10^6 K intervals between 2 and 20×10^6 K. The calculated responses of the set of ionization chambers to a trial multi-thermal plasma are compared with the measured responses of the chambers to solar emission and the trial plasma is then varied using 30000 iterations to make the calculated and measured responses converge.

Data obtained during the solar activity of August 1972 from the ionization chambers carried by the Naval Research Laboratory's Solrad 9 and Solrad 10 satellites were analyzed using this technique. Data obtained prior to and during the initial phases of the large X-ray flare which began at approximately 0300 UT on 2 August were used to produce Figs. 1, 2, and 3. The symbols on the plots represent the emission measure B, which is the square of the electron density integrated over the emitting volume, and the electron temperature T in units of 10^6 K. In all three figures the emitting plasma is shown to be divided into a low temperature component which is associated with the normal X-ray back-

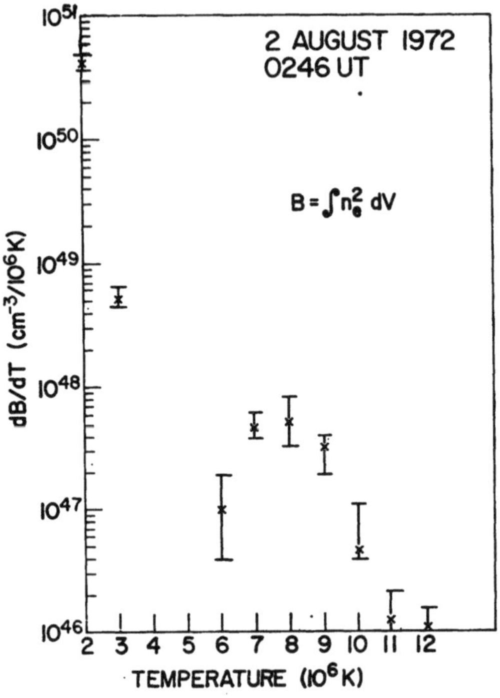

Fig. 1. Plot of the derivative of the emission measure with respect to temperature as a function of temperature based on solar X-ray data of 0246 UT, 2 August 1972.

ground emission at longer wavelengths, and a higher temperature component which is associated with the shorter wavelength X-ray emission observed during solar activity. The higher temperature component was present whenever a measurable response was produced in the 0.5—3 Å ionization chamber.

Fig. 1 is based on X-ray data obtained at 0246 UT. It is typical of results obtained using data acquired over the previous twelve hours in that the higher temperature plasma component had a reasonably constant total emission measure and profile on the dB/dT versus T plot. During these previous twelve hours, there was constant minor flare activity, but nothing compared with the activity which was to follow. Coinciding with the beginning phases of the X-ray flare, the higher temperature plasma component changed between 0255 and 0307 UT by shifting to higher temperatures without otherwise altering its dB/dT profile or total emission measure (Fig. 2). During this twelve minute time period the measured X-ray flux levels increased by a factor of three to four in the 0.5—3 Å and 1—5 Å bands, and by less than a factor of 1.5 in the three lower energy bands. Between 0307 and 0310 UT the higher temperature plasma component accelerated its temperature increase and also changed the shape of its dB/dT profile, but did not change in total emission measure (Fig. 3). During these three minutes, the measured flux for the 0.5—3 Å, 1—5 Å, and 1—8 Å bands increased by a factor between 1.5 and 2, and the flux in the two lower energy bands changed by a smaller factor. The X-ray data show that the flare has clearly begun its

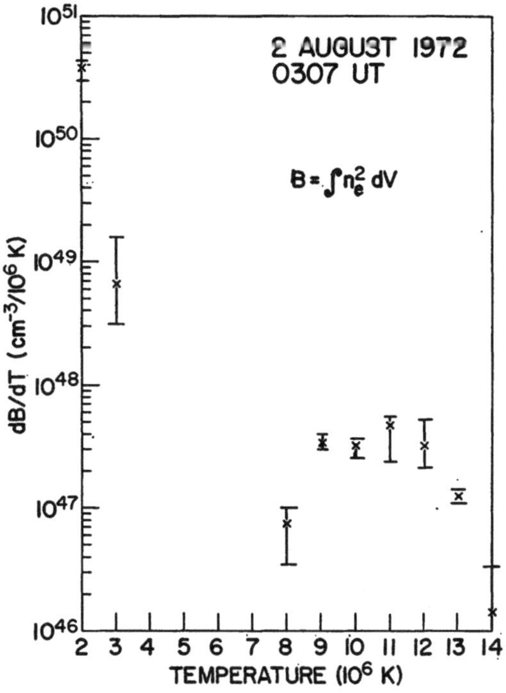

Fig. 2. Plot of the derivative of the emission measure with respect to temperature as a function of temperature based on solar X-ray data of 0307 UT, 2 August 1972.

explosive growth phase by 0310 UT. Application of the multi-thermal plasma analysis to data acquired after 0310 UT indicated that the higher temperature plasma component continued to increase in temperature and began to increase in total emission measure as the X-ray emission continued to increase. However, these results are unreliable because the present form of the analysis technique does not consider temperatures above 20×10^6 K, and this temperature limit was certainly exceeded if the trend exhibited between 0246 and 0310 UT continued. Data acquired at 0731 and 1401 UT, well into the flare's decay phase, show the high temperature plasma component with approximately the same dB/dT profile as at 0246 UT, but larger in total emission measure by an order of magnitude at 0731 and a factor of two at 1401.

The results of application of a multi-thermal plasma technique to X-ray fluxes measured during this large flare show the presence of a higher temperature plasma component associated with the flare. During the initial stages of the flare, this flare plasma increased in temperature before any change in its total emission measure was observed. During the decaying phases of the flare, the temperature of the flare plasma returned to its pre-flare level more rapidly than its total emission measure did. This behavior is compatible with the results of an earlier analysis of large flares which assumed an isothermal plasma and only continuum emission between 0.5 and 8 Å [5]. The multi-thermal plasma technique is presently

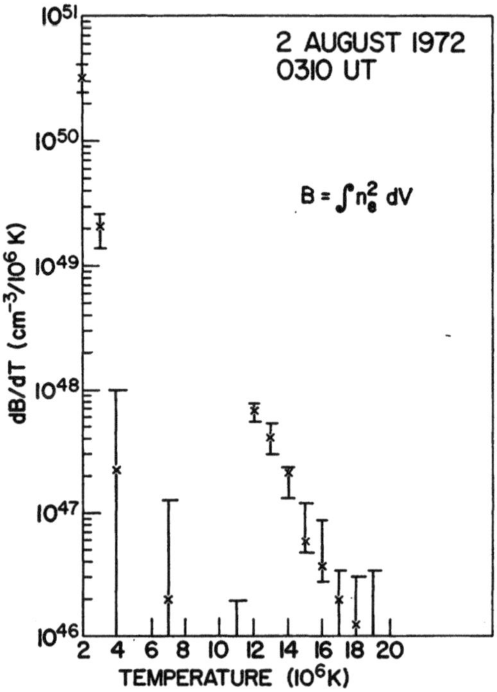

Fig. 3. Plot of the derivative of the emission measure with respect to temperature as a function of temperature based on solar X-ray data of 0310 UT, 2 August 1972.

being modified to consider logarithmically spaced temperatures up to 50×10^6 K so that the analysis can be applied throughout the entire course of flares.

The application of the multi-thermal plasma technique to data obtained several days prior to the onset of the major flare activity also produced interesting results. Fig. 4 is based on X-ray data from the 1—8 Å, 8—16 Å, and 1—16 Å ionization chambers at 1100 and 1720 UT on 26 July 1972. The higher temperature plasma component was not observed and did not begin to appear consistently until early on 31 July, coincident with the commencement of measurable responses in the 0.5—3 Å ionization chamber. Only the low temperature plasma component was present and it showed a significant change in structure during the six-hour interval. At 1100 UT, the observed X-ray emission was coming from a plasma which had no components at 3×10^6 K or above. However, the X-ray emission at 1720 was produced by a plasma which had significant components at 3 and 4×10^6 K. The dB/dT versus T profile obtained using the 1100 UT data was characteristic of such profiles obtained using data collected on 25 July and prior to 1100 UT on 26 July. The dB/dT profile obtained using the 1720 UT data was characteristic of those obtained from data collected until early on 31 July when the higher temperature component began to appear. Comparison of the solar X-ray maps obtained at approximately 1500 UT on 26 and 27 July by the University College London, Leicester University instrument aboard OSO 5 shows that during the 24-hour interval a significant X-ray source appeared on

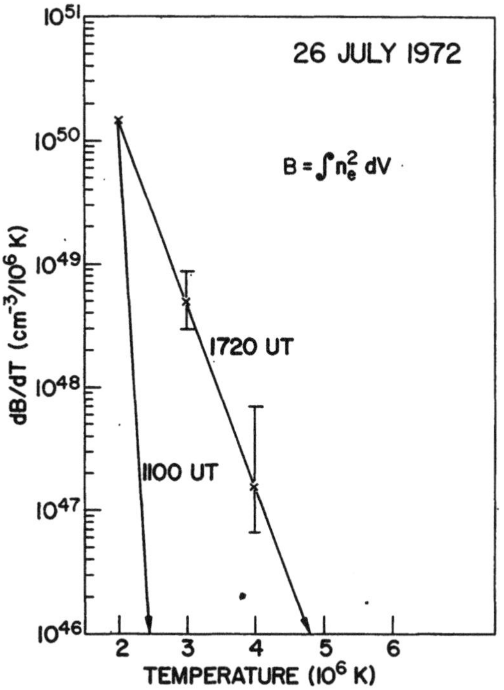

Fig. 4. Plots of the derivative of the emission measure with respect to temperature as a function of temperature based on solar X-ray data of 1100 and 1720 UT, 26 July 1972.

the east limb [6]. Between 1630 UT on 27 July and 1200 UT on 28 July, McMath region 11976 rotated on to the visible disk at the same latitude as the previously observed X-ray source. Therefore, the change in dB/dT profile displayed in Fig. 4 probably occurred as the X-ray source in the corona over region 11976 rotated into view. If this is true, it must be noted that region 11976, the source of the August activity, had a coronal plasma markedly different from those of the eight regions on the disk at 1100 UT on 26 July. It must be considered possible at this time that the presence of coronal plasma at 3 and 4×10^6 K may be related to a region's capability for flare activity.

References

[1] D. M. Horan, Dissertation submitted to the Catholic University of America, Publ. No. 70-22133, University Microfilms, Inc. (1970).
[2] K. P. Dere, D. M. Horan and R. W. Kreplin, Naval Research Lab. Rep. No. 7589 (1973).
[3] J. L. Culhane, Mon. Not. Roy. Astron. Soc. **144**, 375 (1969).
[4] W. H. Tucker and M. Koren, Astrophys. J. **168**, 283 (1971).
[5] D. M. Horan, Solar Phys. **21**, 188 (1971).
[6] Solar-Geophysical Data, 337 Part I, 90—94, September 1972, U.S. Department of Commerce, Boulder, Col.

Space Research XIV — Akademie-Verlag, Berlin 1974

HARD X-RAY OBSERVATIONS OF ELEMENTARY FLARE BURSTS, AND THEIR INTERPRETATION

H. F. VAN BEEK, L. D. DE FEITER and C. DE JAGER

The Astronomical Institute at Utrecht, The Netherlands

Recent observations of solar hard X-ray bursts, performed with high time resolution aboard the ESRO TD-1A satellite, show that often these bursts consist of numerous short-lived spikes with rise and decay times of the order of a few seconds. It is argued that these spikes, which we call "Elementary Flare Bursts" (EFB), are the essential physical phenomena, which, clustered together in large numbers, constitute the conventional high-energy flare. For some cases studied in this paper we find that the rise and decay times can be as low as 1 to 2 seconds; individual points deviate only slightly from this average value. These times do not depend on the photon energy; this seems to indicate that the EFB's occur at a fairly low level in the chromosphere ($\lesssim$ 1500 km above the photosphere). For one particular case studied this conclusion is corroborated by the observation of short-lived flashes as observed optically with an interference filter centred at λ 3835 Å. In this region a considerable part of the ambient electrons are accelerated to energies of several tens of keV's.

1. The Concept "Elementary Flare Burst"

Several studies of hard X-ray bursts with time resolution of about one second or less have been reported in the literature. Kane and Anderson [1] found that the rise and decay times of the impulsive burst, which is the normal component at photon energies above about 20 keV, are 2—5 and 3—10 seconds respectively. In some case [2, 3] indications of fine structure on a time scale less than one second have been reported. Such observations cast doubt on flare models where a sudden release of energy, during the impulsive phase, is followed by a slower relaxation to pre-flare conditions. If these very fast spikes occur in most flare bursts, we should rather think of a repetition of such an instability, at not necessarily the same location in the active region.

In the present communication, we report on observations made with the hard X-ray spectrometer (code name S100) of the Space Research Laboratory of the Utrecht Astronomical Institute [4] aboard the ESRO TD-1A satellite. The instrument performs continuous observations of the solar hard X-radiation in twelve logarithmically spaced energy channels between 24 and 900 keV. The time resolution is 1.2 seconds for the four channels with energies less than 90 keV, and 4.8 seconds for the other channels. Thanks to the good time resolution and the fair sensitivity of the instrument, the observations are very well suited for a study of the temporal fine structure mentioned above. In several events, particularly those referring to weak flares, we have found further evidence for the existence of very short-lived spikes, all with very similar rise and decay times. The apparent similarity of these spikes, with respect to their time profiles as well as their spectra, has led us to the hypothesis that most hard X-ray bursts consist of these spikes, which we propose to call "Elementary Flare Bursts" (EFB).

2. The Observations

The best example to illustrate the concept of an EFB is presented in Fig. 1; it refers to one of the small flare events that occurred during the period of extraordinary high solar activity in August 1972. The flare started at 1939 UT on 2 August 1972, and was classified as importance 1B. It was accompanied by radiobursts at many frequencies; in particular the spectral observations indicate the

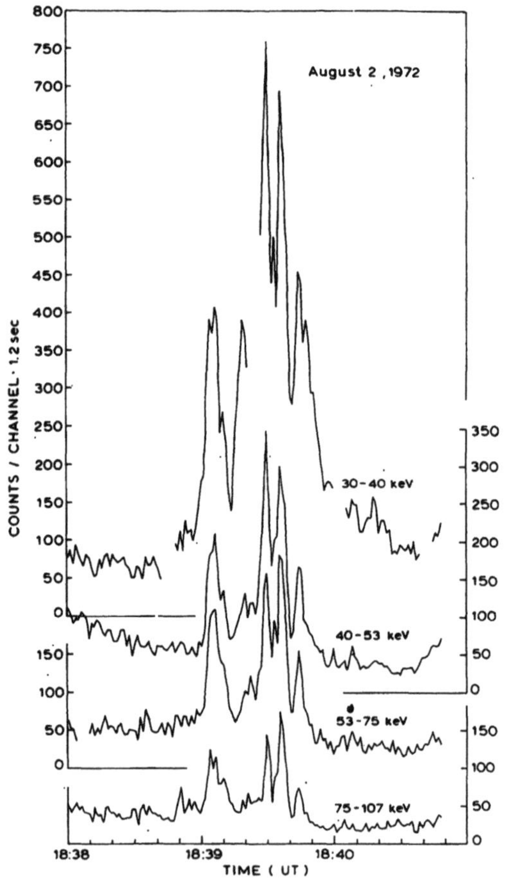

Fig. 1. Time profiles of the hard X-ray burst of 2 August 1972, 1839 UT. The intensity peaks are very close in time at the four energy channels displayed. Note the rapid variations on all four channels and the apparent differences in the spectra of the individual peaks.

occurrence of faint type III bursts. Whereas this activity lasted for some 15 minutes, the initial hard X-ray burst had a duration of no more than one minute. Inspection of the recordings of Fig. 1 indicates the presence of a number of spikes having almost the same time profile. Some of these are relatively isolated, whereas some others occur so closely in time that their time profiles overlap considerably. It is worth noting that the characteristic times of the individual spikes seem to be close to the time resolution of the instrument: 1.2 seconds. Future observations with larger instruments and a better time resolution may show that the EFB's have even shorter durations than those we derive from our observations. As a matter of fact, the observations of Beigman et al. [3] have given indications for time-scales as short as 0.05 second. Also the time profiles do not show any variation with energy, which indicates that these profiles reflect the time profile of the acceleration (or injection) mechanism rather than a relaxation towards thermal energies of energetic electrons that were impulsively injected at the start of the EFB. Qualitatively one can see some variation of the spectral characteristics of the individual EFB's; such spectral variations may distort slightly the correspond-

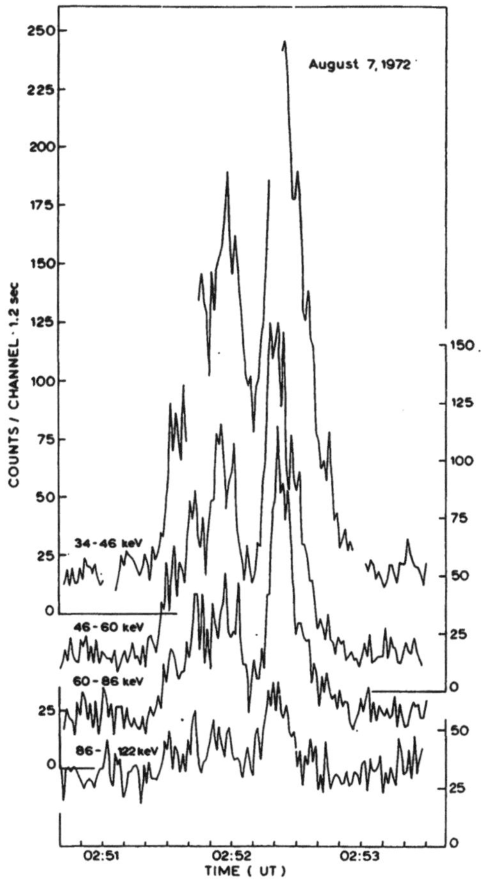

Fig. 2. Time profiles for the hard X-ray burst of 7 August 1972, 0251 UT.

ence of the time profiles in different energy channels. Some of these EFB's may be related to the light flashes that were observed by Zirin and Tanaka [5] with the 15 Å filter centred at 3835 Å, for the same event.

Once the concept of the EFB has been illustrated for a clear-cut case, it becomes interesting to study more complex events. Such an event is presented in Fig. 2. It also refers to a small flare, optical importance SN, of the August 1972 period. Here it may be difficult to accept the idea that the event is composed of a super-

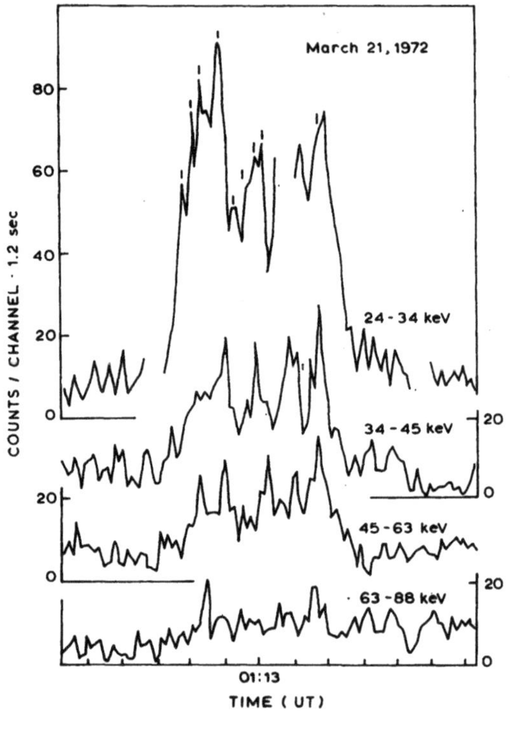

Fig. 3. Time profiles for the hard X-ray burst of 12 March 1972, 0113 UT.

position of a large number of EFB's with durations similar to those found for the previous case. However, the existence of three components can hardly be denied. A closer inspection of some individual peaks, and in particular the correspondence of such peaks on the four energy channels displayed, lends some support to the hypothesis that also this type of burst is composed of a large number of EFB's which overlap in time.

A third example is given in Fig. 3, which in fact was the first case studied in relation to the subject of this paper. Here we deal with a still smaller event and as a consequence of this, the signal-to-noise ratio is rather poor. Nevertheless, the number of peaks that appear simultaneously on at least three energy channels seems to be considerably higher than can be accounted for on the assumption that they are due to statistical fluctuations. Hence, also in this case we find evidence for the existence of EFB's.

It will be clear from the discussion of the three examples that, in many cases, the detection of EFB's is at the limit of the possibilities of the present instrumentation. In particular the limits are set by the signal-to-noise ratio and by the time resolution. In order to make further progress along the lines indicated in this paper the sensitive area of the detectors should be considerably increased, coupled with a reduction of the integration times. Typical values to aim for are 50—100 cm² and 0.2 second respectively.

3. Discussion

Typical values of the rise and decay times found for the EFB's are 2 seconds with almost equal values for τ_r and τ_d, and no dependence upon the photon energy. Assuming that the decay is due to collisions of excited electrons with the ambient electron gas, for which

$$N_e = \frac{1.5 \times 10^8 \bar{E}^{3/2}_{\mathrm{keV}}}{\tau_d} (2^{3/(2\gamma+1)} - 1) \ \mathrm{cm}^{-3} \tag{1}$$

one would find, with $\bar{E} = 50\,\mathrm{keV}$, $\gamma = 3.6$, and $\tau_d = 2$ s, a value of $N_e = 6 \times 10$ cm^{-3}. However, this result cannot be considered correct, since *if* the decay were collision dominated, there should be a clear energy dependence: $\tau_d \sim E^{3/2}$, according to Eq. (1). Since such a dependence is *not* observed, the observed EFB profiles must represent virtually the pure time profile of the injection process. In other words the observed complete independence of τ_d on E shows that the collision time τ_{coll} must be fairly *short* compared with τ_d. It seems a conservative assumption to take $\tau_{\mathrm{coll}} \lesssim 0.5$ s, also for the channel with the highest energy ($E = 90$ keV), which would then yield

$$N_e \gtrsim 5 \times 10^{10} \ \mathrm{cm}^{-3}. \tag{2}$$

Since this N_e value refers to the *ambient* electron density, this result allows us to define the chromospheric height level where the EFB's seem to occur. With a model of the low chromosphere [6] one finds that an electron density $\log N_e$ (in cm^{-3}) $= 10.7$ occurs at a height level about 1400 km above the limb ($\tau_0 = 0.003$), where the quiet-sun temperature takes a value of about 8000 K. According to (2) this height is an *upper limit*. We conclude therefore that the EFB's seem to occur at fairly low height levels in the solar chromosphere, which is in agreement with the conclusions drawn by Zirin and Tanaka for the event of 2 August 1972, 1938 UT.

For a better understanding of the phenomenon of the EFB's it is important to estimate the total number of accelerated electrons. Since we concluded before that the observed time profiles reflect the time variation of the injection of energetic electrons into the radiation producing region, such an estimate should be based upon model calculations where continuous injection rather than impulsive injection is assumed. The relation between the energy distribution function $F(E)$ of the number of electrons injected per second and the spectral distribution of the photon flux $F(\varepsilon)$ as observed at a distance of 1 AU has been given by Brown [7]. For a power law distribution of $F(\varepsilon)$ Brown finds:

$$F(E) = 2.68 \times 10^{33} \gamma^2 (\gamma - 1)^3 B(\gamma - {}^1/_2, {}^3/_2) \frac{I(\varepsilon_1)}{\varepsilon_1^2} \left(\frac{\varepsilon_1}{E}\right)^{\gamma+1} \mathrm{s}^{-1} \tag{3}$$

where γ is the spectral index of the photon flux, B is the Beta function and $I(\varepsilon_1)$ is the integral photon flux above photon energy ε_1 keV. Using (3) we find for the peaks of the event displayed in Fig. 1, values for $F(>10 \text{ keV})$, the number of electrons with energies above 10 keV injected per second at the times of peak intensity, ranging between 10^{35} and 10^{37}. Since the duration of the individual EFB's is a few seconds, we estimate the total number of energetic electrons that produce these spikes to range between about the same values. They represent a total kinetic energy of $10^{27}-10^{29}$ ergs.

The correlation between the hard X-ray EFB's and the flashes observed at λ 3835 Å that is suggested by Zirin and Tanaka [5], and the high densities found from the near independence of the decay times on photon energy, might be used as an indication of the spatial extent of the X-ray emitting region. The bright spots at 3835 Å have an angular size of about one arcsecond, and the scale height at 1400 km is smaller than 1000 km. It would, therefore, be surprising if the volume V were larger than that of a sphere with a radius of 2000 km, or 10^{26} cm^3. Thus the energetic electrons represent a large fraction of the total number of electrons present in this volume. Of course, we do not have any indication that the electrons are accelerated in this emitting region. If they are accelerated higher up in the corona, the magnetic field must provide a very efficient guiding for the electrons to penetrate down into the fairly deep chromospheric layers. In order to study this matter in further detail we will need spatially resolved X-ray observations together with optical observations with a time resolution of about one second.

References

[1] S. R. Kane and K. A. Anderson, Astrophys. J. **162**, 1003 (1970).
[2] K. J. Frost, Astrophys. J. **158**, L 159 (1969).
[3] I. L. Beigman, L. A. Vainshtein, B. H. Vasilyev, B. N. Zhitnik, V. D. Ivanov, V. V. Korneyev, V. V. Krutoo, S. L. Mandelshtam, I. P. Tindo and A. I. Shurygin, Kosm. Issled. **9**, 123 (1971).
[4] H. F. van Beek, Thesis, Utrecht 1973.
[5] H. Zirin and K. Tanaka, UAG Rep. No. 28, Vol. 1, 121 (1973).
[6] R. W. Noyes and W. Kalkoven, Solar Phys. **15**, 120 (1970).
[7] J. C. Brown, Solar Phys. **18**, 489 (1971).

Space Research XIV — Akademie-Verlag, Berlin 1974

INVESTIGATION OF THE SOLAR X-RAY FLARE SPECTRA BY THE INTERCOSMOS 4 AND INTERCOSMOS 7 SATELLITES

YU. I. GRINEVA, V. I. KAREV, V. V. KORNEEV, V. V. KRUTOV, S. L. MANDEL'STAM, U. I. SAFRONOVA, A. M. URNOV, L. A. VAINSHTEIN, B. N. VASILYEV and I. A. ZHITNIK

Lebedev Physical Institute, Moscow, USSR

Results are given of the further experimental study of solar X-ray flare spectra of the Fe XXV line region in the vicinity of 1.85 Å. From theoretical analysis it is known that the spectral range 1.85—1.95 Å includes the 1s—2p and 1s—2s Fe XXV lines and the satellite lines due to 1s—2p transitions from the auto-ionizing levels of the Fe XXIV—XVIII ions. The spectrum contains a few hundred satellite lines and is characterized by a quasi-continuous form. Energy distribution in the spectrum depends strongly on the temperature; electron temperatures of 16×10^6 °K and 18×10^6 °K were estimated for two flares.

1. In preceding papers [1, 2] we have given the results of the study of solar X-ray flare spectra in the range 1.85—1.87 Å with high resolution $\sim 4 \times 10^{-4}$ Å, which were performed aboard the Intercosmos 4 satellite. Here results over the extended region 1.85—1.95 Å are given.

Observations were made by both the Intercosmos 4 and Intercosmos 7 satellites (launched 30 June 1972; programme finished 5 October 1972). Bragg spectrometers, which were similar to the device described in [1, 2], were installed aboard the satellites. Quartz crystals with atomic plane spacing $d = 4.246$ kX, and half width of the rocking curve $\sim 10''$ were used. The effective area of the crystal on Intercosmos 4 was 1.1 cm^2 and on the Intercosmos 7 about 8 cm^2. The detectors were argon-filled Geiger counters with Be windows (density = 160 mg cm^{-2}).

The spectra were obtained by scanning the solar disk with the satellite axis within an angle of about 80′. This angle defined the range of wavelengths registered by each spectrometer. The resolution depended on the angular size of flares and the time constant of the electronic circuit (0.7 sec). The scanning velocity for the spectra given below was about one angular minute per second, and the resolution was $\sim 2 \times 10^{-3}$ Å. The wavelength determinations were carried out using recordings of an X-ray heliograph with a Soller type slit collimator as well as a special optical sensor. The error of the wavelength determinations did not exceed $\sim 2 \times 10^{-3}$ Å. Thus the spectra shown below had both a resolution and an accuracy of the wavelength determinations approximately five times better than those of [1, 2].

Fig. 1 shows two spectra I(b) and II(b) which were obtained at 0101 and 0104 respectively during the 2B flare (16 November 1970). For comparison spectrum II(a) registered simultaneously with II(b) but with higher resolution of 4×10^{-4} Å is given [1, 2]. Spectrum III(b) of the 2B class flare recorded at 0706 (4 August 1972) is given in Fig. 2.

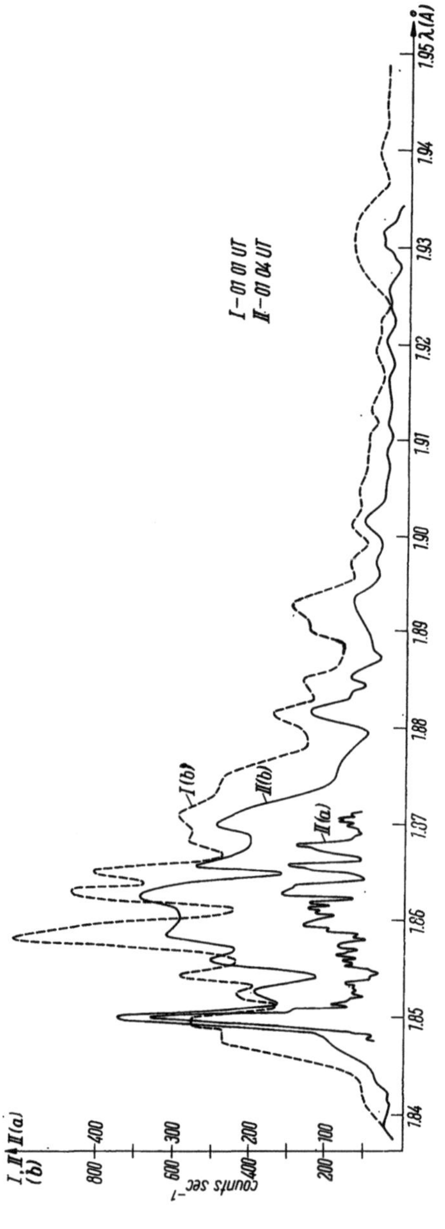

Fig. 1. The spectra of the 2B flare obtained by the Intercosmos 4 satellite with resolutions (a) 0.0004 Å; (b) 0.002 Å.

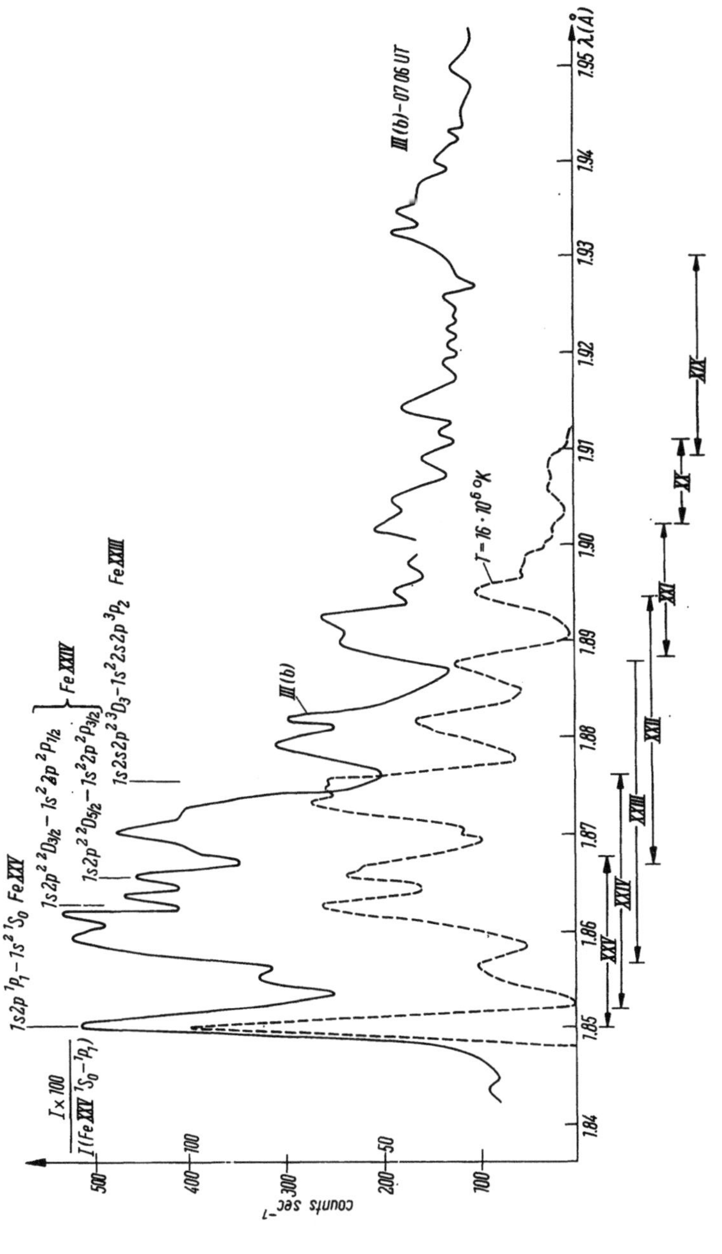

Fig. 2. Comparison of the 2B flare spectrum III (b) obtained by the Intercosmos 7 satellite with the calculated one which consists of a few hundred lines of highly ionized iron. Lines with relative intensities of >3% and the ranges of wavelengths for each ion, which include the lines with intensities of >0.1%, are shown.

The analysis of the curves in Figs. 1—2 reveals that the three spectra of type (b) have similar intensity distributions but with essential differences in their features. This may be attributed partly to changes of the geometrical size of the flares and to macroscopic plasma motions. There is a great difference in the spectra II(a) and II(b) in the overlapping spectral range; it should be noted that they were obtained almost simultaneously. Also the relative intensity of the resonance line at 1.850 Å in the high resolution spectrum always proves to be greater in comparison with that of the low resolution spectra.

The discrepancies between the spectra obtained with the high and low resolution instruments may be partly explained by the superposition of a large number of unresolved weak lines and also by the complex structure of the peaks in the type (a) spectra.

The difference between the spectra I(b) and II(b) at 1.860 Å, however, is difficult to explain. The line at 1.866 Å is of special importance and has been used for derivation of the temperature [1, 2].

2. The spectrum in the range 1.85—1.95 Å includes lines due to 1s—2p and 1s—2s transitions in the Fe XXV ions as well as many satellite lines due to 1s—2p transitions from the auto-ionizing levels in the Fe XXIV—XVIII ions [3]. The excitation of the satellite lines is mainly due to dielectronic recombination. The intensity of the satellite line I_z relative to that of the Fe XXV resonance lines I_{XXV} is given by the formula [4, 5]

$$i'_{\gamma\gamma_0} = \frac{I_z}{I_{\mathrm{XXV}}} = c' \cdot g_\gamma \Gamma_\gamma Q_\gamma$$

$$c' = 5.9 \times 10^{-4} \frac{N_{z+1}}{N_{\mathrm{XXV}}} \frac{E_{10}}{\delta E} \delta\beta e^{\delta\beta} \tag{1}$$

$$Q = \frac{A_{\gamma\gamma_0}}{\Gamma_\gamma + A_\gamma}; \qquad \delta\beta = \frac{\delta E}{\theta}$$

where N_z and I_z are the density and the $\gamma - \gamma_0$ line intensity of the ion with charge $z - 1$ respectively; E_{10} is the energy of the Fe XXV 1^1S—2^1P transition and δE is the electron-binding energy (in Rydberg (ryd) units); $\theta = kT/\mathrm{ryd}$; Γ_γ and A_γ are the probabilities of auto-ionization and radiation decay (without the factor of 10^{13} s^{-1}) of level γ with statistical weight g_γ.

Attempts to attribute the separate peaks in the observed spectra to apparent satellite lines, excluding the three lines marked in Fig. 2, lead to discrepancies between experimental and theoretical values of wavelengths and line intensities greater than the errors in experiments and calculations. Therefore we have made complete calculations of all wavelengths for transitions from the levels $1s2s^k2p^q$ in the Fe XXIV—XVIII ions within the given spectral range (see Table 1).

Values of λ, Γ_γ and A_γ were calculated by means of the method described in [6] to find the total probability of auto-ionization. On the other hand the probability of dielectronic capture, and consequently the satellite intensity, is evidently directly proportional to the probability of decay into the ground state. The use of the total value may lead to a slight overestimate of some satellite intensities. The role of direct excitation of the $1s^2$-shell electrons may be of some importance n the formation of satellites. The intensity of the satellite lines due to this effect

Table 1
Transition for the Fe XXV—XVIII lines in the range 1.85—1.95 Å

Fe ion	Transition	Range of wavelengths	No. of lines calculated	No. of lines with relative intensity* >10%	>1%
XXV	1s2s → $1s^2$ 1s2p → $1s^2$	1.850—1.868	1 3	4	4
XXIV	1s2s2p → $1s^22s$ $1s2p^2$ → $1s^22p$	1.852—1.876	22	7	9
XXIII	$1s2s^22p$ ↗ $1s^22s^2$, ↘ $1s^22p^2$ $1s2s2p^2$ → $1s^22s2p$	1.857—1.888	63	8	22
XXII	$1s2s^22p^2$ ↗ $1s^22s^22p$, ↘ $1s^22p^3$ $1s2s2p^3$ → $1s^22s2p^2$	1.867—1.895	170	4	43
XXI	$1s2s^22p^3$ ↗ $1s^22s^22p^2$, ↘ $1s^22p^4$ $1s2s2p^4$ → $1s^22s2p^3$	1.888—1.902	175	—	27
XX	$1s2s^22p^4$ ↗ $1s^22s^22p^3$, ↘ $1s^22p^5$ $1s2s2p^5$ → $1s^22s2p^4$	1.902—1.911	110	—	6
XIX	$1s2s^22p^5$ ↗ $1s^22s^22p^4$, ↘ $1s^22p^6$ $1s2s2p^6$ → $1s^22s2p^5$	1.909—1.930	24	—	—
XVIII	$1s2s^22p^6$ → $1s^22s^22p^5$	1.925—1.929	2	—	—

* No. of lines with intensity relative to Fe XXV resonance line ($T = 16 \times 10^6$ °K).

was estimated [5] by the expression

$$i''_{\gamma\gamma_0} = \frac{I_z}{I_{\text{XXV}}} = \frac{N_z}{N_{\text{XXV}}} \frac{\langle v\sigma_{\gamma_0\gamma}\rangle}{\langle v\sigma_{01}\rangle} \frac{A_{\gamma\gamma_0}}{\Gamma_\gamma + A_\gamma} \simeq c'' g_\gamma A_\gamma Q_\gamma$$
$$c'' \simeq 3.3 \times 10^{-3} \frac{N_z}{N_{\text{XXV}}}. \quad (2)$$

It was assumed in this that

$$\frac{\langle v\sigma_{\gamma_0\gamma}\rangle}{\langle v\sigma_{01}\rangle} \simeq \frac{g_\gamma A_\gamma}{g(2^1\text{P})\, g(2^1\text{P}-1^1\text{S})} \simeq 3.3 \times 10^{-3} g_\gamma A_\gamma .$$

The estimation of cross sections of this kind is not accurate enough and may lead to more errors. This process is essential only for a few lines, i.e. for Fe XXIV (1.8606 Å, 1.8631 Å) and Fe XXIII (1.8700 Å). The results of the calculations of ionization equilibrium [7] used differ from those of Jordan [8] because more accurate elementary process cross sections were adopted.

The calculated spectra for the given range of wavelengths are shown in Fig. 3 for the three values $T = 14 \times 10^6$, 16×10^6 and 18×10^6 °K. The instrumental

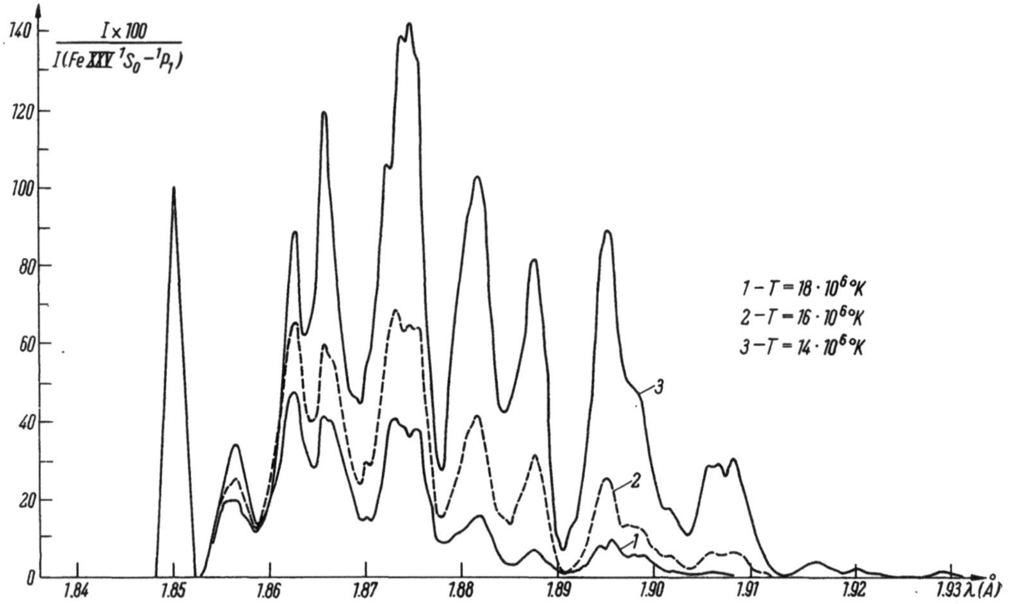

Fig. 3. The theoretical spectra of highly ionized iron with a resolution ~0.002 Å.

function was assumed to be a triangle with half width 2×10^{-3} Å. As shown in Fig. 2, the spectrum for $T = 16 \times 10^6$ °K is the closest to the observed data.

3. The comparison of the experimental and theoretical spectra shows that the thermal model used above gives the general appearance of the spectra in the range 1.85–1.92 Å. The discrepancies in the vicinity of 1.93 Å will be discussed later. In the range 1.89–1.92 Å the group of lines related to different ionization stages of Fe XIX, XX, XXI are separate (respective ranges are 1.909–1.930, 1.902 to 1.911, 1.888–1.902 Å). The resolution achieved makes it practically impossible to identify spectral lines. Many peaks, e.g. some maxima in the vicinity of 1.88 Å, are the sum of a great number of lines with comparable intensities. Therefore it is impossible to make detailed identification of the separate maxima. More extensive comparison of experimental and theoretical spectra, however, shows many differences.

The position of the experimental and theoretical maxima as a rule differ by ~0.001–0.004 Å. This disagreement is of special importance for the maxima at 1.873 Å and 1.875 Å. (It must be mentioned that when high resolution spectra were studied a considerably smaller discrepancy between the theoretical and experimental data was observed [1, 2].) A great difference is found to be close to 1.89 Å.

It is reasonable to estimate the temperature by comparison of the average intensity distributions in the experimental and theoretical spectra taking into account the strong dependence of the calculated spectra intensity on the temperature. Thus for the spectra III we have $T = 16 \times 10^6$ °K. The emission measure obtained from the absolute intensity of the resonances line calculated in [7] was $\sim 10^{49}$ cm^{-3}. The spectra I, II provide a temperature of $\sim 18 \times 10^6$ °K. To obtain the temperature from the intensities of the Fe XXIV satellite lines relative to the Fe XXV resonance line it is necessary to subtract the "background" in the vicinity of $\lambda = 1.850$ Å which may be due to transitions from the $1s2s^k2p^qnl$ levels with $n \geqq 3$. The temperature thus estimated from the satellite line at 1.866 Å proves to be in accordance with the obtained value of $\sim 18 \times 10^6$ °K.

The features in the vicinity of 1.93 Å in the experimental spectra I—III agree in wavelength with the Kα lines [9] emitted by Fe ions of low stages of ionization which were observed in laboratory spectra [10]. It is difficult, however, to explain the observed intensity in this region of the spectrum because the densities of low ionization stages of ions are too small. Presumably the Kα lines are formed due to the penetration of fast electrons or hard X-ray radiation into cold dense regions. It should be noticed that the spectra may be influenced by the effect of Compton scattering in the solar photosphere [11]; the features at 1.93 Å for example might be due to the scattered photons with wavelengths 1.85—1.87 Å. This effect, however, depends essentially on the position of the flare on the solar disk [11].

In conclusion, we can notice that for problems of the identification of spectral lines and their physical interpretations it is necessary to obtain spectra with resolution better than $\sim 5 \times 10^{-4}$ in the range 1.85—1.95 Å. Further increase of resolution is limited by the Doppler width of the lines. It is also important to improve theoretical atomic data, ionization equilibria etc. As for the physical parameters of the flare region, e.g. the temperature, they may be obtained from low resolution spectra. A comparison of the spectra observed simultaneously with different resolutions also provides essential information.

References

[1] B. N. Vasilyev, Yu. I. Grineva, I. A. Žitnik, V. I. Karev, V. V. Korneev, V. V. Krutov and S. L. Mandelstam, Kratkie soob. po fiz., P. N. Lebedev Physical Institute, No. 3, p. 29 (1972).

[2] Yu. I. Grineva, V. I. Karev, V. V. Korneev, V. V. Krutov, S. L. Mandelstam, L. A. Vainshtein, B. N. Vasilyev and I. A. Žitnik, Solar Phys., **29**, 441 (1973).

[3] W. N. Neupert, in: Physics of the Solar Corona, D. Reidel Publ. Co., Dordrecht 1971 (p. 237).

[4] A. H. Gabriel, Mon. Not. Roy. Astron. Soc. **160**, 99 (1972).

[5] L. A. Vainshtein, I. A. Žitnik, V. V. Korneev and S. L. Mandelstam, Kratkie soob. po fiz., P. N. Lebedev Physical Institute, No. 3, p. 35 (1972).

[6] L. A. Vainshtein and U. I. Safronova, Kratkie soob. po fiz., P. N. Lebedev Physical Institute, No. 3, p. 40 (1972).

[7] I. L. Beigman, L. A. Vainshtein and A. M. Urnov, preprint P. N. Lebedev Physical Institute, No. 28 (1971).

[8] C. Jordan, Mon. Not. Roy. Astron. Soc. **148**, 17 (1970).

[9] T. Bearden, NSRDS-NBS 14.

[10] I. L. Schwob and B. S. Fraenkel, Space Sci. Revs. **13**, 589 (1972).

[11] F. F. Tomblin, Astrophys. J. **171**, 377 (1972).

Space Research XIV — Akademie-Verlag, Berlin 1974

A STUDY OF SOLAR X-RAY FLARES

P. R. SENGUPTA*

Tripura Engineering College, Tripura, India

X-ray solar flares have been classified into three types. The spectrum in type I flares is essentially thermal with only one peak between 6 and 10 Å. The spectrum in type II flares may have two peaks, one below 3 Å corresponding to hard X-ray emission temperature and the other between 6 and 10 Å corresponding to soft X-ray emission temperature. In type III flares the spectrum in general has two peaks, one between 1 and 2 Å due, primarily, to non-thermal emission, and the other between 6 and 10 Å due to thermal emission. In both type II and type III flares the soft X-ray peak flux is much higher than the hard X-ray peak flux. Ionospheric effects of these flares, such as short-wave fade-outs (SWF) and sudden phase anomalies (SPA) are mentioned.

From a detailed study of 3500 solar X-ray flares in the 0.5—3 Å, 1—8 Å, 2—12 Å and 8—20 Å wavelength ranges, with peak 2—12 Å fluxes greater than 4 m erg cm^{-2} s^{-1}, or peak 1—8 Å fluxes greater than 2 m erg cm^{-2} s^{-1} recorded by NRL and UI detectors during the six years from July 1966 to June 1972 it is concluded that the X-ray flares may be divided into three types depending on their nature and origin [1].

(i) **Type I** flares are of purely thermal origin. These are *not* associated with hard X-ray bursts and are rarely associated with microwave bursts. These are caused by the heating of the coronal plasma in the flare region to temperatures of about $6-9 \times 10^6$ °K. They have, in general, smooth and small rise, the time of growth being equal in all the wavelength ranges. The spectral hardness of the flare emission is compatible with thermal emission at an emission temperature deduced from the ratio of emissions in the 2—12 Å and 1—8 Å ranges, indicating that the flare X-rays in all the wavelength ranges come from the same region. The enhanced emission during the flare is primarily quasi-thermal bremsstrahlung emission. A typical emission measure for this type of flare is about 10^{48} cm^{-3}. Flares belonging to this type have a better correlation with SWF than with SPA.

(ii) **Type II** flares have a harder spectral distribution than type I. In this case the flare emission in the 0.5—3 Å range is more than that given by thermal emission at the soft X-ray emission temperature. This suggests that the flare X-rays in all the wavelength ranges do not originate in the same region. The harder component (0.5—3 Å) in general has a sharp rise and reaches the maximum value

* Now at The Institute of Applied Manpower Research, Indraprastha Estate, New Delhi, India.

earlier than the softer components. Flares belonging to this type are fairly well correlated with microwave bursts and are associated with hard X-ray emission in the 22 keV range. From the analysis of decay times of 0.5—3 Å flares, it is suggested that the harder component of the flare emission in this type of flare originates at lower heights, i.e. within about 10^4 km above the photosphere. The emission in all the wavelength ranges can be explained assuming quasi-thermal bremsstrahlung emission at different temperatures originating at different levels of the solar atmosphere. The soft X-ray emission temperature is typically about $7-9 \times 10^6$ °K while the hard X rays are emitted at $20-30 \times 10^6$ °K. Non-thermal emission at the initial phase is, however, not ruled out. Type II flares are better correlated with SPA than with SWF.

(iii) **Type III** flares have the hardest spectrum, and are very well correlated with microwave bursts. They are associated with the hardest X-ray bursts ($\geqq 80$ keV) and with high-energy solar electron emission. The harder component has a very sharp rise, steep slope and reaches its maximum value earlier than the softer component. Type III flares are, in general, impulse flares. The time profile often clearly shows two or three impulses. In some cases these impulses are found to be roughly periodic. The impulsive hard X-ray emission in this type of flare is in good agreement with a model of non-thermal bremsstrahlung emission by non-relativistic electrons accelerated during the eruptive phase of the flare. The soft X-ray emission is quasi-thermal in nature, the emission temperature being typically about $8-11 \times 10^6$ °K. Flares belonging to this type are better correlated with SPAs than with SWF.

The flare spectrum in all types of flares is harder during the growth than during the decay. The intensity I of a single impulse soft X-ray (2—12 Å) flare at time t during growth is of the form

$$I(t) = I_{\mathrm{m}}[1 - \exp(-t/T_{\mathrm{r}})]$$

where T_{r} is the rise time of the flare and I_{m} the peak intensity. The decay of the flare is approximately exponential, the e-folding time being smaller the harder the component. The rise time varies from flare to flare lying typically between 2 and 8 minutes in different wavelength ranges.

A complex multi-impulse flare has as many rise times as there are peaks. An X-ray flare has more than 80% probability of being associated with a reported optical flare but the intensity of an X-ray flare is *not* correlated with the intensity of the associated optical flare.

Most of the X-ray flares originate in active regions. The flare emission and the hardness of the flare spectrum are found to be consistent with Sengupta's model of the solar active region [2] with maximum flare emission occurring at the final phase of the growth of the active region. While the frequency of type I and type II flares increases with increases in the temperature and emission measure of the active region, most of the type III flares occur at the final phase of the growth.

References

[1] P. R. SENGUPTA, in: Solar Radiations and the Earth, Hindustan Publ. Agency, Delhi 1973.

[2] P. R. SENGUPTA, D. Sc. Thesis, Dept. of Electronics and Electrical Communication Engineering, I. I. T., Kharagpur 1971.

Space Research XIV — Akademie-Verlag, Berlin 1974

GAMMA-RAY AND NEUTRON MEASUREMENTS AND THEIR RELATION TO THE SOLAR FLARE PROBLEM

E. L. Chupp*, D. J. Forrest and A. N. Suri

Department of Physics, University of New Hampshire, Durham, New Hampshire, USA

In order to interpret the recent observations of solar gamma-ray lines from OSO 7 we consider the possible nuclear reactions which can occur and assess their relative importance in the solar environment. This analysis shows that $\sim 3 \times 10^{32}$ protons ($E_p > 30$ MeV) must have been produced during the initial phase of the flare on 4 August 1972. The ability to determine the temperature in the solar region where the positrons annihilate has been estimated from the 0.51 MeV line width, giving a temperature there of $< 7 \times 10^6$ °K. The major conclusions from these observations are discussed and some outstanding problems mentioned. Finally, requirements for the next generation of solar gamma-ray and neutron experiments are discussed.

1. Introduction

The role of nuclear reactions in association with solar flares has long been of interest, from at least two points of view. For example, Severny et al. [1] have hypothesized that a significant number of thermonuclear reactions can occur in the high temperature region behind shock fronts developed during a solar flare. More recently Lingenfelter and Ramaty [2] have made detailed calculations of the nuclear reaction rates for solar cosmic rays interacting with the ambient solar atmosphere. In either case the consequence of the nuclear reactions is to produce, by a variety of processes, gamma rays and neutrons. In addition the possibility that energetic particles and their interaction with matter play a more fundamental role in the overall flare phenomena, has received attention from many directions especially by Elliot [3] and Najita and Orrall [4], Švestka [5] and others.

In view of the fact that the charged solar cosmic rays are strongly affected by coronal and interplanetary magnetic fields, it is difficult to relate their properties directly to the phenomena in a transient flare. For this reason it has been assumed that the time history of solar gamma rays and neutrons observed at the earth could be directly related to the time structure of flare emissions, such as optical, X-ray and radio.

The gamma-ray monitor on the OSO 7 satellite has provided [6], for the first time, direct observations of gamma-ray lines during the solar flares on 4 August and 7 August 1972. These observations also give indirect evidence for the pro-

* Alexander von Humboldt Senior Fellow and Honorary Fulbright Senior Fellow at the Max-Planck-Institut für extraterrestrische Physik, Garching, FRG.

duction of neutrons, so that the early theoretical predictions can now be tested. The discussion presented here is taken partly from a final report on the OSO 7 August flare observations [7].

2. Flare Observations

The satellite gamma-ray monitor has been previously described [8, 9]. The instrument, designed to measure the gamma-ray spectrum in 377 channels from 0.3 to 10 MeV, consists of a 3″ × 3″ NaI(Tl) crystal surrounded by a CsI(Na)

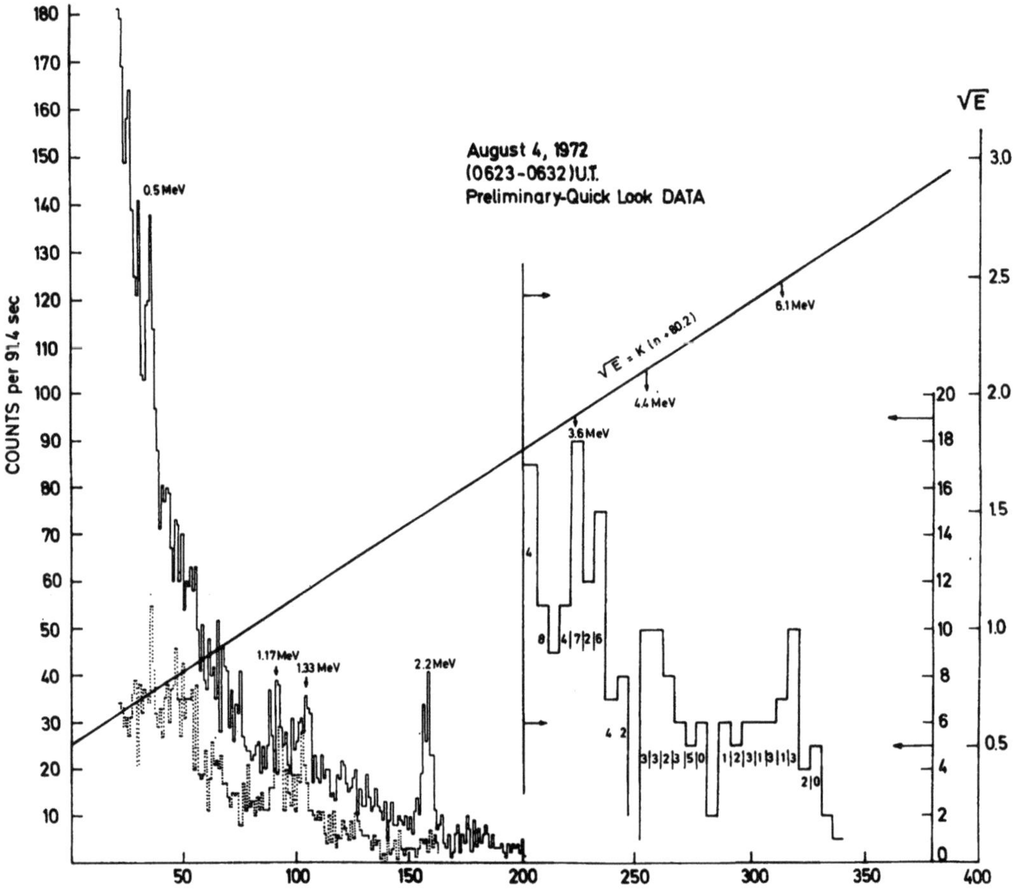

Fig. 1. The time-integrated gamma-ray spectrum is shown plotted for the solar and background directions against the analyzed channel number. The left ordinate is the total counts observed in a channel for 91.4 seconds of solar live time (upper curve) or 101 seconds of background live time (lower curve). The inner ordinate on the right is the total counts in five consecutive channels and applies to channels $n > 200$. The outer ordinate is the square root of the gamma energy, $E^{1/2}$, corresponding to a given channel number as given by the straight line $E^{1/2} = k(n + 80.2)$ in the figure. The constants in the calibration curve are obtained from the in-flight calibrations preceding and following the flare event, which is identified by the UT time and date shown.

anticoincidence shield. An auxiliary 4-channel X-ray detector measures the photon spectrum in the energy range 7.5—120 keV.

The 3B Hα flare on 4 August began at ~0621 UT reaching a maximum at 0638 UT and ending at ~0852 UT [10]. Before the satellite went into eclipse at ~0633.5 UT excess gamma ray line and photon continuum radiation was observed. Additional evidence for gamma rays was obtained in the 3B flare of 7 August 1972. This 3B flare began in Hα at ~1500 UT during satellite nighttime. As the satellite emerged into the sunlight, excess gamma ray line radiation was observed. We will only review here briefly the 4 August 1972 observations previously reported [6].

Fig. 1 shows the time integrated solar (upper histogram) and background (lower histogram) gamma-ray spectrum accumulated between 0623 UT and 0632 UT. The left ordinate gives the counts observed in each channel for the live time indicated in the caption. Above channel 200, the ordinate scale is shifted as shown on the right and the total number of counts in five consecutive channels is shown. On the far right ordinate the square root of the photon energy E (MeV) is given, corresponding to a given channel n according to the relation $E^{1/2} = K(n + n_0)$, which is the straight line in the figure. The constants K and n_0 are found from in flight calibration data.

Only the solar quadrant gamma-ray flux is strongly enhanced at 0.5 and 2.2 MeV during the rise time of the optical flare. In addition, the solar continuum is enhanced over the background spectrum to above 2 MeV. Above channel 200, there are three or more significant peaks, at 3.4—3.8 MeV, 4.2—4.6 MeV and 5.9—6.3 MeV respectively. The leakage ^{60}Co calibration lines at 1.7 and 1.33 MeV are also clearly evident on both the solar and the background spectra. To identify other possible photopeaks in the flare spectrum, it is necessary to separate out the hard X-ray burst spectrum before treating small features [11]. The time history of the 0.5 and 2.2 MeV lines has been shown [6] to follow approximately the rise time of the hard X-ray (60—120 keV) emission. Table 1 gives a summary of the line identifications and tentative flux values.

Table 1

Associated flare and the time of observations	Tentative Designations and Preliminary Solar Flux at 1 AU (photons cm^{-2} s^{-1}) 0.5 MeV	2.2 MeV	4.4 MeV	6.1 MeV
2B (Hα) 4 Aug. 1972 0626—0632 UT (before Hα max)	$(7 \pm 1.5) \times 10^{-2}$	$(2.2 \pm 0.2) \times 10^{-1}$	$(3 \pm 1) \times 10^{-2}$	$(3 \pm 1) \times 10^{-2}$
3B (Hα) 7 Aug. 1972 1538—1547 UT (after Hα max)	$(3.4 \pm 1.1) \times 10^{-2}$	$(4.8 \pm 1) \times 10^{-2}$	$< 2 \times 10^{-2}$	$< 2 \times 10^{-2}$

3. Origin of Gamma-Ray Lines

We shall first discuss the processes that can lead to positrons and neutrons, assuming the two strongest lines are due to the capture of these particles. Production of positrons and neutrons can take place through exoergic or endoergic nuclear reactions. The former are usually called thermonuclear reactions since the reactions can proceed at zero relative velocity between the interacting particles, but particles in the tail of a Maxwellian distribution of velocities can penetrate the Coulomb barrier more easily. Therefore the reaction rate is strongly temperature dependent. The latter are usually referred to as non-thermal reactions, since particles with energies above the reaction threshold are required. Table 2 gives

Table 2

Exoergic Reactions

$^{1}H\ (p, \beta^{+}\nu)\ ^{2}D$	$Q = 1.442$ MeV	$T\ (d, n)\ ^{4}He$	$Q = 17.6$ MeV
$^{2}D\ (p, \gamma)\ ^{3}He$	$Q = 5.493$ MeV	$D\ (d, p)\ T$	$Q = 4.0$ MeV
$^{3}He\ (\alpha, \gamma)\ ^{7}Be$	$Q = 1.587$ MeV	$D\ (d, n)\ ^{3}He$	$Q = 3.25$ MeV

important exoergic reactions yielding positrons, neutrons and gamma rays. The reactions in Table 2 can in principle be induced by low-energy ions in a thermal or non-thermal setting. The reaction rate in a medium consisting of particle types x and y, with particle densities n_x (cm^{-3}) and n_y (cm^{-3}), may be expressed as:

$$r_{xy} = n_x n_y \langle \sigma v \rangle_{\mathrm{Th,NT}}\ \mathrm{cm^{-3}\ s^{-1}}$$

where the average value of the product of cross section σ (cm^2) and velocity v (cm s^{-1}) is taken over a Maxwellian distribution (Th), or some non-thermal distribution (NT).

3.1. Positron Production

Considering the first reaction for a thermal distribution, and assuming the β^+ annihilate instantly into two photons, we can obtain an upper limit to the 0.5 MeV flux at the earth. Using $r_{\mathrm{pp}} = 2.3 n_{\mathrm{p}}^2 \times 10^{-42}$ cm^{-3} s^{-1} at a temperature of 5×10^7 °K and a hydrogen density of $n_{\mathrm{p}} = 10^{13}$ cm^{-3} gives $r_{\mathrm{pp}} \simeq 2.3 \times 10^{-16}$ cm^{-3} s^{-1}. The 0.51 MeV gamma-ray flux at the earth is then $F_{\oplus}$ (0.51) $\simeq 2.3 \times 10^{-43}$ V cm^{-2} s^{-1} for a nuclear reaction region of volume V (cm^3). Since flare volumes are $V \sim 10^{28}$ to 10^{29} cm^3 the 0.51 MeV flux observed on 4 August 1972 could not be due to this process. A similar calculation shows that a non-thermal distribution of ion velocities will not produce the required reaction rate under any reasonable circumstances.

Table 3 shows endoergic positron producing processes such as the decay of π^+ mesons and several positron emitting nuclei. All of these sources have thresholds

Table 3

Endoergic Reactions for Positron Emitters

From $^{1}H\ (p, n\pi^{+})\ X$; $^{4}He\ (p, n\pi^{+})\ Y$; $\pi^{+}(\nu_{\mu})\ \mu^{+}(\beta^{+})\ \nu_{e}\nu_{\mu}$; $E_{p,\mathrm{thres}} = 285$ MeV, $E_{\alpha,\mathrm{thres}} = 180$ MeV	
From protons on C, N, O	
$^{10}C\ (\beta^{+})\ ^{10}B$; $T_{1/2} = 19$ sec	$^{13}N\ (\beta^{+})\ ^{13}C$; $T_{1/2} = 10.0$ min
$^{11}C\ (\beta^{+})\ ^{11}B$; $T_{1/2} = 20.5$ min	$^{14}O\ (\beta^{+})\ ^{14}N$; $T_{1/2} = 71$ sec
$^{12}N\ (\beta^{+})\ ^{12}C$; $T_{1/2} = 0.011$ sec	$^{15}O\ (\beta^{+})\ ^{15}N$; $T_{1/2} = 2.06$ min

above several MeV and in the case of π^+ production the energy $E_p \gtreqq 285$ MeV for production in pp collisions or $E_p \gtreqq 180$ MeV for production in p, ^{4}He collisions. Lingenfelter and Ramaty [2] have shown that these processes can easily produce sufficient numbers of positrons to give a detectable annihilation line in large solar flares. The major unknown factor is the relative contribution from the π^+ or the low z positron emitters, since the former is more important the higher the average proton energy. The presence or absence of π^0 gamma rays could help settle this question experimentally. The only high-energy solar gamma-ray experiment which was in operation at the time of the flare in question was the TD-1 Milan experiment, but a preliminary analysis has not yielded any positive results [12]. Pomerantz and Duggal [13] have given arguments that there was a ground-level cosmic-ray effect associated with the 4 August flare, but delayed $\sim$8 hours after the flare. If this is true then protons of several hundred MeV kinetic energy are required and significant meson production could have taken place.

Model calculations [2, 14] give the yield of annihilation photons from proton and alpha reactions on C, N, O in terms of the characteristic rigidity P_0 of an assumed exponential rigidity spectrum for the accelerated particles. Using the recent work of Ramaty and Lingenfelter [14] for a thick target geometry and $P_0 = 70$ MeV, we find that to give the observed 0.51 MeV integrated flux, a total number $N(P > 0) \simeq 1.6 \times 10^{34}$ protons is required; this corresponds to 5×10^{32} protons with energies greater than 30 MeV ($P \simeq 240$ MeV) for the thick target geometry assuming a 10^3 second burst. For the thin target geometry we find that $nN(P > 0) = 1.4 \times 10^{45}$ cm^{-3}, where n(atoms cm^{-3}) is the hydrogen number density. From a requirement [7] on the positron annihilation time we find $n \gtreqq 2 \times 10^{11}$ cm^{-3}, and therefore $N(P > 0) \cong 10^{33}$ protons or $N(P > 240$ MeV) $\cong 3 \times 10^{31}$ protons.

The last source we shall consider is the production of positrons by pair production from the 2.2 MeV photons. The pair production cross section in hydrogen at 2.2 MeV is $\sigma(2.2) \sim 2 \times 10^{-4}$ barns/H nucleus so the β^+ production rate is $R(\beta^+) \simeq 6 \times 10^{23} \times F(2.2) \times (\varrho t) \times \sigma(2.2)$, where $F(2.2)$ is the total rate of 2.2 photons (s^{-1}) incident on (ϱt) g cm^{-2} of hydrogen. Using $F(2.2) = 6 \times 10^{26}$ s^{-1} from the measured [6] line flux given in Table 1 and $(\varrho t) = 10$ g cm^{-2}, we arrive at $R(\beta^+) \cong 10^{23}$ s^{-1}; which gives a source strength of 2×10^{23} (0.5 MeV photons s^{-1}) if all the positrons annihilated instantly. The required source strength is 2×10^{26} s^{-1}, so this source fails by three orders of magnitude.

3.2. Neutron Production

Table 4 gives endoergic neutron producing reactions and their thresholds, and in Table 2 exoergic reactions were given. The solar abundances of deuterium and tritium are not known, but there have been suggestions for significant thermal D-D reactions during flares [1]. For example, in this case if the D-abundance is 5×10^{-5} times that of hydrogen in a region of ion density 10^{13} cm^{-3} then the D-D reaction rate at 10^7 °K is about 10^{-1} cm^{-3} s^{-1}, each reaction giving a neutron of energy 2.45 MeV. For comparison the reaction rate for production of neutrons via endoergic reactions (e.g. protons on helium) is about 10 cm^{-3} s^{-1} for 100 MeV protons. The density in the flare plasma is probably much lower [7] than the 10^{13} cm^{-3} value used for the D-D estimate, so again it appears very difficult to get sufficient neutrons from thermonuclear reactions, considering other factors. It is shown in a separate study by Reppin et al. [15] that the endoergic reactions can

Table 4
Endoergic Reactions

Reaction	Threshold (MeV/nucleon)	Reaction	Threshold (MeV/nucleon)
^{1}H (p, nπ^{+}) ^{1}H	292.3	^{12}C (p, n ...	19.8
^{4}He (p, pn) ^{3}He	25.9	^{14}N (p, n ...	6.3
^{4}He (p, 2pn) ^{2}H	32.8	^{16}O (p, pn ...	16.5
^{4}He (p, 2p2n) ^{1}H	35.6	^{20}Ne (p, pn ...	17.7

produce the required neutrons if a total number of protons $N(P > 0) \simeq 10^{34}$ are accelerated with a characteristic rigidity $P_0 = 60$ MV.

Following the discussion above concerning positron production, it can also be shown that exoergic reactions with light isotopes of H all fail to provide the neutrons even with a large group of non-thermal ions.

4. Sources of Other Gamma-Ray Lines

The strong spectral lines at 0.51 and 2.22 MeV are most probably from the sources just discussed. The line widths show no evidence for broadening and the photopeak energy positions are in excellent agreement with positron annihilation and neutron capture occurring effectively at rest relative to the observer. We must now consider other sources of gamma-ray lines to explain the weaker features at 3.6, 4.4 and 6.1 MeV and determine if there are any possible contributions to the stronger lines from other sources. On the assumption that there are no Doppler shifts of lines at 4.4 MeV and 6.1 MeV we can determine their production rate from the processes $^{12}C\,(p, p'\gamma)\,^{12}C$ and $^{16}O\,(p, p'\gamma)\,^{16}O$ which have thresholds at their respective excitation energies in the center of mass system. Since it has been established that positron and neutron production is from the endoergic reactions discussed above, the yield of inelastic scattering gamma rays is found directly from the calculations of Ramaty and Lingenfelter [14]. The observed fluxes shown in Table 1 are obtained, in the thick target case, with a total time-integrated proton flux of $N(P > 0) = 10^{34}$ protons, which is essentially the same as that required for the positron and neutron production. In the thin target case we find $nN(P > 0) = 10^{45}$ protons cm^{-3}. Using the target density $n \sim 1.6 \times 10^{11}$ cm^{-3} from the annihilation time [7] we find $N(P > 0) \sim 6 \times 10^{33}$ protons, again nearly the same as that required for the thick target case. Preliminary indications are that the total number of protons released to the earth [14] is at least two orders of magnitude larger than the number $N(P > 0) \sim 10^{34}$ needed for the gamma rays.

5. Temperature of Positron Annihilation Region

By study of the line width of the 0.5 MeV line it is possible to estimate the temperature of the plasma where annihilation is occurring as proposed by Stecker [16]. The fractional line width is $\Delta E/E = 2[2kT(\ln 2)/mc^2]^{1/2}$ where k is Boltzmann's constant, T the absolute temperature and mc^2 the rest energy of the electron. Even though the total observed counts under the photopeak was $\sim$150,

Dunphy et al. [17] were able to show that there is no measurable broadening of the 0.51 MeV line and therefore the temperature must have been no larger than 7×10^6 °K. This measurement is limited by the resolution of the detector which is ~9% FWHM at 0.51 MeV. Future instruments using cooled solid state detectors could achieve a resolution of much less than 1%, therefore reducing the temperature limit by a factor of over 100 to ~10^4 °K.

6. Summary

1. Gamma-ray lines were observed during the impulsive phase of the 4 August 1972 flare for ~10 minutes. It is likely that this was also the case on 7 August 1972.

2. The number of protons required to produce the time integrated gamma-ray flux is 3×10^{32} with E_p (>30 MeV) on 4 August 1972.

3. The source of β^+ and neutron production and (p, p'γ) reactions is due to non-thermal reactions of accelerated solar ions in the ambient solar atmosphere. Thermonuclear reactions do not appear to play any significant role.

4. A preliminary analysis of the continuum emission on 4 August 1972 [11] indicates that ~5×10^{36} electrons (>100 keV) were accelerated over ~600 seconds.

5. No white light observations are yet available during the initial phase of the 4 August event, so the time relation of these emissions to the protons is not verifiable.

6. The temperature and the density of the plasma in the positron-electron annihilation region is $<7 \times 10^6$ °K and $n_e \geqq 1.6 \times 10^{11}$ cm^{-3} respectively.

Two cautionary remarks should be made here:

First, the calculations on gamma-ray and neutron production by charged particles in the solar atmosphere have not taken into account the directivity of the proton motion in solar magnetic fields. This will influence at least the neutron and positron production and the resulting gamma-ray yields.

Second, we must not generalize about flares and particle acceleration from these meager gamma-ray data. As observed before and even during the August activity, there are cases of hard electron acceleration (e.g. 60—120 keV burst, 2 August, 2135 UT) [18] with no observable optical flare and no detectable gamma-ray lines.

7. Future Requirements

In order to increase our understanding of nuclear reactions associated with solar activity many more observations are needed on all possible gamma ray and neutron emissions. The most sophisticated instruments must probably await free-flying platforms launched during the post-Apollo Shuttle program.

Acknowledgments

This work was supported by NASA under Contract Nas 5-11054. We are grateful for the contributions of several persons in the fabrication and data analysis phase of the experiment, particularly Dr. P. R. Higbie, A. A. Sarkady, I. U. Gleske,

S. Foss, S. Croteau, M. Simmons, K. and D. Patel, R. Adams, P. Dunphy, C. Tsai, and P. Ferguson. We also express our appreciation to M. M. Chupp for editing the manuscript.

References

[1] A. B. Severny et al., Soviet Astron. AJ **4**, 583 (1961).
[2] R. E. Lingenfelter and R. Ramaty, in: High Energy Nuclear Reactions in Astrophysics, Benjamin, New York 1967.
[3] H. Elliot, Planet. Space Sci. **12**, 657 (1964).
[4] K. Najita and F. Q. Orrall, Solar Phys. **15**, 176 (1970).
[5] Z. Švestka, Solar Phys. **19**, 202 (1971).
[6] E. L. Chupp et al., Nature **241**, 333 (1973).
[7] E. L. Chupp, D. J. Forrest, P. R. Higbie, A. N. Suri, C. Tsai and P. P. Dunphy, to be published.
[8] P. R. Higbie, E. L. Chupp, D. J. Forrest and I. U. Gleske, IEEE Trans. Nucl. Sci. NS-**19**, No. 1, 606 (1972).
[9] D. J. Forrest, P. R. Higbie, L. E. Orwig and E. L. Chupp, Nucl. Inst. and Mtds, **101**, 567 (1972).
[10] J. V. Lincoln and H. I. Leighton, Rep. UAG-21, November 1972.
[11] A. N. Suri, D. J. Forrest, E. L. Chupp and C. Reppin, Bull. Am. Phys. Soc. **18**, No. 4, paper JH1 (1973).
[12] G. Bignami, private communication.
[13] M. Pomerantz and S. P. Duggal, Nature **241**, 33 (1973).
[14] R. Ramaty and R. E. Lingenfelter in: Proc. Symp. on High Energy Phenomena on the Sun, Goddard Space Flight Center, Greenbelt, Md, 1973 (p. 301).
[15] C. Reppin, E. L. Chupp, D. J. Forrest and A. N. Suri, in: Proc. Int. Conf. on Cosmic Rays, Denver, Col., Vol. **2**, p. 1577 (1973).
[16] F. W. Stecker, NASA SP-249, 49 (1971).
[17] P. Dunphy, E. L. Chupp, D. J. Forrest and A. N. Suri, Bull. Am. Phys. Soc. **18**, No. **4**, paper JH 2 (1973).
[18] E. Chupp, D. Forrest, C. Reppin, A. Sori and C. Tsai, World Data Center A, Rep. UAG-28, Pt. II, p. 325 (1973).

Space Research XIV — Akademie-Verlag, Berlin 1974

SIGNIFICANT PROBLEMS IN IDENTIFYING SOLAR PARTICLE INCREASES

D. F. SMART and M. A. SHEA

Air Force Cambridge Research Laboratories, Bedford, Mass., USA

The identification of solar particle increases and the association of these increases with specific solar phenomena is beset with many difficulties, particularly as the energy of the measurement is decreased. Some of the problems encountered in attempting to catalog solar particle events detected by satellite sensors during the 20th solar cycle are: (i) different sensitivity levels of various detectors that result in different onset detection times even for sensors on the same spacecraft such as on IMP 3 and IMP 4; (ii) the resolution of multiple injections from several flares in a short period of time: unique separation of these multiple events is necessary to identify their source on the sun; (iii) the differentiation between a discrete low energy particle increase and an enhancement associated with the rotation of active regions across the solar disk; (iv) sensor contamination by either significant fluxes of higher energy particles or different particle species. Specific examples of each of these problems are given utilizing data obtained by different sensors on the IMP and Pioneer satellites. This analysis shows that most of these problems can be resolved using data from multiple spacecraft observations or multi-sensor observations on the same spacecraft.

1. Introduction

The detection and identification of solar particle increases are often considered to be a relatively simple matter as typified by the documentation of many classic cases in the literature. The significant events are reported by a number of researchers; this indicates that the detection of large and interesting solar particle events is the impetus for many detailed analyses. In an attempt to identify and catalog solar particle events during this solar cycle by an inspection of the various original satellite records, several persistent, significant and annoying problems have been encountered; some of these are discussed here.

2. Different Onset Times

In our cataloging of each obvious increase discernible on the individual detector data records, we attempted to define the onset time of the detection of particles, the time of maximum intensity, and the maximum flux. As these data were assembled it became obvious that there were times when a specific detector on one satellite was recording an increased intensity several hours prior to detection of this increase by a detector with a similar energy threshold but on a different spacecraft. An example of this is illustrated in Fig. 1a and b where the 0.6—13.9 MeV proton flux on Pioneer 6 is shown together with the 0.9—190.0 MeV proton flux on IMP 3 for 16—20 March 1966. At this time the Pioneer 6 space probe was relatively close to the earth at 0.9 AU with an earth-sun-probe (ESP) angle of 3 degrees; it was close enough to the earth-orbiting IMP 3 to make a meaningful

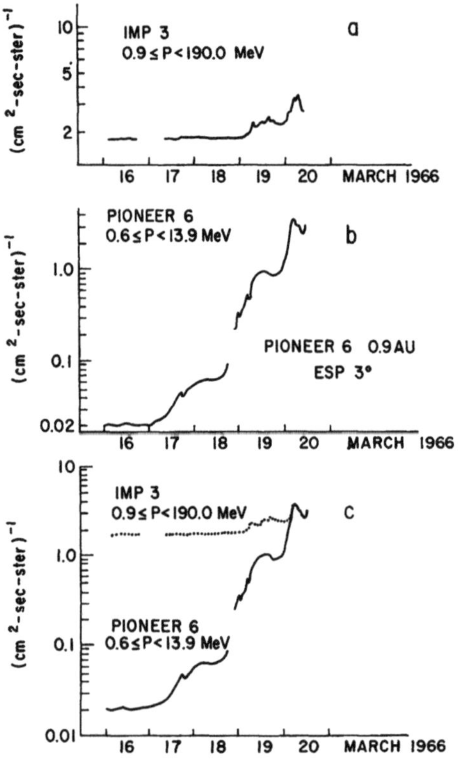

Fig. 1. Illustration of the relative sensitivities of the IMP 3 and Pioneer 6 detectors.

data comparison. An inspection of the Pioneer 6 record indicates the following times of enhancements: 17 March, 18 March (20 UT) and 20 March (00 UT). An inspection of the IMP 3 record indicates two discrete enhancements during this same period: 19 March (06 UT) and 20 March (00 UT). It could be concluded that the increase on Pioneer 6 on 17 March was not observed by the IMP 3 detector; however, comparison of the sensitivities of these two detectors reveals that the Pioneer 6 detector will respond to a flux greater than the background of 0.02 particle $cm^{-2}\,s^{-1}\,sr^{-1}$ while the IMP 3 detector will not respond until the flux is greater than a back ground of 1.6 particles $cm^{-2}\,s^{-1}\,sr^{-1}$. When these increases are plotted on the same flux scale (Fig. 1c) it is obvious that the low energy enhancement detected by the Pioneer spacecraft on 17 March could not have been detected by the IMP 3 instrument, simply because of its lower sensitivity. The same comparison can be made for the event detected by the Pioneer 6 sensor at 20 UT on

18 March. This event is first sensed on the IMP 3 satellite at 16 UT on 19 March when the flux rises above the background level of the IMP 3 detector. As further shown in this figure, both instruments respond to the increase at 00 UT on 20 March.

An example of the difference in sensitivities of similar detectors on the same satellite is shown in Fig. 2: the University of Chicago 29.5—94.2 MeV proton

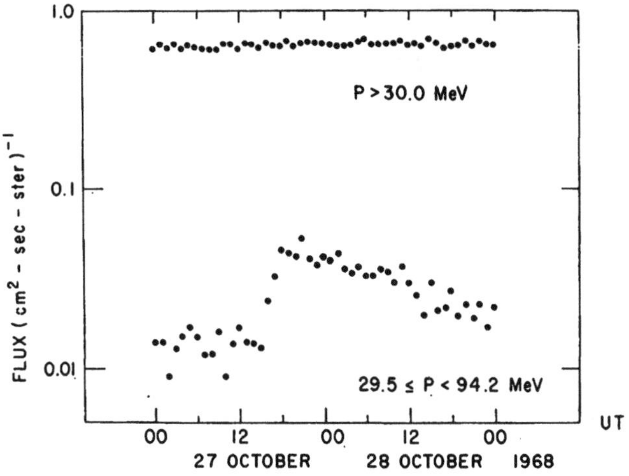

Fig. 2. Comparison of the relative sensitivities of two approximately 30 MeV detectors on the same spacecraft IMP 4. The background flux for the $P > 30$ MeV detector is about 0.65 particle cm^{-2} s^{-1} sr^{-1} (upper points).

sensor on IMP 4 detects a factor of three increase in intensity between 16—18 UT on 27 October 1968 which is not observed on the >30 MeV proton channel of the solar proton monitoring experiment on the same satellite.

3. The Problem of Multiple Injections

A period of enhanced solar activity occurred between 22 October and 5 November 1968 with many discrete particle injections superimposed upon an active region transit. An inspection of Fig. 3 shows that the 0.8—9.6 MeV proton channel on IMP 4 recorded an enhanced intensity (active region transit) before the first discrete particle event at 02 UT on 24 October. The next discrete particle event is at 09 UT on 26 October which is detected on all three energy channels. After the maximum intensity is reached on the 0.8—9.6 MeV channel at 20 UT on 26 October, the intensity slowly decays throughout the next two days. However, an inspection of the intensities on two higher energy channels shows a discrete increase on 27 October. The 0.8—9.6 MeV channel data alone would not have enabled the identification of this discrete particle injection associated with a flare at 1232 UT, at S17 E17, McMath Plage 9740 [1], the comprehensive flare index [2] being 12.

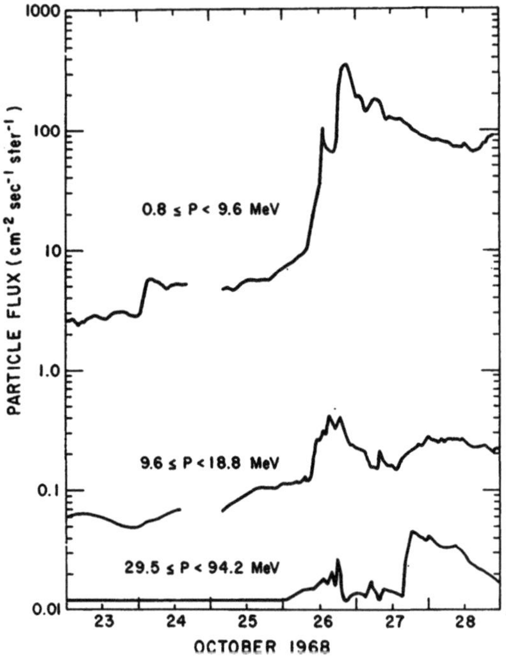

Fig. 3. An example of multiple discrete solar particle injections observed using three different energy channels in October 1968. IMP 4.

4. Differentiation of Type of Enhancement

It has been noted [3] that there is a correlation between central meridian passage of some active regions and the enhancement of the low energy particle flux. Small low-energy particle flux increases associated with active regions and co-rotating with the sun become more numerous and more difficult to identify with enhanced solar activity. Therefore the danger exists of erroneously identifying a low energy proton event when it is really this co-rotating type of phenomenon. An example of this is given in Fig. 4 where the structure observable on the

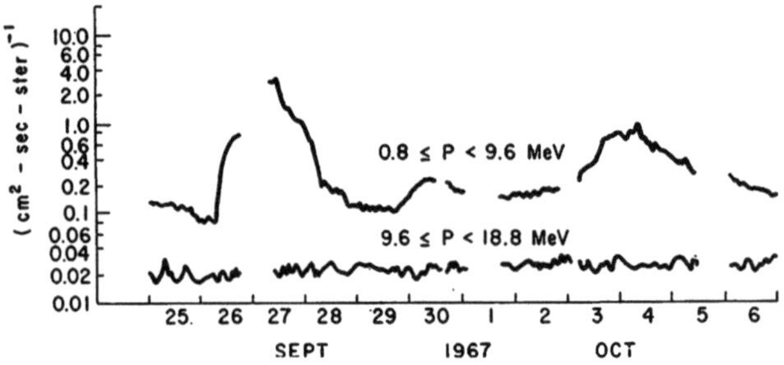

Fig. 4. Examples of particle enhancements observable at low energies, and co-rotating with the sun. IMP 4.

0.8—9.6 MeV proton channel on IMP 4 was identified [4] as solar proton events at 10 UT on 26 September, 07 UT on 27 September and 0330 UT on 3 October 1967 associated with type 3 radio bursts. While the first increase may be flare associated, it is our opinion that the other enhancements are convective type structures co-rotating with the sun which happen to be observable because of the relatively low level of solar activity at this time.

5. Sensor Contamination Problems

Fig. 5 illustrates the response of two similar energy proton detectors on IMP 4 between 11 and 15 July 1968. This figure shows an apparent event on 13 July on the >30 MeV proton detector which is not obvious on the 29.5—94.2 MeV proton detector. This event has been identified [5] as containing a significant flux of relativistic (0.3—0.9 MeV) electrons. The 29.5—94.2 MeV channel of the Uni-

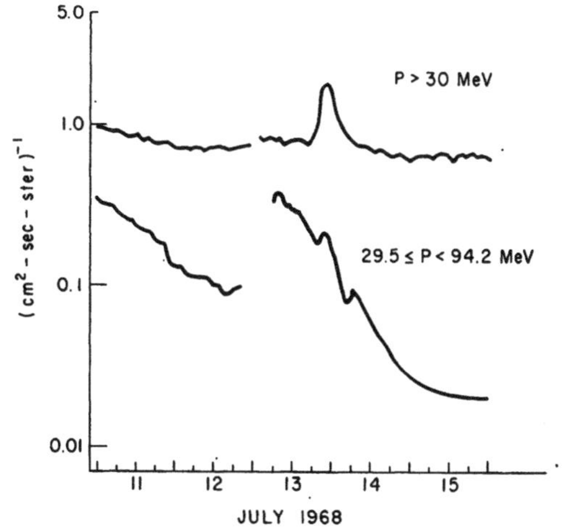

Fig. 5. Illustration of sensor contamination of the $P > 30$ MeV detector on IMP 4, which has a background flux of about 0.65 particle cm^{-2} s^{-1} sr^{-1}, while the 29.5—94.2 MeV detector has a much lower background.

versity of Chicago experiment is relatively insensitive to electron contamination since it requires not only a triple coincidence of three detector elements but also no triggering of the guard counter surrounding the sides of the detector assembly. The >30 MeV solar proton monitoring experiment channel will record any particle that penetrates the 1.6 mm copper shield and deposits more than 150 keV in a thick detector and so is quite sensitive to contamination by relativistic electrons.

The problem of electron contamination can become quite serious when trying to select accurate onset times for various proton energy channels. Fig. 6 illustrates the response of the 0.8—9.6 MeV and the 1—10 MeV proton detectors on IMP 4 for the event on 3 December 1967. In this case the 0.8—9.6 MeV proton channel is contaminated by a large >40 keV electron event [6] at 0918 UT, whereas the 1—10 MeV channel has a very thin detector designed so that it is unlikely that the electrons will deposit sufficient energy to trigger the counter.

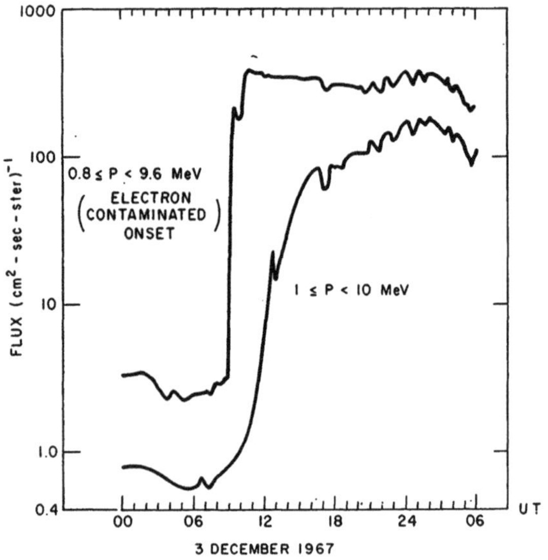

Fig. 6. Illustration of an electron contaminated onset on the 0.8—9.6 MeV proton channel on IMP 4.

6. Summary

We have discussed some of the problems related to the identification of solar particle increases as detected by various satellite instruments. These problems, once recognized, can usually be resolved by multi-spacecraft or multi-detector measurements.

Acknowledgments

We wish to thank the principal investigators (especially Drs. J. A. Simpson and C. O. Bostrom) and the US National Space Science Data Center for providing the satellite data used in this analysis.

References

[1] H. W. DODSON and E. R. HEDEMAN, private communication, 1972.
[2] H. W. DODSON and E. R. HEDEMAN, UAG-14, NOAA, US Dept. of Commerce (July 1971).
[3] F. B. McDONALD and U. D. DESAI, J. Geophys. Res. **76**, 808 (1971).
[4] T. E. GRAEDEL and L. J. LANZEROTTI, J. Geophys. Res. **76**, 6932 (1971).
[5] G. M. SIMNETT, T. L. CLINE, S. S. HOLT and F. B. McDONALD, Acta Physica Acad. Scient. Hung. **29**, Suppl. 2, 649 (1970).
[6] R. P. LIN, Solar Phys. **12**, 266 (1970).

Space Research XIV — Akademie-Verlag, Berlin 1974

THE INFLUENCE OF CORONAL MAGNETIC STRUCTURE ON LOW-ENERGY SOLAR PROTON EVENTS

R. E. Gold[a], J. T. Nolte[a], E. C. Roelof[a] and R. Reinhard[b]

[a] University of New Hampshire, Durham, New Hampshire, USA
[b] Institut für Kernphysik, Kiel, FRG

A technique has been developed for estimating the coronal connection longitude of interplanetary field lines using observed solar wind velocities. This has been used to "map" intensity histories of ~ 1 MeV solar protons back to the corona. These "injection profiles" are compared with equatorial chromospheric magnetic neutral lines deduced from absorption features in Hα filtergrams. When the profiles derived from the separated spacecraft are mapped on to Hα synoptic charts for Carrington rotations 1521-7 (May—December 1967), the neutral lines are found to delineate polarity cells which repeatedly act as release sites for protons on successive rotations. Although the proton acceleration may usually be identified with a given active region, that region is often not contained in the polarity cell from which the particles were injected into the interplanetary medium. This implies a large-scale ordered transport of protons in the outer corona, since it has been shown independently that there can be extended acceleration of $\lesssim 1$ MeV protons without the necessity of extended storage.

1. Introduction

A technique has been developed for the study of low-energy solar charged particles that yields their coronal profiles and the influence of magnetic structures on their coronal propagation. The method is based on the following four points which have been justified elsewhere [1—3]:

(i) low-energy solar charged particles follow interplanetary magnetic field lines closely and undergo negligible diffusion perpendicular to the field; (ii) the large scale structure of the interplanetary field is determined by the bulk motion of the solar wind; the solar wind plasma observed at earth originates near the solar equator and the source longitude of the plasma may be deduced from the observed solar wind velocity; (iii) coronal structure can be inferred from strong chromospheric features deduced from Hα filtergrams; (iv) points (i) and (ii) applied to data from multiple, well separated spacecraft clearly separate spatial and temporal features in the observed fluxes of low-energy solar charged particles.

Carrington solar rotations 1521—1527 (May—December 1967) have been analyzed by this method; and the results of rotations 1523—1525 are presented in [1]. Here, only the highlights of rotation 1521 will be considered, but all the important features of the rotation are summarized in Table 1.

Table 1
Analysis of Rotation 1521

Time (UT)	Connection longitude (Carrington)	Particle flux history	Hα structure	Solar wind	Conclusions
1800–2000 24 May	317		3 flares (2N, 2B, 2B) in 8818		
25 May	290–250	Very high flux ($F_{E34} = 25000$) event. Initial decay constant $\tau \sim 1$ hr.	Connected to large negative cell containing 8818		Temporal event with strong western boundary
0000 26 May to 1800 27 May	250	Change in decay slope, $\tau \sim 10$ hr.	Enter dwell around small positive cell 240–250°	Fast stream rapidly shifts connection point to dwell	Decay is purely temporal since in dwell
0000 28 May	232			Start 2-day dwell	
0550 28 May		Event onset at 0700 ($F_{E34} \sim 1300$)	3B flare in 8818 at ~225°		Temporal event in solar wind dwell
1400 30 May		Shock associated increase at 1426		In dwell until shock passage	
1700 30 May	190–200	Sharp drop in flux at all energies	Strong neutral line 190–200°		Spatial structure associated with neutral line crossing. Particles excluded from positive cell 180–200°
1 June to 0800 3 June				No data	
0800 3 June		3 June particle event ($F_{E34} \sim 28$) high electron flux $F_{>70keV} = 250$	Probable cause is flare in 8818 on invisible hemisphere		

1800 3 June to 2000 5 June	165 until 1200, 5 June; then rapid shift	Continual rise in flux during dwell. $F_{P7} = 11$ at 1200, 2 June (153°); $F_{E34} = 22$ at 0700, 5 June (153°); $F_{P6} > 50$ at 1200, 11 June (153°). Strong field-aligned anisotropy (10 to 40%) seen continuously by Explorer 34	Connected to negative cell containing 8824, 8818	2-day dwell. Continuous increase in solar wind speed (km s^{-1}) at 153°, $V_{P7} = 349$, $V_{E34} = 365$, $V_{P6} = 415$	Continuous acceleration of particles
1200 5 June	145	Rise in flux	Connection longitude crosses into poorly defined positive cell 125 to 150° containing part of 8824		Positive cell 125—150°. Better connection to acceleration region
0000 6 June	125	Sharp drop in flux	Connection longitude enters well-defined, isolated negative cell ~120°		Strong neutral lines are barriers to particle propagation
1200 6 June to 0000 8 June				No data	
0800 6 June		6 June particle event ($F_{E34} \sim 100$) high electron flux $F_{>70keV} > 400$	Probable cause is flare in 8818 on invisible hemisphere		
9 June to 12 June	84—76	Smooth decay of proton flux. Electron flux nearly constant		Solar wind dwell at ~80°	Temporal decay of proton flux. Continual production of electrons
12 June to 24 June	· 60 (rotation 1521) to 270 (rotation 1522)	Slow rise and fall of flux seen by all spacecraft. Explorer 34 proton flux $F_{E34} = 8$. Explorer 34 electron flux $F_{>70keV} = 180$. Event shape similar for all spacecraft. Electron sensitivity increases observed Pioneer fluxes	Large weak positive cell in negative network connected to 8818, 8824. Only small flux changes corresponding to weak neutral line crossings	Uniform except 2.5 day dwell at 335°	Long lived spatial event with temporal evolution. Spatial structure spans 150°. Electron and proton fluxes evolve separately

2. Analysis of Rotation 1521

The solar activity chart shown in Fig. 1 displays the solar connection longitude as a function of time for the spacecraft used in this study and the position, McMath plage region, time and intensity of Hα flares and X-ray bursts. Symbols along a horizontal line represent simultaneous events. As an example, consider 28 May 1967, which is day 148. At 00 UT Explorer 34 is at 194.3° Carrington longitude.

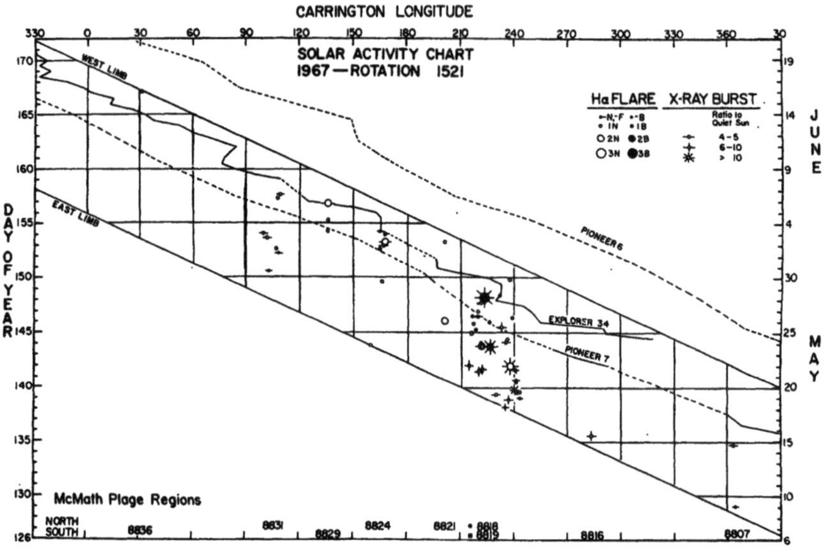

Fig. 1. Solar activity chart for Carrington rotation 1521. The connection longitude for Explorer 34 is dashed during data gaps greater than a day. The Pioneer lines are dashed for gaps greater than 2 days. Flares from active regions at the same longitude are indicated by different symbols (squares and circles).

The visible disk of the sun therefore extended from 104.3° at the east limb to 284.3° at the west limb. The solar wind velocity is 680 km s⁻¹, so the coronal connection longitude for the field line through Explorer 34 is

$$194.3^\circ + \frac{(14.3^\circ/\text{day})\,(1.5 \times 10^8\ \text{km AU}^{-1})}{(680\ \text{km s}^{-1})\,(8.64 \times 10^4\ \text{s d}^{-1})} = 230.8^\circ\ \text{Carrington}.$$

On day 148 there is also a large solar flare of importance 3B with X-ray intensity 94 times that of the quiet sun, which is the large 8-spiked circle at 225° on day 148.

The Hα synoptic chart for rotation 1521, constructed by P. S. McIntosh of NOAA/ERL, is the upper panel in Fig. 2. The techniques for the construction of synoptic charts of chromospheric magnetic neutral lines have been reviewed by McIntosh [4]. The Hα structures from which the charts are constructed (filaments, filament channels, plage corridors, arch-filament systems, and fibril patterns) undergo considerable evolution during their transit across the disk.

In order to give a better indication of their influence on particle fluxes received at earth, the features are mapped from their appearance at $\sim$W 60° rather than at central meridian. The lower panel of Fig. 2 shows the observed proton flux mapped back to its solar connection longitude. The data are from the 1—10 MeV channel of the solar proton monitor experiment on Explorer 34 (F_{E34}), the 0.3 to 12 MeV proton channel on Mariner 5, the 0.6—13.9 MeV and 0.6—12.7 MeV

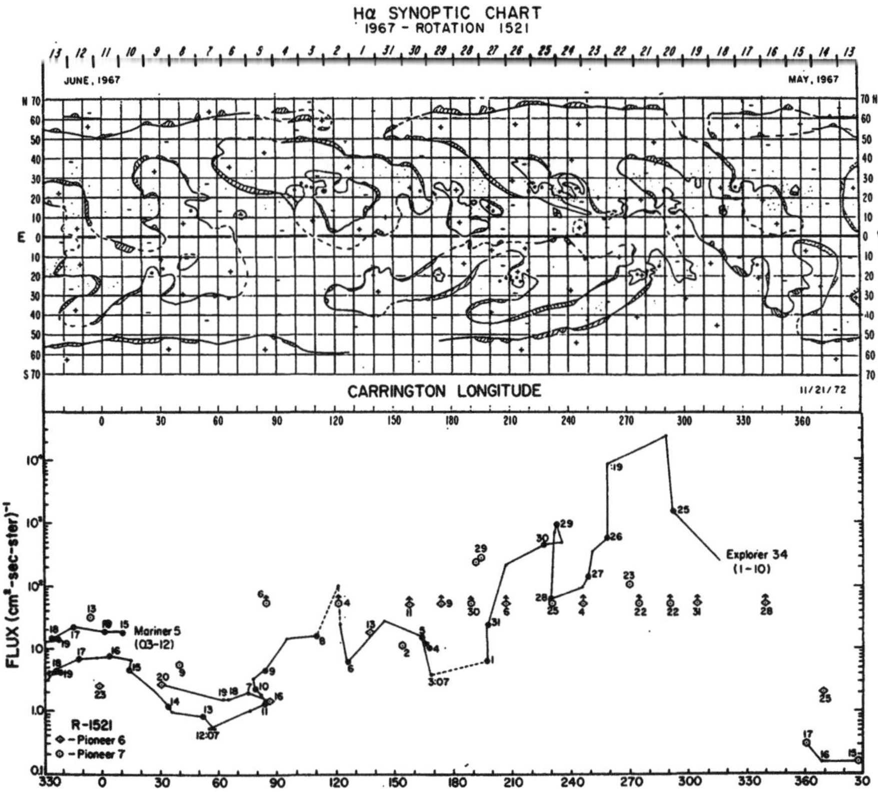

Fig. 2. Upper panel: synoptic chart of chromospheric magnetic field features derived from Hα filtergrams for rotation 1521 prepared by P. S. McIntosh. Filaments, hatched; definite neutral lines, solid; weak or inferred neutral lines, dashed. Sunspots, large dots; plage areas, stippled. Magnetic polarity: +, outward; —, inward.
Lower panel: observed particle intensities plottedat their connection longitudes. Numbers mark day of observation.

channels of the University of Chicago experiments on Pioneers 6 and 7 respectively (F_{P6}, F_{P7}). The presentation of data from multiple, well-separated spacecraft enables spatial and temporal events to be easily distinguished. A temporal event is characterized by simultaneous flux changes, regardless of connection longitude; whereas for a spatial event, the fluxes change at the same connection longitude regardless of the time.

3. Intercomparison of Coronal and Interplanetary Data

The entire rotation was dominated by McMath plage region 8818 which was responsible for the five major low-energy particle events during the rotation. While on the visible disk 8818 produced several large flares, three of which were accompanied by strong related disturbances in X-rays, radio bursts and energetic particles. Four other active regions (8821, 8824, 8829, and 8831) also produced strong Hα flares or X-ray bursts and are responsible for some of the smaller particle flux features during the rotation. 8818 was a rich electron-producing region and all the events associated with it show large electron fluxes.

Explorer 34 was launched on 24 May into the particle flux from the 23 May flares. The solar connection during this event was in the large negative cell stretching from Carrington 190° to 290°. Pioneer 6 which was connected at 10° saw only a very small flux increase indicating that the particles could not cross the intervening structures. Late on 25 May Explorer 34 intersected a fast solar wind stream which rapidly shifted the connection point to a solar wind "dwell" (connection longitude nearly constant for more than a day) near 250°. With this shift in connection point, the decay time constant changed from $\tau \simeq 1$ hr to $\tau \simeq 10$ hrs. This high speed solar wind stream and dwell appear to be the result of an open magnetic field configuration related to the small positive region around 245°. There was a large flare in 8818 at 0550 UT on 28 May while Explorer 34 was in a solar wind dwell at ~230°. Since the spacecraft was in a dwell, there was little coronal distortion of the event profile and the observed time history is purely temporal. An interplanetary shock passed the spacecraft at 1426 UT on 30 May and produced a sharp spike in the proton flux [5, 6] and began a rapid shift in the connection longitude. At 1700 UT, when the fast stream had shifted the connection point across the filament at ~195°, there was a sharp drop in the observed flux at Explorer 34 indicating that the filament corresponds to a coronal propagation barrier. This implies that the particles were confined to the negative cell that contained region 8818, and were excluded from the positive cell at 180° to 200°. It has been suggested that the drop in flux was caused by the shock sweeping the particles ahead of it [7]. However, the 2.5-hour time lapse between the shock passage and the flux change is inconsistent with this explanation.

The only flare on the visible disk that can be associated with the 3 June event was a 1 N flare in region 8831. However, the high electron flux (250 $\text{cm}^{-2}\,\text{s}^{-1}\,\text{sr}^{-1}$ for $E > 70$ keV) seen during this event [8] and the continued low level of activity of 8831 throughout its disk transit suggest that the source of these particles was a flare on the invisible hemisphere in 8818. From 1800 on 3 June until 1200 on 5 June Explorer 34 was in a solar wind dwell and connected to the sun at ~165° Carrington in the negative cell that contains part of 8824 and is connected to 8818. During this time there was a continuous rise in the flux seen at Explorer 34 and the anisotropy remained large (10–40%) and field-aligned. Both the solar wind velocity and the particle flux from this region show progressive increases as they are sampled by the traversal of Pioneer 7 (2 June), Explorer 34 (5 June) and Pioneer 6 (11 June). We believe this increase to be an example of the continuous acceleration of low-energy particles. Region 8824 produced numerous subflares during this period but 8818 could not be observed since it was on the invisible hemisphere. However, the large electron flux favors 8818 as the acceleration region. Extended acceleration was deduced on rotations 1523-5 [1].

The most striking feature of rotations 1521 and 1522 is the long-lived event that produced high fluxes of both electrons and protons that were seen by Explorer 34 for 12 days (12—24 June). The event is spatial, but over the 10-day span which separates comparable Pioneer 6 and Pioneer 7 observations, there is considerable temporal evolution. Pioneer 7 was the first spacecraft to traverse the region, which extends from 60° to 270° on rotation 1522, and observed a flux increase of the same general shape as Explorer 34, Mariner 5 and Pioneer 6 which followed. The peak fluxes recorded by Explorer 34 were 8 $cm^{-2} s^{-1} sr^{-1}$ on 16 June by the 1—10 MeV proton channel and 180 $cm^{-2} s^{-1} sr^{-1}$ on 15 June by the >70 keV electron channel. This event covered a large range of coronal source longitudes which include the positive closed cell that stretches from 345° to 65°. The equatorial features of this cell were the remains of old filaments and other weak and inferred neutral line structures. The observed fluxes were only weakly ordered by this cell with slightly higher fluxes being observed when the spacecraft was connected to the surrounding large negative region as it traversed from 25° to 355° and after it crossed 340°. Thus, it appears that the negative region was well connected to the source of these particles. McMath plage 8818 seems to be the only active region capable of accelerating enough particles, especially electrons, to qualify as the source of these particles. The previous 8818 events on 21, 23, 28 May and 3, 6 June have all shown that the large negative network that dominated the solar equatorial regions during rotation 1521 was relatively open to particles produced in 8818.

The Pioneer detectors have a significant electron sensitivity [9]. Allowing for this, it appears that the temporal development of the flux reached its peak about the time of Explorer 34 transit and had significantly decayed by the time Pioneer 6 passed through. The electron and proton temporal decays are not simultaneous. Explorer 34 is in a 2.5-day solar wind dwell from the middle of 17 June until the beginning of 20 June. During this period the observed 1—10 MeV proton flux is constant showing purely spatial and no temporal development. However, the >70 keV electron flux during the same period on the same spacecraft shows a purely temporal decay from 50 to 15 $cm^{-2} s^{-1} s$-r^{-1}.

4. Summary

The four points of analysis have been applied to the low-energy solar charged particle data from several spacecraft for rotation 1521 with the requirement of consistency with solar wind, anisotropy and magnetic field direction. This demanding intercomparison of the data shows that they are strikingly ordered by coronal features inferred from the Hα synoptic charts. Rotation 1521 is dominated by two structures, region 8818 and the large negative polarity network that covers more than half of the equatorial area of the sun. The active region was directly observed to produce large fluxes of low-energy electrons and protons on 23 and 28 May, while all the subsequent events show the same electron-rich characteristic. The release of these particles into the interplanetary medium is not from the active region, but from areas of the large negative network. This implies large-scale ordered transport of low-energy particles in the outer corona.

Acknowledgments

This work has been supported by Air Force Cambridge Research Laboratories contract F19628-73-C-0070 and National Oceanic and Atmospheric Administration contract 03-3-022-41.

References

[1] E. C. ROELOF and S. M. KRIMIGIS, J. Geophys. Res. **78**, 5375 (1973).
[2] E. C. ROELOF, Proc. Symp. on High Energy Phenomena on the Sun (Goddard Space Flight Center), NASA/Goddard preprint X-693-73-193, p. 486 (1973).
[3] J. T. NOLTE and E. C. ROELOF, Solar Phys. **33**, 241 (1973).
[4] P. S. MCINTOSH, Rev. Geophys. **10**, 837 (1972).
[5] K. W. OGILVIE and J. F. ARENS, J. Geophys. Res. **76**, 13 (1971).
[6] R. A. R. PALMEIRA et al., Solar Phys. **21**, 204 (1971).
[7] I. D. PALMER, Solar Phys. **27**, 466 (1972).
[8] F. R. ALLUM et al., Solar Phys. **17**, 241 (1971).
[9] J. RETZLER and J. A. SIMPSON, J. Geophys. Res. **74**, 2149 (1969).

Astronomy

Space Research XIV — Akademie-Verlag, Berlin 1974

PROPERTIES OF COSMIC X-RAY SOURCES

S. Hayakawa

Department of Physics, Nagoya University, Nagoya, Japan

Recent progress in space observation of cosmic X-ray sources is reviewed. Observational bases of the classification of X-ray sources are given referring mainly to UHURU results. Galactic X-ray sources associated with compact stars are discussed. X-rays from radio pulsars are described. Companions of six close binary systems are found to be X-ray sources. Some of them are probably neutron stars, whereas Cyg X-1 is likely to be a black hole. X-ray and optical variations of Sco X-1 can be understood by the emission from an optically thick plasma.

1. Introduction

Recent progress in X-ray astronomy was extensively discussed at the IAU/COSPAR Symposium [1] at Madrid in May, 1972. Many fruitful results have been made available by the UHURU spacecraft, in orbit since December 1970. About 120 sources have been catalogued, and the spectra, time variations and brightness distributions studied for some of them in the energy range 2—10 keV. Important information was also obtained by means of other spacecraft, rockets and balloons, covering wider energy ranges, attaining better time resolutions and achieving finer angular resolutions.

It is not possible to discuss all the results in the present review, and we restrict ourselves to galactic X-ray sources, paying particular attention to compact X-ray sources associated with degenerate stars. The existence of compact stars has been predicted theoretically, but their detailed study has been developed only after they were found to be relevant to astronomical observations. It has thus been realized that these peculiar objects play a fundamental part in the understanding of Nature.

The white dwarf is a star in which gravitational attraction is balanced by the quantum mechanical pressure of degenerate electrons. If the mass of a star exceeds a limit, called the Chandrasekhar limit, the star can no longer be supported by the Fermi pressure of degenerate electrons but begins to collapse. The collapse may cease owing to the quantum mechanical pressure of degenerate neutrons. A star in such a stable stage is called a neutron star. If the mass is still larger, the pressure of any constituent cannot prevent the star from collapsing, and the radius of the star becomes smaller than the Schwarzschild radius, at which the gravitational red shift is infinite. Since no signal can come out, the star is called a black hole. The masses and radii of these stars are theoretically predicted as shown in Table 1.

White dwarfs have been observed optically and radio pulsars have been identified with neutron stars, although the identification of X-ray sources with neutron stars was suggested as early as in 1963. The existence of black holes has been

Table 1
Mass Limits of Compact Stars

	Initial mass	Events	Final mass	Radius
White dwarf	$1-3\,M_\odot$	Mass ejection at RG	$\leqq 1.4\,M_\odot$	$\sim 10^9$ cm
	$3-8\,M_\odot$[a]	C detonation SN	none	
Neutron star	$8-30\,M_\odot$[a]	Fe dissociation SN	$\lesssim 3\,M_\odot$[b]	$\sim 10^6$ cm
Black hole	$30-100\,M_\odot$	Fe dissociation SN	$\gtrsim 3\,M_\odot$[b]	$\leqq 3\left(\frac{M}{M_\odot}\right) \times 10^5$ cm

[a] These mass values are based on a theoretical study of implosion and subsequent explosion processes taking place at the final stage of stellar evolution. If, however, the number of pulsars and the spatial distribution with respect to the distance from the galactic plane are incorporated with the mass function of stars, the lower limit of initial mass for the neutron star is estimated as about $4\,M_\odot$.

[b] The mass limit of the neutron stars depends on the equation of state of highly condensed nuclear matter. The value of about $3\,M_\odot$ is obtained for a nuclear force with a hard core plus a weak repulsive force.

advocated through analysis of a few binary systems, and a pulsating X-ray source, Cyg X-1, is considered to be a more likely source of the black hole. The compact star has been suggested to be an X-ray source, since matter accreting on to the star gains energy in a strong gravitational field. The accreting matter forms a hot plasma at or near the surface of the star by the conversion of infalling kinetic energy into thermal energy. The accreting matter is supplied by a companion star of a close binary system.

The above theoretical suggestions had been advocated before X-ray pulsars and X-ray binaries were discovered. However, the observed results are not mere confirmation of the theoretical predictions, but reveal a wealth of properties that were not predicted. The real nature of highly condensed matter and the behaviour of the peculiar plasma surrounding the compact star are beginning to be understood as observational X-ray astronomy develops.

2. Classification of X-Ray Sources

Among about 120 X-ray sources in the UHURU Catalogue about two thirds near the galactic plane ($|b| < 20°$) are galactic, whereas the rest at high galactic latitudes ($|b| > 20°$) are mostly extragalactic. The classification of X-ray sources into these two classes is based on the identification with known objects for about one fifth of the galactic sources and for about one third of the extragalactic sources, the concentration toward the galactic plane for the former and isotropy for the latter, and number-intensity distributions that are significantly different from each other.

The number of sources N per steradian with intensities greater than S in units of counts/sec is obtained, taking into account the sky coverage correction for the observed $N - S$ relation [2]. The functions $N(S)$ expressed on a log-log scale for these two classes are reproduced in Fig. 1, in which two sources identified with

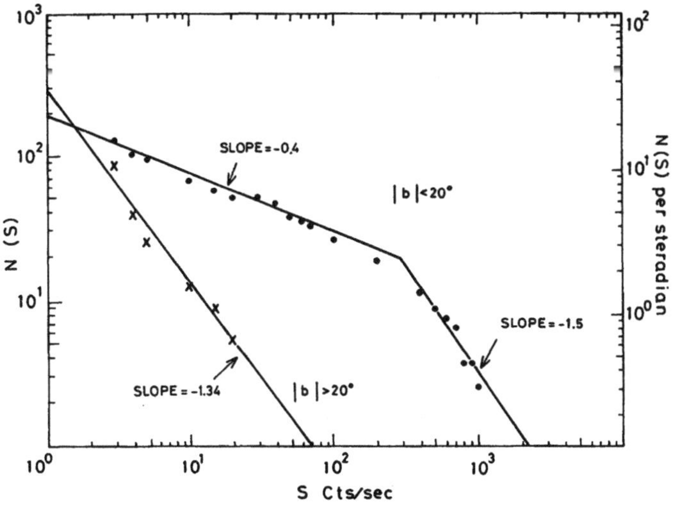

Fig. 1. Number-intensity distribution of X-ray sources. The numbers N per steradian of X-ray sources with intensities greater than S counts/sec are plotted separately for sources with $|b| < 20°$ and $|b| > 20°$. $S = 1$ corresponds to the energy flux of 1.7×10^{-11} erg cm^{-2}s^{-1} for the energy range 2—10 keV. After [2].

galactic objects with $|b| > 20°$ are excluded from the extragalactic class. If sources are uniformly distributed in space, and if their absolute luminosities in a given energy range are independent of distance, the slope of the log N — log S relation should be -1.5. The slope observed for the extragalactic class is -1.34 ± 0.2, consistent with the slope expected for uniformly distributed extragalactic sources.

The log N — log S relation for the galactic sources cannot be represented by a single straight line. The slope for $S < 200$ is consistent with the value of -0.5 that is expected for sources distributed along a line. These sources are fairly uniformly distributed over the galactic longitude except for clustering in spiral arm regions, such as in the Cygnus region. They belong to the disk population, and their luminosities are of the order of 10^{36} erg/sec. Stronger sources with $S > 200$ are concentrated toward the galactic centre and in a lower latitude region. They seem to be located in the galactic bulge, and their luminosities are of the order of 10^{38} erg/sec. In addition, several sources are visible only in the soft X-ray region. Their average luminosity may be of the order of 10^{34} erg/sec.

Among galactic X-ray sources six have been identified with supernova remnants. Two of them, Crab nebula and Vela X, are associated with pulsars of the first and second shortest periods, respectively. They are extended as large as the optical nebulae and contain pulsating components. These X-ray pulsars are identified with radio pulsars NP 0531 and PSR 0833-45, respectively. Their detailed properties will be discussed in § 3.

X-ray emissions have not yet been observed from other radio pulsars. On the other hand, there are several X-ray pulsars that are not associated with radio pulsars. Among them Cen X-3 and Her X-1 show regular pulsations and also eclipsing characteristic of close binary systems. There are some other sources that show similar binary features, but their pulsations have no regular periods. Cyg X-1 is pulsating but with varying periods. Although it does not show eclipsing, its optical counterpart shows binary features and coincides with a radio source whose radio emission correlates with the X-ray emission. These binary sources will be further discussed in § 4.

The strongest X-ray source Sco X-1 has been observed in more detail than other sources, but the mechanism of X-ray emission is yet unclear. § 5 will be devoted to a discussion of this.

Most galactic sources are found to be variable. Variability seems pronounced in the soft X-ray region, in view of the fact that some soft X-ray sources are detected at one time and are not always detectable at other times. Dramatic variability has been observed for some sources whose time variations are similar to those of optical novae. Only about one third of the galactic sources are known to belong to either of the above classes. Further studies of other sources will disclose whether or not there are new classes other than those discussed above. Among extragalactic sources four are located in Magellanic Clouds. SMC X-1 in Small Magellanic Cloud is an eclipsing binary. The X-ray emission of Andromeda Nebula is as strong as that of our galaxy. Several extended sources are identified with clusters of galaxies, whereas several point sources are identified with radio galaxies, a Seyfert galaxy and a quasi-stellar object. Sources in the latter classes as well as the source at the nucleus of our galaxy are probably associated with the compact galaxies, and they may share the same features with the compact stars in some respects, although the masses, sizes and energy generation rates are significantly different. In the present review we do not discuss extragalactic sources but concentrate on three classes of galactic sources.

3. X-Rays from Radio Pulsars

Among about 60 radio pulsars thus far found, the Crab pulsar NP 0531 and the Vela pulsar PSR 0833-45 have periods shorter than 0.1 sec, and the periods increase with e-folding times comparable with the ages of the supernova remnants (SNR). They are regarded as remnant neutron stars left after supernova explosions and are still active to supply radiation energy at the expense of rotational energy. Their characteristic properties are given in Table 2.

They are also the only pulsars from which X-ray emissions have been detected. Most X-rays are emitted from nebulous regions that approximately coincide with optical nebulae. A fraction of the X-rays are pulsed with the same periods as the radio pulsations. Detailed properties of the Crab pulsar are extensively discussed in [1] and will not be repeated here.

Pulsed X-rays from the Vela pulsar were first indicated by Harnden et al. [3] in the hard X-ray range 23—80 keV. Recently Harnden and Gorenstein [4] have detected pulsed X-rays in the soft X-ray range 0.5—1.5 keV. According to the

Table 2
X-ray Sources associated with Radio Pulsars

Radio pulsar		NP 0531	PSR 0833-45
Period P (ms)		33	89
Period Change $P/\dot{P}$ (year)		3×10^3	2×10^4
Age (year)		9×10^2	2×10^4
$\frac{\text{Pulsed comp.}}{\text{Nebula comp.}}$ (%)	Radio	0.1	0.03
	Optical	0.3	$< 10^{-5}$
	Soft X	3	2
	Hard X	20	6

latter, about 6% of the intensity of the entire nebula comes from the position of the pulsar, and about one third of it is pulsed. The period of X-ray pulsation agrees with that of radio pulsation, whereas the phase of the former is shifted more than expected from a delay of radio propagation in the interstellar medium. This may be interpreted as that X-rays are mainly emitted between the pulses of the radio pulsation, analogous to the case of the Crab pulsar. The spectrum of the pulsed component is harder than that of the nebula component [5]. A harder spectrum of the pulsed component has been observed also for the Crab nebula.

Although the Vela SNR is similar to the Crab SNR in many respects, they are different in the following two points: The pulsating optical emission from the Vela pulsar has not been detected, whereas the central star of the Crab nebula is optically pulsating in phase with the X-ray pulsation. X-rays from the Crab nebula are emitted by synchrotron radiation of relativistic electrons, as indicated by their linear polarization and power-law spectrum, whereas the Vela-X nebula emits X-rays, at least in part, from a hot plasma produced by the propagation of a shock wave in the interstellar medium.

Apart from minor differences, the similarity between the Crab and Vela pulsars indicates a common mechanism of X-ray emission associated with rapidly rotating neutron stars. Although the detailed mechanism is still to be investigated, the release of rotational energy as indicated by an increase of the pulse period maintains the activity of the SNR. In fact, the energy emitted as X-rays is of the order of 10% of the loss of rotational energy. The pulsed component may be associated with beamed emission in the vicinity of the neutron star. Since the X-ray luminosity is supposed to be proportional to the loss rate of rotational energy, other pulsars with small loss rates can hardly emit detectable X-rays.

4. Binary X-Ray Sources

4.1. Cyg X-1

A short period pulsation was discovered for Cyg X-1, from which a significant part of the energy is emitted as pulses. The pulse period was, however, found to vary; the pulses form a train that lasts for about 10 sec. Although the pulsation of Cyg X-1 is different from that observed for radio pulsars, a very short pulse width indicates that the source size is comparable with or smaller than the radius of the white dwarf.

Within an error box of the position of Cyg X-1 an early type supergiant was found that showed a sinusoidal variation of the wavelength of an emission line, which is characteristic of a binary system. Although eclipsing of the X-ray source was not observed, the identification of Cyg X-1 with the binary system was established by a correlation between the spectral variation of Cyg X-1 and the flux variation of a radio source that appeared in April 1971 at the position of the binary system.

An emission line of He II λ 4686 is variable and appears to be associated with matter infalling on to the X-ray source [6, 7]. This also supports the mass accreting binary nature of Cyg X-1. The absence of X-ray eclipse may be attributed to a small inclination angle of the binary orbit.

Assuming that Cyg X-1 is a companion of the supergiant star of mass 22 $M_\odot$, the mass of the X-ray source is determined as greater than 5.5 $M_\odot$, the minimum corresponding to the inclination angle of 90° [8]. For an inclination angle of 30° the mass is doubled. Since the mass of Cyg X-1 is significantly larger than the mass limit of the neutron star given in Table 1, the X-ray emitting star is inferred to be a black hole.

The black hole is a physical state that results from a solution of the Einstein equation. In the spherically symmetric case one obtains the well-known Schwarzschild solution. At the surface of radius 2.7 $(M/M_\odot)$ km the red shift is infinity, and events taking place inside the surface are not observable. The rotating spherical system is described by the Kerr solution. In this case the surface of event horizon lies inside the surface of infinite red shift. The rotating non-spherical system is complicated as recently shown by Tomimatsu and Sato [9]. According to their solution a singularity appears outside the surface of event horizon, and its effect may be observable. The black hole nature of Cyg X-1 will be established if such general relativistic effects are observed. However, X-ray emission probably takes place in a region rather far from the stellar surface, in which general relativistic effects may not be important.

4.2. Cen X-3

A regular pulsation of X-ray emission was found for Cen X-3. The pulsation period of about 4.84 sec varies sinusoidally with the same period (2.087 days) as the X-ray eclipsing. The transition state between the high and low intensity states is as long as 50 min, which is interpreted as due to the absorption of X-rays by the extended atmosphere of a companion star. From these orbital elements one can obtain the orbit radius of about 0.1 AU and the mass of the X-ray source smaller than 1 $M_\odot$. The optical counterpart is not observable because of large extinction. Both the pulsation period and the mass of Cen X-3 are consistent with those expected for a neutron star, but may also be consistent with a vibrating white dwarf because of the rather long pulsation period and the small mass. However, the latter alternative is not favourable for the soft X-ray spectrum, as will be discussed below. The rotating beam model proposed for the radio pulsar favours the presence of a possible long low-intensity state before the discovery of Cen X-3, since this takes place when the beam misses the earth as the rotation axis changes.

The binary nature of X-ray sources is favoured, since matter in the atmosphere of a primary star flows through the Roche point on to a secondary star. The infalling matter gains energy in the strong gravitational field of the compact star, and the kinetic energy of the infalling matter is converted to heat at a distance R from the centre of the secondary compact star. The luminosity is estimated as

$$L = GM\dot{M}_a/R,$$

where G is the gravitational constant, M the mass of the compact star and $\dot{M}_a$ the accretion rate. For a given luminosity the accretion rate required is proportional to the radius of the X-ray emitting region. If R is as large as the radius of a light white dwarf, say 10^9 cm, the accretion rate required is large, and the infalling

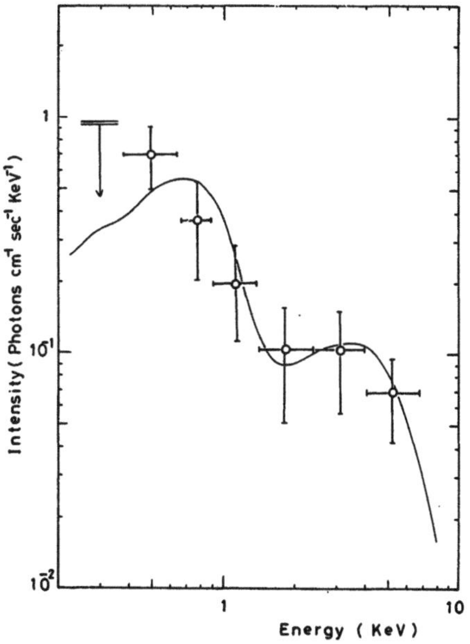

Fig. 2. Soft X-ray spectrum of Cen X-3. The solid line represents the spectrum predicted for the thermal bremsstrahlung of temperature 30 keV partially absorbed by the accreting matter of 3.5 g cm^{-2} thickness and also by the interstellar matter of the hydrogen column density of 5×10^{20} atoms cm^{-2}. The efficiency and energy resolution of the counter are taken into account. After [11—13].

velocity is small. Consequently the thickness of the accreting matter is so large as to absorb X-rays. In order that X-rays are not appreciably absorbed at several keV, the radius of the star should be as small as 10^8 cm or even smaller. Even for such a compact star the absorption is appreciable at low energies, as observed for Cen X-3 [10]. However, the absorption is different from that by neutral gases, since the accreting matter is strongly ionized by X-rays from the central source. Oxygen is almost completely ionized, and the accreting shell is practically transparent at energies lower than the K-edge of O VIII, 0.78 keV. Hence soft X-rays come out without appreciable absorption. The spectrum of Cen X-3 turns up toward low energy, as shown in Fig. 2 [11]. The observed spectrum is in essential agreement with the theoretical prediction taking into account the partial absorption of ionized accreting matter [12, 13].

4.3. Her X-1

Her X-1 is similar to Cen X-3 in many respects, e.g. a regular pulsation with a period of 1.2 sec, a sinusoidal variation of the pulsation period characteristic of the binary system and eclipsing with a period of 1.700 days [14]. However, there are differences in the following respects: There exists a large period of 35 days, in which Her X-1 is observable during about 11 days. This may be interpreted in terms of the precession of an oblique rotator. If X-rays are emitted in the

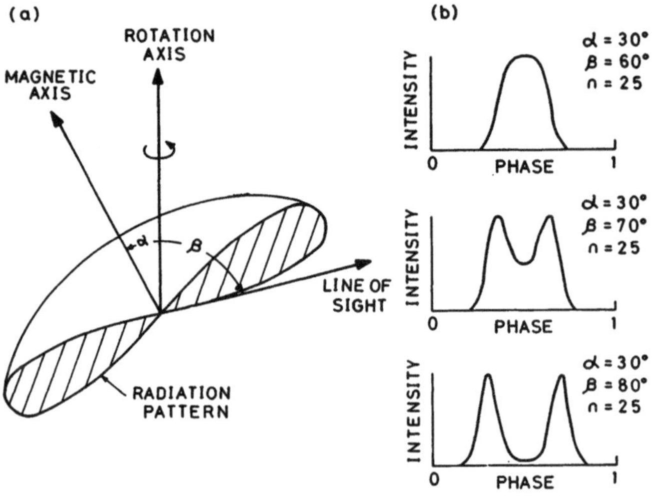

Fig. 3. Oblique rotator model of pulsar. The pulse profiles expected for selective sets of α and β are shown. After [15].

direction perpendicular to an axis, say the magnetic axis, and this symmetry axis is inclined with respect to the rotation axis, the intensity of X-rays forms pulses with the rotational period, as shown in Fig. 3. The pulse profile depends on the angle of these two axes α and the angle between the rotation axis and the line of sight β. If $\alpha + \beta < 90°$, the X-ray beam cannot reach the observer. According to the precession motion, the X-ray intensity waxes and wanes. The oblique rotator model should reveal itself in the pulse profile, as simulated for the radiation pattern of $\sin^n\theta$ in Fig. 3, where θ is the polar angle from the symmetry axis. The pulse profile observed in the energy range 1.2—11 keV is given in Fig. 4 [15], resembling a pattern expected from the model, although the asymmetry and other details may require further elaboration. It seems necessary to observe the phase dependence of the pulse profile in the large period before establishing the model.

The X-ray spectrum of Her X-1 has been extended to the hard X-ray region by satellite [16] and balloon [17] observations, as shown in Fig. 5. The spectrum in the energy range 2—70 keV may be represented by a power law of E^{-2}, but its slope increases with energy so that $6.3\, E^{-2.38}$ cm^{-2} s^{-1} keV^{-1} gives better fit in the range 20—70 keV [17]. The spectrum may also be represented by an exponential law with the apparent temperature of 19 ± 10 keV. It should be remarked that the spectral shape resembles that of Cyg X-1, thus suggesting a common origin of X-rays for the two sources.

The optical counterpart of Her X-1 has been identified with a variable star HZ Her by several authors. In the X-ray eclipsing phase the brightness is minimum and the spectral type based on the Ca II K line is late A. The brightness varies

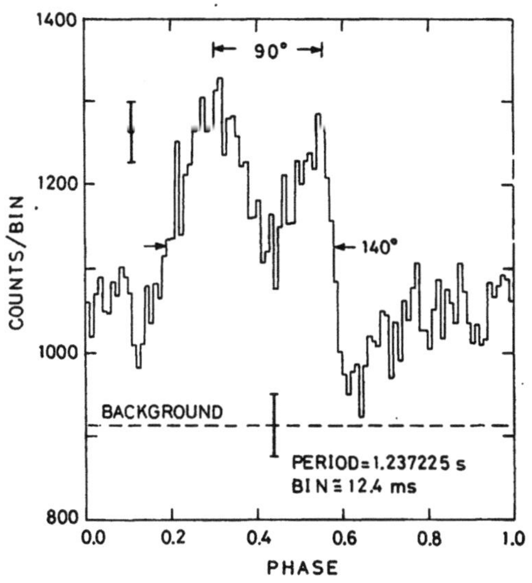

Fig. 4. X-ray pulse profile of the Her X-1 pulsar. After [15].

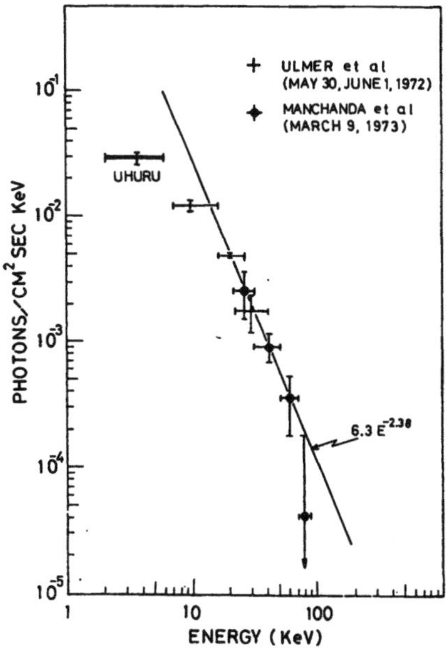

Fig. 5. Hard X-ray spectrum of Her X-1. After [17].

with phase of the orbital motion of the binary system derived from the X-ray observation, and the spectral type shift to B near the maximum brightness. The optical variation persists even when the X-ray source is turned off in the 35-day cycle. The optical variation is interpreted in terms of irradiation of the stellar surface by X-rays from the orbiting X-ray source. Detailed observation of the light curve shows a deviation from symmetry with respect to the X-ray

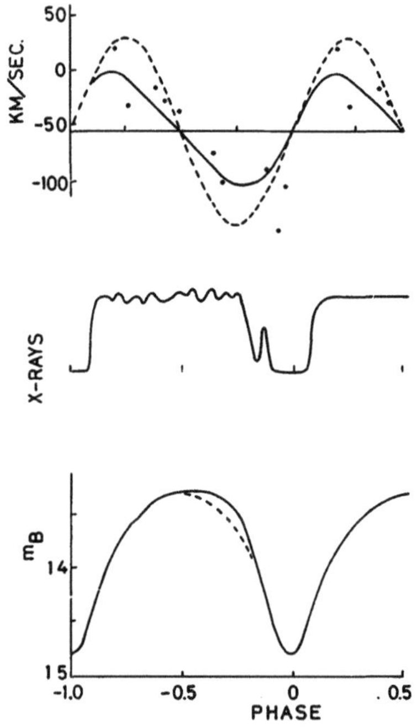

Fig. 6. Light curve of HZ Her. Top: Balmer-line radial velocities. ---- expected variation, —— line smoothly connecting observed values (●).
Middle: Typical variation of the X-ray intensity.
Bottom: —— B light curve. ---- symmetric variation.

phase, as shown in Fig. 6 [18]. The deviation is accounted for in terms of heating and distortion of the atmosphere of HZ Her by the companion X-ray star. The amplitude and shape of the light curve are found to change with time in the 35-day cycle [19].

Further observation of Her X-1 X-rays shows more detailed features [20]: (i) The pulsation period decreased by about one part in 5×10^5 from February to July 1972. This may be explained by the angular momentum transfer from the accreting matter. (ii) The 35-day period also decreased by about one day in March 1972 and seems to decrease further. (iii) The turn-on of X-rays in the 35-day cycle is fairly sharp, the intensity reaches the maximum after about three days, and it declines gradually until turn-off after about 11 days. (iv) The turn-on epochs are concentrated to two regions of the binary phase. (v) In the binary period the X-ray intensity shows a dip a few hours before the eclipse, and the dip occurs progressively earlier as one approaches the end of the "on" state of the 35-day cycle.

Table 3
Binary X-ray Sources

Source	Period (day)	Pulsation period (s)	Long-term variation	Spectrum[a] $E^{-\gamma}\exp(-E/E_a)$ γ	E_a (keV)	Mass $(M_x/M_\odot)$	Luminosity (10^{37} erg/s)	Distance (kpc)	Optical object
Cyg X-1	5.60	$\gtrsim 0.05$ (quasi)	Single transition	1.8	$\leqq 1.5$	>6	1	2	HDE 226868
Cen X-3	2.09	4.84	Extended low	0.4	1.5	<1			
Her X-1	1.70	1.24	35 d	1	1.5	~ 1	1	5.8	HZ Her
SMC X-1	3.89	$\sim 10^2$ (non-per.)	Extended low	0	1.5		30	61	Sanduleak No. 160
GX 263 + 3	8.95	~ 1 (non-per.)	No	0.2	2.5	>3	0.4[b]	1.3	HD 77581
2U 1700-37	3.41	$\gtrsim 0.1$ (non-per.)	No	0.4	2.1		0.3	1.7	HD 153919

[a] in units of keV/keV. The values of γ and E_a are variable; the values at the high state are given here.
[b] An increase by an order of magnitude at flares.

4.4. Other Binary Sources

Three more sources, SMC X-1 [14], GX263 + 3 [21] and 2U1700-37 [22], have thus far been found to be binaries. Their properties are summarized in Table 3 along with those of three that have been discussed above. None of these three show periodic pulsations. The pulsation amplitude of 2U1700-37 is quite large, whereas the pulsations of other two are barely observable. Their energy spectra show low-energy cut-off, and the cut-off energies are variable.

The spectral variation of the optical counterpart (a supergiant star) of GX 263 + 3 has been recently observed, and the mass of the X-ray source is derived from the Doppler shift of a spectral line and the binary period may be greater than $3M_\odot$. Hence GX263 + 3 is another likely source of black hole.

To these sources may be added Cyg X-3 that flares in the radio region [23]. The intensity variation with a period of 4.8 hours could be attributed to its binary

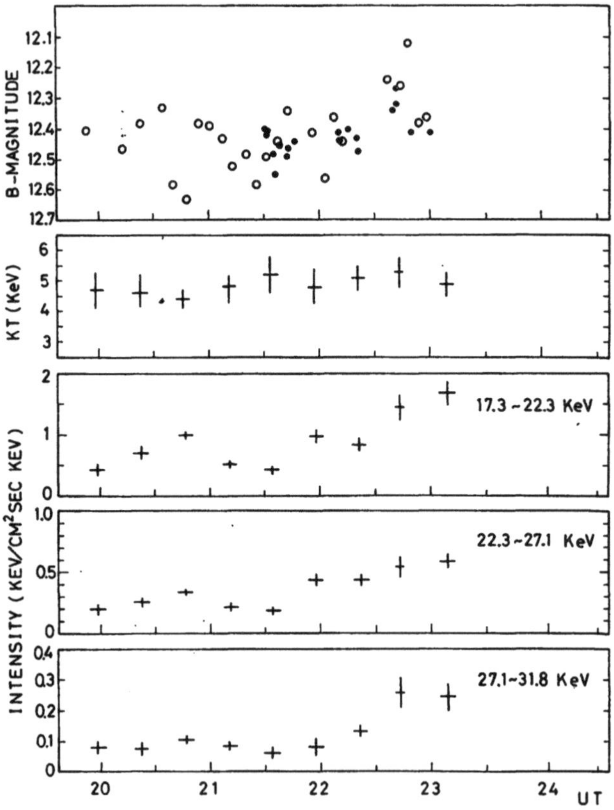

Fig. 7. X-ray and optical variations of Sco X-1 on 16 April 1972. The open and filled circles in the top figure represent the B-magnitudes obtained by photographic and photoelectric observations respectively. The temperatures are derived assuming simple exponential spectra. X-ray enhancements in three energy ranges took place simultaneously with an optical flare at about 2245 UT. After [29].

nature, but the period is much shorter than the periods of other close binary systems, and the time variation is sinusoidal and does not show sharp eclipses [24—26].

5. Sco X-1

The brightest X-ray source Sco X-1 has been observed most extensively not only in the X-ray band but also in the optical and radio bands. However, its nature is not yet fully understood. Cyg X-2 shows similar features in many respects. These two have spectra of the thermal bremsstrahlung type with high energy tails, and their intensities are variable by factors of two to three. As presented in [1], particularly on the basis of a long-term observation over 25 days by OSO 7, the X-ray emission from Sco X-1 fluctuates occasionally even in the quiescent period and continually in the active period. The presence of pulsations has been searched for, but without positive result [27]. Although no apparent correlation between X-ray and radio activities has been found, a positive correlation

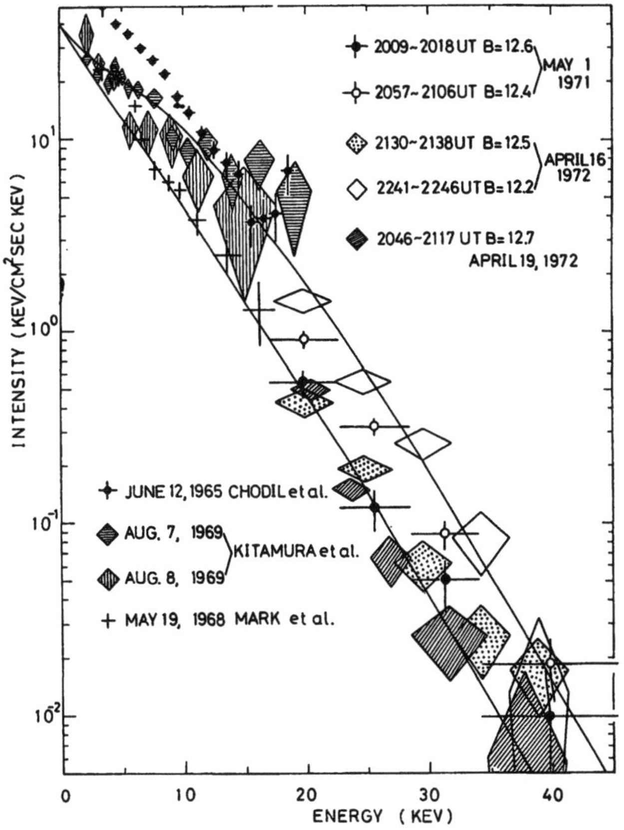

Fig. 8. Hard X-ray spectra of Sco X-1 at different B-magnitudes. The solid lines represent theoretical spectra for the plasma temperature of 3 keV and for the optical depths for electron scattering of 9 and 11 respectively. After [29].

of optical emission with the intensity and temperature of X-ray emission has been observed [28]. Since the apparent temperature derived from the observed spectrum depends rather sensitively on the optical depth of an X-ray emitting plasma due to Compton scattering of X-ray photons by electrons, it is desirable to separate the effect of scattering from that of free-free emission. This can be achieved by observing hard X-rays, for which the spectral shape depends only weakly on the optical depth. In Fig. 7 is shown the result of simultaneous hard X-ray and optical observations [29]. The intensity of hard X-rays positively correlates with B-magnitude, whereas the apparent temperature obtained from the spectrum in the energy range 20—40 keV does not significantly change. The spectra at different B-magnitudes are given in Fig. 8. Theoretical spectra [30] for a plasma temperature of 3 keV and for the optical depths of 9 and 11 are compared. The theoretical spectra change slope as energy decreases and may fit some spectra obtained by rockets in this energy range.

The value of optical depth is consistent with the enhancement of X-ray emission relative to that of optical emission. For example, the X-ray intensity is doubled when the optical brightness increases by 0.2 magnitude. A smaller value of optical enhancement is due to free-free absorption that is appreciable in the optical region for a thick plasma. The enhancement can be accounted for in terms of an

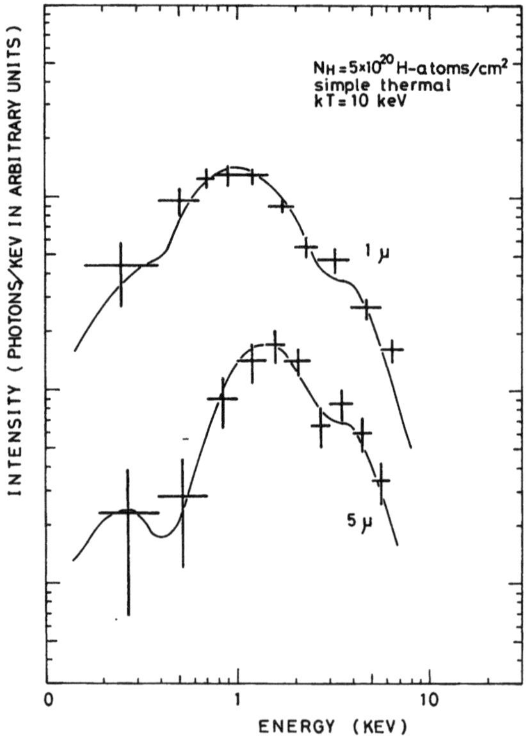

Fig. 9. Soft X-ray spectrum of Sco X-1. The solid curve represents the exponential spectrum of temperature 10 keV modified by the interstellar absorption for the hydrogen column density of 5×10^{20} atoms cm^{-2}. The effeciency and energy resolution of the counter are taken into account. After [32].

increase of the plasma mass without appreciable change of the plasma temperature [31]. The supply of matter and energy may not be attributed to the accretion as in the case of binary sources, since the matter accreting on to a rather large source whose thickness is estimated to be as large as 10^9 cm would have to absorb low energy X-rays.

The soft X-ray spectrum shown in Fig. 9 [32] does not show an absorption effect such as the one observed for Cen X-3, in spite of the fact that a stronger absorption effect would be expected if the X-ray emission were caused by accretion. The low energy cut-off of the spectrum varies from one observation to another, and that in Fig. 9 can be explained by the interstellar absorption with a hydrogen column density of 6×10^{20} atoms cm^{-2}. If this is compared with Lyman α absorption data for nearby stars that are imbedded in a rather dense cloud, Sco X-1 seems to be located on our side of the periphery of the cloud at a distance of about 180 pc.

Although several new features of Sco X-1 have been found in the last year, the sources of energy and matter that maintain X-ray emission and cause enhanced radiation still remain to be investigated.

References*

[1] Proceedings of IAU Symposium No. 55 on X- and Gamma-Ray Astronomy, D. Reidel Publ. Co. Dordrecht, Holland 1973.

[2] T. Matilsky, H. Gursky, E. Kellogg, H. Tananbaum, S. Murray and R. Giacconi, Astrophys. J. **181**, 753 (1973).

[3] F. R. Harnden, W. N. Johnson and R. C. Haymes, Astrophys. J. Lett. **172**, L91 (1972).

[4] F. R. Harnden and P. Gorenstein, Nature Phys. Sci. **241**, 107 (1973).

[5] E. Kellogg, H. Tananbaum, F. R. Harnden, H. Gursky, R. Giacconi and J. Grindlay, Astrophys. J. Lett. **183**, 935.

[6] N. R. Walborn, Astrophys. J. Lett. **179**, L 123 (1973).

[7] H. E. Smith, B. Morgan and P. S. Conti, Astrophys. J. Lett. **179**, L 125 (1973).

[8] R. Brucato and J. Kristian, Astrophys. J. Lett. **179**, L 129 (1973).

[9] A. Tomimatsu and H. Sato, Phys. Rev. Lett. **29**, 1344 (1972).

[10] E. Schreier, R. Giacconi, H. Gursky, E. Kellogg and H. Tananbaum, Astrophys. J. Lett. **178**, L 71 (1972).

[11] J. A. M. Bleeker, A. J. M. Deerenberg, K. Yamashita, S. Hayakawa and Y. Tanaka, Astrophys. J. Lett. **183**, L1 (1973).

[12] S. Hayakawa and Y. Tanaka, Astrophys. J. Lett. **183**, L5 (1973).

[13] S. Hayakawa, Progr. Theor. Phys. **50**, 459 (1973).

[14] H. Tananbaum, H. Gursky, E. Kellogg, R. Levinson, E. Schreier and G. Giacconi, Astrophys. J. Lett. **174**, L 143 (1972).

[15] R. Doxy, H. V. Bradt, A. Levin, G. T. Murthy, S. Rappaport and G. Spada, Astrophys. J. Lett. **182**, L25 (1973).

[16] M. P. Ulmer, W. A. Baity, W. A. Wheaton and L. E. Peterson, Astrophys. J. Lett. **178**, L 61 (1962); **181**, L 33 (1973).

[17] R. K. Manchanda, V. S. Iyengar, N. Durgaprasad, G. S. Gokhale, P. K. Kunte and B. V. Sreekantan, Nature, Phys. Sci. **244**, 59 (1973).

[18] D. Crampton and J. B. Hutchings, Astrophys. J. Lett. **178**, L 65 (1972).

[19] L. Petro and W. A. Hiltner, Astrophys. J. **181**, L 39 (1973).

* References published earlier than and discussed at the Symposium [1] are not given here, unless direct comparisons with recent results are made.

[20] R. GIACCONI, H. GURSKY, E. KELLOGG, R. LEVINSON, E. SCHREIER and H. TANANBAUM, Astrophys. J. **184**, 227 (1973).
[21] W. FORMAN, C. JONES, H. TANANBAUM, H. GURSKY, E. KELLOGG and R. GIACCONI, Astrophys. J. Lett. **182**, L 103 (1973).
[22] C. JONES, W. FORMAN, H. TANANBAUM, E. SCHREIER, H. GURSKY, E. KELLOOG and R. GIACCONI, Astrophys. J. Lett. **181**, L 43 (1973).
[23] Special Issue on Cygnus X-3 of Nature Phys. Sci. **239**, No. 95 (1972).
[24] P. W. SANFORD and F. H. HAWKINS, Nature Phys. Sci. **239**, 135 (1972).
[25] C. R. CANIZARES, J. E. MCLINTOCK, G. W. CLARK, W. H. G. LEWIN, H. W. SCHNOPPER and G. F. SPROTT, Nature Phys. Sci. **241**, 28 (1973).
[26] D. R. PARSIGNAULT, H. GURSKY, E. M. KELLOGG, T. MATILSKY, S. MURRAY, E. SCHREIER, H. TANANBAUM, R. GIACCONI and A. C. BRINKMAN, Nature Phys. Sci. **239**, 123 (1972).
[27] S. S. HOLT, E. A. BOLDT, P. J. SERLEMTSON and A. F. BRISKEN, Astrophys. J. Lett. **180**, L 69 (1972).
[28] C. R. CANIZARES, G. W. CLARK, W. H. G. LEWIN, H. W. SCHNOPPER, G. F. SPROTT, R. M. HJELLMING and C. M. WADE, Astrophys. J. Lett. **179**, L 1 (1973).
[29] M. MATSUOKA et al., private communication.
[30] J. NISHIMURA, private communication.
[31] S. HAYAKAWA, I. KASAHARA and M. MATSUOKA, Publ. Astr. Soc. Japan **26**, in press (1974).
[32] A. J. M. DEERENBERG, J. A. M. BLEEKER, P. A. J. DE KORTE, K. YAMASHITA, Y. TANAKA and S. HAYAKAWA, Nature Phys. Sci. **244**, 4 (1973).

Space Research XIV — Akademie-Verlag, Berlin 1974

REVIEW OF ULTRAVIOLET INTERSTELLAR EXTINCTION AND INTERSTELLAR LINES

L. F. Smith

Max-Planck-Institut für Radioastronomie, Bonn, FRG

The interstellar matter is composed of atoms, molecules and dust grains. Both density and temperature are highly variable from place to place. At optical wavelengths, one observes resonance lines of neutral and ionized atoms and molecules in absorption against the continuum spectrum of bright hot stars; most of these lines fall in the ultraviolet. One also observes extinction due to the interstellar dust grains; the most important wavelength range is again the ultraviolet. Observation of resonance lines yields the column density of the absorbing atom or molecule. In this way, one determines the distribution of a given element over its different ionization stages, and its total abundance. These parameters are needed to understand what mechanism is responsible for heating parts of the interstellar medium to temperatures of several thousand degrees. Observation of lines from low-lying excited levels yields an excitation temperature which may provide an indication of the kinetic temperature and density of the gas. Measurement of the extinction due to dust grains (especially if the contributions due to scattering and absorption can be measured separately) may indicate the chemical composition of these grains. Ultraviolet observations, combined with ground-based radio observations, should ultimately be able to answer the principal questions regarding the physics of the interstellar matter.

1. Introduction

Optical observations have, until recently, been limited by lack of lines of important substances in the visible range and by lack of sufficient frequency resolution to resolve the structure of the lines. The frequency resolution has recently been improved by an order of magnitude, to 0.01 Å, by the use of Fabry-Perot scanners. The advent of rocket photographic spectroscopy brought identification of interstellar lines of C, N, O, Si, S, HI and H_2; resolutions were characteristically 0.5—2 Å, and photometric accuracy was not generally high enough for abundance determinations, except for the strongest lines. OAO-A2 gave us photometric accuracy, but without the resolution to observe any but the strongest lines, namely Lyman alpha. Copernicus brings us now to the stage of 0.05 Å resolution and high photometric accuracy at the same time. TD1 has 1.8 Å resolution and photometric accuracy, and rocket and balloon observations are also achieving accurate results at resolutions of $\sim$0.1 Å. Thus, the amount of information from optical observations of the interstellar gas is likely to increase by an order of magnitude in the immediate future.

Until now, most of our knowledge of the interstellar gas has been derived from radio observations. The advantages of radio observations are: that neutral (as well as ionized) gas may be observed in emission, so that one is not dependent on the presence of background sources; lines due to the presence of HI, DI, ^{3}HeII, ^{4}HeII, CII, and many molecules are observed, and the observations are not limited by dust absorption. Thus, most of our knowledge of the large scale properties of the interstellar gas will always depend on radio observations. At optical wavelengths, most of the observations depend on the attenuation of star light by the dust and gas. Thus, one requires the presence of bright background stars and one is limited to distances up to 1 kpc from the sun. The advantages of optical observations are: very high angular resolution (the angular size of the star compared with 9′—35′ arc (21 cm, single dish) at radio wavelengths), and the presence, in the ultraviolet, of lines of the most important heavy element constituents of the interstellar gas, C, N, O, Si, S, in several stages of ionization, and HI and H_2. It must be firmly stressed that the radio and optical observations do not generally refer to the same substances or to the same regions of space, and it cannot be assumed that parameters derived from radio observations apply to regions observed optically and vice versa.

It is believed that the sun is located on the inner edge of the 'Orion' arm. This is believed to be a 'material' arm, as distinct from 'density wave' arms such as the 'Perseus' and 'Sagittarius' arms are believed to be. Matter flows through a density wave arm, and shock fronts probably exist along the inner edges where the matter enters. Matter does not flow through a material arm, and there are no associated shock fronts. Thus, one may anticipate that the physical state of the interstellar medium may be different in a material arm than in a density wave arm. The arms are characterized by the formation of stars, the massive ones ionizing the remaining hydrogen in their vicinity forming HII regions. The majority of spiral arm HII regions are found within 8 kpc of the galactic centre, i.e. the sun is outside the region of maximum activity. Thus, optical observations apply to a very small region of the interstellar medium, fairly far from the galactic centre, and outside the region of active star formation, which contains a piece of a material arm and a piece of interarm. A high proportion of the radio observations, on the contrary, apply to the region within 8 kpc of the galactic centre. Density wave arms, and greater rates of star formation are found in these regions than in the solar neighbourhood. Obviously one may not assume that the results obtained apply, in detail, to the regions observed optically.

2. Clouds

On a scale that is small compared with the spiral arms etc., the interstellar medium is not homogeneous. The idea of 'clouds' is old, dating from the realization that an optical interstellar line can have several sharp components with different velocities. There was casual speculation that there may be a hot tenuous gas between the clouds that maintains approximate pressure equilibrium. The temperature required for such a gas was obviously high, and it was not until Clark in 1965 [1] observed the intercloud gas at 21 cm that the presence of the 'second component' of the interstellar medium was firmly established. An explanation for the clouds was put forward by Field, Goldsmith and Habing [2]. The explanation depends on the assumption that gas outside HII regions is partly ionized, and thereby heated, by some agent other than photons from stars. The assumption is easy to justify. The photons from stars with wavelengths less than 912 Å are able to ionize hydrogen, and the cross section for this process is high. The photons are therefore absorbed in the vicinity of the stars, forming fully ionized 'HII' regions. The radiation at wavelengths longer 912 Å penetrates the HI regions, but is capable of ionizing only C, Mg, Si, S, Fe and some other heavy elements of lower abundance. Because the number of such atoms is small, the energy input to the HI gas is small, and the temperature would be very low, about 11 °K [3, 4, p. 127]. Since we observe HI (at 21 cm) that has temperatures from 10 °K to greater than 1000 °K, there must be another energy source capable of ionizing hydrogen, and thereby supplying energy to the interstellar neutral gas at a relatively high rate. Field et al. [2] assumed that the source of energy is cosmic rays.

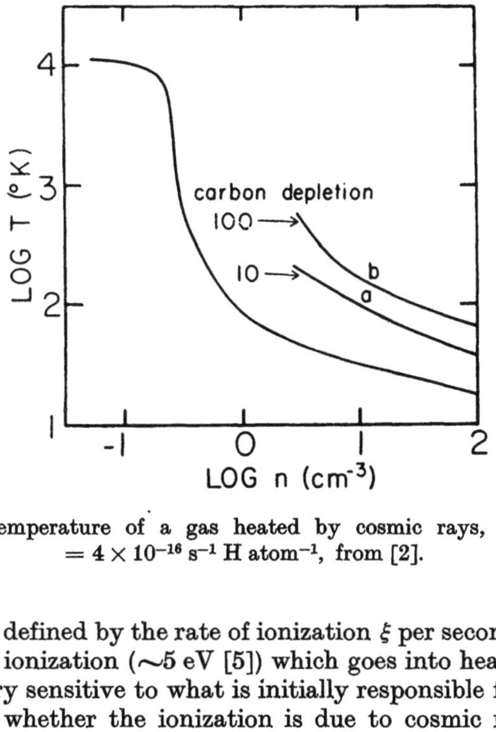

Fig. 1. Equilibrium temperature of a gas heated by cosmic rays, with ionization rate $= 4 \times 10^{-16}$ s^{-1} H atom^{-1}, from [2].

The energy input is defined by the rate of ionization ξ per second and per H atom, and the energy per ionization ($\sim$5 eV [5]) which goes into heating the gas. Since the latter is not very sensitive to what is initially responsible for the ionizations, the theory applies whether the ionization is due to cosmic rays or to X-rays. An ionization rate of the order of 4×10^{-16} s^{-1} is required to heat the interstellar medium. It appears, at the present time, that the observed cosmic and X-ray fluxes in the solar neighbourhood are not sufficient to account for this ionization and heating rate.

Basically, the production of a two-phase interstellar medium is due to the fact that the heating is due to ionizations and is proportional to the density, while the cooling is due to collisions and is proportional to the square of the density. Thus, there is a range of density in which compression causes increased cooling, which causes a temperature drop, and this causes compression. Thus the medium is unstable to density perturbations. For $\xi = 4 \times 10^{-16}$ s^{-1} the region of instability is at about 0.2 cm^{-3}. Figs. 1 and 2 (taken from [2]) show the temperature, pressure and electron density as functions of the total density, and show the possibility of pressure equilibrium between a diffuse gas, $n < 10^{-1}$ cm^{-3}, $T = 10^4$ °K, and clouds with n between 10 and 100 cm^{-3} and T between 100° and 10°K respectively. The time scale of the instability is $\sim 10^6$ years [6, 7]. Thus, the presence of two phases in any region of neutral gas is a reasonable expectation. The presence of detailed equilibrium between the phases is, however, not assured. On the contrary, the presence of disturbances by cloud collisions, formation of stars and H II regions, formation of supernovae and again H II regions will cause continual disturbance of the interstellar equilibrium. Dalgarno and his associates (see [5]) have investigated the consequences of allowing the ionization rate to be variable in time also.

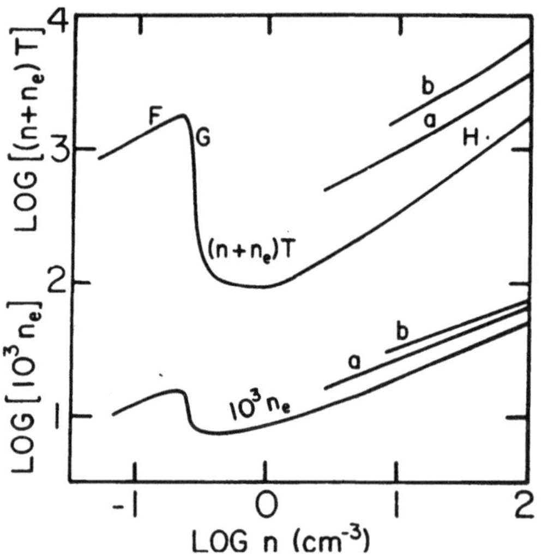

Fig. 2. Equilibrium pressure $(n + n_e)T$ and electron density n_e of a gas heated by cosmic rays, with ionization rate $= 4 \times 10^{-16}$ s^{-1} H $atom^{-1}$, from [2].

Observations indicate that the physical parameters of clouds cover a wide range. Neckel [8] has compiled observations pertinent to the distribution of absorbing matter. Scheffler's [9] analysis of these observations indicates that the absorbing matter is concentrated into clouds of two types, with the following dimensions:

mean cloud extinction	$\langle A_1 \rangle = 0^m.26$	$\langle A_2 \rangle = 1^m.6$
mean cloud frequency	$\langle \nu_1 \rangle = 5$ kpc^{-1}	$\langle \nu_2 \rangle = 0.5$ kpc^{-1}
mean cloud diameter	$\langle D_1 \rangle = 3$ pc	$\langle D_2 \rangle = 70$ pc

Similar analyses by Munch [10] and by Schatzmann [11] have yielded similar results. The smaller of Scheffler's clouds probably achieve the closest approximation to the theory outlined above. However, there are many much denser, larger clouds which are not included in Scheffler's clouds—mainly because their visual absorption is so high that few, if any, stars are observed through them, and a survey based on stellar extinction excludes them almost completely. These larger/denser clouds include the 'dark clouds' studied at radio wavelengths [12] and found to abound in OH and formaldehyde and (by induction from the lack of

HI) in molecular hydrogen. These dark clouds include objects such as the dust clouds in Taurus and Ophiuchus; the extinction is greater than five magnitudes and may be much greater. The 'molecular clouds' [4, p. 121] that are conspicuous at radio wavelengths because of emission in the molecular lines, are probably even more opaque, and optical observations are completely lacking. Thus, once again, the radio and optical observations generally refer to different objects. It is not, however, always the case. Davies [13] concluded that OH is seen only in clouds whose surface densities $N_H > 2 \times 10^{20}$ H atoms cm^{-2} and whose volume density $n_H > 30$ cm^{-3}. From the relationship derived by Savage and Jenkins [14] (see § 9) between N_H and E_{B-V}, this column density corresponds to $E_{B-V} \sim 0.04$ mag., extinction $A_V \approx 0.12$, of the same order as that derived by Scheffler for his small clouds.

3. HII Regions

When an O or B star is formed within a dense cloud of gas, it will, in a short time, create a sphere of ionized gas around itself. The density is still the same as that of the neutral gas, and probably high. The temperature is, however, about 10^4 °K, not 10 °K like the neutral cloud. Clearly, the pressure equilibrium (or quasi-equilibrium) is destroyed and the cloud expands.

Consider a single star with excitation parameter U ($= R_S n_e^{2/3}$ pc cm^{-3} where R_S is the radius of the Stromgren sphere produced in a gas of uniform density n_e cm^{-3}); U $\approx$ 167 for an O4 star and $\approx$4 for a B1V star [18]. In a region with density 10^{-1} cm^{-3}, these stars will produce HII regions with emission measures (EM $= \int n_e^2\, dr \approx 2Un_e^{4/3}$) $\approx$ 17 (O4Ia) and 0.4 (B1V), respectively. The practical limit for detection is EM $\approx 10^3$ for continuum radio scans and for the Palomar Atlas. Photography with narrow-band interference filters can detect objects about 60 times fainter than the Palomar plates [15], i.e. EM $\approx$ 15. Fabry-Perot interference photography and photometry can reach an EM of 5–50 [16, 17], but have not been widely applied. Thus, probably no observed nebula is in pressure equilibrium with the intercloud medium.

However, many B stars are not in HII regions bright enough to appear on the Palomar prints; these include a high proportion of the stars that have been intensively observed for interstellar lines. While EM $= 10^3$ is not a very stringent limit, it may not be unrealistic to assume that the HII regions around these (comparatively old) stars have expanded to the point of pressure equilibrium with the surrounding intercloud gas. In intercloud gas with $n_H = 0.1$ cm^{-3}, $T = 10^4$ °K, this means an HII region with density 5×10^{-2} cm^{-3}, $T = 10^4$ °K. At such low density even a B1V star, $U = 4.4$ pc cm^{-2} [18], creates an HII region of radius 32 pc. Hence, a large portion of the intercloud medium may be fully ionized. Of course, since hot HI intercloud gas is observed at 21 cm [1, 19] we know that it is not *all* ionized.

Excitation parameters are calculated by Churchwell and Walmsley [18] for O4 to B1 stars, using the non-LTE model atmosphere calculations of Auer and Mihalas [20] and absolute visual magnitudes from Conti and Alschuler [21]. For stars later than B1, published model atmospheres do not give fluxes at wavelengths below 912 Å. In order to get an estimate of the excitation parameters for these cooler stars, a linear extrapolation in the log $(N_L/\pi F_v)$ vs T_{eff} diagram given by Churchwell and Walmsley has been made. N_L is the number of Lyman photons per second emitted by the star; πF_v is the emergent flux at the surface of the star in the visual filter band. T_{eff} values for the spectral classes are taken from Morton and Adams [22] and M_V's from Landolt-Börnstein [23]. The results are given in Table 1 with the notation used in [18] and [22].

Table 1

Excitation Parameter Spectra Type Relationship for B Stars

Sp	T_{eff} (°K)	$\log_{10} \frac{N_L}{\pi F_V}$	$\log_{10} L_{UV}$ (photons s^{-1}) V	III	Ia	U (pc cm^{-2}) V	III	Ia
B 1	22600	9.13	45.42	45.78	46.94	4.4	5.8	13.6
B 2	20500	8.68	44.81	45.25	46.53	2.7	3.8	10.2
B 3	17900	8.13	43.94	44.50	45.98	1.4	2.2	6.7
B 4	16800	7.89	43.58	44.10	45.78	1.1	1.6	5.8
B 5	15600	7.64	43.17	43.65	45.57	0.8	1.1	4.9

4. The Observations: n_e

Given this general description of the physical state of the interstellar matter, let us consider the situation that obtains in a normal optical observation of interstellar absorption lines. Consider an early B star, excitation parameter about 10 (about B0.5 to B1) at a distance of 500 pc. We assume that its HII region is in pressure equilibrium with the intercloud gas. Along the lines of sight we expect 2.5 small clouds. We take one with a density of 10 cm^{-3} and one with density 100 cm^{-3}. The densities, temperatures, electron densities (for $\xi = 4 \times 10^{-16}$ s^{-1}), n_e/n_H and dimensions are given in the first six columns of Table 2.

Table 2

Contribution of Components along the Line of Sight

Component	n_{tot} (cm^{-3})	T_e (°K)	n_e[a] (cm^{-3})	$\frac{n_e}{n_{tot}}$	$\int dr$ (pc)	$\int n_{tot}\, dr$ (cm^{-3} pc)	$\int n_e\, dr$ (cm^{-3} pc)	$\frac{n(X)}{n(X^+)}$
HII	5×10^{-2}	10^4	5×10^{-2}	1	74[b]	3.7[c]	3.7[c]	4
Hot HI	10^{-1}	10^4	1.3×10^{-2}	1.3×10^{-1}	420	42	5.5	1
Cold HI	10	10^2	2×10^{-2}	2×10^{-3}	3	30	0.06	46
Cold HI	10^2	10	5×10^{-2}	5×10^{-4}	3	300	0.15	622

[a] For $\xi = 4 \times 10^{-16}$ s^{-1} H $atom^{-1}$.
[b] For the HII region $\int dr = U n_e^{-2/3}$.
[c] For the HII region $\int n_{tot}\, dr = \int n_e\, dr = U n_e^{1/3}$.

Note that, at this low density, the HII region has a radius of 74 pc, a significant fraction of the line of sight. More important, note the general feature of the two-component model, that the degree of ionization of the neutral intercloud medium is very high, 0.13. The relative contribution of the various components to the

$\int n_{tot}\,dr$ and $\int n_e\,dr$ along the line of sight are given in columns 7 and 8. Most of the matter is in the densest cloud. Most of the electrons are in the H II region and the hot intercloud gas. Thus, any attempt to derive n_e for the neutral gas, or n_e/n_H, from such integral quantities is meaningless. As pointed out by Grewing and Walmsley [24], this includes, in particular, the comparison of pulsar dispersion measures giving $\int n_e\,dr$ with 21 cm observations giving $\int n_H\,dr$.

The other method of deriving the electron density from observations involves the ionization balance. The number ratio of atoms in two successive ionization states is controlled by the simple equation

$$n(\mathrm{X}) \cdot \Gamma = n(\mathrm{X}^+) \cdot n_e \cdot \alpha(T_e)$$

where $n(\mathrm{X})$, $n(\mathrm{X}^+)$ are the volume densities of the atoms in the two states, n_e is the electron density, Γ is the ionization rate, and α is the recombination coefficient which is temperature-sensitive with an exponent in the vicinity of -0.7 (neglecting dielectronic recombination). We consider the case where ionization can be by radiation significantly beyond the Lyman limit. Then we may assume that Γ is a constant along the line of sight; this is justified below. Then the ionization balance

$$\frac{n(\mathrm{X})}{n(\mathrm{X}^+)} \propto n_e T_e^{-0.7}.$$

This quantity (relative to the value for the intercloud gas) is tabulated in column 9. Most of the ions will be in the state X^+ and the strength of the line will reflect $\int n_{tot}\,dr$. The proportion of atoms in the state X varies enormously among the various components and is much the highest in the dense clouds. Note that, if the H II region is more than two orders of magnitude denser, i.e. $n_e > 10\ \mathrm{cm}^{-3}$ (less if di-electronic recombination is important) its ionization balance will be the same as for the denser cloud; to affect seriously the neutral X line strength, the density must be even higher. Thus, the ionization ratio derived will usually be close to the true value for the denser cloud. In the event that there are only clouds of $10\ \mathrm{cm}^{-3}$ along the line of sight, an overestimate of the degree of ionization will result from the fact that the hot intercloud medium contributes significantly to the X^+ line but not to the X line. Of the elements that are mentioned in this paper, dielectronic recombination is important only for Ca I and Mg I. In practice, Ca at high temperatures will be mostly Ca III and the lines of both Ca I and Ca II from hot regions will probably be negligible. For Mg I, at $10^4\,°\mathrm{K}$, dielectronic recombination increases α by a factor of 75 [25].

Observations of the ionization balance have been made by Boksenberg et al. [26] using Mg I/Mg II. They derive $n_e = 2 \times 10^{-3}\ \mathrm{cm}^{-3}$ (for $T_e = 60\,°\mathrm{K}$) with comparatively little scatter for six stars. Hobbs' results for Na I/H [30] suggest about $5 \times 10^{-3}\ \mathrm{cm}^{-3}$ if a cosmic abundance for Na and $T_e = 60\,°\mathrm{K}$ are assumed. These values are rather lower than the values 10^{-2} to $10^{-1}\ \mathrm{cm}^{-3}$ predicted by Field et al. [2] for $\xi = 4 \times 10^{-16}\ \mathrm{s}^{-1}$. Further, if as we suppose, the n_e refers to the cool clouds, $n_{tot} = 10$—$100\ \mathrm{cm}^{-3}$ and $n_e/n_H = 3 \times 10^{-4}$—$3 \times 10^{-5}$. This is equal to or less than the number of electrons expected from photoionization of carbon and other heavy elements, i.e. it implies a very low ionization rate due to cosmic rays and X-rays. On the other hand, observations of Ca I/Ca II ratios [28, 29] give results for n_e between 0.4 and $0.04\ \mathrm{cm}^{-3}$ for the few stars where Ca I is strong enough to be observed. This would appear to be in better accord with the theory.

Some caution is in order in interpreting these results. There may be something about the ionization balance in the cool clouds that we do not yet understand. This is demonstrated graphically by Hobbs [27] who finds the interstellar line profiles in the spectrum of ζ Oph due to Ca II, Na I, K I and CH^+ to be all different. In particular, the Na I line appears to have four components, while the K I line corresponds to only one of these. Thus K I appears to be absent in three clouds along the line of sight. Since K I and Na I have similar ionization potentials, we would not have expected a very different ionization balance; one is equally hesitant to propose massive abundance differences. The CH^+ line peaks at about $-13\,km\,s^{-1}$; this corresponds well with the peak of the H 21 cm profile [30] which is very weak at the more negative velocities corresponding to the other components of the Na I profile. Weak 21 cm emission implies low temperatures for the H or the presence of H_2. Thus, one would have expected CH^+ to be strongest in the components where 21 cm emission was weakest, not vice versa, as observed.

5. Ionization Rate of Heavy Elements

We assumed above that the ionization rate of X atoms was constant along the line of sight. This is a good assumption so long as the radiation field is dominated by the diffuse galactic background, and so long as the ionization potential of X is significantly less than 13.6 eV. Most of the ions concerned are ionized by radiation between 1000 and 2000 Å, i.e. between 12.4 and 6.2 eV. In this wavelength range, the diffuse galactic background [43] is approximately 8×10^{-17} ergs cm^{-3} Å^{-1} and corresponds to an O5V star at 65 pc, a B0V star at 21 pc or a B4V star at 3 pc. In the example above, one needs to be well within the H II region before the star begins to have a significant effect on Γ. The presence of photons <912 Å will (except for the hotter stars) have little effect on the ionization since their number is rather less than the number >912 Å and the ionization cross section decreases with wavelength. Likewise, the rate of ionization of X by cosmic rays or X-rays is expected to be negligible.

The ionization of ions with ionization potentials greater than 13.6 eV follows the same pattern as for hydrogen. Thus, as for hydrogen, the integral along the line of sight of the number of ionized species will be completely dominated by the H II region and the hot neutral gas, while the integral for the un-ionized species will be dominated by the cool clouds. The ratio of the two integrals would be hard to interpret.

6. Abundances

To determine reliable abundances, one needs observations of the most abundant ion of the species. In ground-based observations, the ions of greatest importance were Na I and Ca II and neither is the dominant ion. It was, however, generally believed that both were under-abundant with respect to H. The most recent observations by Hobbs [30], who compares Na I with 21 cm H I observations, and by Savage and Jenkins [14], who compare Na I with Lyman alpha observations indicate N(Na I)/N(H) $\approx 3.5 \times 10^{-9}$. This is about three orders of magnitude less than the cosmic value, but can be rationalized with it if $n_e \approx 5 \times 10^{-3}\,cm^{-3}$, a number which agrees with the Mg I/Mg II ratio but which seems low. Assuming the same value of n_e applies to Na and Ca, it follows that Ca must be more under-abundant than Na by several orders of magnitude (see e.g. [31]). However, since again the dominant ion is not observed, and Hobbs' [27] observations indicate radically different distributions of Na I and Ca II, this conclusion should be viewed with caution.

de Boer et al. [32] have observed Mg, Fe and Mn in the spectrum of ζ Pup and find a ratio between the elements which is solar but a ratio to H which is ten times less than solar. ζ Pup is an O5 star and most of the interstellar medium between here and there is ionized. The observed by Boksenberg et al. [26] were all later than O9.5 and the lines are assumed to be formed in cool H I. This leads to different choices between possible values of the recombination coefficients, line widths etc.; one wonders if an incorrect choice may be responsible for the difference in the results of the two groups of authors.

7. Temperatures

Clark [1] demonstrated the presence of two components in the interstellar medium. He observed broad components of the emission profiles which have no counterparts in the absorption profile obtained when the gas was seen in absorption against a hot continuum source. For the cool clouds seen in emission and absorption, excitation temperatures T_{ex} can be derived and were, characteristically, less than 100 °K. For the hot component, which does not appear in absorption, a lower limit to the excitation temperature can be determined, and was about 10^3 °K. Further observations by Radhakrishnan et al. [33] and by Hughes et al. [34] refined these numbers to:

$$\langle T_{ex} \rangle \leqq 70\,°\mathrm{K}\ \text{clouds}$$
$$\geqq 700\,°\mathrm{K}\ \text{hot component.}$$

The excitation temperature is probably equal to the kinetic temperature T_{kin} in the clouds, but may be less than the kinetic temperature in the low density intercloud medium (see below, and next section). Mebold [19, 35] and, independently, Baker [36] have analysed the broad emission lines and indicate kinetic temperatures for the hot component of about 5000 °K. Thus, the value of 10^4 °K predicted by the theory for the intercloud medium would appear to be too high.

The comparison, by Savage and Jenkins [14], of hydrogen column densities derived from Lyman alpha and from 21 cm observations, indicates that the value from 21 cm falls below that from Lyman alpha at a surface density of about 10^{21} cm^{-2}, indicating that the gas is becoming optically thick at 21 cm and, therefore, that the temperature of the emitting gas is less than 180 °K ([4], p. 105), where the 'less than' allows for the contribution of turbulence to the line width.

The kinetic temperature can, in principle, be derived from observations of lines from levels which are collisionally excited. The observed excitation temperature is a lower limit to the kinetic temperature, *assuming* that the upper levels are populated only by black-body background radiation and by collisions, and that the collisions tend to produce a Boltzmann distribution at the kinetic temperature of the gas. However, radio observations have shown tahat this is often not the case. The observed excitation temperatures can be less than 2.7 °K (formaldehyde), or negative (OH), indicating population of the upper level is even more than in LTE at infinite temperature.

Observations of CN [37] in the spectra of ζ Oph and ξ Per indicated $\sim 3\,°K$, i.e. no collisional excitation. At low densities, the level populations are in equilibrium with the radiation field, the 2.7 °K black-body background. (At frequencies involved in the excitation of low levels, <0.01 eV, the black-body background totally dominates the radiation field, except in the immediate vicinity, <0.1 pc, of a star.) Smith and Stecker [38] did not resolve the structure in the CO lines in the spectrum of ζ Oph, but found their strengths were consistent with 2.7 °K excitation temperature. Smith [39] derived 47 °K from the ratio of $J = 1$ to $J = 0$ levels of H_2 in the spectrum of ζ Oph; but collisional transitions between these levels are forbidden and thus the excitation temperature is not related to the kinetic temperature (see [49]).

8. Density

The collisional excitation of molecules discussed above is generally more useful for determination of densities than of temperatures. The effect of collisional excitation on the population of levels depends on whether collisions can populate the levels at a rate which is competitive with the rate of depopulation by spontaneous radiative decay. Since the 21 cm line of H has very low transition probability, collisions control the population and $T_{ex} = T_{kin}$ except in the very low densities of the intercloud gas. For most of the other molecules, the lifetimes of the excited levels are much shorter and $T_{ex} < T_{kin}$. It needs to be noted that the excitation rate depends not only on T_{kin} but also on the nature of the colliding particle; electrons are more efficient excitors than are atoms or molecules. Thus, the degree of ionization must also be known or assumed. In clouds, it may be fairly safe to assume $n_e/n_H = 4 \times 10^{-4}$, the ratio due to ionization of C and other heavy elements. Given the collisional cross sections and the radiative decay rates, and assuming a temperature, say 70 °K, for the clouds, we may calculate a density for the colliding particles required to produce an enhanced population in an upper level. Collisional cross sections are not well known. However, using $\sigma = 4 \times 10^{-15}\,cm^2$, the geometric cross section for H atoms, one may show (following [4], p. 111) that, in order to see collisional excitation of CO to $T_{ex} = 5\,°K$, one needs a density of the order of 50 cm^{-3}; to get $T_{ex} = 20\,°K$ one requires 500 cm^{-3}. The failure to observe collisional excitation of CN places an upper limit of the order of $10^4\,cm^{-3}$ on the densities in those regions. Similar considerations apply to the observation of excited fine structure levels in ions such as C I and C II. By contrast with the low densities implied for optically observed clouds, the densities needed for collisional excitation of the molecules observed in emission at radio wavelengths range from 10^3 to $10^7\,cm^{-3}$. Much the simplest way to get an estimate of the density is to divide the column density of H atoms by the distance to the star. The problem, of course, is that one would, in general, get an average value for both clouds and intercloud. By taking the nearby stars, one may obtain an estimate for intercloud gas only. Savage and Jenkins [14] derive 0.25 cm^{-3}. For distant stars where the line of sight crosses one or more clouds, we may obtain an estimate of the cloud density by assuming the cloud parameters derived by Scheffler [9]. The number of clouds, $n = E_{B-V}/0.09$ and the path length through clouds is $3n$ pc. The contribution of the intercloud medium can be ignored. The result is 50 cm^{-3}, in surprisingly good agreement with expectations considering the roughness of the estimate.

9. Dust

We have, so far, assumed that the dust and gas are fairly well mixed. The first convincing demonstration of this fact came from the OAO observations of Savage and Jenkins [14]. They showed a good correlation between the colour excess of the star and the column density of H derived from the Lyman alpha absorption in its spectrum. They give

$$\frac{N_{\rm H}}{E_{B-V}} = 5 \times 10^{21} \text{ atoms cm}^{-2} \text{ mag}^{-1}$$

and they suggest that some of the scatter is probably due to the presence of H_2. This would increase the number ratio above by, probably, a factor of two.

The extinction of starlight by interstellar dust is wavelength dependent and therefore causes 'reddening'. The dependence of the extinction on wavelength, the 'extinction curve', has been observed from 1 μm down to 1200 Å. The curve shows a maximum at about 2200 Å and continued increase at very short wavelengths. The shape of the curve appears to vary from place to place [40], and Laget [41] has found differences that appear to be correlated with luminosity class. Witt and Lillie [42], from observations of the diffuse galactic background with OAO-A2, have derived the scattering and absorbing parts of the extinction. Absorption is a maximum at 2200 Å, in the vicinity of the extinction maximum. In the far UV, the extinction is mainly due to scattering; this causes a peak in the diffuse galactic background at short wavelengths [43]. The composition and size of particles causing the extinction is still unknown. No model is currently able to fit the UV observations. Silicates and water are indicated by IR observations [44]. Carbon is suggested to account for the maximum at 2200 Å [45].

10. Concluding Remarks

It is clear that high resolution, high accuracy observations of ultraviolet spectra provide new and powerful ways of studying the interstellar medium. The new results from Copernicus [46—51] already begin to exploit the possibilities; this is only a beginning. By combining the new information obtained with the complementary information provided by radio observations, our understanding of the interstellar medium should improve rapidly in the very near future.

Acknowledgments

It is a pleasure to thank P. G. Mezger for extensive discussions and M. Grewing and M. Walmsley for numerous helpful comments. G. Winnewisser provided the spontaneous transition probability values required for § 8. I thank the Princeton group for providing me with copies of their papers prior to publication.

References

[1] B. G. Clark, Astrophys. J. **142**, 1398 (1965).
[2] G. B. Field, D. G. Goldsmith and H. J. Habing, Astrophys. J. Lett. **155**, L 149 (1969).
[3] E. Churchwell, Thesis, Indiana University, 1970.

[4] P. G. Mezger, in: Interstellar Matter, Geneva Observatory, Sauverny, Switzerland 1972.
[5] A. Dalgarno and R. A. McCray, Ann. Rev. Astron. Astrophys. **10**, 375 (1972).
[6] G. B. Field, Astrophys. J. **142**, 531 (1965).
[7] P. Biermann et al., Astron. Astrophys. **19**, 113 (1972).
[8] Th. Neckel, Veröff. Landessternwarte Heidelberg—Königstuhl **19** (1967).
[9] H. Scheffler, Z. f. Astrophys. **63**, 207 (1966); **65**, 60 (1967).
[10] G. Munch, Astrophys. J. **116**, 575 (1952).
[11] E. Schatzman, Ann. d'Astrophys. **13**, 367 (1950).
[12] C. Heiles, Ann. Rev. Astron. Astrophys. **9**, 293 (1971).
[13] R. D. Davies, in: Highlights of Astronomy, D. Reidel Publ. Co., Dordrecht, Holland 1971 (p. 402).
[14] B. D. Savage and E. B. Jenkins, Astrophys. J. **172**, 491 (1972).
[15] G. Courtes and P. Cruvellier, Ann. d'Astrophys. **28**, 683 (1965).
[16] G. Monnet, Astron. Astrophys. **12**, 379 (1971).
[17] R. J. Reynolds et al., Astrophys. J. Lett. **181**, L 79 (1973).
[18] E. Churchwell and C. M. Walmsley, Astron. Astrophys. **23**, 117 (1973).
[19] U. Mebold, Astron. Astrophys. **19**, 13 (1972).
[20] L. H. Auer and D. Mihalas, Astrophys. J. Suppl. **24**, 205 (1972).
[21] P. S. Conti and W. R. Alschuler, Astrophys. J. **170**, 325 (1971).
[22] D. C. Morton and T. F. Adams, Astrophys. J. **151**, 611 (1968).
[23] Th. Schmidt-Kaler, in: Landolt-Börnstein, New Series Group VI, Vol. 1, Springer, Berlin 1965 (p. 301).
[24] M. Grewing and M. Walmsley, Astron. Astrophys. **11**, 65 (1971).
[25] M. Grewing et al., Astron. Astrophys. **27**, 115 (1973).
[26] A. Boksenberg et al., Nature Phys. Sci. **240**, 127 (1972).
[27] L. M. Hobbs, Astrophys. J. Lett. **180**, L 79 (1973).
[28] L. M. Hobbs, Astrophys. J. Lett. **170**, L 85 (1971).
[29] R. E. White, Bull. Am. Astron. Soc. **4**, 327 (1972).
[30] L. M. Hobbs, Astrophys. J. **166**, 333 (1971).
[31] G. H. Herbig, Z. f. Astrophys. **68**, 243 (1968).
[32] K. S. de Boer et al., Astron. Astrophys. **21**, 447 (1972).
[33] V. Radhakrishnan et al., Astrophys. J. Suppl. **24**, 15 (1972).
[34] M. P. Hughes, A. R. Thompson and R. S. Colvin, Astrophys. J. Suppl. **23**, 323 (1972).
[35] O. Hachenberg and U. Mebold, Proc. IAU Meeting in Athens, IAU, in press (1972).
[36] P. L. Baker, Astron. Astrophys. **23**, 81 (1973).
[37] G. Field and J. L. Hitchcock, Astrophys. J. **146**, 1 (1966).
[38] A. M. Smith and T. P. Stecker, Astrophys. J. Lett. **164**, L 43 (1971).
[39] A. M. Smith, Astrophys. J. Letters **179**, L 11 (1973).
[40] R. C. Bless and B. D. Savage, Astrophys. J. **171**, 293 (1972).
[41] M. Laget, Astrophys. J. **180**, 61 (1973).
[42] A. N. Witt and C. F. Lillie, in: Scientific Results from the OAO-2, NASASP-310, Washington, D.C., 1972 (p. 199).
[43] A. N. Witt and M. W. Johnson, Astrophys. J. **181**, 363 (1973).
[44] N. J. Woolf, in: IAU Symposium 52 "Interstellar Dust and Related Topics", Reidel Publ. Co. Ltd, Dordrecht, 1973 (p. 485).
[45] T. P. Stecher and B. Donn, Astrophys. J. **142**, 1681 (1965).
[46] J. B. Rogerson et al., Astrophys. J. Lett. **181**, L 97 (1973).
[47] D. C. Morton et al., Astrophys. J. Lett. **181**, L 103 (1973).
[48] J. B. Rogerson et al., Astrophys. J. Lett. **181**, L 110 (1973).
[49] L. Spitzer et al., Astrophys. J. Lett. **181**, L 116 (1973).
[50] E. B. Jenkins et al., Astrophys. J. Lett. **181**, L 122 (1973).
[51] D. G. York et al., Astrophys. J. Lett. **182**, L 1 (1973).

Space Research XIV — Akademie-Verlag, Berlin 1974

REPORT ON THE CELESCOPE ULTRAVIOLET OBSERVATIONS FROM THE OAO 2 SATELLITE AND ASSOCIATED RESEARCH AT THE SMITHSONIAN ASTROPHYSICAL OBSERVATORY

E. Avrett, R. Davis, W. Deutschman, K. Haramundanis, R. Kurucz, C. Payne-Gaposchkin, E. Peytremann* and R. Schild

Smithsonian Astrophysical Observatory, Cambridge, Mass., USA

We briefly survey two recent publications, the *Celescope Catalog of Ultraviolet Stellar Observations* and a companion volume, *Blanketed Model Atmospheres for Early-Type Stars*, and then describe current research at SAO on the analysis of Celescope data, utilizing ground-based observations and the theoretical models.

The Orbiting Astronomical Observatory (OAO 2) containing the four Schwarzschild telescope Celescope Experiment was launched by the National Aeronautics and Space Administration on 7 December 1968. During 16 months of normal operation, Celescope observed over 10% of the sky, including approximately half the region along the Milky Way where the majority of ultraviolet stars are found.

Each telescope images a star field approximately 2.8° square on to the ultraviolet-sensitive photocathode of a television image tube, or Uvicon. Each field is optically split into two areas of different spectral sensitivity by a pair of optical filters mounted in front of each Uvicon. Further spectral selectivity is achieved by using two types of Uvicons, each with a different photocathode material. There are thus four photometric passbands. The U1 filter extends from 3200 to 2100 Å, while U2 covers a broader band from 3200 to 1600 Å. The U3 filter extends from 2100 to 1300 Å and U4 from 2100 to 1100 Å. Most of the final published magnitudes correspond to U2 and U3. Relatively few stars were observed in the U1 band, partly because U1 is narrower than U2 while the observed stellar flux tends to be smaller in the U1 band, and partly because of some degradation with time of the U1 filter material. Stellar photometry in the U4 band was strongly limited by hydrogen Lyman-α radiation from the geocorona.

The *Celescope Catalog of Ultraviolet Stellar Observations* [1] has been compiled from over 13000 observations of 5068 stars in the passbands described above. The spatial distribution of the observed stars in galactic coordinates is shown in Fig. 1. The Catalog lists, in addition to the ultraviolet magnitudes, each stellar position, spectral type and other known characteristics, available UBV photometry, and references to other observations. Most of the stars observed fall in the spectral classes B and A. A companion volume to the Celescope Catalog, *Blanketed Model Atmospheres for Early-Type Stars* [2] provides the most realistic available prediction of the intrinsic ultraviolet flux for B and A stars observed by Celescope. Fig. 2 shows the theoretical flux distribution for two of the models with an

* Now at Dept. d'Astrophysique, Université de Mons, Av. Maistriau 19, 7000 Mons, Belgium.

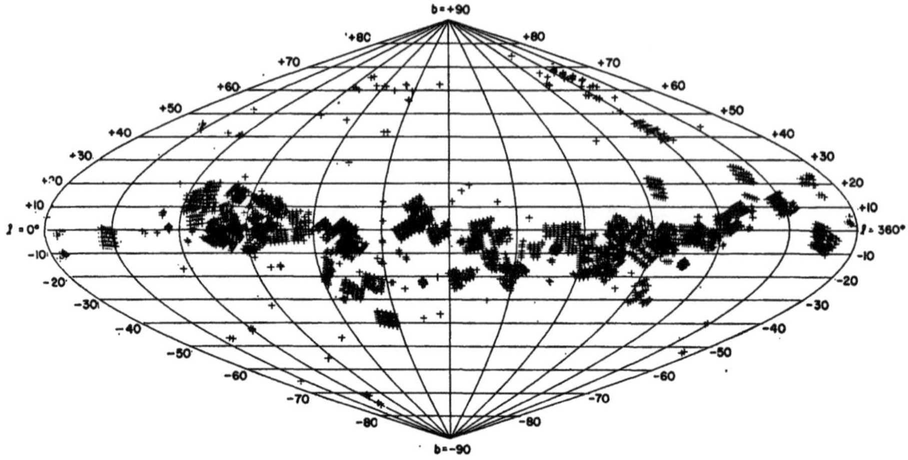

Fig. 1. Plot in galactic coordinates of the exposures taken by Celescope.

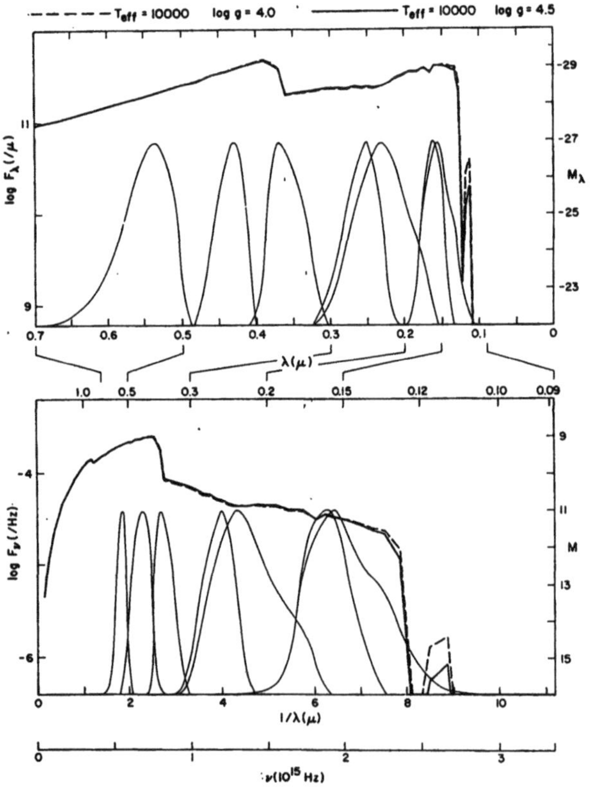

Fig. 2. The emergent flux for models with 10000 K effective temperature and log gravity of 4.0 and 4.5.

effective temperature of 10000 K and log gravity (g) of 4.0 and 4.5. The lower panel shows the absolute flux-per-unit frequency at the stellar surface and the corresponding magnitude plotted against both frequency and reciprocal wavelength. The upper panel shows the flux-per-unit wavelength and the corresponding magnitude plotted against wavelength. In both graphs, the sensitivity curves, reading from right to left, correspond to U4, U3, U2, U1, U, B, and V.

The calculated models range in effective temperature from 8000 to 50000 K and in log gravity from 2 to 5. Solar abundances and a microturbulent velocity of 2 km s^{-1} have been assumed. A major part of this research effort was devoted

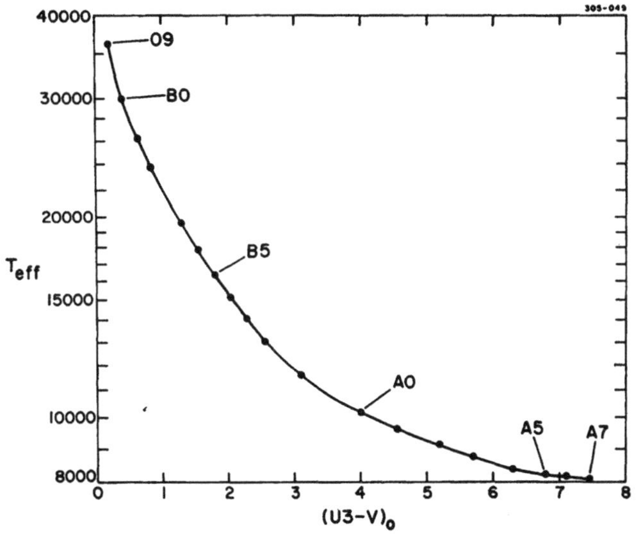

Fig. 3. Effective temperature and spectral type as a function of the intrinsic U3 − V color.

to the inclusion of realistic models of opacity, which determines how radiation escapes from the star. The line opacity was included in the form of distribution functions based on a list of 1760000 lines obtained from the literature and computed by Kurucz and Peytremann with the help of an atomic-structure program written by Dr. Robert D. Cowan at Los Alamos Scientific Laboratory. Tables are presented in this volume of the models, fluxes, UVB colors, Celescope colors, bolometric corrections, and Balmer line profiles.

A general study of the *Celescope Catalog* data has been made by Peytremann and Davis with emphasis on the statistical properties of various groups of stars. Such statistical studies are possible because of the large number of Celescope observations. Over 700 normal stars that have available UBV colors and magnitudes and MK spectral types have been analyzed. A basic part of this work consists of deriving reference sequences of intrinsic ultraviolet colors for normal stars. (By intrinsic color, we mean the magnitude difference of the star itself, unaffected by interstellar extinction.) These intrinsic colors have been compared with the theoretical colors from the grid of model atmospheres to determine the effective temperature scale. Fig. 3 shows the effective temperature for normal main-sequence stars derived from model atmospheres plotted against the intrinsic ultraviolet color $(U3-V)_0$. Spectral type is also indicated. This diagram shows that, for example, a main-sequence A0 star of effective temperature 10000 K should have an intrinsic $U3-V$ color of 4.0. Fig. 4 shows the relationship between the intrinsic $U2-V$ and $U3-V$ Celescope colors and spectral type for normal main-sequence and giant stars (luminosity classes IV, V and II, III respectively). Two methods are used to derive intrinsic colors: (i) we can choose stars that have very little visual extinction and consider that their ultraviolet colors are intrinsic; or (ii) for a group of stars of the same spectral type and luminosity class, we can construct a linear relationship between the ultraviolet color U_i-V and the visual

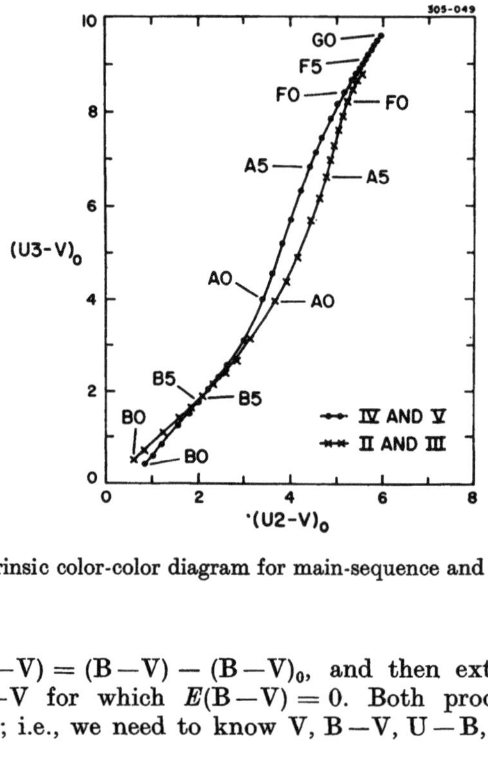

Fig. 4. Intrinsic color-color diagram for main-sequence and giant stars.

color excess $E(B-V) = (B-V) - (B-V)_0$, and then extrapolate to obtain the value of U_i-V for which $E(B-V) = 0$. Both procedures depend on ground-based data; i.e., we need to know V, $B-V$, $U-B$, and the luminosity and spectral class.

Various groups of stars have been compared with the reference sequences of normal intrinsic colors. For example, normal supergiants are brighter than giants having the same intrinsic visual color by 0.3 magnitude in U2 and by 0.5 mag in U3. Emission-line supergiants are fainter than normal supergiants by about 0.4 mag in both U2 and U3. Other emission-line stars do not appear to have peculiar colors. Metallic-line A stars have normal ultraviolet colors, but peculiar A stars are too faint in U3 by about 0.6 mag.

Differences between observed and intrinsic colors, i.e. the color excess, can be used to derive ultraviolet extinction properties of the interstellar medium. We now consider the average observed interstellar extinction in U2 and U3 determined from all normal stars in the *Celescope Catalog* for which there are adequate ground-based observations and for which $E(B-V)$ exceeds a minimum value of about 0.1 (which is several times the standard deviation in measurements of $E(B-V)$. Peytremann and Davis have used the Celescope data to evaluate average values of the ratio $X_i = E(U_i-V)/E(B-V)$ for U2 and U3, where $E(U_i-V)$ is the observed ultraviolet color minus the intrinsic one. They find that $X_2 = 3.6 \pm 1.1$ and $X_3 = 5.0 \pm 0.7$ for the U2 and U3 filters, respectively.

These results are plotted as filled squares in Fig. 5. The solid curve in this figure is the average interstellar reddening law derived by Bless and Savage [3] from an analysis of 17 stars selected on the basis of their extinction properties. The curve shows a pronounced maximum at about 2200 Å. The triangles in the

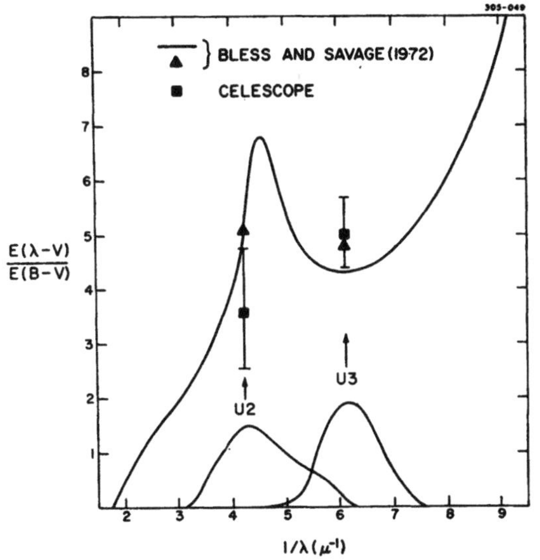

Fig. 5. The wavelength dependence of the ratio of ultraviolet color excess to visual excess.

figure represent the Celescope U2 and U3 extinction values based on the Bless and Savage reddening curve together with the theoretical spectral-energy distribution of a B0 star ($T_{eff} = 30000$ K, $\log g = 4.0$) that is typical of the stars used by Bless and Savage to obtain the average reddening law shown in the figure. The observed Celescope extinction value for U3 is in good agreement with the corresponding value based on the Bless and Savage reddening law. However, the observed U2 value is definitely smaller than that corresponding to the indicated reddening law.

The Celescope U2 and U3 extinction values shown here are based on an average of over 700 stars in each filter, well distributed in galactic coordinates, and not selected for their extinction properties in any way except for the requirement that $E(B-V)$ be above a minimum threshold value. On the other hand, the 17 stars selected by Bless and Savage all exhibit significant amounts of extinction. (This is a requirement since their reddening law is obtained by observed differences in the spectra of reddened and unreddened stars of similar type.)

Seven of the stars observed by Bless and Savage were also observed by Celescope. Peytremann and Davis have compared the two sets of observations in detail and have found no systematic differences. In most cases, when Bless and Savage observe an extinction bump at 2200 Å, Celescope is also able to detect it. On the other hand, the average Celescope U2 extinction value based on observations of over 700 stars does not indicate the presence of an extinction bump at 2200 Å. This result may lead to a change in current theories dealing with properties of the interstellar medium: the extinction bump may be a characteristic of high-extinction interstellar clouds but not a characteristic of the general intercloud medium. These results and others are discussed at length by Peytremann and Davis [4].

Another extensive study of the Celescope data has been carried out by Haramundanis and Payne-Gaposchkin [5]. They have derived intrinsic colors for main-sequence stars from B0 to G3 and have compared observations and predictions for stars in the range B0 to A2. They show that there is good agreement for $-0.20 \leqq (B-V)_0 \leqq 0.00$ but that earlier type stars usually appear slightly reddened and later ones do not seem well represented by the theoretical models. Their examination of the zero-point setting of the Celescope system shows that the component calibration and zero point appear to be correctly established, and that the difficulty in setting the zero point on a standard star arises from the fact that reddening in the ultraviolet is not simply correlated with that in the visual. Certain stars that have unusually blue ultraviolet colors appear to be unresolved multiples. These include a large group of stars in the Orion complex and the striking object HD 4810.

Owing to the large number of Celescope observations, well distributed along the galactic plane, it is possible to detect some variation of the extinction law with galactic longitude l^{II}. A study by Haramundanis [6] of 185 stars averaged within regions of 10° square along the galactic equator, gives values of the extinction ratio $E(U3-V)/E(B-V)$ for stars having a visual excess $E(B-V)$ greater than 0.14. The observed stars range in spectral class from O4 to A0, and about half of these are in the range B0 to B3. The results clearly indicate a maximum in the 60° to 100° region, a second broad maximum around 280°, and minima at 240° and toward the galactic center (located at 0° or 360°). So far, no attempt has been made to associate these extinction features with particular structures located at various distances in the galaxy. Instead, a more detailed analysis will be carried out that uses a much larger number of observations.

All the studies mentioned here require ground-based observations of the stars observed by Celescope. Davis, Deutschman and Schild are compiling UBV and Hβ data for all stars not already having UBV entries in the *Celescope Catalog*. All such stars without published photometry are being measured in an observing program at Kitt Peak National Observatory and at Cerro Tololo Interamerican Observatory. Schild and Deutschman have successfully observed 2300 southern hemisphere stars and are currently reducing the observations to UBV magnitudes and luminosities. Davis has completed over half the observations and magnitude reductions in a program involving 1200 stars in the northern sky. These observations will allow more detailed studies of (i) the intrinsic ultraviolet colors and magnitudes of different types of stars, (ii) particular stars found to be unexpectedly bright or faint in the ultraviolet, (iii) the wavelength dependence of the interstellar reddening law determined by U2 and U3 measurements, and finally (iv) interstellar extinction as a function of galactic longitude, latitude, and radial distance from the sun.

Acknowledgments

This research has been supported by contract NAS 5-1535 from the National Aeronautics and Space Administration and by NASA Grants NGR 09-015-198, NGR 09-015-200, and NGR 09-015-201. The support of a Smithsonian Research Foundation Fellowship (to Kurucz) and an International Fellowship (to Peytremann) from the European Space Research Organization is also gratefully acknowledged.

References

[1] R. J. Davis, W. A. Deutschman and K. L. Haramundanis, Celescope Catalog of ultraviolet Stellar Observations, Smithsonian Institution Press, Government Printing Office, Washington, D.C., 1973.
[2] R. L. Kurucz, E. Peytremann and E. H. Avrett, Blanketed-Model Atmospheres for Early-Type Stars, Smithsonian Institution Press, Government Printing Office, Washington, D.C., 1973.
[3] R. C. Bless and B. D. Savage, Astrophys. J. **171**, 293 (1972).
[4] E. Peytremann and R. J. Davis, Astrophys. J. Suppl. (1974) (in press).
[5] K. L. Haramundanis and C. Payne-Gaposchkin, Astron. J. **78**, 395 (1973).
[6] K. L. Haramundanis, Proc. Advanced Study Inst. on the Interstellar Medium, Max Planck Institute (1973) (in press).

Space Research XIV — Akademie-Verlag, Berlin 1974

OBSERVATIONS OF ULTRAVIOLET STELLAR SPECTRA WITH THE UTRECHT ORBITING STELLAR SPECTROPHOTOMETER S59

K. A. VAN DER HUCHT and H. J. LAMERS

Astronomical Institute, Utrecht, The Netherlands

This paper reports on the analysis of spectral scans of early type stars, obtained with the Utrecht Orbiting Stellar Spectrophotometer S59 on board the ESRO satellite TD-1A. The spectral resolution is 1.8 Å and the observed wavelength ranges are 2060—2160 Å, 2495 to 2595 Å and 2775—2875 Å. Pre-flight calibration values are compared with model atmosphere calculations and observations by others. The spectra of ζ Ori, ζ Pup and γ^2 Vel are briefly discussed. The equivalent widths of the Mg II lines in 31 stars are compared with model atmosphere calculations. In early B-type stars the lines are stronger than expected. The Mg II resonance lines in α Cyg are shifted to shorter wavelengths; a velocity of 140 km s^{-1} indicates a rate of mass loss of 3×10^{-7} to 3×10^{-8} $M_\odot$ yr^{-1}.

1. Behaviour of the Instrument

The Utrecht Orbiting Stellar Spectrophotometer S59 [1] is one of seven major experiments which were launched into a sun-synchronous polar orbit as a part of the ESRO satellite TD-1 A [2] in March 1972. Ever since, the experiment has been operating continuously, only interrupted by a hibernation period from October 1972 to February 1973 when the northern part of the satellite's orbit lay in the earth's shadow. In orbit the gimballed telescope-spectrometer, which has a pointing accuracy of 1 arcsec, scans the celestial sky along ecliptic meridians. It is automatically locked on each star brighter than $m_V = 4.2$ and having an UV flux of more than 30 photons cm^{-2} s^{-1} Å^{-1}, for 260 seconds of time, while counter-balancing the satellite's rotation. During that time the three wavelength regions (2060—2160 Å (the 2100 channel), 2495—2595 Å (2500) and 2775—2875 Å (2800)) are scanned simultaneously with a spectral resolution of 1.8 Å. There are about 225 stars which meet the above-mentioned criteria.

The photometric sensitivity S of S59, defined as the output in counts per 1.05 sec divided by the incoming flux of photons in cm^{-2} s^{-1} Å^{-1} was measured before launch [3]. The values $S_{2100} = 1.49$, $S_{2500} = 2.78$ and $S_{2800} = 1.30$ were found. We realize that these values are more reliable relative to each other than in an absolute sense. Table 1 gives the relative values scaled to $S_{2800} = 1$. An in-flight calibration was performed by comparing non-LTE (local thermodynamic equilibrium) model atmospheres for O and B stars of Mihalas [4] with the S59 spectra of four nearby unreddened stars earlier than B6 of which the continuum could be determined accurately, viz. β CMa, σ Sgr, α Pav and α Gru (S' in Table 1).

Table 1

Relative Sensitivities of the S 59 Channels scaled to the 2800 Channel

Channel	2100	2500	2800
S	1.15	2.14	1
S'	1.35	2.95	1
S''	1.29	2.57	1

S, pre-flight calibration; S', comparison with models; S'', comparison with Stecher [5].

We also compared our observations of ζ Pup with those by Stecher [5] (S'' in Table 1). The values S'' have to be multiplied by a factor 1.69 to get the absolute sensitivity of S 59 with Stecher's measurement of ζ Pup as standard.

In the first four months after launch 765 spectra were obtained of 105 stars of spectral types ranging from WC 8 to F 5. On 4 August 1972, however, the increased density of charged particles in the earth's magnetosphere injected by some strong solar flares interfered with the electronics of the fine-pointing control. As a result only one or two useful spectra per star were obtained till the end of October 1972. Fortunately the electronics recovered during the hibernation and since 14 February 1973, S 59 has been working normally again.

2. Results obtained until now

Classification. One of the major research topics concerning the S 59 spectra is the classification, both in temperature and luminosity, of the early type stars. The absorption lines of singly and doubly ionized metals allow such, especially those of Fe, because line strengths in this wavelength region are very sensitive to temperature and pressure changes. This will be discussed in the next paper.

WC and O type stars. Figs. 1, 2 and 3 show the spectra of γ^2 Vel, ζ Pup and ζ Ori, brought to about the same relative intensity level. We have tentatively identified most spectral features, although there are some broad and shallow lines for which no absorption agent has yet been found. The atomic data are lacking for many of the third and fourth ions of metals. Also the determination of the continuum level is difficult, especially in the 2100 channel. In ζ Pup, 04f [6, 7] we observe ions of HeI and HeII with excitations up to 48.2 eV, of CIII and CIV to 55.7 eV, of NIII, NIV, NV to 84.1 eV, of OIII and OIV to 54.4 eV, of SiIII and SiIV to 31.5 eV, of NeIII to 48.2 eV, of MgIII to 66.2 eV, of CrIII and CrIV from 6.2 to 20.0 eV and of FeIII from 5.1 to 11.1 eV. The interstellar absorption lines in ζ Pup of MgI, MgII, MnII and FeII are discussed in [8]. No emission of nitrogen ions is found in ζ Pup, so there is no evidence here of a WN component. In a recent paper on infrared measurements of ζ Pup Barnes et al. [9] conclude from the absence of HeI λ 9231 that a potential WN companion has $M_v \geqq -2.3$ while WN stars have $M_v = -3.9$ to -6.8. The equivalent width of HeII λ 2511 agrees within 3% with model atmosphere calculations by Mihalas [10].

The spectrum of ζ Ori, O 9.5 I [6], has many lines in common with ζ Pup. Of the differences we mention the absence of NV lines, the much stronger appearance of FeIII and a slight evidence of FeIV at an excitation level of about 20 eV (FeIV lines computed from levels calculated by Warner and Kirkpatrick [11]).

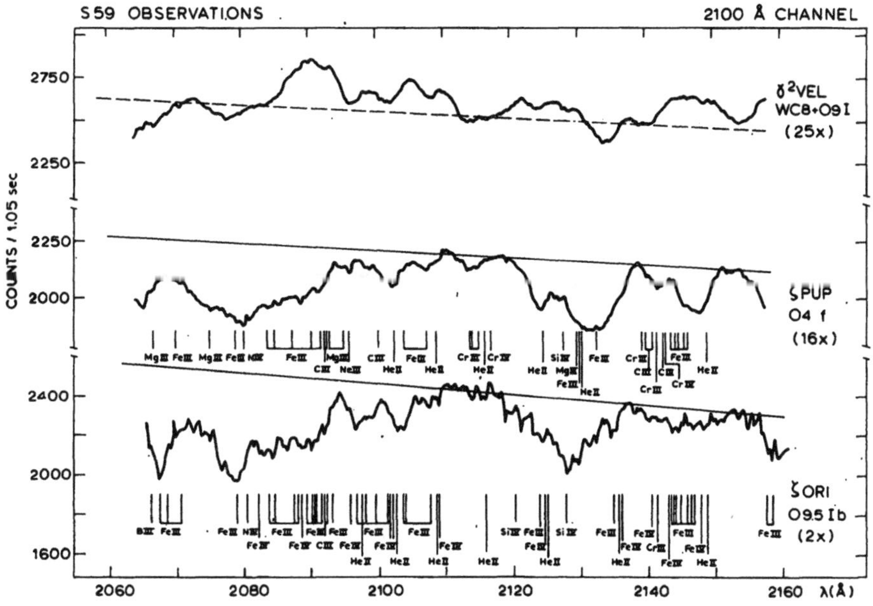

Fig. 1. S 59 spectra. 2100 Å channel.

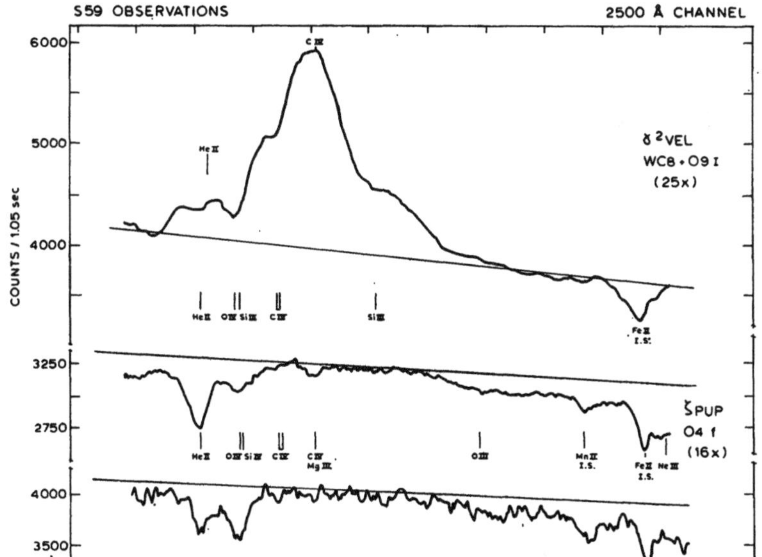

Fig. 2. S 59 spectra. 2500 Å channel.

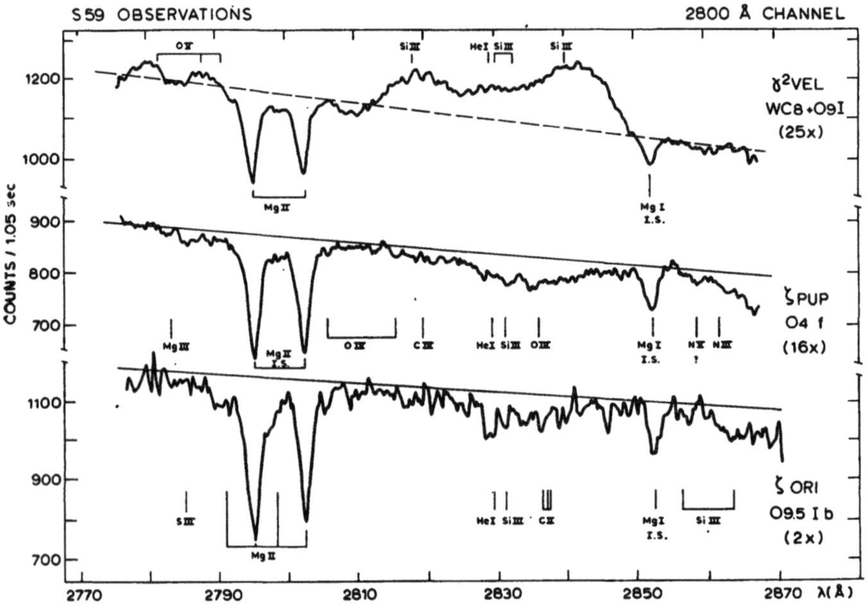

Fig. 3. S59 spectra. 2800 Å channel.

At 2518 Å SiIV dominates in ζ Ori while in ζ Pup OIV is the main contributor. In γ^2 Vel, WC8 + O9I [12], the continua were determined by matching the absorption features with those of ζ Ori, which has about the same spectral type as the O component of γ^2 Vel. As usual the emission lines are attributed to the WC star and the absorption lines to the O component. In the 2100 Å region no absorption lines are indicated because they correspond with ζ Ori. However no emission lines have so far been identified in this region. The 2500 Å region is discussed by van der Hucht and Lamers [13]. The newly established continuum gives a smaller value of the possible red wing of the CIV 2530 emission line, which now extends to 22 Å (2600 km s^{-1}) from the line centre. The observations of γ^2 Vel of May 1973, obtained at a different phase of the binary system, are not yet available. The interstellar MgI, MgII and FeII absorption lines in γ^2 Vel are discussed in [14].

The MgII lines. The MgII lines near 2800 Å are strong over a wide spectral range. Owing to the relatively large separation of 4.5 eV between the 3s and 3p levels and the 3p and 3d levels the MgII lines are sensitive to non-LTE effects and therefore to the structure of stellar atmospheres and their radiation fields. This property makes the MgII resonance lines suited for study of early type stars where deviations from LTE in the level populations of H and He have large effects on the atmospheric structure. A study by Lamers et al. [15] gives the following results: for stars of types A and F the equivalent width depends strongly on the effective temperature while for B type stars there is no temperature dependence. Stars of class II and III show about the same equivalent widths as main sequence stars. The observed equivalent widths agree reasonably well with non-LTE predictions by Mihalas [16] if an abundance of 7.48 is adopted. In stars earlier than B3 the lines are stronger than predicted. The difference increases with temperature, reaching a factor 1.8 or 0.7 Å at B1. The most likely explanation for this discrepancy is the influence of far-UV line blanketing which reduces the ionization rate of the 3p level. For the peculiar A star α^2 CVn Burger et al. [17] found a abundance for Mg of about 7.2. The MgII resonance lines in α Cyg (A2Ia) show a radial velocity of -140 km s^{-1}. This value is higher than the velocity of the Balmer lines in A type supergiants, which is usually -40 to -80 km s^{-1}. In order to estimate the rate of mass loss, we computed the MgII lines in a model atmosphere of $T_{eff} = 9170$ K, $\log g = 1.13$ and variable microturbulent velocity according to Groth [18]. The core of the lines is formed in a layer where the density is $10^{-15}-10^{-14}$ g cm^{-3}. Using these densities and assuming a radius of 45 $R_\odot$ [19], we derive a rate of mass loss between 3×10^{-7} and 3×10^{-8} $M_\odot$ yr^{-1}. This is approximately two orders of magnitudes smaller than in OB supergiants, but it is consistent with the rate estimated from the velocity of the Balmer lines [20].

References

[1] R. Hoekstra, K. A. van der Hucht, C. de Jager, T. M. Kamperman, H. J. Lamers, A. Hammerschlag and W. Werner, Nature Phys. Sci. **236**, 121 (1972).
[2] B. Tilgner, ELDO/ESRO Sci. Tech. Rev. **3**, 567 (1971).
[3] R. Hoekstra, K. A. van der Hucht, T. M. Kamperman and H. J. Lamers, in: Proc. IXth Congr. of the I. C. O. at Santa Monica (1972).
[4] D. Mihalas, NCAR-TN/STR-76 (1972).
[5] T. Stecher, Astrophys. J. **159**, 543 (1970).
[6] P. S. Conti and W. R. Alschuler, Astrophys. J. **170**, 325 (1971).
[7] S. R. Heap, Astrophys. Lett. **10**, 49 (1972).
[8] K. S. de Boer, Space Research XIV, 549 (1974).
[9] T. G. Barnes, D. L. Lambert and A. E. Potter, Astrophys. J. **187**, 73 (1974).
[10] D. Mihalas, private communication (1972).
[11] B. Warner and R. C. Kirkpatrick, Mon. Not. Roy. Astron. Soc. **147**, 115 (1970).
[12] P. Conti and L. F. Smith, Astrophys. J. **172**, 623 (1972).
[13] K. A. van der Hucht and H. J. Lamers, Astrophys. J. **181**, 537 (1973).
[14] M. Grewing, C. Wulf-Mathies, H. J. Lamers and C. M. Walmsley, Space Research XIV, 543 (1974).
[15] H. J. Lamers, K. A. van der Hucht, M. A. J. Snijders and N. Sakhibullin, Astron. Astrophys. **25**, 105 (1973).
[16] D. Mihalas, Astrophys. J. **177**, 115 (1972).
[17] M. Burger, K. A. van der Hucht and H. J. Lamers, Astron. Astrophys. **26**, 149 (1973).
[18] H. G. Groth, Z. Astrofiz. **51**, 206 (1961).
[19] C. W. Allen, Astrophysical Quantities, The Athlone Press, University of London, London (1955).
[20] J. B. Hutchings, in: Colloq. on Supergiant Stars, Trieste 1971 (p. 38).

Space Research XIV — Akademie-Verlag, Berlin 1974

THE NEAR ULTRAVIOLET SPECTRUM OF EARLY TYPE STARS OBTAINED WITH THE S59 SPECTROPHOTOMETER

H. J. LAMERS[a], K. A. VAN DER HUCHT[a], R. HOEKSTRA[a], R. FARAGGIANA[b] and M. HACK[b]

[a] Astronomical Institute, Utrecht, The Netherlands
[b] Osservatorio Astronomico, Trieste, Italy

The near ultraviolet spectra of early type stars from O4 to F5 are shown. The lines are identified and the general characteristics are discussed. A classification criterion for B-type stars, based on the Fe II/Fe III ratio is proposed. The line blocking has been measured and the consequences for the effective temperature scale of late B and A stars is discusssed.

1. Introduction

From the material obtained by S59 so far, we have selected 31 stars which represent a grid of spectral types from O4 to F5 and all luminosity classes. The programme stars are listed in Table 1. The MK-spectral types are from the *Catalogue of Bright Stars* [1]. The values of $B - V$ and V are from Johnson [2]. The intrinsic colours are from Johnson [3], except in those cases where $B - V$ is smaller than $(B - V)_0$. For these stars, the observed $B - V$ was adopted.

The spectra of a sample of 13 stars are shown in Figs. 1, 2 and 3, all on approximately the same scale relative to the continuum. The highest points in the spectra were assumed to represent the position of the continuum. In some cases, especially in A and F stars, the run of the continuum could only be guessed by comparison with stars of similar type. This procedure will result in a possible underestimate of the height of the continuum, in particular in stars of types B8 and later. The model atmosphere computations by Burger and van der Hucht [4] have shown that the continuum should be increased by 8% in the A0 stars.

2. The Identifications

We have tried to identify the spectral features in the stars ζ Pup (O4f), ζ Ori (O9.5 Ib), o^2 CMa (B3 Ia), and α Cyg (A2 Ia). The O-type stars are discussed in [5]. The identifications of the major spectral features are shown in Figs. 1, 2 and 3. The following conclusions can be drawn from these figures: (i) Supergiant stars show considerably stronger lines than main-sequence stars. This is known from visual spectra as well and it is usually attributed to a large microturbulent velocity. (ii) The spectrum of early-B stars is dominated by lines of Fe III and Cr III, especially in the 2100 and 2500 channels. The absorption feature at 2077—2092 Å

Table 1
The Programme Stars

Star	Type	BSC	V	$B-V$	$(B-V)_0$	$E(B-V)$	No.	Remarks
γ^2 Vel	WC8+O9	3207	1.82	−0.27	−0.30	0.03	15	
ζ Pup	O4f[2]	3165	2.25	−0.29	−0.32	0.03	12	
ξ Per	O7.5I[2]	1228	4.06	−0.01	−0.30	0.29	1	
ι Ori	O9III	1899	2.76	−0.23	−0.31	0.08	1	
θ Car	O9.5V	4199	2.76[3]	−0.22[3]	−0.30	0.08	5	B0.5p[4]
ζ OriA	O9.5Ib	1948/49	1.74	−0.21	−0.30	0.09	1	
τ Sco	B0V	6165	2.82	−0.25	−0.30	0.05	1	
η Ori	B0.5V	1788	3.32	−0.18	−0.28	0.10	2	
β Cru	B0.5IV	4853	1.24[3]	−0.24	−0.28	0.04	1	B0.5III[4]
$\varkappa$ Ori	B0.5Ie	2004	2.06	−0.18	−0.25	0.07	1	B0.5V[5]
θ Ara	B0.5II	6743	3.66[3]	−0.09[3]	−0.26	0.17	10	B2Ib[4]
β CMa	B1II	2294	1.98	−0.24	−0.24	0.00	9	
σ Sgr	B2V	7121	2.07	−0.22	−0.24	0.02	8	B3IV[4]
$\varkappa$ CMa	B2Ve	2538	3.96	−0.24	−0.24	0.00	8	
α Pav	B3IV	7790	1.93[3]	−0.20[3]	−0.20	0.00	12	B2.5V[4]
o^2 CMa	B3Ia	2653	3.01	−0.11	−0.14	0.03	9	
α Gru	B5V	8425	1.73	−0.17	−0.17[6]	0.00	10	B7IV[4]
η CMa	B5Ia	2827	2.41	−0.07	−0.10	0.03	7	
ζ Dra	B6III	6396	3.17	−0.11	−0.14	0.03	83	
ε Del	B6III	7852	4.04	−0.12	−0.14	0.02	5	
φ Sgr	B8III	7039	3.17	−0.11	−0.11[6]	0.00	6	
α^2 CVn	B9.5pv	4915/14	2.84	−0.10	−0.10[6]	0.00	6	
α Lyr	A0V	7001	0.00	0.00	0.00	0.00	19	
γ Gem	A0IV	2421	1.96	0.00	0.00	0.00	8	
γ UMa	A0V	4554	2.44	0.00	0.00	0.00	11	
ε UMa	A0pv	4905	1.78	−0.03	0.00	0.00	8	
β Aur	A2V	2088	1.90	0.02	0.02[6]	0.00	10	
α Cyg	A2Ia	7924	1.25	0.09	0.05	0.04	10	
α Aql	A7V	7557	0.74	0.23	0.19	0.04	8	
α Car	F0Ib	2326	−0.73	0.16	0.16[6]	0.00	15	
α CMi	F5IV	2943	0.35	0.43	0.43	0.00	8	

Remarks:

[1] Grygar [11].
[2] Conti and Alschuler [8].
[3] Catalogue of bright stars [1].
[4] Hiltner, Garrison and Schild [9].
[5] Underhill [10].
[6] $(B-V) < (B-V)_0$ Johnson [2].

is representative. This blend is visible in the spectrum of α Gru (B5 V) and hotter stars, but absent in α Lyr (A0 V). (iii) The spectrum of types B5 and later are dominated by lines of Fe II and Cr II, especially in the 2500 channel. The features at 2528, 2538, 2549 and 2563 Å are representative. These lines are visible in the spectrum of α Pav (B3 IV) and later, but absent in β Cru (B0.5 IV). The spectrum of β CMa (B1 II—III) shows some faint lines at these wavelengths. Another typical Fe II and Cr II feature is visible in the 2100 channel at 2124—2140 Å.

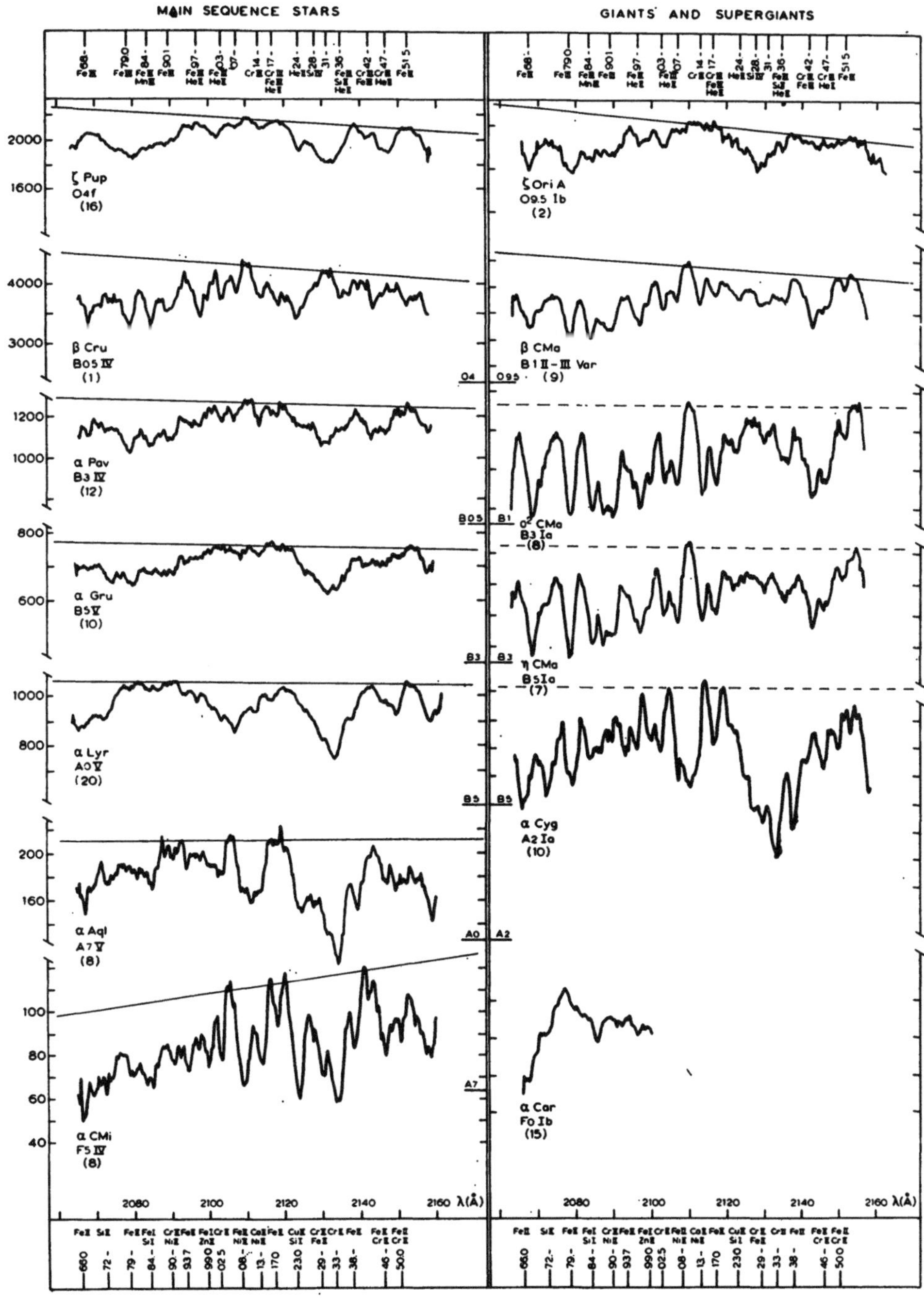

Fig. 1. The spectra of several stars, observed by S 59, 2100 Å channel. The position of the continuum is described in § 1. The less reliable continua are indicated by a broken line. The zero-level of each spectrum is indicated near the middle of the figure. The major spectral features are identified in the upper (for hot stars) and the lower (for cool stars) part of the figure.

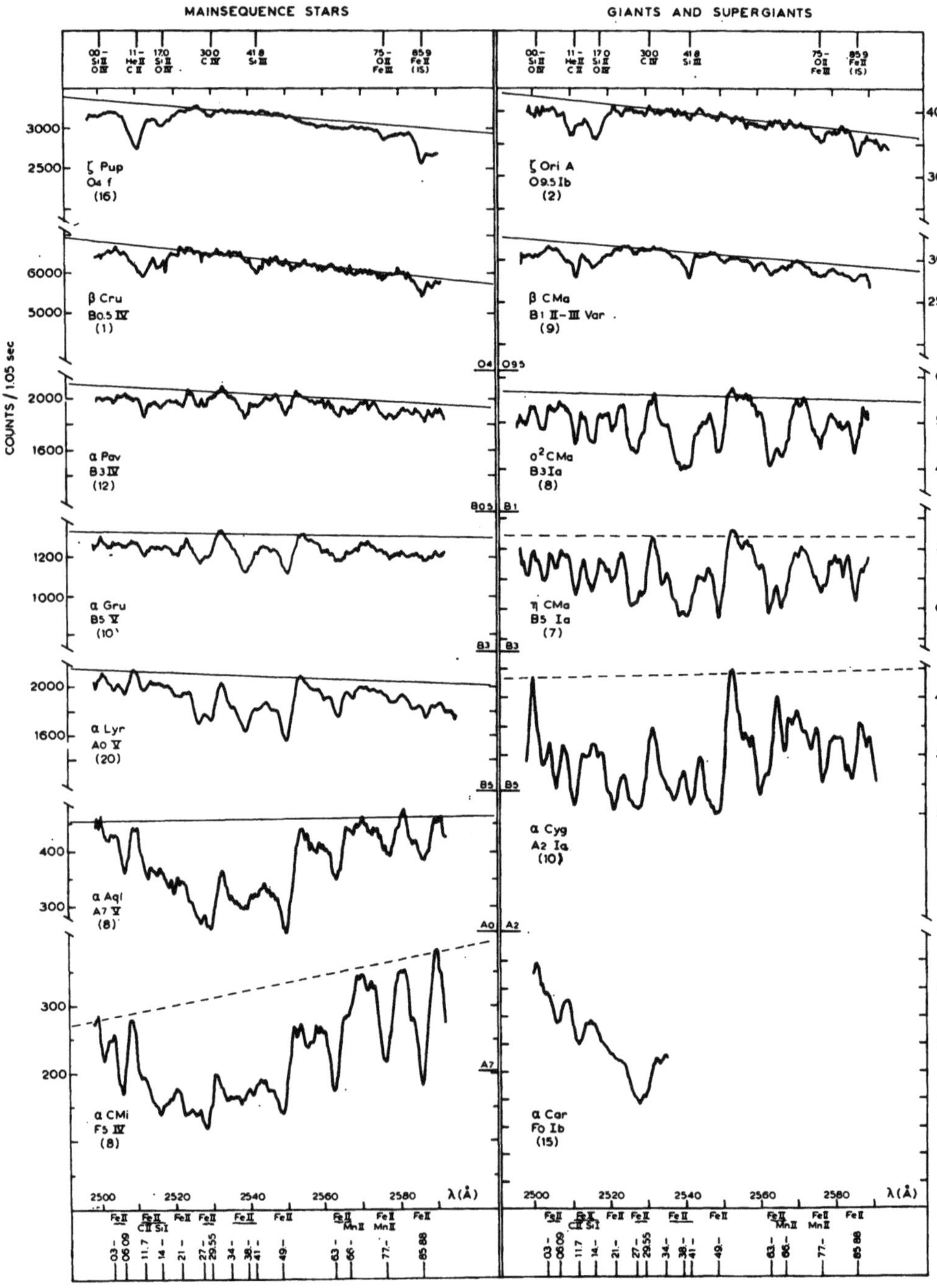

Fig. 2. As Fig. 1, for 2500 Å channel.

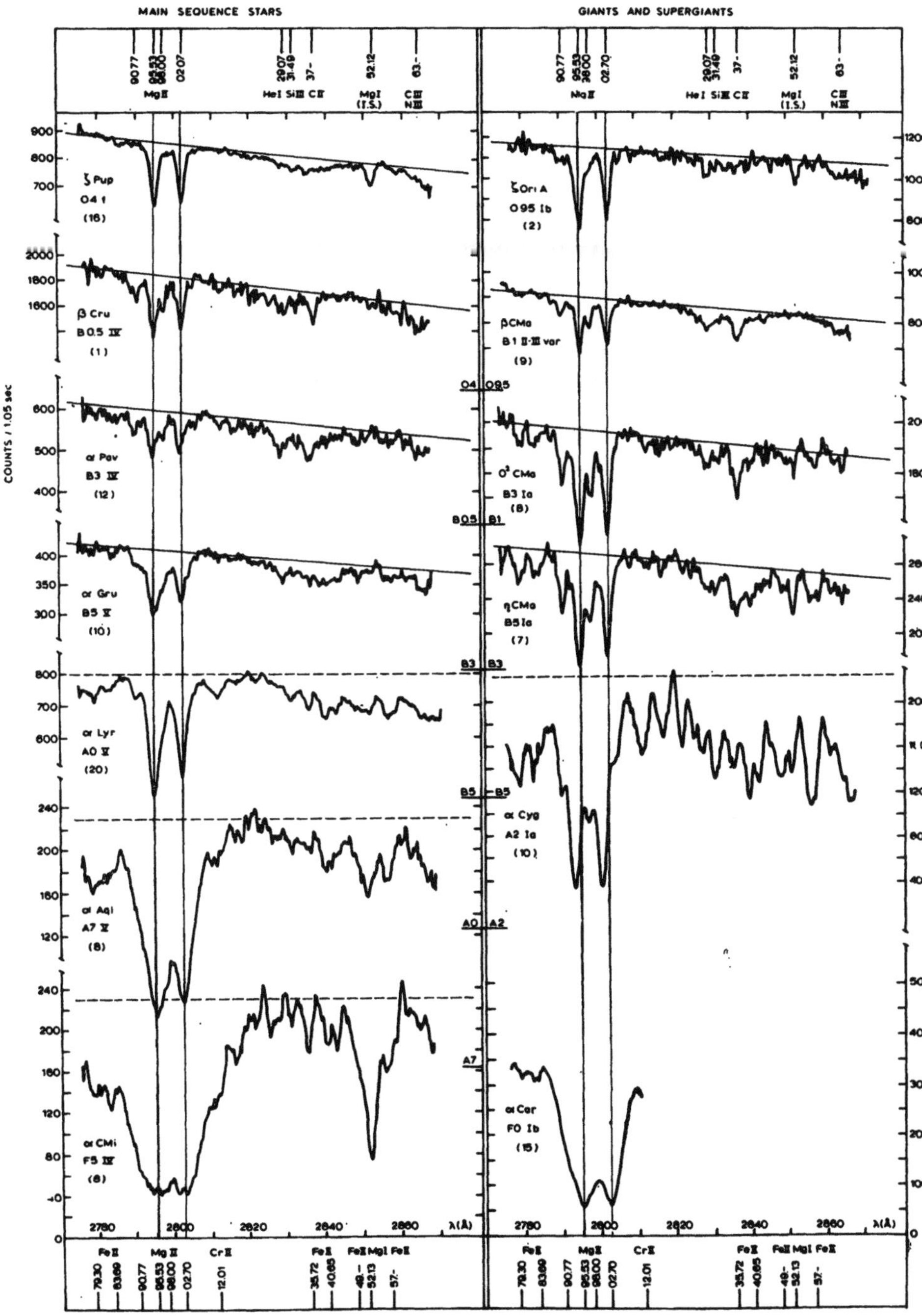

Fig. 3. As Fig. 1, for 2800 Å channel.

In main-sequence stars this blend is present in the types B3 and later, whereas in supergiants it only begins in A type stars. This difference reflects the influence of the electron pressure on the Fe^{++}/Fe^{+} ionization equilibrium. (iv) In the 2800 channel, four Mg II lines are prominent in all stars. In the hottest stars ζ Pup and ζ Ori, part of these lines are of interstellar origin. This is confirmed by the presence of the interstellar Mg I line at 2852 Å. In the spectra of α Aql (A7 V) and α CMi (F5 IV) the stellar Mg I lines are very strong.

3. A Classification Criterion

The Fe II and Fe III lines are visible over a large spectral range and hence their ratio might serve for classification purposes. As a measure of the Fe II line strength we used the average equivalent width of the lines at 2549 and 2563 Å, while the feature at 2077—2092 Å was used as a measure of the Fe III lines.

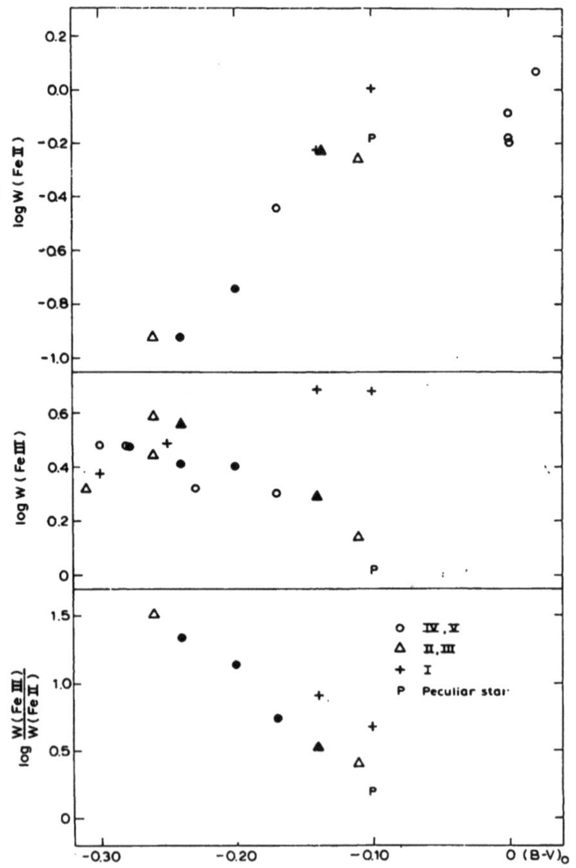

Fig. 4. The equivalent width of the Fe II and Fe III lines, and their ratio. W(Fe II) = (W(2549) + W(2563))/2, W(Fe III) = W(2077 — 2092). The filled symbols refer to the most reliable values.

The equivalent widths of these features and their ratio are plotted against $(B - V)_0$ in Fig. 4. The ratio Fe III/Fe II is strongly dependent on the intrinsic colour and the limited number of observations suggests that this criterion can be used from B0.5 to B6 with a total range of a factor 10. This would enable a classification of the B stars within half a subclass. An accurate classification of this type, based on line-ratios, should be compared with a classification based on the continuum properties, e.g. those proposed by Humphries et al. [6], in order to sort out the present differences between photometric and spectroscopic classification.

4. Line Blocking

The fraction of the stellar flux blocked by the spectral lines has been measured in the S59 spectra of the program stars. The result is shown in Fig 5. The blocking fraction increases from B3 to later types and reaches a typical value of about 15% near A0. The supergiants show values about twice as high as those for main-sequence stars. Owing to our continuum determination, these values are lower limits, especially for stars of types B5 and later. Burger and van der Hucht [4]

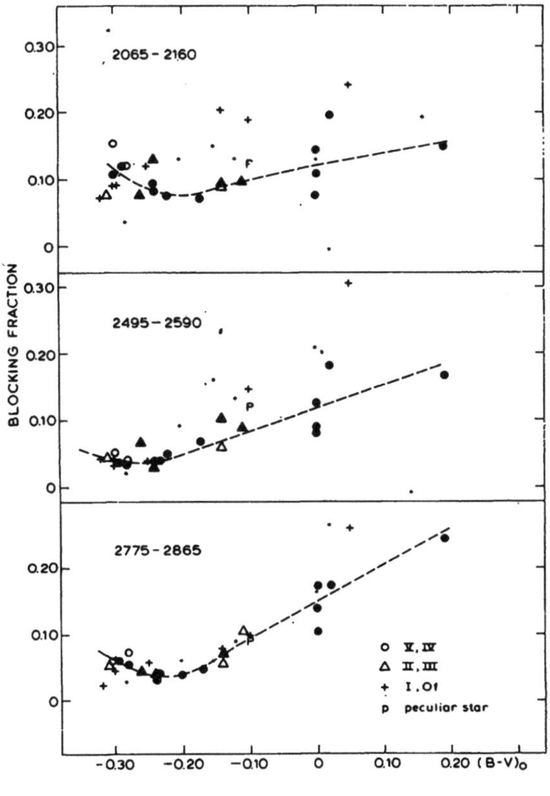

Fig. 5. The line-blocking fractions in the S59 spectra. The filled symbols refer to the most reliable values. The values derived by Underhill [7] are indicated by points. The points just above the dividing lines between the graphs refer to the graph below in each case.

propose an increase of 0.08 for A0 stars. The blocking fractions derived by Underhill [7] from the OAO-II observations are shown for comparison. They agree reasonably with our values of early-B stars but are higher for later types. If the derived line-blocking fractions persists over the whole near ultra-violet wavelength range, then the bolometric corrections for late-B and A stars should be increased (Δ B.C.) and the effective temperatures decrease relative to the values obtained by comparison of the visual spectrum with unblanketed model atmospheres:

$$\Delta \log (T_{\mathrm{eff}}) = -0.1 \Delta \,\mathrm{B.C.} = 0.25 \log (1 - \mathrm{BF})$$

where BF = blocking fraction. A blocking fraction of 0.10 results in $\Delta \log (T_{\mathrm{eff}}) = -0.013$, and BF = 0.30 in $\Delta \log (T_{\mathrm{eff}}) = -0.039$.

The scatter in the blocking fractions of stars around A0 is larger than the uncertainty of the measurements. This scatter might reflect the errors in the present classification or differences in metal abundances or rotation velocities. A comparison with theoretical spectra will solve this problem. The intrinsic scatter in the ultraviolet colour-indices found by Humphries et al. [6] probably originates from the same effect.

References

[1] D. HOFFLEIT, Catalogue of Bright Stars, Yale Univ. Obs., New Haven 1964.
[2] H. L. JOHNSON, Sky Telesc., **30**, 21 (1965).
[3] H. L. JOHNSON, Ann. Rev. Astron. Astrophys. **4**, 193 (1966).
[4] M. BURGER and K. A. VAN DER HUCHT, Space Research XIV, 537 (1974).
[5] K. A. VAN DER HUCHT and H. J. LAMERS, Space Research XIV, 523 (1974).
[6] C. M. HUMPHRIES et al., Space Research XIV, 565 (1974).
[7] A. B. UNDERHILL, in: Scientific Results from OAO-II, 1972 (p. 367).
[8] P. S. CONTI and W. R. ALSCHULER, Astrophys. J. **170**, 325 (1971).
[9] W. A. HILTNER et al., Astrophys. J. **157**, 313 (1969).
[10] A. B. UNDERHILL, The Early Type Stars, D. Reidel Publ. Co., Dordrecht, Holland 1966 (p. 213).
[11] J. GRYGAR, Bull. Astron. Inst. Czechosl. **16**, 196 (1965).

Space Research XIV — Akademie-Verlag, Berlin 1974

LINE IDENTIFICATIONS IN THE NEAR ULTRAVIOLET SPECTRUM OF THE PECULIAR A STAR ε URSAE MAJORIS

M. Burger[a] and K. A. van der Hucht[b]

[a]Astrophysical Institute, Brussels, Belgium
[b]The Astronomical Institute, Utrecht, the Netherlands

A synthetic spectrum using about 250 spectral lines is computed in the wavelength region 2495—2590 Å and is compared with the observed ultraviolet spectrum of the peculiar A star ε UMa. The position of the continuum is indicated and line identifications are given. The spectrum of ε UMa suggests an overabundance of Mn by a factor of 10 to 20.

1. Introduction

The Orbiting Stellar Spectrophotometer S59 on board the ESRO TD-1A Satellite yields observations of ultraviolet stellar spectra [1]. The spectra are scanned simultaneously in three wavelength bands centred around 2110, 2550 and 2820 Å, each 90 Å wide. The spectral resolution of the spectrograms is about 1.8 Å. Reduction of the data provides several difficulties. A resolution of 1.8 Å is insufficient to observe all spectral lines, particularly as the ultraviolet region of the spectrum is extremely rich in lines. It will be difficult to establish the height of the continuum because of the narrowness of the observed spectral region and the multitude of lines.

As a first attempt to solve this problem we computed a synthetic line spectrum which will provide a more reliable determination of the continuum and will facilitate the identification of spectral lines. In the present study we have restricted ourselves to one wavelength region only: 2495—2590 Å. We shall apply the result of the computations to the spectrum of the peculiar A star ε Ursae Majoris.

2. The Spectrum of ε UMa

ε Ursae Majoris is known as a peculiar star of spectral type A0 V. As many other peculiar A stars ε UMa shows light and spectrum variations. Guthnick [2, 3] established a period of 5.0887 days, based on the variation of the K line of Ca II. Fourteen scans of ε UMa were obtained by S59 between JD 2441467.147 and JD 2441468.338. This time interval covers the phases between 0.63 and 0.87 of the period of ε UMa. In this paper we consider the average of the fourteen scans only, although there are indications of variations of spectral lines. The averaged spectrogram is shown in the upper part of Fig. 1. The wavelengths of the most important absorption features are indicated. For comparison Fig. 1 shows the spectrum of a normal main-sequence star, γ Ursae Majoris, which is also classified as A0 V. The most prominent differences between the two spectrograms are indicated by arrows.

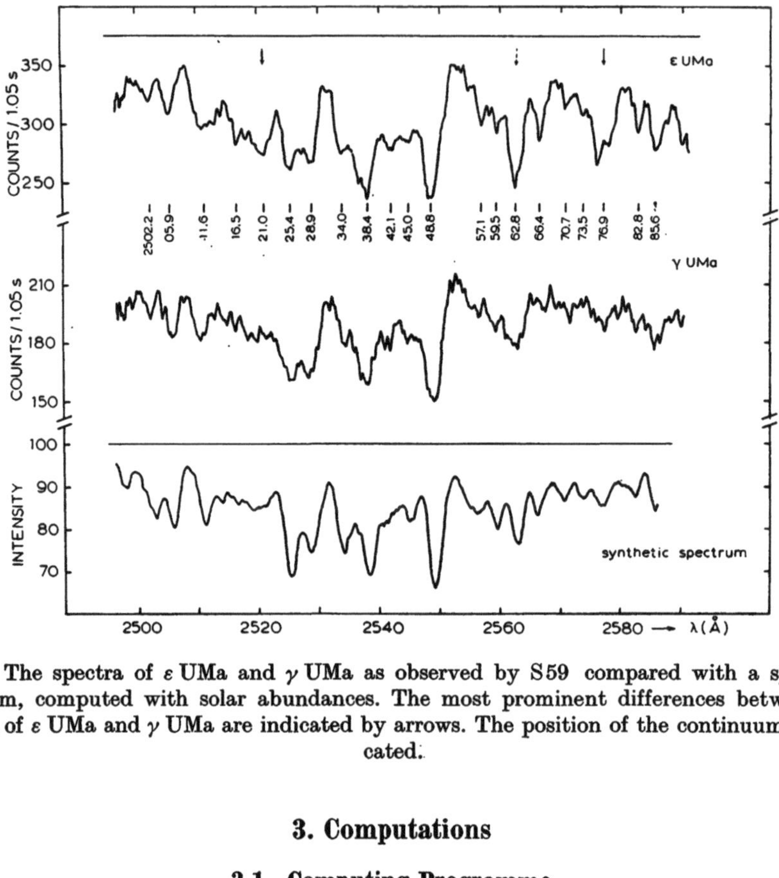

Fig. 1. The spectra of ε UMa and γ UMa as observed by S59 compared with a synthetic spectrum, computed with solar abundances. The most prominent differences between the spectra of ε UMa and γ UMa are indicated by arrows. The position of the continuum is indicated.

3. Computations

3.1. Computing Programme

For the computation of the line spectrum we used a hydrogen-line blanketed model atmosphere of Klinglesmith [4]. The lines were computed asssuming LTE and a Voigt-profile for the absorption coefficient. To save computing time we only included lines with such absorption coefficients that the resulting equivalent width is greater than 0.01 Å. The emergent intensity is computed every 0.02 Å. Finally, the calculated spectrum is convoluted with a triangular instrumental profile with a halfwidth of 1.8 Å, which agrees with the observed instrumental profile [5].

3.2. The Model Atmosphere

The $B - V$ and $U - B$ colour indexes of ε UMa are -0.01 and $+0.01$ respectively. According to Schild et al. [6] the effective temperature of ε UMa is 9400 K. The relation given by Wolff et al. [1] which uses the $B - V$ value only yields an effective temperature of 10200 K. The hydrogen spectrum suggests a value of 10000 K [8]. The curve of growth analysis of ε UMa by Casati and Hack [9] results in $\log g = 4$ and a microturbulent velocity of 1.4 km s^{-1}. From these data we adopted a model atmosphere of Klinglesmith with an effective temperature of 10000 K, $\log g = 4$ and a microturbulent velocity of 1.5 km s^{-1} throughout the atmosphere. However, it should be kept in mind that ultraviolet observations by rockets and satellites indicate that the emerging flux in the ultraviolet is lower than predicted by model atmospheres that fit the observations in the visible region of the spectrum.

The abundances of the elements in the atmosphere of ε UMa are poorly known. Therefore we used solar abundances as a starting point in this computation. We adopted the values compiled by Withbroe [10].

3.3. Atomic Line Data

We considered lines of elements up to $Z = 71$. Between 2495 and 2590 Å we recorded about 700 lines of which *gf* values either are available in the literature or could be computed by means of the scaled Thomas-Fermi method (cf. [11]).

Table 1

Lines included in the Computations for the Model Atmosphere with $T_{\rm eff} = 10000$ K and $\log g = 4.0$

Ion	Abundance	No. of lines	Ref. for *gf* values
Al I	6.40	1	[13]
Si I	7.55	6	[13]
Si II	7.55	2	[13]
Sc II	3.07	5	[14]
Ti II	4.74	8	[14]
V II	4.10	11	[14]
Cr II	5.70	1	[14]
Mn II	5.20	27	[14][a]
Fe I	7.40	36	[14, 15]
Fe II	7.40	124	[14, 16][a]
Co II	4.50	19	[14][a]
Ni II	6.28	15	[17]
Nb II	2.30	2	[17]

[a] also calculated, scaled Thomas-Fermi method.

As a first approximation for the damping constant we took into account two broadening mechanisms: classical radiation damping: $\gamma = 0.221/\lambda^2$ (λ in Å), and, collisions with hydrogen and helium atoms in the van der Waals approximation [12]. The restriction to exclude the weakest lines (see § 3.1) reduced the number of lines included in the computation to about 250.

The references concerning the *gf* values and the number of lines of each ion included are given in Table 1.

4. Results

The computed spectrum is represented in the bottom part of Fig. 1, drawn on about the same scale as the observed spectrograms. It is clear that the position of the real continuum is higher than the highest points in the observed spectra. Therefore computations like these may be useful for determination of the height of the continuum, especially in line-rich regions. Fig. 1 shows a good correspondence between the observed spectrum of γ UMa and the computed one. This indicates that the computed spectrum is a fair representation of the spectrum of an A0 V star. The differences between the computed and the observed spectrum may be caused by differences in the atmospheric parameters, contributions of lines not taken into account in the synthetic spectrum, unreliable gf values, unreliable damping parameters, differences in abundances.

Line identifications based on the result of the computations are given in Table 2. Only the lines contributing the most to the various absorption features of ε UMa are indicated. The identifications are most uncertain at those wavelengths where

Table 2

Line Identifications of ε UMa

Measured Wavelength (Å)	Ion	Laboratory Wavelength (Å)
2502.2*	Fe II	2502.39
	Fe II	03.33
	Fe II	03.56
	Fe II	03.87
2505.9	Fe II	0.522
	Ni II	05.84
	Fe II	06.09
	Co II	06.46
	Fe II	06.80
	Si I	06.90
	Fe II	07.02
2511.6	Ni II	10.87
	Fe II	11.37
	Fe II	11.76
2516.5*	Si I	16.11
	Fe II	17.12
	Fe II	17.21
2521.0*	Fe II	19.05
	Si I	19.20
	Fe II	19.41
	Fe II	20.68
	Fe II	21.09
	Fe II	21.81
	Fe II	22.20
	Fe I	22.85
2525.4	Si I	2524.11
	Co II	25.02
	Fe II	25.11
	Fe II	25.39
	Ti II	25.60
	Fe II	25.87
	Fe II	26.08
	Fe II	26.30
2528.9	Si I	28.51
	Co II	28.62
	Fe II	29.08
	Fe II	29.55
2534.0	Fe II	33.62
	Cr II	34.34
	Fe II	34.42
	Ti II	34.62
2538.4	Fe II	38.20
	Fe II	38.50
	Fe II	38.58
	Fe II	38.81
	Fe II	38.90
2542.1*	Fe II	40.53
	Fe II	40.67
	Fe II	41.10
	Fe II	41.84
	Fe II	42.73

Table 2 (cont'd.)
Line Identifications of ε UMa

Measured Wavelength (Å)	Ion	Laboratory Wavelength (Å)
2542.1*	Fe II	2543.38
	Fe II	43.42
	Mn II	43.45
2545.0	Fe II	44.97
	Fe II	45.22
	Fe II	45.43
	Ni II	45.90
2548.8	Fe II	48.33
	Fe II	48.59
	Fe II	48.74
	Fe II	48.92
	Fe II	49.08
	Fe II	49.40
	Fe II	49.46
	Fe II	49.77
	Fe II	50.02
	Fe II	50.58
	Fe II	50.68
	Fe II	51.21
2557.1	Fe II	55.07
	Fe II	55.44
	Mn II	56.67
	Fe II	57.08
	Fe II	57.50
2559.5	Fe II	2559.24
	Fe II	59.77
	Fe II	59.93
	Fe II	60.27
	Fe II	60.44
2562.8	Fe II	62.10
	Fe II	62.54
	Fe II.	63.47
	Mn II	63.65
	Fe II	63.84
2566.4	Fe II	66.21
	Fe II	66.62
	Fe II	66.91
2570.7	Fe II	69.76
	Fe II	70.84
	Fe II	71.55
2573.5	Fe II	72.97
	Fe II	73.21
	Fe II	74.37
2576.9	Mn II	76.10
	Fe II	76.87
	Fe II	77.92
2582.8	Fe II	82.58
2585.6	Fe II	85.88

* Line identifications uncertain.

conspicuous differences occur between the synthetic spectrum and that of ε UMa (see Fig. 1). We expect these differences to be due to differences in abundances. For instance, an overabundance of Mn can be expected in most of the peculiar stars. Therefore as an example we recomputed a part of the spectrum with Mn abundances of respectively 10 and 20 times the solar value. With these higher abundances a better agreement with the spectrum of ε UMa is obtained (Fig. 2) at 2576.1 Å, where a strong resonance line of Mn II occurs.

Acknowledgments

One of us (M. B.) thanks Professor C. de Jager and his colleagues for making available to her data from the Orbiting Stellar Spectrophotometer S59. She wishes to acknowledge the support of an ESRO fellowship.

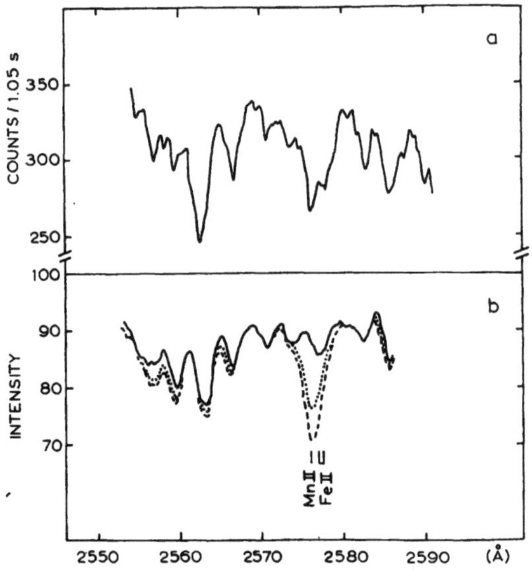

Fig. 2. a, spectrum of ε UMa; b, computed spectrum: —— using solar abundances, ····· 10 times the solar Mn abundance, ---- 20 times the solar Mn abundance.

References

[1] R. Hoekstra, K. A. van der Hucht, Th. Kamperman, H. J. Lamers, A. Hammerschlag and W. Werner, Nature Phys. Sci. **236**, 121 (1972).

[2] P. Guthnick, Sitz. Preuss. Akad. Berlin, No. 27 (1931).

[3] P. Guthnick, Sitz. Preuss. Akad. Berlin, No. 30 (1934).

[4] D. A. Klinglesmith, NASA SP-3065 (1971).

[5] R. Hoekstra, K. A. van der Hucht, Th. Kamperman and H. J. Lamers, Proc. 9th Congr. of the ICO, Santa Monica 1972.

[6] R. Schild, D. M. Peterson and J. B. Oke, Astrophys. J. **166**, 95 (1971).

[7] S. C. Wolff, L. V. Kuhi and D. S. Hayes, Astrophys. J. **152**, 871 (1968).

[8] C. J. Durrant, Mon. Not. Roy. Astron. Soc. **147**, 75 (1970).

[9] B. Casati and M. Hack, Mem. Soc. Astron. Ital. **28**, 53 (1957).

[10] G. L. Withbroe, in: Menzel Symp. on Solar Physics, Atomic Spectra and Gaseous Nebulae, Nat. Bur. Stds Spec. Publ. 353, 1971 (p. 127).

[11] W. van Rensbergen, Solar Phys. **11**, 11 (1970).

[12] A. Unsöld and V. Weidemann, Vistas in Astronomy **1**, 249 (1955).

[13] W. L. Wiese, M. W. Smith and B. M. Miles, Nat. Bur. Stds Ref. Dat. Ser. NBS 22 (1969).

[14] R. J. Takens, private communication 1970.

[15] C. H. Corliss and J. L. Tech, NBS Monograph 108, Nat. Bur. Stds, Washington 1968.

[16] B. Warner, Mem. Roy. Astron. Soc. **70**, 165 (1967).

[17] S. E. Strom and R. L. Kurucz, J. Quant. Spectrosc. and Rad. Transf. **6**, 591 (1966).

Space Research XIV — Akademie-Verlag, Berlin 1974

MAGNESIUM AND IRON ABUNDANCES IN THE GUM NEBULA REGION DETERMINED FROM INTERSTELLAR ABSORPTION LINES IN THE SPECTRUM OF γ^2 VEL

M. Grewing[a], C. Wulf-Mathies[a], H. J. Lamers[b] and C. M. Walmsley[c]

[a]Institut für Astrophysik und extraterrestrische Forschung der Universität Bonn, Bonn, FRG
[b]Astronomical Institute, Utrecht, The Netherlands
[c]Max-Planck-Institut für Radioastronomie, Bonn, FRG

Interstellar absorption lines of Mg I, Mg II and Fe II have been detected in the spectrum of γ^2 Vel, observed with the Orbiting Stellar Spectrophotometer on board the ESRO TD-1 A satellite. No straightforward explanation for the derived column densities of these and other observed ions is available. Possible alternatives are discussed.

1. Introduction

Analysing the emission, dispersion and hydrogen measure of objects located in the Gum Nebula complex, Brandt et al. [1] were led to the conclusion that the Gum Nebula is not ionized by ζ Pup and γ^2 Vel, but was produced by the Vela supernova outburst. Just the opposite conclusion was reached by Beuermann [2], who studied the extent of the Hα emission 21 cm data on the contiguous neutral medium, and low-frequency absorption measurements in the non-thermal radio continuum. The Brandt et al. [1] model depends critically on the total volume that is ionized, which they took to be $\sim 1.3 \times 10^{63}$ cm^3. This is based mainly on the very small hydrogen measure observed against ζ Pup and γ^2 Vel. With $\int_0^l n_\mathrm{H}\, ds = 7 \times 10^{19}$ cm^{-2} for ζ Pup [3] and *assuming* $\langle n_\mathrm{H} \rangle = 0.25$ cm^{-3} for the solar neighbourhood, Brandt et al. obtained $l = 100$ pc for the distance of the ionization front from the sun. As the distance of the centre of the Gum Nebula is estimated to be ~ 450 pc, this leads to a radius of 350 pc for the ionized volume (in deriving the above volume cyclindrical symmetry was assumed). The angular diameter of the Gum Nebula should then be roughly 90°.

Both Bok [4] and Johnson [5] pointed out that so large an angular diameter is contradiceted by observations, since most of the Hα emission seems to be concentrated within an angular diameter of 40°. Indeed, Beuermann [2] claims that a ridge of enhanced 21 cm emission of *neutral* hydrogen closely follows the nearly circular envelope of the strongest Hα contours for almost 360°. We feel that the observations do not justify so strong a statement; however, Beuermann's further claim that low frequency absorption measurements of the non-thermal radio continuum also indicate an angular diameter of the ionized region of 40—50° seems to be valid.

In view of this evidence we feel that the argument applied by Brandt et al.

should be reversed, i.e. starting from the observed angular diameter of the Gum Nebula, the distance of the ionization front from the sun should be calculated (again assuming cylindrical symmetry) and the hydrogen measure of ζ Pup and γ^2 Vel should be used to derive $\langle n_H \rangle$. Following this scheme, we adopt a value of 45° for the angular diameter of the Gum Nebula. This leads to a radius of 170 pc for the ionized region. In this model the ionization front would be at a distance of 280 pc from the sun. The observed hydrogen measures observed against ζ Pup and γ^2 Vel, 23 and 10 cm^{-3} pc respectively, then yield $\langle n_H \rangle = 0.08$ and 0.04 cm^{-3}, where we have assumed a distance of 450 pc for both stars. Since γ^2 Vel is further away from the galactic plane a slightly smaller value is expected, though not by a factor of two. Mean hydrogen densities as low as 0.05 ± 0.02 cm^{-3} are entirely consistent with the values found by Rogerson et al. [6] varying between 0.22 cm^{-3} and 0.02 cm^{-3} from OAO-C data for unreddened stars [$E(\mathrm{B}-\mathrm{V}) \leqq 0.03$]. This case is applicable since the Gum Nebula region shows very low extinction. We therefore assume in what follows that the line of sight to the centre of the Gum Nebula traverses 280 pc of intercloud medium with $n_H = 0.06 \pm 0.02$ cm^{-3} before entering the H II gas. Even if the hydrogen is 20% ionized in the intercloud gas, the contribution from this region to the dispersion measure (DM) of the Vela pulsar (PSR 0833-45) will be negligible. We therefore conclude that

$$\mathrm{DM} = 69\ \mathrm{cm}^{-3}\ \mathrm{pc} = \int_{280}^{500} n_e\, ds \quad ([ds]\ \text{in pc})$$

where we have used a distance of 500 pc for the pulsar. From this relation we obtain $\langle n_e \rangle = 0.3$ cm^{-3} inside the H II region.

An independent estimate of the electron density in the H II region can be derived from the emission measure (EM). However, a reasonable average EM over the 40—50° region observed in Hα is difficult to assess. Whereas EM values as high as 3000 cm^{-6} pc have been observed close to γ^2 Vel [4], the faintest parts are close to the detection threshold which is estimated to be at $\sim$300 cm^{-6} pc. From

$$\mathrm{EM} = \int_{280}^{620} n_e^2\, ds \geqq 300\ \mathrm{cm}^{-6}\ \mathrm{pc} \quad ([ds]\ \text{in pc})$$

we obtain $\langle n_e^2 \rangle \geqq 0.9$ cm^{-6}. This, together with the above result, indicates that clumping occurs. For the filling factor f we get

$$f = \frac{\langle n_e \rangle^2}{\langle n_e^2 \rangle} \leqq 0.1.$$

If we use $\langle \mathrm{EM} \rangle = 1300$ cm^{-6} pc [1], we find $\langle n_e^2 \rangle = 3.8$ and $f = 0.024$. This value is slightly larger than the one obtained by Brandt et al. [1] who found $f = 0.015$. The difference results from the modified distance to the ionization front used here.

2. Results from Interstellar Absorption Lines in the Spectrum of γ^2 Vel

The main purpose of the above discussion was to provide the necessary background for the interpretation of interstellar absorption lines in the spectrum of γ^2 Vel. We have analysed interstellar absorption lines seen in the 2060—2160 Å,

2495—2595 Å and 2775—2875 Å bands of the spectrum of γ^2 Vel, observed by the Orbiting Stellar Spectrophotometer S59 on board the ESRO TD-1A satellite. The resolution of this instrument is 1.8 Å. This excludes the study of the absorption profile in order to determine a possible multi-component structure. Our basic results are displayed in Table 1. Details of their derivation are given in [7].

Table 1
Observed Column Densities

Ion	λ Å	width W_λ(mÅ)	Ref.	N_i*(cm^{-2})
H I	1215.7	4100	[8]	3×10^{19}
Na I	5889.9/5895.9	97/<50	[9]	5×10^{11}
Mg I	2852.1	120	[7]	1.1×10^{12}
Mg II	2295.5/2802.7	340/265	[7]	2.5×10^{13}
Ca II	3333.7	36	[9]	4×10^{11}
Fe II	2285.9	105	[7]	5×10^{13}

* these results are based on a velocity dispersion of 13 km s^{-1}.

It should be emphasized that the observed doublet ratio of 1.28 for the Mg II-lines $\lambda\lambda$ 2795.53, 2802.70 Å allows two fits to the curve of growth, one for a velocity dispersion of 3 km s^{-1} in the region between the flat and the damping part of the curve, and one with a velocity dispersion of 13 km s^{-1} in the region between the linear and the flat part. The arguments in favour of the second solution which we have adopted here are summarized in [7]. The high velocity dispersion solution agrees with that chosen by de Boer [10] in a similar analysis of interstellar absorption lines in the spectrum of ζ Pup. As can be seen from Table 2, which we will discuss later, the largest contribution to the observed column densities will come from the ionized hydrogen region. Such regions usually have turbulent velocities of at least 10 km s^{-1}.

In order to interpret the above results we assume that the line of sight to γ^2 Vel passes through two physically different regions characterized by the following parameters: (1) intercloud H I gas with $\langle n_{\rm H} \rangle = 0.04$ cm^{-3}, $\langle n_e \rangle \leqq 10^{-2}$ cm^{-3}, $l = 280$ pc, (ii) H II gas, $\langle n_{\rm H} \rangle = \langle n_e \rangle = 0.3$ cm^{-3}, $l = 170$ pc. On the basis of this model we have computed the expected column densities for the elements Na, Mg, Ca, Fe. We have done this for two sets of abundances, called Set A and Set B, which are normal solar abundances and abundances based on OAO-C results, respectively (see also [11]). The OAO-C results were obtained by analysing the interstellar absorption lines seen in the spectrum of α Leo. This is a star with very little reddening, and the data indicate that metal depletion by factors of 10—100 may be found also in those regions of interstellar space which contain low density gas.

Table 2 also gives the total expected column densities which were obtained by adding the contributions from regions 1 and 2. As can be seen from a comparison of these results with those of Table 1, the expected column densities for Na, Mg, Ca and Fe are all larger than the observed ones. This holds true even for Set B abundances. Since the total expected column densities are essentially determined by the contribution from the H II region, we face the alternative that either the

Table 2
Expected Column Densities (cm^{-2})

Element	Region 1 Set A[a]	Set B[b]	Region 2 Set A	Set B	Total Set A	Set B
H[c]	3×10^{19}	3×10^{19}	0	0	3×10^{19}	3×10^{19}
Na	1.3×10^{14}	6.7×10^{13}	6.1×10^{14}	3.1×10^{14}	7×10^{14}	4×10^{14}
Mg	1.3×10^{15}	8.4×10^{13}	6.1×10^{15}	3.8×10^{14}	7×10^{15}	5×10^{14}
Ca	1.0×10^{14}	3.4×10^{12}	4.6×10^{14}	1.5×10^{13}	6×10^{14}	2×10^{13}
Fe	1.0×10^{15}	2.7×10^{13}	4.6×10^{15}	1.2×10^{14}	6×10^{15}	1×10^{14}

[a] normal abundances H:Na:Mg:Ca:Fe $= 1:4 \times 10^{-6}:4 \times 10^{-5}:3 \times 10^{-6}:3 \times 10^{-5}$ (Unsöld, private communication).
[b] abundances based on OAO-C results H:Na:Mg:Ca:Fe $= 1:2 \times 10^{-6}:2.5 \times 10^{-6}:1 \times 10^{-7}:8 \times 10^{-7}$ [6].
[c] fitted to observations.

depletion of metals is still larger inside the H II region than found by the Copernicus satellite for intercloud gas, or else the ionization must be such that most of the Na, Mg, Ca and Fe atoms are in higher stages of ionization than those observed.

Since we have estimated the electron density inside the H II region, we know the recombination rate. Consequently, it is easy to compute the ionization rate necessary to maintain, for example, $\geqq 99\%$ of the Mg at least doubly ionized. If we assume that the recombinations are balanced by photoionizations, we find that $\Gamma(\text{Mg II})/\Gamma(\text{Mg I}) \geqq 15$, where Γ denotes the ionization rate, is required to satisfy this condition. The ionization of Mg II needs photons with $\lambda < 826$ Å. We therefore conclude that in order to explain the small Mg I and Mg II column densities as being due to ionization effects, the number of photons with $\lambda < 826$ Å must greatly exceed the number of photons with $\lambda < 1632$ Å capable of ionizing Mg I. Such a photon spectrum cannot be produced by any of the stars in the area, nor can the requirement be met by the X-ray source Vela X observed in this region [12].

In conclusion one finds that either (a) the metal depletion in the Gum Nebula is considerably larger than that observed by the Copernicus satellite for intercloud gas, or (b) the Nebula is indeed a fossil H II region which was ionized by the Vela X supernova outburst and has not yet reached an equilibrium state again. An examination of the results of Schwarz [13] suggests that if the Gum Nebula is indeed a fossil H II region resulting from the Vela X supernova, heavy elements such as magnesium and iron may have less than 10% of their atoms in the neutral and singly ionized states. For example, for electron temperatures $T > 30000$ K, magnesium is maintained in the doubly ionized state due to collisional ionization. Observations of the Vela pulsar suggest an age for the supernova remnant of $\sim 10^4$ years whereas the cooling time for the hot central portion of a fossil H II region is $2000/n_H$ years. It follows that in regions with densities $\leqq 0.2$ cm^{-3}, magnesium may be mostly doubly ionized. This cannot be the case in the clumped gas which produces the observed Hα emission. However a model seems just feasible in which the main contribution to pulsar dispersion and hence to the total magnesium column density comes from the low density gas between the clumps. For example, clumps of density 100 cm^{-3} and filling factor 3×10^{-4}

within the ionized region would approximately account for both the observed emission measure and our observed equivalent widths. In the low density interclump medium, which gives rise to the pulsar dispersion ($n_e \sim 0.3$ cm^{-3}), heavy elements would be highly ionized. Such a model seems to us possible but not very plausible. It is in any case open to test when Copernicus reaches the Gum Nebula.

References

[1] J. C. Brandt, T. P. Stecher, D. L. Crawford and S. P. Maran, Astrophys. J. **163**, L 99 (1971).
[2] K. P. Beuermann, Astrophys. Space Sci. **17**, 123 (1972).
[3] B. D. Savage and E. B. Jenkins, Astrophys. J. **172**, 491 (1972).
[4] B. J. Bok, in: Proc. Conf. at Goddard Space Flight Center, X-683-71-375, p. 158 (1971).
[5] H. M. Johnson, in: Proc. Conf. at Goddard Space Flight Center, X-683-71-375, p. 12 (1971).
[6] J. B. Rogerson, D. G. York, J. F. Drake, E. B. Jenkins, D. C. Morton and L. Spitzer, Astrophys. J. **181**, L 110 (1973).
[7] M. Grewing, H. J. Lamers, C. M. Walmsley and C. Wulf-Mathies, Astron. Astrophys. **27**, 115 (1973).
[8] G. R. Carruthers, Astrophys. J. **151**, 269 (1968).
[9] G. Wallerstein and J. Silk, Astrophys. J. **170**, 289 (1971).
[10] K. S. de Boer, Space Research XIV, 549 (1974).
[11] M. Grewing and C. M. Walmsley, Astron. Astrophys. **30**, 281 (1974).
[12] F. D. Seward, D. A. Burginyon, R. J. Grader, R. W. Hill, T. M. Palmieri and J. P. Stoering, Astrophys. J. **169**, 515 (1971).
[13] J. Schwarz, Astrophys. J. **182**, 449 (1973).

Space Research XIV — Akademie-Verlag, Berlin 1974

INTERSTELLAR ABSORPTION LINES OBSERVED WITH THE ORBITING SPECTROPHOTOMETER S59

K. S. de Boer

Kapteyn Astronomical Institute, Groningen, The Netherlands

Interstellar absorption lines of Mg^0, Mg^+, Fe^+ and Mn^+ have been detected. The Mg ionization equilibrium and the interstellar gas in the Gum Nebula are discussed as well as aspects of interstellar abundances.

On board the ESRO satellite TD-1A, the S59 spectrophotometer has recorded spectral scans with about 2 Å resolution in three bands 100 Å wide between 2000 and 3000 Å [1]. Interstellar absorption lines of Mg I, Mg II, Mn II and Fe II have been detected in the spectra of many OB-stars (θ Ara, θ Car, ι Ori, ζ Ori, ζ Pup, τ Sco and γ^2 Vel) but only for a few stars has the averaging of the available scans been performed, thus limiting the accurate interpretation to only ζ Pup and γ^2 Vel. With later type stars the interstellar lines vanish into the strong stellar components. The resolution of the S59 does not allow the study of the absorption profile (multicomponent structure), so in general the analysis of the interstellar spectra has to be restricted to those cases where other information on the structure of the line of sight to the star is available. Here the results for ζ Pup are presented, together with a discussion of the Gum Nebula gas.

In the preliminary account on the detection of these interstellar lines, de Boer et al. [2] based their results on an average of three ζ Pup scans. In Table 1 the

Table 1
Observational Results

Spectrum	λ (Å)	W_λ(mÅ)	N(cm^{-2}) 12 km s^{-1}	N(cm^{-2}) 8 km s^{-1}
Mg I	2852	160	2×10^{12}	2.5×10^{12}
Mg II	2803	405	2.5×10^{14}	3×10^{16}
Mg II	2796	445		
Mn II	2576	40	2×10^{12}	2×10^{12}
Fe II	2586	115	5×10^{13}	5×10^{13}

results from the average of 12 spectral scans are presented. Neutral hydrogen is seen by Ly α absorption in the spectrum [3] giving a column density of $N_{H^0} = 7.5 \times 10^{19}$ cm^{-2}. The amount of ionized hydrogen in this region of the galaxy may be estimated from the column density of free electrons: pulsar dispersion measures in this region of the sky give $\int n_e\,dl \approx 70-120\ \mathrm{cm^{-3}\,pc}$ or $N_{H^+} \approx 3 \times 10^{20}\ \mathrm{cm^{-2}}$ and in particular for the Vela pulsar $\int n_e\,dl = 69\ \mathrm{cm^{-3}\,pc}$ [4] or $N_{H^+} = 2.1 \times 10^{20}\ \mathrm{cm^{-2}}$; though the pulsar is separated by about 10° from ζ Pup, its distance is also about 500 pc [5]. Thus we find that about three quarters of the material on the line of sight to ζ Pup is the ionized gas of the Gum Nebula, the other quarter is the neutral, probably local, gas.

Assuming for the moment that all Mg II is due to the ionized component, we find that the doublet ratio of 1.1 can be interpreted in two ways. One is an absorbing medium with a velocity dispersion of 12 km s^{-1} and $N_{Mg^+} = 2.5 \times 10^{14}\ \mathrm{cm^{-2}}$ and the other a medium with 8 km s^{-1} and $N_{Mg^+} = 3 \times 10^{16}\ \mathrm{cm^{-2}}$. The ion ratio of Mg^+/Mg^0 has been calculated for a few physical conditions, shown in Table 2.

Table 2

Calculated Mg^+/Mg^0 Ratio

Mg^+/Mg^0	Local neutral gas $n_e \approx 0.05\ \mathrm{cm^{-3}}$ $\Gamma \approx 100 \times 10^{-12}\ \mathrm{s^{-1}}$*	Ionized gas $n_e \approx 0.15\ \mathrm{cm^{-3}}$ $\Gamma = 2 \times \Gamma_{local}$
α (100 K) $= 6 \times 10^{-12}\ \mathrm{cm^3\,s^{-1}}$	330	
α (6000 K) $= 0.5 \times 10^{-12}\ \mathrm{cm^3\,s^{-1}}$	4000	2800
α (10^4 K) $= 10 \times 10^{12}\ \mathrm{cm^3\,s^{-1}}$	200	140
α (2×10^4 K) $= 60 \times 10^{12}\ \mathrm{cm^3\,s^{-1}}$		20

Local UV-radiation density from [7]. Photoionization cross sections from [8] and [9]. Recombination coefficients from [6] and [10].

* Contribution of radiation with $\nu > 3.3 \times 10^{15}$ Hz is negligible if its radiation density is less than $10^{-32}\ \mathrm{erg\,cm^{-3}\,Hz^{-1}}$.

Assuming ionization by UV radiation density which is two times larger than in the local gas, the maximum possible Mg^+/Mg^0 ratio in the ionized gas is about 3000, with $T = 6000$ K. In the low velocity case this means that there should be more Mg^0 than we see, or $N_{Mg^0} \geqq 10^{13}\ \mathrm{cm^{-2}}$. The ion ratio is smaller for other temperatures, especially at higher temperatures due to the effect of dielectronic recombination [10]. If the H II and the H I component have the same chemical composition also the local gas then produces additional absorption in the Mg I line, which increases the noted discrepancy. In the high velocity case we can match the observed Mg^+/Mg^0 ratio to the calculated one, assuming a temperature of about 1×10^4 K. Here the Mg I line can be treated as being due to the ionized gas indeed, because with equal chemical composition the local gas can produce now only about 20 mÅ additional absorption.

We also assumed that the Mg II lines are due to the H II gas. If this is so, the local gas can contribute about 150 mÅ, but this has to be absorbed from the remaining continuum in the saturated lines. This contribution is negligible with respect to the stellar continuum. Only a radial velocity difference of about 30 km s^{-1} between both gas components would make this argument invalid.

Grewing et al. [11] analysed the S59 absorption lines towards γ^2 Vel, a WR star a few degrees off ζ Pup at about the same distance as ζ Pup. Because the H II gas in the Gum Nebula may be of about the same density and temperature over large volumes of space, the results of the two stars can be compared. Also with γ^2 Vel two velocity dispersion solutions are found from the Mg II lines, 13 km s^{-1} and 3 km s^{-1}. The high velocity case coincides with the ζ Pup results, the low velocity case clearly is in disagreement. From these arguments we conclude that the high velocity dispersion solution has to be preferred. Mg^{++} may also be present. The calculation of the photoionization rate of Mg^+ requires knowledge of the radiation density below the Lyman limit. Observations of the X-ray flux from Vela X by Seward et al. [12]. give only information at photon energies higher than 100 eV. If we take a large constant energy density of 10^{-32} erg cm^{-3} Hz^{-1} between 15 and 600 eV and if the photoionization cross section [13] is approximated with constant $\sigma = 3 \times 10^{-19}$ cm^2, then $\Gamma \leqq 5 \times 10^{-14}$ s^{-1} giving $Mg^{++}/Mg^+ \leqq 0.1$. The amount of Mg^{++} seems to be negligible and so the total magnesium on the line of sight in H II is N_{Mg} (total) $= 3 \times 10^{14}$ cm^{-2}. The Fe II line strength indicates that we have (in the high dispersion velocity solution) $N_{Fe^+} = 5 \times 10^{13}$ cm^{-2} and from the Mn II line we find $N_{Mn^+} \approx 2 \times 10^{12}$ cm^{-2}. In both cases the calculated ion ratio shows that the + state dominates in the interstellar gas.

Baschek and Scholz [14] studied ζ Pup spectrographically and from their material the NaD and CaH + K interstellar line absorption was obtained (private communication). Both spectra have doublet ratios of 1.9 ± 0.1 and with the respective small equivalent widths, we find for the velocity dispersion (if the lines are due to one gas component) the lower limit of 5 km s^{-1}. Unfortunately with both Na and Ca the ion stage next higher than the observed one dominates in the interstellar gas. Calculation of the ion ratios of Na^+/Na^0 and Ca^{++}/Ca^+ at the ζ Pup line of sight, indicates that both the H II and the H I region may contribute equally to the D, H + K absorption. With the assumption that this total absorption is due to the H II region only, we find $N_{Na^0} \approx 5 \times 10^{11}$ cm^{-2} and $N_{Ca^+} \approx 7 \times 10^{11}$ cm^{-2}.

In the spectra of Smith [3] many narrow absorption lines are seen and four of them may be due to interstellar Si^+. Because the derived equivalent widths are rather uncertain, they can only be used with an existing curve of growth for the absorbing medium. With the 12 km s^{-1} derived from the S59 lines, we find a column density of about $N_{Si^+} \approx 7 \times 10^{13}$ cm^{-2}. The photoionization cross section of Si^0 is so large that only about 1 in 10^4 atoms will be neutral so the Si I line at 2514 Å is undetectably weak for the S59 experiment. Comparing our column densities with those of Grewing et al. [11] (Table 3) we see that there is about a

Table 3

Comparison of Column Densities log N (cm^{-2})

Ion	ζ Pup present analysis	γ^2 Vel [11]
H^0	19.9	19.5
Na^0	11.7	11.7
Mg^0	12.3	12.0
Mg^+	14.4	13.4
Ca^+	11.9	11.6
Mn^+	12.3	<12.1
Fe^+	13.7	13.7
Si^+	13.8	—

factor of two less material in the direction of γ^2 Vel as seen from the weaker line ions. With Mg^+ the difference is a factor of ten, but this value is less certain because both column densities are found from the flat portion of the curve of growth. The overall difference may be due to (i) a more dilute gas to γ^2 Vel (intrinsically, or due to a z-distribution effect because γ^2 Vel is at $b \approx 10^0$) or (ii) the distance of γ^2 Vel being less than that of ζ Pup.

The element abundances, to be compared with N_{H^+}, are lower by factors of 10 to 100 than solar values. In this region of the sky this cannot be explained by locking up metals into dust, because all stars in the Gum Nebula show very little extinction [15]. The derived column densities are comparable with those found by the OAO-C [16], but smaller than those derived by Boksenberg et al. [6] because in their analysis they give the total value for all, possibly denser, components in one line of sight.

References

[1] R. Hoekstra, K. A. van der Hucht, C. de Jager, Th. Kamperman and H. J. Lamers, Space Research XIII, 871 (1973).
[2] K. S. de Boer, R. Hoekstra, K. A. van der Hucht, Th. Kamperman, H. J. Lamers and S. R. Pottasch, Astron. Astrophys. **21**, 447 (1972).
[3] A. M. Smith, Astrophys. J. **160**, 595 (1970).
[4] R. N. Manchester, J. D. Murray and V. Radhakrishnan, Astrophys. Lett. **4**, 229 (1969).
[5] D. K. Milne, Austr. J. Phys. **21**, 201 (1968).
[6] A. Boksenberg, B. Kirkham, W. A. Towlson, T. E. Venis, B. Bates, G. R. Courts and P. P. D. Carson, Nature Phys. Sci. **240**, 127 (1972).
[7] A. N. Witt and M. W. Johnson, Astrophys. J. **181**, 363 (1973).
[8] R. W. Ditchburn and G. V. Marr, Proc. Phys. Soc. A **66**, 655 (1953).
[9] E. J. McGuire, Phys. Rev. **175**, 21 (1968).
[10] L. Goldberg in: Interstellar Ionized Hydrogen, Benjamin, New York 1968 (p. 396).
[11] M. Grewing, H. J. Lamers, C. M. Walmsley and C. Wulf-Mathies, Astron. Astrophys. **27**, 115 (1973).
[12] F. D. Seward, G. A. Burginyon, R. J. Grader, R. W. Hill, T. M. Palmieri and J. P. Stoering, Astrophys. J. **169**, 515 (1971).
[13] J. H. Black, J. C. Weisheit and E. Laviana, Astrophys. J. **177**, 567 (1972).
[14] B. Baschek and M. Scholz, Astron. Astrophys. **15**, 285 (1971).
[15] G. Wallerstein and J. Silk, Astrophys. J. **170**, 289 (1971).
[16] D. C. Morton, J. F. Drake, E. B. Jenkins, S. B. Rogerson, A. Spitzer and D. G. York, Astrophys. J. **181**, L 103 (1973).

Space Research XIV — Akademie-Verlag, Berlin 1974

SPECTRAL OBSERVATIONS OF STARS AND INTERSTELLAR GAS IN THE BALLOON ULTRAVIOLET

A. Boksenberg[a], B. Kirkham[a], W. A. Towlson[a], T. E. Venis[a], B. Bates[b], P. P. D. Carson[b] and G. R. Courts[b]

[a]Department of Physics and Astronomy, University College London, London, UK
[b]Department of Pure and Applied Physics, The Queen's University of Belfast, Belfast, N. Ireland

Spectral observations of several stars at a resolution of 0.1 Å covering the region 2870 to 2740 Å have been made with an objective grating spectrograph mounted on a balloon-borne star-stabilized platform. Of particular interest are the interstellar lines of Mg II and Mg I near 2800 Å and 2850 Å respectively, from which values for the interstellar Mg column density and interstellar electron density are derived. Consideration is given briefly to the strengths of the Mg II lines originating in the background stars.

1. Introduction

The observations reported here are primarily of the interstellar lines of Mg II at 2795.53 Å and 2802.70 Å and Mg I at 2852.13 Å in the directions of stars in Orion and Cassiopeia. There are two important astrophysical reasons for observing these interstellar lines. First, in the general interstellar medium Mg is predominantly in the Mg^+ state, so the interstellar abundance of Mg can be determined directly from the measured equivalent widths of the Mg II lines. This avoids the difficult problem of allowing for atoms in unobserved states by consideration of the ionization balance, which is encountered in the optical region for Na and Ca. Second, observation of both Mg^+ and Mg^0 gives information on the ionization balance, and values for the interstellar electron density can be derived using estimates of the interstellar radiation density.

2. The Payload

The payload consists of an altazimuth mounted primary star pointing platform [1] carrying a high resolution objective grating spectrograph with an associated secondary, fine guidance system. The latter stabilizes the spectrum image to a few arcseconds in the direction of dispersion. First, coarse stabilization of the platform to about 1 arcmin peak to peak is achieved by use of an attached star sensor which is directed to the target stars by ground command using a magnetometer for sensing the azimuth direction, and by reference to the local vertical. A second star sensor, in the fine guidance system, provides an error signal for a secondary platform actuated in one axis and containing two flat

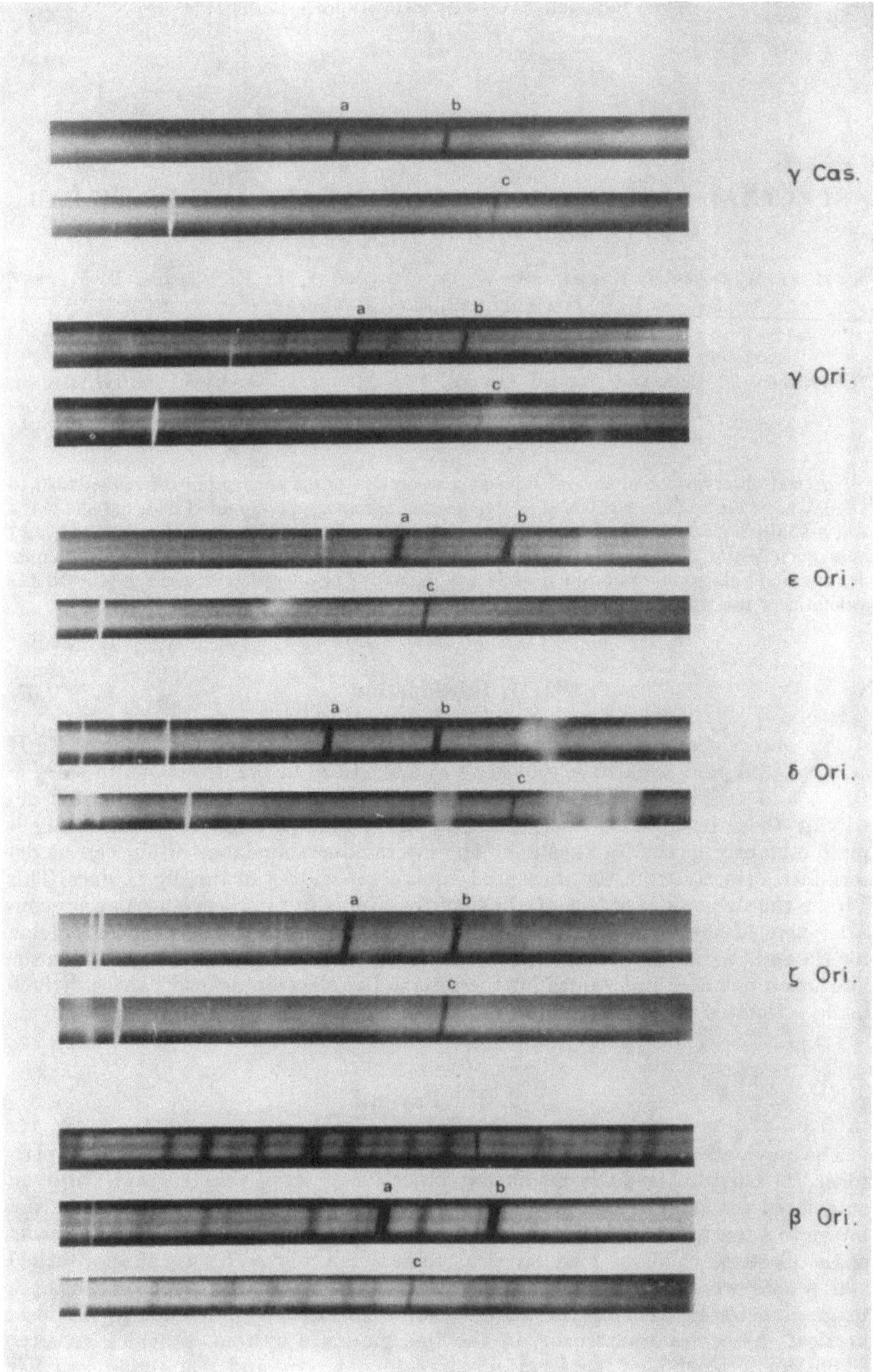
a
b
γ Cas.
c
a
b
γ Ori.
c
a
b
ε Ori.
c
a
b
δ Ori.
c
a
b
ζ Ori.
c
a
b
β Ori.
c

folding mirrors which pass the incident starlight both into the fine star sensor and into the spectrograph mounted side by side. The spectrograph has an objective plane grating of 2160 rulings mm^{-1} and ruled area 102 mm × 128 mm, used in second order. A small-angle quartz prism preceding the servo mirror is used to separate out the unwanted orders. The dispersed light from the grating is reflected from a plane slotted folding mirror to a spherical camera mirror of focal length 70 cm, used on-axis, bringing the spectrum to a focus through the slot in the folding mirror. The spectral recordings were made on 103a-0 roll film, caused to be wound on by ground command. The instrument covers the spectral region 2870—2740 Å with a resolution better than 0.1 Å, governed almost equally by the inherent optical performance and the fine pointing error.

3. Flight Data

The payload was launched on a 20×10^6 ft^3 balloon from the National Center for Atmospheric Research (NCAR) Balloon Flight Station, Palestine, Texas, at 1830 on 4 October 1972. The balloon reached a float altitude of 42 km, decreasing gradually to about 40 km until cut-down at 0650 on 5 October. Over this altitude range the zenith atmospheric transmission is in the region of 50% for the wavelength range of interest. Sample portions of the photographic spectra observed are shown in Fig. 1. The spectra are widened by the motion of the pointing system, which is fine-stabilized only in the direction of dispersion. A microdensitometer tracing of the regions near to and including the Mg II and Mg I lines in β Ori is reproduced in Fig. 2. Apart from the bright fiducial mark to the far left of the Mg I line, the bright trails and smudges present in Fig. 1 are undispersed images of nearby field stars; they are easily recognized and caused no severe problems in the reduction of the interstellar data. The hotter of the six stars studied, viz. γ Cas and γ, δ, ε and ζ Ori, show very little inherent spectral structure in the wavelength range covered and the interstellar Mg II and Mg I lines stand out clearly. The coolest star, β Ori, does show very strong inherent spectral features; nevertheless, the interstellar components are easily recognized by the relative narrowness of the Mg I line, and the relative strengths and shapes of the H and K Mg II lines. The stellar spectrum of β Ori has also evidence of an expanding atmosphere.

To date, γ, δ, ε and ζ Ori, have been subjected to a more complete analysis of the interstellar features than for the preliminary publication of results [2]. The revised data for these stars are presented here. In determining the equivalent widths of the interstellar lines, particularly the Mg II lines, extreme care had to be taken in accounting for the stellar contributions to the observed spectra. For all stars except δ Ori, the stellar components could be recognized easily by their relatively greater breadth and shallower profile than the interstellar components. The stellar line profiles could then be drawn in by eye, based on the profiles of the nearby subordinate lines. As an alternative means of defining the Mg II stellar contributions, an attempt was made to use the non-LTE Mg II

Fig. 1. Segments of sample spectra of the six stars observed, including the Mg II and Mg I interstellar lines. a, Mg II K; b, Mg II H; c, Mg I.

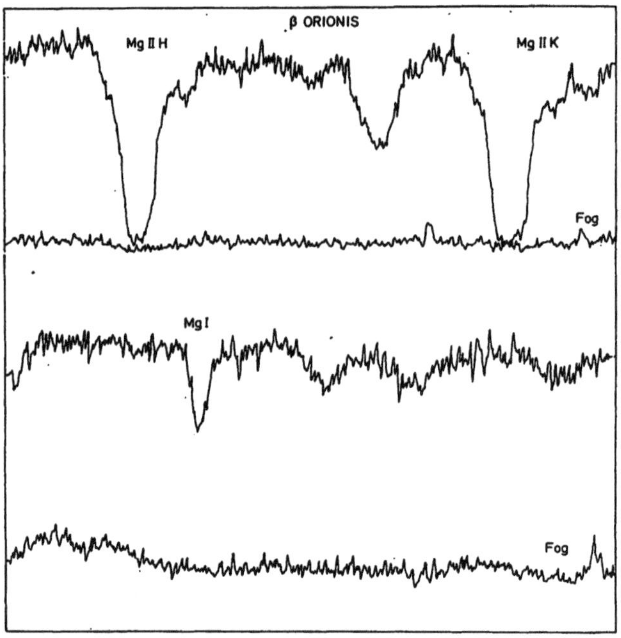

Fig. 2. Microdensitometer tracing of the regions around the Mg II and Mg I lines in β Ori. The stellar subordinate line of Mg II at 2798 Å appears between the Mg II H and K lines.

line results of Mihalas [3, 4] (actually, these are for main-sequence stars, not strictly for stars of the luminosity classes of those observed) to aid in the d finition of the stellar resonance lines by appropriate scaling of the observed nearby subordinate lines. This was done in the case of ζ Ori, for which the strengths of the observed stellar lines seem to accord well with those computed (see below); an eye-estimated determination gave similar results. The method could not be used for γ or ε Ori, because for these there was significant disagreement between the observed and computed stellar values.

The individual equivalent widths determined for the Mg II and Mg I interstellar lines are listed in columns (5) and (6) of Table 1. The indicated possible errors reflect systematic uncertainties present in interpretation rather than errors of measurement.

4. Column Densities

Individual curves of growth, including radiation damping, were computed for the regions of interstellar medium in line with each star. The multicloud velocity profiles found by Hobbs [5] and Marschall and Hobbs [6] for Na^0 and Ca^+ were used as a basis for defining the Mg^0 and Mg^+ profiles, assuming a gaussian distribution of velocities in each of the several cloud components evident. The oscillator strengths were taken from [7]. Strictly, the use of the radial velocity parameters obtained from the Na^0 and Ca^+ data is justified only if the velocity distribution of these and Mg^0 and Mg^+ are similar. According to Hobbs [8], generally good agreement exists in the kinematics of the various interstellar species, a notable

Table 1
Interstellar Magnesium Abundance and Electron Density

(1) Star	(2) S. T.	(3) V	(4) E(B−V)	(5)[1] Equivalent Width MgII, K (mÅ)	(6)[1] Equivalent Width MgII, H (mÅ)	Equivalent Width MgI (mÅ)	(7)[2] Column Density $N(Mg^+)$ (10^{16} cm^{-2})	(8) Column Density $N(Mg^0)$ (10^{12} cm^{-2})	(9)[3] Column Density N(H) (10^{20} cm^{-2})	(10)[4] N(Mg)/N(H) 10^{-5}	(11)[4] $N(Mg^+)/N(Mg^0)$ 10^3	(12)[5,4] Electron Density n(e) 10^{-3} cm^{-3}
γ Ori	B2III	1.64	0.01	202 (±20)	178 (±20)	60 (±10)	0.17 $\binom{+0.15}{-0.11}$	0.63 $\binom{+0.17}{-0.13}$	0.65 $\binom{+0.25}{-0.20}$	2.7 $\binom{+4.4}{-2.0}$	2.7 $\binom{+3.3}{-1.9}$	2.2 $\binom{+5.2}{-1.2}$
ε Ori	B0Ia	1.70	0.05	550 (±30)	480 (±30)	275 (±15)	1.8 $\binom{+0.7}{-0.6}$	7.2 $\binom{+1.8}{-1.2}$	1.75 $\binom{+0.70}{-0.50}$	10 $\binom{+10}{-5}$	2.5 $\binom{+1.7}{-1.1}$	2.4 $\binom{+1.7}{-1.0}$
δ Ori	09.5II	2.20	0.09	546 (±20)	479 (±20)	192 (±20)	2.1 $\binom{+0.7}{-0.5}$	3.8 $\binom{+1.6}{-1.0}$	1.3 $\binom{+0.5}{-0.4}$	16 $\binom{+15}{-7}$	5.5 $\binom{+4.5}{-1.9}$	1.1 $\binom{+0.6}{-0.5}$
ζ Ori	09.5Ib	1.74	0.06	570 (±25)	477 (±25)	250 (±20)	2.0 $\binom{+0.5}{-0.5}$	8.0 $\binom{+2.5}{-2.0}$	1.75 $\binom{+0.70}{-0.50}$	11 $\binom{+9}{-5}$	2.5 $\binom{+1.7}{-1.1}$	2.4 $\binom{+1.8}{-1.0}$

[1] The possible errors are largely systematic.
[2] The errors are obtained by separately applying the respective H and K upper limits and lower limits quoted in column (5), to appropriate curves of growth which give good agreement with the individual doublet ratios.
[3] These are the OAO values of Savage and Jenkins [9].
[4] The errors are the most pessimistic extremes obtained by compounding the errors quoted in columns (7), (8) and (9).
[5] These values are appropriate for the cloud components. See text for the assumptions made in the derivation.

exception being in the gas towards ζ Oph which is quite atypical in this respect. However, for the stars observed here, the Na^0 and Ca^+ profiles do not exactly correspond and computational tests were made to identify the sensitivity of the results to the parameters of the velocity models.

As an example, in Fig. 3 two cases are shown for ε Ori, corresponding approximately to the observed Na^0 and Ca^+ data. The corresponding curves of

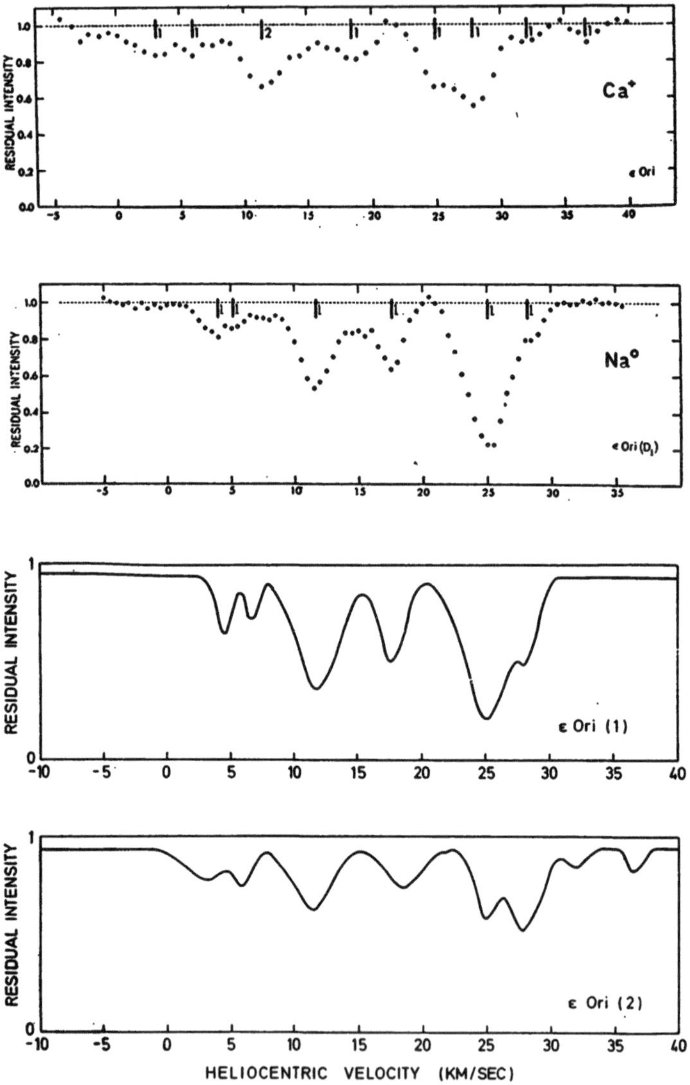

Fig. 3. Interstellar cloud absorption profiles in the direction of ε Ori. The observed Na^0 D_1 and Ca^+ K profiles of Hobbs [5] and Marschall and Hobbs [6], are reproduced at the top. Corresponding model profiles (actually inverted velocity profiles, shown in this way for purposes of comparison, and representing optically thin absorption) are shown respectively as (2) and (1). The parameters of these models are listed in Table 2.

growth are reproduced in Fig. 4. In spite of the large differences between the two sets of cloud profiles, the dispersion between the corresponding results for Mg II is only about a factor two. The results of Mg I, however, are dispersed by an order of magnitude: for case 1 the equivalent width of the Mg I line lies on the saturated part of the curve, but for case 2 it is on the linear part. From the fact that the Mg I line is observed actually not to be saturated the true situation must be closer to case 2 than case 1. Also, the doublet ratio for the observed Mg II H and K lines conforms with case 2 and not case 1. For these reasons we adopt

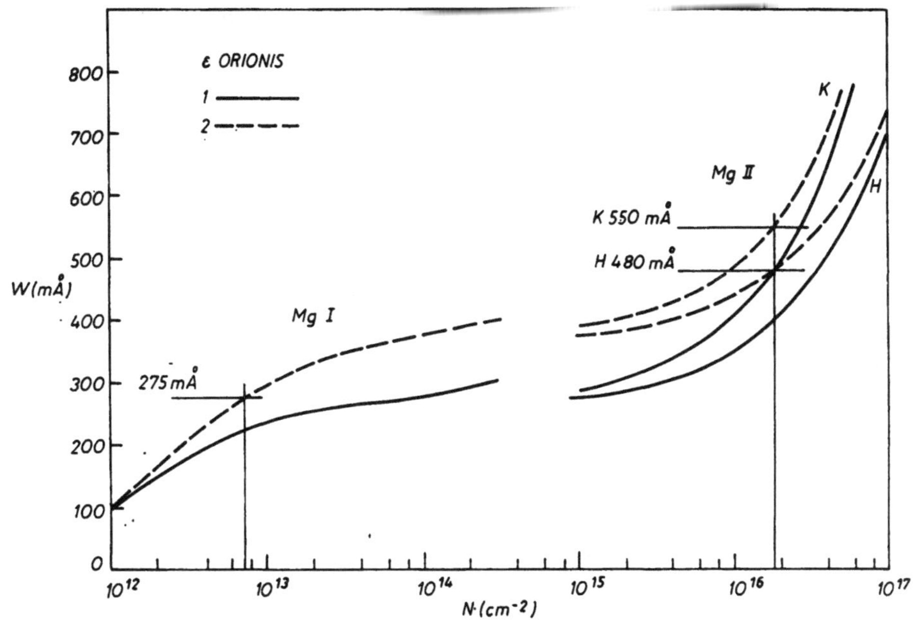

Fig. 4. Curves of growth for the interstellar gas in the direction of ε Ori corresponding to the model cases (1) and (2) illustrated in Fig. 3.

the curve of growth 2 (based on the Ca^+ 8-cloud velocity profile) as being closely representative of the interstellar gas in the direction of ε Ori. Results for all four stars are given in Table 1, and details of the cloud parameters in Table 2. In all cases but γ Ori it was found that the Mg II lines lay close to the damping portions of the curves of growth, and without exception the Mg I lines lay close to the Doppler portions. In each case the actual curves of growth used to derive the listed column densities were chosen on the basis of conformity with the Mg II H and K doublet ratio. The possible errors are difficult to assign; the indicated uncertainties were judged for each result on the basis of the systematic uncertainties in measured equivalent width. Also listed in Table 1 are the values for the H column densities obtained by Savage and Jenkins [9]. The resulting values of derived interstellar Mg abundance (assumed to be predominantly Mg^+) relative to H are listed in column (10). These values do not depend significantly on the ionization state of the H I medium and fundamentally are average values along the lines of sight. For comparison, the cosmic abundance of Mg relative to H is

Table 2
Parameters for the Interstellar Cloud Models

Star	Cloud	b^a (km s^{-1})	Heliocentric radial velocity (km s^{-1})	Relative weight	Remarks
γ Ori	1	2.1	+17.2	0.32	The value listed in Table 1 are derived from this model.
	2	2.2	+22.0	0.37	
	3	1.4	+25.6	0.31	
ε Ori (1)	1	0.85	+4.5	0.06	Case 1 in Figs. 3 and 4.
	2	0.72	+6.7	0.04	
	3	2.30	+11.8	0.30	
	4	1.40	+17.6	0.14	
	5	2.50	+25.1	0.41	
	6	0.93	+28.3	0.06	
ε Ori (2)	1	2.22	+3.20	0.12	Case 2 in Figs. 3 and 4. The values listed in Table 1 are derived from this model.
	2	0.97	+6.14	0.06	
	3	2.05	+11.5	0.22	
	4	2.02	+18.6	0.13	
	5	1.18	+25.0	0.14	
	6	1.73	+28.0	0.25	
	7	1.05	+32.1	0.04	
	8	0.74	+36.7	0.03	
δ Ori (1)	1	1.5	+4.34	0.19	
	2	2.0	+10.0	0.16	
	3	1.5	+16.0	0.09	
	4	1.0	+21.1	0.03	
	5	1.0	+23.0	0.10	
	6	2.0	+25.6	0.43	
δ Ori (2)	1	2.5	+4.34	0.13	
	2	2.5	+10.0	0.19	
	3	2.0	+16.0	0.11	
	4	1.0	+21.2	0.03	
	5	1.2	+23.0	0.10	
	6	2.0	+25.6	0.44	
δ Ori (3)	1	2.5	+4.34	0.13	
	2	2.5	+10.0	0.19	
	3	2.0	+16.0	0.11	
	4	1.0	+21.2	0.03	
	5	1.2	+23.0	0.10	
	6	2.0	+25.6	0.44	
	7	1.5	+35.0	0.00005	
δ Ori (4)	1	2.5	+4.34	0.13	The values listed in Table 1 are derived from this model.
	2	2.5	+10.0	0.19	
	3	2.0	+16.0	0.11	
	4	1.0	+21.2	0.03	
	5	1.2	+23.0	0.10	
	6	2.0	+25.6	0.44	
	7	1.5	+35.0	0.00025	

Table 2 (contd.)

Star	Cloud	b^a (km s^{-1})	Heliocentric radial velocity (km s^{-1})	Relative weight	Remarks
ζ Ori	1	0.80	−5.93	0.01	The values listed in Table 1 are derived from this model.
	2	1.22	−3.40	0.07	
	3	1.85	−1.00	0.15	
	4	4.50	+10.8	0.13	
	5	1.10	+18.9	0.06	
	6	1.56	+23.4	0.31	
	7	1.39	+26.3	0.25	

[a] For each cloud component the parameter b is defined by:

$$\psi(u) = \frac{1}{b\pi^{1/2}} \exp\{-(u - u_0)/b^2\}$$

where $\psi(u)$ is the radial velocity distribution function, u_0 is the central velocity of the cloud.

about 3×10^{-5} by number [10]. For the Belt stars, the derived values of relative Mg abundance are a few times higher than the cosmic value, but for γ Ori the value is about normal.

5. Electron Density

Making the assumptions that ionization is due to the general radiation field and that recombination occurs by collisions between ions and free electrons, the relative numbers of singly ionized and neutral Mg atoms are given by

$$\frac{n(\mathrm{Mg}^+)}{n(\mathrm{Mg}^0)} = \frac{\Gamma(\mathrm{Mg}^0)}{\alpha(\mathrm{Mg}^0)\, n(\mathrm{e})} \quad (1)$$

where $n(\mathrm{Mg}^+)$ and $n(\mathrm{Mg}^0)$ are the respective number densities, $n(\mathrm{e})$ is the number density of electrons, $\Gamma(\mathrm{Mg}^0)$ is the photoionization rate per atom in units s^{-1}, and $\alpha(\mathrm{Mg}^0)$ is the radiative recombination coefficient in units cm^3 s^{-1}. The value of $\Gamma(\mathrm{Mg}^0)$ is given by

$$\Gamma(\mathrm{Mg}^0) = 10^{-8}\, h^{-1} \int_{\lambda_0}^{0} a(\lambda)\, u(\lambda)\, \lambda\, d\lambda \quad (2)$$

where h is Planck's constant, λ is a wavelength expressed in Å, $a(\lambda)$ is the photoionization cross section for Mg^0, $u(\lambda)$ is the interstellar radiation density and λ_0 is the photoionization threshold, at 1618.4 Å. We assume for HI regions $u(\lambda) = 0$ at $\lambda < 912$ Å. To obtain $\Gamma(\mathrm{Mg}^0)$, we use the data of Witt and Johnson [11] for $u(\lambda)$ appropriate for the solar vicinity, and of Dubau and Wells [12] for $a(\lambda)$. The radiation density calculations by Witt and Johnson are based on recent observations and give values about a factor two higher than earlier calculations by Habing [13] over the wavelength range of interest here. The result terminates before 912 Å, at 1250 Å, and here was extrapolated linearly to zero at 912 Å. The photoionization cross section was computed by Dubau and Wells using atomic wave functions obtained in Seaton and Wilson's frozen cores approximation [14]; a similar method was previously employed by Burke and McVicar [15]. The resulting value of $\Gamma(\mathrm{Mg}^0)$ is $5.0 \times 10^{-11}\ \mathrm{s}^{-1}$. In this we assume that most of the absorption by Mg^+ and Mg^0 occurs sufficiently far from the background stars for their individual contributions to the radiation field here to be insignificant. Values of $\alpha(\mathrm{Mg}^0)$, depending on the assumed temperatures, were obtained from [16, 17].

For the ratio $n(\mathrm{Mg}^+)/n(\mathrm{Mg}^0)$ we take $N(\mathrm{Mg}^+)/N(\mathrm{Mg}^0)$, the ratio of column densities. The implicit assumption is that the space distributions of neutral and ionized Mg largely correspond. Further, as is generally accepted, we assume that for H I regions the interstellar gas is predominantly concentrated into cool clouds. For a cloud temperature of 60 °K the value of $\alpha(\mathrm{Mg}^0)$ is $8.5 \times 10^{-12}\ \mathrm{cm}^3\ \mathrm{s}^{-1}$. Values of $N(\mathrm{Mg}^+)/N(\mathrm{Mg}^0)$ for the six stars are listed in column (11) of Table 1, and corresponding values of $n(\mathrm{e})$ assuming a cloud temperature of 60 °K, in column (12). Note that the values of $N(\mathrm{Mg}^+)/N(\mathrm{Mg}^0)$ and the corresponding values of $n(\mathrm{e})$ are similar for all the stars, while the values of $N(\mathrm{Mg})/N(\mathrm{H})$ vary rather widely. The average of the mean values of $n(\mathrm{e})$ derived for the cool clouds in the directions of the four stars is $2 \times 10^{-3}\ \mathrm{cm}^{-3}$.

6. The Stellar Spectra

The equivalent widths of the Mg II line at 2797.99 Å, originating only in the spectra of the six background stars not in the interstellar medium, are plotted against effective temperature in Fig. 5. A curve giving results of Mihalas [3, 4],

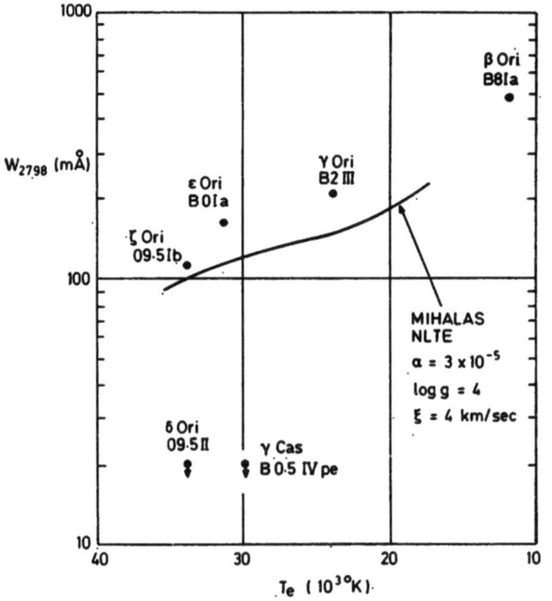

Fig. 5. The equivalent widths of the stellar Mg II subordinate line at 2798 Å in the six stars observed (points), and a curve of computed non-LTE results by Mihalas [3,4] scaled up for $\alpha = N(\mathrm{Mg})/N(\mathrm{H}) = 3 \times 10^{-5}$.

scaled up by a factor 1.3 as described by Lamers et al. [18] to conform with $N(\mathrm{Mg})/N(\mathrm{H}) = 3 \times 10^{-5}$, is shown as the continuous line. The temperature scale is the one adopted by Mihalas [3]. Similar comparisons were made by Lamers et al. but they included the four resonance and subordinate lines near 2800 Å as a sum and made corrections for the interstellar components of the resonance lines. For luminosity classes II—V they found that the lines were stronger than predicted in stars earlier than B3, reaching a factor 1.8 at B1—B2. Two supergiant stars showed still stronger lines but strictly cannot be compared with Mihalas' predictions because of their much lower gravity. The present results, which need no interstellar correction, do not completely follow the general systematic trend found by Lamers et al.

Acknowledgments

It is a pleasure to acknowledge the support and encouragement of Professor Sir Harrie Massey and Professor D. L. Bradley, and the work of our colleagues in the two establishments in helping to make this project possible. We are grateful to Dr. Yoji Kondo of NASA Manned Spacecraft Center, and Professor R. Wilson, late of the SRC Astrophysics Research Unit, Culham, and now of University College London, for helpful discussions and the use of the facilities of their laboratories, to The National Center for Atmospheric Research at Palestine, Texas, for launch and recovery operations and general hospitality, and to the Science Research Council for financial and other support.

References

[1] I. Furniss, R. E. Jennings and A. F. M. Moorwood, Astrophys. J. Lett. **176**, L 105 (1972).
[2] A. Boksenberg, B. Kirkham, W. A. Towlson, T. E. Venis, B. Bates, G. R. Courts and P. P. D. Carson, Nature Phys. Sci. **240**, 127 (1972).
[3] D. Mihalas, Astrophys. J. **177**, 115 (1972).
[4] D. Mihalas, private communication.
[5] L. M. Hobbs, Astrophys. J. **157**, 135 (1969).
[6] L. A. Marschall and L. M. Hobbs, Astrophys. J. **173**, 43 (1972).
[7] W. L. Wiese, M. W. Smith and B. M. Miles, NSRDS-NBS22, **2**, 1 (1969).
[8] L. M. Hobbs, Astrophys. J. **180**, L79 (1973).
[9] B. D. Savage and E. B. Jenkins, Astrophys. J. **172**, 491 (1972).
[10] C. W. Allen, Astrophysical Quantities, Athlone Press, University of London 1963.
[11] A. N. Witt and M. W. Johnson, Astrophys. J. **181**, 363 (1973).
[12] J. Dubau and J. B. Wells, private communication.
[13] H. J. Habing, Bull. Astr. Inst. Netherlands **19**, 421 (1968).
[14] M. J. Seaton and P. M. H. Wilson, J. Phys. B. **5**, 11 (1972).
[15] P. G. Burke and D. D. McVicar, Proc. Phys. Soc. **86**, 989 (1965).
[16] M. J. Seaton, Mon. Not. R. Astron. Soc. **111**, 368 (1951).
[17] C. B. Tartar, Astrophys. J. **168**, 313 (1971).
[18] H. J. Lamers, K. A. van der Hucht, M. A. J. Snijders and N. Sakhibulin, Astron. Astrophys., in press (1973).

Space Research XIV — Akademie-Verlag, Berlin 1974

OBSERVATIONAL RESULTS FROM THE ULTRAVIOLET SKY-SURVEY TELESCOPE IN THE TD-1 SATELLITE: ULTRAVIOLET PHOTOMETRY AND SPECTRAL TYPE CLASSIFICATION OF EARLY TYPE MAIN SEQUENCE STARS

C. M. Humphries, K. Nandy and G. I. Thompson

Royal Observatory, Edinburgh, UK

Analysis of data obtained from the TD-1 satellite Sky-Survey experiment (S2/68) has yielded an ultraviolet photometric system which permits determination of (i) the amount of interstellar reddening and (ii) the spectral type classification. The proposed system is based on colour indices derived from the measured ultraviolet magnitudes at 2750 Å, 2500 Å, and 2190 Å, together with the magnitude V in the visible region. The colour index derived from the 2500 Å and 2190 Å magnitudes is shown to be a sensitive reddening parameter and approximately independent of spectral type. A colour difference method is used to derive a reddening free parameter which enables determination of spectral type to within ± 2 sub-classes for main sequence types in the range from 0 to A5.

The work is published in full in *Monthly Notices of the Royal Astronomical Society* **163**, 1 (1973).

Space Research XIV — Akademie-Verlag, Berlin 1974

SOLAR RADIATION ASYMMETRIES AND HELIOSPHERIC GAS HEATING INFLUENCING EXTRATERRESTRIAL UV DATA

H. J. Fahr and G. Lay

Institut für Astrophysik und extraterrestrische Forschung der Universität Bonn, Bonn, FRG

Up to now the interpretation of extraterrestrial Lyman α data of the satellite OGO 5 has been based on the assumption of spherically symmetric solar radiation fields. This assumption, however, turns out to be very crude and it is shown that appreciable improvement in the interpretation of OGO 5 data can be achieved if actual solar radiation asymmetries both in the corpuscular and in the electromagnetic fluxes are taken into account. Methods of correcting for asymmetries in heliographic latitude and longitude are developed and facilitate the deduction of interstellar parameters. The galactic Lyman α background is found to be below $10R$; the density n_0 and temperature T_0 of the nearby interstellar gas have been obtained as 0.1 cm^{-3} and $(4-6) \times 10^3$ K. The temperatures of $(6-8) \times 10^3$ K indicated by the OGO 5 data are shown to be different from T_0 due to the fact that the interstellar hydrogen penetrating into the solar system is subjected to a heliospheric heating process caused by elastic collisions with solar protons.

1. Introduction

The extraterrestrial Lyman α isophotes obtained by OGO 5 [1, 2] have been interpreted by Blum and Fahr [3], Bertaux et al. [4] and Thomas [5] as being due to backscatter of solar Lyman α radiation from interstellar hydrogen in interplanetary space. Whereas the interpretations of Bertaux et al. [4] and Thomas [5] are based on the assumption that the solar radiation fields are spherically symmetric, Blum and Fahr [3] have tried to give an estimate of how much the extraterrestrial Lyman α sky is influenced by asymmetries in the solar Lyman α emission pattern. They showed that such asymmetries due to enhanced emissions from solar activity centres might give rise to a shift in the celestial position of the extraterrestrial Lyman α radiation maximum by about 40°. This indicates clearly that the deduction of the celestial direction from which interstellar hydrogen approaches the solar system, without any knowledge of the actual solar radiation asymmetry, may be uncertain by about this value, at least if only one Lyman α sky map for one specific date is interpreted. Since Bertaux et al. [4] and Thomas [5] have deduced an average direction of approach obtained from three of their Lyman α sky maps, each of them spaced by three months with respect to the other, this uncertainty due to rotating solar radiation asymmetries may have been cancelled out for its most part.

Therefore the directional coordinates $\alpha = 263°$, $\delta = -22°$ given by these authors seem to be reliable. In order to fit by theory specific OGO 5 Lyman α

data given in any one of these sky maps for selected celestial directions, it seems, however, to be essential to correct for the actual radiation asymmetries given at that time when OGO 5 registered this sky map.

2. Heliographic latitude effect

We shall first investigate the effect of solar radiation asymmetries in heliographic latitude. From the monitoring of the solar Lyman α radiation $I_{s,Ly}$ by the satellite OSO 5 [6] it has become evident that enhanced radiation intensities are emitted from solar activity centres into a relatively narrow cone. Since activity centres are located at moderate heliographic latitudes at a position often stable on the solar disk for several solar rotations this leads to a latitudinal variation of the solar emission not only in the Lyman α range, but also in the range of ionizing EUV wavelengths and possibly also in the range of solar wind proton fluxes. Both the EUV fluxes and the solar wind flux contribute to the ionization of interstellar hydrogen advancing towards the sun from interstellar space.

While latitudinal asymmetries in the $I_{s,Ly}$ are immediately reflected by the extraterrestrial Lyman α background radiation, such asymmetries in the ionizing solar fluxes $I_{s,i}$ influence the resulting interplanetary hydrogen densities only by their averages over several solar rotations. This is due to the fact that a hydrogen atom sees several solar rotations while it advances from infinity towards a specific space point in interplanetary space where it is to scatter solar Lyman α photons.

Therefore only an average latitude effect smoothed over some solar rotations is reflected in the density. The extinction function $E_3(a, \Theta)$ (a being the solar distance, Θ the angle with the velocity $\boldsymbol{v}_0$ of approach) that is given by Fahr [7] and describes the atomic losses due to EUV and charge exchange ionization along the Keplerian orbits of the entering interstellar particles should contain corrections for the average latitudinal variation of $I_{s,i}$. Such averaged variations are described as the mean emission characteristic of a solar activity centre at moderate latitude smoothed over all heliographic longitudes. For the emission pattern of an activity centre we adopt the characteristic emission function given by Priester and Cattani [8] for the activity-correlated EUV fluxes. Therefore in addition to an isotropic solar radiation part I_{s0} characteristic for the quiet solar disk emission we add an activity-dominated part that is modulated with respect to heliographic latitudes B. The total emission $I_{s,i}$ of solar EUV and proton fluxes may therefore be represented by:

$$I_{s,i} = I_{s0}\left(1 + P\left[\exp\left(-\left(\frac{B - B_0}{D}\right)^2\right) + \exp\left(-\left(\frac{B + B_0}{D}\right)^2\right)\right]\right) \tag{1}$$

where $\pm B_0$ is the mean latitudinal position of northern and southern activity centres, D is the angular width of the emission characteristic of a centre, and P is the ratio of the activity-modulated radiation to the radiation of the quiet solar disk. $I_{s,0}$ is chosen so that for $B = B_E$ (B_E = latitude of the earth) $I_{s,i}$ is the mean flux measured at the earth. The average emission $I_{s,i}$ for a smoothed-out activity centre at $B_0 = 14$ and a width of $D = 15°$ is shown in Fig. 1.

The latitudinal correction of the solar Lyman α radiation $I_{s,Ly}$ has been carried out as follows: The intensity variation of $I_{s,Ly}$ as registered by OSO 5 [6] is taken

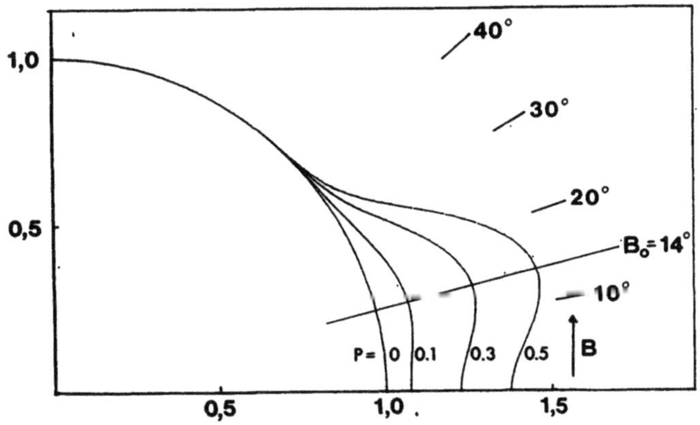

Fig. 1. The normalized emission $I_{s,i}$ as a function of heliographic latitude B according to Eq. (1) for various ratios P of the activity-modulated radiation to the isotropic radiation. The Gaussian has been taken to be centred around $B_0 = 14°$ with a width $D = 15°$.

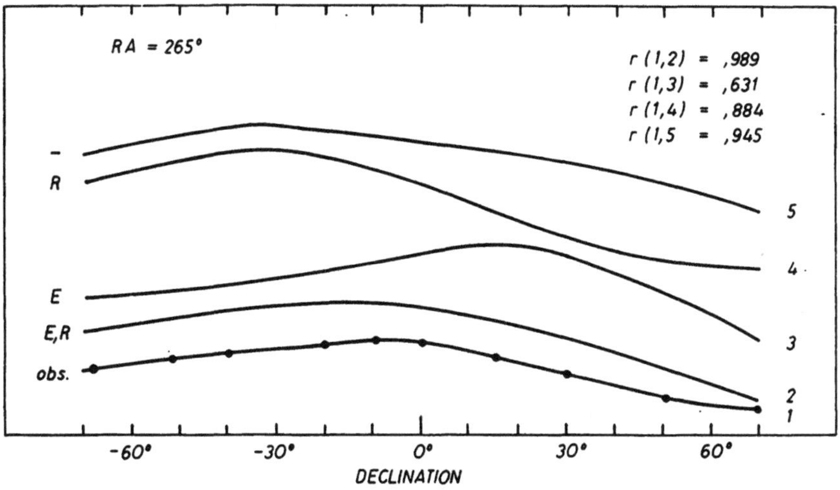

Fig. 2. Theoretical and observed Lyman α resonance intensities in arbitrary units from directions with a constant right ascension RA = 265° as a function of declination. The observed intensities (curve 1) have been extracted from the OGO 5 sky map for September 1969 [1]. The theoretical intensities have been calculated with the use of a solar Lyman α radiation field that has been corrected for longitudinal solar anisotropies as deduced from the solar Lyman α monitoring by OSO 5 [6]. In the intensity curves the position of the vector v_0 has been assumed to point from the direction $\alpha = 265°$; $\delta = -23°$. The calculated intensities have been corrected for asymmetries of the solar radiation fields in heliographic latitude in the curves R, E and E,R. In curve R the latitudinal asymmetry has only been taken into account in the exciting solar Lyman α field; in curve E only the latitudinal asymmetries of the solar fluxes affecting the density of heliospheric neutral hydrogen have been taken into account. In curve E,R effects of solar latitudinal asymmetries upon both the exciting Lyman α field and on the density have been considered. For curve 5, marked —, no latitudinal corrections have been applied. It is found that the curve E,R gives by far the best fit to the observed intensities.

to be due to the rotation of an activity-correlated emission pattern synchronous with the rotation of the solar disk. Therefore the monitoring of the $I_{s,Ly}$ by OSO 5 can be used as an indication of the spatial solar emission pattern in heliographic longitude L, at least for $L_E - 90° \leqq L \leqq L_E + 90°$, i.e. a quarter of a solar rotation (L_E = longitude of the earth). Following this assumption, $I_{s,Ly}$ at a specific value of L can be taken as an indication of the locally inherent activity-correlated part that should be latitudinally modulated corresponding to Eq. (1), i.e. we can obtain an indication of $P(L)$. Hence at each longitude L a value $P(L)$ can be determined that describes the latitudinal variation of $I_{s,Ly}$ when introduced into Eq. (1).

Fig. 2 now shows the effect of latitudinal corrections in the ionizing solar flows $I_{s,i}$ and the solar Lyman α flux $I_{s,Ly}$. For a constant right ascension α intensities of scattered Lyman α radiation from selected directions are plotted versus declination δ. It is evident by comparison with the observational intensities (see curve 1) that an appreciable improvement of the theoretical fit is achieved when latitudinal corrections are applied to both $I_{s,i}$ and $I_{s,Ly}$, whereas corrections in only one of these fluxes is not advisable. This is also pointed out by the correlation coefficients r between the different curves shown.

3. The Heliographic Longitude Effect

Asymmetries of the solar radiation field in heliographic longitude L are only effective as far as they concern the solar Lyman α flux $I_{s,Ly}$. This is due to the fact that only the longitudinal asymmetries of $I_{s,Ly}$ are reflected in the resonantly scattered Lyman α intensity. Contrary to this, the response of the interplanetary hydrogen density to longitudinal variations in the ionizing solar fluxes $I_{s,i}$ only occurs with respect to an average over several solar rotations. Therefore it is unnecessary to correct the interplanetary hydrogen densities for longitudinal variations of $I_{s,i}$. An attempt to include the real longitudinal variations of $I_{s,Ly}$ into the calculation of scattered intensities is shown in Fig. 3. Here observed intensities I_{ob} have been extracted from the OGO 5 sky map of Bertaux and Blamont [1] for specific celestial directions and have been compared with theoretically determined intensities of the identical directions. In Fig. 3 we have assumed that I_{ob} is composed of an isotropic part I_{gal} from the galaxy and a heliospheric part $n_o I_h$ that is proportional to the density n_o of nearby interstellar hydrogen. This leads to the equation

$$I_{ob}(\alpha, \delta) = I_{gal} + n_o I_h(\alpha, \delta). \tag{2}$$

Thus, if this assumption is correct, a plot I_{ob} versus I_h should yield I_{gal} as the intersection of a straight line defined by (2) with the ordinate and n_o as the slope of this line. In the upper half of Fig. 3 it is, however, evident that the selected intensity points (I_{ob}, I_h) are only fitted by a straight line very badly with a correlation coefficient of 0.7. As is shown in the lower part of the figure this correlation coefficient can be improved to a value of 0.97 if the intensities I_h that in the upper half have been calculated for an isotropic solar Lyman α emission are corrected for the real longitudinal variations in $I_{s,Ly}$ that have been observed by OSO 5. Here again the longitudinal anisotropy of $I_{s,Ly}$ is deduced from the monitoring of $I_{s,Ly}$ from the position of the earth with the help of the assumption that observed temporal changes can be interpreted as being due to the rotation of a spatial emission pattern. The success of this procedure is evident in the lower half of Fig. 3 where the selected points now are nearly perfectly aligned on a straight line.

For the calculation of I_h some *a priori* assumptions are required. For the velocity of approach of interstellar matter we have adopted a value of $v_0 = 20$ km s^{-1}. The deduction of low velocities of the order of $v_0 = 5$ km s^{-1} [4, 5] does not seem convincing to us, since these authors, in obtaining such values v_0, try to fit their intensities with the relatively large penetration depths of $r_c \approx 5$ AU indicated in the Lyman α sky maps. This could, however, be reached by taking higher ionization rates. We prefer to assume a plausible ionization rate of $\nu_i = 5.6 \times 10^{-7}$ s^{-1}

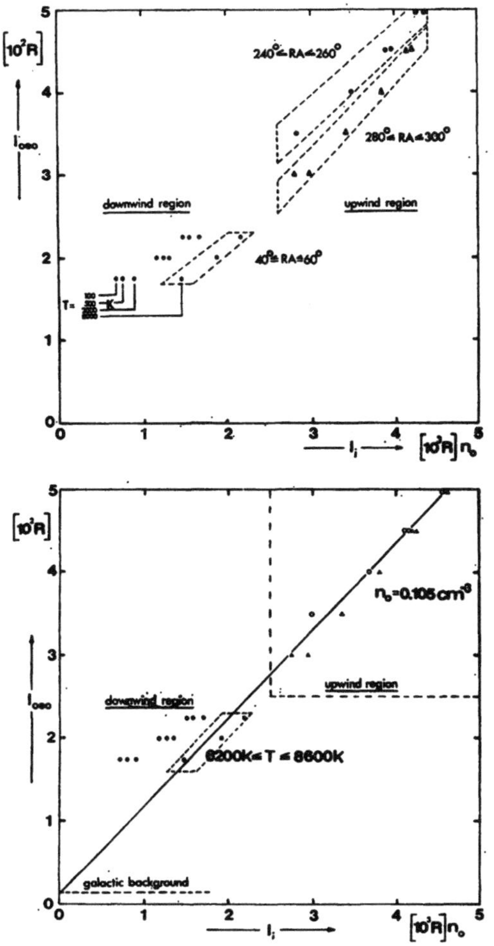

Fig. 3. Intensities I_{OGO} from selected celestial directions from the September 1969 sky map [1] plotted against theoretically calculated intensities I_i from identical directions. The upper figure shows this plot for intensities I_i based on the assumption of an isotropic solar Lyman α emission; in the lower figure I_i has been corrected for longitudinal solar anisotropies deduced from the solar Lyman α monitoring by OSO 5 [6].

and can then fit the data with a velocity $v_0 = 20\ \mathrm{km\ s^{-1}}$, which permits the neglect of contributions to I_{ob} from hot hydrogen of transheliospheric origin [9]. Since upwind interplanetary hydrogen densities that have been used for the calculation of the intensities I_h (dashed square of lower half of Fig. 3) can be determined without *a priori* assumption on the temperature T_0 of the nearby interstellar gas, we can determine the straight line of Eq. (2) with knowledge of the vector $\boldsymbol{v}_0$ only (the direction of $\boldsymbol{v}_0$ is assumed to be $\alpha = 263°$, $\delta = -23°$ in accordance with the above-mentioned authors). From Fig. 3 we can then determine n_0 to be $0.1\ \mathrm{cm^{-3}}$ and I_{gal} to be below $10R$. Using this information on interstellar parameters we can now calculate downwind intensities which are temperature-sensitive. For various temperatures T_0 we obtain values of I_{ob} and I_h. The best-fitting temperature T_0 is then indicated by that point which is closest to the straight line in Fig. 3. From this method we find a temperature interval $6200 \leqq T_0 \leqq 8600$ K.

4. Heliospheric Gas Heating

The temperature T_0 deduced from Fig. 3 now has turned out to be different from the real temperature of the interstellar gas outside the solar system, since appreciable heating of the interstellar gas advancing towards the sun occurs via elastic collisions of the incoming atoms with solar wind protons. Inspired by estimates given by Wallis [10], Fahr [11] has calculated the temperature of the interstellar gas within the heliosphere. The results of these calculations are shown in Fig. 4 where the heliospheric temperature of the interstellar gas for various solar distances is plotted against the inclination Θ to the velocity $\boldsymbol{v}_0$.

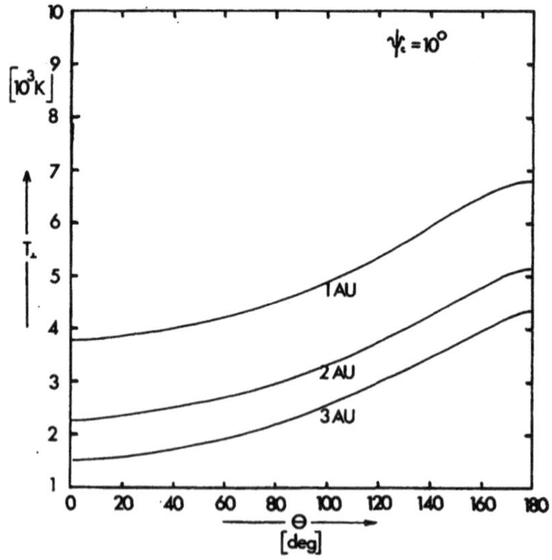

Fig. 4. Heliospheric temperatures $T_\perp$ of neutral interstellar hydrogen penetrating into the heliosphere with $T_0 = 0$ K are shown as a function of inclination Θ against the velocity v_0 of approach for various solar distances (1, 2, 3 AU). The critical deflection angle ψ_c (impact parameter) has been assumed as 10° [11].

It is shown that an interstellar hydrogen gas approaching the solar system with a vanishing temperature T_0 is heated up to 3800 K in upwind, and to 6800 K in downwind positions for $r = 1$ AU. This, however, seems to mean that an appreciable part of the temperature indicated by Fig. 3 is due to the heliospheric heating of the interstellar gas caused by elastic collisions. On the other hand one has to be very careful with this interpretation, since the temperature of between 6 and 8×10^3 K indicated in Fig. 3 is connected with that downwind density from which the maximum of the scattered Lyman α intensity is coming to the observer. The position of this radiation maximum as pointed out by the OGO 5 sky maps is to be adopted at about 15 AU in downwind regions. That means that only the hydrogen densities at solar distances about 15 AU are responsible for the scattered Lyman α intensity. At such distances the heliospheric heating of downwind interstellar hydrogen causes the temperature to rise to only 2000 K. That means that the temperatures $6200 \leqq T_0 \leqq 8600$ K may be reduced by at most 2000 K in order to reveal the temperature T_0 of the nearby interstellar gas to be between 4200 K and 6600 K. Even this result, however, would require the undisturbed interstellar gas outside the solar system to be a hot H I intercloud medium.

References

[1] J. L. Bertaux and J. E. Blamont, Astron. Astrophys. **11**, 200 (1971).
[2] G. E. Thomas and R. F. Krassa, Astron. Astrophys. **11**, 218 (1971).
[3] P. W. Blum and H. J. Fahr, Space Research XII, 1569 (1972).
[4] J. L. Bertaux, A. Ammar and J. E. Blamont, Space Research XII, 1559 (1972).
[5] G. E. Thomas, Solar wind conference, Asilomar, March 1971, NASA, SP-308 (1972).
[6] J. E. Blamont and A. V. Vidal-Madjar, J. Geophys. Res. **76**, 4311 (1972).
[7] H. J. Fahr, Astron. Astrophys. **14**, 263 (1971).
[8] W. Priester and D. Cattani, J. Atmos. Sci. **19**, 121 (1962).
[9] H. J. Fahr, Preprint, Astron. Instit. Univ. of Bonn (May 1973).
[10] M. K. Wallis, Preprint S-100, Stockholm (1972).
[11] H. J. Fahr, Space Sci. Rev. **15**, 483 (1974).

Space Research XIV — Akademie-Verlag, Berlin 1974

POSSIBLE LYMAN ALPHA AND COSMIC RAY ANISOTROPIES EXPECTED IN AN ASPHERICAL HELIOSPHERE

M. K. Bird

Institute for Astrophysics and Extraterrestrial Research, University of Bonn, Bonn, FRG

The boundary of the heliosphere or "heliopause" (determined by the pressure balance between the radially outward flowing solar wind plasma and the interstellar magnetic field) can be shown to differ appreciably from sphericity because of the anisotropic interstellar field pressure along the interface. The heliosphere tends to extend itself along the local galactic field, the exact form and size depending on parameters such as the strength of the interstellar field, the ionic particle pressure in interstellar space, the density of interstellar hydrogen, and the solar wind flux. Observable consequences of an aspherical heliosphere are (a) an additional contribution to the anisotropy of galactic cosmic rays, and (b) an anisotropic distribution of Lyman-α background radiation due to scattering from "hot" hydrogen atoms originating from charge exchanges in the region just outside the heliopause. The intensity maxima of this radiation would be expected to be observed along the direction of the local galactic field. The available satellite data will be reviewed for possible confirmation of these effects.

1. Introduction

The interaction of the solar wind with the interstellar medium has been treated by a number of authors [1—5], and a recent review by Axford [6] presents a comprehensive documentation of the theoretical and experimental progress in the field. The solar wind plasma flows radially outward from the sun until its dynamic pressure is exceeded by the combined interstellar magnetic field and interstellar ionized particle pressures. The interstellar field cannot penetrate into the diamagnetic solar wind, but the solar wind also cannot diffuse into interstellar space (except possibly along the interstellar field lines), so a transition interface is formed which bounds a volume in interstellar space roughly centred on the sun. This volume has been designated the "heliosphere", and the boundary of this volume, upon which dynamic solar wind pressure equals interstellar pressure, is termed the "heliopause".

The solar wind undoubtedly suffers a collisionless shock shortly ahead of the heliopause, in analogy with the magnetosphere of the earth, causing a "heliosheath", or region of subsonic solar wind flow, within which considerable turbulence may exist. Whether or not the solar wind is shocked is immaterial for the form of the heliosphere, which will be controlled entirely by the strength of the interstellar field at all points along the heliopause. The field will consist not only of the assumed constant galactic field, but also will contain contributions from surface currents arising naturally on the heliopause from the almost discontinuous pressure gradients there. The net magnetic field attains its maximum on the heliopause at those points where the surface normal is perpendicular to the undisturbed interstellar field, and thus these points will be "squeezed" harder by the galactic field and should lie closer to the sun than other points on the heliopause. It has recently been shown [7] that the heliosphere reacts to this external anisotropic magnetic pressure by assuming a cigar-like shape oriented roughly along the local galactic field. The elongation of the heliosphere is determined to be primarily dependent upon the ratio of interstellar magnetic pressure to effective interstellar ionic particle pressure.

As *in situ* observations of the heliospheric boundary will not be available for a while [8], we must resort to more indirect methods of inferring the shape and size of the heliosphere experimentally. Since the solar wind dominated heliospheric space is known to be effective in the modulation of galactic cosmic rays [9], it is possible that this modulation will be stronger where the heliosphere is extended, thus influencing the directional distribution of cosmic rays as measured in the vicinity of the earth. A more promising possibility of detecting asphericity of the heliosphere would be a high resolution Lyman-α sky map from which one could unambiguously determine the various sources present. In general one expects the Lyman-α radiation to arise from: (i) a dominant contribution from resonant scattering of solar Lyman-α from "cold" interstellar hydrogen ($v \simeq 20$ km s^{-1}) that penetrates deep into interplanetary space [10]; (ii) a smaller contribution from resonant scattering from "hot" hydrogen atoms ($v \simeq 400$ km s^{-1}) in interplanetary space that are produced by charge exchange in the region just outside the heliopause [11, 12]; (iii) resonant scattering from "hot" hydrogen atoms ($v \simeq 400$ km s^{-1}, directed radially away from sun) which have undergone charge exchange processes in interplanetary space [13]; and (iv) an isotropic contribution from the Galaxy [14]. Contribution (iii) is expected to be negligible, since the scattering agent has such a high associated Doppler shift that it will not be appreciably excited by the solar Lyman-α line. The galactic intensity, contribution (iv), has also recently been estimated to be negligible [15]. The purpose of this paper is to examine more closely contribution (ii), which would be isotropic in the case of a spherical heliosphere. The intensity of Lyman-α radiation will show a double peaked distribution, however, which would become more pronounced with enhanced elongation of the heliosphere. The expected Lyman-α intensities from these hydrogen atoms will be computed and compared with the available Lyman-α data [16—18]. A brief description of the calculation of these intensities, as well as an outline of the method for determining the heliospheric shape, are presented in the following sections.

2. Heliospheric Deformation by the Local Galactic Field

The heliopause is defined as that surface upon which the total interstellar pressure is equal to the dynamic solar wind pressure. This may be expressed as

$$\frac{|\hat{\boldsymbol{n}}_s \times \boldsymbol{B}|^2}{8\pi} + P_s = {}^1/_2\, n_E (r_E/r)^2\, m_p v_p^2 \cos^2 \psi\, W(r) \tag{1}$$

where $\hat{\boldsymbol{n}}_s$ is a unit vector pointing perpendicularly out from the surface, P_s is the interstellar ionic particle pressure (assumed isotropic), ψ is the angle between $\hat{\boldsymbol{n}}_s$

and the radial direction, and $W(r)$ is a reducing factor that represents the effect of the neutral interstellar wind on the solar wind pressure [19]. Since the interstellar field contains only a tangential component on the heliopause, only this component is contained in (1). Knowing the field along the heliopause, one could use (1) to determine the shape and size of the boundary. The field, however, contains a significant contribution from surface currents, and is thus critically dependent upon the surface shape. Both surface and field configuration can be computed using the self-consistent iteration procedure of Beard [20], in which one starts with a good first approximation magnetic field and then uses this field to obtain a first approximation surface from (1). The currents flowing on this surface are given by

$$\boldsymbol{j} = \frac{c}{4\pi} \hat{\boldsymbol{n}}_s \times \boldsymbol{B} \tag{2}$$

and the magnetic field due to these currents is computed from the Biot-Savart law, so that the total magnetic field (the second approximation) at points $\boldsymbol{x}$ just outside the heliosphere is given by

$$\boldsymbol{B}_2(\boldsymbol{x}) = \frac{1}{4\pi} \int \frac{(\hat{\boldsymbol{n}}_s \times \boldsymbol{B}_1) \times (\boldsymbol{x} - \boldsymbol{x}')}{|\boldsymbol{x} - \boldsymbol{x}'|^3} \, ds' + \boldsymbol{B}_0 \tag{3}$$

where $\boldsymbol{B}_0$ is the undisturbed constant galactic field and the subscripts 1 and 2 denote first and second approximations respectively. The equation for the heliopause is given by $r = R(\theta)$, where $\theta = 0$ defines the direction of $\boldsymbol{B}_0$. The field $\boldsymbol{B}_2$ is now used again in (1) to get a new surface, and this process is repeated until no significant change is made between successive iterations. The resultant field and surface configuration is termed "self-consistent".

A series of these self-consistent surfaces is presented in Fig. 1 for an interstellar field strength of 0.4γ and an interstellar hydrogen density of 0.1 cm^{-3}. The assumed

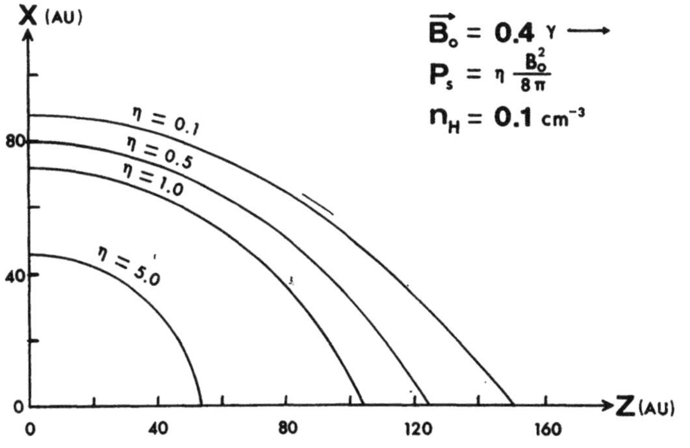

Fig. 1. Self-consistent heliospheric boundary surfaces for various amounts of effective interstellar ionic particle pressure. The constant interstellar field is directed along the Z-axis, and only one fourth of the picture is drawn due to assumed symmetry (rotationally symmetric about Z, mirror symmetric about plane $Z = 0$).

isotropic particle pressure is varied according to the relation

$$P_s = \eta \frac{B_0^2}{8\pi} \tag{4}$$

so that it is a given fraction η of the undisturbed magnetic field pressure in interstellar space. For large η the heliosphere is practically a sphere, but for smaller η (a more probable condition) the heliosphere becomes appreciably elongated.

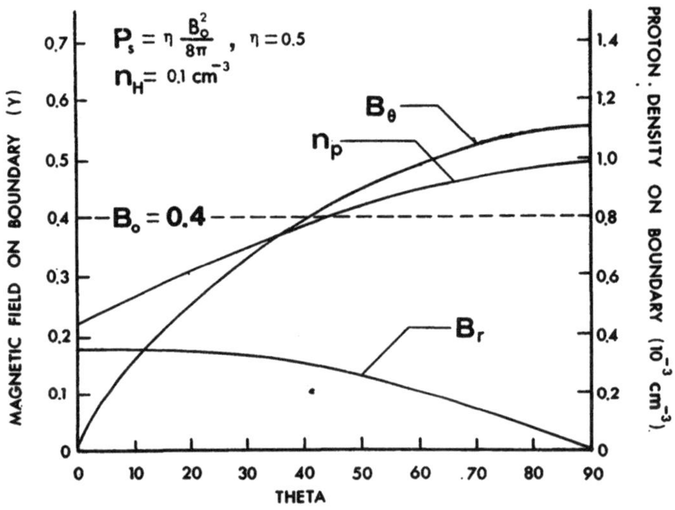

Fig. 2. Magnetic field and solar proton number density along the heliopause. The magnetic field and number density have their maxima in the plane perpendicular to $\boldsymbol{B}_0$ ($\theta = \pi/2$).

The behaviour of $\boldsymbol{B}$ and the proton density along the interface is displayed in Fig. 2 for the case $\eta = 0.5$, $B_0 = 0.4\gamma$ and $n_H = 0.1$ cm^{-3}. The boundary field is seen to exceed the undisturbed field for $40° < \theta < 90°$, the maximum being attained as expected at $\theta = \pi/2$. Solar wind proton density, being a monotonically decreasing function of distance from the sun, also reaches its maximum at $\pi/2$.

3. Effects of the Aspherical Heliosphere on the Distribution of Lyman-α Radiation

Solar wind protons arriving at the heliopause penetrate typically one gyroradius into the interstellar field and are diverted essentially along magnetic field lines into thin channels flowing along the heliopause and eventually along the magnetic field directly away from the sun. The protons impinging on the heliospheric boundary at $\theta = \pi/2$ will be virtually brought to stagnation, while those coming into the boundary at about $\theta = 0°$ will flow freely through. An intermediate situation will arise for other less singular points on the heliopause.

Since an ample supply of atomic hydrogen is present in this region of space, every solar wind proton which has survived up to the heliospheric boundary will undergo charge exchange within a thin sheet approximately coinciding with the heliopause. If the former solar wind protons are transformed at random orientations into "hot" hydrogen atoms, and if the thin transition sheet has a thickness that is very small in comparison with the heliospheric dimensions, approximately one half of the solar wind flux outward at the heliopause will re-enter the heliosphere as kilovolt neutral hydrogen atoms. The trajectories of these particles will be considered to be straight lines — a good approximation in view of the high injection velocities. One may thus write for the source function at the heliopause

$$Q(\theta) = {}^1/_2 n_p(R) V_0 \text{ atoms cm}^{-2}\,\text{s}^{-1}. \tag{5}$$

The density at the position $\boldsymbol{r}'$ inside the heliosphere is obtained by integrating over the source function Q. Therefore we can write

$$n(\boldsymbol{r}') = \frac{1}{4\pi} \int \frac{E_3(\boldsymbol{r}', \boldsymbol{R}) n_p(\boldsymbol{R}) R^2 \sin\theta \, d\theta \, d\phi}{|\boldsymbol{r}' - \boldsymbol{R}|^2} \tag{6}$$

where $E_3(\boldsymbol{r}', \boldsymbol{R})$ is an extinction function that accounts for losses to the density $n(\boldsymbol{r}')$ due to charge exchange with solar wind protons or photoionization by solar EUV radiation before the particle can arrive at $\boldsymbol{r}'$ [21].

The Lyman-α intensity observed at earth from the direction θ when viewed radially outward from the sun is

$$I_{ob}(\theta) = \frac{\Omega}{4\pi} \int_0^\infty dr' u_s B_{12} n(r') \left[\frac{r_E}{r_E + r'}\right]^2 \tag{7}$$

where B_{12} is the Lyman-α transition coefficient, Ω is the solid angle of the detector, and u_s is the solar Lyman-α radiation density at the earth's orbit. The transition coefficient and radiation density are taken as constant over a range of frequencies centred on the Lyman-α line, which is assumed to have a square profile of width 1 Å. If the solar radiation outside the Lyman-α line is very small, then only those "hot" hydrogen atoms with Doppler shifts less than 0.5 Å will actually contribute to $n(\boldsymbol{r}')$ in (6). This restricts the velocities of the scattering agents to be such that

$$\lambda_0 \frac{\boldsymbol{v}_H \cdot \boldsymbol{r}'}{cr'} < \Delta\lambda_0 = 0.5\,\text{Å}. \tag{8}$$

For $v_H \simeq 400$ km s^{-1} this means that only atoms that are moving with velocities with small projections on to $\boldsymbol{r}'$ will not be Doppler-shifted out of the solar Lyman-α emission line. For this reason only those hydrogen atoms arriving from sources in the plane perpendicular to $\boldsymbol{r}'$ through the point $\boldsymbol{r}'$ need be considered in evaluating (6). For a point $\boldsymbol{r}'$ along the direction of $\boldsymbol{B}_0$ ($\theta = 0$), the integral (6) may be shown to yield the approximate result

$$n(r', \theta = 0) = {}^1/_4 n_p(R_1) \tag{9}$$

where r' and R_1 are related by

$$r' = R_1(\theta) \cos\theta, \tag{10}$$

i.e. R_1 is a point on the heliopause that has the same value of z as r' (see Fig. 1). For a point $\boldsymbol{r}'$ perpendicular to the direction of $\boldsymbol{B}_0$ ($\theta = \pi/2$), the density $n(r')$ is again controlled by those sources that project their hydrogen atoms on to $\boldsymbol{r}'$ with velocities perpendicular to $\boldsymbol{r}'$. For the case $\theta = \pi/2$ this requirement means only sources with the same value of x (Fig. 1) need be considered in computing $n(r')$. Therefore to a good approximation

$$n(r', \theta = \pi/2) = {}^1/_4 n_p(R_2) \tag{11}$$

where R_2 is some average point on the heliosphere having the same x value as $\boldsymbol{r}'$

$$r' = R_2(\theta) \sin \theta . \tag{12}$$

For values of r' of the order of a few AU, the density along $\theta = 0°$ will be greater than the density along $\theta = \pi/2$. The function $n(r', 0)$ is a decreasing function of r', whereas $n(r', \pi/2)$ increases with increasing r'.

Using the densities (9) and (11) in the expression for the resonantly scattered Lyman-α intensity (7), one obtains the expected maximum and minimum pattern for the two directions $\theta = 0$ and $\theta = \pi/2$. Since the "hot" hydrogen densities are proportional to the solar wind proton densities at the heliopause, it is clear that the Lyman-α intensity from this component will increase for smaller heliospheres. The size of the heliosphere, but not the shape, will decrease for increasing values of the interstellar field, provided η in (4) is held constant. One thus expects that the Lyman-α intensity computed here should increase with increasing B_0. This trend is shown in Fig. 3, where both $I_{max}(\theta = 0)$ and $I_{min}(\theta = \pi/2)$ are seen to become quite large for small R_{min}. The value of η was taken as 0.5 in Fig. 3, for which the ratio I_{max}/I_{min} remains a constant 1.4 over the entire range of B_0. The Lyman-α intensity would be isotropic for a spherical heliosphere ($I_{max}/I_{min} = 1$) and thus this ratio may be viewed as a measure of the elongation of the heliosphere.

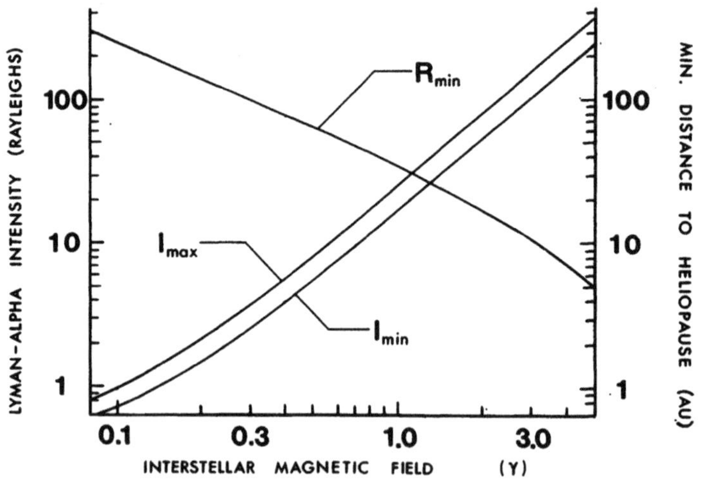

Fig. 3. Minimum and maximum Lyman-α intensities due to 'hot' component of interplanetary hydrogen for various interstellar field strengths and heliosphere sizes. The minimum distance R_{min} is seen to decrease for increasing B_0, but the heliosphere shape is assumed constant ($\eta = 0.5$). The ratio I_{max}/I_{min} increases for more elongated heliospheric shapes.

4. Summary

The asphericity of the heliosphere is shown to lead to an anisotropic Lyman-α intensity distribution from resonantly scattered keV hydrogen atoms injected into the heliosphere from thin subsonic plasma zones located at the heliopause. The distribution shows two maxima located in the directions along the interstellar magnetic field and a minimum intensity in the plane perpendicular to the undisturbed galactic field. The ratio of maximum to minimum intensity increases for greater elongation of the heliosphere along the interstellar field.

A bi-directional sidereal anisotropy in the galactic cosmic-ray distribution is believed to have been observed [22, 23] with maxima located at about 6 hours R. A. in the southern hemisphere and 18 hours R. A. for northern declinations. The degree of the anisotropy is uncertain but is estimated to be less than 0.0002, and may in fact not be an intrinsic characteristic of the galactic cosmic-ray distribution. These maxima deviate only slightly from the directional orientation believed assumed by the local galactic field [24], but no definite evidence of asymmetric modulation of the cosmic-ray distribution has as yet been presented.

If one were to assume that the interstellar hydrogen gas is moving exactly along the local galactic field, then since the solar apex velocity (20 km s^{-1} toward $\alpha = 271°$, $\delta = 30°$) is directed almost along $\boldsymbol{B}_0$, one would expect the arrival direction of the interstellar wind to be close to $\boldsymbol{B}_0$ also. The approach velocity of the neutral wind, as taken from the maximum in the total Lyman-α intensity distribution, is $\alpha = 263°$, $\delta = -20°$, or about 50° inclined to $\boldsymbol{B}_0$. The Lyman-α intensity distribution is dominated by contributions from cold interstellar neutrals in interplanetary space, so that the problem of determining heliospheric size and shape is reduced to attempting to establish (a) the indication of relative maxima in the directions of the local galactic field, and (b) the superimposed amount of the intensity contributed by "hot" hydrogen atoms. The OGO 5 sky maps do not show any structure suggesting a small relative maximum about 50° north of the large 500 rayleigh absolute maximum, and the suspected position of the conjugate southern relative maximum was always hidden behind the geocorona during measurement. Since the minimum Lyman-α intensity produced by this "hot" component is 200 rayleighs for $R_{\min} = 7$ AU ($B_0 = 4.0\gamma$), it may be concluded from the OGO 5 intensity isophotes that the heliopause is certainly farther than 7 AU from the sun. More definite conclusions about the elongation and actual size of the heliosphere cannot be formulated from the available Lyman-α background maps.

References

[1] E. N. PARKER, Interplanetary Dynamical Processes, Interscience, New York 1963.
[2] W. I. AXFORD, A. J. DESSLER and B. GOTTLIEB, Astrophys. J. **137**, 1268 (1963).
[3] P. W. BLUM and H. J. FAHR, Astron. Astrophys. **4**, 280 (1970).
[4] T. E. HOLZER, J. Geophys. Res. **77**, 5407 (1972).
[5] M. WALLIS, Nature **233**, 23 (1971).
[6] W. I. AXFORD, in: NASA SP-308, U. S. Government Printing Office, Washington D. C. 1972 (p. 609).
[7] M. K. BIRD and H. J. FAHR, Mitteil. Astron. Inst. Universität Bonn, FRG (1973).
[8] W. I. AXFORD, Space Sci. Rev. **14**, 582 (1973).

[9] U. RAO, Space Sci. Rev. **12**, 719 (1972).
[10] H. J. FAHR and P. W. BLUM, Space Research XI, 1239 (1971).
[11] T. N. L. PATTERSON, F. S. JOHNSON and W. B. HANSON, Planet. Space Sci. **11**, 767 (1963).
[12] A. J. HUNDHAUSEN, Planet. Space Sci. **16**, 783 (1968).
[13] C. T. GREGORY, Planet. Space Sci. **20**, 841 (1972).
[14] T. F. ADAMS, Astron. Àstrophys. **12**, 280 (1971).
[15] H. J. FAHR and G. LAY, Mitteil. Astron. Inst. Universität Bonn, FRG (1972).
[16] W. H. CHAMBERS, P. E. FEHLAU, J. C. FULLER and W. E. KUNZ, Nature **225**, 713 (1970).
[17] J. L. BERTAUX and J. E. BLAMONT, Astron. Astrophys..**11**, 200 (1971).
[18] G. E. THOMAS and R. F. KRASSA, Astron. Astrophys. **11**, 218 (1971).
[19] H. J. FAHR, Planet. Space Sci. **19**, 1121 (1971).
[20] D. B. BEARD, Rev. Geophys. **2**, 335 (1964).
[21] H. J. FAHR, Astron. Astrophys. **14**, 263 (1971).
[22] D. B. SWINSON, J. Geophys. Res. **76**, 4217 (1971).
[23] R. M. JACKLYN and A. VRANA, Proc. Astron. Soc. Aust. **1**, 278 (1969).
[24] D. S. MATHEWSON and D. C. NICHOLLS, Astrophys. J. **154**, L11 (1968).

Space Research XIV — Akademie-Verlag, Berlin 1974

SPECTROMETRY OF HEAVY NUCLEI IN THE PRIMARY COSMIC RAYS: EXPERIMENT S 67 ON SATELLITE TD-1

C. JULLIOT, L. KOCH, N. PETROU, A. SOUTOUL and J. ENGELMANN

Service d'Electronique Physique, CEN-Saclay, Gif-sur-Yvette, France

The experiment S 67 on the astronomy satellite ESRO TD-1 determines, for $2 \leqq Z \leqq 28$, the charge composition of primary cosmic rays in the range and for energies greater than 200 MeV/nucleon. The experiment has been working correctly since 17 March 1972. With the data at present available we consider the absolute flux of M nuclei for energies greater than 1 GeV/nucleon and the abundances with respect to oxygen for $5 \leqq Z \leqq 28$, again for energies greater than 1 GeV/nucleon.

1. Introduction

Recently much effort has been devoted to the measurement of the chemical composition of cosmic nuclei. The principal motivation for this study arises from the fact that the relative abundance of cosmic nuclei, and the variation of the abundances as a function of energy, contain information about the sources of the relativistic nuclei and about their propagation through space. In the relativistic region, most of the results have been obtained from balloons [1, 2]. Measurements from satellites only exist for light nuclei, $Z \leqq 14$ [3, 4]. A synthesis of all the results can be found in [5]. Now measurements obtained from balloons at a residual atmosphere of several g cm^{-2} must be corrected for nuclear interactions suffered by the incident nuclei before they reach the detector. In particular for the elements between silicon and iron, this correction can cause errors in the results: first of all the abundances of these elements are small, and secondly the fragmentation coefficients are not yet known with precision in this domain.

In this article we present preliminary results on the abundances of nuclei from boron to iron at energies greater than 1 GeV/nucleon. The results have been obtained since 17 March 1972, with a telescope carried on the satellite TD-1.

2. Experimental Apparatus

The identification of each nucleus in the telescope is made by an energy-loss measurement in two solid state detectors J_1 and J_2, and by measuring the Cerenkov radiation in a piece of sapphire C. This detector allows an unambiguous identification for all the nuclei having charges between $Z = 2$ and $Z = 30$, and energies greater than 210 MeV/nucleon.

Since the initial switch on of the experiment on 17 March 1972 it has been working correctly. Meanwhile we have noticed evidence for a gain decrease of 3% in the Cerenkov detector following a very intense solar eruption that occurred at the beginning of August 1972.

3. Results

The data are presented in the form of a diagram in two parameters. The first parameter is the mean response of the two solid state detectors, the second parameter is the response of the Cerenkov counter. In certain cases we have selected the nuclei detected during the time that the geomagnetic cut-off was greater than a given value P_0. Fig. 1 shows a photograph of the three-dimensional model of that part of the charge spectrum from oxygen to argon, with $P_0 = 3$ GV. When $P \geqq 5$ GV (1.8 GeV/nucleon) the detector response is practically independent of energy. For this case, Fig. 2 shows the spectrum of the M-nuclei (C, N, O) as a function of $(J_1 + J_2)/2$.

4. Analysis of the Results

4.1. Abundance of the M-Nuclei (C, N, O) and of Boron as a Function of the Rigidity

Taking into account on the one hand the variation of the cut-off during the time of the measurement and on the other hand the fact that the integral flux of the nuclei as a function of rigidity R is of the form $\phi = AR^{-\gamma}$, we have calculated

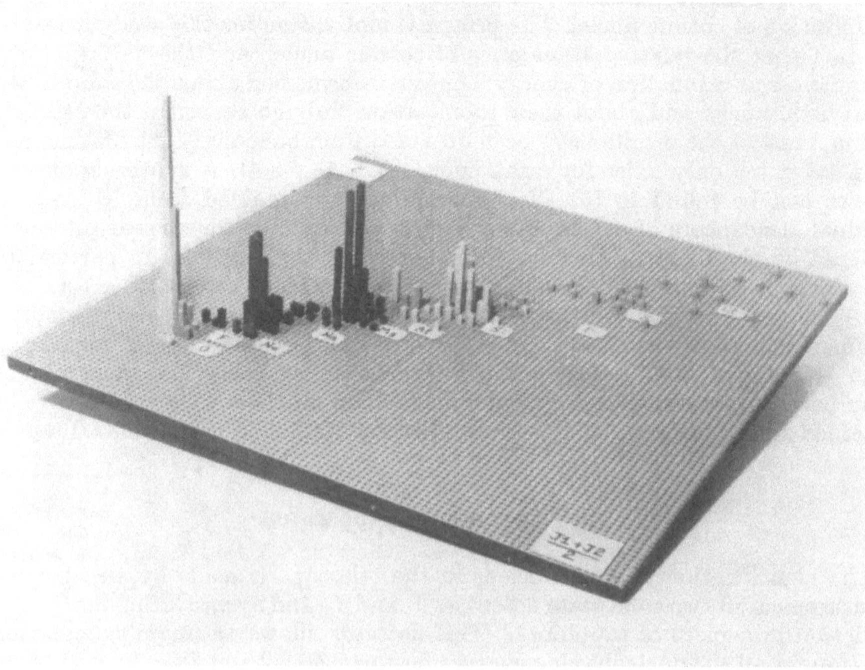

Fig. 1. Heavy ions spectrum representation.

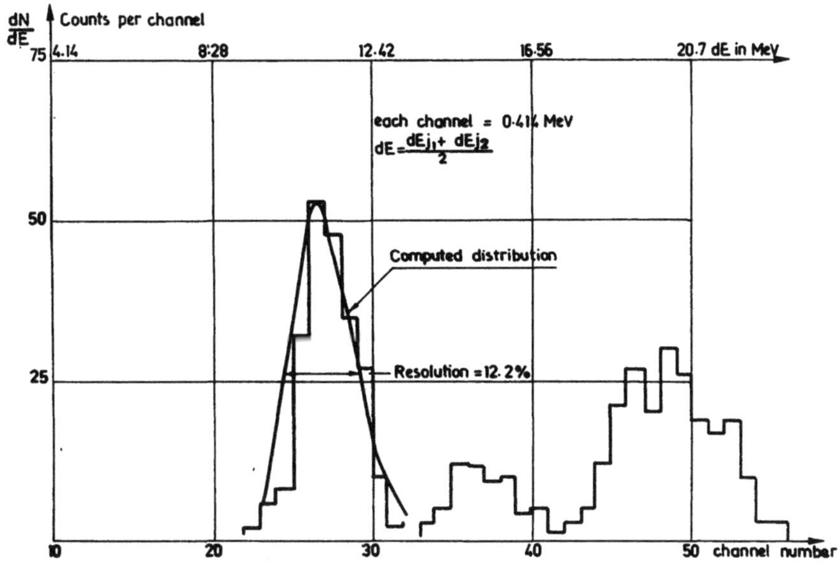

Fig. 2. dN/dE energy loss of the ions C, N, O in the two solid state detectors.

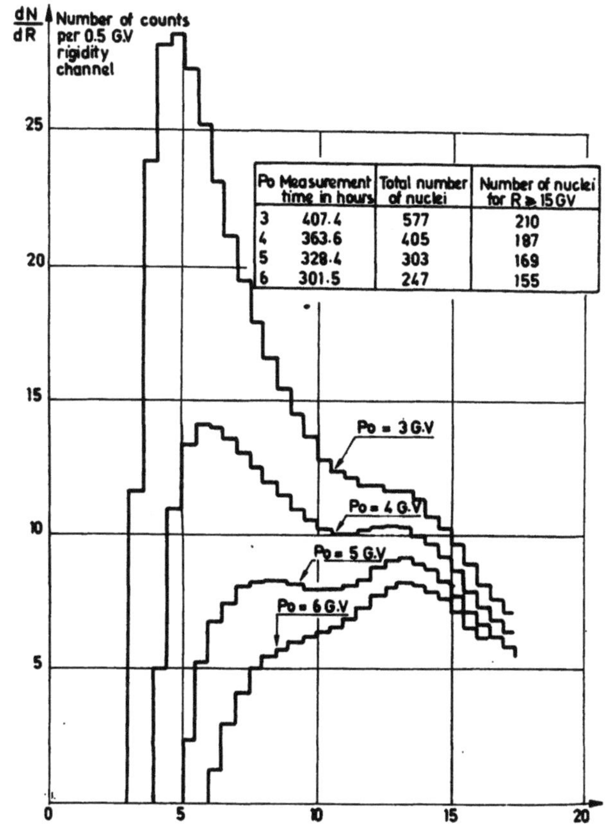

Fig. 3. Computed spectrum of the M-nuclei as a function of rigidity R.

the spectrum of the M-nuclei as a function of R for given values of P_0. The calculated results have been compared with the experimental results.

The calculation of P along the vertical direction has been accomplished by ESRO's Computation Centre using the results of Shea and Smart [6]. P is given in each telemetry format which transmits the data at intervals of 4.8 seconds. The spectra have been calculated (Fig. 3) for $P_0 = 3$, 4, 5 and 6 GV, with the

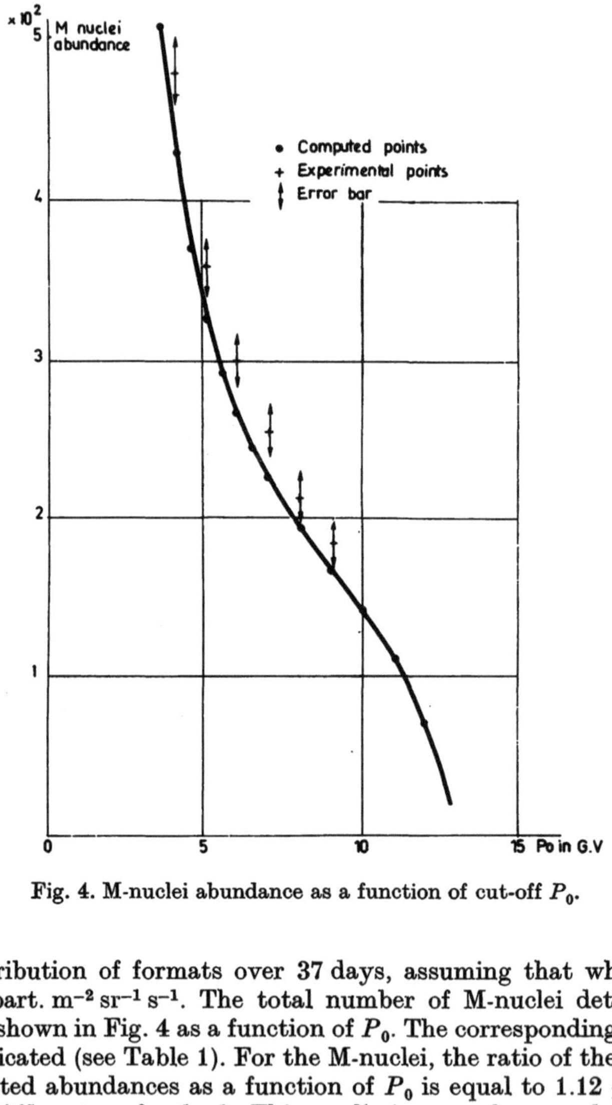

Fig. 4. M-nuclei abundance as a function of cut-off P_0.

observed distribution of formats over 37 days, assuming that when $R \geqq 3$ GV $\phi = 43R^{-1.43}$ part. m^{-2} sr^{-1} s^{-1}. The total number of M-nuclei determined from the spectra is shown in Fig. 4 as a function of P_0. The corresponding experimental values are indicated (see Table 1). For the M-nuclei, the ratio of the experimental to the calculated abundances as a function of P_0 is equal to 1.12 ± 0.04, where $\phi = 48R^{-1.43\pm0.05}$ part. m^{-2}sr^{-1} s^{-1}. This preliminary value can be improved in the future by reducing the error due to statistical fluctuations, by making more precise calculations of the corrections due to nuclear interactions, and by making a correction to the form of the spectra in rigidity: at present calculations are in drogress to determine the value of the cut-off P, taking into account the inclination of the axis of the experiment with respect to the vertical (±10°), and the solid angle of the telescope.

For each value of P_0, Table 1 contains: for each element, the experimental abundance corrected and referenced to that of oxygen; for the M-nuclei, the corrected experimental abundance and the calculated abundance.

Table 1

Abundances of M-Nuclei

P_0 (GV)		B	C	N	O	M-nuclei exp.	calc.
3	Nc[a]	98 ±10 0.292 ±0.05	290 ±19 0.88 ±0.12	75 ±9 0.225 ±0.04	336 ±20 1	702 ±27	614
4	Nc	70 ±9 0.315 ±0.06	205 ±15 0.923 ±0.13	53 ±8 0.225 ±0.05	222 ±17 1	480 ±22	429
5	Nc	56 ±8 0.35 ±0.08	157 ±14 0.985 ±0.17	41 ±7 0.256 ±0.06	160 ±14 1	300 ±17	267
6	Nc	47 ±7 0.345 ±0.08	130 ±12 0.96 ±0.17	33 ±6 0.24 ±0.06	136 ±12 1	256 ±16	227

[a] Nc = Number of nuclei after correction for nuclear interactions.

The value of the ratio C/O, slightly less than 1, is a little smaller than those published until now. The others range from 1.06 to 1.11 ([1—4] and especially [5]). The same remark can be made for the ratio N/O, where the presently available values are between 0.24 and 0.29. On the other hand, the ratio B/O is rather higher than the usual values: 0.27 to 0.31. Taking account meanwhile of the indicated errors, the data are comparable with those given in the above references, and they are independent of the rigidity in the range from 3 to 10 GeV.

4.2. Abundances of Nuclei with $Z = 9$ to $Z = 26$

In the two parts of Table 2, $P_0 = 3$ GV, there are presented for each element the measured value of the abundance, and the value corrected and compared with that of oxygen. In the present state of the data available, only the relative values of the most abundant elements Ne, Mg, Si and Fe are quantitatively significant; these agree with the most recent data [5]. With respect to the other elements in Table 2 (F, Na, Al), we note that aluminium is more abundant than sodium, which is in turn more abundant than fluorine. We also note that, after

Table 2

	O	F	Ne	Na	Mg	Al	Si	Sum of Ne, Na, Mg, Al, Si
Nb[a]	590 ±24	6	103 ±10	13	132 ±12	21	100 ±10	
Nc[a]	680	7	121	15	157	25	120	438 ±21
	1	0.01	0.17 ±0.024	0.02	0.23 ±0.03	0.036	0.17 ±0.024	

	P	S	Cl	Ar	K	Ca	Sc	Ti	V	Cr	Mn	Fe
Nb[a]	2	8	5	3	1	13	1	6	2	5	4	71 ±8
Nc[a]		10	6	4		16		8		6	5	89
		0.014				0.02		0.07				0.13 ±0.02

[a] Nb, raw values; Nc, values after correction for nuclear interactions.

silicon and iron, sulphur and calcium are the most abundant. There is some disagreement on these last two points [5].

Our data are still preliminary. They represent about 20% of the data which will be available to us by the end of 1973.

Acknowledgments

We would like to express our appreciation to those who contributed to the success of this project: Mrs C. Jehanno and Mrs A. Robin who fabricated the solid-state detectors, Mr M. Cantin, Mr Labade and Mr Vigot who participated in the study and in the preparation of the telescope, Mr Pagnier who collaborated in the data analysis and Mr Keirle who directed the construction of the flight models.

We would also like to thank the ESRO team who participated in the integration of the satellite, in the data recovery, and in the preliminary treatment of the data.

References

[1] J. A. LEZNIAK, T. T. VON ROSENVINGE and W. R. WEBBER, Act. Phys. Hung. **29** suppl. 1 (1970).
[2] M. CASSÉ, L. KOCH, N. LUND, J. P. MEYER, B. PETERS, A. SOUTOUL and S. N. TANDON, Proc. 12th Int. Conf. on Cosmic Rays, Hobart, **1**, 215 (1971).
[3] B. G. CARTWRIGHT, M. GARCIA-MUNOZ and J. A. SIMPSON, Proc. 12th Int. Conf. on Cosmic Rays, Hobart, **1**, 295 (1971).
[4] W. R. WEBBER, S. V. DAMLE and J. M. KISH, Proc. 12th. Int. Conf. on Cosmic Rays, Hobart, **1**, 229 (1971).
[5] W. R. WEBBER, Cosmic Rays: Charge Measurements 1972, Summary (preprint).
[6] M. A. SHEA and D. F. SMART, Canad. J. Phys. **46**, S 1098 (1968).

Comets

Space Research XIV — Akademie-Verlag, Berlin 1974

RECENT ADVANCES IN COMETARY PHYSICS AND CHEMISTRY

L. BIERMANN

Max-Planck-Institute for Physics and Astrophysics, Munich, FRG

This review discusses mainly the observations of some recent comets by OAO 2 and OGO 5 and their interpretation in terms of the physical processes and the chemical constitution of comets. The most important finding is that hydrogen atoms are much more abundant in cometary atmospheres than those molecules which make up the ordinary coma; from the intensity of the OH emissions it is concluded that ordinary water is a main constituent. The brightness in Lyman α is at least as large as that due to emission bands in the ordinary optical range, and particularly the size of the hydrogen atmosphere greatly exceeds that of the ordinary coma. Finally, some plans for fly-by missions to comets are dealt with, and their potential and promise is discussed.

To a considerable degree recent advances in our understanding of the physics and especially the chemistry of comets are due to observations from space vehicles. It has been known for a long time that the selection of those molecules which are visible from the ground is essentially due to the fact that they emit light mainly in resonance transitions at wavelengths above 3000 Å, while for the most important constituents these are at wavelengths below the atmospheric cut-off near 3000 Å. A first indication of the true amount of the gas production of comets became available with the proof that the forbidden line of oxygen near 6300 Å, which is one of the most prominent night sky emissions, is present also in the spectra of bright comets. This permitted an estimate of the total gas production, and led, in conjunction with Whipple's picture of the chemical constitution of comets, to the prediction that comets should be very bright in the Lyman α line of atomic hydrogen (1216 Å), and that the hydrogen atmosphere should be very much larger than the ordinary coma of comets [1, 2].

After comet Tago-Sato-Kosaka (1969g) had become accessible to observation from the US Orbiting Astronomical Observatory OAO 2, the first such observations were obtained on 14 January 1970 by Code, Houck and Lillie (IAU Announcement Card 2201). These observations revealed at once the large brightness of the comet in Lyman α (equivalent to several 10 kR in the centre) and the great extent of the hydrogen atmosphere, the true size of which, however, was revealed only by the observations in Lyman α from another space vehicle, OGO 5, by Bertaux and Blamont [3]. These latter observations referred to the second bright comet of 1970, comet Bennett, which fortunately appeared a few months after comet Tago-Sato-Kosaka and which was easily observable, especially in April 1970. The large brightness in Lyman α was independently confirmed by observation of comet Tago-Sato-Kosaka by means of a sounding rocket [4].

The spectrometric observations from OAO 2 revealed also the presence in comparable amounts of OH, the hydroxyl radical, which suggest that at least a considerable fraction of the observed hydrogen originates from molecules of ordinary water. The UV line of atomic oxygen at 1302 Å was observed also with an intensity which is consistent with the assumption that the atomic oxygen observed in comets is derived largely from ordinary water molecules. There are some indications of the presence of molecular hydrogen again in large quantity, but owing to the small excitation probability of the transitions in question the

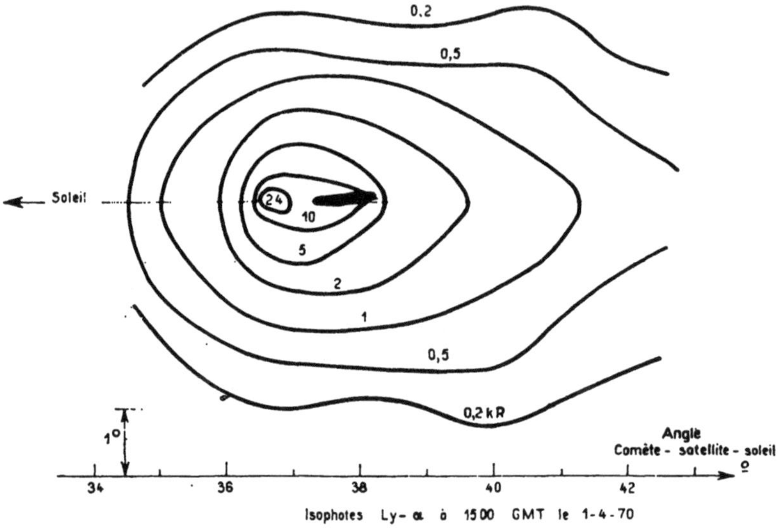

Fig. 1. Comet Bennett. The picture shows the hydrogen atmosphere and the ordinary coma and tail on the same scale. Units, minutes of arc.

problem of the presence of molecular hydrogen in cometary atmospheres has not yet been settled. There are still other molecules which could be present in considerable quantity which, however, would be observed best, not by their transitions in the UV, but by the ones in the cm- or mm-wavelength range, or in the infrared. Positive results of such attempts cannot yet be reported. The present review is concerned mainly with the observations by OAO 2 and OGO 5 and their interpretation in terms of the physical process and the chemical constitution of comets. Some plans for fly-by missions to comets, with the potential and promise of such missions are also briefly discussed.

Fig. 1 shows a composite picture giving a superposition of an ordinary photograph of comet Bennett on the Lyman α contours as measured from OGO 5. The observations from OGO 5 were actually carried out by two groups, Bertaux and Blamont in Paris [5] who worked with a spatial resolution of 40 arc minutes, and Barth and Thomas from the University of Colorado who obtained a still higher sensitivity while having an even lower spatial resolution. Both instruments on board OGO 5 were essentially UV photometers operating in a well defined rather narrow spectral range; on the other hand, the instrument on board OAO 2 was a spectrophotometer described in detail in [6]; see also [3]. A full bibliography is given by Keller [7], who has worked in Munich and in Boulder, Colorado, extensively on the interpretation of the results obtained with OAO 2 and OGO 5.

The figure shows that the extent of the hydrogen atmosphere is more than ten times larger than that of the coma, the atmosphere defined by the neutral constituents visible from the ground (mainly CN and C_2). The difference in size is on one hand due to the larger speed with which the hydrogen atoms stream out from the central parts (8 km s^{-1} compared with about 0.8 km s^{-1}), on the other hand to

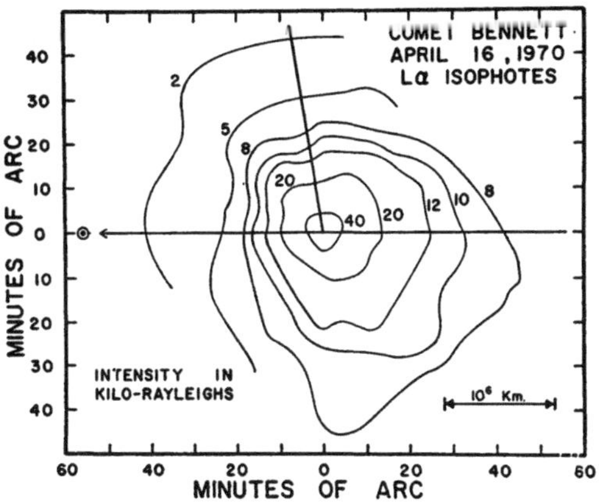

Fig. 2. Comet Bennett. Intensity distribution in Lyman α based on the OAO 2 observations (inner part of the hydrogen atmosphere).

the fact that in Lyman α we observe the true size against the fainter background, whereas for the coma as visible from the ground the outer parts actually merge with the sky background before the real outer boundary due to ionization or dissociation is reached.

Fig. 2 shows the intensity distribution in Lyman α as measured from OAO 2. Three parts can be distinguished in the hydrogen atmosphere: an inner part extending to about half a million kilometres, in which the optical thickness in Lyman α is obviously comparable with or even larger than unity. The spread in radial velocity has been derived from spectrophotometric observations (Fig. 3) at times during which the relative radial velocity between the comet and the earth went through zero, such that the residual absorption by geo-coronal hydrogen atoms (producing an optical depth of about 0.6) could be seen superposed on the emission from the comet. From such observations it became clear that the spread of velocities of the cometary hydrogen atoms was certainly larger than that due to thermal motion in the geo-corona, but not by an order of magnitude. The outer parts of the hydrogen atmospheres are those in which the surface brightness decreases much faster than with the inverse first power of the distance from the nucleus, which characterizes approximately the intermediate parts (cf. Fig. 1). The shape of the isophotes reveals furthermore the influence of the radiation pressure in Lyman α, which produces an acceleration away from the sun compensating or even slightly over-compensating solar gravity; in addition inaccuracy of the measurements or local or temporal variations in the properties of the hydrogen outflow seem to play some role.

For the interpretation we follow essentially the ideas of Bertaux et al. [5], Keller [7] and Delsemme [8]. It is obvious that in a system moving with a comet the influence of light pressure in Lyman α and the distance towards the sun until deceleration is reached determine essentially the kinetic energy of the hydrogen atoms, whereas the lateral extent of the contour (perpendicular to the solar vector) indicates the distance from the nucleus reached before ionization. A combi-

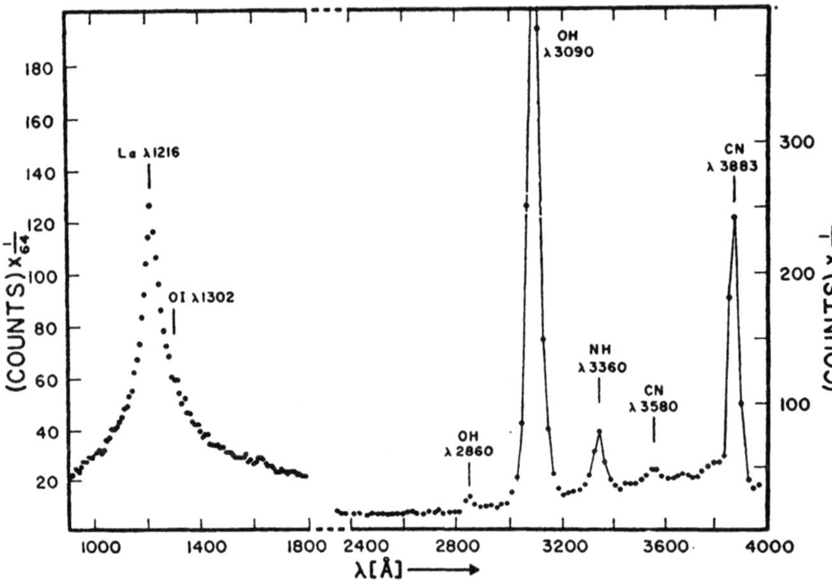

Fig. 3. Spectrum of Comet Bennett (OAO 2).

nation of the two figures may lead to an estimate of the time scale of ionization, which of course can be compared as long as the comet is not too far from the ecliptic with the value expected from the flux of solar wind protons (which ionize by charge transfer reactions with the high cross section of 20 $Å^2$), and by photoionization. The Lyman α contours can in principle be used to determine the solar wind flux off the ecliptic. The surface brightness at a given distance, especially that at intermediate distance, gives evidence on the local density of hydrogen atoms and, more important, on the production rate of hydrogen atoms by the comet; for mid-April 1970 this was found to be close to 10^{30} hydrogen atoms per second for comet Bennett, which at that time was at a solar distance of slightly more than 0.8 AU. The velocity of outflow was found to be approximately 8 km s^{-1}. and the time scale of ionization close to 10^6 sec. The quantitative results rest in fact on the comparison of rather detailed models and of the resulting contours (Fig. 4) with the observed ones. The production rate of hydrogen atoms is, as had been expected, two or three powers of ten larger than that of the molecules observed from ground-based instruments for similar comets, varying with solar distance with the power -2.3 [8] (Fig. 4).

Especially interesting are the observations of the OH resonance transition at 3090 Å. This transition is just at the limit for ground-based work, and therefore its true intensity was in fact not known though it could have been at least roughly determined also from the ground. The spectrophotometric observations revealed that within the relatively large errors of measurement (due partly to finite optical thickness) most of the hydrogen atoms could have been derived from ordinary ice. Another approach to the interesting question whether ordinary ice is a major or even the main constituent of cometary nuclei was provided by the observation (with OAO 2) that the surface brightness in the central parts both of OH and H

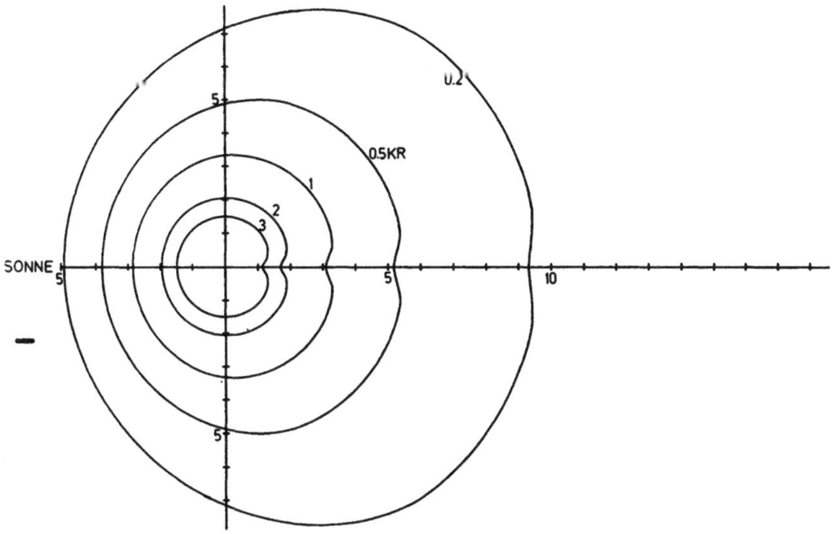

Fig. 4. Computed model of the intensity distribution in Lyman α with distance (in 10^6 km), giving optimum agreement with the observations of Comet Bennett on 15 April 1970. Derived quantities are: source strength 7×10^{28} s^{-1} sr^{-1}, production rate of H atoms by comet 8.8×10^{29} s^{-1}, probable velocity of the radial Boltzmann distribution 8 km s^{-1}, lifetime of H atoms 9×10^5 s.

varied with almost the -6th power of solar distance when determined through a fixed diaphragm. A recent analysis by Delsemme [8] has shown that more than half of this exponent is due to this particular instrumentation, and that the true exponent which characterizes the dependence of the gas production on solar distance is only about -2.8. When comparing this with what on the basis of physical chemistry would be expected for various ices, it is seen that ordinary ice gives just about the right value (cf., however, Keller's aforementioned result). This is one of several indications which led Delsemme to suggest that ordinary ice makes up roughly 90% of the ices from which the gases observed in cometary atmospheres originate. This, however, says so far only that a consistent picture may be reached in this way, but further evidence on components which would not contribute to the UV or the ordinary spectrum of a comet has to be awaited until this question can be regarded as settled.

It is also important to realize that differences between young and old comets, in the sense of Oort's theory of the origin of comets, would be expected. Comets like Humason of 1962, which show a copious gas production already at the distance of Jupiter's orbit, can certainly not consist predominantly of ordinary ice, because the evaporation rate of H_2O at the temperature in question is far too low. This rather complicated question cannot be discussed in more detail here.

In connection with the last mentioned problem observations of molecular hydrogen near 1100 Å and of CO near 1800 Å as well as further observations of the 1302 Å line of atomic oxygen would be very valuable because large amounts of H_2 and CO would tend to indicate, though not necessarily prove, that chemical processes take place when a comet approaches the sun. Also this question can and will probably be attacked by other means; but especially for H_2 UV observations are probably the only feasible way of obtaining quantitative information. The interest of these questions lies in the fact that they may provide clues to the conditions and the processes in the outer solar nebula, in which cometary nuclei formed in very large numbers $4^1/_2$ billion years ago; since their total mass should have been of the same order as that of the outer major planets [9], and since the nuclei of new comets have been less affected by solar heat than any planet, such information may be expected to be of more general interest than one might expect at first sight.

We consider finally the possibility and the scientific potential of future space missions towards a comet. With present techniques a fly-by mission could be carried out with moderate cost. The next such possibility, studied in detail by Ness and coworkers at the Goddard Space Flight Center, would be a mission to comet Grigg-Skjellerup in 1977; another would be a mission to comet Encke in 1980 to be carried out by a third version of the US-German project Helios. The scientific goals to be pursued with such missions with a fly-by relative velocity of the order of 10 km s^{-1}, include the following: with a distance of closest approach of around 10000 km, which seems attainable, the cometary nucleus, if the most widely held view of its constitution, Whipple's icy-conglomerate model, is correct, should subtend an angle of about 1 ′ (roughly) at that distance and thus be measurable in some way. Comet Encke fortunately produces little dust, and might there-

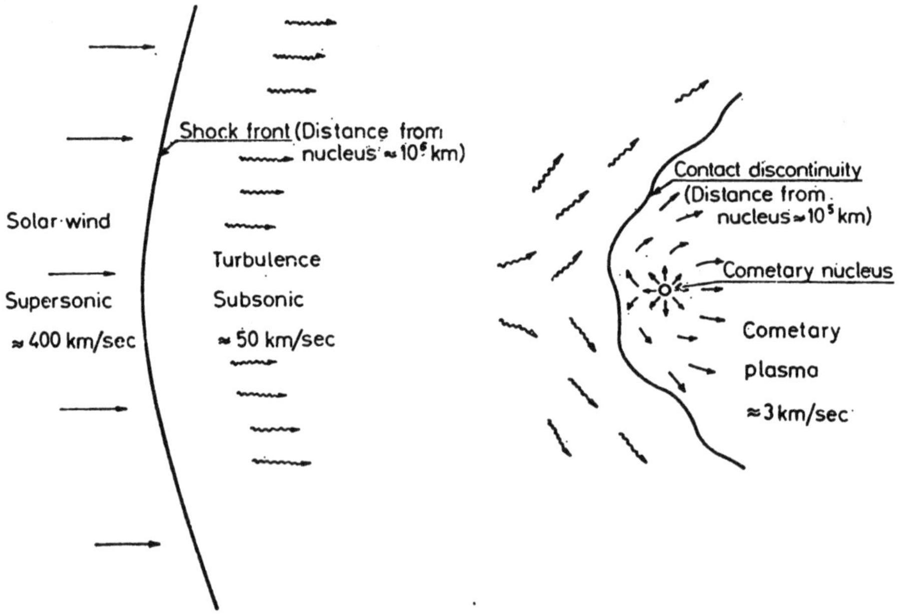

Fig. 5. Plasma flow near a comet.

fore be particularly promising for such an attempt; also its orbit is better known than that of almost any other comet.

Our knowledge of the chemical composition of comets would gain very much by mass-spectrometric measurements of the molecules in the comet's atmosphere. It is practically unknown whether there are as many hydrocarbons as would correspond to the cosmic abundance ratio of C:O (roughly 1:2), whether the C is present mainly in the form of CO or H_2CO, or whether it is less abundant in comets than elsewhere in the solar system.

An interesting problem of cosmical plasma physics is offered by the interaction of the solar wind with a comet. Numerical predictions have been made [10, 11] by means of magnetohydrodynamical numerical computations (Fig. 5). Since this is an example of a large class of such problems, and since very few opportunities for direct comparisons between theory and observation exist, a measurement of the plasma flow around the comet, which could be made with the same instruments as are used for plasma measurements in interplanetary space, would have an interest far beyond cometary physics. This, however, has been a feature of cometary physics for at least some decades. Questions of more general interest like those connected with the origin of the solar system or with the properties of the solar wind are so intimately related to modern cometary research that it should no longer remain the domain of a few specialists; the tools of space research seem particularly suited to the attack of some of the open questions in this area from new angles.

References

[1] L. Biermann and E. Trefftz, Z. f. Astrophys. **59**, 1 (1964).
[2] L. Biermann, JILA Rep. No. 93 (1968).
[3] J. L. Bertaux and J. Blamont, C. R. Acad. Sci. Paris B **270**, 1581 (1970).
[4] E. Jenkins and D. Wingert, Astrophys. J. **174**, 697 (1972).
[5] J. L. Bertaux, J. Blamont and M. Festou, Astron. Astrophys. **25**, 415 (1973).
[6] A. Code, T. Houck and C. Lillie, Proc. NASA Symp. Amherst 1971, NASA-SP-310, 1972 (p. 109).
[7] H. U. Keller, Astron. Astrophys. **23**, 269 (1973).
[8] A. Delsemme, Astrophys. Lett. **14**, 163 (1973).
[9] L. Biermann, Proc. Nice Symp. on the Origin of the Solar System, CNRS 1972 (p. 311).
[10] L. Biermann, B. Brosowski and H. U. Schmidt, Sitz. Ber. Bayer. Akad. d. Wiss., Sonderdruck No. 7 (1967).
[11] B. Brosowski and R. Wegmann, Max-Planck-Inst./PAE-Astro 46 (1972).

The Moon

Space Research XIV — Akademie-Verlag, Berlin 1974

PRELIMINARY RESULTS OBTAINED WITH AN ASTROPHOTOMETER INSTALLED ON LUNOKHOD 2

A. B. SEVERNY, E. I. TEREZ and A. M. ZVEREVA

Crimean Astrophysical Observatory, p/o Nauchny, Crimea, USSR

The astrophotometer was used for measurements of lunar sky brightness during lunar day and night. Preliminary results showed unexpectedly high values both in the near ultraviolet region and, particularly, in the visible light region. The preliminary conclusion from these results is that this is due to the scattering of solar radiation by dust particles above the surface of the moon.

1. Instrumentation

An astrophotometer was installed on Lunokhod 2 for measurements of sky brightness during lunar day and lunar night. This instrument comprised a double lensless telescope one tube of which was sensitive to visible light ($\lambda_{max} = 5400$ Å, $\Delta\lambda = 700$ Å)*, while the other tube was sensitive to near ultraviolet ($\lambda_{max} = 2700$ Å, $\Delta\lambda = 730$ Å.) Special blinds against scattered and outside light restrict the field of view to 12.2° for visible light and 17.6° for ultraviolet light (the outside light beam inclined at the angle $\geqq 40°$ to the axis of the tube does not have any effect on the measurement of the faintest light sources). The measured and calculated rejection for off-axis angles $\geqq 40°$ is greater than 10^5.

The detection system of each tube uses a spectral filter and sensitive photomultiplier preceded by a rotating-disk light chopper. A 4-position programme disk which can be stepped every 3 seconds provides a choice of: (i) a large stop, (ii) the first artificial calibrated light source (radioactive luminofor), (iii) a small stop and (iv) the second luminofor to which the corresponding photomultiplier-filter set is insensitive. The last of these is used to measure the photomultiplier dark current.

The astrophotometer was calibrated by means of a source calibrated by a black-body radiator [1] and using the luminescence of a solution of rhodamine B in ethylene glycol. For visual light the absolute sensitivity is 4.3×10^{-6} erg s^{-1} cm^{-2} per volt (accuracy $\pm 5\%$); for ultraviolet light the sensitivity is 4.2×10^{-6} erg s^{-1} cm^{-2} with accuracy $\pm 10\%$. These accuracies are determined by calibration limitations. The channel for visual light was also calibrated by measuring the night-sky brightness from the ground; the brightness of the radioactive artificial source was estimated as 267 stars 10^m per square degree.

---

* The band pass of the filters is defined as twice the decrease of transmission (of maximal)

The astrophotometer was looking at local zenith on the moon and the measurements which were used for the present work were all obtained in the shadow of the panel of the solar cell.

2. Preliminary Results

From the landing of Lunokhod 2 on 15 January 1973 up to 20 March 1973 13 telemetric records were obtained, ten of which were made during lunar day, two during lunar night and one during "lunar twilight", when the upper edge of the sun was 1.0° below the local horizon. For all lunar day measurements the readings in the visible channel were so large that they were outside the scale while for ultraviolet light there were quite distinct readings comparable with the readings of the luminofor. This means that even when measured in shadow the visible sky brightness is very large, far above 4400 units (1 unit = brightness of 1 star of 10^m per square degree).

More precise determination of sky brightness in visible light was realized for the "twilight" lunar sky, after the sun was 1° below the local horizon. The visible brightness appeared to be ten times brighter than the brightest part of the night sky observed from the earth (9260 units). The upper boundary of the shadow was 260 metres above the point of observation. (This measurement had been made 30 hours after Lunokhod was stopped for night rest.) This phenomenon of very bright "twilight" radiation is probably similar to the phenomenon observed by Surveyor in January 1968 [2]: the aureole or halo around the dark lunar border after the sunset, bright and narrow, but wide in azimuth.

Our preliminary conclusion is that in the visible light the lunar sky is bright during lunar day and immediately after sunset. This bright radiation originates not only near the surface of the moon, but probably extends up to the heights of at least 260 metres.

Ultraviolet brightness was recorded in seven cases (one corresponding to lunar twilight) and it appeared to be several (from 2 to 15) times brighter than would be expected from the stars and zodiacal light corresponding to the local zenith regions of the sky (assuming the mean spectral distribution the same as that of the sun). In lunar "twilight" conditions (24 January 1973) the *ultraviolet* brightness appeared to be the same as one would expect for the brightness of the sky outside the earth's atmosphere, and very near to the value determined from our measurements on the satellites Cosmos 51 and 213 [3, 4]. It should be especially noted that the excess of ultraviolet brightness over the expected values showed the dependence on zenith distance decrease with $\cos z$ wich is characteristic of some cases of scattering in planetary atmospheres.

Comparison of visual and ultraviolet brightness in twilight conditions showed a ratio of 62:1 while the same ratio for lunar surface is 70:1 if we take for the lunar albedo in our ultraviolet region the value 1% according to the rocket data obtained by Australian scientists [5] (7% was assumed for the albedo in the visual region, and the decrease of solar flux towards the ultraviolet is $\simeq 10$). This leads to a preliminary conclusion that the source of enhanced radiation is the scattering of solar radiation at the cloud of dust particles above the surface of the moon with albedoes similar to that of the lunar surface.

The dependence on zenith distance and the values of excess of ultraviolet radiation has enabled us to determine the optical depth of the scattering layer ($\tau_0 = 10^{-6}$). Taking the height of this layer as 260 metres we can estimate the concentration of dust particles for the case of our twilight measurements, assuming the size of these particles to be from 10 to 70 μm. The concentration is found to be about ten particles per 100 cubic metres (10^{-9} cm^{-3}). This is 10^4 times larger than the mean concentration of dust particles in interstellar space. This concentration agrees quite well with Gault's [6] estimate of concentration of dust particles expelled from the surface of the moon due to the bombardment by micrometeorites with velocities higher than 10 km s^{-1}.

Two telemetric records during the lunar night were not quite successful (due to the very low temperatures and to the slow rotation of the programme disk only a small amount of information was obtained). The visible brightness of the night sky in the region of Leo Minoris was found to be 220 units instead of 107 as expected, which is practically the same and also coincides with the minimal values found in the earth's shadow with the satellites Cosmos 51 and 213.

Our preliminary conclusion is that the lunar sky is not appropriate for astronomical observations during *lunar day* in visible light, while it is quite dark during the night. At the same time there is scarcely any gain in penetrating power of telescopes working on the moon as compared with those working in the earth's satellites.

In ultraviolet light the lunar sky is quite dark even during the lunar day, if screened from solar radiation and stray light.

References

[1] E. I. Terez, Izv. Krimsk. Astrofiz. Obs. **46**, 144 (1972).
[2] L. D. Jaffe, in: Symp. on Physics of the Moon and Planets 1968 (p. 8), and references therein.
[3] N. A. Dimov and A. B. Severny, Izv. Krimsk. Astrofiz. Obs. **45**, 53 (1972).
[4] N. A. Dimov, A. M. Zvereva and A. B. Severny, Izv. Krimsk. Astrofiz. Obs. **45**, 67 (1972).
[5] J. H. Carver and B. H. Norton, Proc. Astr. Soc. of Austr. **1**, 11 (1967).
[6] D. E. Gault et al., NASA SP-173, May 1968 (p. 233).

Space Research XIV — Akademie-Verlag, Berlin 1974

COMPOSITION AND PHYSICS OF THE LUNAR ATMOSPHERE

J. H. HOFFMAN[a], R. R. HODGES, JR.[a], F. S. JOHNSON[a] and D. E. EVANS[b]

[a] The University of Texas at Dallas, Dallas, Texas, USA
[b] NASA/Johnson Space Center, Houston, Texas, USA

The existence in the lunar atmosphere of helium, neon, argon and possibly molecular hydrogen has been confirmed by the Apollo 17 mass spectrometer. The observed helium concentrations and distribution agree closely with model predictions for a non-condensable gas based on a solar wind source, thermal escape and a longitudinal distribution based on a Monte Carlo random walk calculation. Heavier gases are lost by photoionization and subsequent sweeping away by the solar wind electric field. The observed night-time neon concentration of 8×10^4 molecules cm^{-3} is consistent with expected amounts. Argon, however, is adsorbed on the lunar surface late at night when the surface temperature is lowest. It shows the expected pre-dawn enhancement exhibited by condensable gases released into the atmosphere at the sunrise terminator. Hydrogen appears to exist in the molecular rather than the atomic state. Its observed concentration is less than a factor of 3 higher than that predicted by a model similar to that used for helium. The total night-time concentration of known species (H_2, He, Ne, Ar) in the lunar atmosphere is 2×10^5 molecules cm^{-3}.

1. Introduction

The moon does indeed have an atmosphere, but it is so tenuous that it acts as a collisionless gas, the molecules travelling in ballistic trajectories between encounters with the surface of the moon. Trajectory heights and horizontal travel are determined by the surface temperature and molecular mass.

The major source of known gases in the lunar atmosphere is the solar wind [1]. Its ions impinge directly on the lunar surface and become imbedded in the surface materials, but once the surface is saturated with a given constituent of the solar wind, that constituent must be released from the surface at the same average rate as it is accrued. When released, the neutralized molecules form an atmosphere. Those molecules released with sufficiently high velocity will escape from the moon. This is the dominant loss process for hydrogen and helium. Heavier gases, for which thermal escape is slow, are distributed around the moon by diffusion, their local concentration being a function of temperature, $T^{-5/2}$ [2], unless they condense or are adsorbed on the cold nightside. A non-condensable gas will then have a night-to-day concentration ratio of about 30. For a condensable gas, a rapid release occurs at the sunrise terminator forming a locally enhanced concentration. Gases from this region can travel eastward throughout the dayside region or westward into the nightside region several scale heights before being readsorbed on the surface, thus giving rise to a pre-dawn enhancement.

The light gases do not follow the $T^{-5/2}$ concentration law, but their distribution about the moon can be handled by a Monte Carlo calculation. Assuming a symmetric daytime temperature distribution about the subsolar point and a gradually decreasing night temperature to 90 K at the sunrise terminator, the diurnal ratio for helium and hydrogen has been shown to be 24 and 6.5 respectively [3] with the maximum for both occurring at night.

Heavy gas molecules, not lost by thermal escape, are photoionized and accelerated by the solar wind electric field [4]. As their speeds increase, the ions begin to react to the solar wind magnetic field, eventually travelling in cycloidal paths. Since the radius of gyration is large compared with the moon, about half of the ions escape from the moon and half impinge on the surface to be re-released into the atmosphere as neutral gas molecules. Using these concepts and measured solar wind fluxes, expected gas concentrations have been derived for both the subsolar and antisolar points. These are summarized in Table 1.

Table 1
Summary of Gases in the Lunar Atmosphere

Gas	Observation Molecules cm^{-3}		Model [3, 10] Molecules cm^{-3}	
	Day	Night	Day	Night
H_2	—	6.5×10^4	3.6×10^3	2.3×10^4
^4He	2×10^3	4×10^4	1.7×10^3	4.1×10^4
^{20}Ne	—	8×10^4	4.7×10^3	1.2×10^5
^{36}Ar	—	3×10^3 b	—	—
^{40}Ar	—	3.5×10^4 b	—	—
	—	8×10^3 a	—	—
O_2	—	$<2 \times 10^2$	—	—
CO_2	—	$<3 \times 10^3$	—	—
CO	—	$<3 \times 10^3$	—	—
Total (Night-time)				
Mass spectrometer		2×10^5		
CCGE		2×10^5		

a, sunset terminator; b, sunrise terminator.

Apollo 17 carried a miniature mass spectrometer to the Taurus Littrow site to identify the gas species in the lunar atmosphere and determine their concentrations. Because of the very tenuous nature of the lunar atmosphere, the instrument was designed [5] with an extremely high sensitivity, of the order of 100 molecules cm^{-3} for most gases, such as nitrogen and argon. It features an electron bombardment ion source, simultaneous scanning of three mass ranges (1—4, 12—48, and 27—110 amu), ion counting techniques, and a unique floating point data compression scheme which maintains 7-bit (1%) accuracy throughout a 21-bit (2×10^6) range of data counts. Calibration was done using a molecular beam system at NASA Langley Research Center [6].

Operation of the instrument began on 27 December 1972. Data reported herein have been taken during the first five lunations since deployment.

2. Results

Quick-look strip chart type records have been used for data reduction to date. An example of such a record is given in [5]. The accuracy of data from this type of record (not the final data form) is considered to be $\pm 30\%$, the smaller amplitude peaks having a factor of 2 uncertainty.

Because of the very high sensitivity of the instrument, daytime spectra are characterized by large amplitude peaks at nearly every mass number. These gases result from outgassing of the instrument, lunar module and landing site, and, except for helium, are all considered to be artifact at this time. Night-time spectra are generally quite clear of contaminant peaks but exhibit dominant peaks at 19, 35 and 37 amu. These are believed to be from fluorine and chlorine which are probably artifacts, but their origins are unknown. Hydrogenated halogen peaks occurring at masses 20, 36 and 38 are probably formed in the ion source by a reaction with hydrogen as their abundance is a strong function of ion source temperature.

In order to be assured that any given mass peak in the spectrum is not an artifact (produced by outgassing of the ion source electrodes) a cool-down test of the ion source is performed periodically. The normal night-time ion source temperature is 270 K; removing filament power for nearly an hour causes the ion source temperature to drop to approximately 195 K. Subsequent turn-on of the filament produces a relatively clean spectrum for the first spectrum scan: that is, peaks due to outgassing have decreased markedly, some to zero amplitude, while those gases not originating in the ion source produce essentially unchanged peak amplitudes.

There exists positive identification of at least three gases native to the lunar atmosphere, helium, neon and argon. In addition, recent ion source cool-down tests show hydrogen to exist in the molecular form at a concentration of 6.5×10^4 molecules cm^{-3}. While this may still be only an upper limit on lunar molecular hydrogen, since the mass 2 peak was dominated by hydrogen outgassing from the ion source early in the operation of the instrument, recent measurements have shown a stable peak amplitude which is only slightly depressed during the cool-down tests, suggesting that the ambient level is being sampled. The existence of hydrogen in molecular form is consistent with the result from the far UV spectrometer experiment flown on Apollo 17, which set a daytime upper limit for atomic hydrogen of 10 atoms cm^{-3} [7]. The results also showed the daytime upper limit for H_2 to be less than the instrument detection limit 6×10^3 molecules cm^{-3}. This value combined with the Monte Carlo model diurnal ratio of 6.5 [3] gives an inferred night-time upper limit of 4×10^4 cm^{-3}, which is close to the presently measured value. Using a solar wind flux of 3×10^8 molecules cm^{-2} s^{-1}, Hodges [3] obtained a daytime concentration of 3.6×10^3 and a night-time value of 2.3×10^4, a factor of nearly 3 lower than the observed value. Earlier predictions of the hydrogen concentration in the lunar atmosphere [1] assumed no recombination at the surface and thus an atmosphere of H. The existence of H_2 has not been predicted.

The mass 1 peak is dominated by dissociatively ionized (in the instrument ion source) hydrogenated molecules and no measurement of its ambient lunar concentration can be expected from the instrument.

Fig. 1 shows the helium concentration plotted as a function of solar zenith angle during the first five lunations after deployment of the instrument. The abscissa scale begins at the subsolar point and progresses through the lunation. Sunset, midnight and sunrise are identified. The scatter of the night-time data is indicative of the variations that have been measured [8] in the solar wind flux of helium. Residence time is of the order of 7.6×10^4 s [3] indicating a response time to fluctuations in the solar wind of a few tens of hours. The frequencies of the observed variations are consistent with this time scale.

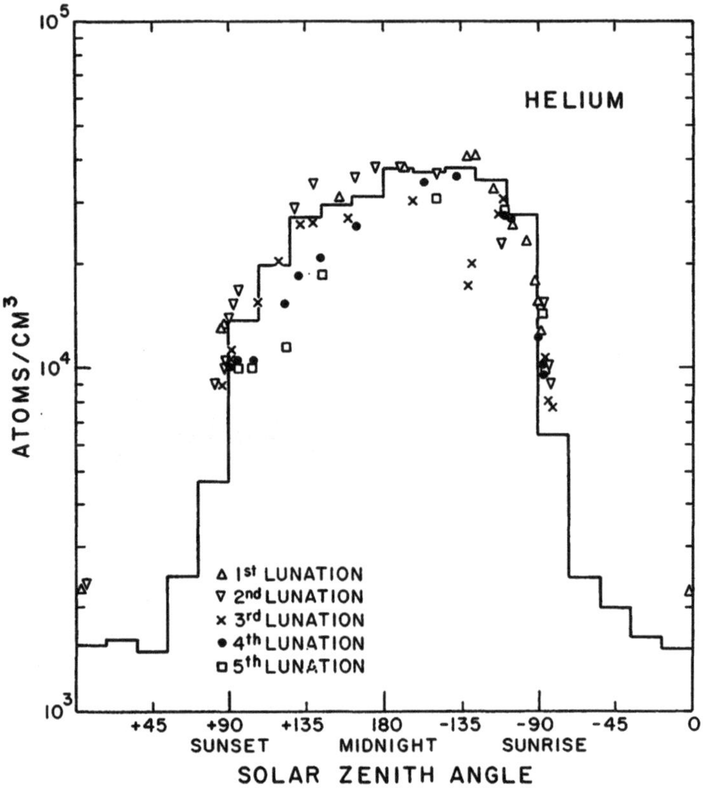

Fig. 1. Diurnal variation of ^{4}He concentration as a function of solar zenith angle starting at subsolar point. Data points are from the first five lunations. Solid curve is theoretical distribution based on a Monte Carlo calculation.

The solid curve in Fig. 1 is the helium distribution at 18° latitude from an updated Monte Carlo calculation based on a solar wind flux of 1.3×10^7 He ions $cm^{-2}\,s^{-1}$ observed by Geiss et al. [9]. The agreement with the data is seen to be excellent. The maximum calculated diurnal ratio is 24; the measured value is approximately 20. The asymmetry about the antisolar point results from the temperature decrease throughout the night and the rotation of the moon.

During ion source cool-down tests occurring in the fourth and fifth lunar nights, the level of the artifact peak (HF) at mass 20 has been reduced sufficiently to give valid neon measurements. The average ^{20}Ne concentration is 8×10^4 cm^{-3} which agrees favorably with the atmospheric model value of 1.2×10^5 cm^{-3} [10]

The observed isotopic ratio, $^{20}Ne/^{22}Ne$, is 14.2, close to the ratio of 13.7 from the solar wind composition experiment [9].

The diurnal distribution of ^{40}Ar during the third, fourth and fifth lunar nights is shown in Fig. 2. Coordinates are similar to those of Fig. 1. Daytime data are masked by hydrocarbon peaks from outgassing of the instrument and site. By

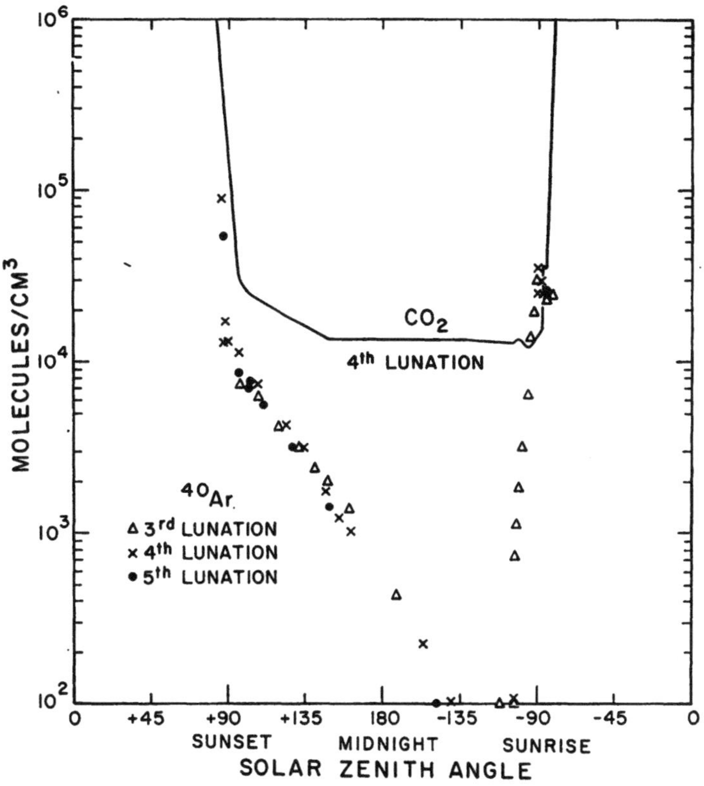

Fig. 2. ^{40}Ar and CO_2 diurnal variation. Coordinates similar to Fig. 1.

the time the sunset terminator (solar zenith angle of 90°) has crossed the landing site meridian (6 hours after the sun has dipped below the mountains to the west), the site has cooled sufficiently for contaminant peaks no longer to interfere with the ^{40}Ar measurement. The ^{40}Ar concentration at sunset is about 8×10^3 cm^{-3}. This steadily decreases through the night as the surface cools, reaching a value of 10^2 cm^{-3} (the instrument detection limit) at $-145°$ solar zenith angle. It appears that the argon is adsorbed on the cold night-time surface. At about 20° before sunrise, the ^{40}Ar concentration begins to increase, reaching a value of 3.5×10^4 at the terminator crossing of the site, a factor of nearly 5 larger than at sunset. Sunrise at the site is delayed 8 hours from the time the sun is at a zenith angle of 90° by shadowing from the mountains to the east, precluding significant local heating before this time. This behavior of argon matches closely that of a condensable gas as described above. The concentration of an artifact gas, CO_2, which does not exhibit a pre-dawn enhancement, is shown for comparison. After sunrise, rapid heating occurs and hydrocarbon peaks again dominate the spectrum.

A more detailed picture of the pre-dawn and sunrise conditions is given in Fig. 3. Gas concentrations are plotted as a function of solar zenith angle from $-110°$ to $-80°$. After terminator crossing there appears to be a slight decrease in ^{40}Ar concentration as though the local source were being somewhat depleted of argon as would be expected after the terminator had passed. Mass 36 also exhibits a pre-dawn enhancement, about 10% of that of ^{40}Ar. Because of the presence of

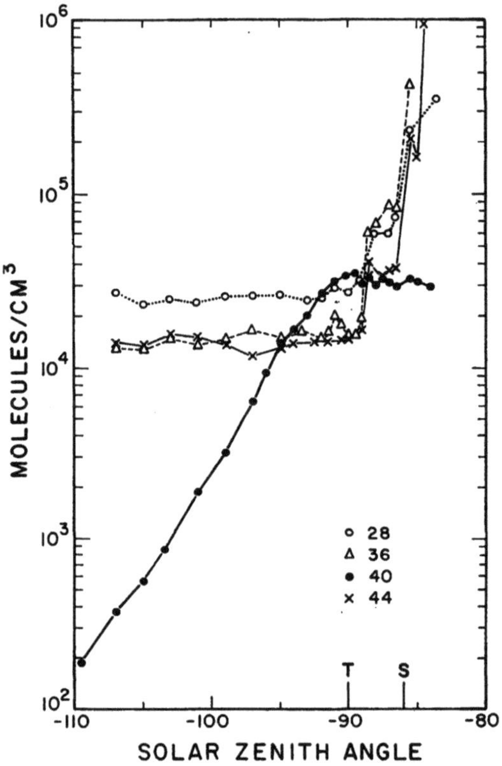

Fig. 3. Concentrations of masses 28, 36, 40 and 44 as a function of solar zenith angle from $-110°$ to $-80°$. T is time of terminator crossing landing site longitude. S is sunrise at landing site and the time of onset of local heating.

HCl, only the enhanced portion of the peak is believed to be ^{36}Ar, yielding a concentration of 3×10^3 cm^{-3} at the terminator. The ^{36}Ar source is the solar wind, whereas ^{40}Ar is believed to come from degassing of the regolith. From the near equality of the ^{36}Ar and excess ^{40}Ar trapped in the soil [11], the atmospheric ratio would be expected to be closer to unity. Perhaps the soil is not saturated with ^{36}Ar and hence most of the impinging solar wind flux of ^{36}Ar is permanently trapped.

Other gases shown in Fig. 3 (masses 28 and 44) do not appear to exhibit a pre-dawn enhancement. Mass 28, if it is N_2, would probably not be adsorbed at the night lunar temperature and would therefore not show a pre-dawn enhancement.

If it is CO, it would probably be adsorbed and would be expected to follow the argon pattern. Native CO_2 (mass 44) would also be expected to show a pre-dawn enhancement. The negative result places an upper limit on both native CO and CO_2 at the sunrise terminator of 3×10^3 cm^{-3}. Mass 32, O_2, also exhibits no pre-dawn enhancement, and several ion source cool-down tests have shown a zero amplitude peak for O_2. An upper limit is in the low 10^2 cm^{-3} range.

The total gas concentration measured near the antisolar point on the fourth lunation is 2×10^5 molecules cm^{-3} excluding the mass 19, 35 and 37 peaks. All other peaks in the spectrum are of significantly lower amplitude. This value is the same as that obtained by the Cold Cathode Gauge Experiment (CCGE) on Apollos 14 and 15 [1]. However, addition of the 19, 35 and 37 peaks (F and Cl) doubles the night-time gas concentration. It is not clear whether these gases were present at the Apollo 14 and 15 site and were being measured as part of the night-time gas concentration by the cold cathode gauges. Daytime comparisons are not meaningful at this time owing to the high outgassing rate still prevailing at the Apollo 17 site. CCGE daytime readings were not made until the ninth lunation.

3. Conclusion

The existence of three gases in the lunar atmosphere is confirmed by the Apollo 17 mass spectrometer. Helium and neon follow the expected behavior of non-condensable gases, and have concentrations close to those predicted by lunar atmospheric models considering the solar wind as their source. Argon behaves on the nightside as a condensable gas exhibiting a pre-dawn enhancement, with the $^{40}Ar/^{36}Ar$ ratio approximately 10. In addition, night-time molecular hydrogen measurements are within a factor of 3 of a Monte Carlo model prediction, and close to an upper limit inferred from application of the model's night-to-day ratio to the daytime value of the far UV spectrometer experiment. The existence of hydrogen in the molecular form has not been predicted.

The sum of all the known gases in the night-time lunar atmosphere (H_2, He, Ne, Ar) equals that measured by the cold cathode gauges at the Apollo 14 and 15 sites: 2×10^5 molecules cm^{-3}. Table 1 lists a summary of these gas concentrations. All other peaks in the spectrum are at present considered to be artifacts. As the degassing of the site continues other gas species may eventually be identified as native lunar gases.

Acknowledgments

This work was supported by NASA Contract NAS9-12074 and by Bendix Aerospace Subcontract SC-830 of NASA Contract NAS9-5829. Special thanks are due the crew of Apollo 17 who so successfully deployed the instrument on the moon.

References

[1] F. S. Johnson, J. M. Carroll and D. E. Evans, Proc. 3rd Lunar Sci. Conf., Geochim. et Cosmochim. Acta, Suppl. 3, **3**, 2231 (1972).
[2] R. R. Hodges and F. S. Johnson, J. Geophys. Res. **73**, 7307 (1968).
[3] R. R. Hodges, Jr, J. Geophys. Res. **78**, 8055 (1973).

[4] R. H. MANKA and F. C. MICHEL, Proc. 2nd Lunar Sci. Conf., Geochim. et Cosmochim. Acta Suppl. 2, **2**, 1717 (1971).
[5] J. H. HOFFMAN, R. R. HODGES, JR., F. S. JOHNSON and D. E. EVANS, Apollo 17 Prelim. Sci. Rep., NASA Rep., SP-330 (1973).
[6] J. H. HOFFMAN, R. R. HODGES, JR., F. S. JOHNSON and D. E. EVANS, Apollo 15 Prelim. Sci. Rep., NASA Rep. SP-289 (1972).
[7] W. G. FASTIE, private communication.
[8] A. J. HUNDHAUSEN, S. J. BAME, J. R. ASBRIDGE and S. J. SYDORIAK, J. Geophys. Res. **75**, 4643 (1970).
[9] J. GEISS, F. BUEHLER, H. CERUTTI, P. EBERHARDT and C. FILLEUX, Apollo 16 Prelim. Sci. Rep., NASA Rep. SP-315 (1972).
[10] R. R. HODGES, JR., J. H. HOFFMAN, F. S. JOHNSON and D. E. EVANS, Proc. 4th Lunar Sci. Conf., Geochim. et Cosmochim. Acta, Suppl. 4, **3**, 2855 (1973).
[11] A. YANIV and D. HEYMANN, Proc. 3rd Lunar Sci. Conf., Geochim. et Cosmochim. Acta, Suppl. 3, **2**, 1967 (1972).

Space Research XIV — Akademie-Verlag, Berlin 1974

MAIN DIFFERENCES BETWEEN LUNAR REGOLITH OF MARE AND HIGHLAND ORIGIN ACCORDING TO DATA OF MÖSSBAUER SPECTROSCOPY

T. V. Malysheva

V. I. Vernadsky Institute of Geochemistry and Analytical Chemistry, USSR Academy of Sciences, Moscow, USSR

Comparison of regolith samples returned by the automatic stations Luna 16 and Luna 20 has shown that regolith from the highland region contains ~1.85 times less iron than regolith from the Sea of Fertility; metallic iron contains more nickel and less of the finely dispersed component; about 5% of the iron is in the trivalent state; there is more olivine and it is less ferruginous; the pyroxene component contains less glasses and is more homogeneous; instead of ilmenite ulvospinel appears. It follows from the comparison given that the minerals contained in regolith returned by the automatic station Luna 20 from the highland region of the moon had been crystallized from the less reduced magma than the minerals of regolith returned by the automatic station Luna 16 from the Sea of Fertility.

The automactic station Luna 20 landing place is a typical highland region [1] which is located 120 km to the north of the Luna 16 landing place in the region of Mare Fecunditatis [2] and is above it by approximately 1 km. It is interesting to carry out a comparative Mössbauer study of regolith from two surfaces of so different lunar regions. The measurements were carried out with a $^{57}Co(Cr)$ source, of 50 mC activity and a Mössbauer spectrometer of the type described in [3] at both room temperature and liquid nitrogen temperature. Regolith fractions of grain sizes 0.250—0.400; 0.127—0.250; 0.083—0.127 and $\leqq$ 0.083 mm were measured. The resulting data were analysed by means of electronic computers BESM-3m and BESM-4. The isomer shifts were determined with reference to stainless steel.

The total iron content in the regolith samples returned by the automatic station Luna 20 is found to be less than in the regolith samples of Luna 16. Fig. 1 shows the Mössbauer spectra of the two types of regolith fractions with grain size $\leqq$ 0.083 mm, which were measured under identical conditions. In the spectra of the finest regolith fractions the peaks of metallic iron become most distinctly apparent so that its nature can be determined with greater precision.

Three peaks of six of the Zeeman magnetic hyperfine splitting belonging to metallic iron are seen in Fig. 1. The high-velocity peaks of the pure metallic iron superfine splitting are also shown for comparison. The numbers of analyser channels for the extreme peaks of Luna 16 metallic iron (112:9 $\pm$ 0.17 and 450.48 $\pm$ 0.18) coincide well with the numbers of channels for pure metallic iron peaks (112.48 $\pm$ 0.14 and 449.96 $\pm$ 0.15) and differ from the values for the Luna 20 (115.50 $\pm$ 0.36 and 447.24 $\pm$ 0.85). This suggest that the metallic iron in the Luna 20 regolith is present in an alloy with nickel [4]. The peaks of lunar metallic iron are broadened in comparison with peaks for pure metallic iron, this broadening being greater for the Luna 20 regolith. The widths of low-velocity peaks are (0.317 ± 0.02) mm s^{-1}; (0.417 ± 0.03) mm s^{-1} and (0.65 ± 0.08) mm s^{-1} for pure metallic iron and the Luna 16 and Luna 20 samples respectively,

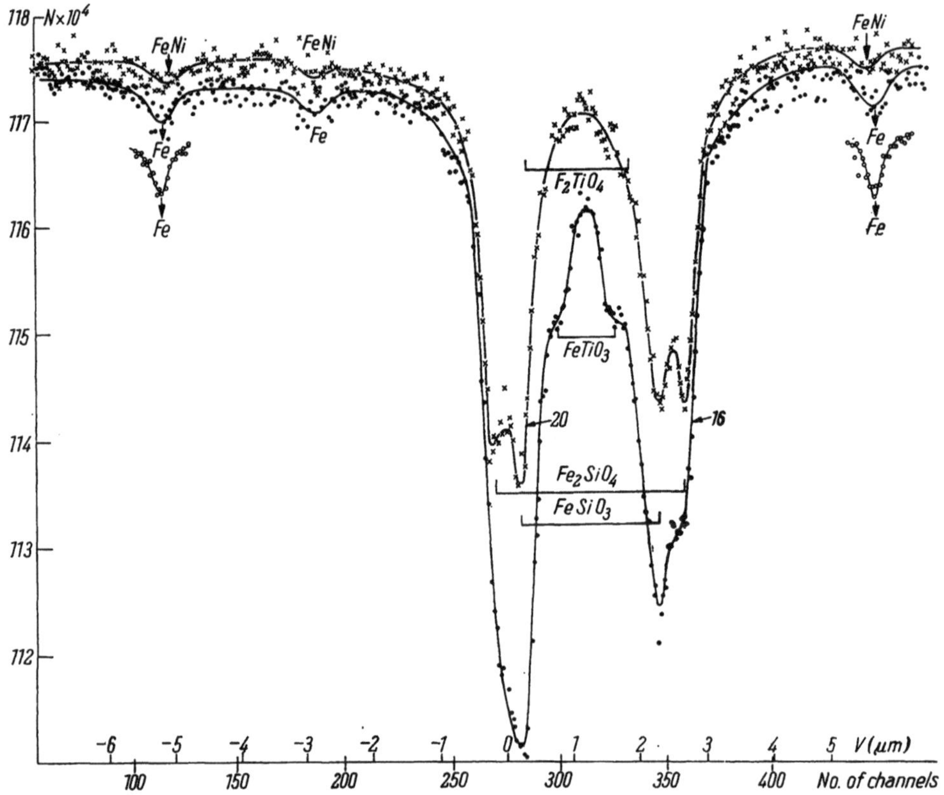

Fig. 1. Mössbauer spectrum of small regolith fractions (≦0.085 mm), returned by the automatic stations Luna 16 and Luna 20.

which once more confirms the presence of a noticeable amount of nickel in the alloy with metallic iron in the Luna 20 regolith.

In order to determine the amount of finely dispersed metallic iron in the regolith samples we have carried out measurements of small fractions (≦ 0.083 mm) of Luna 16 and Luna 20 regolith at liquid nitrogen temperatures (~80 °K). According to Housley et al. [5] the transition from the ferromagnetic state, which is characterized in Mössbauer spectra by six lines of the Zeeman hyperfine splitting, to the superparamagnetic state giving only a quadrupole doublet for pure iron occurs at 285 °K for 134 Å diameter particles and at 77 °K for particles of 85 Å diameter. Hence, in the case of 85—134 Å metallic iron particles the area of the magnetic-hyperfine-splitting peaks at liquid nitrogen temperature is larger than at room temperature.

As a result of measurements it has been found that the area of the metallic iron peaks of the Luna 16 regolith had increased ~ 2 times and the area of the metallic iron peaks of the Luna 20 regolith had only slightly changed (see Table 3). This proves that there is much more finely dispersed iron in the regolith of Mare Fecunditatis than in the highland regolith.

From the comparison of room temperature Mössbauer spectral-peak areas for Luna 16 and Luna 20 regolith we find that the total amount of iron in ferruginous minerals is ~1.85 times less in the Luna 20 regolith. It is clearly seen from Fig. 1 that the structure the spectrum due to ilmenite $FeTiO_3$ in the Luna 16 regolith is not present in the spectrum of Luna 20. This fact is more notable as in the lunar samples of Apollo 11, 12, 14, 15 the ilmenite makes up from 5 to 20% [6]. Better identification of the high-velocity Luna 20 peak can be seen in Fig. 2. By ana-

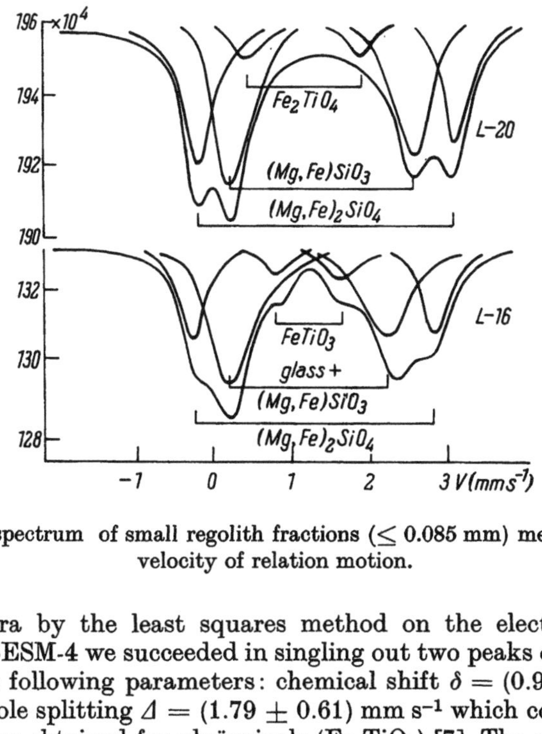

Fig. 2. Mössbauer spectrum of small regolith fractions ($\leq$ 0.085 mm) measured at a smaller velocity of relation motion.

lysing the spectra by the least squares method on the electronic computers BESM-3m and BESM-4 we succeeded in singling out two peaks of the quadrupole doublet with the following parameters: chemical shift $\delta = (0.96 \pm 0.08)$ mm s^{-1} and the quadrupole splitting $\Delta = (1.79 \pm 0.61)$ mm s^{-1} which coincides well with the δ and Δ values obtained for ulvöspinels (Fe_2TiO_4) [7]. The great width of the line $\Gamma = (0.78 \pm 0.03)$ mm s^{-1} indicates that we obviously are dealing with solid solutions of the Fe_2TiO_4—$FeCr_2O_4$ type but not with pure ulvöspinel. In larger grain size regolith fractions we did not succeed in identifying ulvöspinel.

As in the Luna 16 regolith, most of the iron in the Luna 20 samples is in the form of silicates. Peaks of the iron in pyroxene component have, on the average, a high Δ value (~ 2.13 mm s^{-1} instead of ~ 2.05 mm s^{-1} for Luna 16), more ionic bonds (δ = ~ 1.26 mm s^{-1} instead of ~ 1.24 mm s^{-1} for Luna 16) and narrower lines (Γ ~ 0.45 mm s^{-1} and ~ 0.50 mm s^{-1} instead of ~ 0.59 mm s^{-1} and 0.70 mm s^{-1} for the left and right peaks respectively) (Table 1). This indicates that there are less glasses in the pyroxene of the Luna 20 regolith than in Luna 16 regolith.

Olivine peaks are more pronounced in the Luna 20 regolith, which suggests a greater amount of olivine, which is characterized by a slightly greater value of the

Table 1

Parameters of Olivine Mössbauer Spectra from Luna 16 and Luna 20 Lunar Regolith

Grain size	δ (mm s^{-1})		Δ (mm s^{-1})		Γ (mm s^{-1})			
					Luna 16		Luna 20	
	L-16	L-20	L-16	L-20	Low-velocity peak	High-velocity peak	Low-velocity peak	High-velocity peak
0.200 – 0.450	1.27 ± 0.02	1.29 ± 0.01	2.94 ± 0.01	3.00 ± 0.005	0.310 ± 0.003	0.422 ± 0.008	0.290 ± 0.005	0.360 ± 0.005
0.127 – 0.200	1.26 ± 0.02	1.275 ± 0.002	2.92 ± 0.01	2.998 ± 0.002	0.364 ± 0.005	0.426 ± 0.005	0.336 ± 0.003	0.360 ± 0.004
0.083 – 0.127	1.26 ± 0.02	1.275 ± 0.002	2.90 ± 0.01	2.984 ± 0.004	0.327 ± 0.002	0,417 ± 0.004	0.325 ± 0.002	0.357 ± 0.002
≦0.083	1.25 ± 0.005	1.26 ± 0.005	2.88 ± 0.005	2.93 ± 0.01	0.340 ± 0.005	0.396 ± 0.005	0.335 ± 0.002	0.360 ± 0.002

δ is the chemical shift; Δ is the quadrupole splitting; Γ is the line width

Table 2

Parameters of Mössbauer Spectra of Luna 16 and Luna 20 Lunar Regolith Pyroxene Component

Grain size	δ (mm s^{-1})		Δ (mm s^{-1})		Γ (mm s^{-1})			
					Luna 16		Luna 20	
	L-16	L-20	L-16	L-20	Low-velocity peak	High-velocity peak	Low-velocity peak	High-velocity peak
0.200–0.450	1.21 ± 0.03	1.28 ± 0.005	2.06 ± 0.04	2.130 ± 0.005	0.61 ± 0.01	0.83 ± 0.01	0.47 ± 0.005	0.510 ± 0.005
0.127–0.200	1.24 ± 0.01	1.257 ± 0.002	2.05 ± 0.01	2.126 ± 0.002	0.61 ± 0.01	0.82 ± 0.01	0.440 ± 0.004	0.495 ± 0.006
0.083–0.127	1.24 ± 0.01	1.263 ± 0.002	2.05 ± 0.01	2.140 ± 0.001	0.59 ± 0.01	0.68 ± 0.01	0.442 ± 0.002	0.495 ± 0.005
≦0.83	1.24 ± 0.004	1.25 ± 0.01	2.03 ± 0.003	2.100 ± 0.002	0.580 ± 0.003	0.601 ± 0.008	0.445 ± 0.003	0.503 ± 0.006

Table 3

Iron Distribution over the Main Mineral Phases in Regolith returned by the Automatic Stations Luna 16 and Luna 20 as percentage of total area of the Mössbauer Spectrum

Mineral / Grain size	Luna 16				Luna 20			
	0.250—0.400	0.127—0.200	0.083—0.127	$\leqq 0.083$	0.250—0.400	0.127—0.200	0.083—0.127	$\leqq 0.083$
Olivine	15.8	16.2	22.6	22.0	38.0	35.6	34.8	36.0
Pyroxene + glass	79.3	75.8	71.0	55.0	60.07	64.4	65.5	56.8
Ilmenite	4.9	4.9	5.8	5.72	$\leqq 1$	$\leqq 1$	$\leqq 1$	$\leqq 1$
Ilvöspinel	$\leqq 1$	$\leqq 1$	6.25	2.1	$\leqq 1$	$\leqq 1$	$\leqq 1$	6.6
Metallic iron 300°K	3.5	3.3	3.1	5.85	$\leqq 1$	$\leqq 2$	4.5	5.65
80°K	—	—	—	10.7	—	—	—	6.12
Troilite	$\leqq 1$	$\leqq 1$	2.8	2.0	$\leqq 1$	$\leqq 1$	$\leqq 1$	$\leqq 1$

chemical shift δ, a greater value of the quadrupole splitting Δ and narrower lines in comparison with Luna 16 olivine (Table 2). Δ decreases with decrease of grain size from 3.00 ± 0.005 mm s^{-1} to 2.93 ± 0.01 mm s^{-1}, repeating the tendency noted for olivine from the Luna 16 [8] regolith. The greater Δ proves the lower ferruginity of olivine from Luna 20 regolith, which agrees well with the general lower ferruginity of highland regolith.

The relative iron distribution among mineral phases determined from the area of corresponding superfine splitting peaks is listed in Table 3. The assumption has been made that the values of resonance absorption are equal for all minerals. Data from the Luna 16 regolith with grain size 0.250—0.400, 0.127—0.250 and 0.083—0.127 are taken from [3]. From the comparison of iron distribution data among mineral phases for regolith of mare (Luna 16) and highland (Luna 20) origin it is seen that the types of magma from which the minerals had crystallized are different.

It is interesting to consider the data in the light of regularities reported in [9] on the study of conditions of native iron formation in basalts surrounding the tree molds of the Fujiyama volcano in Japan. The following composition changes of minerals being crystallized from magma corresponding to the increase of its reducing properties have been obtained: titanomagnetite, titanomagnetite + ilmenite, ilmenite, ilmenite + native iron. Olivines, being present everywhere, become more ferruginous as the reducing properties of magma increase. With the augmentation of iron entry into silicate phases, the Ti/Fe ratio in the main basalt mass increases and ilmenite becomes the predominant associated mineral.

In the light of the above the Mössbauer data on iron distribution over mineral phases in regolith from lunar maria and highlands (Table 3) indicate that the minerals of lunar maria have been crystallized from a more reducing magma than minerals of lunar highlands. In the regolith of Luna 16 a more ferruginous olivine, ilmenite and a greater amount of metallic iron are present, while in the regolith of Luna 20 the olivine is less ferruginous, instead of ilmenite ulvöspinel appears, and there is less metallic iron.

Acknowledgments

The author expresser her thanks to the head engineer V. V. Kurash for his aid during the conduct of the experiment.

References

[1] A. P. Vinogradov, Geokhimiya 3, 8 (1971).
[2] A. P. Vinogradov, Geokhimiya 7, 763 (1972).
[3] T. V. Malysheva and V. V. Kurash, Space Research XII, 137 (1972).
[4] C. Johnston, M. Ridout, T. Cranshaw and P. Madsen, in: The Mössbauer Effect, ed. Foreign Liter., Moscow 1962 (p. 367).
[5] R. M. Housley, R. W. Grant and M. Abdel-Gawal, Proc. Third Lunar Sci. Conf., 1, 1065 (1972).
[6] K. Thiel, F. W. Neugebauer, M. Heimann and W. Herr, in: Lunar Science III, Lunar Science Institute Contr. No. 88, 1972 (p. 747) (abstract).
[7] K. Ono, L. Chandler and A. Ito, J. Phys. Soc. Japan, 25, 174 (1968).
[8] T. V. Malysheva, Proc. Third Lunar Sci. Conf., 1, 105 (1972).
[9] K. Kanehira and J. Shimazaki, Neues Jahrb. Mineral. Monatch., No. 3, 124 (1971).

Space Research XIV — Akademie-Verlag, Berlin 1974

THE OPTICAL PARAMETERS OF MARE AND HIGHLAND LUNAR SOILS

I. I. Antipova-Karataeva, M. V. Akhmanova, B. V. Dementyev, M. N. Markov, Yu. I. Stakheyev and L. S. Tarasov

USSR

The diffuse-reflectance spectra (albedo) and variation of intensity with phase angle of mare and highland lunar soils which have been brought back by the Luna 16 and Luna 20 stations and Apollo 15 mission were investigated in the UV, visible and IR regions. The albedo of all specimens was found to increase continuously from blue to red and to have the maximum in the IR region (the reflectance coefficient was up to 20—30% for mare soils, and up to 40% for highland soils). The albedo of highland fines was higher than that of mare fines, and that difference was smaller in the UV than in the IR region. The albedo values of all specimens were rather low in the UV region (6—12%). The backscattering effect was found in all specimens in the spectral region investigated. The scattering coefficients of soils depend on landing site and on spectral region.

The diffuse-reflectance spectra and variations of intensity with phase angle of mare and highland fines were obtained. The samples which had been brought back by the automatic stations Luna 16 and Luna 20 and the Apollo 15 mission were investigated. All optical parameters were measured under near-natural conditions. Such data are the most interesting for astronomers as they can be compared with telescopic investigations of the moon and other objects of the solar system, especially planets and asteroids.

The specimens were prepared by free pouring of powder into the cell without smoothing its surface; the bulk density of samples was 1.2—1.8 g cm^{-3}. The fine samples from the surface area of Luna 16 and Luna 20 core (four fractions: <83 μm, >83 μm, >127 μm and >200 μm) and from Apollo 15 core at different depths were investigated. The diffuse-reflectance spectra allow us to obtain information about quite a number of the characteristics of the samples: presence of metal ions (Ti^{3+}, Fe^{2+} etc.), H_2O, OH^- groups, and surface properties of lunar particles. Such information is obtained by investigation in a wide spectral region. The surface properties of fines were investigated by measurement of the variation of the reflected light intensity with phase angle.

The measurements in the UV and visible regions were carried out relative to a MgO-smoke standard by using an EPS-3T spectrophotometer with an integrating sphere attachment. For the measurements in the IR region the spectral equipment described in [1, 2] was used. Diffuse-reflectance spectra of all specimens were similar and had no evident structure. The reflectance increased slowly from blue to red and to the IR region and had a maximum at 4—5 μm. Complete absorption was observed in the region between 8 and 14 μm. Let us consider the spectra in more detail (Fig. 1).

The diffuse-reflectance spectra of measured samples between 0.23 and 1.8 μm had a marked difference in absolute value of the reflectance coefficient, but were

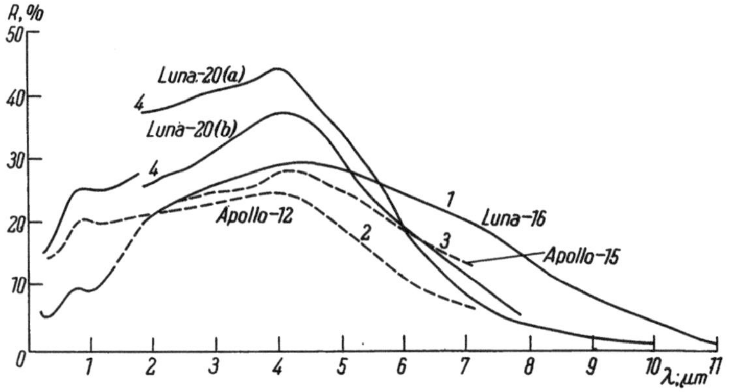

Fig. 1. The diffuse-reflectance spectra of lunar soils samples in the region 0.3—14 μm. 1, Luna 16, No. 1603; 2, Apollo 12, No. 12028; 3, Apollo 15, No. 15001,113; 4, Luna 20, non-fractionated sample, (a) smoothed surface, (b) rough surface.

similar in that the reflectance coefficient increased slowly with wavelength, there was a broad weak absorption band at 1 μm and a minute increase in reflectance Luna 20 fines fractions depended markedly on particle size: the smaller the mean diameter the larger the reflectance coefficient (curves 6—9). The spectrum of a in the UV region 0.23—0.3 μm. The spectra of different Luna 16 fines fractions from the same surface layer of the core were very similar, and the reflectance coefficient of this layer was the lowest (Fig. 2, curves 1—4). The spectra of different non-fractionated sample nearly coincides with the spectrum of the finest fractions

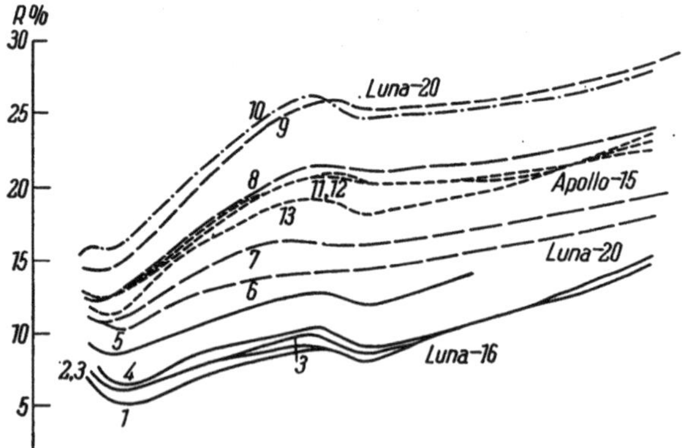

Fig. 2. The diffuse-reflectance spectra of lunar soils samples in the region 0.23—1.8 μm. 1—4, Luna 16, No. 1603-2: 1, > 200 μm; 2, > 127 μm; 3, > 83 μm; 4, < 83 μm; 5, No. 1609-1a; 6—9, Luna 20, No. 2002: 6, > 200 μm; 7, > 127 μm; 8, > 83 μm; 9, < 83 μm; 10, non-fractionated sample; 11—13, Apollo 15: 11, No. 15001,113-1 (h = 220 cm); 12, No. 15005,281 (h = 80 cm); 13, No. 15002,131 (h = 160 cm).

(curve 10). The data obtained indicated that the diffuse-reflectance spectra in the UV and visible regions probably depended not only on the chemical composition of the finest particles, but also on their surface properties. The low reflectance coefficient and the absence of the spectral dependence on particle size is probably connected with the properties of mare soils described in the literature: with quite a number of particles covered with a film of glass, sintered and slaggy particles with a rough light-absorbing surface.

The small spectral variations in the 0.5—1 μm region were probably connected with the difference of chemical and mineralogical composition of different fractions. Unlike Luna 16 soils, the highland Luna 20 soils consist mainly of crystalline material, and the reflectance spectra of these soils is determined essentially by scattering on crystalline particles. The effect of scattering depends on particle size. The great similarity of the parent and the finest-fraction spectra indicated that the concentration of small particles in Luna 20 soils was relatively large. This conclusion was in agreement with the results of granulometric analysis [3]. Probably the optical properties of small particles mainly determined the sample reflectance spectrum. It should be noted that diffuse-reflectance spectra of Luna 16 fine fraction (> 83 μm) from zone B ($h = 20—25$ cm) had a little greater intensity than that for zone A (surface zone); this difference was due to depth variations of chemical composition and concentration of scoriaceous particles (curve 5).

The diffuse-reflectance spectra of Apollo 15 core samples which had been taken from different depths were similar to each other. The reflectance coefficient is a little greater than that of Apollo 15 surface soils [4] and falls within the range of the different fractions of Luna 20 sample.

Examination of the data shows that diffuse-reflectance spectra (albedo) depended on both chemical composition and size and surface properties of particles. The optical characteristics of a non-fractionated sample in the region 0.23—1.8 μm were the most similar to that of the finest-fraction sample. Probably the optical characteristics of only these samples could be compared with data from earth-based telescopic measurements. The position of the Fe^{2+} band in the 1 μm region characterized the mean pyroxenes composition in the samples and did not depend on physical properties of the surface of the reflecting particles.

The IR spectra (2—14 μm) of a non-fractionated sample from the surface zone of Luna 20 core and of samples from an Apollo 15 core were investigated. The data obtained were compared with data from Luna 16, Apollo 11 and Apollo 12 samples. IR reflectance spectra of mare samples were observed to be similar to each other [1]: the reflectance increased continuously in the near IR region, there was a maximum at 4.5 μm with a value of 30% for the reflection coefficient. At 10—14 μm the reflection gradually falls practically to zero. The reflectance spectra of the Apollo 15 samples had a similar character with the maximal reflection ~30% at λ equal to 4.5 μm. The reflectance spectra of highland soils (Luna 20) had a somewhat different character: the maximum was at 4 μm with a greater reflection coefficient value, 40%; the reflection falls steeply in the long wavelength region.

The origin of the reflectance maximum in the region 3—5 μm was analysed in [2]. There it was assumed that the maximum corresponded to the most transparent region of the absorption spectra of the soil. The reflection decrease in the short wavelength region was connected with an increase of absorption of transitive metal ion impurities, with specific structure and with surface properties of soil particles. The reflection decrease in the long wavelength region was connected with absorption inherent for silicate matter Si—O bonds. The greater reflectance of highland soils was connected with a lower concentration of colouring metal ions and probably with the lower number of surface defects. It is important to note the complete absence of H_2O and OH^--group absorption bands in the spectra of both highland and mare samples. The lunar soils investigated not only did not contain water, but also lacked the power to adsorb it. In this property lunar soils are greatly different from earth rocks, which as a rule contain water and are able to adsorb it.

The absolute value of the reflectance coefficient of the measured samples depended on experimental conditions, in particular, on the degree of smoothing of the sample surface. The difference in reported data could be accounted for by the fact that samples were not identical [4].

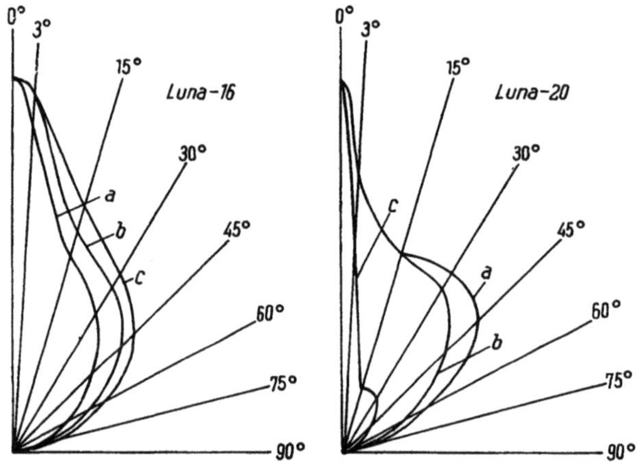

Fig. 3. Scattering coefficient of regolith at different angles of observation and at normal angle of incidence for Luna 16 (No. 1603-2) and Luna 20 (non-fractionated sample): a, $\lambda = 2.2\ \mu m$; b, $\lambda = 4.5\ \mu m$; c, $\lambda = 7.2\ \mu m$.

The scattering coefficient of the regolith was measured with a view to investigating the effect of surface properties on optical parameters. The measurements were made in the IR region because the IR scattering coefficient was the most sensitive to the granulometric composition and to surface properties of the reflecting layer. The value of the scattering coefficient for different angles of viewing and for an angle of illumination equal to 0° at three wavelengths is shown in Fig. 3. It had previously been demonstrated that the light-scattering character of the surface layer of lunar mare samples was different from that of terrestrial rocks; the backscattering effect was much greater in lunar samples. This effect was even greater in highland samples, as can be seen from Fig. 3. In addition the scattering coefficient depended differently on wavelength in Luna 16 and Luna 20 samples. The backscattering effect was most marked in the long wavelength range for Luna 20 samples. Probably this difference was connected with the greater value of absorption coefficient of Luna 20 soils in the region $\lambda \sim 4\ \mu m$. With this provision the "metallic" character of reflection could arise and consequently would be an additional contribution to backscattering at angles of illumination close to angles of observation.

References

[1] M. V. Akhmanova et al., Space Research XII, 87 (1972).
[2] M. V. Akhmanova et al., Kosm. Issled., **10**, 423 (1972).
[3] A. P. Vinogradov, Geochimia No. 3, 261 (1971).
[4] J. B. Adams et al., Geochim. Cosmochim. Acta **37**, 731 (1973).

Space Research XIV — Akademie-Verlag, Berlin 1974

COMPARISON OF APOLLO AND EARTH-BASED OBSERVATIONS: DEFINITION OF SURFACE UNITS IN MARE SERENITATIS

R. W. Shorthill[a]*, T. W. Thompson[b], E. A. Whitaker[c] and S. H. Zisk[d]

[a] Boeing Aerospace Company, Seattle, Washington, USA
[b] Jet Propulsion Laboratory, Pasadena, Calif., USA
[c] University of Arizona, Lunar and Planetary Laboratory, Tucson, Arizona, USA
[d] M. I. T. Haystack Observatory, Westford, Mass., USA

Surface units in Mare Serenitatis and the Montes Haemus region can be identified by remote observations. The earth-based visual, infrared and radar measurements form the data base for this study. An improved surface description can be obtained when the Apollo-orbital measurements are employed.

Observations of Mare Serenitatis in the radar infrared and optical wavelengths have been made with similar resolutions of 1 to 15 km. Physical properties of the lunar surface are investigated by comparing these earth-based measurements [1]. The Apollo orbital experiments such as the gamma-ray spectrometer, X-ray fluorescence, infrared mapper, lunar sounder and pan and metric photography can be compared with the earth-based data set to form a broader basis for a synthesis of lunar data.

A comparison of the earth-based measurements alone reveal that at least five units are to be found on the surface of Mare Serenitatis. Patches of the Montes Haemus highlands, southwest of Mare Serenitatis, are covered with recent mantling. While these conclusions are principally due to the color differences, they are corroborated by the radar and infrared (eclipse) mappings [2]. The dark ring which appears in the white-light optical photographs is comprised of several units. It should be noted that these remote observation data identify some units that are not revealed on the optical photographs. The earth-based data set for these comparisons are the color differences reported by Whitaker [3], the eclipse measurements by Shorthill and Saari [4] at 11 nm, the bistatic radar measurements of Tyler and Howard [5]; and the 70 cm and 4 cm radar backscatter measurements of Thompson and Zisk [6]. Unit I has the strongest blue color difference and lowest radar backscatter. There are nine of these patches indicated by the darkest cross hatching in Fig. 1. A large region in the northwestern part of the Mare surface labeled Unit II is 250 km in diameter. This area has a weaker color difference and a weaker radar backscatter than Unit I. Unit II is not an obvious feature in the "white-light" photographs of Mare Serenitatis. Unit III includes the patches occurring along the northeastern, southwestern ring of Mare Serenitatis and a circular patch south of Unit II approximately 100 km in diameter. This unit has the weakest color difference and radar backscatter. The central region of Serenitatis is denoted as IV, which in color difference is the reddest and

* Now at University of Utah Research Institute, Salt Lake City, Utah, USA.

has the strongest radar backscatter. Sulpicius Gallus Formation, Unit V, has the lowest albedo, has the same color difference as IV, a moderate radar backscatter and is the coolest from the infrared measurements. The infrared response of the Units I, II, III and IV indicates a very slight thermal enhancement.

Thus, based primarily on color differences, radar, and infrared we conclude that the surface of Mare Serenitatis is comprised of at least five units. The color

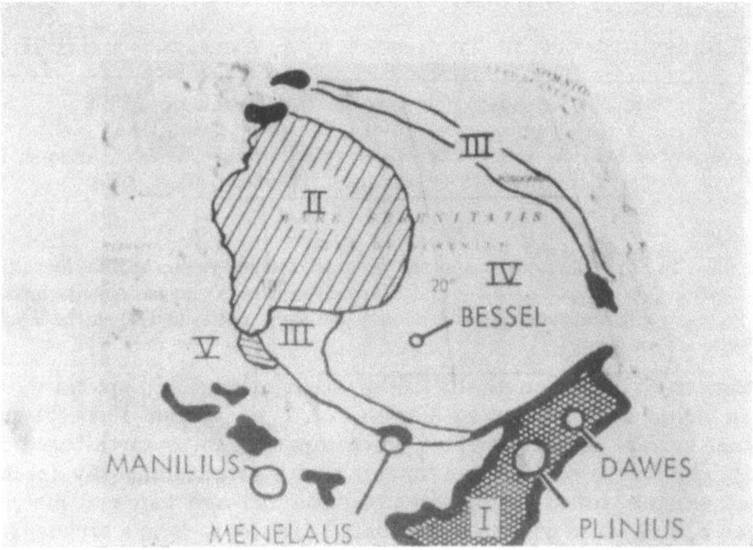

Fig. 1. Sketch map of units and features described in the text. The background in Mare Serenitatis and its environs from the LEM-1 photomosaic (ACIC, USAF, St. Louis, Mo, USA).

differences which vividly separate these units may well be indicative of petrological differences in the surface material. One cannot, however, exclude the possibility that these differences are primarily due to different ages of the units. In support of this interpretation, it should be noted that the Apollo bistatic radar data give differences in slope distributions between these units. For example, the Apollo 15 data show the average slope of the surface changes between the average mare surface and Unit III. The radar topography measurements of Zisk [7] indicate an elevation difference or a change in slope coincident with the boundaries of several of these units. The topography data suggests age differences in that they would form physical boundaries for the emplacement of more recent surface materials. This material would undoubtedly have a slope distribution different from that of the pre-existing surface. A more detailed description of these observations and correlations with other Apollo orbital measurements (i.e. X-ray fluorescence) appear in the Apollo 17 Preliminary Science Report [8].

References

[1] T. W. Thompson, H. Masursky, R. W. Shorthill, S. H. Zisk and G. L. Tyler, Contrib. No. 16, Lunar Science Institute, Houston, Texas, 1970.

[2] T. W. Thompson, R. W. Shorthill, E. A. Whitaker and S. H. Zisk, The Moon, in press (1974).

[3] E. A. Whitaker, The Moon **4**, 348 (1972).
[4] R. W. Shorthill and J. M. Saari, Science **150**, 210 (1965).
[5] G. L. Tyler and H. T. Howard, J. Geophys. Res. **78**, 4852 (1973).
[6] T. W. Thompson and S. H. Zisk, in: Thermal Characteristics of the Moon, M. I. T. Press, Cambridge, Mass. 1972.
[7] S. H. Zisk, The Moon, in press (1974).
[8] T. W. Thompson, K. A. Howard, R. W. Shorthill, G. L. Tyler, S. H. Zisk, E. A. Whitaker, G. G. Schaber and H. J. Moore, Apollo 17 Preliminary Science Report NASA SP-330 (1973).

Space Research XIV — Akademie-Verlag, Berlin 1974

APOLLO 15 AND 16 SUBSATELLITE MAGNETOMETER MEASUREMENTS OF THE LUNAR MAGNETIC FIELD

C. T. Russell[a], P. J. Coleman Jr[a,b], B. R. Lichtenstein[b], G. Schubert[b] and L. R. Sharp[b]

[a] Institute of Geophysics and Planetary Physics, University of California, Los Angeles, Calif., USA
[b] Department of Planetary and Space Science, University of California, Los Angeles, Calif., USA

Measurements with the UCLA fluxgate magnetometers on the Apollo 15 and 16 subsatellites have been used to map the lunar magnetic field. The strongest magnetic fields found to date occur between 165° and 175° east longitude at about 20° south latitude near the crater Van de Graaff. The measurements place an upper limit of $3 \times 10^{18}\ \Gamma\ cm^3$ on the dipole moment in the plane of the subsatellite orbit. Assuming that the total dipole moment lies parallel to the lunar rotation axis, an upper limit of $7 \times 10^{18}\ \Gamma\ cm^3$ results. This is 8×10^{-8} of the earth's dipole moment.

1. Introduction

Lunar magnetic investigations, initiated on the Luna spacecraft, were extended significantly with the Explorer 35 lunar orbiter, and have been continued during the Apollo program both on the surface and in orbit. These measurements have been recently reviewed [1], and we will only briefly summarize previous findings here. During the lunar month the moon is immersed in three distinctly different plasma environments. For about five days around full moon the moon is enveloped in the rather steady field of the geomagnetic tail and shielded from the solar wind flowing from the sun. Mapping of the lunar magnetic field is carried out at this time. Orbital studies reveal a lunar field of up to about 1γ at 100 km [2]. Surface measurements reveal fields of up to 300γ [3]. For about three days on either side of the geomagnetic tail, the moon is immersed in the shocked magnetosheath plasma, a turbulent region with large amplitude fluctuations. These fluctuations are useful in lunar sounding studies from which the electrical conductivity of the lunar interior is inferred [4]. During the rest of the lunar month, the moon is in the comparatively steady solar wind flow. The solar wind, for the most part, directly impacts the lunar surface on the sunward side of the moon, leaving a void behind the moon into which the solar wind expands with increasing distance downstream. When certain lunar regions appear at or near the lunar limbs, compressional disturbances are generated in the solar wind above the terminator. These limb compressions appear to be the result of the deflection of part of the solar wind by the lunar magnetic field [5].

In this paper we discuss first some recent results on mapping the lunar magnetic field using the Apollo 15 and 16 subsatellite magnetometers. Then, we use these measurements to deduce an upper limit on the lunar magnetic dipole and discuss the implication of these results.

2. Instrumentation

The Apollo 15 and 16 particles and fields subsatellites are essentially identical small satellites placed in lunar orbit from their respective command modules at an altitude of about 100 km. Each carried a biaxial magnetometer which returned a measurement of the magnetic field amplitude and phase in the spin plane and the component along the spin axis every 24 seconds at the lowest data rate. Since the orbital period was just under 2 hours, there was approximately one vector sample per orbit of the magnetic field every degree of lunar longitude. Magnetometer observations continued on the Apollo 15 mission for seven months until a telemetry failure terminated transmission of the magnetometer data. Apollo 16 subsatellite operations ceased after only 35 days when the subsatellite crashed into the moon. More extensive descriptions of the subsatellite and the magnetometers can be found in [6, 7].

3. Data Reduction Technique

When the moon is in the lobes of the geomagnetic tail, the non-lunar magnetic contribution to the magnetometer measurements is relatively steady and variations in the data due to the motion of the subsatellite through lunar magnetic features are more readily discernible than in other regions. Furthermore, in the near vacuum conditions of the lobes there is little plasma to distort the lunar magnetic field. In this study we have attempted to separate the contribution of the geomagnetic tail to the measurements from that of the moon in two ways. First, for each orbit we have subtracted the average vector magnetic field in inertial coordinates from the individual measurements and rotated the residual fields into a local lunar coordinate system consisting of the local radial, east and north components. Second, we have removed the average field and the linear slope of the data for each orbit and then proceeded as in the first case. We will refer to these latter data as the detrended data. Both these data processing procedures attenuate the low-order harmonics of the lunar field somewhat but in a readily calculable manner.

In the present analysis we have used 70 orbits of Apollo 15 data in contrast to our previous work with 48 orbits. This increase is due to the processing of additional orbits of tail data. Orbits were selected on the basis of our judging the satellite to be in the lobes of the tail for the entire orbit. This judgment is based on magnetic field direction, magnitude and variability.

In the analysis presented here we have not corrected for any altitude dependence of the data. In part this is because we are studying the body-centered dipole field for which we do not expect the altitude dependence to be discernible over the range of altitudes of the subsatellite orbit. Fig. 1 shows the altitude dependence of the radial component for each of four longitudinal quadrants using the Apollo 15 detrended data and some similarly treated Apollo 16 data at low altitudes. (Positive and negative values have been averaged separately.) The altitude dependence above 80 km where most of our data lie is at best weak. Since lunar magnetic features, such as in the Van de Graaff region, have a measurable altitude dependence, the lack of one here is merely a reflection of the fact that over most of the lunar surface mapped to date, the lunar field is much

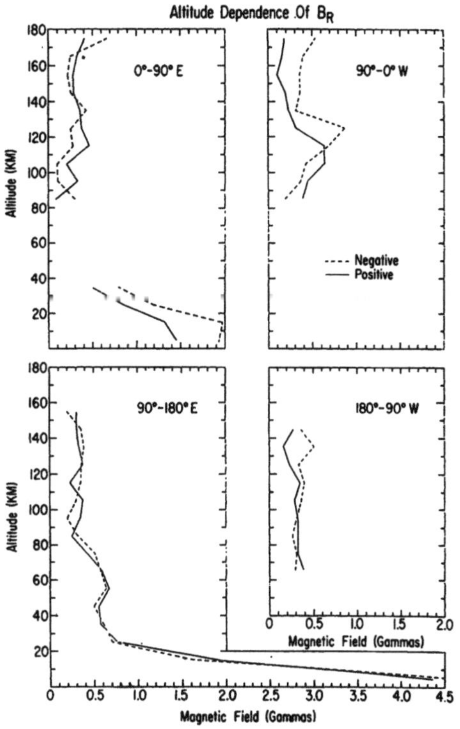

Fig. 1. Altitude dependence of radial component of residual fields for the four lunar quadrants.

weaker than in the Van de Graaff region. It also suggests that the noise level of an individual measurement is about 0.4γ. We note, however, that in our maps, which are constructed from many orbits, the noise level is reduced below this value at some cost of spatial resolution by averaging independent samples over the same region.

4. The Van de Graaff-Aitken Region

We have used the residual detrended data to construct average radial, east, and north components of the lunar field in boxes one degree in latitude by one degree in longitude. Fig. 2 shows contour maps for each of the three components between 130° and 210° east longitude, which includes the Van de Graaff anomaly. Shaded regions denote negative fields. The fields here are the strongest encountered in the regions mapped thus far. In the radial component, the strongest fields are 1γ into the moon in the Van de Graaff region and 1γ out of the moon just west of Van de Graaff. A strong, 1.2γ, eastward field is observed between these two regions suggesting that many of the field lines leaving the positive anomaly enter the Van de Graaff region. The southward component observed in the bottom panel in this same location is consistent with this interpretation. Most of the features which appear to parallel the orbit tracks occur in regions where there is little coverage, and should be viewed with some suspicion.

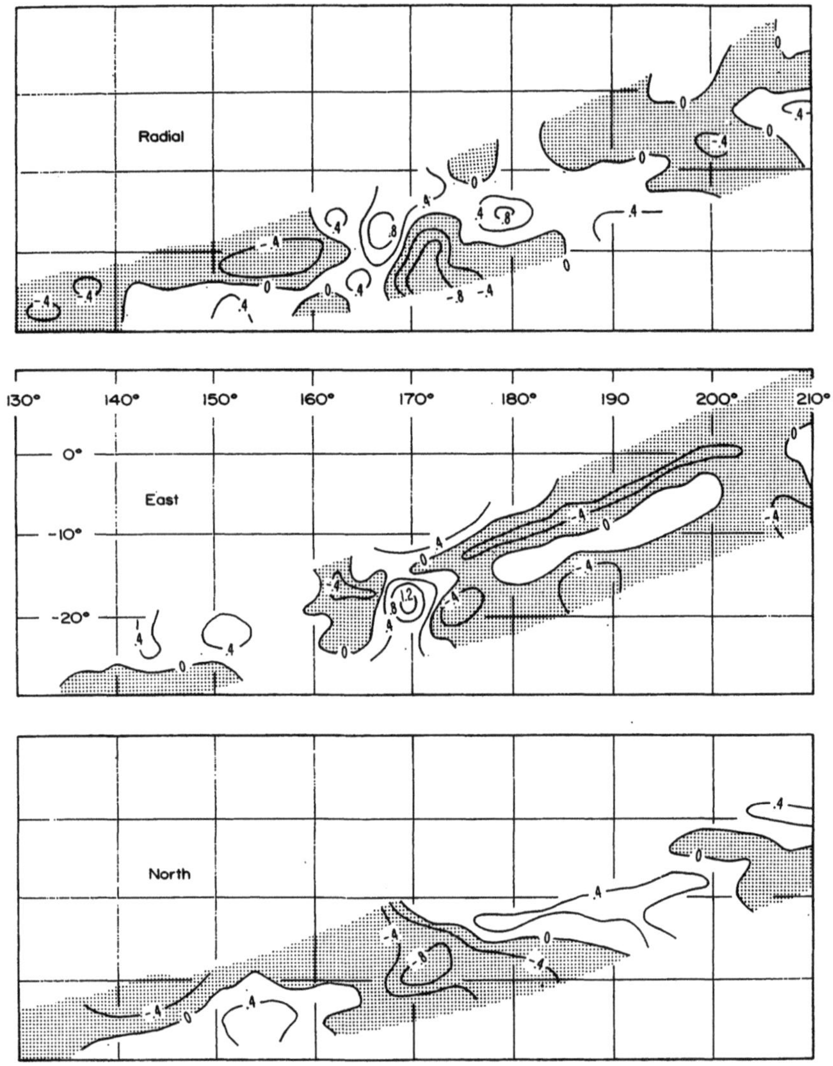

Fig. 2. Contour maps of the radial, east and north components of the lunar field in the Van de Graaff region using the detrended data with no altitude dependence. Positive fields point outward, eastward and northward respectively in the three maps.

5. The Lunar Dipole Field

The magnetic field of a body-centered lunar dipole would be constant and perpendicular to the orbit plane if its dipole moment were perpendicular to the orbit plane, but would have a sinusoidally varying radial component if it lay in the orbit plane. In the inertial coordinate system the former field would be constant and would be removed in our creation of the residual fields. The latter field, however, consists of two parts in the inertial system, one constant and one varying at twice the orbital frequency. The former part is removed but the latter is unattenuated by the subtraction of either the average or the trend.

Fig. 3 shows the radial component averaged as a function of distance along the orbit track from the 0° selenographic meridian using the detrended data. Both

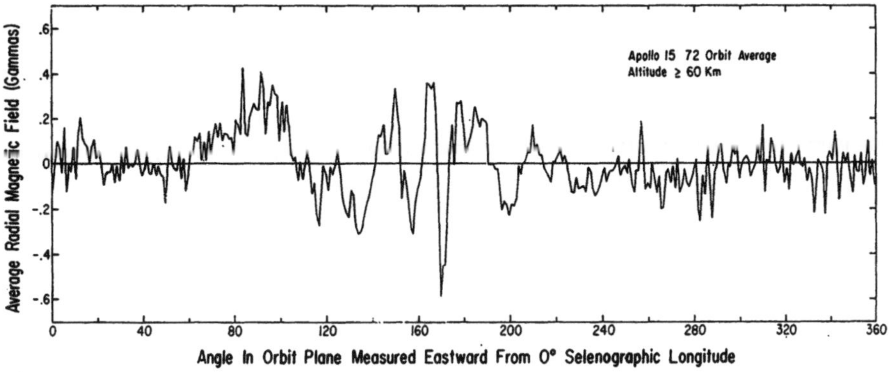

Fig. 3. The average radial component as a function of angle in the orbit plane using the detrended data.

sets of residuals have been averaged in this way and their first 19 Fourier harmonics are shown in Fig. 4. As expected both sets of data give the same dipole field. We note that a constant radial field would be sinusoidally varying at the orbital frequency in the inertial coordinate system and quadrupole variation would have components at both the orbital frequency and three times the orbital frequency, all of which would be attenuated by detrending. As expected, this is observed.

The first harmonic variation of 0.05γ is not much different from some of the neighboring harmonics and thus may represent simply the noise level of our analysis; in other words, we may be measuring only an upper limit to the first

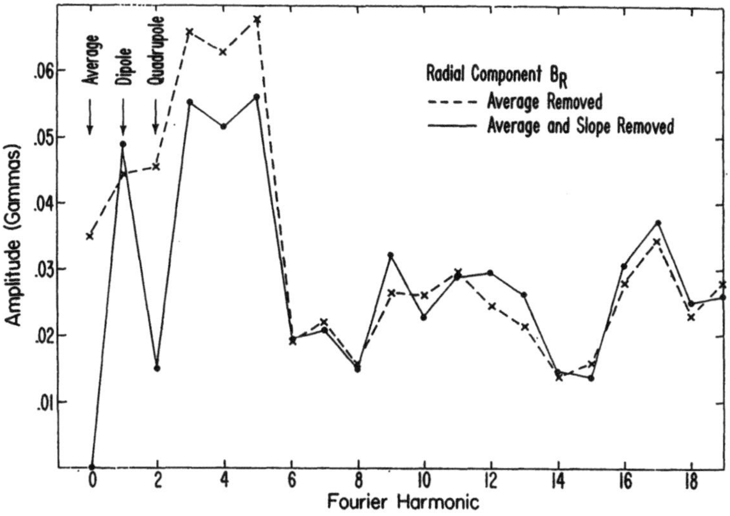

Fig. 4. Fourier amplitudes of the radial components variation for both sets of data.

harmonic variation. A 0.05γ variation would be caused by a magnetic moment in the orbit plane of $3 \times 10^{18}\,\Gamma\,cm^3$. Assuming a remanent magnetization of 10^{-5} emu cm^{-3}, this moment could be produced by a uniformly magnetized lunar surface layer 5 km thick. If we assume that the overall dipole moment is parallel to the lunar rotation axis, then a total dipole moment of $7 \times 10^{18}\,\Gamma\,cm^{-3}$ would be required to give the observed projection in the subsatellite orbit plane. This is 8×10^{-8} of the earth's magnetic moment.

The ancient magnetic fields which magnetized the lunar rocks may have been of external or internal origin. If they were of internal origin a magnetic moment of at least $1.1 \times 10^{23}\,\Gamma\,cm^3$ was once present [8]. The ratio of the present to this hypothesized ancient magnetic moment is less than 6×10^{-5}. If it were due to an overall remanent magnetization of the whole moon this remanent magnetization has disappeared. If this was due to an internal dynamo, this dynamo has stopped. We note if a present-day lunar dynamo were operative we would not have expected such low levels of seismic activity as are presently observed. Our analysis to date does not distinguish between the various theories of the source of this magnetizing field. An ancient lunar dynamo could have stopped as the moon's rotation rate dropped. A remanent field would disappear as the moon warmed up due to radioactive heating. If the source were the earth's geomagnetic field as the moon orbited close to the earth during the often hypothesized capture of the moon, then the field would not so much have disappeared but, rather, the moon had simply moved out of it.

Finally, we note the possibility of an early dynamo in a melted outer layer of the moon [9]. It has been proposed that this mechanism could explain the hypothesized low magnetic field strength at high selenographic latitudes deduced from Explorer 35 observations of limb compression occurrence rates [10]. While the subsatellite measurements indicate that limb compression formation is related to lunar magnetic regions [5], the low latitude orbits of Apollo 15 and 16 do not cover this interesting region. Such measurements must await a polar orbiting lunar satellite.

Acknowledgments

This work was supported by the National Aeronautics and Space Administration under contract NAS 9-12236.

References

[1] C. T. RUSSELL et al., Space Research XIII, 951 (1973).
[2] L. R. SHARP et al., The Moon, **7**, 322 (1973).
[3] P. DYAL et al., in: Apollo 16 Prelim. Sci. Rep. p. 11-1 (1973).
[4] C. P. SONETT et al., Proc. 3rd Lunar Sci. Conf., Geochim. Cosmochim. Acta, Suppl **3**, **3**, 2309 (1972).
[5] B. R. LICHTENSTEIN et al., EOS Trans. AGU, **55**, 413 (1974).
[6] P. J. COLEMAN, JR, et al., in: Apollo 15 Prelim. Sci. Rep., NASA SP-289, p. 22-1 (1972).
[7] P. J. COLEMAN, JR, et al., in: Apollo 16 Prelim. Sci. Rep., NASA SP-315, p. 23-1 (1973).
[8] C. T. RUSSELL et al., Proc. 4th Lunar Sci. Conf., Geochim. Cosmochim. Acta, Suppl **4**, **3**, 2833 (1973).
[9] R. SMOLUCHOWSKI, Nature **242**, 516 (1973).
[10] C. P. SONETT and J. D. MIHALOV, J. Geophys. Res. **77**, 688 (1972).

Space Research XIV — Akademie-Verlag, Berlin 1974

THE NATURE OF CIRCULAR MARIA BASED ON GRAVITY STUDIES

R. J. Phillips, J. E. Conel and W. L. Sjogren

Jet Propulsion Laboratory, California Institute of Technology, Pasadena, Calif., USA

Analyses of Lunar Orbiter, Apollo and Apollo subsatellite data have constrained possible anomalous mass distributions in the circular maria. Perimeters of the masses in Serenitatis and Crisium are uniquely determined, show strong correlation with circular wrinkle ridge structure, and separate regions of shallow and deep mare flooding; a basin-fill origin for mascons is strongly suggested. Photogeologic and radar data indicate these ridges are scarps for arcuate fault structures. A surface mass model for Serenitatis shows a thinning of the excess mass toward the perimeter, further suggesting basin fill. Specifically, Apollo 15 data suggest the Serenitatis mascon is lava fill in a two-ring basin. Recent Apollo 15 subsatellite data show that a mass excess at the surface plus a mass deficiency at some depth provide a better fit to the Humorum and Crisium anomalies than surface mass distributions alone. A constitent model is one of lunar crust faulted into mantle over regions bounded by arcuate fault structures.

The circular maria have thus experienced varying degrees of volcanic flooding; gravity anomalies are roughly proportional to the isostatically excess mass added. Many tectonic features seen are consequences of subsequent isostatic adjustments in the basins.

1. Introduction

Studies of the moon's gravity field from Lunar Orbiter and Apollo, radar and laser profiling from lunar orbit and photogeologic investigations have shed new light on the structure of large lunar circular basins, and on the nature and origin of surface structures observed in the fill deposits. This paper summarizes our current thinking on the evolution of these major lunar morphological features.

The great lunar circular basins must result from high-velocity impact of large bodies with the moon [1]. Mare Orientale provides the detailed surface structure of one such feature that is relatively young. The morphology is basically that of concentric mountainous rings of extremely chaotic physiography, separated by annular depressions. The elevations of mountainous rings and montane basins increases radially outward relative to the center. No adequate theoretical or experimental basis appears to exist for understanding the mechanical details of formation of these enormous structures. For example, there is no basis upon which to decide the "original" depths of excavation, how the surface might have relaxed to its present configuration, or whether the present surface relief was ever much greater than that presently observed (about 6 km for Mare Orientale and Mare

Nectaris). Likewise, the deformation and/or patterns of fracture deep in the interior are not known.

The well-known irregularities in the lunar gravitational field discovered from Lunar Orbiter tracking data [2] indicated detailed inequalities of mass distribution in the lunar interior, and profound differences between circular and irregular maria. In particular, the mass per unit area over the great basins is greater than that over either the irregular maria or lunar highlands. It was immediately speculated that these major inequalities of mass distribution arose from the infalling objects themselves [2] buried beneath the surface as a final result of the impact process. Subsequently it was noted [3] that the gravity data alone gave no unique solution for the distribution of anomalous mass, and that bodies with the geometry of unfilled circular basins (such as Orientale) at the surface would equally well give rise to gravity effects that fitted the data. Wise and Yates [4] further demonstrated the non-uniqueness of interpretation by showing that the anomalies could arise from rebound (accompanying impact) of an interface at depth on the order of 100 km, between low-density lunar crust and higher density mantle. The anomalous body involved (Mare Imbrium in this instance) is virtually a mirror image of the filled-crater model, although with detailed scale adjustments.

Attempting to settle the question of the precise distribution of anomalous mass, and hence provide constraints on mechanisms of mascon origin, Phillips et al. [5] carried out detailed analyses on most available lunar orbital gravity data for Mare Serenitatis. The conclusions of that study were: (i) the body involved has a very small ($<1{:}20$) thickness/width ratio, (ii) its horizontal dimensions are firmly constrained and coincide precisely with boundaries between regions of deep and shallow fill in the basin interior and with concentric wrinkle ridge patterns in the surficial deposits, (iii) if the body is deeper than about 50 km it would require increasingly bizarre and unlikely radial fluctuations in (surface) mass distribution to account for the external field, (iv) that the body, if at the surface, increases in thickness toward the center, in an approximately stepwise fashion. For single-body models it was concluded that all of the gravitational and photogeologic evidence was consistent with a model in which the anomalous mass is at the surface, and possesses the geometry of large circular mare basins; but multiple body models involving some contribution from other sources such as "Moho relief" could not be ruled out on the basis of available evidence.

The question of mass transport is obviously crucial to any mechanism for anomalous mass distribution. Wise and Yates [4] require large stress differences arising from initial surface relief to move material from one place to another. Muller and Sjogren [2] following Urey [6] suggested the initial impacting objects as sources of the observed gravity effects. Kaula [7] imagined thermoelastic stresses arising from global contraction on cooling as the driving force, while Gilvarry [8] and others thought that ancient stream systems had transported sediments from the highlands and filled the low depressions. In this paper we shall use the simple hydrostatic mechanism of Wood [9] for mass transport, although mantle rebound [4] may contribute to, but is not crucial for, the process we will describe.

The hydrostatic mechanism functions as follows: If for a circular impact structure, a laterally pervasive basaltic source magma of density ϱ_L is at some depth H below the surrounding terra, then it will *always* stand at a level h in the central impact basin with $\varrho_L h = \varrho_C H$ if the magma has free upward mobility;

ϱ_C is the crustal (terra) density. The height h represents the limiting height to which solidified basalt may occur. If isostasy occurs totally by mantle rebound, then superisostatic material (i.e. in excess of that mass required for isostasy) may be added to the central basin if h stands higher than the floor of the basin. The excess mass is linearly proportional to the basin depth. If isostasy occurs by basin filling alone, then a basaltic magma at height h in the central basin is in isostatic equilibrium with the surrounding region. After solidification, the basalt of density ϱ_S stands at a height $h' = (\varrho_L/\varrho_S)h$, and if upward mobility is still possible then subsequent magma may still rise to the height h. If the process goes to completion, then the departure from isostasy, in terms of mass per unit area, is given by

$$\Delta m = \varrho_S \left[D \left(1 - \frac{\varrho_C}{\varrho_S} \right) - H \left(1 - \frac{\varrho_C}{\varrho_{-L}} \right) \right] \tag{1}$$

where D is depth of the basin. Δm will be (i) subisostatic ($\Delta m < 0$), (ii) isostatic ($\Delta m = 0$), or (iii) superisostatic ($\Delta m > 0$), depending on the five quantities in Eq. (1). If, for example, the magma source is at the base of the central basin, then $D = H$ and

$$\Delta m = D \varrho_C \varrho_S \left[\frac{1}{\varrho_L} - \frac{1}{\varrho_{-S}} \right],$$

guaranteeing superisostasy. Based on seismic and geochemical evidence it is reasonable to assume that mare basalts are denser than the general lunar crust (i.e. $\varrho_C/\varrho_S < 1$). If $\varrho_C/\varrho_L > 1$ and $\varrho_C/\varrho_S < 1$ then superisostasy is also guaranteed independently of the ratio H/D. Without constraining assumptions on ϱ_C/ϱ_L, and/or ϱ_C/ϱ_S, and/or H/D, then any of the three isostatic conditions may prevail. This perhaps makes the mantle rebound theory more attractive, because *any* magma that reaches the floor of the basin creates a superisostatic mass. Note that if $\varrho_C/\varrho_S < 1$, and if superisostasy occurs by the mechanism of Eq. (1), then the excess mass is linearly proportional to the basin depth. Thus, independent of *which* mechanism achieves isostasy, we would expect the amount of superisostatic mass to be a direct function of the basin depth. The amount of excess mass required to achieve the required gravity anomaly is not exceedingly great; the thickest part of the Serenitatis mascon needs only be about $2^1/_2$ km if ϱ_S is taken as 3.3 g cm^{-3} [5]. It should be pointed out that the rebound and basin fill mechanisms are "end members" of an infinite number of combinations of *both* mechanisms. Both processes have probably operated. We have discussed the "end members" only for simplification. Eq. (1) can be generalized to include mantle rebound, introducing two more variables. This mechanism of superisostatic fill may be self-perpetuating, since accumulation of excess mass on the surface gives rise to an excess pressure. Downward crustal deflections may thus be expected which in turn could give rise to further potential excess mass accumulation and further deformation.

Following the hydrostatic mechanism outlined, we present a working hypothesis for the evolution of lunar circular basins subsequent to their origin by impact. We shall appeal in the main to Doppler gravity data to support our hypothesis, although photographic and altimetric information will also be used. A summary of the argument is: All large ringed circular basins follow a common evolutionary path of superisostatic volcanic flooding followed by partial and variable isostatic adjustment. The *difference* between basins is the *amount* of flooding, which in turn may be related to the center of figure—center of mass offset of the moon. Many of the surface structural features in the circular basins are the result of yielding and partial isostatic adjustment to superisostatic loads.

2. Stage I—Isostatic Adjustment of the Central Basin

The central impact basin comes to approximate isostatic equilibrium with its immediate surroundings, in a time period indeterminate from present observations, but short with respect to the time increment between basin origin and filling (Fig. 1). The mechanism of adjustment is perhaps a combination of mantle

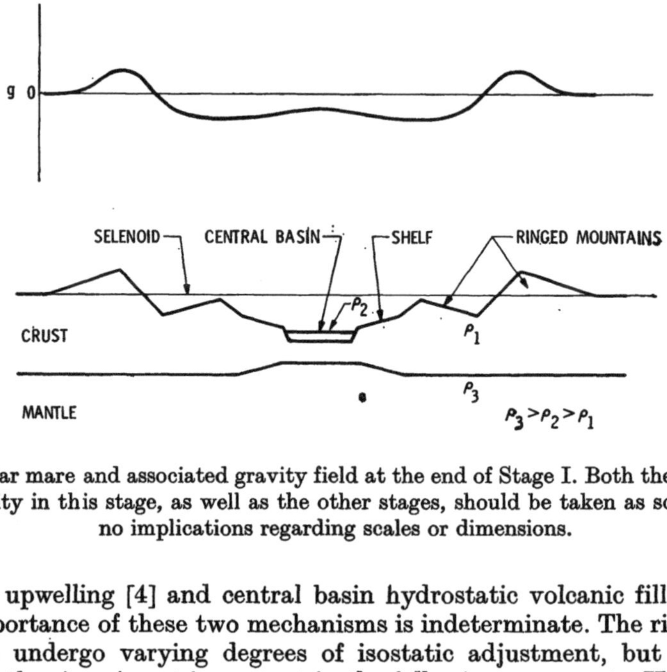

Fig. 1. Circular mare and associated gravity field at the end of Stage I. Both the cross section and the gravity in this stage, as well as the other stages, should be taken as schematic with no implications regarding scales or dimensions.

rebound or upwelling [4] and central basin hydrostatic volcanic filling [9]. The relative importance of these two mechanisms is indeterminate. The ringed mountain chains undergo varying degrees of isostatic adjustment, but the precise behavior of the rings is not important in the following arguments. Here we must speculate that backside circular basins may have never reached isostatic equilibrium. The supposed thicker backside crust [10] may have precluded both a hydrostatic mechanism and mantle rebound. The evolution would be arrested without completing Stage I. The gravity anomaly at perhaps 50 km altitude *following* isostatic adjustment is also shown schematically in Fig. 1. Note that isostasy does not imply a zero free air gravity anomaly.

3. Stage II—Superisostatic Filling of the Central Basin

Assume that, as discussed above, the hydrostatic mechanism fills the central basin to a level beyond that required for isostasy [4, 9]. The dashed line of Fig. 2 indicates schematically the amount of superisostatic fill. A positive gravity anomaly will result from this excess. The crucial difference between the approach taken here from that of Wise and Yates [4] is that mantle rebound is not necessarily required, and that in any event we take the mare basalts as denser than the surrounding highlands (a supposition consistent with current measurements and speculation about lunar surface material), and it is mare material that gives rise to positive gravity anomalies.

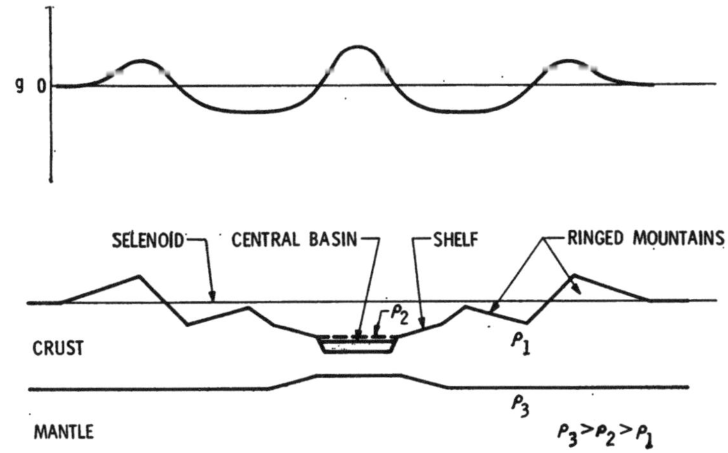

Fig. 2. Circular mare and associated gravity field at the end of Stage II.

Positive gravity anomalies are not unique to circular basins. For example, the more recent lava flows in Oceanus Procellarum are regions of positive gravity anomaly [11]. The Marius Hills, thought to be of volcanic origin, show a free air anomaly of about 60 mgal [12]. The question is, why are the free air anomalies of the circular basins so much larger than the free air anomalies in the irregular basins? The hydrostatic arguments given earlier show that the mass of super-isostatic fill is linearly proportional to the depth of the basin, regardless of which isostatic adjustment mechanism operated. Thus, if we accept that the circular maria were deeper than the irregular maria before flooding, then a simple explanation for the large free air anomalies over the circular basins is immediately apparent.

Mare Orientale, with its filled central basin, shelf region and prominent rings, represents a circular basin at the end of Stage II except for adjustment to a super-isostatic state (Stage IV). Fig. 3 shows the gravity measured by the Apollo 15 subsatellite 50—90 km over the Orientale basin. Reliable data are available for only the eastern half of the basin. Also shown is the topographic profile derived from Lunar Orbiter photography; the elevations shown are relative and approximate only, but the location of features is fairly accurate. There is very clearly a free air gravity high over the central basin as well as over the dominant ring, Montes Cordillera. Orientale has the best preserved ring structure of all circular maria, and the most pronounced gravity correlation with ring structure of the front side circular maria. There are, however, correlations of gravity with the less well preserved ring structure of other circular basins [12, 13].

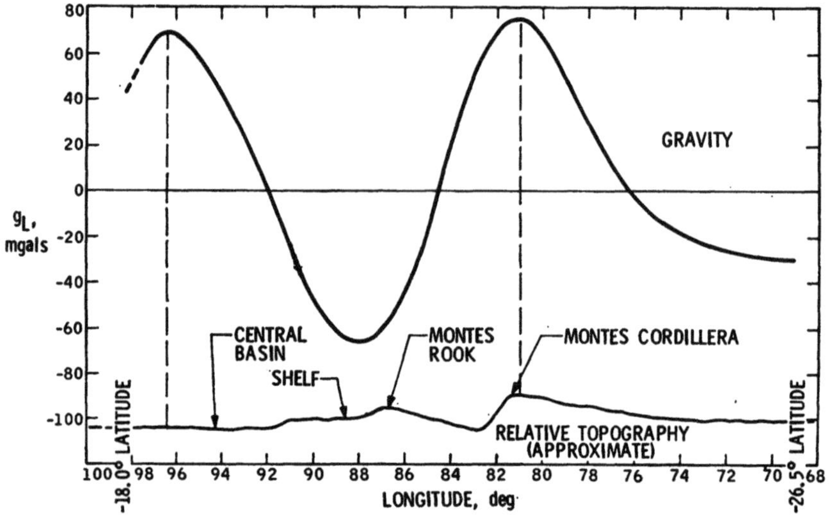

Fig. 3. The line-of-sight gravity g_L measured over Mare Orientale by the Apollo 15 subsatellite. Also shown is the relative topographic profile derived from Lunar Orbiter photography. No vertical scale is implied in the topographic profile, but the location of features is fairly accurate; they show a high correlation with the gravity field.

4. Stage III—Overflow into the Shelf Region

Within limitations of the hydrostatic theory, the central basin will continue to fill, with an accompanying increase in the free air anomaly. If the hydrostatic limiting height of fill is above the rim of the central basin (the height is controlled by relative values of ϱ_L and ϱ_C), then volcanic material overflows into the shelf region (Fig. 4). This overflow condition appears to be the exception rather than the rule. It has been definitely detected for Mare Serenitatis, but is ruled out for

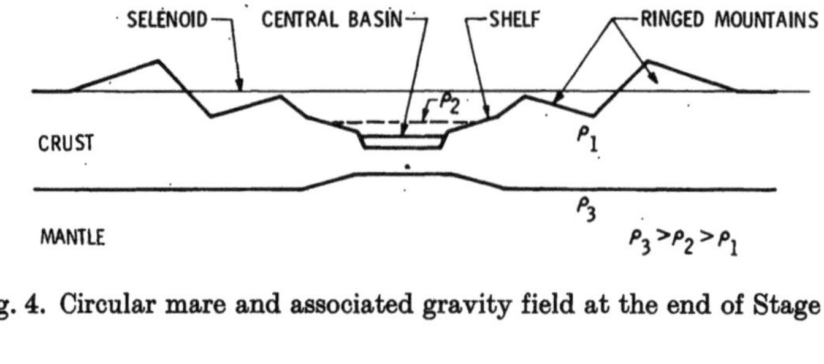

Fig. 4. Circular mare and associated gravity field at the end of Stage III.

Mare Crisium, Mare Orientale, and probably Mare Humorum. We include this stage for the sake of completeness. Fig. 5 shows the Apollo 15 gravity data taken 15—20 km over Mare Serenitatis. The accompanying model is required to satisfy

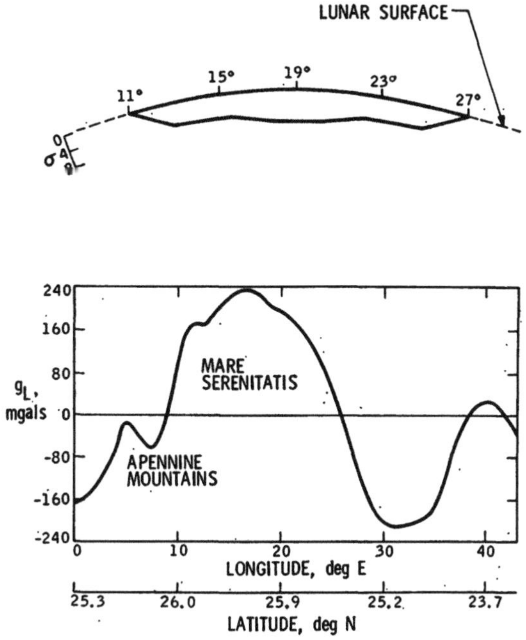

Fig. 5. Line-of-sight gravity g_L measured by the Apollo 15 CSM over Mare Serenitatis (bottom) and the surface density σ of a lunar surface model (top) required to satisfy the gravity data.

the gravity data *if* the model is placed at or near the lunar surface [5]. There is strong suggestion in the model of fill in the central basin as well as some fraction of a shelf region.

5. Stage IV—Isostatic Adjustment to Superisostasy

The final stage in the evolution of filled circular basins is an incomplete isostatic adjustment to the superisostatic state (Fig. 6). *All* circular basins that have substantiated positive free air anomalies show concentric normal faulting near the basin perimeter, with the basin side downthrown. The normal faulting is often associated with wrinkle ridge structure. Some of the circular basins (e.g. Serenitatis, Humorum) show concentric rilles. We hypothesize that the normal faulting represents isostatic adjustment to the superisostatic loads. The concentric rilles may be tensional features also associated with attempts of the moon to achieve an isostatic condition.

Fig. 7 shows the uniquely determined [5] horizontal boundaries for the Serenitatis and Crisium mascons. In each case, the boundary is associated with wrinkle ridge structure; the downthrown inner basin is particularly apparent for Crisium. It was also found [5] that the horizontal boundaries marked the outer extent of deep basin flooding, tending to substantiate that the mascon is associated with (near surface) fill of a central impact basin.

Fig. 8 shows a radar profile-image, acquired by the Apollo 17 Lunar Sounder, of the eastern part of Crisium basin. The mascon boundary is associated with a 400 m surface displacement near the outer margin of the basin. The abrupt change

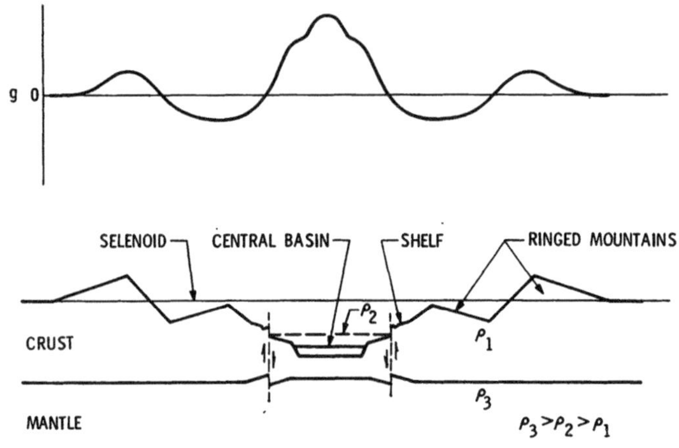

Fig. 6. Circular mare and associated gravity field during Stage IV.

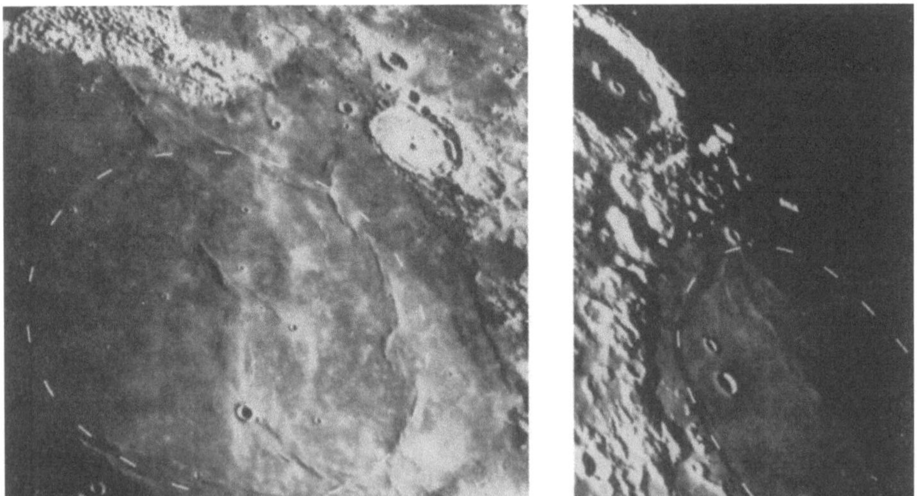

Fig. 7. Mare Serenitatis (left) and Mare Crisium (right) showing the uniquely determined [5] horizontal boundaries of the mascons.

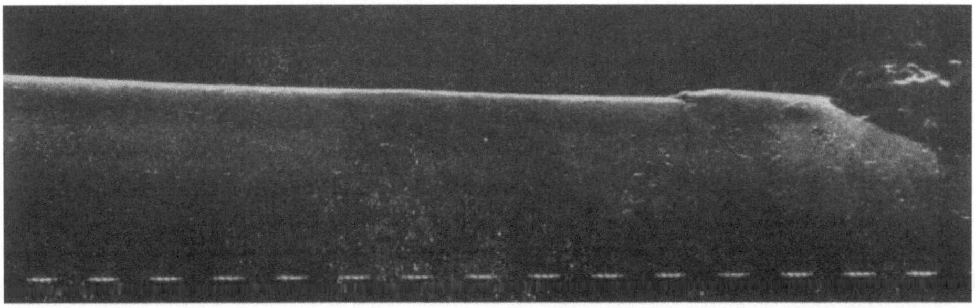

Fig. 8. Lunar Sounder profile-image of eastern Mare Crisium. In this image, east is to the right and the scene is imaged to the north. Time marks shown at the bottom of Lunar Sounder images represent a ground distance of approximately 15 km. The general linear trend in surface elevation over each image is the uncorrected effect of the elliptical orbit of the spacecraft.

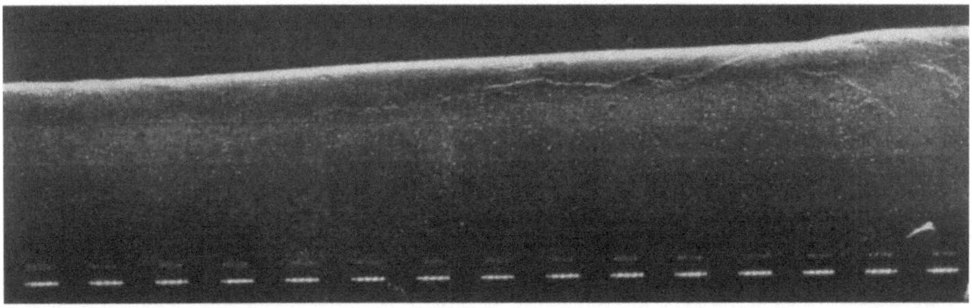

Fig. 9. Lunar Sounder profile-image of western Mare Serenitatis. Here east is to the left and the scene is imaged to the north.

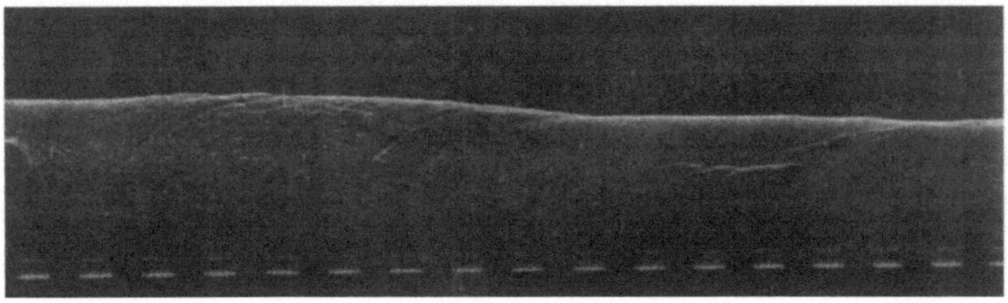

Fig. 10. Lunar Sounder profile-image of the Tranquillitatis-Serenitatis boundary showing the crater Dawes (extreme left) and the Plinius rilles (center left). Here east is to the left and the scene is imaged to the north. The cross-cutting rille is actually an image superposition from the south.

of slope to the right of the displacement suggests elastic but unrecovered bending before structural failure. The same geometry appears at the mascon boundary in western Mare Serenitatis (Fig. 9); here a wrinkle ridge is clearly associated with a 300 m displacement. A Lunar Sounder profile-image of the boundary between Mare Serenitatis and Mare Tranquillitatis is shown in Fig. 10. The concentric Plinius rille system in the left center of the image is a possible tension feature associated with the isostatic adjustment of the Serenitatis mascon. The floor of Tranquillitatis stands 900 m above the outer bench of Serenitatis. We interpret this difference in elevation as reflecting the greater depth of Serenitatis basin before flooding. The 300 m displacement and associated wrinkle ridge at the mascon boundary may also be seen.

Recent Apollo data show that a mass excess at the surface plus a mass deficiency at some depth provide a better fit to the Humorum and Crisium gravity anomalies than a surface mass distribution alone. If it is assumed that the deficiency is caused by crustal displacement into the mantle due to isostatic adjustment, then the horizontal dimensions of the deficiency may be taken as those of the surface mass. Under this assumption, a least squares solution for the depth of the crust-mantle interface under Humorum yields 55 km, a reasonable number based on seismic evidence. While this result is not unique, it is a composite model that fits the data better than a surface model alone.

6. Summary

We have presented a working hypothesis, subject to much further test, for the evolution of lunar circular maria following their impact formation. In brief, we believe their evolution can be explained by simple hydrostatic arguments. The front-side circular basins filled superisostatically; the superisostatic idea is supported by tectonic features that appear to represent subsequent isostatic adjustment. The isostatic adjustment is far from complete and the question of supporting excess loads for the order of 4×10^9 years must be raised. Phillips et al. [5] point out that even with the small amount of adjustment that has taken place in Serenitatis, material in the lunar interior at depths of the maximum shear stress could have remained at temperatures on the order of 1000° to 1300 °K during the time of adjustment, temperatures consistent with plausible thermal histories of the moon.

The maximum shear stress for the Serenitatis mascon occurs at a depth of approximately 100 km beneath the region of maximum anomaly [3]. The *apparent* coincidence of 600—1000 km source depth moonquakes with the margins of circular basins [14] would thus appear to be unrelated to the isostatic adjustment of the mascons. The coincidence, if true, might be explained as failure along original zones of weakness imparted at the time of impact formation.

Acknowledgments

This paper represents the results of one phase of research carried out at the Jet Propulsion Laboratory, California Institute of Technology, under Contract No. NAS 7-100, sponsored by the National Aeronautics and Space Administration.

References

[1] R. S. SAUNDERS et al., Polarforschung **7**, Jg. 40, 33 (1970).
[2] P. M. MULLER and W. L. SJOGREN, Science **161**, 680 (1968).
[3] J. E. CONEL and G. B. HOLSTROM, Science **162**, 1403 (1968).
[4] D. U. WISE and M. T. YATES, J. Geophys. Res. **75**, 261 (1970).
[5] R. J. PHILLIPS et al., J. Geophys. Res. **77**, 7106 (1972).
[6] H. C. UREY, Science, **162**, 1408 (1968).
[7] W. M. KAULA, Phys. Earth Planet. Interiors **2**, 123 (1969).
[8] J. J. GILVARRY, Nature **221**, 732 (1969)
[9] J. A. WOOD, J. Geophys. Res. **75**, 6497 (1970).
[10] G. RANSFORD and W. SJOGREN, Nature **238**, 260 (1972).
[11] MISSION EVALUATION TEAM, Apollo 17 Mission Report, 5-1 (1973).
[12] W. L. SJOGREN et al., in: Apollo 15 Prelim. Sci. Rep. NASA SP-289, 1972 (p. 20-1).
[13] W. L. SJOGREN et al., in: Apollo 16 Prelim. Sci. Rep., NASA SP-315, 1972 (p. 24-1).
[14] G. LATHAM et al., in: Abstracts, 4th Lunar Sci. Conf., Lunar Science Institute, Houston 1973 (p. 457).

Space Research XIV — Akademie-Verlag, Berlin 1974

ON THE ACCURACY OF THE MOON'S GRAVITY FIELD AS DERIVED FROM LUNAR ORBITERS

W. Köhnlein

Astronomical Institutes of the University of Bonn, Bonn, FRG

The standard deviations of the moon's potential, gravity and gravity gradient fields are determined from the discrepancies between different gravity models at the elevations 0, 250, 1000, and 10000 km. For the zonal field we find a strong correlation between the uncertainty and the selenographic latitude. The relative accuracies of the potential, its gradient and gravity gradient fields appear lowest on the moon's surface and increase with higher elevations. Compared with the zonal part, the non-zonal field structure ($\leqq 8$, 8) is less accurate; this is particularly true for the far side of the moon where the error is larger than on the near side. The relative uncertainty is lowest for the potential, increasing towards gravity and gravity gradient; most of the error amplitude has, however, vanished at 1000 km elevation and can be neglected in most cases at still higher elevations.

1. Introduction

The moon's harmonic coefficients are, compared with the earth's field, still estimated with great uncertainty; this is due mainly to the interruption of the line-of-sight observation as orbiters move behind the moon, and to the limited variety of the orbital parameters available for numerical analysis. The estimates of the uncertainty of the field structure obtained by least-squares methods are, in general, much too small; they reflect only an internal uncertainty assuming that not systematic errors are present [1]. To obtain a more realistic picture we consider the actual discrepancies between the numerical results of different gravity models and compute the average uncertainty of the field structure at a certain elevation.

Let ε be the field discrepancy at latitude ϕ and longitude λ on a sphere at an elevation h above the moon. We define a standard deviation:

$$\hat{\sigma} = \pm \left[\frac{1}{4\pi} \int\limits_{\lambda=0}^{2\pi} \int\limits_{\varphi=-\pi/2}^{\pi/2} \varepsilon^2(\phi, \lambda) \cos \phi \, d\phi \, d\lambda \right]^{1/2} \tag{1}$$

from which we obtain the scatter or uncertainty of the field

$$\sigma = \hat{\sigma}/2^{1/2}, \tag{2}$$

or equivalently the relative standard deviation

$$m = \sigma / f \tag{3}$$

where f is the average of the field value at elevation h.

In the following we restrict our investigation to the accuracies of:

(i) *the potential field*

$$U = \frac{\mu}{r}\left[1 + \sum_{n=2}^{\infty}\sum_{m=0}^{n}\left(\frac{a}{r}\right)^n (C_{nm}\cos m\lambda + S_{nm}\sin m\lambda)P_{nm}(\sin\phi)\right] \tag{4}$$

(standard notation), or equivalently to the radius

$$r = r(U = \text{const}); \tag{5}$$

(ii) *the gravity field*

$$g = |\nabla U|, \quad \text{and} \tag{6}$$

(iii) *the gravity gradient field*

$$\nabla g = -|\nabla|\nabla U|| \tag{7}$$

(at elevations 0, 250, 1000 and 10000 km above the moon's surface)

using as residuals the spread among the gravity models of Liu and Laing [2], Blackshear et al. [3], and Michael et al. [4] for the error analysis. (For brevity we refer to them as L, B and M models.)

2. Accuracy of the Zonal Field

Fig. 1 shows the systematic deviation of the gradient line of the moon's zonal potential field [5] using different sets of harmonic coefficients. There are huge discrepancies near the moon's surface, i.e. up to 210 m compared with the earth's

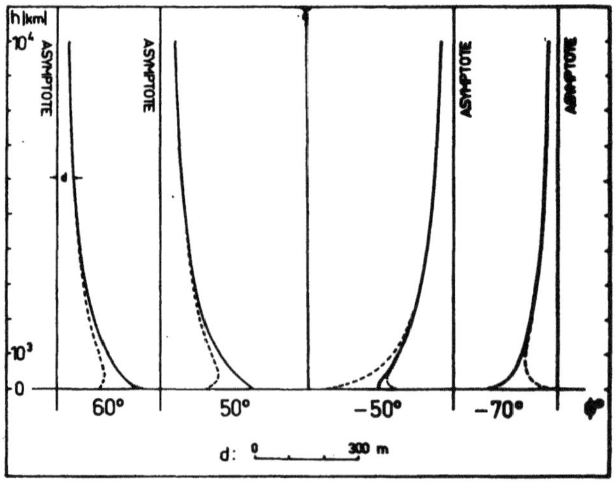

Fig. 1. Shape of the zonal gradient line. ϕ = latitude of the gradient line on the moon's surface, d = distance of the gradient line from its asymptote at the elevation h.
—— Liu and Laing, --- Blackshear et al., ··· Michael et al.

field of only a few metres [6]; the spread, however, decreases at 1000 km to $\sim$20 m, and becomes negligible from 5000 km upwards. An analogous picture is obtained for the zonal gravity $\bar{g}$ along a meridian circle at different elevations (see Fig. 2 for 0 km elevation). Near the equator we have a fairly good agreement between the gravity models, but towards the poles the discrepancy $\delta\bar{g}$ increases very strongly. This correlation of uncertainty with latitude is equally present in the potential field and in the gravity gradient field.

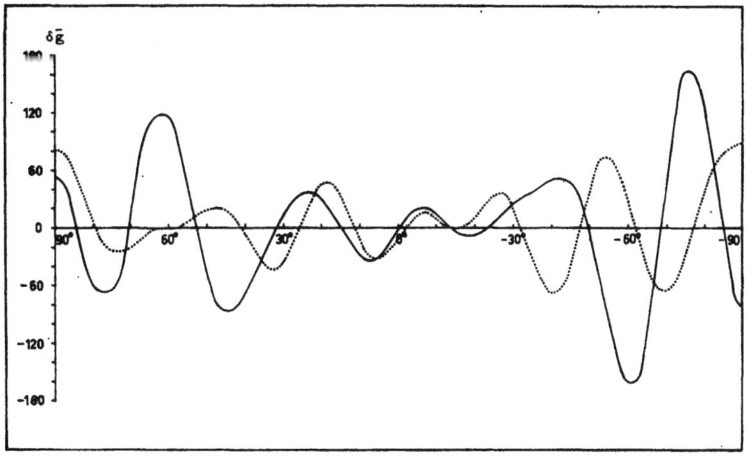

Fig. 2. Discrepancies $\delta\bar{g}$ of the zonal gravity field along a meridian. ($90° \geqq \phi \geqq -90°$, at $h = 0$ km elevation). $\delta\bar{g} \times 10^{-3}$ (cm s^{-2}) —— B — L, ··· M — L.

Using the harmonic coefficients of Liu and Blackshear we obtain from the discrepancies ε (Eqs. (1)—(7)) the relative standard deviations of the potential (or its radial analogue $\sigma_{\bar{r}}/\bar{r} \sim \sigma_{\bar{U}}/\bar{U}$), the gravity and the gravity gradient fields as shown in Fig. 4 (broken lines). The relative uncertainty of the field structure is lowest for the potential, increasing toward gravity and gravity gradient:

$$\sigma_{\bar{r}}/\bar{r} < \sigma_{\bar{g}}/\bar{g} < \sigma_{\nabla\bar{g}}/|\nabla\bar{g}|.$$

At 1000 km elevation most of the error amplitudes have vanished and the relative standard deviation becomes very small.

3. Accuracy of the Total Field

Similarly to the zonal part, we treat the total field according to the Eqs. (1) to (7). Fig. 3 shows the discrepancies of the gravity field g between the Liu and Blackshear coefficients (non-zonal $\leqq 8,8$) on a sphere at zero elevation: the front side of the moon ($-90° \lesssim \lambda \lesssim 90°$) has much smaller discrepancies than the far side. This result was expected from the unequal distribution of observations, i.e. the direct line-of-sight is interrupted as soon as the lunar orbiter moves behind the moon. The gap between the gravity models of L and B is greater than that between L and M; hance the uncertainty was computed from L and B, reflecting the systematic distortions between both models also. The total error for the

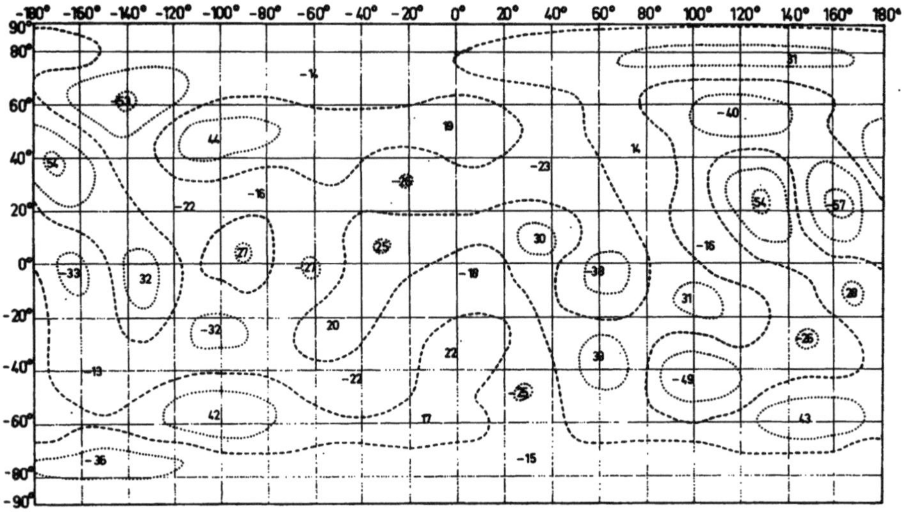

Fig. 3. Discrepancies δg of the total gravity field along a sphere at $h = 0$ km elevation L — B. ($90° \geqq \phi \geqq -90°$; $180° \geqq \lambda \geqq -180°$) $\delta g \times 10^{-2}$ (cm s^{-2}); interval: 25×10^{-2} (cm s^{-2}).

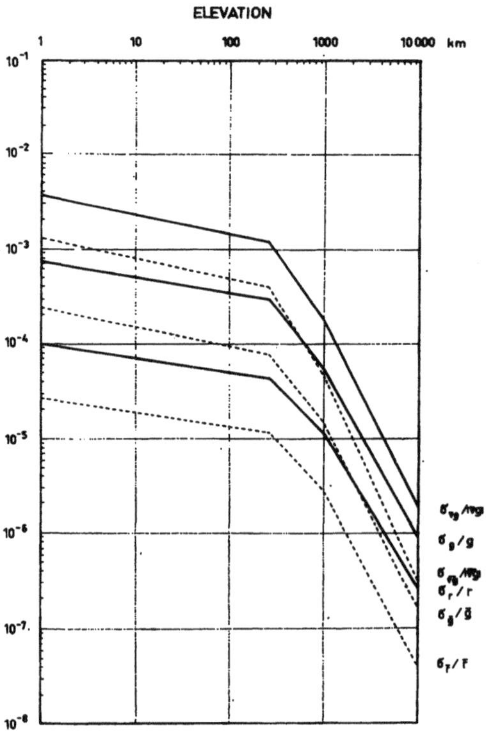

Fig. 4. Relative standard deviations computed from the discrepancies between L and B (with non-zonal coefficients $\leqq 8,8$; see text). --- zonal, —— total.

different fields is about 30% (and more) of the amount of the extreme amplitude assumed by the non-zonal field portion.

Again using relative standard deviations, the solid lines in Fig. 4 represent the relative accuracies of the potential field (or its radial analogue), the gravity field and the gravity gradient field at different elevations:

$$\sigma_r/r < \sigma_g/g < \sigma_{\nabla g}/|\nabla g| \quad \text{with} \quad \sigma_U/U \sim \sigma_r/r .$$

Compared with the zonal part, the non-zonal field discrepancy is larger, but beyond 1000 km elevation the error amplitude again has strongly decreased as in the zonal case.

A comparison of the gravity field, of L and M, with local accelerations of space probes near the moon's surface [7] favours slightly the Liu coefficients.

4. Summary

(i) L and M agree more closely than L and B or M and B;

(ii) there are strong systematic discrepancies between L and B (Fig. 1);

(iii) the zonal field uncertainty increases from the equator towards the poles. This latitudinal correlation appears in the potential, gravity and gravity gradient fields at all elevations (Fig. 2);

(iv) the non-zonal field is less accurate than the zonal field;

(v) the uncertainty of the non-zonal field is larger on the far side of the moon than on the near side (see Fig. 3);

(vi) the relative standard deviations for both the zonal and the total fields are, according to Fig. 4:

$$\sigma_r/r \sim \sigma_U/U < \sigma_g/g < \sigma_{\nabla g}/|\nabla g| ;$$

(vii) at the moon's surface the relative uncertainty is largest. Most of the errors have, however, disappeared at 1000 km and become negligible in most cases at still higher elevations;

(viii) local accelerations of space probes near the moon's surface (Muller and Sjøgren [7]) favour the Liu coefficients slightly.

As an example, the standard deviations at 0 km elevation, following from Fig. 4, are given in Table 1.

Table 1

Standard Deviations at 0 km

Field	Standard deviation Zonal	Standard deviation Total	Average field value
potential [$cm^2\,s^{-2}$]	0.76×10^6	0.28×10^7	0.282×10^{11}
radial analogue (m)	50	170	1738×10^3
Gravity ($cm\,s^{-2}$)	0.04	0.12	162
Gravity gradient (s^{-2})	0.25×10^{-8}	0.70×10^{-8}	-0.187×10^{-5}

Acknowledgments

This work was supported by grant WRK 226 from the Bundesministerium für Bildung und Wissenschaft.

References

[1] W. Köhnlein, Smithson. Astrophys. Obs. Spec. Rep. No. 200, **3** (1966) (p. 103).
[2] A. S. Liu and P. A. Laing, Science **173**, 1017 (1971).
[3] W. T. Blackshear, E. F. Daniels and S. G. Anderson, NASA-TM-X-2260 (1971).
[4] W. H. Michael, W. T. Blackshear and J. P. Gapcynski, paper presented at Plenary Meeting of COSPAR, Prague, May 1969.
[5] W. Köhnlein, The Moon **6** (1973).
[6] W. Köhnlein, Smithson. Astrophys. Obs. Spec. Rep. No. 216 (1966).
[7] P. M. Muller and W. L. Sjøgren, Science **161**, 680 (1968).

Space Research XIV — Akademie-Verlag, Berlin 1974

RADON EMANATION FROM THE MOON: SPATIAL AND TEMPORAL VARIABILITY

P. GORENSTEIN*, L. GOLUB and P. BJORKHOLM

American Science and Engineering, Cambridge, Mass., USA

Observations of ^{222}Rn and ^{210}Po on the lunar surface with the orbiting Apollo Alpha Particle Spectrometer reveal a number of features in their spatial distribution and indicate the existence of time variations in lunar radon emission. Localized ^{222}Rn or ^{210}Po around the craters Aristarchus and Grimaldi and the edges of virtually all maria indicates time varying radon emission and suggests a correlation between alpha "hot spots" and sites of transient optical events observed from the earth. In a gross sense, the slower variations of ^{222}Rn seem to correlate with the distribution of gamma activity.

1. Introduction

Radon-222 is a radioactive noble gas which is produced during the decay of uranium. Under certain conditions it may diffuse through the lunar regolith and reach the surface before decaying by alpha particle emission. The alpha particle spectrometer aboard the Apollo 15 and 16 spacecrafts was designed to detect and identify the alpha particles emitted by ^{222}Rn and her daughter products during decay [1]. A variation in the ^{222}Rn signal over the lunar surface indicates either a variation in the parent uranium concentration and/or the ability of the radon to migrate to the lunar surface. Since the uranium concentration in the regolith has been determined by other orbital measurements [2], lunar regions with high radon mobility can be identified.

2. Observations

We will discuss general trends and then the specific correlation with lunar features. The important point to bear in mind is that because of its short half-life, 3.8 days, the decay of ^{222}Rn represents activity that was occurring at the time of observation. However, its descendent ^{210}Po is the daughter of ^{210}Pb which has a half-life of 21 years. Therefore, ^{210}Po activity is the result of ^{222}Rn reaching the lunar surface prior to the observation (from about 10 days to 60 years). Hence, the observed ^{222}Rn to ^{210}Po ratio is an indication of the secular history of the radon emanation.

---

* Now at the Center for Astrophysics, Harvard College Observatory/Smithsonian Astrophysical Observatory, Cambridge, Mass., USA.

2.1 Global Trends

Fig. 1 shows the observed count rate for ^{222}Rn (plus her two prompt daughters) as a function of longitude for the Apollo 15 groundtrack. While the ability to observe detailed structure in these data is limited by the counting statistics, there is an increase in the amount of ^{222}Rn observed over the western front side (Oceanus Procellarum and Mare Imbium) and to a lesser extent over the eastern front side (Maria Serenitatis, Tranquillitatis, Fecunditatis and Crisium). The single bin labelled "Arist." (for Aristarchus) in Fig. 1 has been excluded in the

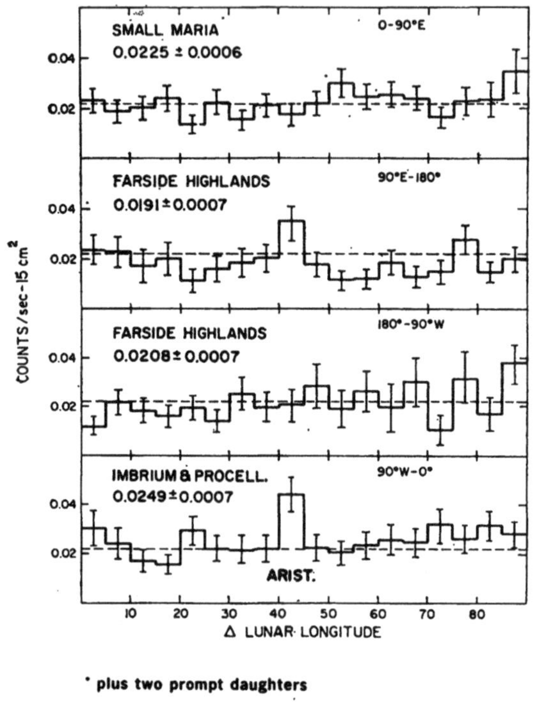

Fig. 1. Distribution of the observed ^{222}Rn (plus two prompt daughters) count rate as a function of longitude for the Apollo 15 ground track. The dashed line represents the average decay rate for the total observed data (including background). The average count rate for each quadrant is indicated in each panel.

average. This general trend is in agreement with the uranium and thorium concentrations over the same region as observed with the gamma ray spectrometer [2]. The highest levels are seen in the quadrant containing Mare Imbrium and Oceanus Procellarum and the lowest over the farside highlands. The ordering of the alpha particle and gamma-ray data is similar. Both depend on the variation of uranium so that it is not surprising that a correlation exists. However, the rate of alpha emission also depends on the local radon conductance through the regolith. As we have noted, this varies as a function of time. If there were sufficient counting statistics, a detailed comparison of the alpha and gamma results would locate the regions of high conductance.

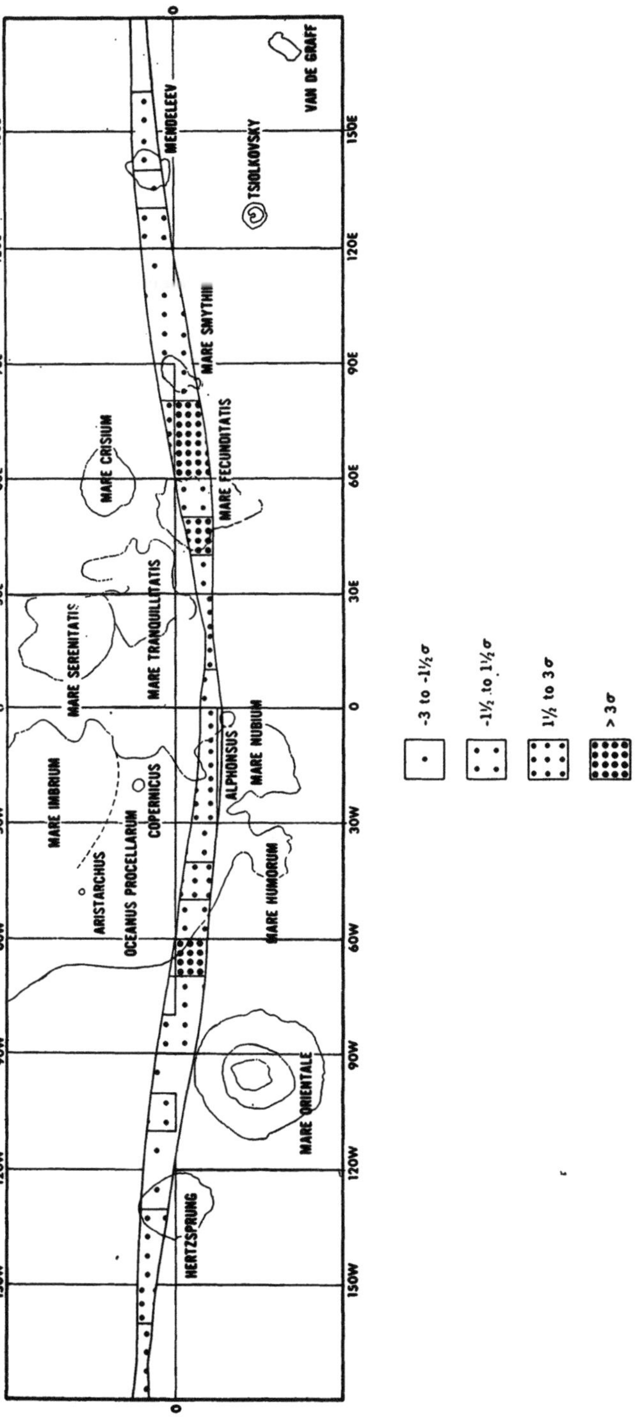

Fig. 2. Apollo 16 count rate of lunar ^{210}Po displayed on a schematic lunar map. The data are grouped in 10 × 10° bins. The symbols are given on the figure and refer to standard deviations (counting statistics only) above (or below) the measured background level.

The ^{210}Po distributions show a considerably larger spatial variation. The Apollo 15 ^{210}Po distribution (including a cosmic ray background) shows an increase in signal between 40° E and 180° E. The average count rate between 40° E and 180° E is 0.72 ± 0.002 counts/sec and 0.062 ± 0.001 counts/sec elsewhere. The excess count rate between 40° E and 180° E corresponds to a ^{210}Po decay rate of $(4.6 \pm 1.4) \times 10^{-3}$ disintegration $cm^{-2}\ s^{-1}$ (dis $cm^2\ s^{-1}$). The Apollo 16 ^{210}Po distribution shows an even more striking spatial variation. Because the Apollo 16 mission had a fairly low inclination orbit, the statistics are sufficient to generate a two-dimensional display. This is shown in Fig. 2. Here the data are grouped into 10° × 10° bins and displayed relative to the background counting

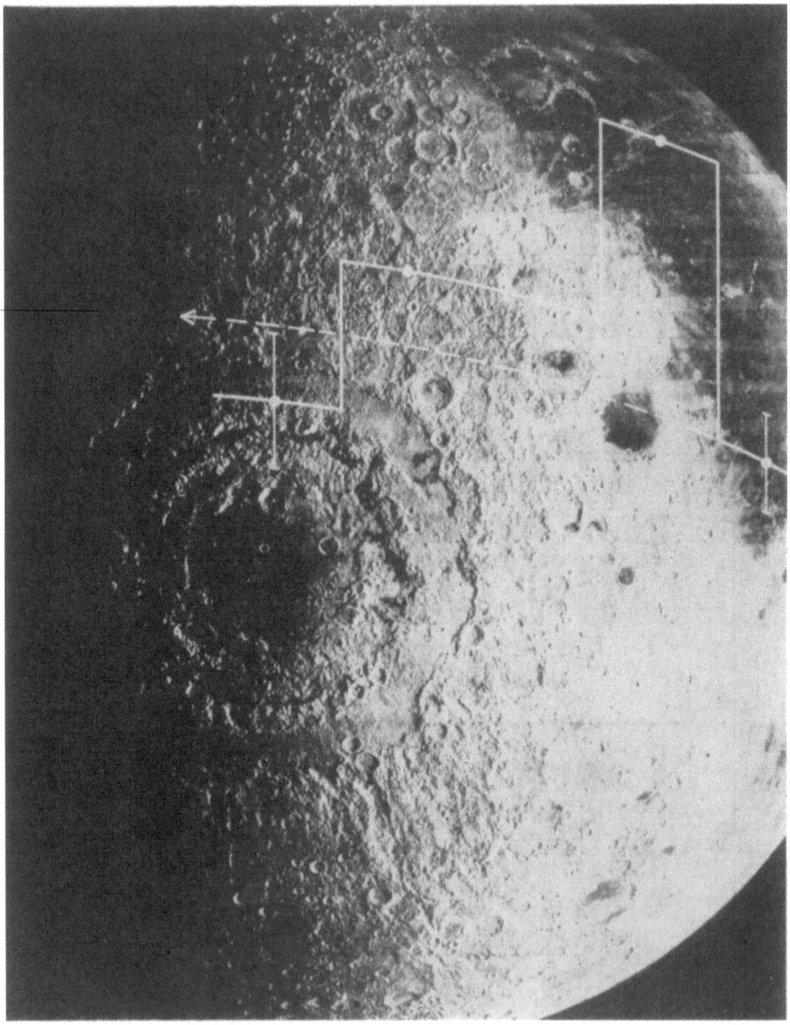

Fig. 3. The observed ^{210}Po count rate superimposed on a photograph of the moon. The dashed line represents the average ground track and is used to represent the background counting level. The crater directly beneath the local increase in count rate is the crater Grimaldi.

rate obtained from data taken with the instrument in a non-lunar orientation. These data are discussed in more detail in [3].

The observed amounts of ^{222}Rn and ^{210}Po are not in radioactive equilibrium. This implies that the process causing the distribution is time varying.

2.2. Correlations with Topological Features

Craters. We have already reported on the basis of Apollo 15 data an increase in ^{222}Rn activity in the region of the crater Aristarchus [4]. The excess over background is $(10.1 \pm 3.3) \times 10^{-3}$ dis cm^{-2} s^{-1}. A similar feature was seen in the Apollo 16 ^{210}Po data, a very strong increase over the crater Grimaldi (Fig. 3). The ^{210}Po decay rate directly over the crater Grimaldi exceeds the background rate by $(3.3 \pm 1.1) \times 10^{-3}$ dis cm^{-2} s^{-1}.

Maria Edges. Another example of a spatially localized concentration of ^{210}Po is shown in Fig. 4. This count rate for ^{210}Po decays is shown as the ground track crosses Mare Fecunditatis. The highlands to either side of the Mare, the edges of the Mare, and the central region are indicated on the figure. The observed rate falls to background levels on either side of the Mare and in the center. The increase is associated strictly with the edges.

While Mare Fecunditatis shows this effect most dramatically, it is not unique. Table 1 gives the count rate at all observed Mare edges relative to background. The data are divided into five categories based on the distance from the Mare edge. The terms "In" and "Out" are used to refer to data taken over the Mare

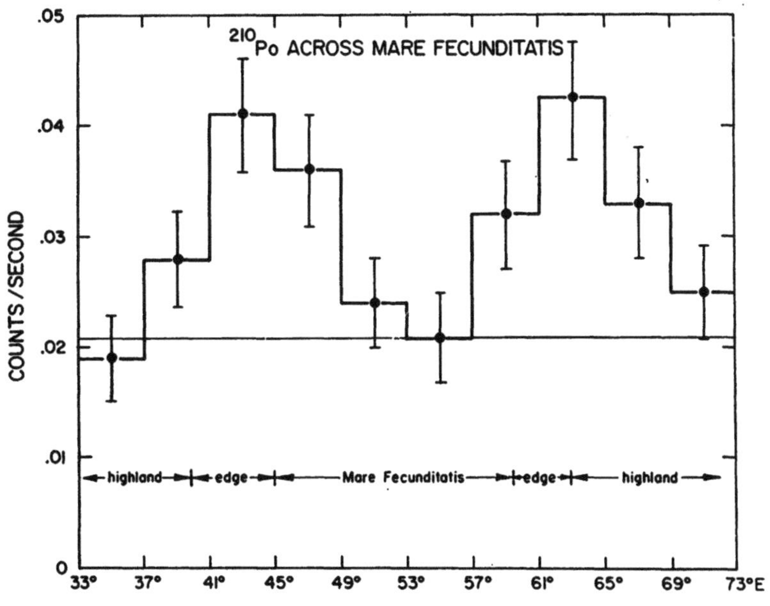

Fig. 4. The observed ^{210}Po count rate as a function of longitude as the ground track Mare Fecunditatis (Apollo 16). The solid line is the observed background level. The highland, mare, and edge regions are indicated. The finite width to the "edge" is caused by the oblique crossing of the mare-highland interface by the ground track.

Table 1

^{210}Po Decay Rate at Maria Edges[a]

Mare	In 200–350 km	In 75–200 km	Edge ±75 km	Out 75–200 km	Out 200–350 km
A. *Apollo 15*					
Fecunditatis	−0.1 ± 4.3	11.1 ± 4.4	12.5 ± 3.5	8.1 ± 5.8	no data
Crisium	no data	no data	20.1 ± 5.2	4.1 ± 5.6	−5.6 ± 8.0
Smythii	no data	−5.6 ± 10.1	14.1 ± 4.2	9.0 ± 6.3	6.2 ± 7.5
Procellarum	−7.7 ± 5.3	1.5 ± 3.8	5.4 ± 4.1	4.0 ± 4.2	−6.7 ± 6.2
Imbrium	3.4 ± 4.5	2.3 ± 4.2	4.6 ± 3.1	no data	no data
Tranquillitatis	2.1 ± 6.3	−7.7 ± 2.8	8.0 ± 2.6	no data	no data
Serenitatis	no data	5.1 ± 4.4	4.0 ± 2.9	4.5 ± 3.4	no data
Total Apollo 15	−0.1 ± 2.5	1.0 ± 1.6	8.4 ± 1.3	5.3 ± 2.1	−1.9 ± 4.1
B. *Apollo 16*					
Fecunditatis	no data	6.8 ± 2.1	19.5 ± 3.1	7.1 ± 2.8	8.3 ± 7.1
Smythii	no data	3.4 ± 4.7	9.5 ± 3.1	10.9 ± 3.0	7.5 ± 5.3
Tranquillitatis	no data	no data	11.6 ± 3.8	4.7 ± 2.4	9.3 ± 2.7
Nubium	no data	no data	11.5 ± 3.6	8.1 ± 2.8	3.9 ± 3.4
Cognitum	no data	6.6 ± 2.8	11.2 ± 2.7	7.3 ± 3.7	no data
Total Apollo 16	no data	6.5 ± 1.7	13.3 ± 1.5	7.7 ± 1.3	7.6 ± 1.9
Total A 15 & A 16	−0.1 ± 2.5	2.9 ± 1.3	9.8 ± 1.0	6.7 ± 1.2	4.9 ± 1.8

[a] Rates are $\times 10^{-3}$ cts s^{-1} = 0.36×10^{-3} dis cm^{-2} s^{-1} at the lunar surface.
"In" and "Out" refer to mare surface and surrounding highlands respectively.

surface and over the surrounding highlands, respectively. The data from the two missions for the same Mare need not agree since different portions of the Mare were observed. All of the observed Maria, with the exception of Serenitatis, show an enhanced decay rate for ^{210}Po at their edges. In particular, Maria Fecunditatis and Crisium show this effect most dramatically.

3. Discussion

3.1. Summary of Observational Results

The level of ^{222}Rn is finite, about 10^{-3} of terrestrial values. ^{222}Rn is one of perhaps only two gaseous species that have been identified as having a lunar origin, the other being ^{40}Ar [5]. It is reasonable to expect that the rather short half-life of ^{222}Rn gas (3.8 d) could lead to the preservation of features which appear initially in the spatial distribution of ^{222}Rn and her daughter product ^{210}Po. The features include localized emission that is associated with certain well-known lunar craters and maria edges, in addition to a diffuse component of ^{222}Rn and ^{210}Po that exists more or less over the entire moon.

The following statements summarize our results.

(1) The process of radon evolution occurs over a large portion of the moon as evidenced by the widespread distribution of ^{222}Rn and ^{210}Po. However, it need not act simultaneously in many places.

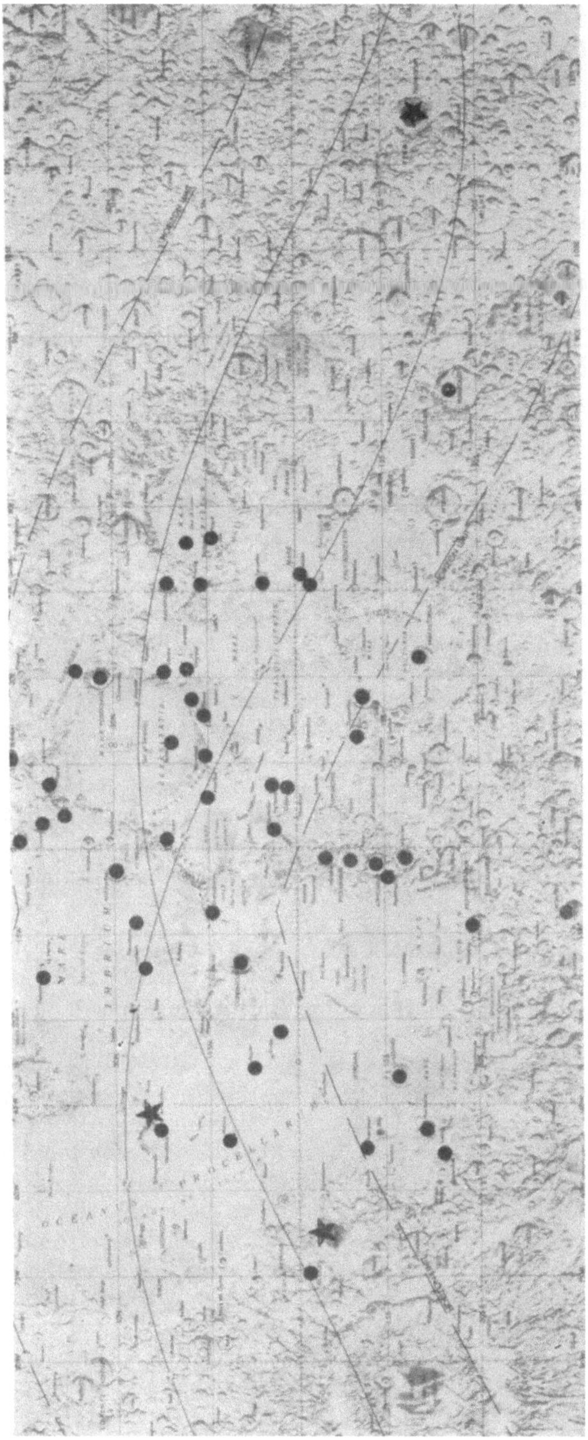

Fig. 5. Black dots indicate sites where the more reliable transient events have been reported. Stars indicate localzed concentrations of ^{222}Rn and ^{210}Po detected in either Apollo 15 or Apollo 16. Reports of transient events have been summarized by Middlehurst [6].

(2) There exist variations of ^{222}Rn and ^{210}Po activity across specific regions and across maria edges which are rapid compared with the variation of the concentration of uranium, the source of radon.

(3) The process is time dependent and varying on a time scale between $\sim$3.8 days and $\sim$21 years which are the half lives of ^{222}Rn and ^{210}Po. The evidence for time dependence is the detection of ^{210}Po activity which exceeds that of its progenitor ^{222}Rn at various locations and on the moon as a whole.

These results appear to rule out external processes, e.g., meteor bombardment or mass slumping, as a means of releasing radon. An external process would require a mass of several tenths g cm^{-2} of lunar material *per year* to produce the radon. In view of the persistence of lunar surface features on a time scale of 10^9 years, external processes require more mechanical churning of the surface than actually occurs. Furthermore, an external process could not explain items (2) and (3). Therefore, the mechanism responsible for the emission of radon must be a time-varying, internal process.

3.2. Correlation with Sites of Transient Lunar Phenomena

Two of the brightest alpha particle "hot spots" occur in the vicinity of the well-known craters Aristarchus and Grimaldi. The significance of these two craters is that over a long period they have been reported repeatedly by ground-based observers to be sites of transient optical events. In the case of Aristarchus the reports go back several hundred years. Middlehurst and co-workers [6] have summarized these reports. Along with the discrete repeating sites of transient optical events such as Aristarchus, Grimaldi, and Alphonsus there are a large number of reports of smaller events at non-specific sites. Fig. 5 illustrates the distribution of the events which are judged to be most reliable. It is clear that the distribution correlates with the edges of the maria. In light of the fact that we observe a rather striking ^{210}Po effect at the edges of the maria, it appears that there is a correlation between radon hot spots and sites of transient lunar phenomena.

3.3. Enhancement near Lunar Terminator

Since radon is a gas, the solar heating effects are expected to influence its distribution on the lunar surface. Heymann and Yaniv [7] have described a model of radon transport which predicts that the peak activity will occur at the sunrise terminator. The model is based on the hypothesis that free radon atoms will diffuse rapidly on the sunlit (warm) side and condense as soon as they enter on the dark (cold) side. Fig. 6 shows the observed radon activity for Apollo 15/16 as a function of angle from the sunrise terminator. There is indeed a peak in the vicinity of the terminator; however, it seems to be somewhat past the sunrise terminator. Any attempt to correlate ^{222}Rn data with the terminator is complicated by the fact that there are other spatial features such as the concentration at Aristarchus and a diffuse component. Fig. 6 also shows some preliminary results from another model of radon outgassing that is presently being considered by us. In this model the radon concentration in the regolith just below the surface builds up while the moon is cold. Little radon leaves the surface until the sun heats it to a certain critical temperature. At that temperature the layer is suddenly purged of all the radon accumulated during the entire time it has been cold. As

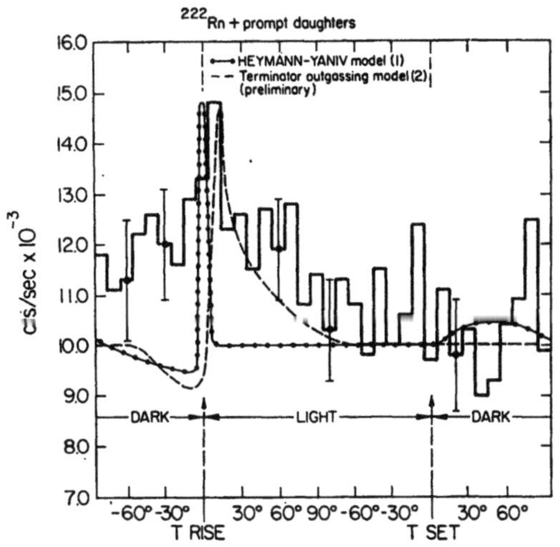

Fig. 6. Distribution of ^{222}Rn during Apollo 15 and Apollo 16 with respect to terminator. Two theoretical models are shown, that of Heymann and Yaniv [7] and another being considered by us.

it is heated to still higher temperatures, there is no storage; radon escapes rapidly and the observed activity should be low once more. This model predicts that peak activity will be seen in the region where the temperature has reached the critical value. This should be somewhat past the sunrise terminator. Due to the complication mentioned above and the fact that our counting statistics are limited, it will be difficult to discriminate between the two models on the basis of these data.

Acknowledgment

This paper was supported by the Johnson Spacecraft Center of the National Aeronautics and Space Administration under Contract NAS 9-9982.

References

[1] P. Gorenstein and P. Bjorkholm, Proc. 3rd Lunar Sci. Conf., Geochim. Cosmochim. Acta Suppl. 3, **3**, 2179 (1972).
[2] A. E. Metzger et al., Science **179**, 800 (1973).
[3] P. Bjorkholm, L. Golub and P. Gorenstein, Science **180**, 957 (1973).
[4] P. Gorenstein and P. Bjorkholm, Science **179**, 792 (1973).
[5] J. H. Hoffman, R. R. Hodges, Jr, F. S. Johnson and D. E. Evans, Proc. 4th Lunar Sci. Conf. Geochim. Cosmochim. Acta Suppl. 4, **3**, 2865 (1973).
[6] B. M. Middlehurst, Rev. Geophys. **5**, 173 (1967).
[7] D. Heymann and A. Yaniv, Nature Phys. Sci. **233**, 37 (1971).

Space Research XIV — Akademie-Verlag, Berlin 1974

RADIOACTIVITY OF LUNAR SOIL SUPPLIED BY THE AUTOMATIC STATION LUNA 20

YU. A. SURKOV, G. A. FEDOSEYEV, O. P. SOBORNOV and L. S. TARASOV

V. I. Vernadsky Institute of Geochemistry and Analytical Chemistry, Academy of Sciences of the USSR, Moscow, USSR

The data on the radioactivity of lunar soil supplied by Luna 20 from the highland region near the crater Apollonium C were obtained by the low-background gamma-spectrometry method. On the basis of the data obtained in this experiment and the results achieved by other authors a conclusion is drawn about some peculiarities of regolith formation in the above region of the moon. Proceeding from the natural radio-element content in lunar samples and from its comparison with the data on the radioactivity of basic and ultrabasic terrestrial rocks, the hypothesis has been made that there is similarity between the primary matter of he earth and that of the moon.

1. Introduction

Investigations of the moon carried out over the last several years with the aid of Soviet and American space vehicles have allowed the return of valuable material on its nature. In particular a vast quantity of information was obtained as a result of laboratory investigations of the radioactivity of lunar samples selected by the first missions from "sea" regions of the moon. At the same time it became apparent that the elucidation of some aspects of the moon's evolution, of the formation peculiarities of its rocks and the regolith, of the radiation conditions etc. would require the extension of these investigations to new regions of the moon, particularly to the highlands.

In the present paper results of radioactivity measurements of the lunar soil supplied by the automatic station Luna 20 from highland regions near the crater Apollonium C are considered and their analysis is carried out in the light of up-to-date ideas about the radioactivity of terrestrial and lunar rocks as well as of stony meteorites.

2. The Experimental Technique and the Results achieved

As is known [1], the lunar soil sampled with the aid of a drilling device of the station Luna 20 is in the form of small granular material containing fragments of various rocks, chiefly of anorthosite type. A sample weighing 10 g, taken from the upper part of the soil column, was used for investigation. The determination of the content of natural radio-elements and cosmogenic radioisotopes in it was carried out in the same way as during analysis of the regolith sample of Luna 16 [2], by the method of low-background gamma-spectrometry. The description of the gamma-spectrometric installation with anticoincidence protection used by us for the measurement of low activities in samples with a relatively small weight and volume and the methods of the analysis of lunar sample radioactivity are given in [3, 4].

The radio-element contents of the sample under investigation were determined by the resolution of recorded gamma-spectra with a computer by the method of least squares. The mean-square errors of content determination and spectrum approximation were also calculated assuming a six-component radioisotopic composition of the sample being measured (^{40}K, ^{238}U, ^{232}Th, ^{22}Na, ^{26}Al, ^{54}Mn). At the time of the lunar sample investigation the other cosmogenic radioisotopes were either disintegrated, or their content was beyond the detection limit of the apparatus. As a result of the analysis carried out the following data on the content of natural radio-elements and cosmogenic radioisotopes in the regolith sample of Luna 20 have been obtained: $K = 0.075 \pm 0.025\%$; $U = (0.22 \pm 0.08) \times 10^{-4}\%$, $Th = (0.73 \pm 0.12) \times 10^{-4}\%$, $^{26}Al = 46 \pm 32$ dpm kg^{-1}, $^{22}Na = 80 \pm 39$ dpm kg^{-1}; $^{54}Mn = 21 \pm 16$ dpm kg^{-1}.

3. Discussion of Results

3.1. Cosmogenic Radioisotopes in the Regolith of Luna 20

As is known, the study of the final distribution of products of nuclear reactions induced by solar and galactic cosmic radiation makes it possible to obtain some information about the history of the regolith formation in the region of the moon being investigated. However, since the sample used for investigation had been homogenized (from 0 to 22 cm depth) during selection from the lunar soil column the data obtained on the content of cosmogenic radioisotopes are of somewhat limited value. It is nevertheless of some interest to consider the data on the radioactivity of cosmogenic radioisotopes averaged over the column depth and to compare them with known data on the ^{26}Al and ^{22}Na content of samples taken from other regions of the moon (Table 1).

Table 1

^{26}Al and ^{22}Na Content (dpm kg^{-1}) in Regolith Samples from Different Regions of the Moon

Isotopes	Region of the Apollonium C crater	Mare Tranquillitatis [5]	Oceanus Procellarum [6]	Mare Fecunditatis [2]	Region of the Fra Mauro crater [7]	Region of the Hadley-Apennines [8]	Descartes region [9]
^{26}Al	46 ± 32	86—137	58—140	63	78—220	38—175	84—225
^{22}Na	80 ± 39	51— 61	27— 80	45	45— 84	25— 57	27— 86

The relatively high ^{22}Na content of the regolith of Luna 20 in comparison with Luna 16 is, in our opinion, connected with a difference in the chemical composition and, in particular, an appreciably (almost 2 times) greater Na_2O content in the former.

The ratio between the ^{26}Al and ^{22}Na contents of lunar samples is, as a rule, 2—3 times that of meteorites, which is caused by the prevailing ^{26}Al formation under the action of low-energy solar protons [10]. However, for the sample of Luna 20 such a regularity is not observed. This can be accounted for by assuming that the sample had not been on the moon's surface long (i.e. that it has a low exposure age [1], and hence the activity of cosmogenic radioisotopes in the Luna 20 sample is, to a considerable extent, caused by the action of the primary galactic component and the secondary nuclear active particles.

As is known, the influence of solar wind is almost entirely limited to the top few millimetres of the regolith. If in the Luna 16 regolith signs of the solar wind and in particular a great amount of solar rare gases [11] were revealed from the top of the column downwards and in olivine samples a high density of tracks ($\sim 7 \times 10^7$ tracks cm^{-3}) has been fixed, then in the Luna 20 olivine the track density and the content of the inert gases is substantially less [1].

Thus it must be supposed that recently there has been either a formation of Luna 20 regolith (young ejection) or a removal of its upper layer. Ablation of the most radioactive upper regolith layers from the highest areas of the moon's surface under the action of various external factors, especially of meteorite impacts, is possible and obviously leads to a rejuvenation of such areas. For instance, very distinct traces of erosion were detected by participants of the Apollo 15 mission in the Hadley Rille region, where on a large platform at the edge of the inspected terrace the loose regolith layer was practically lacking. In order to determine the relative importance of the proposed mechanisms (ejection and ablation) it is necessary to study the distribution of radioactivity in thin layers of regolith from various depths.

3.2. Natural Radio-elements in Lunar Samples

In Table 2 data on K, U and Th contents of regolith samples of various regions of the moon are listed. It follows from the table that regolith of the highland region near the Apollonium C crater has the lowest concentration of natural radio-elements, which is primarily due to the fact that the material of Luna 20 consisted of 50—60% of anorthositic rocks which usually are depleted in these elements. In fact, in the analysis of data on the natural radioactive element distribution of natural objects including lunar rocks, the potassium-uranium systematics are of definite interest. If the ratios between the thorium and uranium contents for samples from various regions of the moon lie in the narrow range of values Th/U 3.2—4.0 then the K/U ratios of samples given in Table 2 and Fig. 1 substantially differ. The wide range of the listed K/U values may be indicative of differences in the character of the magmatic differentiation process for different regions of the moon. At the same time the constancy of K/U ratios for different samples of one region (crystalline rocks, breccia, regolith fines) attests the local character of regolith formation in the investigated regions of the moon [13].

At the present time the difference between the K/U ratios of moon rocks and earth rocks is of great interest. In a number of publications explanations of the

Table 2

Contents of Natural Radioelements in Regolith Samples from Different Regions of the Moon

Elements	Region of the Apollonium C crater [5]	Mare Tranquilitatis [5]	Oceanus Procellarum [6]	Mare Fecunditatis [2, 12]	Region of the Fra Mauro crater [7]	Region of the Hadley-Apennines [8]	Descartes region [9]
K %	0.075	0.11	0.21—0.44	0.095—0.14	0.40—0.48	0.12—0.19	0.055—0.11
U, 10^{-4} %	0.22	0.59	1.5 —3.4	0.25 —0.3	3.8 —4.1	0.4 —1.3	0.32 —0.74
Th, 10^{-4} %	0.73	2.24	6.0 —13.2	0.80	13.4—14.9	1.9 —5.1	1.12 —2.63
Th/U	3.3	3.8	4.0	3.2	3.6	3.8	3.7
K/U	3400	2000	1450	3800—4700	1100	1230—1850	1400—1800

difference between the K/U ratios for the earth and the moon have been presented [8, 14, 15]. Indeed, if we proceed from data on K/U ratios of rocks of the earth (1×10^4—1.2×10^4) and the moon (1.2×10^3—4×10^3) which are given in these papers, the conclusion drawn by the authors seems to be correct. However, we think it expedient to analyse more critically the listed values of K/U ratios according to new data of lunar and terrestrial rocks.

In Vinogradov's opinion [1] the formation of the superficial anorthosite layer on the moon occurred during fast cooling of the erupted high-temperature gabbroid magma. In this case the character of the melt cooling on the surface of the moon is greatly influenced by the high vacuum. As the vapour pressure over the melt surface was practically equal to zero and increases rather slowly with depth, there must have been a boiling of the melt resulting in its depletion in volatile elements. Under these conditions the greatest loss was experienced by those elements with the greatest vapour pressure at the melt temparature. In particular, the pressure of K_2O vapours in the 1250—1500 °C temperature range is much higher than for most oxides of other alkaline and alkaline-earth elements. The selective loss of volatile components in the surface layer of the melt, having been confirmed in a model experiment [16], must be one of the reasons for the difference in the chemical composition between rocks from most superficial and most deep-seated layers of the moon. This peculiarity of the formation of the superficial rocks of the moon applies equally as well to lava outflows of basalt type and to all other effusive rocks. Thus, the potassium content and hence, the K/U ratio in the superficial layers of basalt covers may apparently not correspond to that of abyssal rocks.

Brown and Peckett [17] came to an analogous conclusion when they directly investigated samples of lunar basalt (Fra Mauro) enriched in plagioclase. Based on results of chemical analysis of different zones of the phenocryst plagioclase which was compared with the contents of modal plagioclase, the authors made an estimate of the initial contents of alkaline components in the magma and came to the conclusion that K_2O losses through volatilization reach 78%.

However, it must be noted that till now there has been no clarity in the notions about the mechanism of the above process of the superficial-cover formation of the moon. Thus for instance in the latest papers [18], based on the experimental study of the sequence of basalt phase crystallization from melts with different Na_2O and K_2O content (the experiment was carried out at 1 atm pressure) the authors argue against any role of selective volatilization on the lunar surface.

As is known, regolith is formed from rocks which crop out on the surface of the moon. It may be that both effused and more abyssal rocks appeared on the surface as a result of a long erosion process. Regolith formation consists of the crushing, refusing and sintering of all these rock fragments. During this process a certain redistribution of elements may take place. Potassium or uranium loss in the course of regolith formation may lead to a K/U ratio displacement with respect to the primary rocks. On the other hand some averaging of its composition as a result of a mixing of rock fragments of different origin and history is not ruled out. Thus a comparison of compositions of only abyssal rocks, which may represent the mantle matter (i.e. such rocks the compositions of which are not distorted by secondary processes) will allow us to answer correctly the question about the genetic relationship between the matter of the earth and of the moon.

Taking into account all the above, we shall try to carry out a comparison of K/U ratio in terrestrial and lunar rocks which are listed in Fig. 1. Fisher [19] has noted that the scheme used by Fanale and Nash [13] during analysis of K/U ratios of terrestrial rocks is very simplified and far from being based on complete data. In particular, data on the ratios of natural radio-elements in basic and ultrabasic rocks have been substantially widened and more precisely defined, primarily as a result of intensive investigations of bedrocks in the rift zone of the oceans. As is

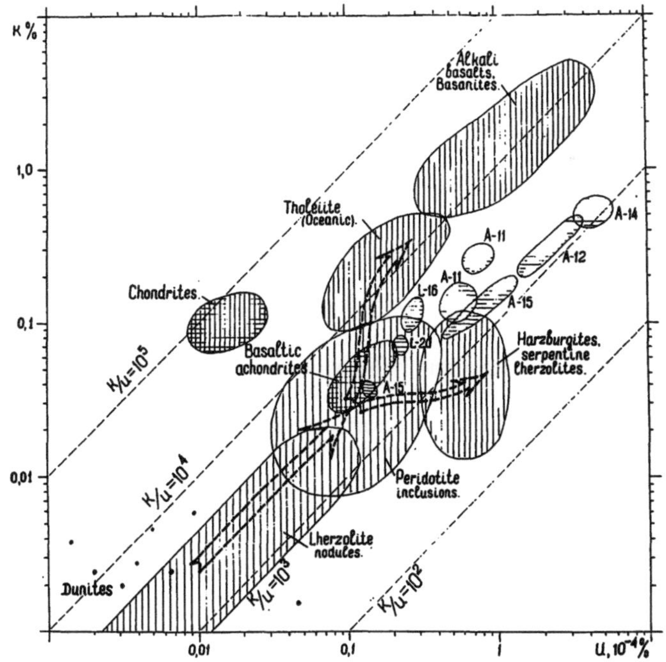

Fig. 1. Potassium-uranium systematics for lunar, chondritic, achondritic and terrestrial samples.

known, these investigations are concerned with the processes of the formation of the primary earth crust which is undoubtedly related to lunar problems.

From the point of view of studyin the mantle matter the analysis of ultrabasic rocks of the middle oceanic ridges is also of great interest. These rocks are subdivided into two main groups: the group of lherzolites and the group of harzburgites. The petrographical and petrochemical investigation of oceanic rock samples allowed the authors [20, 21] to consider the genetic interconnection between ultrabasites and basalts. As a result of these investigations the complementarity of lherzolite, harzburgite and basalt compositions was shown: lherzolites were classified by the authors as the initial matter of the upper mantle and harzburgites as the residuals after tholeiitic basalt smelting.

Recently determinations were made of the uranium and potassium content of some samples of rocks from the rift zone of the Indian ocean. These also widen the series of data [13, 19, 22] on lherzolites, dunites, tholeiitic basalts and basalts of other types. From [19, 20] it follows that the region of average values of K/U ratios for lherzolite nodules of oceanic regions occupies a narrow zone in the $(2-3) \times 10^3$ range, i.e. it is much below the zone of K/U ratios for the succession of tholeiitic basalts, both oceanic and continental ($K/U = 7 \times 10^3$) and the main mass of alkaline basalts ($K/U = (1-2) \times 10^4$).

When the data on K/U ratios listed in Fig. 1 are considered, attention is drawn to the branching of the lherzolite zone of K/U ratois, $(1-5) \times 10^3$, into two series substantially differing from one another. The upper series, which begins with oceanic tholeiitic basalts, is characterized by a higher K/U ratio ($5 \times 10^3 - 2 \times 10^4$). It is natural to assume that the K/U ratio increase in these rocks must be accompanied by its decrease in other formations being connected with the first by a single genetic process. Among rocks and formations within the bounds of the continental crust, local decreases of the K/U ratios are not rare and, as a rule, are connected with heightened uranium concentrations which have arisen in the course of its geochemical cycle. Among oceanic formations, the group of rocks with lowered K/U ratio is of special interest. In this sense the second series, outlined by data on harzburgites and serpentinous lherzolites and located below the continuation of the zone of K/U ratios which are characteristic of lherzolites of the mantle, is thought of as being additional to the series of oceanic tholeiitic basalts.

The K/U ratio decrease in the indicated harzburgitic rocks of the rift zone is caused by the uranium content which is very high for ultrabasic rocks. Based on a number of facts it may be supposed that such high uranium contents are the result of a secondary enrichment, which occurred in the course of the reworking of these rocks by the products of mantle degassing, by which uranium was supplied. In the given case it is of importance that rocks being directly connected with the mantle are enriched in uranium. There can be no influence of the continental crust in rift regions and all variations of K/U ratio are caused only by differentiation in the mantle matter and its products. The comparison of K/U ratios for terrestrial and lunar rocks gives grounds to assume that the "lherzolite" zone, being characteristic of terrestrial mantle formations, must also be characteristic of the primary lunar matter, i.e. K/U ratio $(1-5) \times 10^3$ is also peculiar to it. Indeed, the regions of the K/U ratio of rocks from various places of the moon are within the range of the values of K/U ratio $(1-5) \times 10^3$ continuing by this the zone of lherzolitic mantle rocks of the earth. As they already are a differentiated material of the lunar mantle, i.e. apparently being enriched in uranium and potassium to some extent, the primary undifferentiated material must have a lower concentration of them. The slight variations of the K/U ratio within the bounds of lunar rocks of basalt and anorthosite types allows as to assume that their abyssal parent rocks must also possess similar K/U ratios.

As has been noted in [13], the difference in the inclinations in correlation lines characterizing in K-U coordinates some "families" of lunar samples, apparently argues for the existence of complementary parts of these families. The available experimental data on the radioactivity of lunar samples are still not numerous and cannot yet give a direct proof of the genetic relationship between lunar rocks corresponding to the different correlation lines. The same difficulty is encountered in the determination of the region of the intersection of correlation lines which corresponds to the primary matter. Nevertheless if it is thought that the measured limiting values of the whole range of K/U ratios (1.0×10^3—4.7×10^3) characterize the differentiation degree of the lunar matter, then the average value ($\sim 2.8 \times 10^3$) apparently must in all probability correspond to the initial undifferentiated matter. It is necessary to note that the K/U $\sim 2.8 \times 10^3$ value practically coincides not only with the average K/U value for the lherzolite mantle, but also corresponds with the region of K/U ratios which characterizes calcium-rich achondrites (see Fig. 1). This correspondence supports the achondritic model of the formation of protoplanetic bodies of the solar system.

References

[1] A.P. Vinogradov, Geokhimia, No. 7, 763 (1972).
[2] Yu. A. Surkov, G. A. Fedoseyev and O. P. Sobornov, Kosm. Issled. **10**, 938 (1972).
[3] Yu. A. Surkov, A. S. Stan, F. F. Kirnozov, I. N. Ivanov and L. P. Moskaleva, Space Research XII, 33 (1972).
[4] Yu. A. Surkov and O. P. Sobornov, Atomnaya Energiya **34**, 125 (1973).
[5] LSPET*, Science **165**, 1211 (1969).
[6] LSPET*, Science **167**, 1325 (1970).
[7] LSPET*, Science **173**, 681 (1971).
[8] LSPET*, Science **175**, 363 (1972).
[9] LSPET* Science **179**, 23 (1973).
[10] R. W. Perkins, L. A. Rancitelli, J. A. Cooper, J. H. Kaye and N. A. Wogman, Science **167**, 577 (1970).
[11] A. P. Vinogradov, Geokhimia, No. 3, 261 (1971).
[12] A. L. Albee, A. A. Chodos, A. J. Gancarz, E. L. Haines, D. A. Papanastassiou, L. Ray, F. Tera, G. J. Wasserburg and T. Wen, Earth Planet Sci. Lett. **13**, 353 (1972).
[13] F. P. Fanale and D. B. Nash, Science **171**, 282 (1971).
[14] G. D. O'Kelley, I. S. Eldridge, E. Schonfeld and K. I. Northcutt, Science **175**, 440 (1972).
[15] N. W. Hinners, Revs. Geophys. Space Phys. **9**, 447 (1971).
[16] A. M. Gutkin, M. S. Markov, C. M. Raytburd and M. V. Slonimskaya, in: Physics of the Moon and the Planets, Nauka, Moscow 1972 (p. 189).
[17] G. M. Brown and A. Peckett, Nature **234**, 262 (1971).
[18] D. H. Green, N. G. Ware and W. O. Hibberson, Nature **238**, 450 (1972).
[19] D. E. Fisher, Science **172**, 1166 (1971).
[20] L. V. Dmitriev, in: Investigation on Problem of World Ocean Rift Zones **1**, Nauka, Moscow 1972 (p. 115).
[21] L. V. Dmitriev, A. V. Ukhanov and A. Ya. Sharaskin, Geokhimia, No. 10, 1155 (1972).
[22] D. H. Green, J. W. Morgan and K. S. Heier, Earth Planet. Sci. Lett. **4**, 155 (1968).

* LSPET = Lunar Sample Preliminary Examination Team.

Venus

Space Research XIV — Akademie-Verlag, Berlin 1974

COMPOSITION AND STRUCTURE OF THE CLOUD LAYER OF VENUS

YU. A. SURKOV, B. M. ANDREICHIKOV and O. M. KALINKINA

V. I. Vernadsky Institute of Geochemistry and Analytical Chemistry, Academy of Sciences of the USSR, Moscow, USSR

The problem of the Venus cloud layer is considered from the standpoint of geochemistry. Possible ways of nitrogen evolution in the Venus atmosphere are considered and the organization of the experiment for the determination of the ammonia content is substantiated. Results of the automatic station Venera 8 are given. On the grounds of the data obtained as well as of previously known results of cosmic and terrestrial measurements views about the composition and structure of the Venus cloud layers are expressed. A three-layer multicomponent model is proposed and the levels of the formation of layers and their composition are indicated.

As is known, the gas-analytical equipment of the automatic interplanetary station (AIS) Venera 4 has made it possible to determine for the first time the content of the main components of the Venus atmosphere [1]. These data were thereupon more precisely defined and enlarged with the aid of instruments of the AIS Venera 5 and Venera 6 [2]. It was found that the predominant component of the planet's atmosphere is carbon dioxide, 97%; nitrogen content does not exceed 2%, hydrogen content 0.1% and water vapour is present in relatively greater amounts only in the upper layers of the atmosphere.

However, despite the considerable advances in the study of the physico-chemical characteristics of the Venus atmosphere achieved with the aid of space equipment, the problem of the cloud layer of the planet is still not solved. There are a great number of quite diverse assumptions on this subject. In the main they all lead to the same conclusion, that the prevailing components of the upper clouds are condensates. However, what their nature is, and of what the compact Venus cloud cover consists, remains till now an enigma.

Most widely considered is the hypothesis that the cloud layer is of a water type. Quantitative estimates of water vapour content derived from spectroscopic data lead a number of authors to think that the Venus clouds consist of water drops or ice crystals. Results of direct measurements also indicate the presence of relatively great amounts of water vapour in the upper layers of the planet's atmosphere and the possibility of their condensation at a height of about 60 km. Some characteristics of such a cloud layer are listed in papers [3, 4]. At the level of the lower boundary of clouds a fall-out of precipitations is possible. Falling down, the drops evaporate (in the planet's atmosphere the boiling temperature of water is reached at a pressure of about 5 atm) and the water vapour diffuses to the clouds. The high temperature excludes the possibility of the formation of reservoirs on the surface of the Venus and, hence, a water circulation similar to the terrestrial one is lacking. Therefore the lower layers of the atmosphere are depleted of water to a considerable degree. Venus on the whole loses water during the process of photodissociation in the upper layers of the atmosphere and the subsequent hydrogen dissipation.

The hypothesis about the water cloud layer nevertheless meets with definite difficulties when it has to be brought into agreement with results of terrestrial measurements. One would think that the low relative content of water vapour derived from some spectroscopic data [5] excluded the possibility of H_2O condensation. However, some explanation of this discrepancy has been given in the analysis of Mariner 5 measurements carried out by the authors of [6]; in particular, a sharp decrease of the temperature gradient was noted by them at heights above 58 km. Thus, for instance, in the interval between 60 and 70 km it is about 4°/km, i.e. similar to the value obtained by us [3, 4]. At a height between 80 and 90 km a deep temperature inversion with a minimum about 180 °K has been noted [6]. At this temperature a sublimation of water is possible even at a P_{H_2O}/P_{CO_2} ratio equal to 10^{-6}, which may also co-ordinate estimations of H_2O content derived from data from cosmic and terrestrial measurements, if they can be related to different levels of the Venus atmosphere.

However, the absence of ice absorption bands in infrared spectra of Venus, the sharp decrease of the absorption coefficient in the ultraviolet part of the spectrum and, finally, results of polarization measurements do not agree with the water model of the cloud layer.

In [7] about 500 results of polarization measurements are given, on the grounds of which Loffeen [8] comes to the conclusion that the Venus clouds consist of particles with a 2.5 ± 0.5 μm diameter and that their refractive index lies between 1.43 and 1.55. The careful analysis of these data carried out by the authors of [9] permitted them to draw the conclusion that the particles of the cloud layer must be spherical with a 1 μm radius and refractive index 1.45 ± 0.02. It is known that water in the liquid and the solid phases has a lower refractive index. Particles of mineral dust explain the results of experiments best of all. In the presence of such particles the refractive index of water (or rather of aerosol) increases.

There also exist other hypotheses about the nature of the matter of which the compact cloud cover of Venus is composed. These suppose the existence of formaldehyde [10] as well as carbon suboxide [11] and hydrocarbon [12] and carbon sulphur oxide [13] and ferric chloride and even mercury and its halides [14]. The possibility of the presence of these compounds in the planet's atmosphere is for some reasons problematic; also, it has not be confirmed by spectroscopy.

Of definite interest is paper by Lewis [14] considering an aqueous solution of hydrochloric acid as a possible component of the Venus cloud layer. He has considered in detail the equilibrium of the $HCl-H_2O$ system. Under the conditions of the Venus cloud layer he has shown that it may consist of a 25% solution of hydrochloric acid. However, solid $HCl \cdot 3H_2O$ can only exist at a temperature below 196 °K [14], which does not agree with the infrared Venus temperature [15]. The refractive index of such a solution does not correspond to results of polarization observations. Also it must be noted that the possibility of the presence of such great amounts of HCl in the planet's atmosphere is very doubtful. Hydrochloric acid is a product of endogenic processes and must activity bind with rocks.

Finally it has to be noted that all the above considered assumptions about the nature of the matter forming the Venus clouds cannot explain a number of experimental data obtained with the aid of interplanetary automatic stations. In the first place this applies to the measurements of the density of the planet's atmosphere with an ionization type device installed on Venera 4. As is known, in this experiment a non-monotonic change of atmospheric density in the range of 2—7 atm pressures has been recorded [16]. As one of the possible causes of this anomaly, the authors consider the inconstancy of the atmosphere's chemical composition in the region of measurement which allowed them to estimate the possible maximum moisture content assuming that the atmosphere consists here of a mixture of H_2O and CO_2 and that, below, the partial pressure of water vapour is practically equal to zero in comparison with the CO_2 pressure. However the results obtained by them do not agree with data of the measurements of moisture content at that level carried out by Vinogradov et al. [1, 2].

This lack of correspondence can only be explained by the presence of some phase transition at a height about 50 km over the average surface level. Other results of direct measurements also argue in favour of this. Thus, for instance, according to measurement results of the stations Venera 5 and Venera 6 the refractive index for radiowaves changes at a level where the atmospheric pressure is about 1.5 atm [17]. In [6] it is noted that the weakening of signal at a height of about 50 km, which was observed during radiosounding of the Venus atmosphere by Mariner 5, exceeds the calculated value for an atmosphere consisting of a CO_2 and N_2 mixture (i.e. an additional absorbing agent is necessary). This effect cannot be explained by the presence of water because a great amount of it would be needed [18]. Thus, even the whole complex of results of cosmic and terrestrial measurements does not give a definite idea about the nature of the Venus cloud layer.

Let us try to consider the problem of the Venus clouds from geochemical standpoints. The peculiarity of the physico-chemical characteristics of the Venus atmosphere allows us to assume that the chemical evolution of some of her components must differ from the components of the earth's atmosphere. In this respect the evolution of nitrogen is of definite interest. The source of nitrogen, ammonia, was degassed in consequence of volcanic activity, as on the earth. In the presence of a sufficient amount of oxygen NH_3 is oxidized to free nitrogen, as has obviously taken place on the earth after the appearance of oxygen which had formed as a result of photosynthesis. However, in view of the small amount of O_2 and the CO_2 abundance in the Venus atmosphere the presence of nitrogen may be expected partly in a fixed form.

For these reasons we have carried out an experiment on the AIS Venera 8 on the determination of ammonia content in the planet's atmosphere [19]. For the ammonia determination an instrument was mounted which used the differential linear-colouristic method of determination, i.e. the recording of the change of the chemical-indicator colour during influence of ammonia upon it. Tetrabromophenol-sulphaphthalein on calcined silica gel was used as the indicator. The principle of the instrument working is shown in Fig. 1. The instrument represents an evacuated system consisting of two chambers, the measuring and the calibrated ones. In the measuring chamber a translucent vessel with an ammonia indicator was installed which had contact with the light guide and the detector unit. The vessel had two sections, the working and the standard ones. The standard section of the vessel is hermetically isolated from the working section and the outer atmosphere. The working section is connected with the evacuated calibrated chamber. The light source provided a uniform lighting of the vessel with the aid of the light guide. The light, being partly absorbed, passed through a thin layer of a chemical indicator and was received by photoresistors, one of which was in contact with the working section of the vessel, the other with the standard one.

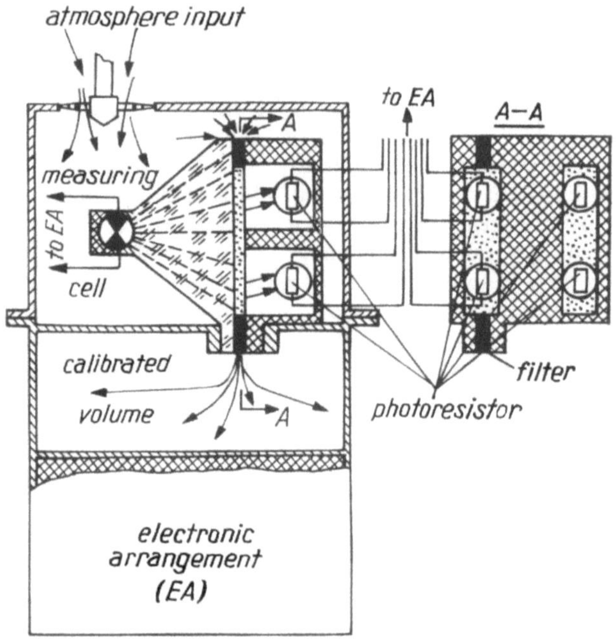

Fig. 1. Schematic diagram of the ammonia measuring equipment.

At the command of the programming-timing unit of the station the instrument was de-hermetized and the atmosphere entered through the working section into the calibrated chamber. During the station's descent the instrument was opened and the gas continuously passed under the action of outer pressure through the working section of the vessel, where the change of chemical indicator colour was taking place. This change of the colour was registered by photoresistors.

The instrument installed on the Venera 8 station carried out measurements at two levels, at 2 atm and 8 atm pressure. Both readings of the instrument are in good agreement and the volumetric gas content in the area of measurement may be estimated as 0.01—0.1%.

As was indicated above we used tetrabromophenol-sulphaphthalein on calcined (anhydrous) silica gel as indicator. Under conditions of the Venus atmosphere it is a sufficiently selective indicator for ammonia. Nevertheless it must be noted that the results obtained may be masked by the presence of alkali vapours. However, neither data of direct and spectroscopic measurements of the composition of the Venus atmosphere [20], nor the considered equilibria of the system atmosphere-lithosphere with the parameters of the Venus atmosphere approaching those measured by the Venera 8 station apparatus [21], indicate the possibility that some other components giving alkaline reactions could be present in the atmosphere. Thus, the results recorded by us with the instrument of the Venera 8 station have to be ascribed to the presence of ammonia in the atmosphere of the planet.

These data together with other results of cosmic and terrestrial measurements allow us to express some views about the structure and composition of the Venus cloud layer. As is known, ammonia easily interacts with carbon dioxide and water vapour forming a number of compounds (carbonate, bicarbonate; ammonium carbamate, etc.). These are white crystalline materials which are easily dissolved in water and sublimated by increase in temperature. The results obtained by instruments of the station Venera 8 indicate the possibility of the presence of such compounds in the upper layers of the Venus atmosphere and of the formation of ammonium clouds. The lower boundary of such a cloud layer must be rather sharp and the ammonia content here a maximum. Probably the measurements with instruments of the station Venera 8 were carried out below the dissociation level of ammonium compounds of the carbonate type. Small amounts of ammonium fluoride and chloride may enter the composition of ammonium clouds which, under conditions of the Venus atmosphere, are dissociated below about the 10 atm pressure level.

Thus, the Venus clouds must be multilayer ones. Their lower level corresponds to a pressure of 10 atm. The prevailing composition of the lower layer of the clouds is ammonium chloride at a very small density of clouds. The lower boundary of the second layer corresponds to the height where the pressure is 1—2 atm. Ammonium carbonates are here the prevailing component of the clouds. Above this level a mixture of ammonium compounds (ammonium chlorides and carbonates, etc.) is present. Finally, depending on the ammonia and water vapour ratio, the presence of ice crystals in the Venus clouds is not excluded. At the dissociation level a precipitation of ammonium salts, their subsequent decomposition and a diffusion of water vapour and ammonia upwards is possible. Clouds of the ammonium type, even if they contain liquid water, must contain a very small amount of it. Probably the out-washing action of rains is not substantial here and, hence, the presence of mineral dust particles in the Venus cloud layer is possible. (It is, for instance, known that volcanic dust on the earth reaches great heights.) Finally, if we take into account the results of Mariner 5 [6], it is possible to assume a formation of carbon dioxide hydrates at the level of the upper boundary of the cloud layer.

Preliminary estimations show the possibility of an intense convection motion of the atmosphere above the 1—2 atm level and a decrease in speed of convective fluxes below this level. In this case the speed of the wind must be relatively slight at a height of less than 30 km over the average surface level.

Thus, one may imagine the following mechanism of the Venus cloud cover formation. The nitrogen source of the atmosphere of Venus, as well as of other terrestrial planets is NH_4^+ which, in a medium saturated with carbon dioxide, had been degassed from the interior parts of the planet in the form of ammonium compounds which are easily dissociated with formation of ammonia. At a relatively low temperature ammonia interacted with carbon dioxide and water forming solid ammonium compounds which led to the decrease of free NH_3 volumetric content in the upper layers of the atmosphere. This fact together with high temperature was the cause of increasing ammonia diffusion towards the cloud layer, where it entered into reaction with carbon dioxide and water vapour. As a result of the action of such a mechanism ammonia and water vapour were transferred into the cloud layer, increasing its thickness. Simultaneously the process of NH_3 oxidation was going on in the presence of free oxygen, the product of the photodissociation of water vapour and carbon dioxide. Probably even if the lower layers of the atmosphere contain ammonia and water vapour, it is in very small amounts, the main mass of them being present in the clouds of the planet in a fixed form. At the level of the upper boundary of the cloud layer ammonia and water vapour are dissociated in the process of NH_3 oxidation and H_2O photodissociation.

Such are our preliminary ideas about the possible chemical nature and structure of the Venus clouds. Obviously further investigation of the planet and in particular obtaining the profile of the chemical composition of her atmosphere, will allow us to construct a model of the Venus cloud layer with greater certainty.

References

[1] A. P. VINOGRADOV, YU. A. SURKOV, K. P. FLORENSKY and B. M. ANDREICHIKOV, Dokl. Akad. Nauk SSSR **179**, 37 (1968).
[2] A. P. VINOGRADOV, YU. A. SURKOV and B. M. ANDREICHIKOV, Dokl. Akad. Nauk SSSR **190**, 552 (1970).
[3] A. P. VINOGRADOV, YU. A. SURKOV, B. M. ANDREICHIKOV, O. M. KALINKINA and I. M. GRECHISHCHEVA, Kosm. Issled. **8**, 578 (1970).
[4] A. P. VINOGRADOV, YU. A. SURKOV, B. M. ANDREICHIKOV, O. M. KALINKINA and I. M. GRECHISHCHEVA, Space Research XI, 129 (1971).
[5] G. H. KUIPER, and F. F. FORBES, Communs. Lunar and Planet Lab. **6**, 177 (1967).
[6] G. FJELDBO, J. KLIORE and R. ESHLEMAN, Astron. J. **76**, 123 (1971).
[7] D. L. COFFEEN and T. GEHERELS, Astron. J. **74**, 433 (1969).
[8] D. L. COFFEEN, Astron. J. **74**, 446 (1969).
[9] J. HANSEN and A. ARKING, Science **171**, 669 (1971).
[10] R. WILDT, Astrophys. J. **96**, 312 (1942).
[11] B. H. KOIPER, On the Threshold of Space, Publ. House of Foreign Lit., Moscow 1969.
[12] F. HOYLE, Frontiers of Astronomy, New York 1955.
[13] J. S. LEWIS, Astrophys. J. **152**, 579 (1968).
[14] J. S. LEWIS, Icarus **11**, 367 (1969).
[15] V. I. MOROZ, Physics of Planets, Nauka, Moscow 1967.
[16] V. V. MIKHNEVICH and V. A. SOKOLOV, Kosm. Issled. **7**, No. 2, 23 (1969).
[17] M. A. KOLOSOV, O. I. YAKOVLEV and A. I. EFIMOV, Dokl. Akad. Nauk SSSR **182**, 93 (1968).
[18] R. J. RASOOL, Radio Sci. **5**, 367 (1970).
[19] YU. A. SURKOV, B. M. ANDREICHIKOV and O. M. KALINKINA, Dokl. Akad. Nauk SSSR **213**, No. 2, 48 (1973).
[20] The Venus Atmosphere, Inst. of Space Res., USSR Acad. Sci., Moscow 1970.
[21] J. S. LEWIS, Earth and Planetary Sci. **10**, 73 (1970).

Space Research XIV — Akademie-Verlag, Berlin 1974

URANIUM, THORIUM AND POTASSIUM CONTENT OF VENUS ROCK MEASURED BY VENERA 8

YU. A. SURKOV, F. F. KIRNOZOV and A. P. VINOGRADOV

V. I. Vernadsky Institute of Geochemistry and Analytical Chemistry, Academy of Sciences of the USSR, Moscow, USSR

The uranium, thorium and potassium contents of Venus rock, which are $2.2 \times 10^{-4}\%$, $6.5 \times 10^{-4}\%$ and 4% respectively have been determined for the first time with the aid of a scintillation gamma spectrometer installed on the automatic interplanetary station Venera 8. A description of the experimental technique and results of gamma-radiation measurements in the atmosphere and on the surface of Venus are given. Data on the content and ratio of natural radioactive elements obtained on the basis of gamma-spectral measurements are compared with data on terrestrial rocks. Views about the character of Venus rocks and the mechanism of their formation are expressed.

1. Introduction

As is known, the first successful *in situ* investigation of Venus was carried out by the automatic station Venera 4 in 1967. This station returned important information about the composition and structure of the atmosphere [1—3]. However, till recently practically no information has been obtained about the surface of Venus. Therefore, besides the further study of the atmosphere parameters, the station Venera 8 was given for the first time, the task of investigating the character of the Venus rock at the landing site of the space vehicle.

As we know, the character of rock may be determined not only by its complete chemical or mineralogical composition, but also by the content of a number of separate chemical elements (even if they are present in very small amounts), which are sufficiently characteristic of one type of rock or another. One such diagnostic indicator is the content of natural radioactive elements: uranium, thorium and potassium. In this connection, a gamma spectrometer was installed on the station Venera 8, which determined the content of natural radioactive elements in the Venus rock.

The use of the gamma spectrometer for the investigation of the nature of Venus rock was decided upon because under the climatic conditions existing on the surface of Venus (in the landing region the temperature and pressure were 470 °C and 90 atm respectively) it is extremely complicated to utilize some of the other methods as they all require the carrying of detectors or a sampling device outside the hermetically thermostated section of the station. In contrast to this the gamma spectrometer was wholly contained in the hermosection of the station and recorded the gamma radiation being emitted by the radio-elements of the rock, which penetrated through the structural material of the station.

The use of the gamma-spectral method was also favoured by the absence in the Venus rock of radiation induced by cosmic rays, which seriously hampers the accomplishment of such measurements on celestial bodies having a small atmosphere or not having one at all. We used this method [4, 5] on the first satellite of the moon, the automatic station Luna 10, when it was first established that on the moon the region of maria is composed of igneous rocks, which according to their content of natural radioactive elements most closely approach basaltic rocks which are widespread in the crust of the earth. Below we describe the chief results of the first experiment carried out on uranium, thorium and potassium determination in Venus rock, which was accomplished by the Soviet automatic station Venera 8. Preliminary data from this experiment are published in [6].

2. Experimental Technique

The gamma spectrometer used on the station Venera 8 consists of a gamma-radiation detection unit (of a scintillation sensor), a multichannel amplitude analyser and an integral count rate meter. A schematic diagram of the spectrometer is shown in Fig. 1. The sensor of the gamma spectrometer consists of a Na I(Tl) crystal of size 63 mm × 63 mm, a spectrometric photomultiplier FEU-93, a high-resistance voltage divider and a high-voltage power source. The high-voltage source provides the power for the photomultiplier which requires 1—2 kV voltage at a load current up to 100 μA. The multichannel amplitude analyser is similar in structure to the equipment which was installed on the automatic station Luna 10 [7]. It ensures the amplification and the digital measurement of the amplitude of sensor impulses, the sorting of impulses, spectrum accumulation and the transfer of the measured spectrum to the telemetric system of the station Venera 8. The medium frequency meter uses the E-I characteristics of diodes in series to approximate a logarithmic scale in the range of 10—10000 Hz. The

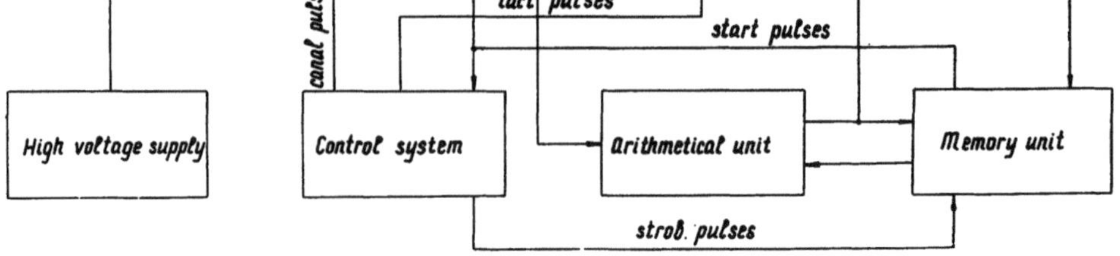

Fig. 1. Schematic diagram of the multichannel scintillation gamma spectrometer installed on the automatic station Venera 8.

output signal of the medium frequency meter is passed to one of the channels of the telemetric switch.

The main characteristics of the gamma spectrometer are: energy range 0.3 to 3.0 MeV; resolution along the line 0.662 MeV of the ^{137}Cs emitter $\sim 11\%$; number of channels 60; channel capacity 2^{15} impulses; integral non-linearity $\pm 1\%$; differential non-linearity $\pm 2\%$; median dead time of the analyser 800 μs; power used 2.5 W. The instrument stands the required linear and vibration overloads and works over a wide range of temperatures.

3. Results of Measuring the Gamma Radiation of Venus

The above described equipment was installed on the descent capsule (DC) of the station Venera 8. In conformity with the working programme of the DC the measurement of the integral counting rate was accomplished both on the descent route in the planet's atmosphere and after landing of the station on the surface. Measurement of the spectral composition of the gamma radiation was carried out only on the planet's surface. Three measurements of the integral counting rate were made during the descent and one after landing on the surface. The counting rate recorded during the descent did not change by an appreciable amount; after the landing of the DC on the surface an increase of the integral counting rate was recorded, which was due to the additional effect of the planet's surface. Thus, the measurement of the integral counting rate of γ-quanta allowed us, on the one hand, to evaluate the effect of gamma radiation induced by cosmic rays in the station material and, on the other hand, to determine the total level of gamma radiation of the Venus rock and thus to judge qualitatively of the content of natural radioactive elements in it.

The measurement of the Venus rock gamma spectrum was started immediately after the station landed and finished in 42 minutes. Transmission of information accumulated by the spectrometer occurred through all channels simultaneously with the spectrum measurement and was carried out twice. The first gamma spectrum was transmitted approximately in the middle of the spectrum measurement, the second at the end. In Fig. 2 and 3 spectra of gamma radiation of the Venus rock in the region of the landing of Venera 8 are shown; both spectra are scaled to 21 minutes. (Owing to a short information gap the second spectrum has no data in several middle channels). The spectra display most conspicuously the 1.46 MeV, 1.76 MeV and 2.62 MeV energy lines characteristic of potassium, uranium and thorium decays.

For the processing of gamma-radiation spectra of the Venus rock it was necessary to obtain the following data: (i) to determine the contribution of the natural gamma radiation of the spectrometer and the DC to the recorded spectra; (ii) to determine the contribution to the recorded spectra of the radiation induced by cosmic rays in the material of the spectrometer and the DC; (iii) to carry out a calibration of the spectrometer on rocks with known content of U, Th and K; (iv) to estimate the effect of scattering and absorption of gamma radiation of the Venus rock by the planet's atmosphere.

For the measurement of the gamma radiation background from the contamination of the DC by natural radioactive elements, the DC, with the spectrometer installed on it, was placed into a dunite rock mass, which, as known, has an extremely low content of uranium, thorium and potassium. In this case one

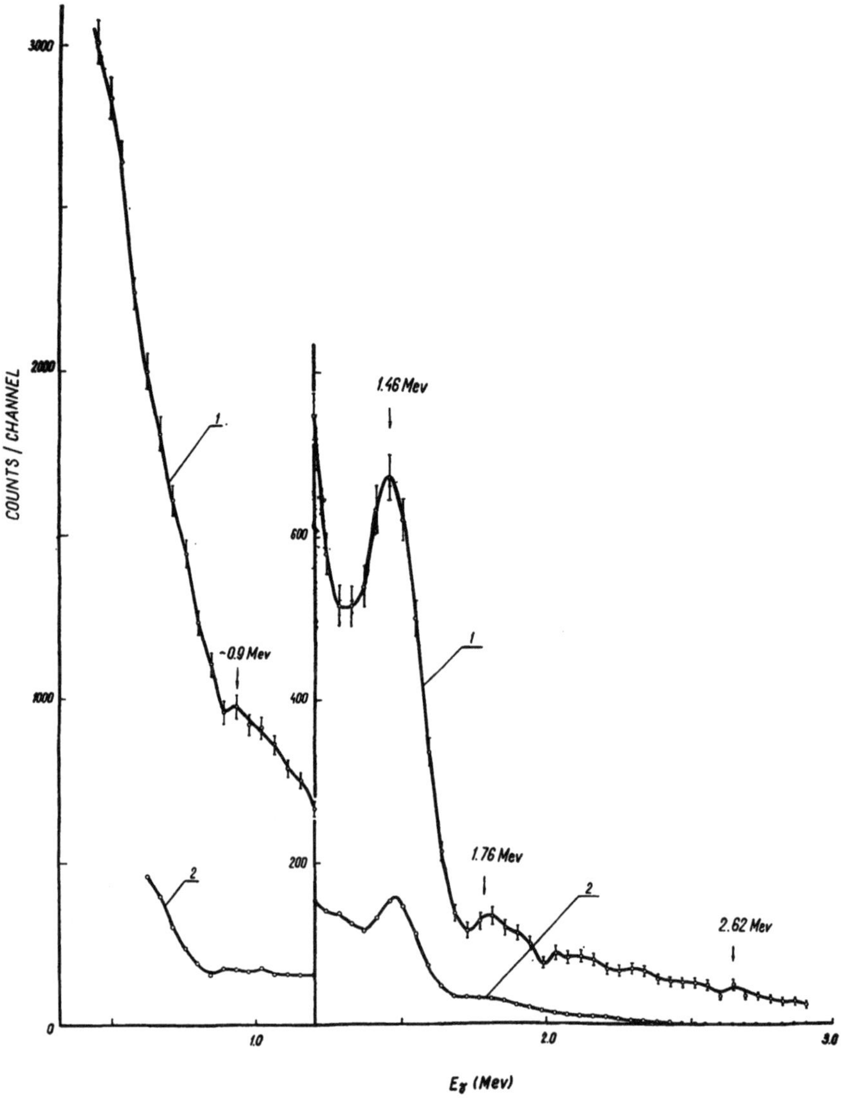

Fig. 2. First gamma-radiation spectrum measured on the Venus surface by the automatic station Venera 8 (measurement time 21 minutes).

may disregard the effect of cosmic radiation in view of the fact that the instrument is shielded by the great rock mass. The measured background of the DC is shown in Figs. 2 and 3 (curves 2). The nature of the background spectrum is consistent with the assumption that it is due to natural radioactive elements (mainly to potassium). The integral counting rate of the background (curves 2, Figs. 2 and 3) in the energy range 0.3—3.0 MeV equals approximately 18% of the value measured for the same time of the Venus surface (curves 1, Figs. 2 and 3).

The fact that the counting rate measured during descent did not change by an appreciable amount proves the practical absence of a contribution from the

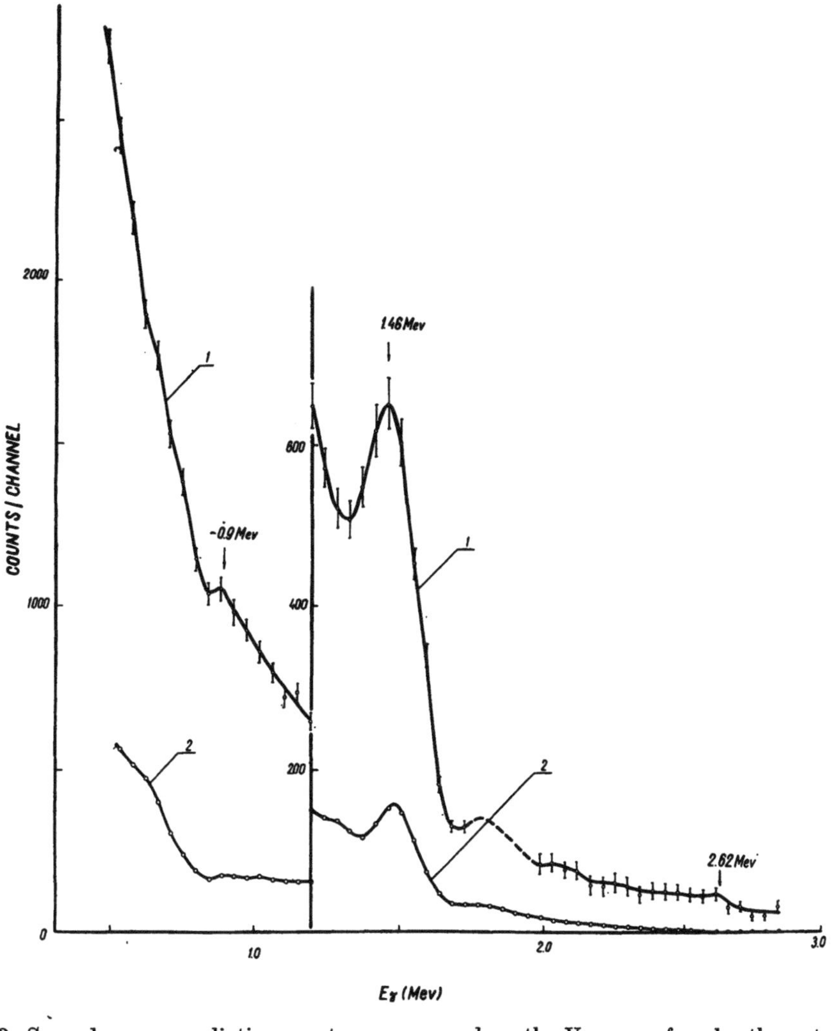

Fig. 3. Second gamma-radiation spectrum measured on the Venus surface by the automatic station Venera 8 (measurement time 21 minutes).

decay of short-lived (minutes to hours) radio-isotopes, which could arise in the construction materials of the station under the action of cosmic radiation. As to the decay of long-lived radio-isotopes (with a half-life of some days and more), their contribution to the DC radioactivity is at least one order less than that of the background from the natural radioactive elements. Thus, only radio-isotopes of Na, I and Tl (with half-lives from hours to days) formed in the detector material during its irradiation by cosmic rays could make any practical contribution to the recorded DC background, which was due to the cosmic-ray irradiation of the station.

Among possible gamma-emitters which are formed during irradiation of these elements, only ^{24}Na($T_{1/2} \sim 15$ h, $E_{\gamma 1} = 1.37$ MeV and $E_{\gamma 2} = 2.75$ MeV) could make an appreciable contribution. The other emitters either have low energy

($E_\gamma < 1$ MeV) or a low yield ($\eta < 5\%$) and thus could not make a noticeable contribution to the background spectrum. For the determination of the possible influence of ^{24}Na, the ^{24}Na spectrum measured on the above instrument was introduced into the Venus spectra and these were resolved into three components (U, Th, K) and four components (U, Th, K and Na); the differences in the contents obtained were small. The potassium content different by 1% from the measured value, that of uranium by 6% and that of thorium by 5%. Also the ^{24}Na contribution, which could be formed from the ^{23}Na(n, γ)^{24}Na reaction was determined by us from model experiments which have confirmed the calculated data.

For calibration of the instrument under terrestrial conditions a number of measurements over outcrops of rocks with high uranium, thorium and potassium content was carried out. In order to approximate the measurement conditions of the actual experiment on the planet the instrument was placed into an analog of the station Venera 8 and the measurements were carried out in $\sim 2\pi$ geometry over strata of granite, granodiorite, andesite-basalt and other rocks. Some of these

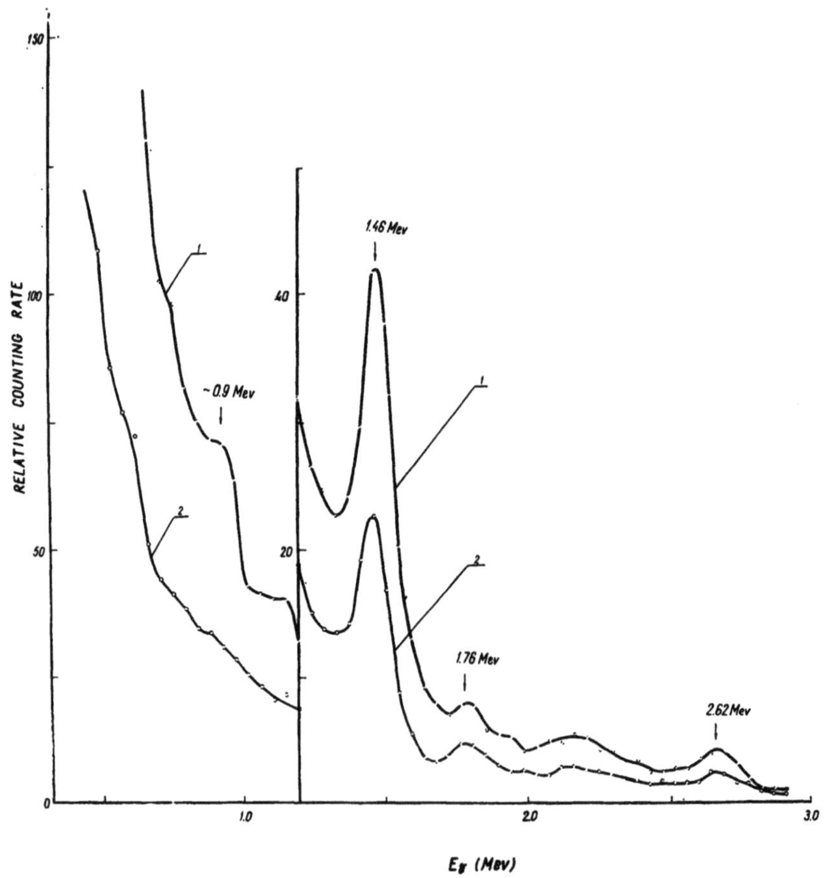

Fig. 4. Calibration gamma-radiation spectra of terrestrial rocks measured by the spectrometer installed on the automatic station Venera 8.
Curve 1, granite; curve 2, granodiorite.

spectra are given in Fig. 4. In order to link these spectra with definite uranium, thorium and potassium contents representative specimens of these rocks were sampled in the regions of measurements. The determination of natural radioactive elements in them was made by methods of radiochemistry and low-background radiometry using international BCR standards. The content of elements has been determined with ~3% error.

Finally, during the processing of the Venus gamma-spectra it was necessary to take into account the scattering and absorption of the radiation emitted from the rock by the planet's atmosphere. This factor is of great importance because the calibration spectra of terrestrial rocks were measured under different conditions.

Calculation by the radial method leads to the following relation between the content in the planet rock of a mono-emitter of the given E_γ energy and the number I of gamma-quanta per time unit crossing the surface of the detector (which with sufficiently good approximation may be considered as spherical):

$$I = \frac{\pi r^2 N}{2\mu_0} E_2(\mu h) \tag{1}$$

where N is the number of gamma-quanta emitted by a unit of rock volume per time unit; μ_0 and μ are linear coefficients of attenuation of gamma-quantum flux in the emitting rock and in the planet's atmosphere respectively; h is the detector height above the rock; r is the detector radius; E_2 is the King function:

$$E_2(x) = \mathrm{e}^{-x} - x \int_x^\infty \frac{\mathrm{e}^{-t}}{t}\, dt. \tag{2}$$

If we use the expression (1) for the determination of gamma-radiation intensity recorded by the detector near the surface of the earth and Venus and assume that μ_0 is approximately the same in both cases, we obtain the following expression for the coefficient, taking into account the influence of the Venus atmosphere:

$$K = \frac{I_V}{I_3} = \frac{E_2(\mu_3 h)}{E_2(\mu_V h)} \tag{3}$$

where I_3 and I_V are gamma-radiation intensities recorded by the detector above the surface of the earth and Venus respectively and μ_3 and μ_V are linear coefficients of the monochromatic radiation attenuation in the atmosphere of the earth and Venus respectively. When the King function $E_2(\mu_3 h)$ and $E_2(\mu_V h)$ is calculated in the expression (3) for each energy, corresponding linear coefficients of attenuation are taken.

It must be noted that in this calculation of correction factors we have disregarded the attenuation of gamma radiation by the material of the station itself which could lead to some over-estimation of the observed uranium, thorium and potassium contents. However, simple calculations show that this kind of over-estimation cannot exceed ~10% of the estimated contents.

Taking into account the above remarks, the content of natural radioactive elements in the Venus rock has been determined. Two known methods have been used: the method of comparing the areas of photopeaks in the Venus and calibration gamma-spectra and the method of least squares.

In the first case, after subtraction of the station background spectrum measured in dunite for each calibration spectrum, the areas of the most characteristic photopeaks, expressed as a number of impulses, were calculated. In this way the intensities in the photopeaks corresponding to the above-mentioned energies could be related to known potassium, uranium and thorium decay constants. After that, the determination of potassium, uranium and thorium contents of the Venus rocks under investigation required a simple recalculation of the areas of corresponding photopeaks to account for the K coefficient. In the second case, after subtraction of the same background, the pure Venus spectrum was resolved by the method of least squares into three components (without taking into account the contribution of gamma radiation induced by cosmic rays). As a result of spectra processing the following contents of natural radioactive elements in the Venus rock were obtained: potassium $4.0 \pm 0.2\%$, uranium $(2.2 \pm 0.4) \times 10^{-4}\%$, thorium $(6.5 \pm 0.3) \times 10^{-4}\%$ (the content is given in weight per cent, the errors are statistical ones).

If we also take into account the errors of content determination in calibration rocks, errors due to the spectrometer scale drift and errors due to the contribution of gamma radiation induced by cosmic rays, the errors in the content determination must be increased. The estimations give ground to think that these errors do not exceed 30%.

4. Discussion of Results

The data obtained by us indicate that the station Venera 8 has carried out measurements in a region with a relatively high content of natural radioactive elements in the rocks under investigation in comparison with their abundance in the earth's crust and all the more in the mantle of the earth. In the level of its uranium, thorium and potassium contents the Venus rock most closely approaches the acid magmatic rocks of the earth (granites).

The ratio of potassium to uranium for the Venus rock proves to be very close to that characteristic of most magmatic rocks of the earth, equalling $\sim 10^4$. In our opinion this fact may prove that, if processes leading to a change of the chemical composition of the primary matter (whatever its nature had been) had taken place, they were, in the main, similar for the earth and Venus.

As is known, the composition and properties of the matter of planets of the solar system are a result both of fundamental processes which occurred in the primary protoplanetic nebula and of large-scale planetary processes, for instance, the differentiation of the planet matter into shells. With respect to specific areas of the surface of the earth and other planets of the solar system, the development of the chemical composition and properties of the rocks composing them was also influenced by physical and climatic conditions, which existed on the surface of the given planet. In particular, on Venus which has a dense and sufficiently active atmosphere, with a high temperature and pressure at the surface, the formation of superficial rocks was apparently determined both by mechanical destruction of the primary crystalline rock as a consequence of atmospheric, temperature and gravitation erosion and by a partial decomposition under the action of the carbon dioxide (in the presence of water) and, finally, probably by the partial melting of easily fusible mineral components. This is also indicated by the fact that the density of the Venus rock measured by the station Venera 8 is 1.5 g cm^{-3}, i.e. a little over half the density of terrestrial crystalline rocks.

The possibility is not ruled out that besides endogenic processes, characterized by the evacuation to the surface of differentiated rocks enriched in uranium, thorium and potassium, some of the above listed factors also have lead to the enrichment in natural radioactive elements of the superficial Venus rock.

Thus, it may be assumed that in the region of the Venera 8 landing we have met a rock which had been smelted from the inner parts of the planet and subsequently was subject to a long influence of the outer environment, which has led to the formation of a metamorphic rock relatively rich in natural radioactive elements in comparison with the main rocks of the earth's crust, the basaltic ones.

For the development of more definite ideas about the formation of the Venus superficial rock it is necessary to carry out further measurements both of the content of natural radioactive elements and of other characteristics of the Venus rocks in a larger number of regions of the planet.

Acknowledgment

In conclusion the authors express their gratitude to V. N. Glazov, O. P. Sobornov, G. A. Fedoseyev, B. N. Kononov and L. N. Myasnikova for their participation in the development of the equipment.

References

[1] A. P. Vinogradov, Yu. A. Surkov and K. P. Florensky, Dokl. Akad. Nauk SSSR **179**, 1 (1968).
[2] A. P. Vinogradov and Yu. A. Surkov, Kosm. Issled. **8**, 578 (1970).
[3] A. P. Vinogradov, Yu. A. Surkov and M. Ya. Marov, Rep. on Int. Congr. on Astronautics, FRG **1970**.
[4] A. P. Vinogradov, Yu. A. Surkov, G. M. Chernov, F. F. Kirnozov and G. B. Nazarkina, Kosm. Issled. **4**, 874 (1966).
[5] A. P. Vinogradov, Yu. A. Surkov, G. M. Chernov, F. F. Kirnozov and G. B. Nazarkina, Kosm. Issled. **5**, 871 (1967).
[6] A. P. Vinogradov, Yu. A. Surkov, F. F. Kirnozov and V. N. Glazov, Dokl. Akad. Nauk SSSR **208**, 723 (1972).
[7] B. G. Egiazarov, B. N. Kononov, S. S. Kurochkin, Yu. A. Surkov and N. A. Shekhovtsov, Kosm. Issled. **6**, 267 (1968).

Symposium on Noctilucent Clouds and Interplanetary Dust

Space Research XIV — Akademie-Verlag, Berlin 1974

SOLID PARTICLES AT THE MESOPAUSE

G. Witt

Institute of Meteorology, University of Stockholm, Sweden

The optical characteristics of noctilucent clouds have been studied by means of rocketborne photoelectric polarization photometers. Combinations of such photometers operating in different wavelength regions have been launched into displays of noctilucent clouds from Esrange in Northern Sweden on repeated occasions. The launchings took place during twilight and the optical measurements were performed at scattering angles varying from 70° to about 105°, depending on the orientation and semi-angle of the rocket's precession cone. It was established that the degree of polarization of the light scattered by the cloud particles was higher than that scattered by the air molecules at scattering angles greater than about 70° and at wavelengths greater than 400 nm. At 90° scattering angle, where the polarization of molecularly scattered light is limited to about 94%, the polarization at 536 nm produced by the clouds was found to exceed 96%. The ratio of the radiance values at 366 nm and 536 nm observed in different cloud displays was found to be only about 60% of the molecular-scattering colour ratio. On all occasions the light-scattering layers was sharply limited from below with the cloud base altitude located at 81 to 85 km.

The light-scattering properties of the cloud particles suggest a size distribution which is peaked near 100 to 120 nm radius. The contribution to the light scattering by particles with radii greater than 120 nm must be exceedingly small in order to account for the observed strong polarization. In order to identify the physical process responsible for the generation of a particle distribution of this kind, the equations describing the growth and decay of particles through deposition and sublimation of water vapour on to suitable nuclei of condensation have been integrated. Based only upon the assumptions that the concentration of water molecules at the mesopause is 1 to 5 parts per million by number and that thermodynamically stable condensation nuclei are generated in the 80—85 km altitude region, the model was found to reproduce faithfully the observed features of noctilucent clouds. The mechanism for production of the necessary condensation nuclei has not yet been definitively clarified. Published observations of an enhanced concentration of heavy ion-water clusters near the cold summertime arctic mesopause seem to support the hypothesis of ion hydration as a possible initial step in the nucleus formation. Details of the rocket-borne observations and of the numerical cloud model are discussed.

Space Research XIV — Akademie-Verlag, Berlin 1974

NOCTILUCENT CLOUDS ON MARS — THE BLUE CLEARING

M. DUBIN

National Aeronautics and Space Administration, Washington, D. C., USA

A new explanation of the blue haze in the Martian atmosphere and the phenomenon of the blue clearing is presented, based on the cosmic dust measurements from Pioneers 8 and 9 and sounding rocket studies of the arctic region noctilucent clouds. In this model, the haze results from scattering of sub-micrometer particles in the Martian atmosphere and the nuclei of the haze particles are of extra-Martian origin. The nuclei are not meteoritic particles, but rather the component of interplanetary dust which has traversed the inner coronal region of the sun and has been accelerated radially out of the solar system by solar radiation pressure. The existence of a significant population of outflowing cosmic dust has been found from the Pioneer measurements and from impact cratering on glassy lunar rocks. A major argument favoring this model is the explanation of the phenomenon of the "blue clearing" which tends to be strongest near opposition. The mechanism of a terrestrial cause of the "blue clearing" is described. In other respects the blue haze on Mars evolves in a manner analogous to that for terrestrial noctilucent clouds, with water vapor as the predominant condensate on the cosmic nuclei.

Space Research XIV — Akademie-Verlag, Berlin 1974

OPTICAL MEASUREMENTS ON NOCTILUCENT CLOUDS BY ROCKETS

F. Rössler

Deutsch-Französisches Forschungsinstitut, Saint-Louis, France

The results of several rocket launchings carried out in 1969 and 1970 at Kiruna are discussed. These experiments aimed at measuring the skylight scattered from noctilucent clouds. A formation of layers was always found at different altitudes near the mesopause. In the case of noctilucent clouds the layer at an altitude of 82 km was only 1 km thick but very distinct. One hour later this layer has disappeared. The results may be illustrated best by representing the light intensity per mass of air as a function of altitude.

For an explanation of the results for different colours Aitken nuclei have to be taken into account as well as Junge nuclei. However, their share is different for fully developed or for disappearing noctilucent clouds.

1. Purpose of the Work

Since 1961 measurements of the diffuse skylight by rockets or balloons have been carried out [1]. The purpose of the experiments is to determine the number of particles of the aerosol. In 1969 and 1970, the experiments were extended to the mesosphere. The four ascents reported here were made at Kiruna in order to study noctilucent clouds. In 1969 no display of noctilucent clouds was observed, whereas in 1970 the rockets could be fired into such clouds. By evaluating the experiments, several stages of the formation of layers could be compared.

2. Evaluation

The evaluation will be described first [2]. The experimental data are transformed by means of two calibration curves into light intensities as a function of altitude; a computer is used, and an attached drawing apparatus produces the curve directly. Fig. 1 shows as an example the smoothed curve of the luminance as a function of altitude for apparatus 14. It can be seen clearly that the formation of a layer at 82 km is characterized by a more horizontal shape of the curve. However, the curve obtained in this way gives only the integral values of all radiation above the given level. The contribution of each layer is determined by differentiation of the curve (or formation of the gradient), in this way yielding physically interesting results. Because of the exponential decrease of air pressure with increasing height, the values extend over a very wide range. This difficulty may be avoided by dividing the values by the air density, thus referring all values to the mass of 1 kg air. In applying the data of the standard atmosphere, the light values in cd kg^{-1} are obtained as a function of altitude as shown in Fig. 2 (blue). Another curve for apparatus 18 (yellow) is shown in [1]. The special advantage of such a

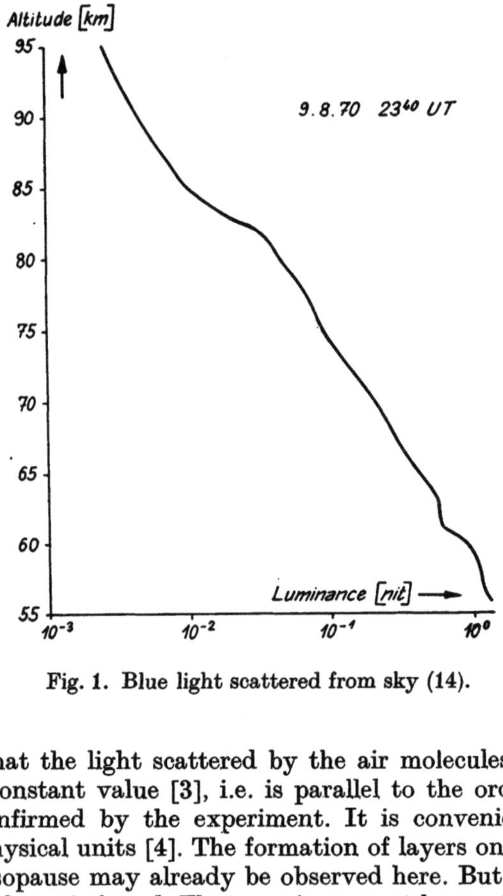

Fig. 1. Blue light scattered from sky (14).

plot is the fact that the light scattered by the air molecules, i.e. the Rayleigh radiation, has a constant value [3], i.e. is parallel to the ordinate axis. This is approximately confirmed by the experiment. It is convenient to express the data of light in physical units [4]. The formation of layers on both sides (82 and 86 km) of the mesopause may already be observed here. But also further down, at 73 km, another layer is found. The most important layer of noctilucent clouds is at an altitude of 82.5 km. It is very small (only 1 km thick) but very distinct. In Fig. 3 we see the result of the measurement made one hour later. The distinct layer has disappeared but the layers at 86 km and 73 km are again present.

3. Number of Particles

By forming the ratio of all the radiation of scattering to that of Rayleigh scattering, the turbidity factor τ is obtained. The value of $\tau - 1$ is proportional to the number of particles. By means of Mie's theory and the values of Bullrich [5] the number of aerosol particles is calculated.

We first assumed that the diameter of the particles is in the range of 0.1—1 μm, which means that they have a Junge distribution. But it has been shown by de Bary and Rössler [6, 7] that, starting from this assumption, the dependence of the diffuse radiation on the angle cannot be explained. In addition, smaller particles of diameter smaller than 0.1 μm, so-called Aitken nuclei, have to be considered. Farlow and Ferry were able to collect such particles [8, 9]. They

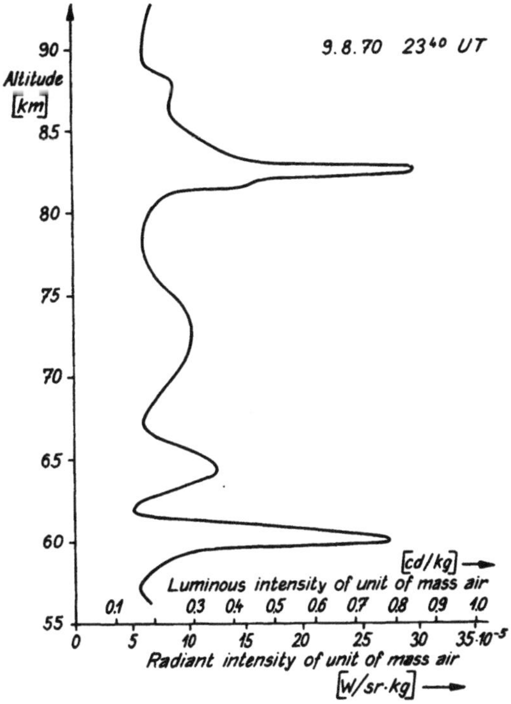

Fig. 2. Scattered blue light of unit of mass air (14).

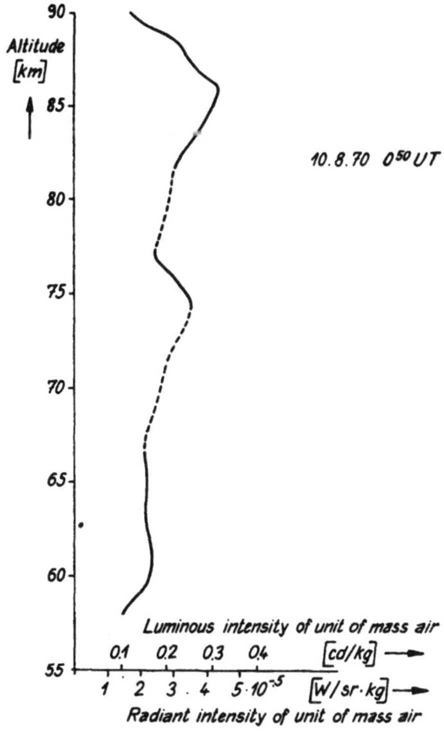

Fig. 3. Scattered blue light of unit of mass air (2).

observe that such nuclei are more numerous if there are no noctilucent clouds. Such small particles are optically less effective but their large number increases their contribution. It is clear, therefore, that many more particles are found by optical means than by Fechtig's method of collection [10]. The scattering power of these particles is highest in the blue while the scattering intensity of larger particles is less dependent on wavelength. The ratio A of the $(\tau - 1)$ values for different wavelengths from yellow to blue gives information about the mutual shares of Aitken and Junge nuclei. It is of interest to note that this ratio is greater than 1 with noctilucent clouds present. One hour later these clouds having disappeared and the ratio is smaller than 1. It may be concluded that in the latter case the Aitken nuclei prevail, in accordance with the observations of Farlow and Ferry. The particle distribution used earlier (Fig. 9 in [1]) with a constant relation between the number of Aitken and Junge nuclei has to be modified by introducing a variable α as a factor characterizing the number of the Aitken nuclei. The unknown parameter α is then determined from the ratio A [2, 7b].

4. Results

Factors of turbidity and the resulting numbers of aerosol particles for different experiments have been plotted in Fig. 4. There were two firings in 1969 without a visible noctilucent cloud display, and two in 1970 into visible clouds. Only the results measured in the blue are represented. There were noctilucent clouds at

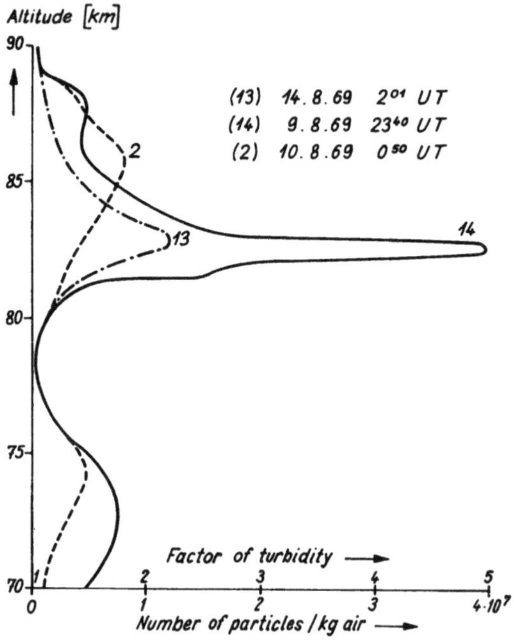

Fig. 4. Several states of the noctilucent clouds.*

82.5 km of average intensity, observed only with apparatus 14; after one hour (apparatus 2) they had faded away. But the layer at 87 km remained. With apparatus 2 it is more distinct than with apparatus 14. A certain number of particles permanently present seem to give the base for the formation of noctilucent clouds. These might be Aitken nuclei originating, after a suggestion of Hemenway et al. [11], directly from the sun. The measurement with apparatus 13 was carried out without visible noctilucent clouds; nevertheless, there is an indication of a layer at 83 km. An accumulation of dust with Aitken particles seems to form in the region of the mesopause, acting as germs for the clouds. According to Hesstvedt [12] this is favoured by air flowing upward adding ambient water vapour. If, according to Witt [13], water clusters attached to ions are formed as has been proved by mass spectrometer by Pollermann et al. [14] during the firing in 1970, the Aitken nuclei increase much in size by formation of ice (from curve 2 to 14) and are optically more efficient (by about r^4); noctilucent clouds then become visible. They may develop without the number of particles increasing.

* According to new calculations the maximum of curve 13 in Fig. 4 is much too high, the real number of particles being only a quarter [2, Fig. 21].

Acknowledgments

I thank Dr. Regula and Dr. Fechtig for their helpful cooperation. The measurements were supported by the Bundesministerium für Bildung und Wissenschaft.

References

[1] F. Rössler, Space Research XII, 423 (1972).
[2] F. Rössler, Bundesministerium für Forschung und Technologie, Forschungsbericht W 73-08 und W 73-09.
[3] F. Rössler, Optik **26**, 242 (1967).
[4] F. Rössler, Optik **34**, 445 (1972).
[5] K. Bullrich, Advanc. Geophys. **10**, 99 (1964).
[6] E. de Bary and F. Rössler, J. Geophys. Res. **71**, 1011 (1966).
[7] E. de Bary and F. Rössler, a, Contr. Atm. Phys. **43**, 87 (1970); b, Contr. Atm. Phys. **47** in press (1974).
[8] N. H. Farlow and G. V. Ferry, Space Research XII, 369 (1972.)
[9] G. V. Ferry and N. H. Farlow, Space Research XII, 381 (1972.)
[10] P. Rauser and H. Fechtig, Space Research XII, 391 (1972).
[11] C. L. Hemenway, D. S. Hallgren and D. C. Schmalberger, Nature **238**, 256 (1972).
[12] E. Hesstvedt, A Meridional Model of the Oxygen, Hydrogen Atmosphere, Aug. 1970, Inst. of Geophys. University of Oslo.
[13] G. Witt, Space Research IX, 157 (1969).
[14] B. Pollermann, J. Kissel, H. Wieder and D. Krankowsky, Verh. Phys. Ges. **8**, 479 (1971).

Space Research XIV — Akademie-Verlag, Berlin 1974

PHOTOGRAPHS OF THE INFRARED AIRGLOW

A. W. Peterson

Department of Physics and Astronomy, The University of New Mexico,
Albuquerque, New Mexico, USA

The discovery last year (Peterson and Kieffaber, Nature, **242**, 321 (1973) that the infrared airglow can be photographed with exposure times of a few minutes began to reveal its detailed and rapidly varying structure. Motions of bright patches are readily measured against the background star trails and yield speeds of 20 to 40 m s^{-1}. Two station parallax measurements during the night of 2 May 1973 yield a height of 89 $\pm$ 3 km for four discrete features. The most frequently observed morphology near the horizon consists of relatively horizontal bands to the east and west with patchy structures to the south. Moving bands are seen; sometimes motion is perpendicular to the band, sometimes parallel to it. During one evening the banded structures faded into a uniform, structureless radiance in an hour. We suggest that the infrared airglow could be a contaminant in infrared studies of noctilucent clouds.

Space Research XIV — Akademie-Verlag, Berlin 1974

MANIFESTATIONS OPTIQUES DES POUSSIÈRES COSMIQUES DANS LA HAUTE ATMOSPHÈRE

F. Link

Institut d'Astrophysique, Paris, France

Suivant notre proposition à la Réunion du COSPAR à Seattle 1971, nous avons inauguré au Pic du Midi les mesures photoélectriques du crépuscule destinées à détecter la présence des poussières cosmiques dans la haute atmosphère. On donne deux exemples des ces mesures obtenues dans la seconde moitié de 1972.

Comme nous en avons annoncé l'intention en 1971 [1], nous avons commencé en juillet 1972 à l'Observatoire du Pic du Midi (2860 m) la photométrie systématique du ciel crépusculaire. Le but genéral de ce programme à long terme est d'arriver à la meilleure compréhension du rôle des influences météorologiques et cosmiques sur les phénomènes crépusculaires. En ce qui concerne les influences cosmiques nous y étions encouragés par les études préliminaires montrant nettement l'influence des essaims météoriques sur la luminance du ciel crépusculaire [2, 3]. Dans la présente Note nous donnerons quelques résultats obtenus au Pic du Midi pendant la seconde moitié de 1972.

Nous avons utilisé un photomètre photoélectrique appartenant à la famille des photomètres dévéloppés au «Service de la luminescence du ciel nocturne» de l'Institut d'Astrophysique spécialement dans sa version [4] pour les vols à bord de ballons. Le photomètre avec quelques simplifications dues à la présence de l'observateur permet de mesurer la luminance du ciel le long d'un cercle à 27° du zénith et en deux couleurs, verte 5100 Å et rouge 7200 Å. Les mesures sont enregistrées en vue d'une réduction ultérieure suivant les critères différents et mal connus d'avance.

La diffusion supplémentaire sur les aérosols méteoriques est plus aisément détectée en rouge qu'en vert. Par contre, l'influence de la nébulosité sur le trajet des rayons éclairants solaires est plus grande dans le rouge que dans le vert. Si donc, nous nous décidons pour la première raison de travailler dans le rouge, il faut trouver la possibilité d'éliminer l'influence néfaste de la nébulosité qui risque de fausser l'interprétation des mesures.

Les résultats de nos mesures sont présentés sous forme de deux courbes crépusculaires $\log b = f(U)$, dont la première (courbe I) a été obtenue lors de l'activité météorique et la seconde (courbe II) en son absence. L'influence des poussières météoriques doit se manifester par un décalage positif de la courbe I pour les dépressions $U > 8°$ approximativement quand seules les couches atmosphériques au dessus de 60 km environ sont directement élairées par le soleil. Par contre, l'influence de la nébulosité s'étend sur toute la courbe en commençant par $U = 5°$. On choisit alors pour la courbe de comparaison (courbe II) celle qui coïncide très approximativement avec la courbe I pour les dépressions solaires $U < 8°$. Dans ce cas, on peut affirmer que l'influence de la nébulosité était la même dans les deux cas.

Nous allons démontrer ce procédé sur deux exemples: Le premier est celui des Géminides avec le maximum de l'activité le 13 décembre 1972. Pour la courbe de comparaison, nous avons choisi celle du 8 octobre 1972. En effet, les deux courbes (Fig. 1) sont sensiblement identiques jusqu'à $U = 8°$ et ce n'est que

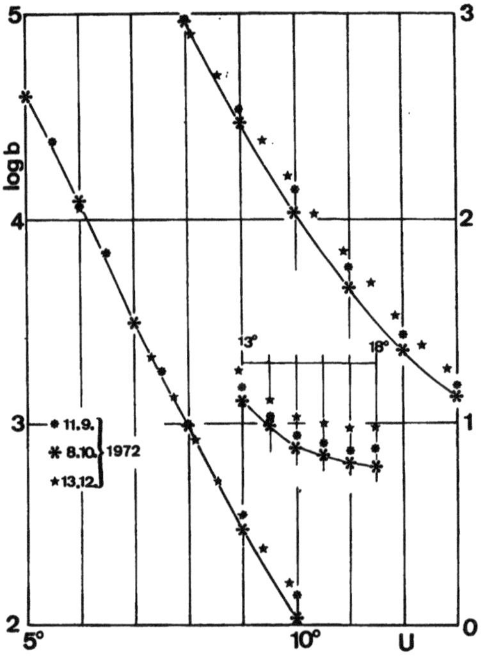

Fig. 1. Courbes crépusculaires $\log b = f(U)$ obtenues au Pic du Midi en rouge 7200 Å, à 27° du zénith, dans l'azimut du Soleil.

plus loin que commence leur séparation qui en $\log b$ va en croissant jusqu'au début de la nuit à $U = 18°$. Le second exemple est celui du 11 septembre 1972 avec la même courbe de comparaison que dans le cas précédent (Fig. 1). Deux essaims météoriques pourraient entrer en ligne de compte à cette date: les Piscides et les γ Aquarides [5]. Sur la courbe des échos de radar, obtenue à Ottawa, on trouve également une pointe bien marquée à cette date [6]. Il est encore à noter que l'examen des cartes météorologiques relatives à ces trois dates, montre que les conditions de nébulosités ont été sensiblement analogues. En vert, le même effet se manifeste également, mais son amplitude est sensiblement plus faible comme le prévoit la théorie.

L'effet météorique s'étend du crépuscule jusqu'au début de la nuit à $U = 18°$, comme nous l'avons dejà signalé [2], à propos des Orionides. Connaissant alors les coordonnées galactiques des plages en question, on peut confronter les luminances observées dans leurs valeurs relatives avec celles du fond stellaire [7] calculées d'après le dénombrement d'étoiles. On obtient ainsi la comparaison suivante:

Dates	Fond stellaire	Mesures en rouge	en vert
8. X. 1972	55 et. $10^{mg}/\square^\circ$ (1,0)	6,2 unités (1,0)	3,1 unités (1,0)
11. IX. 1972	80 et. $10^{mg}/\square^\circ$ (1,4)	7,6 unités (1,2)	3,8 unités (1,2)
13. XII. 1972	30 et. $10^{mg}/\square^\circ$ (0,5)	9,3 unités (1,5)	4,0 unités (1,3)

Il semble donc exister en dehors de la composante stellaire une autre composante qui varie avec l'activité météorique. De toute façon, les effets météoriques paraissent peu différents, dans les deux cas en contraste avec les importances visuelles et radioélectriques fort différentes à la faveur des Géminides.

Bibliographie

[1] F. Link, Space Research XII, 433 (1972).
[2] F. Link et G. Weill, C. R. Acad. Sci. Paris **271** B, 974 (1970).
[3] M. Fehrenbach, D. Frimout, F. Link et C. Lippens, C. R. Acad. Sci. Paris **272** B, 913 (1971).
[4] M. Fehrenbach, D. Frimout, F. Link et C. Lippens, Annls Géophys. **28**, 363 (1972).
[5] Z. Sekanina, Icarus **18**, 253 (1973).
[6] P. M. Millman et B. A. McIntosh, Canad. J. Phys. **42**, 1730 (1964); **44**, 1593 (1966).
[7] F. E. Roach et L. R. Megill, Astrophys. J. **133**, 228 (1961).

Space Research XIV — Akademie-Verlag, Berlin 1974

TIMES, LOCATIONS AND SIGNIFICANCE OF COMETARY MICROMETEOROID INFLUXES IN THE EARTH'S ATMOSPHERE

S. K. POULTNEY

Department of Physics and Astronomy, University of Maryland, College Park, Md, USA

A number of comets during the last seven years have had orbits that make it possible for micrometeoroids released from their nuclei upon approach to the sun to travel directly to the vicinity of the earth under the action of radiation pressure. The micrometeorites collect in the upper atmosphere and fall at a rate determined by the radiation parameter allowed by their orbital elements. One such falling cloud has been reported by the Jamaica laser radar group in May 1970 in an event related to comet Bennett (1970 II). Due to the young age of the micrometeoroids and the small thickness of a comet dust tail emitted before its coma develops, the micrometeoroids impinge on the earth at a definite time and geographical location. Table 1 gives theoretical predictions for the radiation parameters, times of influx, and geographical locations of the micrometeoroid influx for comets 1971 II, 1970 II, 1969 IX, 1969 V, 1968 IV, 1967 VIII, 1967 VII, 1967 IV, and 1967 II.

Table 1

Times and Locations of Possible Cometary Micrometeorite Showers

Comet Name	β	R_0	Time	Hem. Cap Location	Remarks
1971 II	0.85	1.1	1971 23 Feb. 1840	105° W 2.5° S	±1 day
1970 II	0.6	2.1	1970 5 May. 0340	90° E 43° S	Observed
1969 IX	1 (?)	—	1970 2 Jan. 0240	—	?
1969 V	0.5	2.1	1969 15 Nov. 1800	90° W 16° N	Sharp
1968 IV	0.3	5	1968 13 May 0936	52° W 60° S	Sharp
1967 VIII	0.7	1.2	1968 24 Feb. 0100	180° E 1.9° S	±1 day
1967 VII	1 (?)	6	1967 25 Oct. 0800	120° E 50° S	Sharp
1967 IV	0.45	$\geqq 7$	1967 14 Oct. 0230	60° E 54° N	Sharp
1967 II	1 (?)	?	1966 8 Dec.	—	—

On average, about one comet a year injects micrometeoroids equalling the normal one-day influx of interplanetary dust in a period shorter than a day. The contribution of comet micrometeoroids to the total extraterrestrial dust input to the earth is therefore minor. Its permanent contribution to the interplanetary dust population is also minor except possibly in the case of certain short period comets. However, the existence and detectability of these micrometeorite showers are important with respect to comet structure and with respect to studies of aerosol phenomena near the mesopause. The mechanism for these direct cometary micrometeoroid showers to reach the earth is quite sensitive to the micrometeoroid properties and requires a definite release distance from the sun. Detection of this influx to the atmosphere by a means sensitive to particle size (e.g. laser radar) would yield significant information about the activity of a comet nucleus at the release distance, about the size of the micrometeoroids given off there, and about the state of coma development. Most of the influx events considered here are due to emissions before the comet coma develops and so bear a closer resemblance to the "forward spike" of comet Arend-Roland (1957 III) than to the thicker dust tails of comets Arend-Roland and Bennett which have been analysed by Finson and Probstein and Sekinina and Miller respectively.

Injection of the micrometeoroids into the atmosphere at a given time and location(s) in significant amounts (10^2 to 10^4 tons) could yield information on the behavior of the micrometeoroids in near-earth space, information on the behavior of aerosols in the upper atmosphere, and information on competing mechanisms for the origin of the free sodium near 90 km. An indication of the potential in the latter case is the report of Visconti and Fiocco of an increase in Na twilight emission related to a comet P/Encke (1971 II) event (among other comet events). They use these data to support their hypothesis that the Na is produced by sublimation from recently fragmented, incoming dust particles, and indicate that this method is a more sensitive dust detection method than the laser radar method since it is able to detect the lower amount of material quoted above. Their reports suggest the examination of synoptic observations of airglows by the OGO 4 and OGO 6 satellites for correlations of airglow enhancements with the times and locations of cometary showers (as well as meteor showers). A recent study by J. D. Walker of the OGO 4 data spanning the period September 1967 to January 1968 shows promise for a correlation only with the comet 1967 VII event.

Detection of the cometary micrometeoroid showers in space by interplanetary space probes is much less probable due to the much smaller cross sections of their particle detectors. Each passage of a Pioneer 8/9 detector through a shower in space would only be expected to yield one detected particle. These types of direct micrometeoroids are therefore only a minor constituent of the actual Pioneer dust events. However, since the time and trajectory of such an event can be predicted, a correlation with space probe data may still yield identifications with particular comets. It should also be clear that the earth makes a good space probe to one of these comet dust tails due to its large collection cross section. A Pioneer-type sampling mission would collect very few particles unless orbiting in the comet plane itself.

Despite the minor contributions that direct cometary micrometeoroid showers make to both the interplanetary dust population and the dust influx to the earth, their existence and detectability are important with respect to comet structure and with respect to studies of aerosol phenomena near the mesopause.

Acknowledgment

This work has been supported by NASA Grant NGR 21-002-219 and will be reported on more fully in the *Journal of Geophysical Research*.

Space Research XIV — Akademie-Verlag, Berlin 1974

COSMIC DUST INFLUX TO THE UPPER ATMOSPHERE DURING THE MAJOR METEOR SHOWERS

DAVID W. HUGHES

The University of Sheffield, UK

This paper presents the results of measurements made of the increased influx of cosmic dust to the upper atmosphere that occurs during the activity of the Quadrantid, Perseid and Geminid meteor showers. Emphasis is given to the way in which the incident number of dust particles varies as a function of particle mass and to the way in which the mass influx varies as the earth passes through the stream.

In a recent paper Hughes [1] has tabulated the incident flux $\Theta(m)$ to the earth of meteoroids having masses greater than m ($10^{-5} < m < 10^{-3}$ g) at the time of maximum of the Quadrantid, Perseid and Geminid meteor showers. Also given is the mean mass distribution index $\bar{s}_m$ of all the observed shower meteors (when considering the total influx, $\bar{s}_m$ has to be used because s_m can vary from place to place in the stream). The incident flux may be represented by the formula

$$\log \Theta(m) = \log \Theta(1) + (1 - \bar{s}_m) \log m .$$

Fig. 1 shows this incident flux, the $m < 8 \times 10^{-2}$ g lines being drawn through the data points given in [1]. Hughes [2] found that there was a knee in the flux curve at a meteor magnitude of +3 (equivalent to a meteoroid mass of approximately 8×10^{-2} g). The extrapolation from the radar data has therefore been terminated at this mass and the curves continued on the high mass side of this point using mass distribution indices obtained from visual observations of the meteor showers. The values of $\bar{s}_m$ and $\Theta(1)$ used in Fig. 1 are given in Table 1.

From Hughes [6] it can be seen that the mass influx ψ to unit area of the earth's surface per unit time made up of particles having masses between 10^n and 10^{n+1} g is given by

$$\log \psi = n(2 - \bar{s}_m) + \log \left[\frac{(\bar{s}_m - 1)\, \Theta(1)}{(2 - \bar{s}_m)} (10^{2-\bar{s}_m} - 1)\right].$$

ψ has been calculated, using the data given in Table 1, for the maximum of the three showers and is plotted as a function of mass in Fig. 2. These curves refer to the influx at the top of the atmosphere. It can be seen that the majority of the mass of the meteor stream in space is made up of particles in the size range $10^{-2} < m < 10^{1}$ g. The heaviest particles in the meteor stream have masses around 10^{6} g, a meteoroid of this size being just able to escape from the gravitational attraction of the cometary nucleus and move into the stream. According

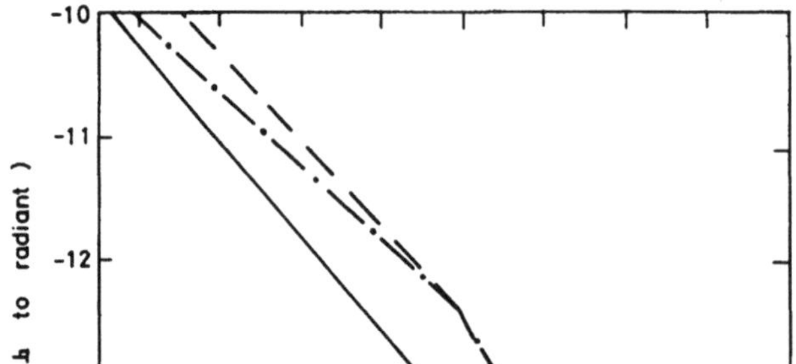

Fig. 1. The incident flux of meteoroids to the earth ($s^{-1}\,m^{-2}$ ⊥ to radiant) having masses greater than m (g) as a function of mass, at the maximum activity of the shower. ——— Perseid meteors; —·—·— Quadrantid meteors; — — — Geminid meteors.

to Verniani [7] all stream meteoroids ($m > 10^{-2}$ g) have low densities of about 0.3 g cm^{-3}, and are porous crumbly objects composed of loosely conglomerate spongelike material. On encountering the earth's atmosphere they ablate and fragment into individual molecules and into particles with masses below the micrometeoroid limit (10^{-6} g); these micrometeoroids slowly sink to the ground under the influence of gravity. Our knowledge of the size and mass distribution of these ablated and fragmented particles is still incomplete.

The total incident mass of meteoric material at shower maximum can be found by summing the masses given in Fig. 2 (see Table 1, line 5). This summation has been carried out between limiting particle masses of 10^{-6} and 10^{+6} g. These limits are not very stringent, for example if the limits were moved to 10^{-10} g and 10^{+10} g the incident mass would increase in the worst case by only 10%. To calculate the total influx throughout the shower, the shower width must be measured. Fig. 3 shows the results of radar measurements carried out at Sheffield which give the incident flux of meteors brighter than 7.6^{m} (masses greater than $\sim 10^{-4}$ g) as a function of solar longitude. Using these results, the equivalent transit width of the shower can be calculated, this width being defined as the

Table 1
Parameters of the Major Meteor Streams

	Quadrantids	Perseids	Geminids
1 Mass distribution index $\bar{s}_m$ ($m < 8.2 \times 10^{-2}$ g)	1.59	1.78	1.71
2 Mass distribution index $\bar{s}_m$ ($m > 8.2 \times 10^{-2}$ g)	2.11 [3]	2.40 [4]	2.30 [5]
3 $\Theta(1)$ ($m < 8.2 \times 10^{-2}$ g)	9.20×10^{-14}	6.05×10^{-15}	7.55×10^{-14}
4 $\Theta(1)$ ($m > 8.2 \times 10^{-2}$ g)	2.58×10^{-14}	1.44×10^{-15}	1.70×10^{-14}
5 Influx of dust at shower maximum (g m^{-2} s^{-1})	3.3×10^{-13}	2.8×10^{-14}	2.4×10^{-13}
6 Equivalent transit width (s)	6.2×10^{4}	76×10^{4}	48×10^{4}
7 Influx to earth each year (g)	2.6×10^{6}	2.6×10^{6}	15×10^{6}
8 Geocentric velocity of meteoroids (km s^{-1})	41.0	59.0	34.2
9 Maximum density of stream (g cm^{-3})	8.0×10^{-24}	4.7×10^{-25}	7.0×10^{-24}
10 Angle between stream and earth's orbit ε (deg)	70.9	60.4	64.8
11 Stream diameter (km)	1.72×10^{6}	19.7×10^{6}	12.9×10^{6}
12 Circumference of stream orbit (km)	2.5×10^{9}	1.39×10^{10}	9.69×10^{8}
13 'Equivalent' volume of stream (km^{3})	5.8×10^{21}	4.2×10^{24}	1.26×10^{23}
14 Mass of dust in stream (g)	4.6×10^{12}	2.0×10^{15}	8.8×10^{12}
15 Mass of originating comet (g)	1.8×10^{13}	7.7×10^{15}	3.4×10^{13}

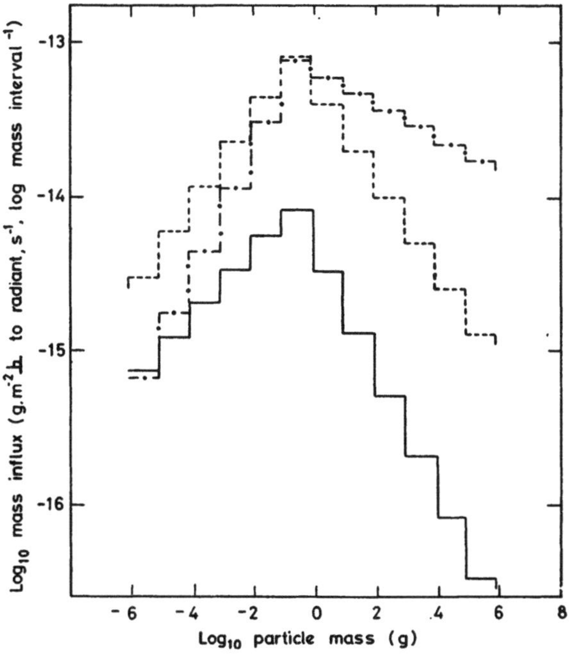

Fig. 2. The mass influx ψ to the earth's surface (g m^{-2} $\perp$ to radiant, s^{-1}, logarithmic mass interval^{-1}) as a function of mass, at the maximum activity of the shower, ——— Perseids, —·—·— Quadrantids, — — — Geminids.

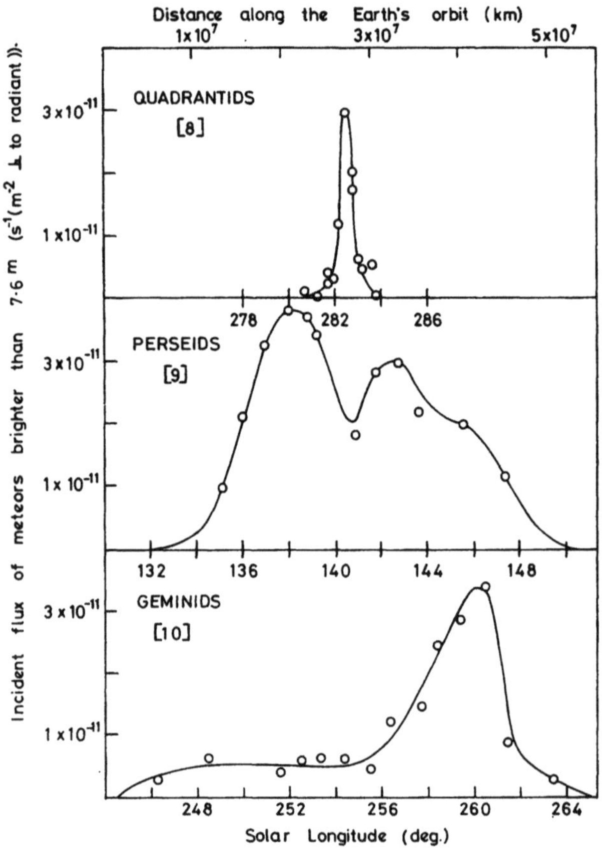

Fig. 3. Meteor shower activity as a function of solar longitude.

time the shower would have to exist at maximum rate to give the same meteoroid influx as normally observed (see Table 1, line 6). The total mass of meteoric material incident on the earth during its passage through the stream can be estimated approximately (i.e. neglecting gravitational enhancement) by multiplying the incident mass of meteoroids at shower maximum (line 5) by the equivalent stream width (line 6) and the cross-sectional area of the earth. This is given in line 7. Notice that the influx from streams is very much less than the normal annual sporadic influx wich has been estimated [11] to be 6×10^9 g.

The maximum spatial density of the stream can be found by dividing the incident mass of meteoroids by the meteoroid geocentric velocity (the relative velocity between the earth and the meteoroid, line 8). These are given in line 9 of the table.

If ε is the angle between the apex of the earth's way and the stream, the equivalent stream diameter (i. e. the diameter it would have if it had a density everywhere equal to the maximum density) is given by

$$D = 2\pi\Delta_{\odot} \sin \varepsilon \cdot 149 \times 10^6/360 \text{ km}$$

where $\Delta_{\odot}$ is the equivalent transit width in degrees of solar longitude. ε and D are given in Table 1, lines 10 and 11 (it has been assumed that the earth goes through the centre of the stream). The circumference of the stream (line 12) can easily be calculated from the orbital parameters and the equivalent volume (line 13) found by multiplying the circumference by the equivalent area of cross section. The total mass of meteoroids in the stream is then found by multiplying the equivalent volume by the maximum spatial density and this is given in line 14.

Assuming that comets contain 26% dust and 74% ice (see [12]) the mass of the originating comet can easily be calculated (line 15).

As the earth moves through the meteor stream the incident flux of particles varies. The instantaneous value of the flux can be found by modifying the maximum rate by the functions given in Fig. 3, care being taken to include a cos χ factor where χ is the radiant zenith angle.

The calculation of the mass deposition rate as a function of height in the earth's atmosphere is a much more difficult problem. Tentatively it can be stated that the maximum deposition rate occurs at about 90 km this becoming negligible at 110 and 70 km.

References

[1] David W. Hughes, Mon. Not. Roy. Astron. Soc. **161**, 113 (1973).
[2] David W. Hughes, Space Research XI, 319 (1971).
[3] K. B. Hindley, J. Brit. Astron. Assoc. **80**, 479 (1970).
[4] B. A. Lindblad, Smithson. Contr. Astrophys. **11**, 171 (1965).
[5] B. J. Levin, in: Meteors, Pergamon Press, London 1955 (p. 131).
[6] David W. Hughes, Planet. Space Sci. **20**, 1949 (1972).
[7] F. Verniani, Space Sci. Rev. **10**, 230 (1969).
[8] L. M. G. Poole, David W. Hughes and T. R. Kaiser, Mon. Not. Roy. Astron. Soc. **156**, 223 (1972).
[9] T. R. Kaiser, L. M. G. Poole and A. R. Webster, Mon. Not. Roy. Astron. Soc. **132**, 225 (1966).
[10] A. R. Webster, T. R. Kaiser and L. M. G. Poole, Mon. Not. Roy. Astron. Soc. **133**, 309 (1966).
[11] David W. Hughes, Mon Not. Roy. Astron. Soc. submitted.
[12] David W. Hughes, J. Brit. Astron. Assoc. in press.

Space Research XIV — Akademie-Verlag, Berlin 1974

MICROMETEOROIDS AND LUNAR ROCKS

F. Hörz[a], D. E. Brownlee[b], J. B. Hartung[c], D. E. Gault[d], D. A. Morrison[a] and J. F. Vedder[d]

[a] NASA Johnson Space Center, Houston, Texas, USA
[b] Dept. of Astronomy, University of Washington, Seattle, Washington, USA
[c] Max-Planck-Institut für Kernphysik, Heidelberg, FRG
[d] NASA-Ames Research Center, Moffett Field, Calif., USA

Microcraters ranging in size from less than 1 μm to 1 cm in diameter have significantly contributed to the understanding of the micrometeoroid complex. The most suitable materials are smooth, glass surfaces because of favorable observational conditions on one hand and the presently available laboratory simulations on the other. The results below are predominantly derived from such materials.

Crater Morphology. Crater morphologies show dependence on central pit diameters; craters larger than 5 μm have a concentric spall, craters between 1 and 5 μm are transitional, and craters less than or equal to 1 μm display no spall. These characteristics are consistent with experimental data and

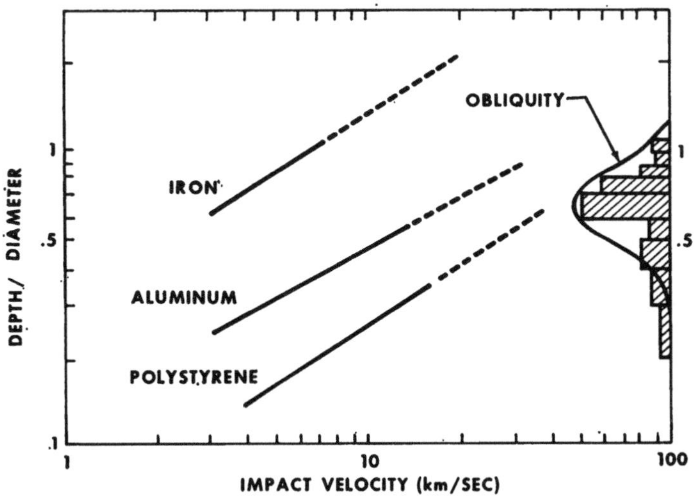

Fig. 1. Experimental depth/diameter ratios using projectiles ranging in density from 1 to 7.8 g cm^{-3} and comparison with lunar microcraters, indicated by the hatched area. An empirical distribution accounting for oblique impact is indicated. The respective surface area under this curve or the hatched histogram corresponds to 100%, i.e. to 70 craters measured. Note the absence of shallow structures. Though not illustrated, there is no correlation of depth/diameter with absolute pit-size.

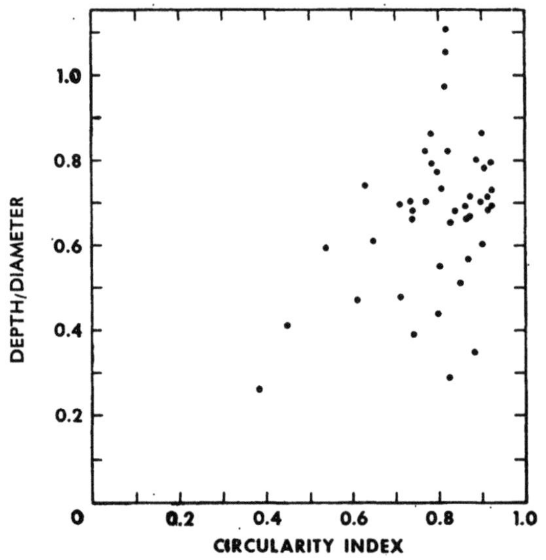

Fig. 2. Correlation of circularity index (CI) and depth/diameter ratio. CI was obtained as the ratio of the measured pit surface area and the area of a perfect circle just enveloping this pit. Note that shallow craters are relatively elongated, indicating that they are due to oblique projectile trajectory; conversely the deep structures are more centro-symmetric, consistent with vertical impact. Micrometeoroids corresponding to the most elongated craters have a ratio of about 2 for longer/shorter axis. These findings are incompatible with elongated particles like needles, etc.

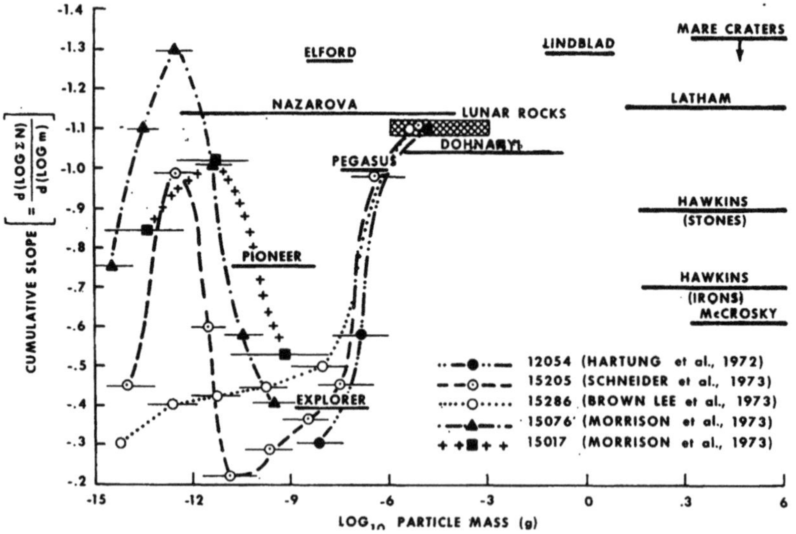

Fig. 3. Comparison of the cumulative mass-frequency slope of a variety of observational techniques but in particular of five well documented lunar glass surfaces. Note that a marked decrease in frequency is common to all surfaces for masses between 10^{-6} and 10^{-9} g but that there are significant differences between individual surfaces for masses smaller than 10^{-9} g. The masses for lunar microcraters are based on a density of 3 g cm^{-3}, an impact velocity of 20 km s^{-1} and a pit/projectile diameter ratio of 2.

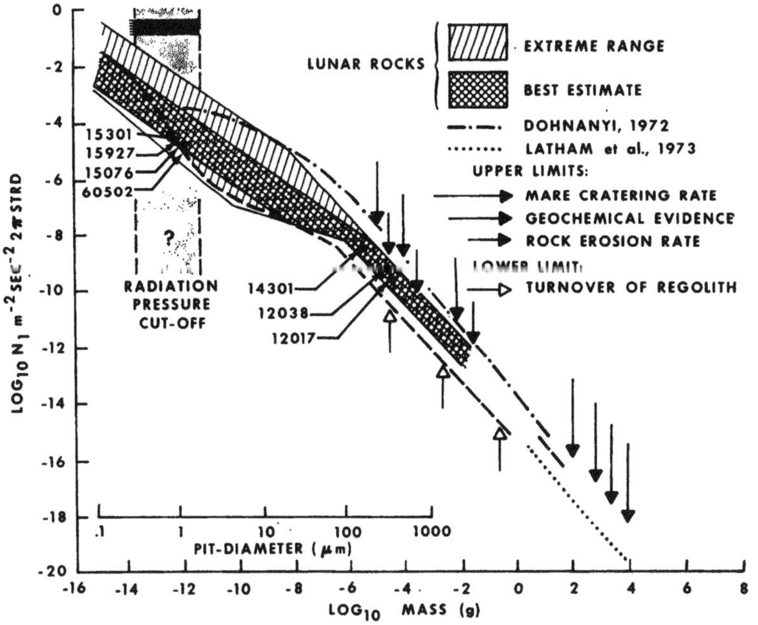

Fig. 4. The flux of micrometeoroids as deduced from lunar rocks in comparison with satellite- and earth-based data (taken from a recent review by Dohnanyi [1]). Upper and lower limits are indicated from entirely independent lunar studies which place serious constraints on the flux in general as well as on the rate of meteorite bombardment on the lunar surface over geological times. Note that all lunar studies are in good agreement for particles $> 10^{-7}$ g in mass.

suggest velocities in excess of 8 km s^{-1}. Comparison of observed and experimental depth/diameter relations for craters 0.1 to 100 μm in pit diameter indicate that micrometeoroids between 10^{-12} and 10^{-8} grams have densities above 1 g cm^{-3}, with 2—4 g cm^{-3} most probable (Figs. 1—2). Circularity indices indicate that micrometeoroids are equant and that rods, needles, platelets and "fluffy" structures are rare or absent.

Size Frequency Distribution. Well established size frequency distributions for craters > 100 μm correspond to a cumulative micrometeoroid mass slope of -1.1 for particles $> 10^{-7}$ g. The size frequency of smaller craters is subject to refinement with a distinct possibility of a bi-modal distribution. Observation of craters of 0.1 μm diameter (corresponding to a particle of 10^{-15} g assuming 20 km s^{-1} impact velocity) suggest no radiation pressure cutoff at $\approx 10^{-12}$ g particle mass (Fig. 3).

Micrometeoroid Flux. Correlation of lunar rock surface exposure ages (determined with cosmic-ray tracks) and absolute crater number densities yield a flux of $\phi = 10^{-15} m^{-1.1}$ per $m^2 s^{-2} 2\pi$ sr in the lunar environment for particles of $m > 10^{-7}$ g. There appears to be uncertainty of ≈ 2 orders of magnitude for significantly smaller objects, i.e. $< 10^{-10}$ g projectile mass. Additional studies not directly related to micrometeoroids such as the turnover-rate of lunar regolith (absolute minimum flux), rock erosion, the passive seismometer, the total amount of meteoritic material in the regolith and the crater formation rate on mare surfaces not only place very serious constraints on everybody's flux data, but they are prefectly consistent with our findings. Furthermore these constraints suggest that the flux for the last 3.2×10^9 years cannot have varied more than a factor of 5 (Fig. 4). The micrometeoroid and meteor fluxes generated in the past appear to be a factor of ≈ 10 too high and we suggest that these data be recalculated using densities > 1 g cm^{-3}, preferably 2.5 g cm^{-3}.

A full paper will be published in [2].

References

[1] J. S. Dohnanyi, Icarus 17, 1 (1972).
[2] F. Hörz, D. E. Brownlee, H. Fechtig, J. B. Hartung, D. A. Morrison, G. Neukum, E. Schneider and J. F. Vedder, Planet. Space Sci., in press.

Space Research XIV — Akademie-Verlag, Berlin 1974

METEOROID MASS DISTRIBUTIONS AND FLUXES FROM MICROCRATERS ON LUNAR SAMPLE 15205

J. B. Hartung and D. Storzer

Max-Planck-Institut für Kernphysik, Heidelberg, FRG

Microcrater populations on lunar samples have been studied by numerous workers to obtain information related to interplanetary dust or meteoroids. Cumulative crater size distributions and rock surface exposure ages have been measured to obtain meteoroid mass distribution and flux data [1—6]. This paper reports on our progress toward obtaining a well-defined *differential* crater size distribution and the formation ages of *individual* microcraters.

A sample of the exposed surface of rock 15205 was selected for this work because it had a smooth glass coating which had not reached equilibrium with respect to the cratering process, but at the same time had a relatively high areal density of microcraters. In addition, the sample had been taken from the top of a metre-sized boulder, thus minimizing the effects of splashes and dust known to exist along the soil line of rocks. Microcrater pit diameters ranging from 0.2 to 200 μm were measured for over 1000 craters on a surface area of less than 0.4 cm^2 using optical and scanning electron microscope techniques. Crater counts were tabulated in equal logarithmic intervals corresponding to a factor of the square root of two. The resulting differential crater size distribution is shown in Fig. 1. Data taken at larger magnifications over smaller areas were joined at conservatively selected end points to obtain a smooth curve over the entire size range. A change in the slope of the distribution is clearly indicated at a pit diameter of 5—10 μm, which corresponds to an impacting particle mass between 10^{-11} and 10^{-10} gramme based on calibration I of Hartung et al. [3]. This result qualitatively confirms an earlier finding of Schneider et al. [5] and suggests a bimodal meteoroid size distribution of interplanetary dust impacting the moon.

Estimates of the exposure ages of 19 microcraters with pits larger than 20 μm in diameter were made using the solar flare track method. In this method the time of exposure is considered proportional to the density of iron-group solar flare particle tracks near the surface of the sample. Particle track records are obtained by repeated polishing and etching of sections through the sample or microcrater, then counting the etch pits using a scanning electron microscope. Corrections up to a factor of 10 to account for the annealing of tracks at lunar temperatures over geologic times are required for glass samples, such as the surface of rock 15205. Calibration to an absolute age is based on an analysis of the Surveyor 3 camera lens [7] which has recently been modified [8].

Using this method the apparent exposure age of the surface of 15205 was found to be about 100000 years. Apparent exposure ages of pit glasses melted during meteoroid impacts are shown in Fig. 2. The figure shows first that impacts are not uniformly distributed in time, i.e. the flux of meteoroids in this size range is not constant for long periods, and second that the flux in the recent past, and probably at the present time, is higher than that for the more distant past (>10000 years).

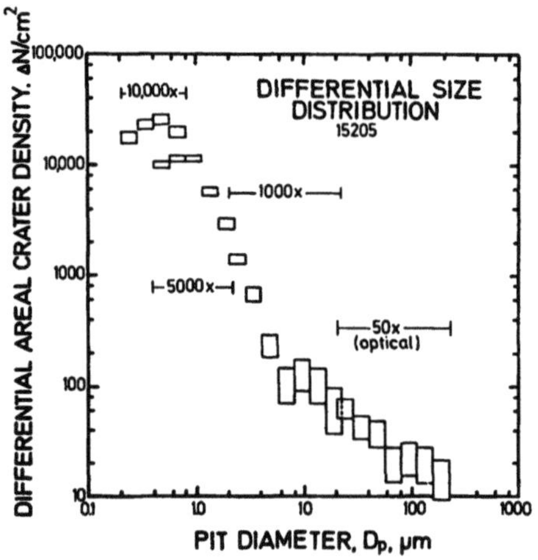

Fig. 1. Differential size distribution of lunar microcraters. Error bars correspond to plus or minus the square root of the number counted in each individual size class. Corresponding impacting particle masses may be obtained based on calibrations reported by Hartung et al. [3].

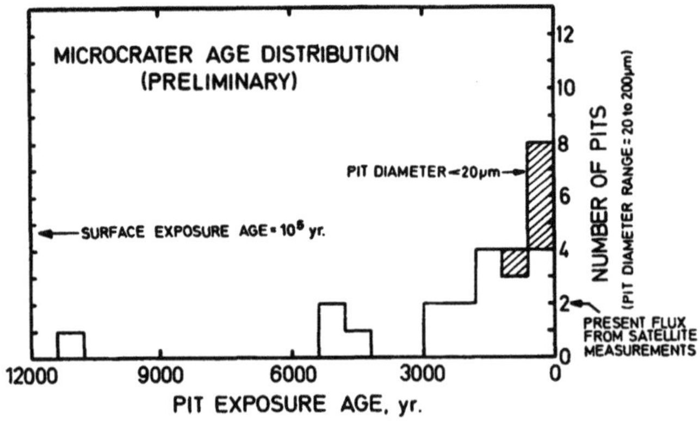

Fig. 2. Preliminary microcrater age distribution. For description see text.

These interpretations rely on two basic assumptions: (i) The iron-group solar flare particle flux is constant over geologic time; (ii) The behavior of the pit glasses and the host glass is the same with respect to their annealing and etching properties. This second assumption is in the process of being tested by exposing a cratered fragment of this same sample to energetic particles having known characteristics.

The average flux of meteoroids over the last few thousand years inferred from these studies agrees with that derived from satellite-borne experiments [9] within the calibration uncertainties for both the satellite-borne and the lunar rock experiments. The meteoroid flux averaged over 10^5 years is shown to be about an order of magnitude lower than the present-day flux, thus confirming earlier results based on analysis of lunar samples.

The present results and conclusions should be considered preliminary, but they suggest how the history of interplanetary dust in the solar system can be studied using microcraters on lunar rocks as natural meteoroid detectors.

References

[1] M. R. Bloch, H. Fechtig, W. Gentner, G. Neukum and E. Schneider, Proc. 2nd Lunar Sci. Conf., Geochim. Cosmochim. Acta, Suppl. 2, **3**, 2639 (1971).
[2] F. Hörz, J. B. Hartung and D. E. Gault, J. Geophys. Res. **76**, 5770 (1971).
[3] J. B. Hartung, F. Hörz and D. E. Gault, Proc. 3rd Lunar Sci. Conf., Geochim. Cosmochim. Acta, Suppl 3, **3**, 2735 (1972).
[4] G. Neukum, E. Schneider, A. Mehl, D. Storzer, G. A. Wagner, H. Fechtig and M. R. Bloch, Proc. 3rd Lunar Sci. Conf., Geochim. Cosmochim. Acta, Suppl. 3, **3**, 2793 (1972).
[5] E. Schneider, D. Storzer and H. Fechtig, in: The Apollo 15 Lunar Samples, Lunar Sci. Inst., Houston 1972 (p. 415).
[6] D. E. Brownlee, F. Hörz, J. B. Hartung and D. E. Gault, in: The Apollo 15 Samples, Lunar Sci. Inst., Houston 1972 (p. 407).
[7] D. J. Barber, R. Cowsik, I. D. Hutcheon, P. B. Price and R. S. Rajan, Proc. 2nd Lunar Sci. Conf., Geochim. Cosmochim. Acta, Supp.l 2, **3**, 2705 (1971).
[8] D. Storzer, G. Poupeau and W. Krätschmer, Proc. 4th Lunar Sci. Conf. Geochim. Cosmochim. Acta, Suppl. 4, **3**, 2363 (1973).
[9] D. E. Gault, F. Hörz and J. B. Hartung, Proc. 3rd Lunar Sci. Conf., Geochim. Cosmochim. Acta, Suppl. 3, **3**, 2713 (1972).

Space Research XIV — Akademie-Verlag, Berlin 1974

MICROMETEORITE INFLUX RATES ON THE LUNAR SURFACE DEDUCED FROM REVISED ESTIMATES OF THE SOLAR WIND SPUTTER RATE AND SURFACE CRATER STATISTICS

D. G. Ashworth and J. A. M. McDonnell

University of Kent at Canterbury, UK

Recent measurements of the sputter rate of simulated solar wind ions on lunar material together with the latest information on the numbers of micrometeorite craters present on samples of lunar material are used to obtain information on the mass spectrum of the micrometeorite flux on the lunar surface. Erosion of lunar material, due to the processes of solar wind sputtering, crater erasure due to the overlapping of small craters by larger ones and crater abrasion by the integrated effect of smaller particles, is included in a computer technique which derives the microparticle influx rate from surface crater statistics. Results show an incompatibility between published lunar production and equilibrium fluxes. Similarly, equilibrium measurements do not lead to agreement with current direct measurements in space over a wide mass range.

1. Introduction

Information on the erosion environment of the lunar surface has been made abundantly available from many lunar sample investigations, particularly from crater studies and cosmic ray track measurements [e.g. 1, 2]. We have made previous attempts [3] to predict equilibrium crater densities by proposing possible interplanetary dust particle flux distributions and computing the crater statistics by assuming appropriate erosion mechanisms. Erosion is considered to be produced by three processes: crater erasure due to the overlapping of small craters by larger ones, crater abrasion by the integrated effect of smaller particles, and solar wind sputtering. In the computer program we take account of the fact that different erosion mechanisms will become more, or less, significant as the size of the craters considered increases. Also included is the time dependent nature of the flux [4]. The program is thus able to predict equilibrium crater statistics for craters with diameter D metres over the whole range of 10^{-7} metre to 10^{3} metres. Alternative analytical techniques [5] are able only to achieve partial solutions covering two size ranges; the continuous solution provided by our computer integration proves very significant. The program requires that the incident flux data be continuous in nature and contain no discontinuities. A similarly continuous prediction of the equilibrium crater statistics can be made over the stated diameter range. A modification to this program has now been made whereby if the equilibrium crater counts are known then the incident particle flux may be calculated. The technique used is an iterative one whereby an estimate of the flux is entered as input data and the equilibrium crater statistics are calculated. These are then compared with the experimentally measured equilibrium crater statistics and the input flux model is adjusted by multiplying it by the ratio of observed crater density to computed crater density. This revised flux model is then used to compute a new set of equilibrium crater statistics which is again compared with the experimental results. The technique is repeated until the computed equilibrium counts converge to the experimental values.

2. Equilibrium Crater Densities

Neukum [2] has examined exposed glass surfaces of lunar samples 60015, 60095, 60135 and 64455 and has plotted the cumulative crater size distributions, together with previous results [6—10], against the central pit diameter. The distributions are arbitrarily normalized (in the 50—150 μm range) to the crater density of sample 60015. Since the exposure age of one of the samples is known, the crater distributions are found to be equivalent to a distribution found on any sample having been exposed for 4.5×10^4 years. Using this age together with the relationship $D = 0.01 k m^{1/\beta}$, where D metres is the central pit diameter and m grammes is the mass of the particle responsible for the crater, Neukum was able to calculate the cumulative flux of interplanetary particles Φ as

$$\Phi_1 = 10^{-8.73}\, m^{-0.38}\,[\mathrm{m^{-2}\,s^{-1}}] \quad (10^{-12}\,\mathrm{g} < m < 10^{-7}\,\mathrm{g})$$

and

$$\Phi_2 = 10^{-15.5}\, m^{-1.35}\,[\mathrm{m^{-2}\,s^{-1}}] \quad (m > 10^{-5}\,\mathrm{g}),$$

valid probably up to the order of grammes). In the calculation a mean impact velocity of 20 km s^{-1} and a density of 3 g cm^{-3} were assumed for the particles, thus leading to values of $k = 6.2$ and $\beta = 2.65$ as a compromise between the laboratory results of Gault [11] and Neukum et al. [10]. In terms of central pit diameter:

$$\Phi_1 = 10^{-9.94}\, D^{-1}\,[\mathrm{m^{-2}\,s^{-1}}] \quad 10^{-6} < D < 10^{-4}\ \text{metres}$$

and

$$\Phi_2 = 10^{-19.8}\, D^{-3.57}\,[\mathrm{m^{-2}\,s^{-1}}] \quad D > 10^{-3}\ \text{metres}.$$

These flux estimates are shown in Fig. 1 (the section between 10^{-5} g and 10^{-7} g is interpolated) together with the best experimentally determined values of the flux of interplanetary particles over the same mass range [12]. The two curves agree very well for masses greater than 10^{-10} gramme but for smaller masses the curve derived from the Pioneer data has a significantly steeper gradient than Neukum's.

Using Neukum's predicted flux as input data for our erosion program results in the equilibrium cumulative crater densities N_E shown in Fig. 2 for four values of the sputter rate. Also shown in Fig. 2 are Neukum's estimates for the equilibrium crater densities. In the region $D > 3 \times 10^{-4}$ m, $N_E = A_1 D^{-2}\,\mathrm{m^{-2}}$, where the areal density $A_1 = 0.4$—0.6 according to Neukum et al. [13] and Morrison et al. [14]. The line DE, which is labelled Neukum, has $A_1 = 0.4$. This line is parallel to the line $N_E = 10^{-1.1} D^{-2}\,\mathrm{m^{-2}}$ which is lettered AB and is an extrapolation of Shoemaker's measurements on large craters [1]. For $D < 10^{-4}$ m Neukum has assumed that sputtering is insignificant and concludes that the equilibrium curve is parallel to the production curve, i.e. $N_E \propto D^{-1}$. The constant of proportionality has been estimated by Neukum who gives $N_E = 10^{2.9} D^{-1} m^{-2}$. This curve is plotted as BC in Fig. 2. From Fig. 2 it can be seen that, for small craters, the equilibrium crater densities N_E we have derived from Neukum's flux estimate do not agree with his values of N_E derived from the above equation. However, several points of interest may be deduced from the curves and from the curves plotted in Fig. 3 which show the relative importance of the three erosion processes.

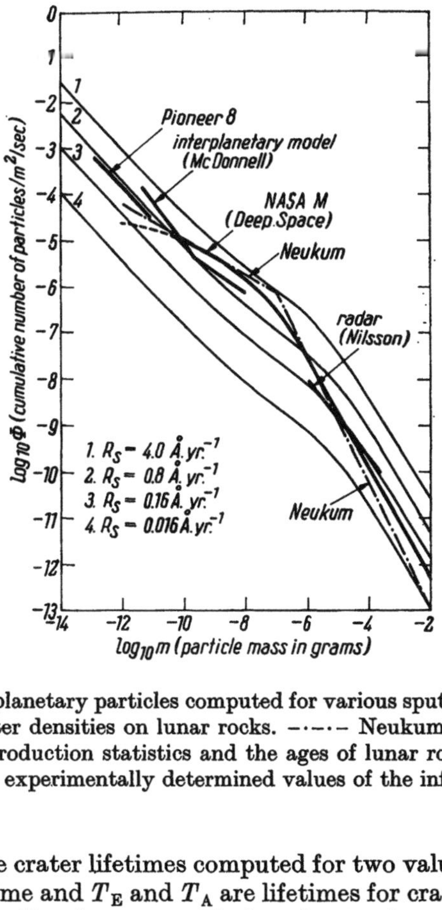

Fig. 1. The flux of interplanetary particles computed for various sputter rates from Neukum's inferred equilibrium crater densities on lunar rocks. –·–·– Neukum's estimate of the influx calculated from crater production statistics and the ages of lunar rocks. Also shown are the best experimentally determined values of the influx.

These curves show the crater lifetimes computed for two values of sputter-erosion. T_S is the sputter lifetime and T_E and T_A are lifetimes for crater erasure and crater abrasion respectively. T_C, the crater lifetime due to the combined effect of the three erosion processes, is calculated at $R_S = 0.016$ Å yr^{-1} only. T_C is evaluated by using the formula

$$T_C = (T_S^{-1} + T_E^{-1} + T_A^{-1})^{-1}.$$

Fig. 3 shows that sputter erosion is important for craters with diameters less than 10^{-5} m *even* at a sputter rate R_S as low as 0.016 Å yr^{-1}. For a sputter rate much less than 0.016 Å yr^{-1} the curves shown in Fig. 2 will approach the dashed line which is parallel to, but greater in magnitude than, Neukum's curve $N_E = 10^{2.9} D^{-1} m^{-2}$. If the sputter rate is not less than 0.016 Å yr^{-1} then Neukum's assumption, that the equilibrium curve and the production curve are parallel, is certainly not valid, and his equilibrium and influx measurements are incompatible as currently understood. Such a low sputter rate has never previously been contemplated; the previous lowest limit, suggested by McDonnell and Ashworth, was 0.02–0.04 Å yr^{-1} [15]. In an accompanying paper [16] experimental data lend support to a sputter rate in the region of 0.04 Å yr^{-1}. Neg-

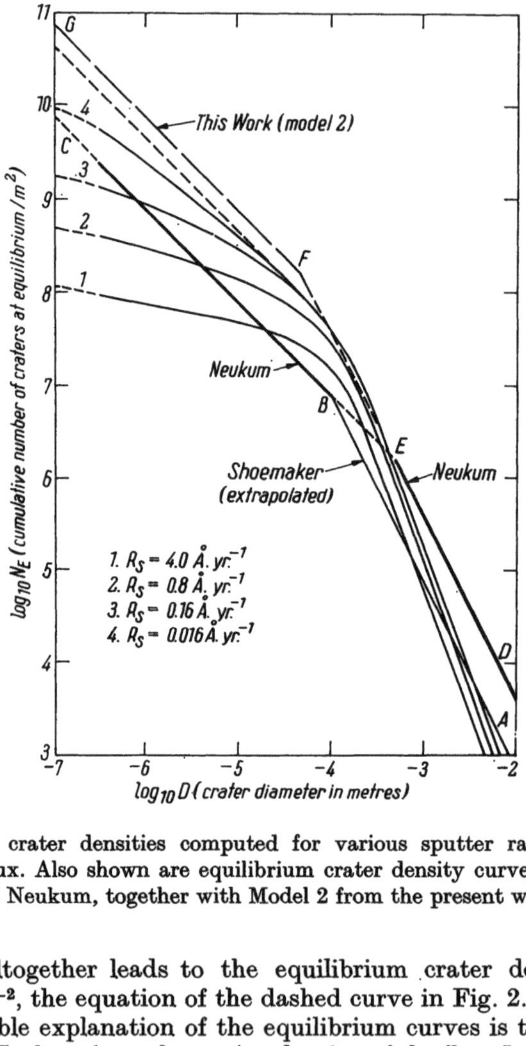

Fig. 2. Equilibrium crater densities computed for various sputter rates from Neukum's estimate of the influx. Also shown are equilibrium crater density curves due to Shoemaker and Neukum, together with Model 2 from the present work.

lect of sputter altogether leads to the equilibrium crater densities given by $N_E = 10^{3.6} D^{-1}$ m^{-2}, the equation of the dashed curve in Fig. 2.

A second possible explanation of the equilibrium curves is that the estimated flux Φ_1 given by Neukum is too large. A reduction of the flux Φ_1 by approximately two orders of magnitude, coupled with a very low sputter rate, would yield Neukum's equilibrium crater count. Unfortunately, such a low flux would pose insurmountable disagreement with current day *in situ* experiments as interpreted by accepted hypervelocity impact parameters, unless the present day micrometeoroid flux could be assumed to be greater (by up to two orders of magnitude) than the long-term average integrated over the past tens of thousands of years, which corresponds to the age of the rocks investigated.

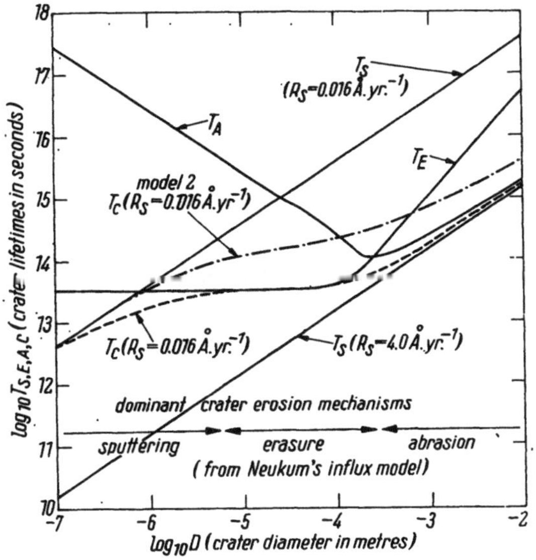

Fig. 3. Crater lifetimes under sputter erosion T_S, erasure T_E, and abrasion T_A, computed from Neukum's estimate of the influx. Also shown is the combined lifetime T_C under the three erosion processes for a sputter rate of 0.016 Å yr^{-1}. T_C is also shown for Model 2 under the same sputter rate.

3. Flux Models

Firstly, we used the computer program to produce a model of the particle flux using Neukum's equilibrium crater density estimates as input data, i.e. the curve DEBC shown in Fig. 2. The flux needed to produce such an equilibrium crater density curve was then evaluated for values of the sputter rate R_S of 4.0, 0.8, 0.16 and 0.016 Å yr^{-1} (Fig. 1). The flux curve which comes closest to the current day experimentally observed flux curves is for $R_S = 0.8$ Å yr^{-1} but, although the agreement is good for particles of mass 10^{-13} g $< m < 10^{-10}$ g and within experimental bounds for particles of mass $m > 10^{-6}$ g where radar measurements are less accurate, the agreement is poor in the intermediate region where the experimental results derived from penetration experiments should yield good agreement with lunar impact data assuming a time invariant flux. Many other models have been tested: the most successful one, in terms of an agreement with current fluxes, is Model 2 (labelled DEFG in Fig. 2). Section EF is an extrapolation of DE and FG satisfies the equation $N_E = 10^{3.9} D^{-1}$ m^{-2}. The resulting flux curves are shown in Fig. 4. The flux curves derived for sputter rates of 0.16 and 0.016 Å yr^{-1} follow very closely the shape of the experimental curves with the 0.16 Å yr^{-1} curve too large in magnitude and the 0.016 Å yr^{-1} curve too small. An accurate fit to current experimental fluxes over the whole range can, therefore, be obtained by using Model 2 at a sputter rate of approximately 0.04 Å yr^{-1}. The combined crater erosion time T_C is shown in Fig. 3 for Model 2 at a sputter rate of 0.016 Å yr^{-1}.

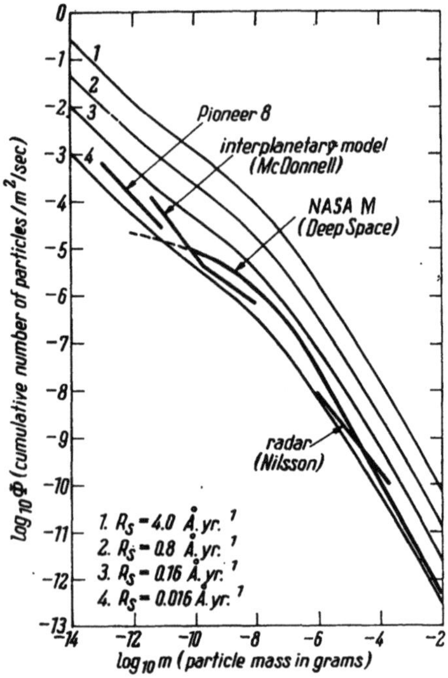

Fig. 4. The flux of interplanetary particles computed from Model 2 for various sputter rates. Also shown are the best experimentally determined values of the influx.

4. Conclusions

Using our computer erosion program it has been found that a basic incompatibility exists between both the magnitude and the slope of the published influx particle distributions and the equilibrium crater distributions of Neukum. Equilibrium distributions also lead to an influx model incompatible with current direct measurements, and for such an agreement we would have to conclude that the published equilibrium values inferred are an order of magnitude lower than we would expect in the range of crater diameter $<10^{-4}$ m. Neukum's equilibrium crater densities for craters of diameter 5×10^{-4} m $< D < 10^{-2}$ m would, however, appear to be compatible. We find that by using equilibrium crater densities an order of magnitude higher than those estimated by Neukum in the range $D < 10^{-4}$ m we need a sputter rate in the region of 0.04 Å yr^{-1} for agreement to be reached between the calculated influx model and current direct measurements. This figure is in agreement with our earlier estimates of the sputter rate for lunar rocks [15] and yet despite this low value sputter erosion still proves to be the most dominant erosion mechanism for micrometre and submicrometre size particles.

Alternatively, using Neukum's equilibrium measurements directly at this sputter rate leads to derived influx curves which are in agreement with current meteor flux measurements for $m > 10^{-4}$ g. This, however, implies that the current micrometeorite flux for $m < 10^{-6}$ g is enhanced by one order of magnitude above average values over the last 10^6 years.

Acknowledgments

The authors would like to thank Mr. D. Bateman for assistance with computer programming and Miss D. Paine for preparation of the manuscript.

References

[1] E. M. Shoemaker et al., Proc. Apollo 11 Lunar Sci. Conf. Vol. 3, Pergamon Press 1970 (p. 2399).
[2] G. Neukum, 4th Ann. Lunar Sci. Conf., Abstract.
[3] D. G. Ashworth and J. A. M. McDonnell, Space Research XIII, 1071 (1973).
[4] M. R. Bloch et al., in: Proc. 2nd Lunar Sci. Conf., Vol. 3, M. I. T. Press 1971 (p. 2639).
[5] G. Neukum and H. Dietzel, Earth Planet. Sci. Lett. **12**, 59 (1971).
[6] J. B. Hartung et al., in: Proc. 3rd Lunar Sci. Conf., Vol. 3. M. I. T. Press 1972 (p. 2735).
[7] F. Horz et al., J. Geophys. Res. **76**, 5770 (1971).
[8] D. E. Brownlee et al., in: The Apollo 15 Lunar Samples, Lunar Sci. Inst., Houston 1972.
[9] G. Neukum, Thesis, University of Heidelberg, 1971 (unpublished).
[10] G. Neukum et al., in: Proc. 3rd Lunar Sci. Conf., Vol. 3, M. I. T. Press 1972 (p. 2793).
[11] D. E. Gault, The Moon (in press).
[12] J. A. M. McDonnell, Space Research XI, 415 (1971).
[13] G. Neukum et al., to be published.
[14] D. A. Morrison et al., in: Proc. 3rd Lunar Sci. Conf., Vol. 3, M. I. T. Press 1972 (p. 2767).
[15] J. A. M. McDonnell and D. G. Ashworth, Space Research XII, 333 (1972).
[16] J. A. M. McDonnell, R. P. Flavill and D. G. Ashworth, Space Research XIV, 733 (1974).

Space Research XIV — Akademie-Verlag, Berlin 1974

LONG TERM LUNAR SURFACE EXPOSURE EXPERIMENT

F. Hörz and D. W. Strangway

NASA Johnson Space Center, Houston, Texas, USA

Surveyor 3 parts returned during the Apollo 12 mission were analyzed by a variety of investigators concerned with the space environment and its interaction with planetary surfaces. These investigations fall in three general categories: (i) solar wind and solar flare studies; (ii) galactic radiation studies; (iii) micrometeoroid studies. Although pre-mission documentation of these materials was not detailed and total surface exposure time was only 31 months, significant and, in part, unexpected results were obtained. As a consequence, significant efforts were undertaken to document some Apollo 17 flight hardware in great detail. It is envisioned that these articles may be returned in some future, though completely undetermined, time and exposure effects ranging for periods of decades may be evaluated.

The actual documentation consisted of three general, though related, activities: (i) Detailed photo documentation was performed with resolutions up to 60 μm of a variety of surfaces, in particular those suitable for micrometeoroid studies, though other investigations were not excluded. (ii) Corresponding or representative parts of the various back-up units (or of other suitable quality, preferably those resembling closest the flight units) were retrieved and are properly stored in the Lunar Sample Curator's facilities. These materials will serve as useful "standards", especially for analytical techniques presently not available. (iii) Detailed design drawings and other technical information were retrieved and are also stored in the Curatorial facilities.

Of particular interest to micrometeoroid studies are mirror surfaces. A total surface area of approximately 0.5 m^2 was photographed with a resolution of 150 μm and about 50 cm^2 with resolutions of 60 μm. The mirror surfaces of the television camera are orthogonally arranged and provide the opportunity to investigate directional anisotropies of micrometeoroid trajectories. A variety of surfaces are of chemically pure materials (SiO_2 glass, Mg, Al and Be surfaces) and therefore the expected microcraters may be successfully searched for the chemical composition of projectile materials.

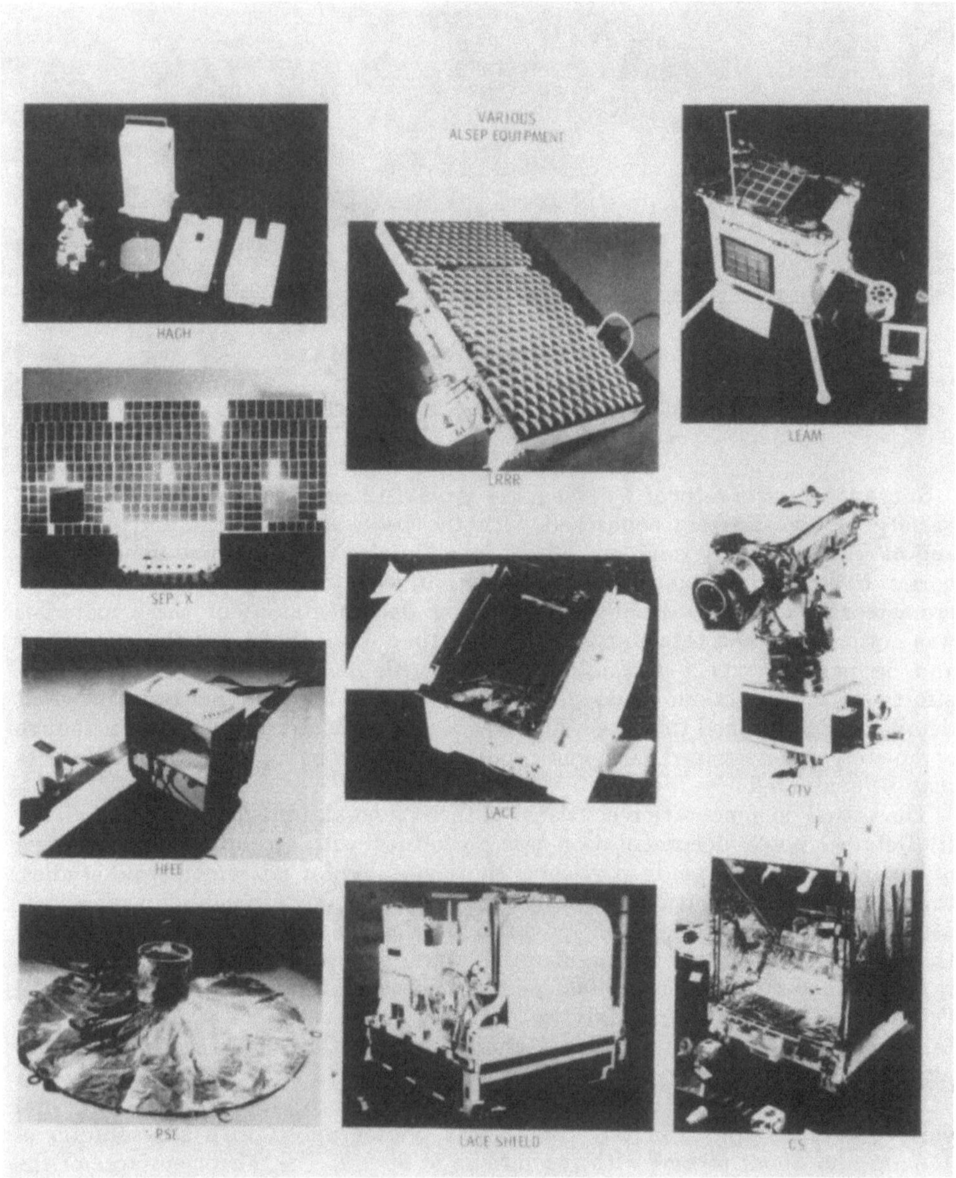

Fig. 1. Various ALSEP equipment documented: HAGH, Helix Antenna Gimbal Housing (glass fibre); LRRR, Laser Ranging Retro Reflector (SiO_2 glass); LEAM, Lunar Ejecta and Meteoroid Experiment (Mirror-surfaces, Mylar foils, etc.); SEP, X, Surface Electrical Properties, Transmitter (solar cells i.e. SiO_2 glass, Ge-back plate); HFEE, Heat Flow Experiment Electronics Box (glass fibre); LACE, Lunar Atmospheric Composition Experiment (glass fibre, SiO_2 glass, Be surfaces); CTV, Color Television Camera (SiO_2 glass); PSE, Passive Seismic Experiment (Mylar-skirt); LACE SHIELD (aluminum); CS, Central Station (Mylar). With the exception of LRRR, PSE and CS all other instruments are photographed in detail. Representative pieces, however, of all of them are stored in the Curatorial facilities.

Space Research XIV — Akademie-Verlag, Berlin 1974

HYPERVELOCITY IMPACT AND SOLAR WIND EROSION PARAMETERS FROM SIMULATED MEASUREMENTS ON APOLLO LUNAR SAMPLES

J. A. M. McDonnell, R. P. Flavill and D. G. Ashworth

University of Kent at Canterbury, UK

A programme of simulated erosion parameter evaluation involving hypervelocity impact, solar wind sputtering and thermal cycling has been conducted using Apollo lunar samples and various reference surfaces. Hypervelocity impacts of iron microparticles yield a primary crater diameter typically three times the particle diameter at 5—15 km s^{-1} and a spall zone of four times the primary crater diameter. A weak velocity dependence is observed and no significant decrease of spall damage at the highest velocities observed. Under simulated solar wind sputtering using helium it has been found necessary to erode a depth of 0.75 μm to remove craters of primary diameter 0.25 μm. Preferential etching of the shock damaged spall zone leads to a sputter lifetime considerably greater than anticipated, and a sputter erosion depth of twice the primary crater diameter is expected for complete extinction of impact craters resulting from regular solid silicate microparticles in the submicrometre range.

1. Introduction

The evaluation of erosion parameters relevant to particulate matter in the zodiacal complex and to the exposed surfaces of planetoids remains an important area of study necessary for the development of evolutionary models of solar systems and for determining current 'equilibrium' states, in particular, the lunar surface. Towards the interpretation of these lunar data recently made abundant, hypervelocity impact parameters have been evaluated at realistic meteoric velocities for solid particles of density $\gtrsim 1$ g cm^{-3} using electrostatic microparticle accelerators, and this can be used to detail inferred values of the interplanetary flux, and probably the time history of the flux, at least in the vicinity of 1 AU heliocentric distance. This derived flux should then be in substantial agreement with currently measured fluxes from *in situ* detectors. To date, this disparity has been considerable [1].

2. Current Measurements

We have extended hypervelocity impacts [2] to several types of lunar samples in the velocity range 2—20 km s^{-1} and to ambient target temperatures of -195 °C to $+120$ °C. Results of the complete study are still under evaluation, but we concentrate here primarily on the subsequent behaviour of these hypervelocity

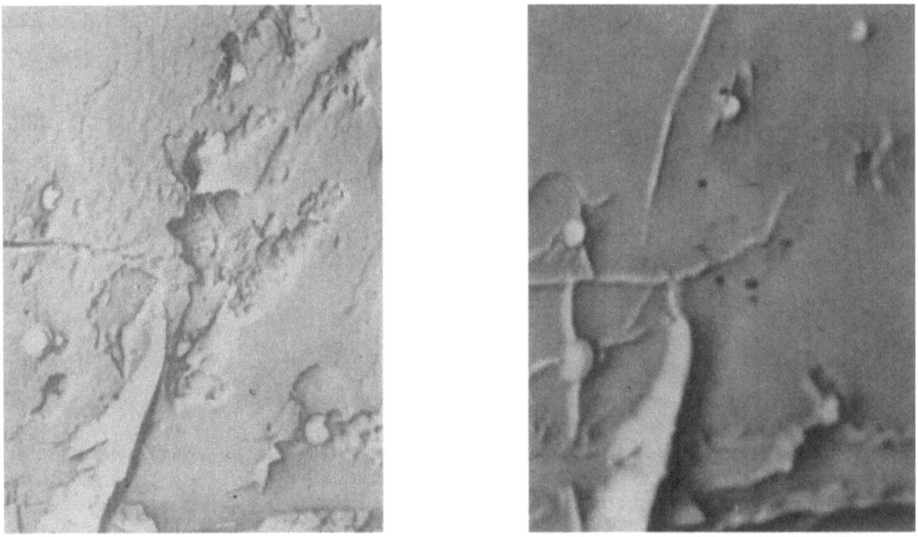

1a 1b

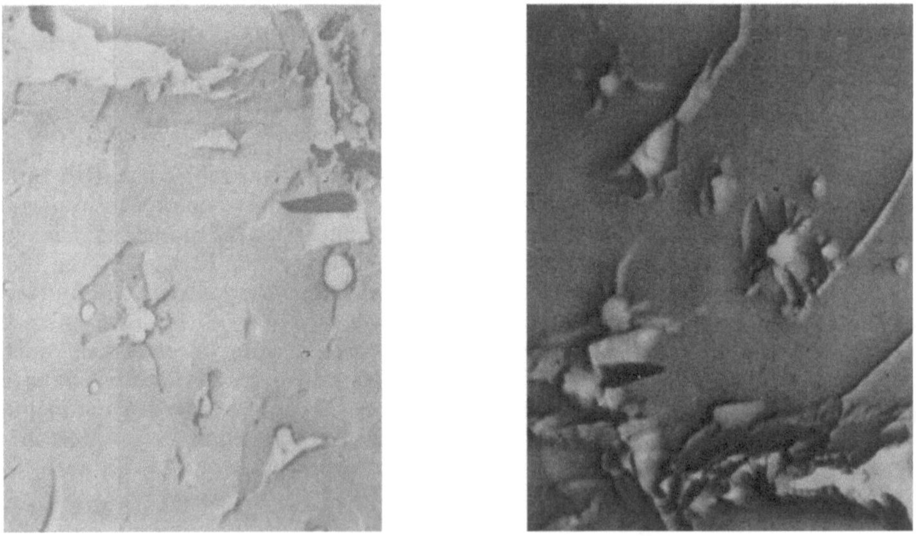

2a 2b

Figs. 1 and 2. Sample areas of Apollo 14, 310.158. Hypervelocity impacts on polished sections (left) are subsequently eroded by helium sputtering (right).

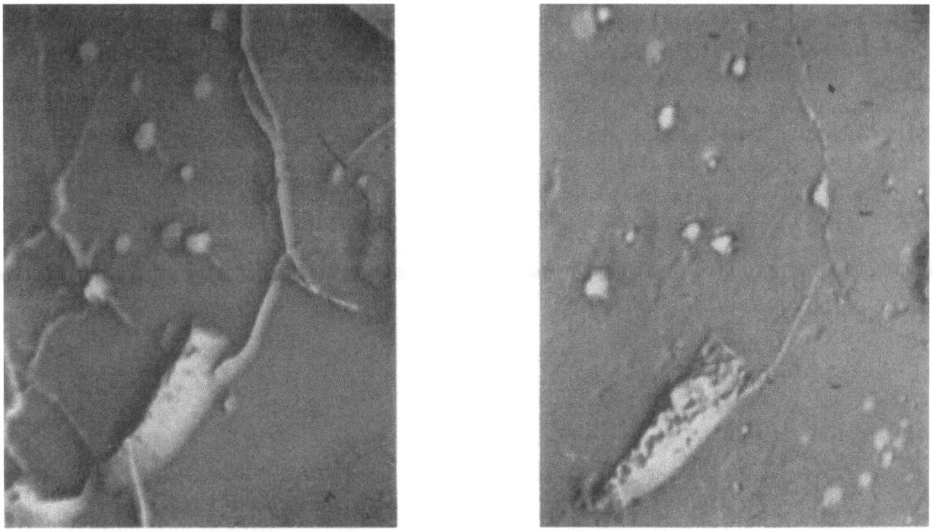

a b

Fig. 3. Hypervelocity impacts before (left) and after (right) simulated solar wind erosion. Preferential etching of grain boundaries and spall faults can be detected and the enlargement of small craters.

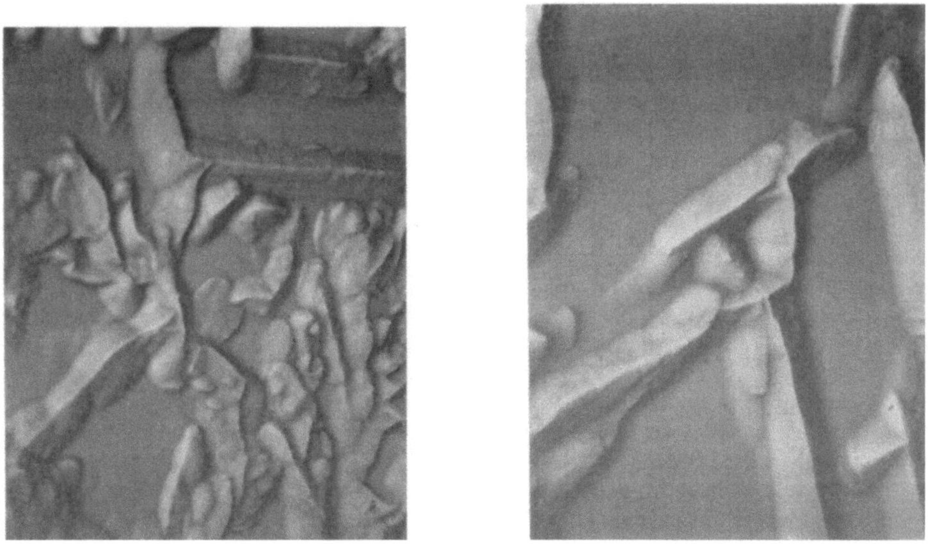

a b

Fig. 4. Etching of grain boundaries by solar wind sputtering on polished section of 14,310.158. These areas contain no hypervelocity impacts.

impact zones under the effects of simulated solar wind sputtering. For this purpose a helium ion source was constructed and used at an energy of 2.5 keV, typical of the solar wind helium component. Ion fluxes of much higher magnitude than the solar wind were necessary to avoid contamination. When insulating samples were used a flood of low energy electrons was provided to neutralize charge repulsion. Reference measurements on copper foil were made to compare with established ion yield ratios.

The results of areas of simulated hypervelocity impact and the subsequent sputter erosion to be anticipated by an equivalent of 2×10^6 years exposure are shown in Figs. 1—3 for Apollo sample 14310.158. A loss of some 0.75 μm of surface material was observed. The larger craters shown in the figures result from

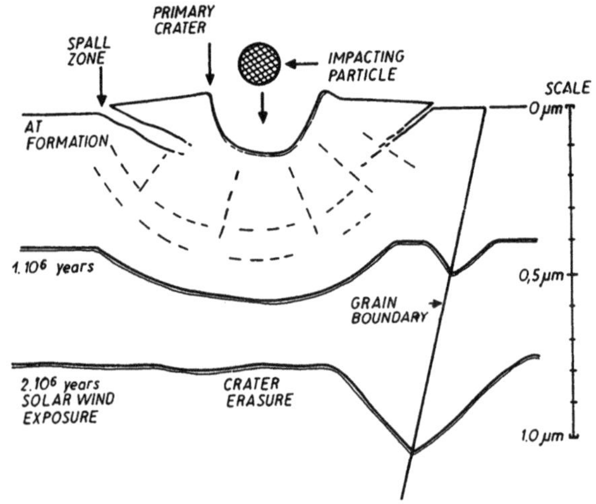

Fig. 5. Schematic illustration of the impact zones created at impacts from 5—15 km s^{-1} and subsequent erasure by solar wind under simulated erosion by helium ions. A sputter erosion of twice the primary crater diameter is estimated for erasure of impacts from silicate microparticles.

low velocity impacts of comparatively large particles of ~1 μm diameter, compared with the smaller crater resulting from particles of diameter ~0.05 μm at 10—15 km s^{-1}. Features observed are a primary crater zone of about three times the impacting particle diameter at above 3 km s^{-1} and a spall diameter of ~13 times the particle diameter. This impact spall zone has particular relevance to solar wind sputter measurements. For low velocity impacts having radial fractures, sputter erosion was noted to occur preferentially along the fractures, and this etched spall pattern could be expected therefore even after sputtering of the primary crater. At the highest velocities observed in this programme to date, the spall zone was still of comparable dimensions, and upon sputtering a shallow, but hemispherical, depression of diameter similar to this spall zone resulted even after removal of the primary crater. In addition to what is probably preferential etching of spall zones, it should be noted that surface features tend to be self-preserving under sputter action, since the sputter depth is almost independent of the angle of incidence for angles up to ~60°.

Significant features of microcrater impact craters and the subsequent solar wind sputtering are (i) microcraters of old exposure age, perhaps $\gtrsim 10^6$ years for micrometre craters, will appear as shallow hemispherical depressions of diameter similar to the extent of spall damage, (ii) a removal of surface material by sputtering of depth considerably greater than the primary crater diameter is required. For impacting iron particles a sputter erosion depth of three times the primary crater diameter is required. A figure of twice the primary crater diameter is estimated for the sputtering of impacts of lunar density silicates. Other features noted under sputter erosion were the appearance of pseudo craters almost indistinguishable from eroded microcraters and the strong preferential etching near both grain boundaries and fractures in spall zones. The preferential sputter erosion of smaller craters near these areas would result in lower equilibrium crater densities, particularly in breccia types.

3. Sputter Rates on the Lunar Surface

Conversion of this observed erosion of 0.75 μm of lunar rock under a simulated helium exposure of 2×10^6 years to a value representative of actual lunar conditions must be cautious at this stage and await further experimental measurements. We would rather confine the use of these data to identify the characteristics of craters and grain boundaries which develop under sputter erosion.

Our earlier estimate based on data prior to the availability of lunar samples [3] had predicted a rate of 0.02 Å yr^{-1} in contrast to the then accepted figures in the region of 1 Å yr^{-1} [4]. The preliminary simulation data presented here would certainly support our rejection of this high figure, and, combined with the need to sputter a depth of twice the primary crater diameter, the effect of solar wind sputter as a crater removal mechanism is certainly weak.

The importance of sputtering relative to other erosion mechanisms is a more critical issue, however, in particular, solid particle impact erosion. The comparison with impact erosion is dependent on the model assumed for mass distribution, but assuming the existence of distributions of similar nature to the current interplanetary distribution we find [5] that at a sputter rate even as low as 0.004 Å yr^{-1} the equilibrium crater density distribution in the micrometre region is still significantly influenced by the sputter rate.

References

[1] G. Neukum and H. Dietzel, Earth Planet. Sci. Lett. **12**, 59 (1971).

[2] J. A. M. McDonnell, D. G. Ashworth, R. P. Flavill and R. C. Jennison, Proc. 3rd Lunar Sci. Conf., Suppl. 3, Geochim. Cosmochim. Acta 2755 (1972).

[3] J. A. M. McDonnell and D. G. Ashworth, Space Research XII, 333 (1972).

[4] G. K. Wehner, C. KenKnight and D. L. Rosenberg, Planet. Space Sci. **11**, 885 (1963).

[5] D. G. Ashworth and J. A. M. McDonnell, Space Research XIV, 723 (1974).

Note added in proof. In a recent extension of sputter measurements we present a sputter rate typical of lunar conditions of 0.043 Å yr^{-1} [J. A. M. McDonnell and R. P. Flavill, Lunar Science V, Abstracts, p. 478, Lunar Science Institute, Houston, Texas 1974].

Space Research XIV — Akademie-Verlag, Berlin 1974

MICROMETEOROID EROSION OF LUNAR ROCKS

G. Neukum

Max-Planck-Institut für Kernphysik, Heidelberg, FRG

Measurements of "production" crater populations on (mainly) lunar glass samples in connection with exposure age determinations give a means of calculating the flux of micrometeoroids on to the moon. Results indicate that the flux of very small particles ($\lesssim 1$ μm diameter) seems to have risen by approximately 1.5 orders of magnitude in the last $\approx 5 \times 10^3$ years (historic time). Thus, we have to distinguish between a historic flux (time span, today to $\approx 5 \times 10^3$ years ago) and a prehistoric flux (time span, $\approx 5 \times 10^3$ to $\approx 5 \times 10^7$ years ago) for this size range. The historic flux is in rough agreement with today's satellite measurements (Fig. 1). It is not yet quite clear if the flux for larger particles has changed in the same way. The prehistoric cumulative flux results from the lunar studies to

$$\Phi = 10^{-16.5} m^{-1.35} [\mathrm{m^{-2}\,s^{-1}}]$$

for the range $10^{-7}\,\mathrm{g} < m < 10^{-3}\,\mathrm{g}$.

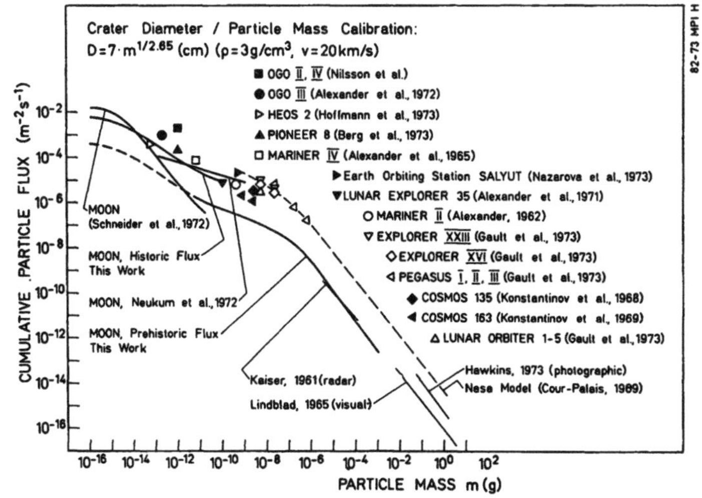

Fig. 1

The erosion by micrometeoroid impacts leads to a gradual destruction of rocks on the lunar surface. The average small scale erosion in the last 10^7 years is of the order of $^1/_{10}$ mm per 10^6 years.

Measurements of crater frequencies on lunar rocks give a means of determining the exposure times of lunar rocks on the lunar surface. Typical exposure ages for Apollo 16 rocks lie between 100000 years and 10 million years.

A full paper will be submitted to *Planetary and Space Science.*

Space Research XIV — Akademie-Verlag, Berlin 1974

INVESTIGATION ON THE CHEMICAL MODIFICATION IN AN IMPACT CRATER ON A LUNAR SAMPLE

A. MEHL*

Max-Planck-Institut für Kernphysik, Heidelberg, FRG

Various methods have been used to obtain information about the chemical composition of meteoroids. Qualitative analysis using an electron microprobe of some sub-μm particles collected at a height of about 115 km was made by Hemenway et al. [1]. Millman [2] identified some elements by spectral analysis of the light emitted during the entry of a meteor into the earth's atmosphere. He was able to assign approximate concentrations to a few of the elements found.

During a high velocity impact of interplanetary dust on the moon a certain amount of the impacting material is left in the crater while the remainder is either evaporated or thrown out as ejecta. Neukum [3] developed a technique for measuring the amount of material left in the crater. Using a dust accelerator he impacted iron particles in the μm-sized range on various metal targets and then used a microprobe to analyse the iron in the crater linings. Scanning over the whole of a crater he determined semiquantitatively the admixture of projectile material to the target. He later applied this method to lunar samples [4].

In this note measurements of the concentrations of various elements in the glass lining of a much larger lunar impact crater are reported. A piece from sample 12070.141 consisting of a glass-lined pit about 2 mm in diameter was imbedded and polished to obtain a vertical cross section through the pit. The concentrations of the elements K, Ca, Na, Fe, Ti, Mg, Ni, Si, Al and O were measured using a 20 keV electron microprobe along radial profiles going from the pit surface into the undisturbed target material. A sketch of the cross section showing the measurement spots can be seen in Fig. 1. K, Ca and Na were measured using a strongly defocused beam (about 10 μm diameter compared with the normal 0.2 μm diameter) to prevent evaporation by excessive local heating. The raw data have been corrected for absorption, atomic number effect, and fluorescence using a computer program by Goldstein and Comella [5]. The concentrations of K and Na were below the detection limit. The distribution of the remaining elements along a path about 15 μm below, but parallel to, the pit surface is shown in Fig. 2. Fig. 3 illustrates the concentrations along a central vertical profile leading from the crater surface into the target material. The concentrations of Mg, Si, Fe and O increase towards the centre of the pit, while the contents of Ca and Al decrease. The concentration of Ti shows no variation with either depth or length. The con-

* Present address: Physikalisch-Technische Bundesanstalt, Braunschweig, FRG

centration of Ni is constant at about 0.06%, which is only slightly above the detection limit.

If one assumes that the elemental variations do not represent the original target composition, which seems improbable because of the marked dependence on depth and length, the variation of elemental concentrations can be explained

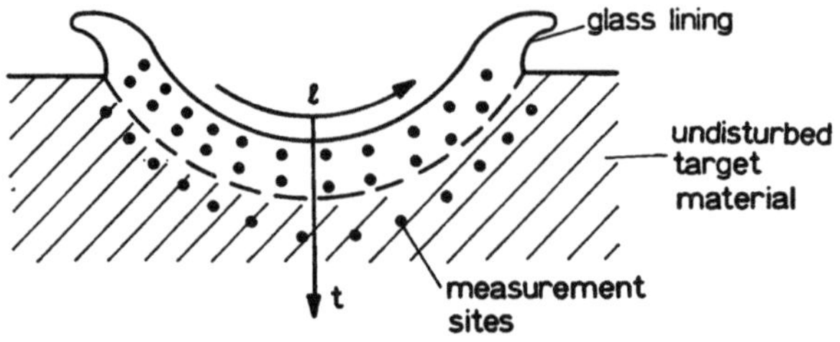

Fig. 1. Sketch of measurement sites on the crater section.

by two different mechanisms. First, the elements depleted in the zone immediately below the crater surface relative to the target material are characterized by comparatively low melting and boiling points, so that differential vaporization is probably one cause for the described phenomenon. Mg, although having relatively low melting and boiling temperatures, may be enriched in the surface zone because of the very high melting and boiling temperatures of its oxide. The second factor which contributes to the chemical modification is the addition of projectile, i.e. meteoritic material, to the surface zone. At present, no information is available about the relative importance of these two processes.

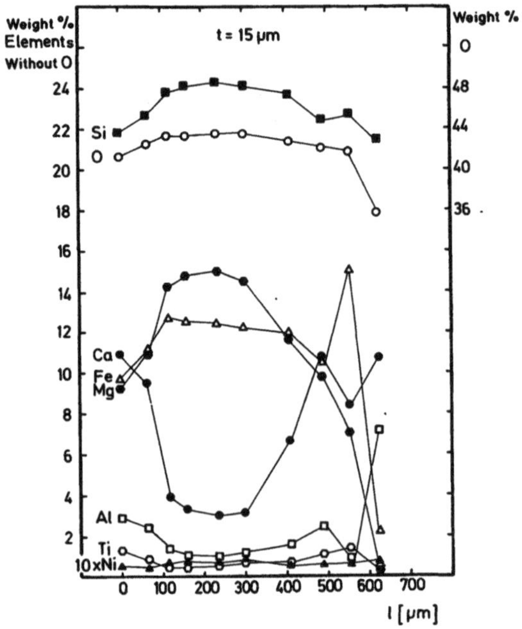

Fig. 2. Variation of elemental abundances with length l.

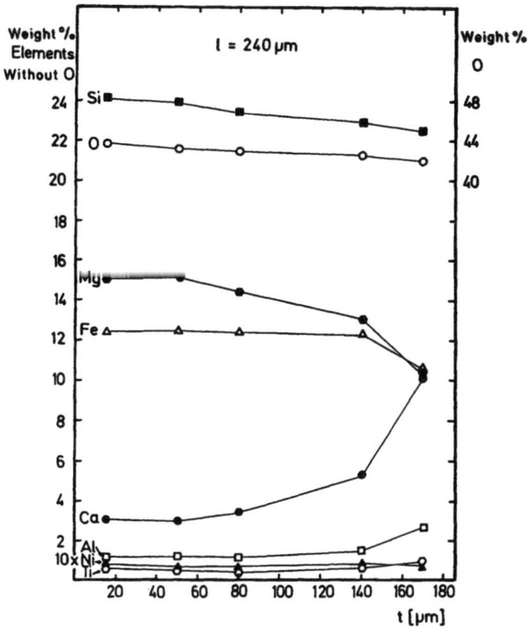

Fig. 3. Variation of elemental abundances with depth *t*.

References

[1] C. L. Hemenway, D. S. Hallgren and D. C. Schmalberger, Nature **238**, 256 (1972).
[2] P. M. Millman, contribution to the 34th Meeting of the Meteoritical Society, Tübingen 1971.
[3] G. Neukum, Diplomarbeit, Universität Heidelberg 1969.
[4] G. Neukum, Doktorarbeit, Universität Heidelberg 1971.
[5] J. I. Goldstein and P. A. Comella, NASA X-642-69-115 (1969).

Space Research XIV — Akademie-Verlag, Berlin 1974

FIRST RESULTS OF THE MICROMETEOROID EXPERIMENT S 215 ON THE HEOS 2 SATELLITE

H.-J. HOFFMANN, H. FECHTIG, E. GRÜN and J. KISSEL

Max-Planck-Institut für Kernphysik, Heidelberg, FRG

The detection of the micrometeoroids is based on the production of a plasma by the impact on the sensor. Their mass and speed are derived from the height and rise time of two coincident pulses originating from the ions and electrons of the plasma. The sensor has a viewing cone half angle of 60° and can be orientated in any direction perpendicular to the sun—earth line. The satellite orbits the earth in a highly eccentric polar orbit with an apogee of 240000 km and a perigee varying between 350 and 5000 km. Within the first year 173 particles have been recorded by the experiment. The data indicate a significant flux enhancement in the near earth region below 60000 km. By analysing the time distribution of the particle impacts, three categories characterized by the period between successive impacts are indicated:

Category (number of events)	Period	Total Number of Particles
1. swarms (three)	less than 15 min.	94
2. groups (seven)	15 min.—12 hours	24
3. random particles	more than 12 hours	55

The swarms are interpreted as being of very recent origin, possibly supplied 1. by meteoroids grazing the earth's atmosphere, 2. ejecta from the moon and 3. collision of particles in space. The groups may be dissipated old swarms or the margin of swarms. The random particles consist of the ancient dust population plus the dissipated groups.

A full paper will be submitted to *Planetary and Space Science.*

Space Research XIV — Akademie-Verlag, Berlin 1974

SPATIAL AND TIME VARIATIONS OF THE INTERPLANETARY MICROPARTICLE FLUX ANALYSED FROM DEEP SPACE PROBES ON PIONEERS 8 AND 9

J. A. M. McDonnell[a], O. E. Berg[b] and F. F. Richardson[b]

[a] University of Kent at Canterbury, England
[b] NASA Goddard Space Flight Center, Maryland, USA

The first results of a comprehensive computer analysis of over 300 front film and grid coincidence events is presented using statistical tests on the observed data. The short-term time dependence of the observed flux is entirely commensurate with a purely random Poisson distribution and any possible contributions from discrete "cometary showers" must certainly be of relatively minor significance compared with the sporadic background for mass $>10^{-13}$ g. The mass spectrum of the flux in the range $10^{-11}-10^{-13}$ g indicates an increasing flux of particles to the lowest limits of mass detected, with a derived flux of $\Phi = 1.4 \times 10^{-12} m^{-0.68}$ g m^{-2} s^{-1} $(2\pi$ sr$)^{-1}$.

A full paper will be submitted to *Planetary and Space Science*.

Space Research XIV — Akademie-Verlag, Berlin 1974

MARINER 4: A STUDY OF THE CUMULATIVE FLUX OF DUST PARTICLES OVER A HELIOCENTRIC RANGE OF 1—1.56 AU 1964—1967

W. M. ALEXANDER[a]* and J. L. BOHN[b]

[a] Max-Planck-Institut für Kernphysik, Heidelberg, FRG
[b] Temple University, Philadelphia, Pennsylvania, USA

Between December 1964 and December 1967, the Mariner 4 dust particle experiment obtained data concerning the distribution of minute zodiacal dust cloud particles over a heliocentric range of 1—1.56 AU. The first measurement was over the complete heliocentric range, while the two additional measurements were made between 1.1 and 1.25 AU in 1966, and between 1.2 and 1.5 AU in 1967. The initial results of these measurements presented the mean cumulative flux for the respective data periods. The results of a detailed study and comparison of the three measurements are presented, with particular emphasis on the variation of the flux as a function of heliocentric range. A small, but statistically significant, increase in the flux is observed between 1.15 and 1.4 AU. The initial reports showed a lower cumulative flux for the latter two measurements. However, a detailed analysis containing corrections for spacecraft attitude indicate that all three measurements yield similar results, and that the particles detected were in low inclination orbits.

1. Introduction

The Mariner 4 spacecraft was launched on 30 November 1964 and reached a near encounter with Mars on 15 July 1965. The cosmic dust experiment detected interplanetary dust particles impacting an exposed sensor plate. Except for the time used for playback of data from the photographic experiment immediately after the encounter, data of the cosmic dust experiment were continuous from launch until 1 October 1965 when the earth—spacecraft distance became too great for successful data transmission. In 1966 and 1967 signals from the spacecraft were re-acquired and information received from the dust experiment. The mean cumulative flux for the heliocentric range of the experiment over the times of data reception have been reported. The purpose of this paper is to give a summary of the statistical analysis of the data, to present the events as a function of heliocentric range, and to show the cumulative flux-mass distribution as derived from the Mariner 4 measurements.

2. Mariner 4 Cosmic Dust Experiment

A description of the experiment was originally reported by Alexander et al. [1]. Further studies concerning calibration of piezoelectric acoustical transducers bonded to large impact plates by McDonnell [2] provided a shift in the threshold

* Permanent address: Baylor University, Waco, Texas, USA.

impulse of the Mariner 4 experiment. This was first reported by McDonnell [3] as a mass threshold of 6×10^{-12} g or a nimpulse of 1.2×10^{-5} dyne s for 20 km s^{-1} particles, which was lower than first reported [1]. The number of events has remained the same since the first report.

3. Noise and Mariner 4

Noise has been an ever-present problem with all of the very low data rate dust particle experiments. Several papers have presented data concerning noise from the standpoint of thermal problems and solar energetic particles [4—7]. From the extensive tests by Bohn et al. [5], there appears to be little reason to suspect that the events in the data are caused by thermal related phenomena. The solar energetic particle events during the mission are shown with the data in Fig. 1. The Mariner 4 instrument does not show any evidence of the kind of

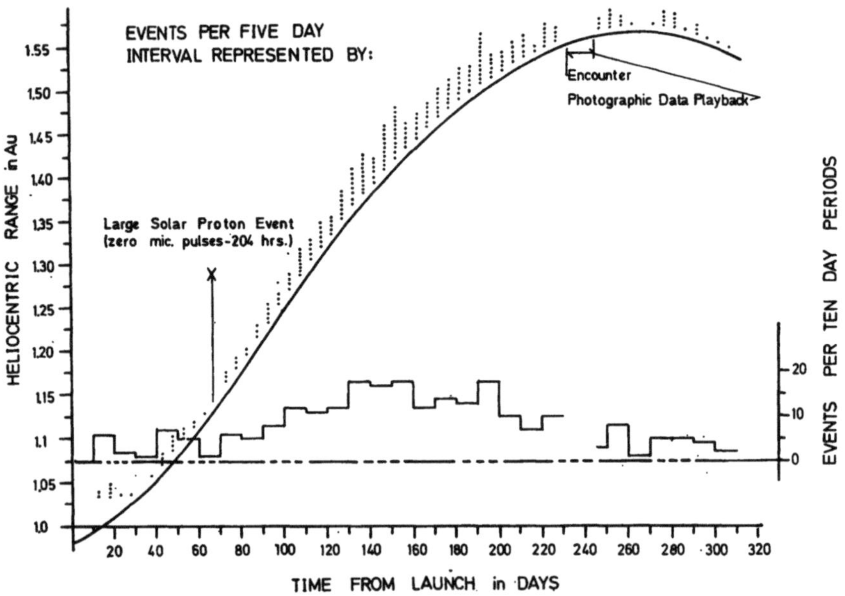

Fig. 1. Mariner 4 events as a function of heliocentric range.

events reported from Pioneer 8 and 9 [6, 7]. One large event occurred during an eight day period of zero counts from the microphone sensor. There does not appear to be any compelling reason or evidence that the Mariner 4 data are subject to noise interference.

4. Data from the Mariner 4 Experiment

Event Information. The cosmic dust experiment detected 241 events during the primary phase of the mission. The data are shown in Fig. 1 as a function of heliocentric range. A pulse height analysis (PHA) was made on each event. The PHA information was erased after each readout of the experiment information.

However, a cumulative counter was never reset and provided a continuous count of the number of events if any short periods occurred where data were not received or processed. Of the 241 events, 23 occurred during periods where in-flight PHA information was not processed. Therefore, the total cumulative flux can be derived from the 241 events, and the slope of the cumulative flux-mass distribution from the remaining 218 events.

Experiment Attitude Information. The attitude of the experiment was such that the normal of the impact plate was in or very close to the ecliptic plane. Also, an angle of 90° with the radius vector to the sun was maintained throughout the mission.

Summary of Statistical Evaluation of the Data. The distribution of the events in time were considered to be represented by a Poisson distribution, and the χ^2 test for variance was employed to test the degree of randomness of the event time intervals. First, all 241 events were used; χ^2 was found to be 22 (5 degrees of freedom), giving a probability p of less than 0.001 that the overall distribution was randomly distributed in time. However, when the distribution is considered in three segments, near earth, near Mars, and between the orbits of the two planets, the following values for p were found: earth to 1.2 AU, $p = 0.7$; 1.2—1.45 AU, $p = 0.6$; 1.45—1.56 AU, $p = 0.95$. The events within each segment are definitely distributed in time in a random manner.

5. Discussion of the Mariner 4 Results

Flux vs Heliocentric Range. The data seen in Fig. 1 shows an increase in flux between the planets. A search was conducted for possible interceptions of nodes of cometary orbits or meteor streams as possible sources of flux increase between the planets. However, only one was found near the Mariner 4 orbit and the increase is much too broad for a single stream. Intercepts of the planes of cometary orbits with the ecliptic plane along the Mariner 4 path were studied. It has been shown in a number of studies [8—11] that the semi-major axis of orbits of small particles, after being created by any process, will be considerably larger (i.e. those not moving into parabolic or hyperbolic orbits). It is not clear that such an increase as seen by Mariner 4 can occur for this reason. A test of this occurred when spacecraft transmission was re-acquired in 1966. The spacecraft was near perihelion, and then moved out again. The mean flux value was essentially the same as for the first part of the mission, and then increased with heliocentric range. However, the solar ecliptic longitude was quite different. Therefore, the probability of a significant increase in one segment of solar ecliptic longitude due to debris from cometary orbits is felt to be low.

The mean flux values derived from the periods in 1966 and 1967 were modulated by shifts in the spacecraft attitude. The normal to the detector plate was as previously stated when the Canopus sensor was locked on Canopus. About 1 September 1966 the Canopus sensor was found to be locked on a star near the ecliptic plane. The plane of the detector plate was then lying in the ecliptic plane with the plate normal pointing to a pole in the celestial hemisphere. The flux was found to be significantly reduced: the ratio of the values was 4:1. This is strong evidence that the orbit inclination angles of the detected particles are quite low. Significant flux range data are given in Table 1.

Table 1

Comparison of Flux from Mariner 4 for 1964—1965 and 1966 over the same Heliocentric Range (from [12])

Range (AU)	Flux (events m^{-2} s^{-1} $\times 10^{-5}$) 1964—1965	1966
1.1—1.08	8	9
1.08—1.12	5	8
1.12—1.19	9	7
1.19—1.22	10	9
1.22—1.25	13	13

The measurement is considered to show depression in flux at the planetary orbits, the most probable cause being planetary scattering. This is not an effective mechanism for particles in these size regimes with fairly short lifetimes. However, the effect may be significant with particles of these sizes when inclinations and eccentricities are low in value [13].

Cumulative Flux-Mass Distribution. The PHA measurements provide a means for determining the cumulative flux-mass distribution for detected events. The data were fitted to a linear curve by applying linear regression curve fitting techniques. The cumulative flux is usually described by using the empirical equation $\Phi = km^{-\alpha}$, where Φ is flux in (particles m^{-2} s^{-1}), m is the mass (g) and k and α are constants.

The log equation

$$\log_{10} \Phi = \log_{10} k - \alpha \log_{10} m$$

is linear with slope of $-\alpha$. The values for α were found for each segment, and the variation fell within statistical bounds. The overall cumulative flux is presented in Fig. 2 since the slope variation was not significant.

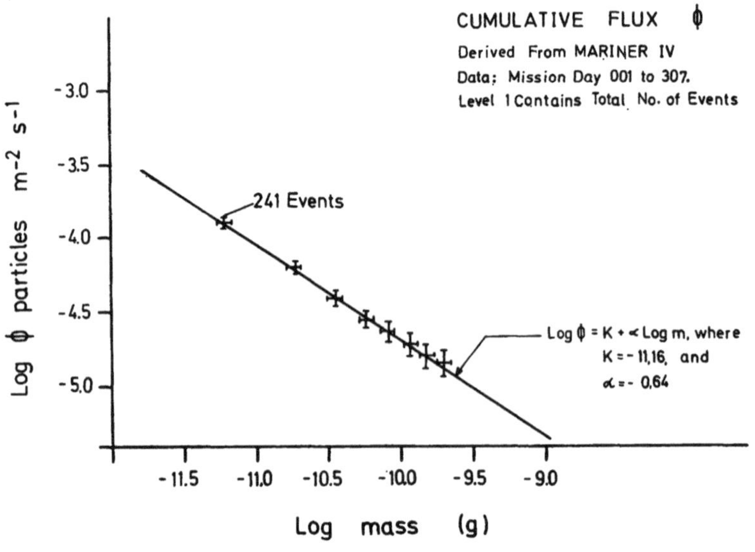

Fig. 2. Cumulative flux-mass distribution derived from the measurement of Mariner 4.

6. Comparison of Mariner 4 Cumulative Flux with other *in situ* Measurements

Selected in situ Measurements. The measurements selected for comparison with Mariner 4 results in Fig. 3 were chosen because they contained either coincidence information or some form of control to enhance the reliability of the data. The data from the penetration experiments of Explorers 16 and 23 and Pegasus are used. The penetration points contain the calibration revisions which have occurred since the review paper of McDonnell [3].

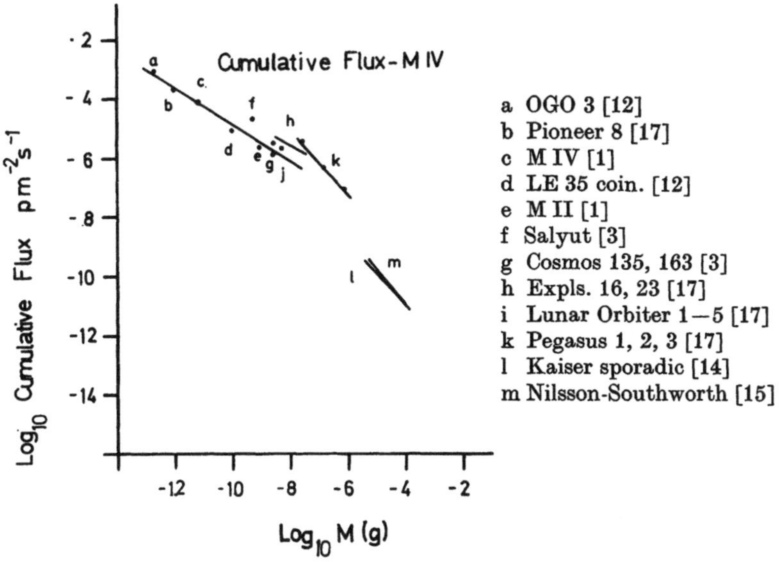

Fig. 3. Comparison of Mariner 4 cumulative flux distribution and *in situ* measurements.

Selected Ground-Based Data. The radar data from two representative sources are used [14, 15], for comparison of the particles just larger than those obtainable from the *in situ* measurements.

Results of Comparison. The cumulative flux-mass distribution from the *in situ* measurements has a slope essentially identical with that determined by the Mariner 4 measurements. The Mariner 4 slope used in Fig. 3 is that between 1 and 1.2 AU. The decrease of the slope from radar through the Pegasus and Explorer points shows an agreement which is quite close. All points of the curve have been corrected for gravitational focusing [16, 17] so the flux is that of interplanetary space near 1 AU. The *in situ* measurements do not, at the present time, show a mass cutoff, i.e. slope going to zero on the cumulative flux-mass distribution curve. In addition, information from the studies of sub-micrometre craters on lunar rocks shows the presence of particle sizes smaller than previously expected [18, 19].

Acknowledgments

The work was partially supported under NASA Research Grant NGR 39-012-001 and the Max-Planck-Institut für Kernphysik. The authors are indebted to Dr. H. Fechtig, Dr. H.-J. Hoffmann, and Dr. E. Grün of the Max-Planck-Institut für Kernphysik for many helpful discussions.

References

[1] W. M. Alexander, C. W. McCracken and J. L. Bohn, Science **149**, 1240 (1965).
[2] J. A. M. McDonnell, J. Phys. E **2**, 1026 (1969).
[3] J. A. M. McDonnell, Space Research XI, 415 (1971).
[4] C. S. Nilsson, Science **153**, 1242 (1966).
[5] J. Lloyd Bohn, W. M. Alexander and W. F. Simmons, Space Research VIII, 588 (1968).
[6] O. E. Berg and U. Gerloff, Space Research XI, 225 (1971).
[7] O. E. Berg and E. Grün, Space Research XIII, 1047 (1973).
[8] F. L. Whipple, Astrophys. J. **121**, 750 (1955).
[9] M. Harwit, J. Geophys. Res. **68**, 2171 (1963).
[10] E. C. Silverberg, Dissertation, University of Maryland, Tech. Rep. No. 70-066 (1970).
[11] J. S. Dohnanyi, Icarus **17**, 1 (1972).
[12] W. M. Alexander, C. W. Arthur, J. D. Corbin and J. L. Bohn, Space Research X, 252 (1970).
[13] T. Schmidt, NASA SP-150, 333 (1967).
[14] T. R. Kaiser, Mon. Not. Roy. Astron. Soc. **123**, 265 (1961).
[15] C. S. Nilsson and R. B. Southworth, Physics and Dynamics of Meteors, Springer-Verlag, New York 1968 (p. 280).
[16] E. J. Öpik, Proc. Roy. Irish Acad., Section A **54**, 165 (1951).
[17] B. G. Cour-Palais, NASA SP-8013 (1969).
[18] G. Neukum, E. Schneider, A. Mehl, D. Storzer, G. A. Wagner, H. Fechtig and M. R. Bloch, paper presented at XVth COSPAR Meeting, Madrid 1972.
[19] J. B. Hartung, F. Hörz and D. E. Gault, Proc. 3rd Lunar Sci. Conf., Geochim. Cosmochim. Acta Suppl. 3, **3**, 2735 (1972).

Space Research XIV — Akademie-Verlag, Berlin 1974

RESULTS OF THE ASTEROID/METEOROID PARTICLE EXPERIMENT ON PIONEER 10 (1.0 – 3.3 AU)

S. L. NESTE and R. K. SOBERMAN

General Electric Space Sciences Laboratory and Drexel University, Philadelphia, Pa, USA

The asteroid/meteoroid detector (Sisyphus), an optical array to measure particle size and orbit, has been performing successfully on Pioneer 10 since it was initially activated on 9 March 1972. During the first ten months of operation over 200 meteoroid and asteroid events were detected between 1.0 and 3.3 AU. These events are being analyzed to determine the spatial concentration as a function of heliocentric distance. The early results of these analyses are presented with particular emphasis on the distribution in the asteroid belt. Preliminary orbital parameters for some particles which have been analyzed are presented with a discussion of instrumental limitations.

1. Introduction

The Pioneer 10 vehicle was launched on 2 March 1972 on a trajectory passing through the asteroid belt to encounter Jupiter on 4 December 1973. Among the eleven experiments being carried by the spacecraft is the asteroid/meteoroid detector or Sisyphus. This instrument was designed to measure the meteoroid and asteroid environment encountered by the spacecraft beyond the earth's orbit. The instrument consists of two subsystems. The optical subsystem contains four Ritchey-Crétien telescopes with 7.5° fields of view that use RCA 7151Q photomultipliers with S20 photocathodes as the sensors. The telescopes are surrounded by a light shield to reduce sunlight reflected or scattered from various parts of the spacecraft. The electronics include four identical analog channels of amplifying and thresholding circuits, coincidence logic, counters and storage registers. The optical subsystem is located behind the three-meter high-gain antenna, pointed at an angle of 45° relative to the spin axis of the spacecraft. As the spacecraft rotates at a rate of approximately 4.8 revolutions per minute, the sensors sweep a viewing annulus on the celestial sphere.

The instrument collects two types of data. The first is the measurement of the sky background for the purpose of zodiacal light determination as a function of heliocentric distance. The data from this portion of the experiment are treated in [1].

The prime purpose of the Sisyphus detector is the measurement of individual particles which reflect sunlight as they cross through the overlapping fields-of-view of the telescopes. The parameters measured by the instrument are the entrance and exit times as the particle enters and leaves each of the fields. The vector equations of motion are solved for the range of the particle from the instrument and its three velocity components relative to an instrument-centered coordinate system. The fourth telescope provides redundancy and the possibility of four independent measurements of a single particle trajectory. From the computed trajectory of the particle with respect to the instrument and the known attitude and orbit of the spacecraft, one can then calculate the orbit of the particle in the solar system.

As a noise rejection technique the instrument requires a three-fold coincidence in order to accept pulses from the telescopes as constituting a real "event". Thus, the instrument is a nearly omnidirectional detector with a "sensitive volume" equal to the overlap region of three fields-of-view. For simplicity this overlap region is also taken as conical. The height of this cone is a function of the particle size. It is expressed as a range-to-radius ratio which is inversely proportional to the first power of the heliocentric distance. From the computed range of the particle and the intensity of the signal as measured by the telescopes we can calculate an "albedo cross section" which is defined as the product of the particle area and its bond albedo.

2. Orbit Determinations

Pioneer 10 and 11 represented the first opportunity to test the Sisyphus detection system in space. Since there was the possibility that few if any particles would be seen during the mission (e.g. [2]), the instrument sensitivity was maximized by operating at comparatively low signal-to-noise ratios. Excellent noise immunity for "events" was built in by the coincidence requirement. However, this noise immunity does not extend to individual orbit determinations. While high confidence can be placed on an "event" as being caused by a particle, the calculated trajectory and orbit may be erroneous due to any of the following causes: (i) premature signal drop-out due to noise, (ii) particle brightness pulse crossing the threshold while the particle is in the instrument field-of-view (such as might

Table 1

Preliminary Orbital Parameters for Five Particles Detected by the AMD near 1.2 AU

Event No.	Semi-major axis (AU)	Eccentricity	Inclination	Longitude of ascending node	Radius of perihelion (AU)	Period (yr.)
0-2-105α	1.91	0.45	10.4°	22.6°	1.05	2.6
0-2-107δ	0.82	2.10	95.0°	213.1°*	0.90	—
0-2-109β	3.88	0.75	4.3°	12.3°	0.99	7.6
0-2-115δ	5.72	0.81	3.6°	248.8°	1.10	13.7
0-2-116β	2.65	0.57	11.4°	229.5°	1.14	4.3

* Descending node.

result from a large axial velocity component), (iii) the particle entering or leaving the shadow of the spacecraft while in the field-of-view of the instrument, (iv) variable particle reflectivity as a function of the sun-particle-telescope angle (primarily for close range particles), (v) noise pulses near the entry or exit times. All but (iii) can be overcome by a good signal-to-noise ratio from the particle. This is a luxury which we did not enjoy for most of the events measured by the Pioneer 10 Sisyphus experiment. In fact, one of the first two orbits which were reduced [3] now appears to be erroneous. After reviewing the instrument performance and simulating the trajectory, it was concluded that premature termination of one of the exit timers resulted in an erroneous reading.

In Table 1 we list some of the orbits which have been calculated to date. All five of these events were recorded in one two-week interval when the spacecraft was near 1.2 AU. No confidence limits are placed on these orbits as yet. In fact, we have reason to distrust the one hyperbolic orbit. Additional effort is required here to determine the limits of error on these orbits as well as many others measured by the Sisyphus experiment on Pioneer 10.

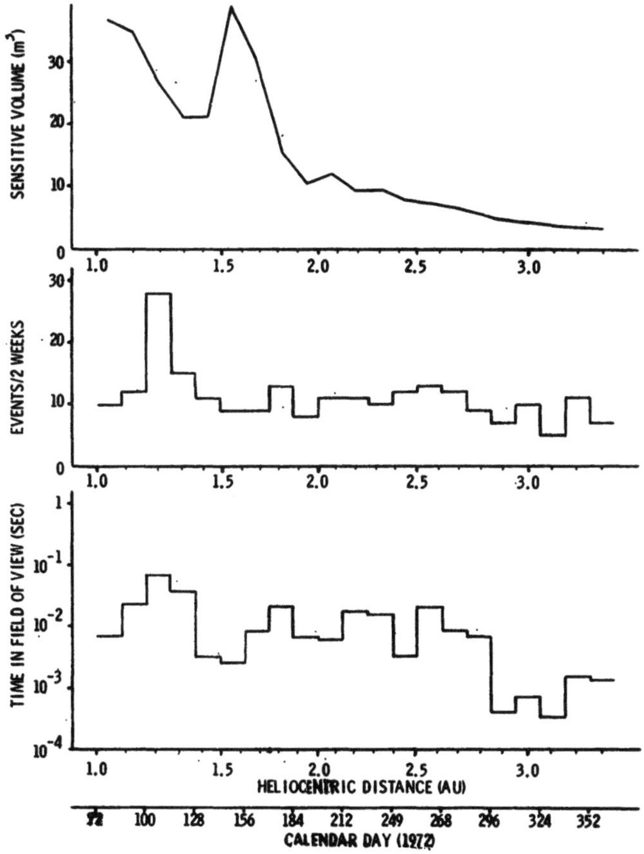

Fig. 1. Top, sensitive volume; middle, event rate; bottom, particle transit time, as functions of solar distance.

3. Spatial Concentration

By measuring the background as the spacecraft rotates we are able to calculate the threshold and consequently the maximum range at which a particle of a given size can be measured as a function of position in the rotation cycle. From this and the known heliocentric distance we can compute the sensitive volume of the Sisyphus instrument. Corrections are made for the time when portions of this volume are in the shadow of the high-gain antenna. The range or height of the cone is determined by the third most sensitive telescope. An average volume is computed for each day from the background maps obtained on that day. The variation of the sensitive volume for a one millimeter particle size as a function of the heliocentric distance is shown in Fig. 1 (top). Also shown in that figure (middle) is the number of events per two-week period and (bottom) the time spent by the particles in the field-of-view.

We have computed a relative concentration of particles as a function of heliocentric distance from 1 to 3.3 AU (Fig. 2) using the data from Fig. 1. All of the particle events not attributable to spacecraft thruster firings, marginal stellar threshold crossings, solar storms, etc. are included in this spatial concentration. It differs from spatial concentration measured by impact detectors in two ways. The first is that there is no abrupt cutoff as far as particle size or mass is concerned. This figure is for a mean particle radius of approximately 1 mm $\pm$ a factor of 10. The scond is that no velocity assumptions are required to compute this spatial concentration. One need note only the amount of time that the particle spent in the sensitive volume. The ordinate is presented in relative units since we expect to refine this figure by presenting the spatical concentration as a function of more specific particle sizes. Putting the concentration in terms of particles per cubic meter at this point might be misleading due to the wide range of particle sizes included.

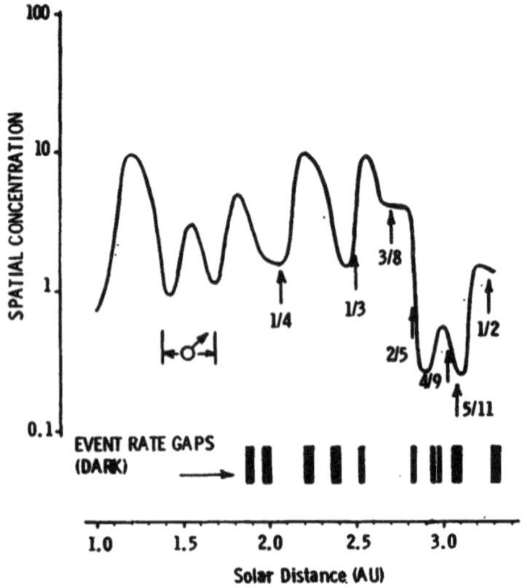

Fig. 2. Relative spatial concentration as a function of solar distance.

Of interest in Fig. 2 is the rising spatial concentration as one goes outward from 1 AU with a peak at approximately 1.2 AU. The concentration then drops off to a minimum in the vicinity of the Mars orbit and then starts rising again as the asteroid belt is approached. It is, of course, possible that the initial rise from 1 AU may be due to the earth's sweeping effect. The variations in spatial concentration within the asteroid belt are seen to coincide with the Kirkwood Gaps. This appears to be more than fortuitous. We have shown under the spatial concentration curve the points at which no events were recorded for intervals of 0.02 AU and longer. The minima may be somewhat displaced as a result of data averaging over two week periods. The average uncertainty on the points in Fig. 2 is approximately $\pm 40\%$. It should be remembered that the Kirkwood Gaps are orbital rather than spatial gaps for the visible asteroids. The presence of heliocentric gaps in the smaller particle concentrations would appear to open new horizons to the solar system theoreticians.

4. Conclusion

The most obvious conclusion to be drawn from the foregoing is that a large amount of work remains to be performed on the analysis of the Pioneer 10 Sisyphus data. We have confidence that we will able to derive a larger number of orbits with reasonable uncertainty limits. Size distributions remain to be determined. The results obtained to date have proven that the Sisyphus particle detection concept works and can realize vital data regarding the particulate environment in our solar system.

Acknowledgments

The authors gratefully acknowledge the cooperation and assistance of the personnel of the Pioneer Project Office, National Aeronautics and Space Administration, Ames Research Center without which this effort would not have been possible. They are also deeply indebted to the many people of the General Electric Space Division who contributed to the Sisyphus instrument and are contributing to the data reduction.

This work was supported in its entirety by the National Aeronautics and Space Administration, under contract numbers NAS 2-5609 and NAS 2-6559.

References

[1] H. A. Zook and R. K. Soberman, Space Research XIV, 763 (1974).
[2] S. Auer, Pioneer File Memo. PAE-0-40-107 (NASA/ARC, 25 November 1970).
[3] S. L. Neste, I. Jurkevich and R. K. Soberman, paper presented at AGU Fall Annual Meeting, San Fransisco 1972.

Space Research XIV — Akademie-Verlag, Berlin 1974

PIONEER 10 METEOROID PENETRATION DATA

W. H. Kinard, R. L. O'Neal, J. M. Alvarez and D. H. Humes

NASA Langley Research Center, Hampton, Va, USA

Penetrations of meteoroids having masses of 10^{-9} gram, or larger, have been recorded by pressurized cell type detectors on the Pioneer 10 spacecraft for the first time in interplanetary space from the earth's orbit out through the region of the Asteroid Belt. The cells have stainless steel walls 25 μm thick. The rate of the detected meteoroid penetration events has remained nearly constant during the first year of the two-year Pioneer 10 Jupiter fly-by mission. Analysis of the history of these penetration events indicates a near-constant spatial density of approximately 9×10^{-9} particle of 10^{-9} gram mass, or larger, per cubic meter. Contrary to most interplanetary dust models, little, if any, decrease in the particulate spatial density was observed with increasing heliocentric distance in the region of space from the earth's orbit out past the Mars orbit and to the edge of the Asteroid Belt. Also, contrary to most dust models, no enhancement of the particulate spatial density was observed in the Asteroid Belt.

Space Research XIV — Akademie-Verlag, Berlin 1974

THE RADIAL DEPENDENCE OF THE ZODIACAL LIGHT

H. A. Zook[a] and R. K. Soberman[b]

[a] NASA Johnson Space Center, Houston, Texas, USA
[b] General Electric Space Sciences Laboratory and Drexel University, Philadelphia, Pa, USA

The asteroid/meteoroid detector (Sisyphus) has, in addition to its primary role of detecting and analyzing individual meteoroids, been used for measuring the brightness of the night sky from its platform on board the Pioneer 10 spacecraft. This spacecraft has traversed the asteroid belt and will fly by Jupiter in December 1973. It was quickly found that the brightness of the night sky in the antisolar hemisphere decreased rapidly with increasing distance from the sun. This decrease is due solely to the decrease in the brightness of the zodiacal light. The heliocentric dependence of the zodiacal light has been deduced from several months of data. Some data are shown to illustrate the sensitivity of the instrument to features in the night sky and the technique used for subtracting out the integrated brightness due to starlight is described. Preliminary results showing the decrease of the zodiacal light with solar distance is not inconsistent with an inverse square depencence.

1. Introduction

On 2 March 1972 the Pioneer spacecraft was launched on an interplanetary trajectory to fly by Jupiter on 4 December 1973. During the interplanetary cruise portion of its journey it traversed the asteroid belt during the latter half of 1972 and early portion of 1973. Among the experiments carried by Pioneer 10 is the asteroid/meteoroid detector or Sisyphus. The Sisyphus instrument collects two types of data. The first deals with the individual particles passing through the fields of view in the vicinity of the spacecraft. The data from this portion of the experiment are treated in [1]. The instrument is also designed to perform photometric mapping of the sky background. The objective of this latter portion of the experiment is to measure the radial dependence of the zodiacal light as the spacecraft traverses the region from 1 to 5 AU. The Sisyphus sensors consist of four 20 cm F 1.0 Ritchey-Crétien telescopes mounted with their optical axes approximately parallel. A light shield surrounds the four telescopes to protect the optical assembly from sunlight reflected or scattered from the spacecraft. The telescopes each have a 7.5° full field of view and are mounted to view the sky at an angle of 45° to the spin axis of the spacecraft. Located in the telescope focal plane (focused for infinity) is a 2.705 cm diameter field stop. The photomultiplier tube, an RCA 7151 Q with an S20 photocathode, is located approximately 2 cm behind the field stop. This placement defocuses regions in the focal plane to minimize the effect of variations in photocathode sensitivity. To avoid light loss between the focal plane and the photomultiplier, a gold coated reflecting cylinder surrounds the intervening space. The lightweight primary and secondary mirrors of the telescope are also coated with electrodeposited gold.

As the spacecraft rotates at a rate of approximately 4.8 revolutions per minute, the four telescopes view an annular region of the celestial sphere. A spoke-wheel generator divides the rotation cycle into 128 sectors. In a semi-random fashion, photometric data from each of the four telescopes as a function of rotation sector are telemetered from the spacecraft. From these data, photometric maps of the viewed annulus can be prepared. The spacecraft spin axis, through the center of the high gain antenna, remains pointing approximately to earth throughout the mission. Thus, the viewed annulus moves across the celestial sphere during the course of the interplanetary cruise. While this yields the possibility of mapping a zone approximately 90° wide and 210° long (approximately 45° above and below the ecliptic plane), there is the disadvantage that only a limited number of points on the celestial sphere can be viewed at different heliocentric distances.

2. Analysis

The two sources used for subtracting the star backgrounds are the Smithsonian Astrophysical Observatory (SAO) Star Catalogue [2] for stars brighter than 8th magnitude and the Roach and Megill [3] analysis for stars of higher magnitude. As we shall note later, there appear to be some significant discrepancies between our calculated stellar background from the Roach and Megill data and that which is actually measured by our instrument. The authors of that paper were well aware of possible disagreements of their data with other types of analyses and, in particular, noted differences between their results and the photometric results of Elsasser and Haug [4].

The photomultiplier used in each telescope not only receives light via the primary optical system but also responds to all light entering the aperture in front of the photomultiplier tube. Light entering by the first route will be called "direct" light and that by the second path "indirect" light. Although the collecting efficiency per unit solid angle for direct light is from 40 to over 100 times greater than that for indirect light. the collecting solid angle of 44.2 deg^2 for direct light is less by a similar factor. Also, the intensity of the light received indirectly by the photomultiplier tubes has not been reduced by partial absorption of the shorter wavelengths as they are reflected from the primary and secondary gold-coated mirrors. Somewhat counterbalancing this latter effect, the indirect light is usually reflected once or more from the gold collimator sleeve located between the aperture and the photomultiplier tube.

The indirect star background is subtracted in nearly the same way as the direct star background. The difference is due to the response curve for the photomultiplier alone which is significantly different from that for the photomultiplier with two gold reflections. Each indirect light ray is reduced by an average gold absorptivity (wavelength independent) for each reflection it makes with the gold collimator. The amount of indirect starlight received is then calculated to three significant figures after exact account has been taken of the shielding afforded by the instrument light shield, the respective secondary mirrors, and the struts supporting the secondary mirrors. The result is that, for a uniform sky, about 65% of the light would come into telescopes Nos. 1 and 4 by the indirect route (35% via direct). For telescopes 2 and 3 the percentage of light arriving indirectly from a uniform sky is about 75%.

Fig. 1 shows the experimental data obtained by telescope 1 and the star background as calculated for this telescope for 28 May 1972. On that day the spacecraft was about 1.57 AU from the sun. The ordinate is in instrument output

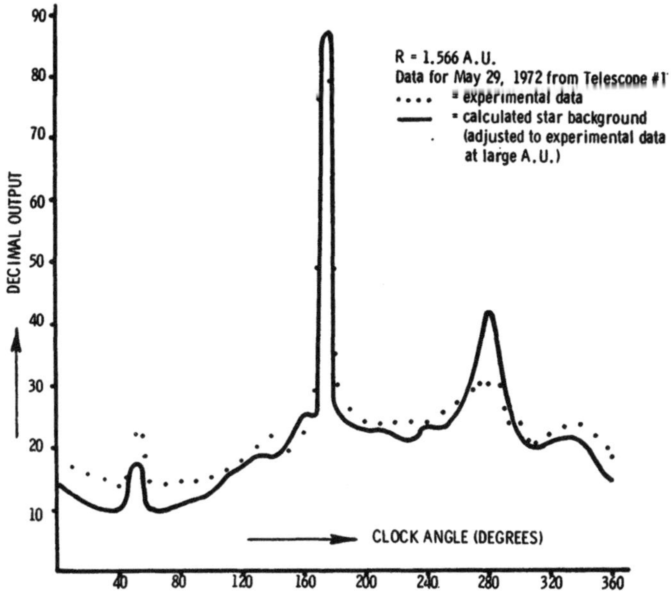

Fig. 1. Experimental data from telescope 1 compared with calculated star background.

units and the abscissa is in degrees of rotation (or clock angle) of our experiment about the spacecraft spin axis. The zero degree point is when the instrument passes north through the ecliptic plane. Apparent in the figure are Arcturus ($M_{PV} \approx 0$) at 55°, Jupiter at 175° and a peak where the pointing axis of the instrument crosses the south galactic plane near the constellation Centaurus at about 280° clock angle. We note that the peak is much smaller experimentally than it was calculated to be from the star catalogues. This indicates that either the SAO catalogue or the Roach and Megill analysis has produced too great a brightness for this portion of the sky. The Roach and Megill analysis contributes considerably more to this portion of the sky than does the SAO catalogue. Also, Elsasser and Haug disagree most strongly with Roach and Megill in this portion of the sky. In short, it becomes clear that the stellar background maps need to be corrected.

An error in the calculated sky background at one point leads to errors at other points. For example, if a variance exists as seen by a telescope in its direct field-of-view, the indirect field-of-view will also give incorrect values when that part of the sky comes into the indirect field (normally about 30 degrees of clock angle before and after it enters the direct field-of-view). The error so propagated will be dependent on clock angle. Note that if a single star is in error, this has little effect on the indirect light. Even Jupiter does not show up in the indirect field, although it could cause the output count to change by, perhaps, one unit. Errors show up in the indirect light when large regions are not correctly catalogued (as appears to be the case when the galactic plane is crossed). Note also in Fig. 1 that Jupiter is not as high experimentally as it is theoretically. This is a designed instrumental effect. The ordinate is linear with respect to incoming light to about 40 units (depending upon the preamplifier temperature) and then becomes non-linear at higher light levels to compress the high amplitude data.

Fig. 2 illustrates the results of subtracting the calculated star backgrounds from the total Sisyphus response as a function of distance from the sun. In the two upper diagrams a clock angle of 0° was chosen for telescopes 1 and 3 and in

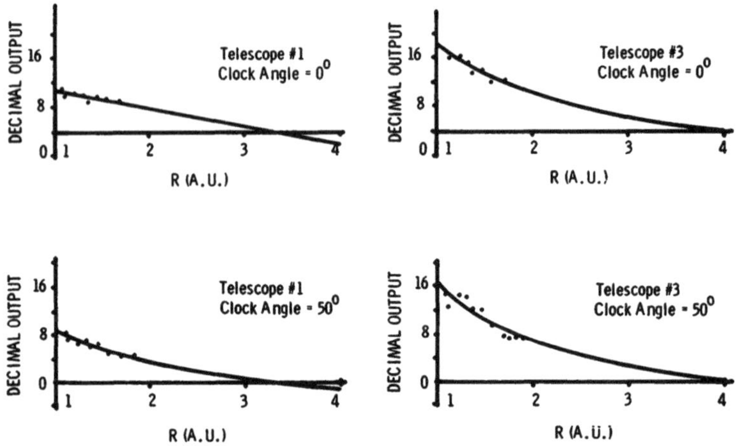

Fig. 2. Weighted average for zodiacal brightness versus heliocentric distance, with preliminary attempts at curve fitting.

the two lower diagrams a clock angle of 50° was chosen for the same telescopes. The correction constants were chosen to make all of the light as nearly as possible equivalent to starlight at large distances from the sun. Once a constant is chosen for a telescope, it is maintained for all clock angles at all distances from the sun. The different curves of Fig. 2 are preliminary attempts to fit the data. In general, it can be seen that the zodiacal light goes to zero within our instrument resolution between 3 and 4 AU from the sun. However, in order to deduce the spatial density of particles as a function of distance from the sun, it is necessary to calculate very precisely the behavior of these zodiacal light curves. The latitude dependence must also be determined.

Because of the uncertainties in the stellar backgrounds derived from SAO and Roach and Megill, we are not, at present, able reliably to derive the radial or latitudinal distributions of the particles responsible for the zodiacal light However, after these catalogues are adjusted to fit the sky as seen by Sisyphus we should be able to determine the zodiacal light brightness radially out from 1 AU with an accuracy to approximately 5% of its value at 1 AU. With the present data, we are able to calculate a weighted average for the zodiacal light brightness at any heliocentric distance traversed by the spacecraft. This is crudely calibrated by comparison with Arcturus as it was measured in the direct field-of-view of the instrument. In telescope 1, Arcturus raises the light level by $7^1/_2$ units and in telescope 3 by $9^1/_2$ units. Extrapolating Fig. 2 back to 1 AU, we find that telescope 1 indicates about 8 units of zodiacal light, while telescope 3 yields 17 units. Taking into account the different indirect fields-of-view and assuming Arcturus to be of magnitude 0 (no color correction), we obtain a weighted average for the zodiacal brightness of approximately 90 S_{10} units at 1 AU.

3. Conclusions

From the foregoing it should be clear that, beyond the extensive effort already expended, a large amount of work remains to be performed on the Pioneer 10 sky background analysis. The data and the techniques we are applying to the analysis of the data give us confidence that we will be able to deduce, in reasonably precise form, the radial distribution of the zodiacal light and consequently the spatial concentration of the zodiacal dust. We believe that the uncertainties in the existing star brightness catalogues remain our largest problem, but the data now being obtained beyond 4 astronomical units will enable us to provide the proper corrections. Variation in solar elongation angle for a given spacecraft reference angle must be examined in greater detail to determine how much this influences our determinations on both the average zodiacal brightness and the elongation angle dependence at a given radial distance. Preliminary estimates indicate that this is a comparatively small correction to the average brightness.

Our results to date indicate that the zodiacal light drops below the sensitivity of our instrument between 3 and 4 astronomical units. Early estimates place the Sisyphus sky background uncertainty limit at about 10 S_{10} units. As shown above, the zodiacal brightness as measured by this instrument at 1 AU was approximately 90 S_{10} units. Thus, our preliminary results are not inconsistent with a gross inverse square fall off with heliocentric distance. Further efforts with the Pioneer 10 and now the Pioneer 11 Sisyphus data should result in a precise definition of this radial dependence.

Acknowledgment

The authors gratefully acknowledge the cooperation and assistance of the personnel of the Pioneer Project Office, National Aeronautics and Space Administration, Ames Research Center without which this effort would not have been possible.

This work was supported in its entirety by the National Aeronautics and Space Administration, partially under contract number NAS 2-6559.

References

[1] S. L. Neste and R. K. Soberman, Space Research XIV, 755 (1974).
[2] SAO Star Catalogue, Smithsonian Institution Astrophysical Observatory, June 1967.
[3] F. E. Roach and L. R. Megill, Astrophys. J. **133**, 228 (1961).
[4] H. Elsasser and U. Haug, Z. f. Astrophys. **50**, 121 (1960).

Space Research XIV — Akademie-Verlag, Berlin 1974

CHANGES IN ZODIACAL LIGHT WITH HELIOCENTRIC DISTANCE: PRELIMINARY RESULTS FROM PIONEER 10

M. S. Hanner* and J. L. Weinberg*

Dudley Observatory, Albany, New York, USA

The Imaging Photopolarimeter on Pioneer 10 is being used during the cruise phase of the mission to map zodiacal light and integrated starlight. Data in the ecliptic plane have been analyzed to derive the absolute brightness of the zodiacal light as a function of spacecraft distance R from the sun. The brightness at $\varepsilon \simeq 140°$ decreases as $1/R^2$ between 1.0 and 1.8 AU, in agreement with the decrease found for the $\varepsilon = 180°$ (Gegenschein) region.

The Imaging Photopolarimeter (IPP) on Pioneer 10 is being used during the cruise phase of the mission to map the brightness and polarization of zodiacal light and background starlight in broad spectral bands in the blue (3900 to 5000 Å) and red (5950—7200 Å). The instrument operation and first observations of the Gegenschein away from the earth are described in [1].

The data at 140° elongation, ε, in the ecliptic plane have been analyzed to derive the absolute brightness as a function of spacecraft distance from the sun. This elongation was chosen because it was the one region in the ecliptic which could be followed consistently throughout the mission at high galactic latitudes and at telescope look angles free of scattered light. Tables of integrated starlight [2] were used to subtract the background starlight for stars of visual magnitude m_v greater than 6.0. The position of every data point was plotted on a star atlas, and fields of view containing stars with $m_v < 6.0$ were eliminated from the averages.

The resulting zodiacal light for the blue spectral band is shown in Fig. 1 as a function of sun—spacecraft distance. The brightness units are equivalent number of 10th magnitude (visual) stars per square degree. Each point represents the average of at least six data numbers in the range of elongation $140 \pm 5°$ and ecliptic latitude $0 \pm 4°$. Only data at galactic latitudes $|b^{\text{I}}| \geqq 40°$ are included.

Fig. 2 shows that the observed variation of zodiacal light between the earth and the asteroid belt fits a $1/R^2$ distribution, in agreement with the rate of decrease in Gegenschein brightness between 1.0 and 1.25 AU [3]. If this distribution continued in and beyond the asteroid belt, it would imply that the particles are distributed as $1/R$. As seen in Fig. 1, however, the slope changes abruptly near the start of the asteroid belt, and the brightness thereafter decreases approximately as $1/R^{0.5}$. The slow rate of decrease and the relatively large residual brightness beyond the asteroid belt imply that the particles increase in number in the

* Current address: State University of New York at Albany.

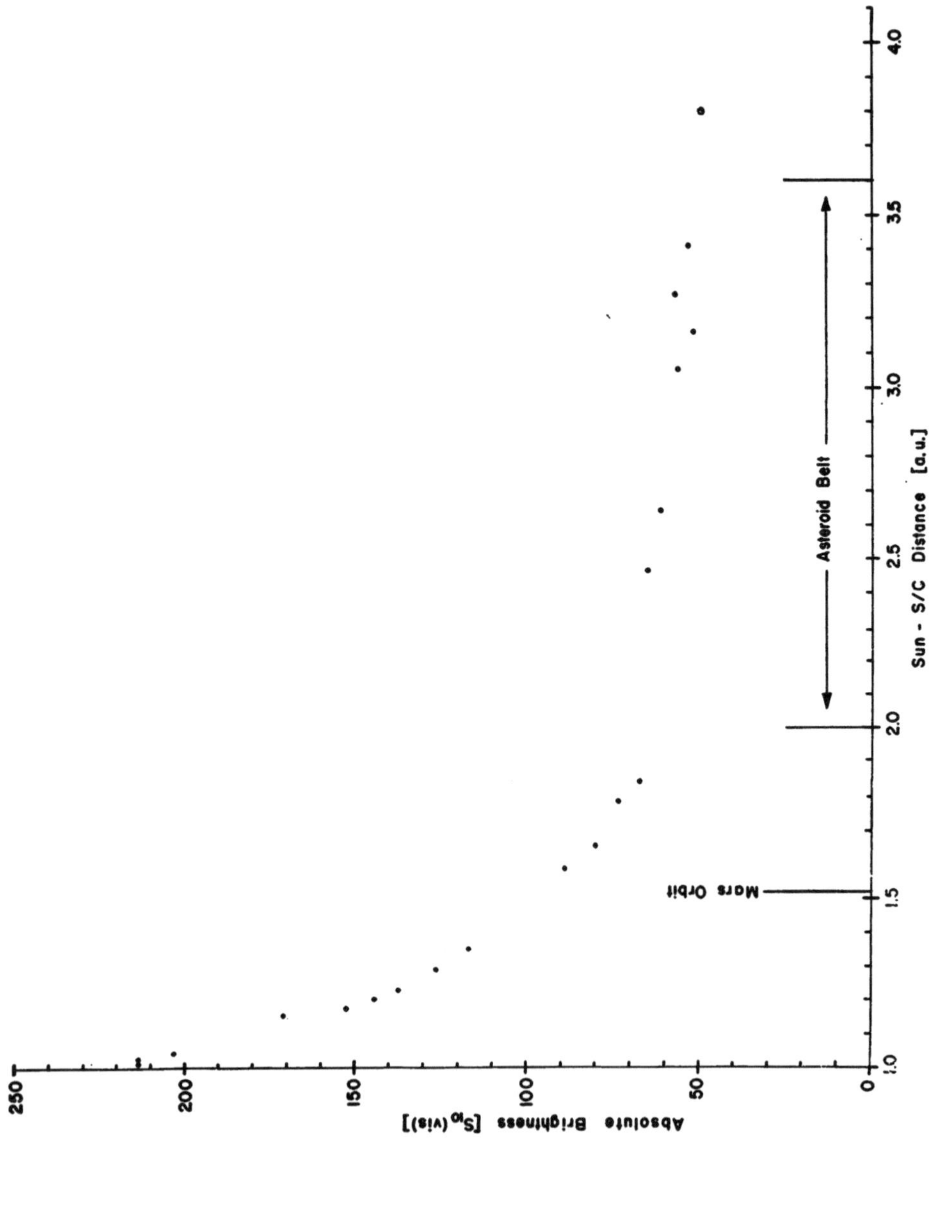

Fig. 1. Preliminary results in the blue on the brightness of zodiacal light as a function of the heliocentric distance R of Pioneer 10.

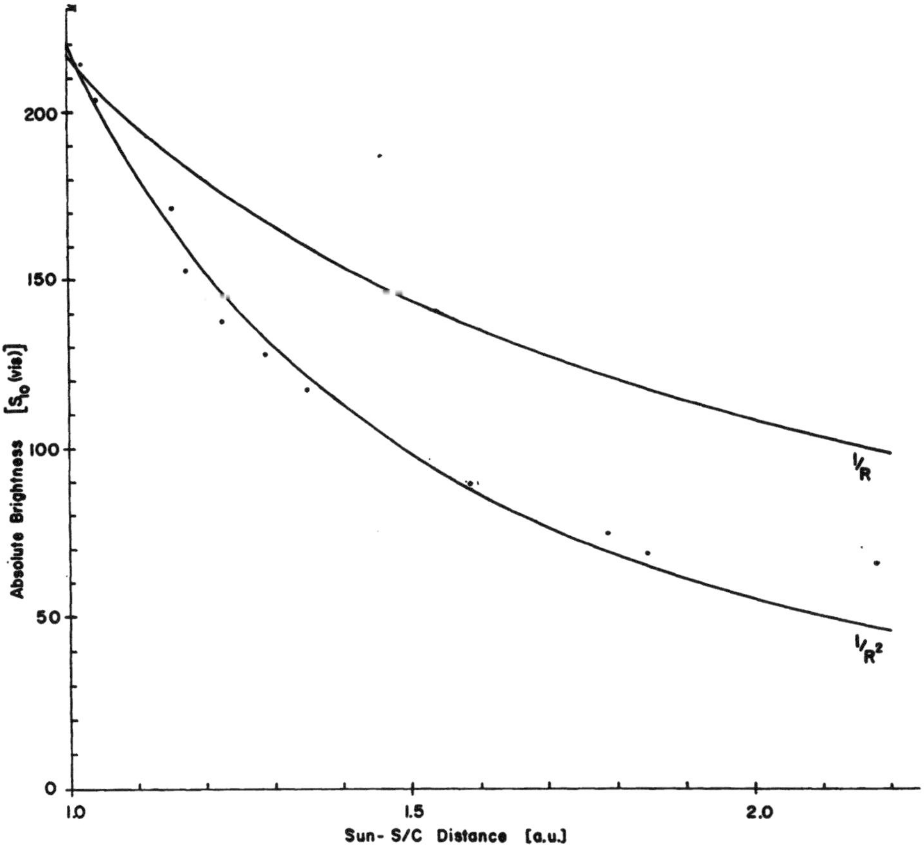

Fig. 2. The observed variation of the zodiacal light between the earth and the asteroid belt. A $1/R$ distribution implies a constant particle density.

belt and beyond. This also requires that the decrease in particle density between the earth and the asteroid belt is considerably steeper than $1/R$. Observations will continue to be made from Pioneer 10 out to approximately 5 AU.

Acknowledgments

This research was supported by the National Aeronautics and Space Administration under contracts NAS 2-5610 and NAS 2-6560. We wish to thank L. De Shields II, B. Green, and J. Hutchison at the Dudley Observatory and staff of the Pioneer Project office at the NASA Ames Research Center for their assistance.

References

[1] J. L. Weinberg, M. S. Hanner, H. M. Mann, P. B. Hutchison and R. Fimmel, Space Research XIII, 1187 (1973).
[2] F. E. Roach and L. R. Megill, Astrophys. J. **133**, 228 (1961).
[3] M. S. Hanner and J. L. Weinberg, Sky Telesc. **45**, 217 (1973).

Note added in proof: Subsequent observations out to 4.8 AU indicate that the background starlight is significantly different from the star count data used here. This accounts for the high residual brightness in Fig. 1. The zodiacal ligth decreases more steeply than $1/R^2$ in the asteroid belt and is absent beyond it [M. S. Hanner, J. L. Weinberg, L. M. DeShields II, B. A. Green and G. N. Toller, *Bull. AAS*, in press].

Space Research XIV — Akademie-Verlag, Berlin 1974

THE METEORIC PARTICLE SPACE DENSITY NEAR THE EARTH AND THE MOON, ACCORDING TO DATA OBTAINED BY SIMULTANEOUS OBSERVATIONS OF SPACE VEHICLES

T. Nazarova and A. Rybakov

USSR

In 1972 a series of simultaneous measurements by means of equal (combined) sensors was carried out near the earth on the satellites Cosmos 470, Intercosmos 6 and Cosmos 502, and near the moon on the lunar satellite Luna 19.

Measurements have shown the presence of heightened spatial density of meteoric particles with masses 10^{-10}—10^{-12} g near the earth compared with the density in the vicinity of the moon and in interplanetary space (where they coincide with one another) Fig. 1.

From study of the whole complex of available experimental data the assumption may be made that the increase in the spatial density of particles near the earth occurs at the expense of particles with $m < 10^{-7}$ g.

The mass index (S parameter) for the range of masses 10^{-12}—10^{-7} g near the moon is equal to $S \sim 0.6$ compared with $S \sim 0.8$ near the earth, i.e. decrease in the number of particles with decrease in their mass occurs to a greater extent in the vicinity of the moon and in the interplanetary space than near the earth.

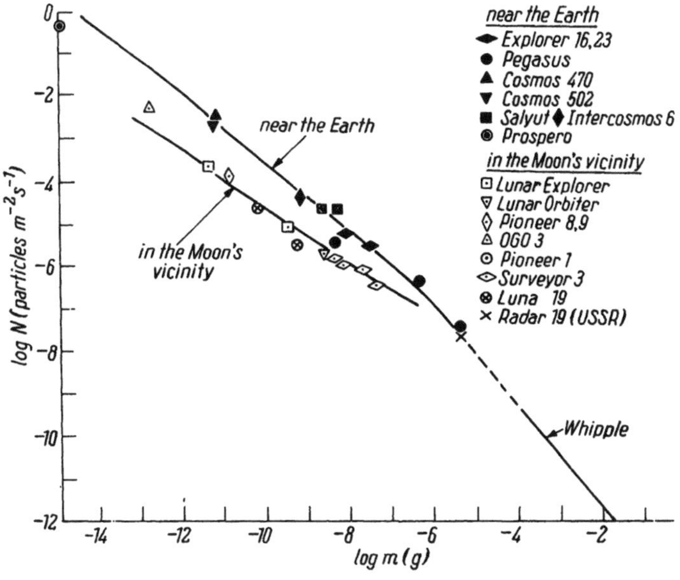

Fig. 1

Space Research XIV — Akademie-Verlag, Berlin 1974

THE ZODIACAL LIGHT AT 2.4 μm

W. HOFMANN and D. LEMKE

Max-Planck-Institut für Astronomie, Heidelberg, FRG

The zodiacal light and the OH airglow have been observed in the infrared by the balloon-borne telescope THISBE from 32 km float altitude. The 15 cm telescope, the photometer and the PbS detector were cooled by dry ice. We found the brightness of the OH airglow at $\lambda = 2.1\ \mu m$ to be 1.6×10^{-9} W cm^{-2} μm^{-1} sr^{-1} at the zenith. Within a gap in the intensive OH airglow spectrum, we were able to measure the zodiacal light at elongations $\varepsilon = 36°$ and $\varepsilon = 26°$ with an accuracy to $\pm 20\%$.

The intensities in the ecliptic plane are 6×10^{-11} and 10×10^{-11} W cm^{-2} μm^{-1} sr^{-1} respectively. These results will be discussed in relation to the temperature of the interplanetary dust.

A full paper has been published in *Nature* Phys. Sci. **243**, 140 (1973).

Space Research XIV — Akademie-Verlag, Berlin 1974

MASS DISTRIBUTION OF INTERPLANETARY PARTICLES RESULTING FROM RANDOM COLLISIONS

S. Grzędzielski and J. Juchniewicz

Astronomical Observatory of the Warsaw University, Warsaw, Poland

The effect of random binary collisions of cube-shaped particles upon their mass distribution has been investigated, assuming that the masses of resulting fragments are simple linear functions of the impact parameter. Both time-independent and time-dependent numerical solutions to the appropriate non-linear, partial integro-differential equations were studied for a number of hypotheses as to the mass input, mode of fragmentation and mass range.

1. Introduction

In recent years several authors have studied the theoretical mass distribution of asteroidal objects in the Solar System. Following the pioneer investigation by Piotrowski [1], time-independent mass distributions were obtained by Dohnanyi [2] and Hellyer [3], and numerical solutions to the time-dependent problem were studied by Hellyer [4] and Bandermann [5]. The last three authors took into account the influence of the debris on the distribution, and, basing their considerations on experimental data, assumed a power law for the distribution of fragments of a disrupted body. It was also tacitly assumed that the effect of collision does not depend upon the impact parameter, i.e. that the effect is the same both for head-on and for grazing collisions. As there seems to be little foundation for such an assumption in the case of large objects with diameters of the order of many metres or kilometres, the present authors attempted to investigate what kind of mass distribution one might expect when the the effect of different impact parameters (as well as of the debris) is taken into account.

2. Model of Collisions

All colliding objects are assumed to be cubes of constant density and their relative velocity $v_{\rm rel}$ in a collision is always the same. The cubes collide with corresponding faces parallel to each other and the impact parameter l is defined in Fig. 1 (i) in which the relative velocity is perpendicular to the plane of paper and the picture is symmetrical with respect to the line defined by l. The differential collision cross section is then $2\pi l\, dl$. The configurations that may result from collision are shown in Fig. 1 (ii). The cuboid 3′ (shaded) combines a part of object 1 with a part of object 2. For $l \leqq {}^1/_2\,(a_2 - a_1)$ the object 2′ does not exist.

Two extremal hypotheses as to the result of collision were investigated:

1. Simple break-up (SB). Upon collision of objects 1 and 2 three objects 1′, 2′ and 3′ are formed.

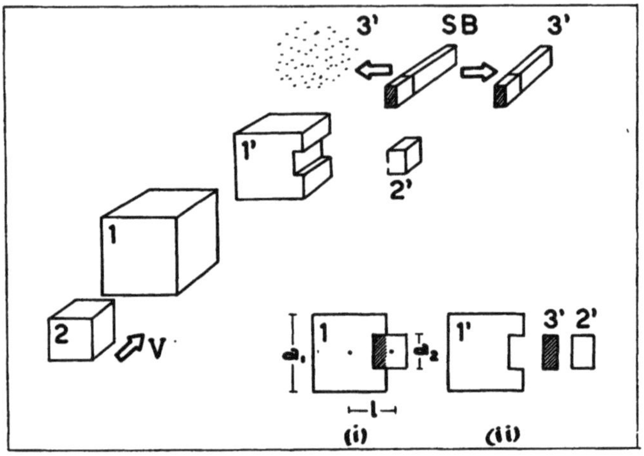

Fig. 1. The model of collision. SB, simple break-up; B + D, break-up with destruction is shown by the swarm to the left of SB.

2. Break-up with destruction (B + D). Those parts of 1 and 2 that collide head-on disintegrate into a swarm of particles with masses smaller than the minimal mass ($m_{\min}$) considered. Hence object 3′ disappears. Objects 1′ and 2′ are defined as previously.

Under these assumptions the fragmentation equation can be written in the form (t = time)

$$\frac{\partial}{\partial t} n(m, t)\, dm = v_{\mathrm{rel}} \iiint\limits_{\Omega} n(m_1, t)\, n(m_2, t)\, \varphi(m; m_1, m_2; l)\, 2\pi l\, dl\, dm_1\, dm_2 + \text{mass source} \tag{1}$$

where $n(m, t)$ denotes the differential mass distribution and where $\Omega = \Omega(m_1, m_2, l)$ is the volume of the (m_1, m_2, l)-space over which the integration has to be performed. The distribution function $\varphi(m; m_1, m_2; l)$ of objects created in a collision of masses m_1 and m_2 can be represented as an appropriate sum of Dirac δ-type functions, since the required mass m will appear only if at least one of the masses m_1', m_2', m_3' is equal to m.

The calculations were performed for a finite mass range ($m_{\min}$, $m_{\max}$) with $m_{\min}/m_{\max} = 10^{-3}$ typically. In some cases a mass source term was introduced into Eq. (1) in order to simulate the effects of mass input from the mass intervals adjacent to the interval considered. Two modes of mass input were tried: (a), mass input only at $m = m_{\max}$, (b), mass input proportional to $m^{-5/3}$. When applied, the output of the mass source term matched the loss of mass from the ($m_{\min}$, $m_{\max}$) interval due to the collisions.

3. Numerical Solutions

Time-dependent solutions were investigated by direct computer integrations of Eq. (1). The time-dependent distributions relaxed on time scales of the order of 3 to 6 collision times towards stationary distributions independent of the initial distributions. Examples of stationary distributions for different $m_{\min}/m_{\max}$ values and for the two modes of mass input are shown in Fig. 2 for case SB. The comparison of stationary distributions obtained for SB and B + D for modes (a) and (b) is shown in Fig. 3.

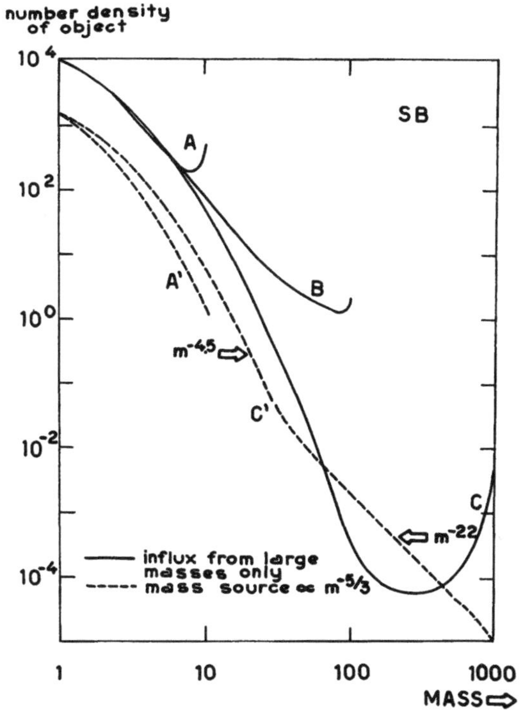

Fig. 2. Dependence of stationary distributions for simple break-up on mass range and mode of mass input. For curves A and A′, $m_{\min}/m_{\max} = 0.1$; for curve B, $m_{\min}/m_{\max} = 0.01$; for curves C and C′, $m_{\min}/m_{\max} = 0.001$.

From the solutions obtained the following conclusions can be drawn: (i) the stationary distributions are, in general, steep; this means that a mechanics of collision which depends on the impact parameter is able to produce by itself a large number of small objects without recurring to cratering and crushing; (ii) for input mode (a) the solutions depend on the ratio $m_{\min}/m_{\max}$ (see curves A, B, C in Fig. 2); this dependence is most pronounced at $m \simeq m_{\max}$, i.e. close to the mass source; on the other hand, for m approaching $m_{\min}$ all solutions look rather similar; this suggests that in the low mass range the internal mechanics of the mill prevails over the effects due to the imposed condition at $m = m_{\max}$; (iii) for a less extreme mode of mass input (mode (b)) the distributions possess almost rectilinear stretches (m^{-s}, cf. Fig. 3) with moderate exponents ($2 \lesssim s \lesssim 3$) encountered in the case of differential mass distributions of objects in the 10^{-3} to 10^3 g mass range.

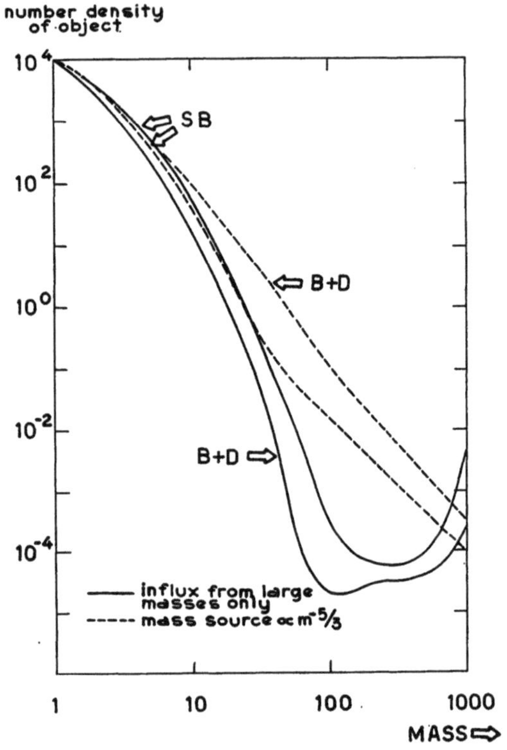

Fig. 3. Comparison of stationary distributions obtained for SB and B + D for the two adopted modes of mass input. (a) ——; (b) ----

4. Debris Production by the Asteroids

In the upper asteroidal mass range the evolution of distribution will occur most probably without external mass input. An evolution without mass input is shown for the case B + D in Fig. 4, where the initial distribution ($t = 0$) was taken quite arbitrarily. If one identifies the distribution labelled $t = 111$ with the *present* distribution of the three upper decades of asteroidal masses, the rate of evolution determines the debris ($m < (1/1000) m_{\max}$) production by the asteroids. Assuming $v_{\text{rel}} = 5$ km s^{-1}, rock density $= 3$ g cm^{-3}, mass of largest colliding asteroid equal to $\beta \times 10^{23}$ g, total mass of the three upper decades of asteroidal matter equal to $\alpha \times 10^{24}$ g, one finds for the rate of debris production

$$\frac{dM}{dt} = \frac{\alpha^2}{\beta^{1/3}} 3 \times 10^5 \text{ g s}^{-1}$$

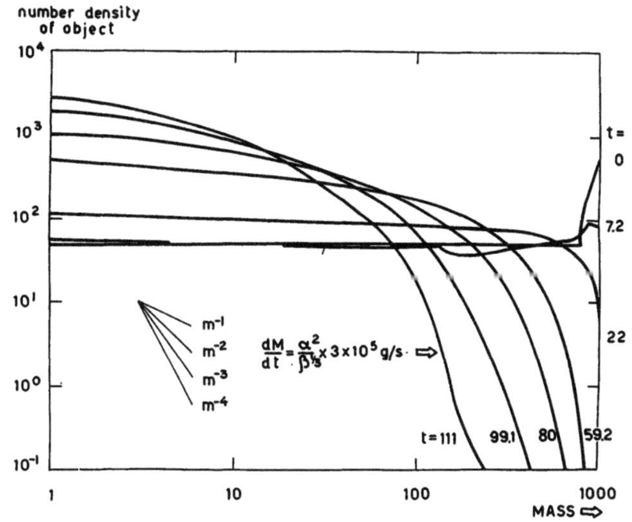

Fig. 4. Evolution of distribution without mass input. The indicated rate of production of masses with $m < m_{\min}$ corresponds to the case when the distribution labelled $t = 111$ is identified with the present distribution of the upper three decades of asteroidal masses. The corresponding time unit is then $(\beta^{1/3}/\alpha) \times 10^{10}$ years (for α and β see text).

Acknowledgment

Financial support from the Polish Academy of Sciences is gratefully acknowledged.

References

[1] S. Piotrowski, Acta Astr. **5**, 115 (1954).
[2] J. S. Dohnanyi, J. Geophys. Res. **74**, 2531 (1969).
[3] B. Hellyer, Mon. Not. Roy. Astron. Soc. **148**, 383 (1970).
[4] B. Hellyer, Mon. Not. Roy. Astron. Soc. **154**, 279 (1971).
[5] L. W. Bandermann, Mon. Not. Roy. Astron. Soc. **160**, 321 (1971).

Space Research XIV — Akademie-Verlag, Berlin 1974

BASIC COLOUR MODELS OF ZODIACAL LIGHT FOR INTERPRETATION OF SPACEBORNE MEASUREMENTS

R. H. GIESE[a], M. S. HANNER[b] and C. LEINERT[c]

[a] Bereich Extraterrestrische Physik, Ruhr-Universität Bochum, FRG
[b] Dudley Observatory, Albany, N.Y., USA
[c] Max-Planck-Institut für Astronomie, Heidelberg, FRG

For interpretation of forthcoming Skylab data colour models of zodiacal light in the ecliptic plane have been calculated for both dielectric and metallic particles in the sub-micrometre and micrometre size range. Two colour ratios were computed, a blue ratio C_b (0.40 μm, 0.53 μm) and a red ratio, either C_r (0.82 μm, 0.53 μm) or C_r (0.71 μm, 0.53 μm). The models with a size distribution $\propto s^{-2.5}\,ds$ generally show a colour close to the solar colour and almost independent of elongation. Especially in the blue colour ratio there is generally no significant dependence on the lower cutoff size (0.1—1 μm). The main feature of absorbing particles is a reddening at small elongations. The models for size distribution $\propto s^{-4}\,ds$ show larger departures from solar colour and more variation with model parameters. This might be a way to distinguish between flat and steep size distributions by colour measurements from Skylab.

A full paper has been published in *Planetary and Space Science* **21**, 513 (1973).

Space Research XIV — Akademie-Verlag, Berlin 1974

COUNTERGLOW FROM THE EARTH-MOON LIBRATION POINTS

J. R. Roach

Ball Brothers Research Corporation, Boulder, Col., and Rutgers, The State University
Newark, N.J., USA

Results from the OSO 6 Rutgers zodiacal light analyzer experiment show photometric perturbations above the background in the anti-sun line of sight. Sixteen successive lunations were examined, and the accumulated perturbations show a maximum value in the direction of the L_4 and L_5 earth-moon libration points. This is interpreted as a counterglow from a cloud of particles at the libration points. The average brightness of these libration clouds is 20 S_{10} Vis. The average angular size of the libration clouds is approximately 6 degrees. Their position varies from one lunation to the next, within an ellipsoidal zone centered on the libration point direction, with its semi-major axis, of approximately 6 degrees, nominally in the ecliptic, and its semi-minor axis, of approximately 2 degrees, perpendicular to the ecliptic. The position of these clouds is towards the moon from the libration point in the summer and away from the moon in the winter.

A full paper has been submitted to *Planetary and Space Science.*

Space Research XIV — Akademie-Verlag, Berlin 1974

ON A COSMIC DUST SOURCE TO SUPPLY THE PIONEER 8 AND 9 PARTICLE FLUX

O. E. Berg[a] and H. A. Zook[b]

[a] Goddard Space Flight Center, Greenbelt, Md, USA
[b] NASA Manned Spacecraft Center, Houston, Texas, USA

One of the authors has previously shown that the bulk of particles intercepted by the Pioneers 8 and 9 cosmic dust experiments emanates generally from the solar direction. It is postulated that the particles are born of comets in perihelion passage ($q < 1$ AU) and expelled quasi-radially from the sun by radiation pressure. The question whether cometary contribution can account for the mass of material inferred by the satellite experiment's measurements can be answered affirmatively by using Whipple's estimate of the daily cometary mass input inside the earth's orbit. Particles introduced near the sun (a few solar radii) with a mass of $10^{-12}-10^{-14}$ grams, may readily attain velocities in excess of 100 km s^{-1} at 1 AU. Lunar rocks should be essentially devoid of microcraters on their north or south sides or any area sheltered from the radially accelerated particles, a possible explanation for the non-uniform distribution of microcraters on lunar samples.

A full paper will be published in *Planetary and Space Sciences* under the title "The Effects of Radiation Pressure on the Orbits of newly created Cosmic Dust" by H. A. Zook and O. E. Berg.

Space Research XIV — Akademie-Verlag, Berlin 1974

INTERPLANETARY DUST AND ITS INFLUX TO THE EARTH'S SURFACE

DAVID W. HUGHES

The University of Sheffield, UK

From the latest cumulative influx rate data the influx of interplanetary dust to the earth's surface in the mass range 10^{-13} — 10^{6} g is found to be 5.7×10^{9} g yr^{-1}; with this value, the solar system dust cloud is found to have a mean density of 5×10^{-23} g cm^{-3} and a mass of 1.6×10^{19} g and to require a mass influx of 1.3×10^{7} g s^{-1} to maintain it in equilibrium. It is concluded that this influx is produced by cometary decay. The mass distribution index s_m of the particles in the meteor stream is changed considerably by the time these particles have decayed into the dust cloud.

The earth on its journey around the sun is continuously colliding with dust particles. Fig. 1 gives the cumulative flux $\Theta(m)$ of these particles (i.e. the incident number having a mass greater than m) as a function of m; the satellite data were provided by W. M. Alexander (private communication) and the radio and visual data from [1]. By an analysis similar to that given in [2] the total mass influx ψ_T to the earth's surface can be calculated. This is found to be 5.7×10^{9} g yr^{-1} (3.5×10^{-17} g cm^{-2} s^{-1}). The incident dust was assumed to lie in the size range 10^{-13}—10^{6} g, 10^{-13} g being the radiation pressure limit and 10^{6} g the mass of the largest meteoroid that can escape from the cometary nucleus [3]. As the mass distribution index s_m (defined by $\Theta(m) \propto m^{1-s_m}$) is a long way from 2.0 at both ends of the mass range the integration limits make very little difference to the total incident mass, 95% of which is made up of particles in the size range $10^{-9} < m < 10^{-5}$ g.

Following Öpik [4] the density ϱ of the solar system dust cloud at 1 AU is given by

$$\varrho = 4u\psi_T(u^2 + v_e^2)^{-1}$$

where u is 5.6×10^{5} cm s^{-1}, the mean relative velocity between the dust particles and the earth and v_e is the escape velocity from the earth. This density is found to be 5.0×10^{-23} g cm^{-3}. Assuming that the cloud has a volume of 3.1×10^{41} cm^{3} and that the spatial density of dust is constant throughout [3], the mass of the cloud is found to be 1.6×10^{19} g equivalent to the mass of dust in about 24000 comets. Particles are being continually removed from the cloud by the combined effects of collisions, radiation pressure and the Poynting-Robertson effect. Using Whipple's [5] values for the life of particles, the loss rate from the cloud has been calculated to be 1.3×10^{7} g s^{-1}. This can be compared with the influx rate to the cloud from decaying comets which has been estimated to be 1.5×10^{6} g s^{-1} [2].

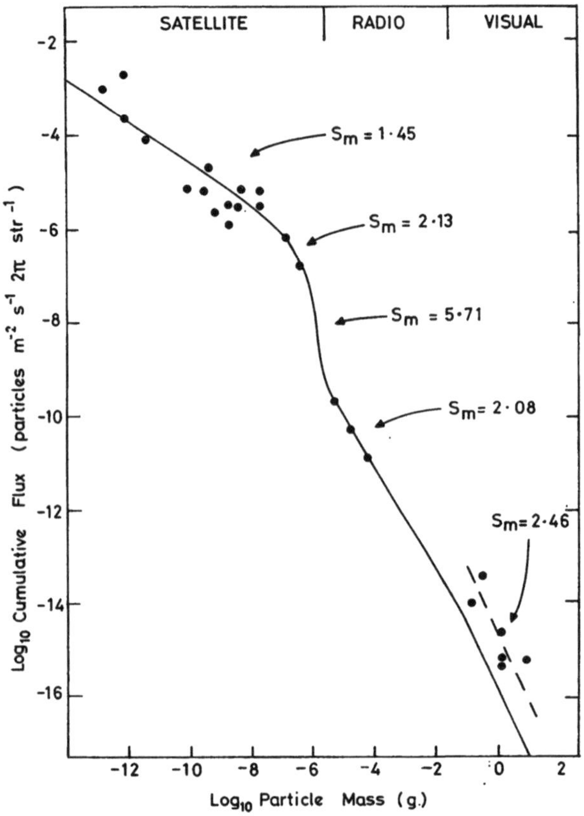

Fig. 1. The cumulative flux of sporadic meteoroids incident on the earth as a function of mass. s_m, the mass distribution index; --- mean of visual data.

This difference is not considered significant. At the above loss rate the complete mass of the cloud will be removed every 4000 years and for equilibrium to occur the cometary decay rate must average 1.3×10^7 g s^{-1} over a period of this order.

Now the mass distribution index of the particles in a meteor stream is decidedly different from the index of the debris of this stream after it has decayed into the dust cloud. The mass distribution indices of the meteoroids in the Quadrantid, Perseid and Geminid streams are given in [6]; if these are typical, mean stream indices are found to be 1.71 for $m < 8 \times 10^{-2}$ g and 2.27 for $m > 8 \times 10^{-2}$ g. Fig. 2 shows how the dust in the meteor stream is distributed between the different logarithmic mass intervals (i.e. the interval $10^n - 10^{n+1}$ g). This has been calculated using the method given in [2] with the following assumptions: (i) the dust in an average meteor stream has a mass of about 10^{15} g [6] and (ii) the upper and lower limits of the particle size in this cometary debris are 10^{-13} g and 10^6 g respectively. Lines (a) and (b) in Fig. 2 both apply to the actual meteor stream, line (b) with the s_m change at 8×10^{-2} g being thought the more probable. From line (b) it can be seen that the majority ($\sim 73\%$) of the mass of a meteor stream is made up of particles with masses in the range $10^{-3} < m < 10^1$ g. Line (c) shows the distribution of this mass when it has been perturbed into the dust cloud and it can be seen that now most of the mass (95%) is made up of much smaller particles ($10^{-9} < m < 10^{-5}$ g).

It is thought that the main process causing meteor stream decay is inter-particle collision between stream and dust cloud particles at relative velocities

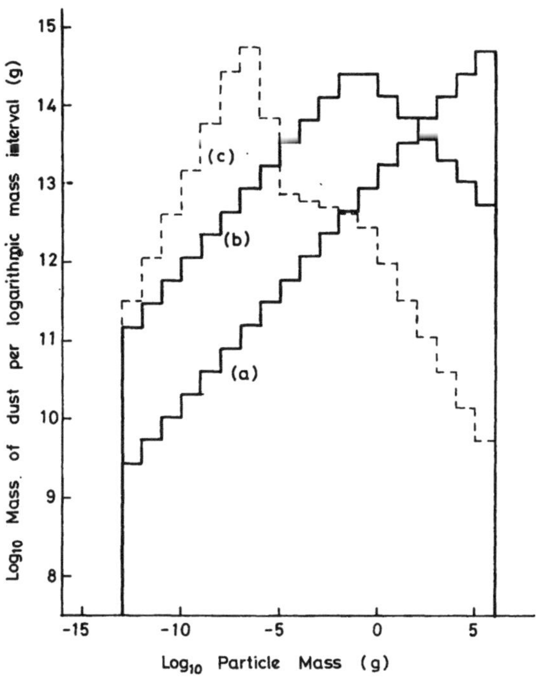

Fig. 2. The mass of dust in each logarithmic mass interval for the 10^{15} g in an average meteor stream: (a) assuming that the stream particles have a mass distribution index of 1.71; (b) assuming that $s_m = 1.71$ ($10^{-13} < m < 10^{-1}$ g) and $s_m = 2.27$ ($10^{-1} < m < 10^{6}$ g); (c) for the dust after it has been perturbed into the solar system dust cloud.

of tens of km s^{-1}. This decay process causes collisional fragmentation which breaks up the fragile sponge-like low density stream meteoroids into much smaller dust cloud particles, the mass distribution index being considerably increased from 1.71 to 2.08 in the radio meteor region and 2.29 to 2.46 in the visual meteor region. This process would take about 10^5 years, the average lifetime of a meteor stream. The problem of why the dirty iceball cometary nucleus breaks up to form meteoroids with s_m values of 1.71 and 2.29 still remains to be solved.

References

[1] David W. Hughes, Mon. Not. Roy. Astron. Soc. **161**, 113 (1973).
[2] David W. Hughes, Planet. and Space Sci. **20**, 1949 (1972).
[3] David W. Hughes, Mon. Not Roy. Astron. Soc. Submitted.
[4] E. J. Öpik, Irish Astron. J. **4**, 84 (1956).
[5] F. L. Whipple, NASA, SP-150, p. 409 (1967).
[6] David W. Hughes, Space Research XIV, 709 (1974).

Space Research XIV — Akademie-Verlag, Berlin 1974

MEASUREMENTS OF THE LIGHT FLASH PRODUCED BY HIGH VELOCITY PARTICLE IMPACT

G. Eichhorn

Max-Planck-Institut für Kernphysik, Heidelberg, FRG

The impact light flash produced by electrostatically accelerated iron particles with diameters ranging from 5 μm to 0.05 μm and velocities lying between 1 km s^{-1} and 30 km s^{-1} has been investigated by means of photomultipliers. As target materials, mainly gold and tungsten were used. The pulse of the multiplier was registered directly and after electronic integration. The pulse height of the multiplier signal, the amplitude of the integrated signal and its rise-time were found to be unique functions of the mass and velocity of the impacting particle. For the pulse height of the differential signal the relation $I = c_1 m^{1.25} v^5$ was obtained, and for the integrated signal the relation $I = c_2 m^{1.25} v^{3.8}$, with only c_1 and c_2 depending on the target material. The rise-time of the integrated signal follows the relation $T = 2.2 \times 10^2 v^{-0.4}$ using gold as target, and for tungsten the relation $T = 9.8 \times 10^2 v^{-1.2}$, where v is given in km s^{-1} and T is in μs. Using the spectral distribution of the light intensity, measured by means of calibrated photomultipliers, the total amount of light energy emitted in the visible range could be calculated. As a result we obtained that for $v = 4$ km s^{-1} and $m = 10^{-11}$g, about 3×10^{-4} of the kinetic energy of the particle was converted into light energy. The variation of the impact flash intensity with the target material and the measured spectral distribution made it possible to estimate the temperature of the crater after the impact as lying between 2000 K and 3000 K.

A full paper will be submitted to *Planetary and Space Science*.

PAPERS PRESENTED AT THE COSPAR MEETING AT CONSTANCE, 1973, BUT PUPLISHED ELSEWHERE

The Earth's Neutral Atmosphere

C. A. Reber and A. E. Hedin, Heating of the high latitude thermosphere during magnetically quiet periods, J. Geophys. Res.

L. G. Jacchia, I. G. Campbell and J. W. Slowey, A study of the diurnal variation in the atmosphere as derived by satellite drag, Planet. Space Sci. **21**, 1825 (1973).

L. G. Jacchia, J. W. Slowey and I. G. Campbell, An analysis of the solar-activity effects in the upper atmosphere, Planet. Space Sci. **21**, 1835 (1973).

L. G. Smith and K. L. Miller, The measurement of O_2 number density by absorption of Lyman-α, J. Geophys. Res. 1974.

F. Barlier, P. Bauer, C. Jaeck and G. Thullier, North-south asymmetry in the thermosphere during the last maximum of the solar cycle, Planet. Space Sci. 1974.

T. G. Scholz and D. Offermann, Measurement of neutral atmospheric composition at 85—115 km by mass spectrometer with cryo-ion source, J. Geophys. Res. **79**, Jan. 1974.

D. Offermann, Seasonal variation of atomic oxygen at altitudes between 85 and 150 km, J. Geophys. Res.

C. S. Weller and R. R. Meier, First satellite observations of He^+ 304 Å radiation and its interpretation, J. Geophys. Res. 1974.

K. Petzoldt, Extreme temperature deviations from the climatological mean in the upper stratosphere—observed by rockets, confirmed by satellites, Meteorologische Abhandlungen, Inst. f. Meteorologie u. Geophysik der Freien Universität Westberlin 1974.

A. D. Belmont and D. G. Dartt, Semi-annual variation in zonal wind from 20 to 65 km at 80° N—10° S, J. Geophys. Res. 78, 6373 (1973).

The Ionosphere

L. G. Smith, M. A. Geller and H. D. Voss, Energetic electrons in the midlatitude nighttime E region, J. Atmos. Terr. Phys. 1974.

K. Hirao and K. Oyama, Electron temperature profilė and its solar activity dependence in the middle latitude region, J. Geomagn. Geoelect. 1973.

B. H. Subbaraya, S. Prakash, S. P. Gupta and A. S. S. Sinha, Measurement of electron temperature in the lower ionosphere by detecting the space potential on the Langmuir probe characteristic, Planet. Space Sci. 1974.

The Magnetosphere

F. A. Morse, E. J. Rogers, P. F. Nelson and P. F. Mizera, Correlation of optical aurora and low energy electrons, J. Geophys. Res.

G. Gustafsson, G. Starkov, V. Vorobyev, Y. Feldstein and N. Shevnina, Dynamics of day and night aurora during substorms, Planet. Space. Sci.

J. C. Cerisier, Ducted and partly ducted propagation of VLF waves through the magnetosphere, J. Atmos. Terr. Phys.

The Sun

J. Vorpahl, 19—20 May 1969, an example of type III emission during the impulsive phase of flares, Solar Phys. **29**, 447 (1973).

E. C. Roelof and S. M. Krimigis, Analysis and synthesis of coronal and interplanetary energetic particle plasma and magnetic field observations over three solar rotations, J. Geophys. Res. **78**, 5375 (1973). (The material presented at the conference is mainly contained in Section 7 of this paper.)

R. C. Henry, P. D. Feldman, W. G. Fastie, H. W. Moos and A. Weinstein, Low resolution ultraviolet spectroscopy of several stars from Apollo 17, Astrophys. J.

J. P. Swings, C. Jamar and J. M. Vreux, Absolute ultraviolet spectrophotometry from the TD-1 satellite: I, Absorption features in the UV spectra of B stars, Astron. Astrophys. **29**, 207 (1973).

J. M. Vreux, D. Malaise and J. P. Swings, Absolute ultraviolet spectrophotometry from the TD-1 satellite: II, Observations of B stars continua shortward of 2500 Å, Astron. Astrophys. **29**, 211 (1973).

D. Malaise, M. Gros and D. Macau, Absolute ultraviolet spectrophotometry from the TD-1 satellite: III, The continuum of A type stars between 1350 and 2800 Å, Astron. Astrophys.

F. Praderie, D. Macau and C. Jamar, Absolute ultraviolet spectrophotometry from the TD-1 satellite: IV, Line features in UV spectra of A stars, Astron. Astrophys. 1974.

F. Beeckmans, D. Macau and D. Malaise, Absolute ultraviolet spectrophotometry from the TD-1 satellite: V, Bolometric corrections for early type stars of luminosity class V, Astron. Astrophys.

C. M. Humphries, K. Nandy and G. I. Thompson, Ultraviolet photometry of early type stars from TD-1 satellite observations, Mon. Not. R. Astr. Soc. **163**, 1 (1973).

J. B. Rogerson, Jr, Observations of interstellar atoms by the Copernicus satellite, this paper was a synthesis of three papers in Astrophys. J. (Letters), **181**, L97, L103 and L110 (1973).

C. S. Weller, G. R. Carruthers and R. C. Henry, Observation of the ultraviolet interstellar radiation field at several wavelengths, Astrophys. J.

The Moon and Planets

J. J. Rennilson and D. R. Criswell, Surveyor observations of Lunar horizon glow, The Moon, Eds. H. Alfvén, Z. Kopal and H. C. Urey.

R. J. Phillips, G. F. Adams, et al., The Apollo 17 Lunar Sounder, Proc. 4th Lunar Sci. Conf. **3**, Geochim. Cosmochim. Acta, Suppl. 4, 2821 (1973).

H. H. Schmitt, Apollo 17 report on the valley of Taurus-Littrow, Science **182**, 681 (1973).

M. Shimizu, Atmospheric mixing in the upper atmospheres of Mars and Venus, J. Geophys. Res. **78**, 6780 (1973).

C. W. Hord, K. E. Simmons and L. K. McLaughlin, Mariner 9 ultraviolet spectrometer experiment: Pressure-altitude measurements on Mars, Icarus 1974.

H. Masursky, An overview of geological results from Mariner 9, J. Geophys. Res. **78**, 4009 (1973).

INDEX OF AUTHORS

Bei Fragen zur Produktsicherheit wenden Sie sich bitte an:
If you have any questions regarding product safety,
please contact:

Walter de Gruyter GmbH
Genthiner Straße 13
10785 Berlin
productsafety@degruyterbrill.com